Mariusz Urbański, Mario Roy, Sara Munday
Non-Invertible Dynamical Systems

De Gruyter Expositions in Mathematics

Volume 69/2

Mariusz Urbański, Mario Roy, Sara Munday

Non-Invertible Dynamical Systems

Volume 2: Finer Thermodynamic Formalism – Distance Expanding Maps and Countable State Subshifts of Finite Type, Conformal GDMSs, Lasota-Yorke Maps and Fractal Geometry

DE GRUYTER

Mathematics Subject Classification 2010
37A05, 37A25, 37A30, 37A35, 37A40, 37B10, 37B25, 37B40, 37B65, 37C05, 37C20, 37C40, 37D20, 37D35, 37E05, 37E10

Authors
Prof. Dr. Mariusz Urbański
University of North Texas
Department of Mathematics
1155 Union Circle #311430
Denton, TX 76203-5017
USA
mariusz@urbanskimath.com

Dr. Sara Munday
John Cabot University
Via della Lungara 233
00165 Rome
Italy
smunday@johncabot.edu

Prof. Dr. Mario Roy
York University
Glendon College
2275 Bayview Avenue
Toronto, M4N 3M6
Canada
mroy@glendon.yorku.ca

ISBN 978-3-11-070061-9
e-ISBN (PDF) 978-3-11-070269-9
e-ISBN (EPUB) 978-3-11-070273-6
ISSN 0938-6572

Library of Congress Control Number: 2021951113

Bibliographic information published by the Deutsche Nationalbibliothek
The Deutsche Nationalbibliothek lists this publication in the Deutsche Nationalbibliografie; detailed bibliographic data are available on the Internet at http://dnb.dnb.de.

Typesetting: VTeX UAB, Lithuania
Printing and binding: CPI books GmbH, Leck

www.degruyter.com

Mariusz Urbański dedicates this book to his wife, Irena.
À mes parents Thérèse et Jean-Guy, à ma famille et à mes amis, sans qui ce livre n'aurait pu voir la vie… du fond du coeur, merci! Mario

Preface

Dynamical systems and ergodic theory is a rapidly evolving field of mathematics with a large variety of subfields, which use advanced methods from virtually all areas of mathematics. These subfields comprise but are by no means limited to: abstract ergodic theory, topological dynamical systems, symbolic dynamical systems, smooth dynamical systems, holomorphic/complex dynamical systems, conformal dynamical systems, one-dimensional dynamical systems, hyperbolic dynamical systems, expanding dynamical systems, thermodynamic formalism, geodesic flows, Hamiltonian systems, KAM theory, billiards, algebraic dynamical systems, iterated function systems, group actions, and random dynamical systems.

All of these branches of dynamical systems are mutually intertwined in many involved ways. Each of these branches nonetheless also has its own unique methods and techniques, in particular embracing methods which arise from the fields of mathematics the branch is closely related to. For example, complex dynamics borrows advanced methods from complex analysis, both of one and several variables; geodesic flows utilize methods from differential geometry; and abstract ergodic theory and thermodynamic formalism rely heavily on measure theory and functional analysis.

Indeed, it is truly fascinating how large the field of dynamical systems is and how many branches of mathematics it overlaps with. In this book, we focus on some selected subfields of dynamical systems, primarily noninvertible ones.

In the first volume, we give introductory accounts of topological dynamical systems acting on compact metrizable spaces, of finite-state symbolic dynamical systems, and of abstract ergodic theory of measure-theoretic dynamical systems acting on probability measure spaces, the latter including the metric entropy theory of Kolmogorov and Sinai. More advanced topics include infinite ergodic theory, general thermodynamic formalism, and topological entropy and pressure. This volume also includes a treatment of several classes of dynamical systems, which are interesting on their own and will be studied at greater length in the second volume: we provide a fairly detailed account of distance expanding maps and discuss Shub expanding endomorphisms, expansive maps, and homeomorphisms and diffeomorphisms of the circle.

The second volume is somewhat more advanced and specialized. It opens with a systematic account of thermodynamic formalism of Hölder continuous potentials for open transitive distance expanding systems. One chapter comprises no dynamics but rather is a concise account of fractal geometry, treated from the point of view of dynamical systems. Both of these accounts are later used to study conformal expanding repellers. Another topic exposed at length is that of thermodynamic formalism of countable state subshifts of finite type. Relying on this latter, the theory of conformal graph directed Markov systems, with their special subclass of conformal iterated function systems, is described. Here, in a similar way to the treatment of conformal expanding repellers, the main focus is on Bowen's formula for the Hausdorff dimension

https://doi.org/10.1515/9783110702699-201

of the limit set and multifractal analysis. A rather short examination of Lasota–Yorke maps of an interval is also included in this second volume.

The third volume is entirely devoted to the study of the dynamics, ergodic theory, thermodynamic formalism, and fractal geometry of rational functions of the Riemann sphere. We present a fairly complete account of classical as well as more advanced topological theory of Fatou and Julia sets. Nevertheless, primary emphasis is placed on measurable dynamics generated by rational functions and fractal geometry of their Julia sets. These include the thermodynamic formalism of Hölder continuous potentials with pressure gaps, the theory of Sullivan's conformal measures, invariant measures and their dimensions, entropy, and Lyapunov exponents. We further examine in detail the classes of expanding, subexpanding, and parabolic rational functions. We also provide, with proofs, several of the fundamental tools from complex analysis that are used in complex dynamics. These comprise Montel's Theorem, Koebe's Distortion Theorems and Riemann–Hurwitz formulas, with their ramifications.

In virtually each chapter of this book, we describe a large number of concrete selected examples illustrating the theory and serving as examples in other chapters. Also, each chapter of the book is supplied with a number of exercises. These vary in difficulty, from very easy ones asking to verify fairly straightforward logical steps to more advanced ones enhancing largely the theory developed in the chapter.

This book originated from the graduate lectures Mariusz Urbański delivered at the University of North Texas in the years 2005–2010 and that Sara Munday took notes of. With the involvement of Mario Roy, the book evolved and grew over many years. The last 2 years (2020 and 2021) of its writing were most dramatic and challenging because of the COVID-19 pandemic. Our book borrows widely from many sources including the books [63, 81, 98]. We nevertheless tried to keep it as self-contained as possible, avoiding to refer the reader too often to specific results from special papers or books. Toward this end, an appendix comprising classical results, mostly from measure theory, functional analysis and complex analysis, is included. The book covers quite a many topics treated with various degrees of completeness, none of which are fully exhausted because of their sheer largeness and their continuous dynamical growth.

List of Figures

https://doi.org/10.1515/9783110702699-202

Introduction to Volume 2

In the first volume, we developed the general theory of thermodynamic formalism, i. e., for general topological dynamical systems (i. e., continuous self-maps of a compact metrizable space) subject to general potentials (i. e., continuous potentials). In this second volume, we carry on with the study of thermodynamic formalism and its dynamical and geometric applications, but this time for specific classes of systems subject to specific classes of potentials.

We now describe in more detail the content of each chapter in this second volume, including their mutual dependence and interrelations.

Chapter 13 – Gibbs states and transfer operators for open, distance expanding systems

In this chapter, we introduce the fundamental concept of Gibbs states. Originating from statistical mechanics, Gibbs states play a prominent role in dynamical systems theory as (1) they describe, from a measure-theoretic viewpoint, the long-term behavior of typical orbits; (2) they are part of the core of thermodynamic formalism; and (3) they are indispensable for studying the fractal geometry of many dynamically-defined invariant sets. We also introduce the transfer operator (sometimes called Perron–Frobenius operator), one of the most powerful tools in thermodynamic formalism. This operator is designed to construct Gibbs states. All of this is done for open, distance expanding maps (cf. Chapter 4) under continuous potentials.

The first five sections are devoted to preparatory material. Section 13.1 concerns potentials. Section 13.2 introduces Gibbs states, specifies their basic properties and reveals that invariant Gibbs measures are equilibrium measures (cf. Chapter 12). Section 13.3 discusses Jacobians and changes of variables in purely measure-theoretic terms. Section 13.4 outlines a general way of constructing invariant measures from quasi-invariant ones. Section 13.5 presents general transfer operators and their fundamental properties.

The existence of Gibbs states generally requires two additional assumptions: (1) that the map be topologically transitive; and (2) that the potential be Hölder continuous. Under these conditions, we establish in Section 13.6 the existence of an ergodic, though not necessarily invariant, Gibbs state. Under these same conditions, we show in Section 13.7 that every system admits exactly one invariant Gibbs state. This measure is also ergodic and is the unique equilibrium measure. In Section 13.8, we study more advanced properties of transfer operators, especially the behavior of the iterates of these operators. Among others, we show that the transfer operator is quasi-compact, i. e., it exhibits a spectral gap. We do this by proving the Ionescu-Tulcea and Marinescu inequality and using the Ionescu-Tulcea and Marinescu theorem (see Appendix B). We further establish the continuity (1) of invariant Gibbs

https://doi.org/10.1515/9783110702699-203

states; (2) of eigenmeasures of dual operators; and (3) of the Radon–Nikodym derivative of the former with respect to the latter; in their dependence on the underlying potential.

Section 13.9 demonstrates that measure-theoretic dynamical systems generated by invariant Gibbs states truly deserve to be called random or chaotic, as they obey stochastic laws such as the exponential decay of correlations, the central limit theorem, the law of the iterated logarithm, K-mixing and the Bernoulli property. They also have asymptotic variances.

Finally, in Section 13.10 we show that if a potential depends real analytically on a parameter, then the corresponding topological pressure also depends real analytically on that parameter. This is proved by complex-analytic methods and by means of the perturbation theory for linear operators applied in our context to transfer operators. This topic will have notable sequels in Chapters 16 and 20.

Chapter 14 – Lasota–Yorke maps

Roughly speaking, Lasota–Yorke maps are piecewise monotone, piecewise differentiable expanding maps of a compact interval. They have finitely many pieces and the reciprocal of the absolute value of their derivative is of bounded variation. After defining them in Section 14.1, we describe the form of their Perron–Frobenius operator in Section 14.2 and use this operator in Section 14.3 to establish the existence of an invariant measure which is absolutely continuous with respect to the Lebesgue measure. In Section 14.4, we demonstrate the exponential decay of correlations with respect to that invariant measure under two additional assumptions on the spectrum of the transfer operator.

Chapter 15 – Fractal measures and dimensions

In this "nondynamical" chapter, we discuss different ways of evaluating the size of sets, especially fractal sets. Classical geometrical objects such as lines, squares or cubes, have integer Euclidean and topological dimensions. In contrast, because of their irregularities at infinitesimal scale, fractal sets cannot be accurately described and distinguished from one another by their Euclidean or topological dimensions. In order to evaluate their size, we need geometric measures (Hausdorff and packing) and the dimensions they induce, to examine the local and global structures of these sets at arbitrarily small scales. We also introduce the box-counting dimensions. In Sections 15.2 and 15.3, we establish the fundamental properties of these measures and dimensions. In Section 15.5, we consider volume lemmas and their geometric consequences. Their main purpose is to tell us when a Hausdorff measure or a packing measure is positive, finite, zero or infinite. We refer to them as Frostman converse theorems. At the end of that section, we formulate Frostman's (direct) lemma and compare it with the Frostman converse theorems. The advantage of the latter is that they provide tools to calculate, or at least estimate, both Hausdorff and packing measures and dimensions. Finally, in Section 15.6, we deal with dimensions of measures.

Chapter 16 – Conformal expanding repellers

In this chapter, we return to the class of expanding repellers that were introduced in Chapter 4. As these maps are open and distance expanding, the theory of Gibbs states developed in Chapter 13 applies. However, this class is too large to achieve many interesting results about the fractal structure of the limit set the repellers act on. So in this chapter, we restrict our attention to the subclass of conformal expanding repellers.

In Section 16.1, we briefly review the notion of conformality. In Section 16.2, we define the class of conformal expanding repellers, provide some examples, and derive some consequences of results presented in Chapters 4 and 13, in particular bounded distortion properties. Beginning with Section 16.3, we engage with the fractal geometry of the repeller's limit set. In Section 16.3, we prove the celebrated Bowen's formula, according to which the unique zero of the pressure function for a particular geometric potential is equal to the Hausdorff dimension of the limit set of the repeller. Bowen's formula is a generalization of Hutchinson's formula, a well-known result in fractal geometry. In Section 16.4, we examine the behavior of the Hausdorff dimension of the limit set of complex repellers under perturbation. This technically-involved study relies on transfer operators and on powerful complex- and functional-analytic methods, including the perturbation theory of bounded linear operators. It necessitates the use of quasiconformal mappings, holomorphic motions and the celebrated λ-lemma, whose complete proof is provided in Theorem 16.4.4. Ultimately, we show in Theorem 16.4.11 that the Hausdorff dimension of the repellers' limit set varies real analytically under analytic perturbations of the repellers. In Section 16.5, we develop a formula expressing the Hausdorff dimension of an ergodic invariant Borel probability measure as the ratio of its measure-theoretic entropy and its Lyapunov exponent. For historical reasons, this formula is frequently referred to as a volume lemma. Finally, in Section 16.6, we perform a multifractal analysis of the Hausdorff dimension of the level sets of Gibbs states for any Hölder continuous potential. This is a fertile field of applications of the theory of conformal expanding repellers and more largely of this book. In addition to its intrinsic interest and importance, the multifractal analysis conducted in this chapter bears witness to the strength and sophistication of the tools and methods developed in this book.

Chapters 17–18 – Countable state thermodynamic formalism

In these two chapters, we present thermodynamic formalism for countable-alphabet subshifts of finite type E_A^∞ generated by transition matrices $A : E\times E \to \{0,1\}$ on countable alphabets E. If the alphabet E is finite, then the symbolic space E_A^∞ is compact and all the knowledge acquired in Chapters 11–13 about pressure, variational principle, equilibrium states, Gibbs states and transfer operators, applies to the shift map $\sigma : E_A^\infty \to E_A^\infty$. If the alphabet E is infinite and the incidence matrix A is finitely irreducible, then the phase space E_A^∞ is not compact and essentially none of the results proved in the compact case holds directly.

The primary goal of Chapter 17 is to introduce the appropriate concept of topological pressure and to prove the existence and uniqueness of equilibrium states and invariant Gibbs states for potentials that are summable and Hölder continuous on cylinders. The definitions of pressure, equilibrium state and Gibbs state reduce to the usual ones in the case of a finite alphabet. In Section 17.1, we give the basic definitions of finite irreducibility and finite primitivity. Unless otherwise stated, the results mentioned in this paragraph assume that the subshift is finitely irreducible. In Section 17.2, we define the pressure. In order to speak of equilibrium states, one needs a variational principle for the pressure and we prove in Section 17.3 three such principles. In Section 17.4, we study Gibbs states and prove the existence of a unique invariant Gibbs state. This state is ergodic, and even totally ergodic when the subshift is finitely primitive. In Section 17.5, we identify a simple condition under which the unique invariant Gibbs state is an equilibrium state. Moreover, we show that equilibrium states are unique. Consequently, we deduce that invariant Gibbs states and equilibrium states exist, are unique, and coincide. In Section 17.6, we examine the transfer operator and we demonstrate that its dual admits at most one eigenmeasure, which turns out to be a (nonnecessarily invariant) Gibbs state. In Section 17.7, the existence of such an eigenmeasure is established. So the dual of the transfer operator has a unique eigenmeasure and its eigenvalue is the exponential of the pressure. The aforementioned facts are summarized in Corollary 17.7.5. Finally, in Section 17.8 we confirm the presence of a pressure gap and, consequently, the positivity of the measure-theoretic entropy of the subshift with respect to the unique equilibrium state.

Chapter 18 treats of refined properties of σ-invariant Gibbs states μ_f for potentials $f : E_A^\infty \to \mathbb{R}$ that are summable and Hölder continuous on cylinders. We show that the measure-theoretic dynamical systems $(\sigma : E_A^\infty \to E_A^\infty, \mu_f)$ deserve to be called random or chaotic by proving in Section 18.3 that they obey classical stochastic laws. This is achieved via a refined study of transfer operators acting on the Banach space of Hölder continuous functions and especially by establishing their spectral properties, such as quasicompactness (spectral gap) and simplicity of the leading eigenvalue $e^{\mathrm{P}(f)}$. To set the table, we prove in Section 18.1 the famous Ionescu-Tulcea and Marinescu inequality. In Section 18.2, we establish the continuity (1) of the pressure function $f \mapsto \mathrm{P}(f)$; (2) of the unique σ-invariant Gibbs states $f \mapsto \mu_f$; (3) of the unique eigenmeasures of the dual operators $f \mapsto m_f$; and (4) of the Radon–Nikodym derivative of μ_f with respect to m_f, i. e. the continuity of $f \mapsto d\mu_f/dm_f$. Finally, in Section 18.4 we characterize, primarily by means of cohomological equations, those potentials that have the same invariant Gibbs state.

Chapter 19 – Conformal graph directed Markov systems (CGDMSs)
Chapter 20 – Real analyticity of topological pressure and Hausdorff dimension for CGDMSs
Chapter 21 – Multifractal analysis for CGDMSs

In these three chapters, we apply the thermodynamic formalism developed in Chapters 17–18 to CGDMSs.

In Chapter 19, we describe a powerful method for constructing and studying the geometric and dynamical properties of fractal sets, including the middle-third Cantor set, the Sierpiński triangle, the Sierpiński carpet, the Koch curve, various kinds of Menger sets and the limit set of real and complex continued fractions. This method arises from the theory of countable-alphabet conformal graph directed Markov systems, which are a generalization of iterated function systems (IFSs). These systems are built from a graph comprising vertices and directed edges and from an edge-transition matrix. In Section 19.1, we start with the introduction of basic notions and notation for graph directed Markov systems (GDMSs): in particular, the underlying symbolic space and the shift map on it, the limit set, and the projection (coding map) from the symbolic space onto the limit set. Section 19.2 consists of properties of conformal maps in Euclidean spaces that will be crucial in deriving in Section 19.3 the core properties of conformal IFSs (CIFSs) and their generalization, conformal GDMSs (CGDMSs). In Section 19.4, we define an appropriate version of the topological pressure, the finiteness parameter, and the Bowen parameter. Based on the behavior of the pressure function, we then classify in Section 19.5 CGDMSs as irregular or regular, and in this latter case a finer splitting into strongly regular and critically regular systems is described. In Section 19.6, a variation of Bowen's formula is derived. Among its many consequences, we get an almost cost-free, effective and non-trivial lower estimate for the Hausdorff dimension of the limit sets of finitely irreducible CGDMSs. We further deduce from it the classical version of Hutchinson's formula. A particularly important tool to study the fractal geometry of limit sets of CGDMSs is the concept of conformal measure. We introduce and study these measures at length in Section 19.8, and we give a formula, called volume lemma, for the Hausdorff dimension of projections of ergodic invariant measures from the underlying symbolic space. This formula is expressed in the familiar form of the ratio of the measure-theoretic entropy and the Lyapunov exponent (cf. Section 16.5). Beforehand, though, we examine in Section 19.7 different separation and geometric conditions: the Strong Open Set Condition (SOSC), the Boundary Separation Condition (BSC), the Strong Separation Condition (SSC) and the Cone Condition (CC). Any one of those, when satisfied, has a profound impact on conformal measures, invariant measures, and the whole dynamics and geometry of CGDMSs. Finally, we provide a variety of examples in Section 19.9.

The first three sections of Chapter 20 are related to our considerations in Section 13.10. They are written in the framework of Chapters 17 and 18. Under various hypotheses, we show that the pressure function $P(\varphi + g)$ depends real-analytically on g (in an appropriately defined way). The potential φ is assumed to be strongly summable and Hölder continuous on cylinders. The methods of proof are of complex- and functional-analytic character. Our strategy consists in complexifying potentials first, and thus Perron–Frobenius operators, too. Then we prove that these complexified operators depend holomorphically on the perturbation g. This is where infinite-dimensional complex analysis (on Banach spaces) is used. Next, enters functional analysis. Indeed, we know that the exponential of the pressure function is a simple isolated eigenvalue of the transfer operator. So the Kato–Rellich perturbation theorem yields holomorphic dependence of appropriate eigenvalues of the complexified transfer operators. Restricting these operators to the real realm gives real-analyticity of the exponential of the pressure function, and hence of the pressure itself. We further identify the pressure's derivative. In Section 20.4, we consider families of CGDMSs in the complex plane $\mathbb{C}$. We assume that these families depend in an appropriate way on a complex parameter λ and we show the real-analytic dependence of the Bowen parameter on λ. We deduce from this the real-analyticity of the Hausdorff dimension of the limit set. Holomorphic motions are once again encountered in this section (cf. Section 16.4).

In Chapter 21, we perform a multifractal analysis of the Hausdorff dimension of the level sets of Gibbs states for any Hölder family of functions of the form $g_e + u \log |\varphi_e'|$, where $\{g_e\}_{e\in E}$ is a bounded Hölder family of functions and $\Phi = \{\varphi_e\}_{e\in E}$ is a CGDMS. In [46, 81], the multifractal analysis used, at each point of the limit set J, the natural filtration generated by the initial cylinders associated to the word that encodes the given point. Aiming to give the multifractal analysis a transparent geometrical meaning, we derive in this chapter an analysis using as the filtration a base of balls centred at the point. We will conduct the analysis of the conformal measure (or, equivalently, of the invariant measure) associated to a family of weights imposed upon a (finite or infinite) CGDMS. We also conduct this analysis over the full set of parameters on which it can be expected to hold. However, we restrict our attention to cofinitely regular, finitely irreducible, conformal-like CGDMSs that satisfy the SOSC and to a large and dynamically significant subset J_0 of the limit set J, as opposed to the entire limit set, though the sets J_0 and J coincide under a mild boundary separation condition. An additional geometric flavor of this analysis results from the fact that we concentrate on a geometrically meaningful family of Hölder weights. The results apply to large classes of CGDMSs, one- and multi-dimensional alike, including real and complex continued fractions.

In Section 21.1, we describe the Hölder families of weights F and $F_{q,t}$ we shall work with and we study the properties of the pressure function $\mathrm{P}(q,t)$ and of the temperature function $T(q)$ they determine. (The general properties of Hölder families of functions and the amalgamated function they induce are studied in Section 19.8.) In Section 21.2, we carry out the multifractal analysis of the conformal measure m_F (or, equivalently, of the invariant measure μ_F) associated to the family of weights F. This analysis is done over a set $J_0 \subseteq J$ of full μ_F-measure and conducted by means of balls. In particular, we show that the Legendre transform of the temperature function $T(q)$

describes the local structure of the set J_0. This is accomplished via the introduction of auxiliary measures that capture subsets of J_0 where the pointwise dimension is constant. Part of the analysis is restricted to a subclass called almost-CIFSs. In Section 21.3, we derive the multifractal analysis of m_F under additional conditions on the almost-CIFSs. Finally, in Section 21.4 we conduct a similar analysis on smaller subsets of points of the limit set J (that have a symbolic coding with at least one infinitely-seen letter from a finite subalphabet) but for all CGDMSs that are cofinitely regular, finitely irreducible, conformal-like and satisfy the SOSC.

Contents

Volume 1

https://doi.org/10.1515/9783110702699-204

Volume 2

Volume 3

13 Gibbs states and transfer operators for open, distance expanding systems

In the first volume, we developed the general theory of thermodynamic formalism, i. e., for general topological dynamical systems (i. e., continuous self-maps of a compact metrizable space) subject to general potentials (i. e., continuous potentials). In this second volume, we carry on with the study of thermodynamic formalism and its dynamical and geometric applications, but this time for specific classes of systems subject to specific classes of potentials.

In this chapter, we introduce the fundamental concept of Gibbs states. Motivated by statistical mechanics, this concept was introduced to dynamical systems theory (primarily to subshifts of finite type with finite alphabets) by David Ruelle, Rufus Bowen and Yakov Sinai, in their foundational works [11, 111–113] and [117], respectively. Gibbs states play a prominent role in dynamical systems theory as (1) they describe from a measure-theoretic viewpoint the long-term unstable behavior of typical orbits; (2) they are part of the core of thermodynamic formalism; and (3) thanks to the seminal paper [12], they are indispensable for studying the fractal geometry of many dynamically defined invariant sets. We also introduce the transfer operator (sometimes called Perron–Frobenius operator), one of the most powerful tools in thermodynamic formalism. This operator is designed to construct Gibbs states. All of this is done for open, distance expanding maps (cf. Chapter 4) under continuous potentials. Our exposition stems and expands the account given in [98]. We also borrowed from [91].

First of all, we prove the existence and uniqueness of (invariant) Gibbs states, along with the existence and uniqueness of equilibrium states. We further show that equilibrium states and invariant Gibbs states coincide. We also establish the basic properties of Gibbs states as well as some refined ones. We also show that the transfer operator is quasi-compact (i. e., it exhibits a spectral gap). This operator, and especially its spectral properties, are essential for establishing fine and transparent stochastic properties of invariant Gibbs states, such as the central limit theorem, the law of the iterated logarithm and the exponential decay of correlations.

Moving on to more detailed content, (X, T) will stand for a topological dynamical system, i. e., a continuous self-map $T : X \to X$ of a compact metrizable space X. We further assume that T is an open, distance expanding map with respect to some metric d which is compatible with the topology on X. Per Definition 4.1.1, the map T is distance expanding if there exist two constants $\delta > 0$ and $\lambda > 1$ such that

$$d(x,y) < 2\delta \quad \Longrightarrow \quad d(T(x), T(y)) \geq \lambda\, d(x,y). \tag{13.1}$$

For such a system, we proved in Chapter 4 that there exists a constant $0 < \xi \leq \delta$ such that T and all its iterates have local inverse branches on all balls of radius ξ. More

https://doi.org/10.1515/9783110702699-013

precisely, for any $x \in X$ and $n \in \mathbb{N}$ the local inverse branch of T^n that maps $T^n(x)$ to x is defined to be (cf. (4.29), (4.30) and (4.31))

$$T_x^{-n} := T_x^{-1} \circ T_{T(x)}^{-1} \circ \cdots \circ T_{T^{n-1}(x)}^{-1} : B(T^n(x), \xi) \to B(x, \lambda^{-n}\xi). \tag{13.2}$$

Obviously, these inverse branches of T^n are contractions with ratio λ^{-n}.

Initially, the system $T : X \to X$ will be submitted to continuous potentials $\varphi : X \to \mathbb{R}$, though we will eventually restrict our attention to Hölder ones.

The first five sections are devoted to preparatory material. Section 13.1 concerns potentials. Section 13.2 introduces Gibbs states, specifies their basic properties and establishes that invariant Gibbs measures are equilibrium measures. Section 13.3 discusses Jacobians and changes of variables in purely measure-theoretic terms. Section 13.4 describes a general way of constructing invariant measures from quasi-invariant ones. Section 13.5 presents general transfer operators and their fundamental properties.

The existence of Gibbs states generally requires two additional assumptions:
(1) An assumption on the map T: we impose that T be topologically transitive; and
(2) An assumption on the potential φ: we demand that φ be Hölder continuous.

As we will observe, both assumptions arise naturally. Under these conditions, we establish in Section 13.6 the existence of an ergodic, though not necessarily invariant, Gibbs state.

Under these same conditions, we show in Section 13.7 that every system admits exactly one invariant Gibbs state. This measure is also ergodic and is the unique equilibrium measure. This section is quite long, with plenty of interesting results. In particular, we prove that Hölder continuous potentials entail a pressure gap, and consequently, the positivity of entropy. We also fully characterize, primarily by means of cohomological equations, those Hölder continuous potentials that have the same invariant Gibbs state.

In Section 13.8, we study more advanced properties of transfer operators, especially the behavior of the iterates of these operators. Among others, the transfer operator is quasi-compact, i. e., it exhibits a spectral gap. We achieve this by proving the Ionescu-Tulcea and Marinescu inequality and using the Ionescu–Tulcea and Marinescu theorem (see Appendix B). We also establish the continuity of invariant Gibbs states, of eigenmeasures of dual operators, and of the Radon–Nikodym derivative of the former with respect to the latter, in their dependence on the potential φ.

Section 13.9 demonstrates that dynamical systems generated by invariant Gibbs states truly deserve to be called random or chaotic. We treat, by means of spectral properties of transfer operators, stochastic laws such as the exponential decay of correlations, the central limit theorem, the law of the iterated logarithm, K-mixing, and the Bernoulli property. Another feature essential for randomness, which we also prove by means of spectral properties of transfer operators, is the existence of asymptotic variances of Hölder continuous functions. We establish two very useful formulas for them.

Finally, in Section 13.10, we prove that if a potential depends real analytically on a parameter (in particular, if it is of the form $t \mapsto t\varphi$) then the corresponding topological pressure also depends real analytically on that parameter. This is proved by complex analytic methods and by means of the perturbation theory for linear operators applied in our context to transfer operators. This topic will have notable sequels: in Chapter 16, where will be shown the real-analytic dependence of the Hausdorff dimension of limit sets of conformal expanding repellers, and in Chapter 20, where the real-analytic dependence of the topological pressure for countable-alphabet symbolic dynamical systems will be substantiated, which will lead to the real-analytic dependence of the Hausdorff dimension of limit sets of conformal graph directed Markov systems. Once it has been ascertained that the topological pressure is real analytic, it is natural to ask what its first and second derivatives are. We give a full answer to this question at the end of Section 13.10. Namely, the first derivative is the integral of the "perturbation" against the appropriate Gibbs state while the second derivative is the asymptotic variance of the "perturbation" with respect to that same state. To our knowledge, the first result on the real-analytic dependence of the topological pressure on a parameter can be found in [111] and [113]. A number of similar results ensued since then; we mention here very few of them, not covered in this book, such as [2, 84, 121, 127, 135, 137].

13.1 Hölder continuous potentials

Let (X, d) be a compact metric space. Recall that a function $\varphi : X \to \mathbb{R}$ is said to be *Hölder continuous with exponent* $\alpha > 0$ if there exists some constant $c \geq 0$ such that

$$|\varphi(x) - \varphi(y)| \leq c\big[d(x, y)\big]^{\alpha}, \quad \forall x, y \in X.$$

The least constant c with this property will be denoted by $v_\alpha(\varphi)$. To simplify notation, we will at times write $d^\alpha(x, y)$ for $[d(x, y)]^\alpha$. Clearly, every Hölder continuous function is continuous. Also, observe that a Hölder continuous function with exponent $\alpha = 1$ is simply a Lipschitz continuous function. (For more, see Exercise 13.11.1.)

For now, we will consider only real-valued functions. Denote by $C(X)$ the Banach space of all continuous functions defined on X. For a given $\alpha > 0$, the space of all Hölder continuous functions defined on X with exponent α will be denoted by $H_\alpha(X)$. For every $\varphi \in H_\alpha(X)$, set

$$\|\varphi\|_\alpha := v_\alpha(\varphi) + \|\varphi\|_\infty. \tag{13.3}$$

Then $(H_\alpha(X), \|\cdot\|_\alpha)$ is a Banach space (see Exercise 13.11.2).

The following result affirms that the nth ergodic sum of any Hölder potential φ along each of the inverse branches of the nth iterate of T is Hölder continuous with Hölder constant independent of the branch and of n.

Lemma 13.1.1 (Local bounded variation principle for ergodic sums). *If $T : X \to X$ is an open, distance expanding system and $\varphi \in H_\alpha(X)$, then for all $w \in X$, all $n \in \mathbb{N}$ and all $x, y \in B(T^n(w), \xi)$ we have*

$$|S_n\varphi(T_w^{-n}(x)) - S_n\varphi(T_w^{-n}(y))| \le C_\alpha(\varphi)\, d^\alpha(x,y), \tag{13.4}$$

where $C_\alpha(\varphi) := \frac{v_\alpha(\varphi)}{\lambda^\alpha - 1}$ while ξ and λ come from (13.1)–(13.2). *Thus*

$$\exp(S_n\varphi(T_w^{-n}(x))) \le D_\alpha(\varphi) \exp(S_n\varphi(T_w^{-n}(y))), \tag{13.5}$$

where $D_\alpha(\varphi) := \exp((2\xi)^\alpha C_\alpha(\varphi))$.

Furthermore, if $d(x,y) < \xi$ then for all $n \in \mathbb{N}$,

$$\sum_{\overline{x} \in T^{-n}(x)} e^{S_n\varphi(\overline{x})} \le D_\alpha(\varphi) \sum_{\overline{y} \in T^{-n}(y)} e^{S_n\varphi(\overline{y})}. \tag{13.6}$$

Proof. Recall that $S_n\varphi := \sum_{j=0}^{n-1} \varphi \circ T^j$. Then

$$\begin{aligned}
|S_n\varphi(T_w^{-n}(x)) - S_n\varphi(T_w^{-n}(y))| &\le \sum_{j=0}^{n-1} |\varphi \circ T^j \circ T_w^{-n}(x) - \varphi \circ T^j \circ T_w^{-n}(y)| \\
&= \sum_{j=0}^{n-1} |\varphi \circ T_{T^j(w)}^{-(n-j)}(x) - \varphi \circ T_{T^j(w)}^{-(n-j)}(y)| \\
&\le \sum_{j=0}^{n-1} v_\alpha(\varphi)\, d^\alpha(T_{T^j(w)}^{-(n-j)}(x), T_{T^j(w)}^{-(n-j)}(y)) \\
&\le v_\alpha(\varphi) \sum_{j=0}^{n-1} \lambda^{-\alpha(n-j)}\, d^\alpha(x,y) \\
&\le v_\alpha(\varphi)\, d^\alpha(x,y) \sum_{k=1}^{\infty} \lambda^{-\alpha k} \\
&= \frac{1}{\lambda^\alpha - 1} v_\alpha(\varphi)\, d^\alpha(x,y).
\end{aligned}$$

The second part of the statement follows from the first one by summing over all branches of T^n. Indeed, by Lemma 4.2.8 we know that $T_u^{-n}(B(z,\xi)) \cap T_v^{-n}(B(z,\xi)) = \emptyset$ for all $z \in X$, $n \in \mathbb{N}$ and $u, v \in T^{-n}(z)$ with $u \neq v$. Moreover, by Lemma 4.2.9 we know that $T^{-n}(A) = \bigcup_{u \in T^{-n}(z)} T_u^{-n}(A)$ for all $z \in X$, all $n \in \mathbb{N}$ and all $A \subseteq B(z,\xi)$. As $d(x,y) < \xi$, there is then a bijection between the sets $T^{-n}(x)$ and $T^{-n}(y)$ defined by

$$\begin{aligned} b : T^{-n}(x) &\longrightarrow T^{-n}(y) \\ \overline{x} &\longmapsto \overline{y} = b(\overline{x}) := T_{\overline{x}}^{-n}(y). \end{aligned}$$

Therefore, using the exponential form of the first part on each branch, we get

$$\begin{aligned} \sum_{\overline{x}\in T^{-n}(x)} e^{S_n\varphi(\overline{x})} &= \sum_{\overline{x}\in T^{-n}(x)} e^{S_n\varphi(T_{\overline{x}}^{-n}(x))} \\ &\leq \sum_{\overline{x}\in T^{-n}(x)} e^{S_n\varphi(T_{\overline{x}}^{-n}(y))} \cdot \exp(C_\alpha(\varphi)\, d^\alpha(x,y)) \\ &\leq D_\alpha(\varphi) \sum_{\overline{y}\in T^{-n}(y)} e^{S_n\varphi(\overline{y})}. \end{aligned}$$

□

Under further assumptions, this local version has a global analogue.

Corollary 13.1.2 (Global bounded ratio of exponentials of ergodic sums). *If, additionally, $T : X \to X$ is topologically transitive or X is connected, then there exists a constant $\overline{D}_\alpha(\varphi) \geq 1$ such that*

$$\sum_{\overline{x}\in T^{-n}(x)} e^{S_n\varphi(\overline{x})} \leq \overline{D}_\alpha(\varphi) \sum_{\overline{y}\in T^{-n}(y)} e^{S_n\varphi(\overline{y})}, \quad \forall x,y \in X,\ \forall n \in \mathbb{N}.$$

Proof. When X is connected, the result follows from (13.6) in Lemma 13.1.1 by constructing a finite chain of balls of radius $\xi/2$ covering X.

If T is topologically transitive, then due to the compactness of X there exists $M \in \mathbb{N}$ such that for all $x, y \in X$ there is $0 \leq m \leq M$ such that $T^m(B(x,\xi)) \cap B(y,\xi) \neq \emptyset$. Indeed, select a point $z \in X$ such that $\omega(z) = X$. Then there are ξ-dense suborbits $T^k(z),\dots,T^{k'}(z)$ and $T^l(z),\dots,T^{l'}(z)$ with $l > k'$. Set $M = l' - k$. There then exists $k \leq k'' \leq k'$ and $l \leq l'' \leq l'$ such that $T^{k''}(z) \in B(x,\xi)$ while $T^{l''}(z) \in B(y,\xi)$. With $m = l'' - k''$, we get $T^m(T^{k''}(z)) = T^{l''}(z) \in T^m(B(x,\xi)) \cap B(y,\xi)$. Thus, there is a point $y' \in B(x,\xi) \cap T^{-m}(B(y,\xi))$. Using (13.6) in Lemma 13.1.1 twice (at the beginning with the pair (x,y') and at the end with the pair (y',y)), we obtain

$$\begin{aligned} \sum_{\overline{x}\in T^{-n}(x)} e^{S_n\varphi(\overline{x})} &\leq D_\alpha(\varphi) \sum_{\overline{y'}\in T^{-n}(y')} e^{S_n\varphi(\overline{y'})} \\ &= D_\alpha(\varphi) \sum_{\overline{y'}\in T^{-n}(y')} e^{S_{m+n}\varphi(\overline{y'})} e^{-S_m\varphi(T^n(\overline{y'}))} \\ &\leq D_\alpha(\varphi) e^{-m\inf\varphi} \sum_{\overline{y'}\in T^{-n}(y')} e^{S_{m+n}\varphi(\overline{y'})} \\ &\leq D_\alpha(\varphi) e^{-m\inf\varphi} \sum_{\overline{y'}\in T^{-(m+n)}(T^m(y'))} e^{S_m\varphi(\overline{y'})} e^{S_n\varphi(T^m(\overline{y'}))} \end{aligned}$$

$$\begin{aligned}
&\leq D_\alpha(\varphi)e^{m(\sup\varphi-\inf\varphi)} \sum_{\overline{y'}\in T^{-(m+n)}(T^m(y'))} e^{S_n\varphi(T^m(\overline{y'}))}\\
&\leq D_\alpha(\varphi)e^{2m\|\varphi\|_\infty}\cdot(\#T^{-1})^m \sum_{\overline{\overline{y'}}\in T^{-n}(T^m(y'))} e^{S_n\varphi(\overline{\overline{y'}})}\\
&\leq (\#T^{-1})^M (D_\alpha(\varphi))^2 e^{2M\|\varphi\|_\infty} \sum_{\overline{y}\in T^{-n}(y)} e^{S_n\varphi(\overline{y})},
\end{aligned}$$

where $\#T^{-1} := \max_{z\in X}\#T^{-1}(z) < \infty$ by the expanding property of T and the compactness of X. Finally, note that M depends only on z, and thus is independent of x, y and n. □

The pressure of an open, distance expanding system can be expressed as a pointwise pressure.

Proposition 13.1.3. *Let $T : X \to X$ be an open, distance expanding system. For any potential $\varphi \in C(X)$, we have*

$$\mathrm{P}(T,\varphi) = \max_{x\in X}\mathrm{P}_x(T,\varphi) \quad \textit{and} \quad \mathrm{P}(T,\varphi) = \mathrm{P}_{x_0}(T,\varphi) \textit{ for some } x_0 \in X,$$

where

$$\mathrm{P}_x(T,\varphi) := \limsup_{n\to\infty}\frac{1}{n}\log\sum_{\overline{x}\in T^{-n}(x)} e^{S_n\varphi(\overline{x})}.$$

If φ is Hölder continuous and, moreover, T is topologically transitive or X is connected, then

$$\mathrm{P}(T,\varphi) = \lim_{n\to\infty}\frac{1}{n}\log\sum_{\overline{x}\in T^{-n}(x)} e^{S_n\varphi(\overline{x})} = \mathrm{P}_x(T,\varphi), \quad \forall x \in X.$$

Proof. Let $x \in X$ and $n \in \mathbb{N}$. Given $\zeta > 0$, recall that a set $E \subseteq X$ is ζ-separated if $d(e_1, e_2) \geq \zeta$ for all $e_1 \neq e_2 \in E$; it is (n,ζ)-separated if $\max_{0\leq j<n} d(T^j(e_1), T^j(e_2)) \geq \zeta$ for all $e_1 \neq e_2 \in E$. By the expanding property of T (cf. (13.1)), the set $T^{-1}(x)$ is δ-separated. Thus the set $T^{-n}(x)$ is (n,δ)-separated, and hence is contained in a maximal (n,ε)-separated set for every $0 < \varepsilon \leq \delta$. By Theorem 11.2.1,

$$\mathrm{P}(T,\varphi) \geq \limsup_{n\to\infty}\frac{1}{n}\log\sum_{\overline{x}\in T^{-n}(x)} e^{S_n\varphi(\overline{x})} = \mathrm{P}_x(T,\varphi). \tag{13.7}$$

On the other hand, let $0 < \varepsilon < \xi/4$ with ξ from (13.2), and pick a maximal ε-separated set $E(\varepsilon) \subseteq X$. Fix $n \in \mathbb{N}$. We claim that $T^{-n}(E(\varepsilon))$ is a maximal $(n+1,\varepsilon)$-separated set. Indeed, let $\overline{x}, \widetilde{x} \in T^{-n}(E(\varepsilon))$ and $\overline{x} \neq \widetilde{x}$. There are two possibilities. If $T^n(\overline{x}) = T^n(\widetilde{x})$, the expanding property of T imposes that $\overline{x}$ and $\widetilde{x}$ be (n,δ)-separated and thereby (n,ε)-separated. On the other hand, if $T^n(\overline{x}) \neq T^n(\widetilde{x})$, then $T^n(\overline{x})$ and $T^n(\widetilde{x})$

are ε-separated since $E(\varepsilon)$ is ε-separated, and hence $\overline{x}$ and $\widetilde{x}$ are $(n+1,\varepsilon)$-separated. So $T^{-n}(E(\varepsilon))$ is $(n+1,\varepsilon)$-separated. Now, let $x \in X$. As $E(\varepsilon)$ is a maximal ε-separated set, there exists $y \in E(\varepsilon)$ such that $d(T^n(x), y) < \varepsilon < \xi$. Since the inverse branches of T on any open ball of radius ξ are contractions with ratio λ^{-1}, we deduce that

$$d(T^j(x), T^j(T_x^{-n}(y))) = d\Big(T_{T^j(x)}^{-(n-j)}(T^n(x)), T_{T^j(x)}^{-(n-j)}(y)\Big) \le \lambda^{-(n-j)}\, d(T^n(x), y) < \varepsilon$$

for all $0 \le j \le n$. As $T_x^{-n}(y) \in T^{-n}(E(\varepsilon))$ and as $x \in X$ was chosen arbitrarily, we conclude that $T^{-n}(E(\varepsilon))$ is a maximal $(n+1,\varepsilon)$-separated set. By Theorem 11.2.8, we then know that

$$\mathrm{P}(T,\varphi) = \lim_{n\to\infty} \frac{1}{n} \log \sum_{\overline{x}\in T^{-n}(E(\varepsilon))} e^{S_n\varphi(\overline{x})} = \lim_{n\to\infty} \frac{1}{n} \log \sum_{x\in E(\varepsilon)} \sum_{\overline{x}\in T^{-n}(x)} e^{S_n\varphi(\overline{x})}. \tag{13.8}$$

Since $\#E(\varepsilon) < \infty$, for each $n \in \mathbb{N}$ there exists $x_n \in E(\varepsilon)$ that maximizes the sum $\sum_{\overline{x}\in T^{-n}(x)} \exp(S_n\varphi(\overline{x}))$ over all $x \in E(\varepsilon)$. Then there is $x_0 \in E(\varepsilon)$ and a strictly increasing sequence $(n_k)_{k=1}^{\infty}$ such that $x_{n_k} = x_0$ for all $k \in \mathbb{N}$ and such that

$$\liminf_{n\to\infty} \frac{1}{n} \log \sum_{\overline{x_n}\in T^{-n}(x_n)} e^{S_n\varphi(\overline{x_n})} = \lim_{k\to\infty} \frac{1}{n_k} \log \sum_{\overline{x_0}\in T^{-n_k}(x_0)} e^{S_{n_k}\varphi(\overline{x_0})}. \tag{13.9}$$

It follows from (13.8) and (13.9) that

$$\begin{aligned}
\mathrm{P}(T,\varphi) &\le \liminf_{n\to\infty} \frac{1}{n} \log\Big[(\#E(\varepsilon)) \sum_{\overline{x_n}\in T^{-n}(x_n)} e^{S_n\varphi(\overline{x_n})}\Big] \\
&= \liminf_{n\to\infty} \frac{1}{n} \log \sum_{\overline{x_n}\in T^{-n}(x_n)} e^{S_n\varphi(\overline{x_n})} \\
&= \lim_{k\to\infty} \frac{1}{n_k} \log \sum_{\overline{x_0}\in T^{-n_k}(x_0)} e^{S_{n_k}\varphi(\overline{x_0})} \\
&\le \limsup_{n\to\infty} \frac{1}{n} \log \sum_{\overline{x_0}\in T^{-n}(x_0)} e^{S_n\varphi(\overline{x_0})} = \mathrm{P}_{x_0}(T,\varphi).
\end{aligned} \tag{13.10}$$

The first part of the proposition ensues from (13.7) and (13.10).

Now, assume additionally that $\varphi \in H_\alpha(X)$ and that T is topologically transitive or X is connected. Let $\varepsilon < 2\xi$, and for every $n \in \mathbb{N}$ let $E_n(\varepsilon)$ be an (n,ε)-separated set. Cover X by finitely many balls of radius $\varepsilon/2$, say $B(z_1, \varepsilon/2), \ldots, B(z_k, \varepsilon/2)$. Then

$$\sum_{z\in E_n(\varepsilon)} e^{S_n\varphi(z)} \le \sum_{j=1}^{k} \sum_{z\in E_n(\varepsilon)\cap T^{-n}(B(z_j,\varepsilon/2))} e^{S_n\varphi(z)}. \tag{13.11}$$

Fix momentarily $1 \le j \le k$. For every $z \in E_n(\varepsilon) \cap T^{-n}(B(z_j, \varepsilon/2))$, define $\overline{z_j}(z) \in T^{-n}(z_j)$ so that $z \in T_{\overline{z_j}(z)}^{-n}(B(z_j, \varepsilon/2))$. The map $z \mapsto \overline{z_j}(z)$ is well and uniquely defined as, according to Lemma 4.2.8, the inverse branches of T^n have mutually disjoint images on open balls of radius ξ. Moreover, this map is injective since $E_n(\varepsilon)$ is an (n,ε)-separated set. By (13.11) and Corollary 13.1.2, we deduce that

$$\sum_{z\in E_n(\varepsilon)} e^{S_n\varphi(z)} \le \sum_{j=1}^{k} \overline{D}_\alpha(\varphi) \sum_{\overline{z_j}\in T^{-n}(z_j)} e^{S_n\varphi(\overline{z_j})} \le k(\overline{D}_\alpha(\varphi))^2 \sum_{\overline{x}\in T^{-n}(x)} e^{S_n\varphi(\overline{x})}.$$

Letting $n\to\infty$ implies $\varepsilon\to 0$, and using Theorem 11.2.1 we obtain that

$$\mathrm{P}(T,\varphi) \le \liminf_{n\to\infty} \frac{1}{n} \log \sum_{\overline{x}\in T^{-n}(x)} e^{S_n\varphi(\overline{x})}. \tag{13.12}$$

The conclusion follows from (13.7) and (13.12). □

13.2 Gibbs measures

13.2.1 Definition and properties

Let us now introduce the concept of Gibbs measures.

Definition 13.2.1. Let $T : X \to X$ be an open, distance expanding system. A Borel probability measure μ on X is called a *Gibbs measure* (or *Gibbs state*) for a potential $\varphi : X \to \mathbb{R}$ provided that there exist a number $P \in \mathbb{R}$ and a constant $C \ge 1$ such that for all $n \in \mathbb{N}$ and all $x \in X$ the following holds:

$$C^{-1} \le \frac{\mu(T_x^{-n}(B(T^n(x),\xi)))}{\exp(S_n\varphi(x) - nP)} \le C, \tag{13.13}$$

where ξ comes from (13.2). Note that P is uniquely defined for any given Gibbs state μ. The set of all Gibbs states for a potential φ will be denoted by $\mathcal{G}_\varphi$.

Remark 13.2.2. At first glance it looks as if $\mathcal{G}_\varphi$ might depend on ξ. Indeed, $\xi \le \delta$ by definition (cf. (4.29)). But if δ delimits a neighborhood of expansion for T, so does any $0 < \delta' \le \delta$. Then the ξ' associated with δ' satisfies $0 < \xi' \le \xi$, and thus $\mu(T_x^{-n}(B(T^n(x),\xi'))) \le \mu(T_x^{-n}(B(T^n(x),\xi)))$. The term $\exp(S_n\varphi(x))$ does not change, though, and therefore either P and/or C depend on ξ or $\mathcal{G}_\varphi$ depends on ξ. In Proposition 13.2.4, we will see that P is independent of ξ. In Proposition 13.6.16, we shall observe that when T is additionally transitive and φ is Hölder continuous, the constant C generally grows with ξ, hence ensuring that $\mathcal{G}_\varphi$ is independent of ξ. In these circumstances, an equivalent definition of a Gibbs measure is the following:

A Borel probability measure μ on X is called a Gibbs state for a potential $\varphi : X \to \mathbb{R}$ provided that there exist $\xi > 0$ and $P \in \mathbb{R}$ such that for every $0 < \xi' \le \xi$ there is a constant $C(\xi') \geq 1$ such that

$$C(\xi')^{-1} \leq \frac{\mu(T_x^{-n}(B(T^n(x), \xi')))}{\exp(S_n\varphi(x) - nP)} \leq C(\xi'), \quad \forall n \in \mathbb{N}, \forall x \in X.$$

Lemma 13.2.3. *Let $T : X \to X$ be an open, distance expanding system. Suppose that μ is a Gibbs state for a potential φ and that a Borel probability measure ν is equivalent to μ, i. e., $\mu \ll \nu$ and $\nu \ll \mu$. Suppose further that the Radon–Nikodym derivative of μ with respect to ν is essentially uniformly bounded away from 0 and ∞, i. e.,*

$$0 < \operatorname{ess\,inf}\Big(\frac{d\mu}{d\nu}\Big) \leq \operatorname{ess\,sup}\Big(\frac{d\mu}{d\nu}\Big) < \infty,$$

where

$$\operatorname{ess\,inf}\Big(\frac{d\mu}{d\nu}\Big) := \sup_{\nu(N)=0} \inf_{x\in X\setminus N} \frac{d\mu}{d\nu}(x) \quad \text{and} \quad \operatorname{ess\,sup}\Big(\frac{d\mu}{d\nu}\Big) := \inf_{\nu(N)=0} \sup_{x\in X\setminus N} \frac{d\mu}{d\nu}(x).$$

Then the measure ν is also a Gibbs state for the same potential φ with the same constant P.

Proof. Since $\frac{d\nu}{d\mu} = \big(\frac{d\mu}{d\nu}\big)^{-1}$ a. e., the Radon–Nikodym derivative of ν with respect to μ is also essentially uniformly bounded away from 0 and ∞. Therefore, there exists a constant $D \geq 1$ such that

$$D^{-1}\mu(B) \leq \nu(B) \leq D\mu(B), \quad \forall B \in \mathcal{B}(X),$$

where $\mathcal{B}(X)$ is the Borel σ-algebra on X. It is then easy to see that ν, like μ, is a Gibbs state with the same constant P: the constant C for μ can be replaced by CD for ν. □

The following proposition states that the converse of Lemma 13.2.3 is also true, and identifies the constant P as $\mathrm{P}(T, \varphi)$, the topological pressure of φ.

Proposition 13.2.4. *Let $T : X \to X$ be an open, distance expanding system. Suppose that μ and ν are both Gibbs states for a potential $\varphi \in C(X)$ and that (C, P) and (D, Q) are the couples of constants arising from Definition 13.2.1 for these respective measures. Then*

(a) *μ and ν are equivalent,*
(b) *$P = Q = \mathrm{P}(T, \varphi)$, and*
(c) *the mutual Radon–Nikodym derivatives of μ and ν are essentially uniformly bounded away from 0 and ∞.*

Proof. Let ξ come from (13.2). Let E be a finite ξ-spanning set for X, i. e., a finite set such that $X = \bigcup_{e\in E} B(e, \xi)$. Observe that for every $n \in \mathbb{N}$,

$$X = T^{-n}(X) = \bigcup_{e\in E} T^{-n}(B(e, \xi)) = \bigcup_{e\in E} \bigcup_{x\in T^{-n}(e)} T_x^{-n}(B(e, \xi)). \tag{13.14}$$

We first claim that for all compact sets $A \subseteq X$, for all $\gamma > 0$ and for all $n \geq N(A,\gamma)$ large enough,

$$\mu(A) \leq CD(\#E)\big(\nu(A) + \gamma\big)e^{n(Q-P)}. \tag{13.15}$$

To prove this claim, note that by the compactness of A and the outer regularity of μ and ν (see Definition A.1.23 and Theorem A.1.24), there exists $\varepsilon > 0$ such that $\mu(B(A,\varepsilon)) \leq \mu(A) + \gamma$ and $\nu(B(A,\varepsilon)) \leq \nu(A) + \gamma$. Fix $n \in \mathbb{N}$ so large that $2\lambda^{-n}\xi < \varepsilon$. For each $e \in E$, define the set $X_n(e)$ by

$$X_n(e) := \{x \in T^{-n}(e) \mid A \cap T_x^{-n}(B(e,\xi)) \neq \emptyset\}.$$

By (13.14) and (13.2) (or, alternatively, (4.31)), we have

$$A \subseteq \bigcup_{e\in E} \bigcup_{x\in X_n(e)} T_x^{-n}(B(e,\xi)) \subseteq B(A, 2\lambda^{-n}\xi) \subseteq B(A,\varepsilon). \tag{13.16}$$

For each $e \in E$ the family $\{T_x^{-n}(B(e,\xi))\}_{x\in T^{-n}(e)}$ consists of preimages under different inverse branches of T^n at e. According to Lemma 4.2.8, these preimages are mutually disjoint. Therefore,

$$\begin{aligned}
\mu(A) &\leq \mu\Big(\bigcup_{e\in E} \bigcup_{x\in X_n(e)} T_x^{-n}(B(e,\xi))\Big)\\
&\leq \sum_{e\in E} \sum_{x\in X_n(e)} \mu(T_x^{-n}(B(T^n(x),\xi)))\\
&\leq C \sum_{e\in E} \sum_{x\in X_n(e)} e^{S_n\varphi(x)-nP}\\
&= Ce^{n(Q-P)} \sum_{e\in E} \sum_{x\in X_n(e)} e^{S_n\varphi(x)-nQ}\\
&\leq CDe^{n(Q-P)} \sum_{e\in E} \sum_{x\in X_n(e)} \nu(T_x^{-n}(B(T^n(x),\xi)))\\
&= CDe^{n(Q-P)} \sum_{e\in E} \nu\Big(\bigcup_{x\in X_n(e)} T_x^{-n}(B(e,\xi))\Big)\\
&\leq CDe^{n(Q-P)} \sum_{e\in E} \nu(B(A,\varepsilon))\\
&\leq CDe^{n(Q-P)}(\#E)\big(\nu(A)+\gamma\big).
\end{aligned} \tag{13.17}$$

This proves the claim.

If $Q < P$, letting $A = X$ in the above inequality and letting $n \to \infty$ would imply that $\mu(X) \leq 0$. This means that $Q \geq P$. Reversing the roles of μ and ν and correspondingly those of P and Q gives $P \geq Q$. So $P = Q$. This is part of (b). Thus, (13.15) reduces to

$$\mu(A) \leq CD(\#E)(\nu(A) + \gamma).$$

Letting $\gamma \to 0$ results in

$$\mu(A) \leq CD(\#E) \cdot \nu(A).$$

By the inner regularity of the measures μ and ν, it follows that

$$\mu(B) = \sup_{\substack{A \subseteq B, \\ A \text{ compact}}} \mu(A) \leq CD(\#E) \cdot \sup_{\substack{A \subseteq B, \\ A \text{ compact}}} \nu(A) = CD(\#E) \cdot \nu(B)$$

for every Borel set $B \subseteq X$. Hence, $\mu \ll \nu$. Exchanging the roles of μ and ν yields

$$\nu(B) \leq CD(\#E) \cdot \mu(B)$$

for every Borel set $B \subseteq X$. Hence, $\nu \ll \mu$. So μ and ν are equivalent. This proves (a).

Moreover, since

$$\int_B \frac{d\mu}{d\nu}\, d\nu = \mu(B) \leq CD(\#E) \cdot \nu(B) = \int_B CD(\#E)\, d\nu, \quad \forall B \in \mathcal{B}(X),$$

we deduce that $\frac{d\mu}{d\nu} \leq CD(\#E)$ ν-a. e. on X. By a similar argument, $\frac{d\mu}{d\nu} \geq (CD(\#E))^{-1}$ ν-a. e. on X. Interchanging μ and ν yields the same lower and upper bounds on $\frac{d\nu}{d\mu}$ (alternatively, recall that $\frac{d\nu}{d\mu} = (\frac{d\mu}{d\nu})^{-1}$ a. e.). So $\frac{d\mu}{d\nu}$ and $\frac{d\nu}{d\mu}$ are essentially uniformly bounded away from 0 and ∞. This demonstrates (c).

We now establish that $P \leq \mathrm{P}(T, \varphi)$. Using (13.17) with $A = X$, we get

$$\begin{aligned} 1 = \mu(X) &\leq C \sum_{e \in E} \sum_{x \in T^{-n}(e)} e^{S_n\varphi(x) - nP} \\ &\leq C(\#E) \sum_{x \in T^{-n}(e_{\max})} e^{S_n\varphi(x) - nP} \\ &\leq C(\#E) \sum_{x \in E_n(\delta)} e^{S_n\varphi(x) - nP}, \end{aligned}$$

where $e_{\max} \in E$ is chosen so that the sum $\sum_{x \in T^{-n}(e)} \exp(S_n\varphi(x) - nP)$ is maximal among all $e \in E$ and where $E_n(\delta)$ is a maximal (n, δ)-separated set containing $T^{-n}(e_{\max})$. Taking logarithms, dividing by n and rearranging, we obtain

$$P \leq \frac{1}{n} \log(C(\#E)) + \frac{1}{n} \log \sum_{x \in E_n(\delta)} e^{S_n\varphi(x)}.$$

Upon letting $n \to \infty$ and $\delta \to 0$, we deduce from Theorem 11.2.1 that $P \leq \mathrm{P}(T, \varphi)$.

It remains to show that $P \geq \mathrm{P}(T,\varphi)$. In virtue of Proposition 13.1.3, let $x_0 \in X$ be such that $\mathrm{P}_{x_0}(T,\varphi) = \mathrm{P}(T,\varphi)$. It follows from the definition of a Gibbs state that

$$\begin{aligned}\sum_{x\in T^{-n}(x_0)} e^{S_n\varphi(x)-nP} &\leq C \sum_{x\in T^{-n}(x_0)} \mu(T_x^{-n}(B(x_0,\xi)))\\ &= C\mu\Big(\bigcup_{x\in T^{-n}(x_0)} T_x^{-n}(B(x_0,\xi))\Big)\\ &= C\mu(T^{-n}(B(x_0,\xi)))\\ &\leq C.\end{aligned}$$

Therefore,

$$e^{nP} \geq C^{-1} \sum_{x\in T^{-n}(x_0)} e^{S_n\varphi(x)},$$

and hence

$$P \geq -\frac{1}{n}\log C + \frac{1}{n}\log \sum_{x\in T^{-n}(x_0)} e^{S_n\varphi(x)}.$$

Letting $n \to \infty$, we conclude that $P \geq \mathrm{P}_{x_0}(T,\varphi) = \mathrm{P}(T,\varphi)$. □

The following is a generalization of Definition 8.2.39.

Definition 13.2.5. Let $\mathbb{R}^X$ be the set of all real-valued functions defined on X, and let $F \subseteq \mathbb{R}^X$. A function $\varphi \in \mathbb{R}^X$ is said to be *cohomologous* in F to a function $\psi \in \mathbb{R}^X$ provided that there exists a function $u \in F$ such that

$$\varphi - \psi = u \circ T - u.$$

Any such function u is called a *coboundary*.

Note that cohomology of functions in a subset F of $\mathbb{R}^X$ is an equivalence relation on $\mathbb{R}^X$ if F is a group with respect to pointwise addition.

We now show that continuous functions that are cohomologous, up to a constant, in the group of bounded functions have (1) pressures that differ by that constant, and (2) the same Gibbs states.

Lemma 13.2.6. *Let $T : X \to X$ be an open, distance expanding system. If two potentials $\varphi, \psi \in C(X)$ are such that $\varphi - \psi$ is cohomologous to a constant c in the additive group of bounded functions on X, then $\mathrm{P}(T,\varphi) = \mathrm{P}(T,\psi) + c$ and $\mathcal{G}_\varphi = \mathcal{G}_\psi$.*

Proof. Temporarily assume that φ is cohomologous to ψ in the additive group of bounded functions on X, i. e., assume that $c = 0$. Let u be a bounded coboundary such that $\varphi - \psi = u \circ T - u$. Then for every $n \in \mathbb{N}$,

$$S_n\varphi - S_n\psi = u \circ T^n - u.$$

Given any family $\{E_n(\varepsilon)\}_{n\in\mathbb{N},\varepsilon>0}$ of maximal (n,ε)-separated sets, by Theorem 11.2.1 we then have

$$\begin{aligned}
\mathrm{P}(T,\varphi) &= \lim_{\varepsilon\to 0}\limsup_{n\to\infty}\frac{1}{n}\log\sum_{x\in E_n(\varepsilon)} e^{S_n\varphi(x)} \\
&= \lim_{\varepsilon\to 0}\limsup_{n\to\infty}\frac{1}{n}\log\sum_{x\in E_n(\varepsilon)} e^{S_n\psi(x)}e^{u\circ T^n(x)-u(x)} \\
&\le \lim_{\varepsilon\to 0}\limsup_{n\to\infty}\frac{1}{n}\log\sum_{x\in E_n(\varepsilon)} e^{S_n\psi(x)}e^{2\|u\|_\infty} \\
&= \mathrm{P}(T,\psi).
\end{aligned}$$

By symmetry (and replacing u by $-u$), we deduce that $\mathrm{P}(T,\psi) = \mathrm{P}(T,\varphi)$.

Suppose now that $\varphi - \psi$ is cohomologous to a constant $c \in \mathbb{R}$ in the additive group of bounded functions on X, and let u be a bounded coboundary so that $\varphi - \psi - c = u \circ T - u$. This is equivalent to saying that φ is cohomologous to $\psi + c$ via u. By the first part and Proposition 11.3.1, we get

$$\mathrm{P}(T,\varphi) = \mathrm{P}(T,\psi + c) = \mathrm{P}(T,\psi) + c.$$

Moreover,

$$\begin{aligned}
S_n\varphi &= S_n(\psi + c + u\circ T - u) \\
&= S_n\psi + S_nc + S_nu\circ T - S_nu \\
&= S_n\psi + nc + u\circ T^n - u.
\end{aligned}$$

Hence,

$$S_n\varphi - n\mathrm{P}(T,\varphi) = S_n\psi - n\mathrm{P}(T,\psi) + u\circ T^n - u.$$

Let ξ come from (13.2). Let $\mu \in \mathcal{G}_\varphi$ and $C \ge 1$ be a Gibbs constant for μ, i. e. such that

$$C^{-1} \le \frac{\mu(T_x^{-n}(B(T^n(x),\xi)))}{\exp(S_n\varphi(x) - n\mathrm{P}(T,\varphi))} \le C, \quad \forall n \in \mathbb{N},\ \forall x \in X.$$

Then

$$C^{-1} \le \frac{\mu(T_x^{-n}(B(T^n(x),\xi)))}{\exp(S_n\psi(x) - n\mathrm{P}(T,\psi) + u\circ T^n(x) - u(x))} \le C, \quad \forall n \in \mathbb{N},\ \forall x \in X.$$

Consequently,

$$(Ce^{2\|u\|_\infty})^{-1} \le \frac{\mu(T_x^{-n}(B(T^n(x),\xi)))}{\exp(S_n\psi(x) - n\mathrm{P}(T,\psi))} \le Ce^{2\|u\|_\infty}, \quad \forall n \in \mathbb{N},\ \forall x \in X.$$

So $\mu \in \mathcal{G}_\psi$, and hence $\mathcal{G}_\varphi \subseteq \mathcal{G}_\psi$. By symmetry, $\mathcal{G}_\varphi \supseteq \mathcal{G}_\psi$. □

We now compare classes of Gibbs states for various Hölder continuous potentials.

Theorem 13.2.7. *If $T : X \to X$ is a topologically transitive, open, distance expanding system and $\varphi, \psi : X \to \mathbb{R}$ are Hölder continuous potentials such that $\mathcal{G}_\varphi \neq \emptyset$ and $\mathcal{G}_\psi \neq \emptyset$, then the following statements are equivalent:*

(a) *There exists a constant $R_1 \in \mathbb{R}$ such that*

$$S_n\varphi(x) - S_n\psi(x) = R_1 n$$

whenever $n \in \mathbb{N}$ and $x \in X$ is a periodic point of T of period n.

(b) *There exist constants $R_2 \in \mathbb{R}$ and $C \ge 0$ such that*

$$|S_n\varphi(x) - S_n\psi(x) - R_2 n| \le C, \quad \forall n \in \mathbb{N},\ \forall x \in X.$$

(c) *$\varphi - \psi$ is cohomologous to a constant $R_3 \in \mathbb{R}$ in the additive group of bounded functions on X.*

(d) *$\varphi - \psi$ is cohomologous to a constant $R_4 \in \mathbb{R}$ in the additive group of continuous functions on X.*

(e) *$\varphi - \psi$ is cohomologous to a constant $R_5 \in \mathbb{R}$ in the additive group of Hölder continuous functions on X.*

(f) $\mathcal{G}_\varphi = \mathcal{G}_\psi$.

(g) $\mathcal{G}_\varphi \cap \mathcal{G}_\psi \neq \emptyset$.

In addition, if any of the above statements holds, then all the constants R_1–R_5 are the same and equal to $\mathrm{P}(T,\varphi) - \mathrm{P}(T,\psi)$.

Proof. Of course (e)⇒(d)⇒(c). It is straightforward to see that (c)⇒(b). To see that (b)⇒(a), take a periodic point x of period n (by Corollary 4.3.10, every transitive, open, distance expanding system has a dense set of periodic points). Write the left-hand side of the relation in (b) with n replaced by kn for an arbitrary $k \in \mathbb{N}$. Let $k \to \infty$ and conclude that (a) holds. It also then follows that $R_5 = R_4 = R_3 = R_2 = R_1$. The implication (c)⇒(f) follows immediately from Lemma 13.2.6.

Let us now show that (f)⇒(a). Suppose that $\mu \in \mathcal{G}_\varphi = \mathcal{G}_\psi$. Let $\zeta = [\varphi - \mathrm{P}(T,\varphi)] - [\psi - \mathrm{P}(T,\psi)] = [\varphi - \psi] - [\mathrm{P}(T,\varphi) - \mathrm{P}(T,\psi)]$. As μ is a Gibbs state for both φ and ψ (cf. Definition 13.2.1), there exists a constant $D \ge 1$ such that

$$D^{-1} \le \exp([S_n\varphi(x) - n\mathrm{P}(T,\varphi)] - [S_n\psi(x) - n\mathrm{P}(T,\psi)]) = e^{S_n\zeta(x)} \le D \tag{13.18}$$

for all $n \in \mathbb{N}$ and all $x \in X$. In particular, if x is a periodic point of period n for T, then $S_n\zeta(x) = 0$. Otherwise, we would have $S_{kn}\zeta(x) = kS_n\zeta(x)$ for all $k \in \mathbb{N}$ and hence (13.18) would not hold. Thus, item (a) holds with $R_1 = \mathrm{P}(T, \varphi) - \mathrm{P}(T, \psi)$.

With this value for R_1, let us move on to showing that (a) implies (e). Keep ζ as above. Since T is transitive, let $x \in X$ be such that $\omega(x) = X$. Define $u(T^n(x)) := S_n\zeta(x)$ for every $n \geq 0$, where we define $S_0\zeta \equiv 0$. This function u is well-defined on $\mathcal{O}_+(x)$. Indeed, if x is periodic then u is well-defined since $S_N\zeta(x) = 0$, where N is the prime period of x. If x is not periodic, then u is well-defined simply because $T^m(x) \neq T^n(x)$ whenever $m \neq n$. Observe further that

$$\begin{aligned}\zeta(T^n(x)) &= S_{n+1}\zeta(x) - S_n\zeta(x) = u(T^{n+1}(x)) - u(T^n(x)) \\ &= u \circ T(T^n(x)) - u(T^n(x)) = (u \circ T - u)(T^n(x))\end{aligned}$$

for all $n \geq 0$, i. e., ζ is cohomologous to 0 on $\mathcal{O}_+(x)$ with u as a coboundary. Equivalently, $\varphi - \psi$ is cohomologous to $\mathrm{P}(T, \varphi) - \mathrm{P}(T, \psi)$ on $\mathcal{O}_+(x)$ with u as a coboundary. If x is periodic, then $\mathcal{O}_+(x) = X$ and u is the sought coboundary. If x is not periodic, then $\mathcal{O}_+(x) \neq X$ and we need to extend u continuously to $\overline{\mathcal{O}_+(x)} = X$. To do this (and more), given that $\omega(x) = X$ let $m < n$ be such that

$$d(T^m(x), T^n(x)) < \frac{1}{2}\frac{\lambda - 1}{\lambda + 1}\xi.$$

To shorten notation, set $\varepsilon := d(T^m(x), T^n(x))$. Let $\beta := \varepsilon(\lambda + 3)/(\lambda - 1) < \xi$. Observe that $\varepsilon < \min\{\xi, \beta(\lambda - 1)/2\}$. According to the closing lemma 4.3.7, there then is a periodic point y of period $n - m$ which β-shadows the suborbit $(T^k(x))_{k=m}^{n-1}$. Moreover,

$$d(T^n(x), y) \leq d(T^n(x), T^m(x)) + d(T^m(x), y) \leq \varepsilon + \beta = \frac{\lambda + 1}{\lambda - 1}(2\varepsilon) < \xi.$$

Using Lemma 13.1.1 with $\varphi = \zeta$, $w = T^m(x)$ and n replaced by $n - m$, we get

$$\begin{aligned}|u(T^n(x)) - u(T^m(x))| &= |S_n\zeta(x) - S_m\zeta(x)| = |S_{n-m}\zeta(T^m(x))| \\ &= |S_{n-m}\zeta(T^m(x)) - S_{n-m}\zeta(y)| \\ &\leq C_\alpha(\zeta)\, d^\alpha(T^n(x), y) \leq C_\alpha(\zeta)\left(\frac{\lambda + 1}{\lambda - 1}\right)^\alpha (2\varepsilon)^\alpha \\ &= 2^\alpha\left(\frac{\lambda + 1}{\lambda - 1}\right)^\alpha C_\alpha(\zeta) d^\alpha(T^n(x), T^m(x)).\end{aligned}$$

This shows that u is locally Hölder continuous and thereby uniformly continuous on $\mathcal{O}_+(x)$. So u extends continuously to $\overline{\mathcal{O}_+(x)} = X$. Moreover, the local Hölder continuity and the boundedness of u on $\mathcal{O}_+(x)$ imply that u is Hölder continuous on $\mathcal{O}_+(x)$. It follows immediately that the extension of u is Hölder continuous on X. Furthermore, thanks to the continuity of ζ and u on X, the fact that ζ is cohomologous to 0 on $\mathcal{O}_+(x)$ with u as a coboundary remains true on X.

Finally, the fact that (f)⇒(g) is trivial while (g)⇒(f) is a direct consequence of Lemma 13.2.3 and Proposition 13.2.4. □

13.2.2 Invariant Gibbs states are equilibrium states

There is a close relationship between invariant Gibbs states and equilibrium states, and this relationship partially explains the interest for Gibbs states and especially invariant ones. Recall from Definition 8.1.5 that a Borel probability measure μ on X is T-invariant if $\mu \circ T^{-1} = \mu$. Recall also that an equilibrium state for φ is a T-invariant Borel probability measure μ that satisfies the variational principle (Theorem 12.1.1), i. e. such that $\mathrm{P}(T,\varphi) = \mathrm{h}_\mu(T) + \int_X \varphi\, d\mu$.

We shall soon show that given an open, distance expanding map $T : X \to X$ and a potential $\varphi : X \to \mathbb{R}$, any T-invariant Gibbs state for φ is an equilibrium state for φ.

First, we observe that given a Borel partition $\mathcal{P}$ of X with diameter less than ξ, each atom of the $(n+1)$th "iterate" $\mathcal{P}^{n+1}$ is contained in a set of the form $T_x^{-n}(B(T^n(x),\xi))$. Recall that Gibbs measures confer to this latter set a weight which is comparable to $\exp(S_n\varphi(x) - n\mathrm{P}(T,\varphi))$.

Lemma 13.2.8. *Let $T : X \to X$ be an open, distance expanding system and $\mathcal{P}$ a Borel partition of X such that* $\operatorname{diam}(\mathcal{P}) < \xi$, *where ξ comes from* (13.2). *Then for all $x \in X$ and all $n \geq 0$,*

$$\mathcal{P}^{n+1}(x) \subseteq T_x^{-n}(B(T^n(x),\xi)).$$

Proof. We prove this lemma by induction. Let $x \in X$. Since $\operatorname{diam}(\mathcal{P}) < \xi$, we have $\mathcal{P}^1(x) = \mathcal{P}(x) \subseteq B(x,\xi) = T_x^{-0}(B(T^0(x),\xi))$. That is, the statement holds when $n = 0$. Suppose now that the statement has been demonstrated for some $n \geq 0$. Then

$$\begin{aligned}
\mathcal{P}^{n+2}(x) &= (\mathcal{P} \vee T^{-1}(\mathcal{P}) \vee \cdots \vee T^{-n}(\mathcal{P}) \vee T^{-(n+1)}(\mathcal{P}))(x) \\
&= \mathcal{P}(x) \cap T^{-1}(\mathcal{P}^{n+1}(T(x))) \\
&\subseteq \mathcal{P}(x) \cap T^{-1}(T_{T(x)}^{-n}(B(T^n(T(x)),\xi))) \\
&= \mathcal{P}(x) \cap \bigcup_{y \in T^{-1}(T(x))} T_y^{-1} \circ T_{T(x)}^{-n}(B(T^{n+1}(x),\xi)) \\
&\subseteq T_x^{-(n+1)}(B(T^{n+1}(x),\xi)).
\end{aligned}$$

□

Proposition 13.2.9. *Let $T : X \to X$ be an open, distance expanding system. Any T-invariant Gibbs state for a potential $\varphi \in C(X)$ is an equilibrium state for φ.*

Proof. Suppose that μ is a T-invariant Gibbs state for φ and let $\mathcal{P}$ be a finite Borel partition of X with $\operatorname{diam}(\mathcal{P}) < \xi$. Since T is ξ-expansive and $\operatorname{diam}(\mathcal{P}) < \xi$, it follows from Theorem 9.4.20, Fatou's lemma (Lemma A.1.36) and Lemma 13.2.8 that

$$\begin{aligned}
\mathrm{h}_\mu(T) &= \mathrm{h}_\mu(T,\mathcal{P})\\
&= \lim_{n\to\infty} \frac{1}{n} \mathrm{H}_\mu(\mathcal{P}^{n+1})\\
&= \lim_{n\to\infty} \frac{1}{n} \int_X -\log \mu(\mathcal{P}^{n+1}(x))\,d\mu(x)\\
&\geq \int_X \liminf_{n\to\infty} -\frac{1}{n} \log \mu(\mathcal{P}^{n+1}(x))\,d\mu(x)\\
&\geq \int_X \liminf_{n\to\infty} -\frac{1}{n} \log \mu(T_x^{-n}(B(T^n(x),\xi)))\,d\mu(x)\\
&\geq \int_X \liminf_{n\to\infty} -\frac{1}{n} \log[C \exp(S_n\varphi(x) - n\mathrm{P}(T,\varphi))]\,d\mu(x)\\
&= \int_X \Big[\mathrm{P}(T,\varphi) - \limsup_{n\to\infty} \frac{1}{n} S_n\varphi(x)\Big]\,d\mu(x)\\
&= \mathrm{P}(T,\varphi) - \int_X \limsup_{n\to\infty} \frac{1}{n} S_n\varphi(x)\,d\mu(x)\\
&= \mathrm{P}(T,\varphi) - \lim_{n\to\infty} \int_X \frac{1}{n} S_n\varphi(x)\,d\mu(x)\\
&= \mathrm{P}(T,\varphi) - \lim_{n\to\infty} \int_X \varphi\,d\mu,
\end{aligned}$$

where the last equality follows from the T-invariance of μ, while the second-to-last equality follows from Lebesgue's dominated convergence theorem (Theorem A.1.38). The variational principle (Theorem 12.1.1) yields the opposite inequality. □

13.3 Jacobians and changes of variables

We momentarily digress from the discussion of Gibbs measures to consider more general measures. Suppose that $T : X \to Y$ is an invertible Borel-measurable transformation between two topological spaces X and Y, meaning that T and T^{-1} are both measurable. Suppose that X is endowed with a finite Borel measure μ while Y is equipped with a finite Borel measure ν such that $\mu \circ T^{-1} \ll \nu$. By the Radon–Nikodym theorem (Theorem A.1.51), there exists an $L^1(\nu)$-function called Radon–Nikodym derivative $\frac{d(\mu\circ T^{-1})}{d\nu} : Y \to [0,\infty)$ such that

$$\mu \circ T^{-1}(B) = \int_B \frac{d(\mu \circ T^{-1})}{d\nu}(y)\,d\nu(y), \quad \forall B \in \mathcal{B}(Y).$$

Define the Borel-measurable function

$$\begin{aligned} \mathfrak{J}_1^{-1} : X &\longrightarrow [0,\infty) \\ x &\longmapsto \mathfrak{J}_1^{-1}(x) := \tfrac{d(\mu\circ T^{-1})}{d\nu}(T(x)). \end{aligned}$$

Then for each Borel set $B \subseteq Y$,

$$\mu\circ T^{-1}(B) = \int_B \frac{d(\mu\circ T^{-1})}{d\nu}(y)\,d\nu(y) = \int_B \frac{d(\mu\circ T^{-1})}{d\nu}(T(T^{-1}(y)))\,d\nu(y) = \int_B \mathfrak{J}_1^{-1}\circ T^{-1}(y)\,d\nu(y).$$

If $\nu\circ T \ll \mu$, then we may define an $L^1(\mu)$-function $\mathfrak{J}_2 : X \to [0,\infty)$ in a similar way by setting

$$\mathfrak{J}_2(x) := \frac{d(\nu\circ T)}{d\mu}(x).$$

By definition, this Radon–Nikodym derivative function satisfies

$$\nu\circ T(A) = \int_A \frac{d(\nu\circ T)}{d\mu}(x)\,d\mu(x) = \int_A \mathfrak{J}_2(x)\,d\mu(x), \quad \forall A \in \mathcal{B}(X).$$

We then have the following change-of-variables formulas.

Proposition 13.3.1 (Change-of-variables formulas). *Let $T : X \to Y$ be an invertible Borel-measurable transformation and μ and ν be finite Borel measures on X and Y, respectively.*

(a) *If $\mu\circ T^{-1} \ll \nu$, then for all Borel-measurable functions $f : X \to \mathbb{R}$ and all Borel sets $A \subseteq X$ we have*

$$\int_A f\,d\mu = \int_{T(A)} (f\cdot \mathfrak{J}_1^{-1})\circ T^{-1}\,d\nu. \tag{13.19}$$

(b) *If $\nu\circ T \ll \mu$, then for all Borel-measurable functions $g : Y \to \mathbb{R}$ and all Borel sets $B \subseteq Y$ we have*

$$\int_B g\,d\nu = \int_{T^{-1}(B)} (g\circ T)\cdot \mathfrak{J}_2\,d\mu. \tag{13.20}$$

Proof. We will derive formula (13.20). First, suppose that g is a characteristic function, i. e. $g = \mathbb{1}_{\widetilde{B}}$ for some Borel set $\widetilde{B} \subseteq Y$. Then

$$\begin{aligned} \int_B \mathbb{1}_{\widetilde{B}}\,d\nu &= \int_Y \mathbb{1}_{\widetilde{B}\cap B}\,d\nu = \nu(\widetilde{B}\cap B) = \nu\circ T(T^{-1}(\widetilde{B}\cap B)) \\ &= \int_{T^{-1}(\widetilde{B})\cap T^{-1}(B)} \mathfrak{J}_2\,d\mu = \int_{T^{-1}(B)} \mathbb{1}_{T^{-1}(\widetilde{B})}\cdot \mathfrak{J}_2\,d\mu = \int_{T^{-1}(B)} (\mathbb{1}_{\widetilde{B}}\circ T)\cdot \mathfrak{J}_2\,d\mu. \end{aligned}$$

So the formula holds for characteristic functions. By linearity of the integral, the formula applies to any measurable simple function g, i. e., to any finite linear combination of characteristic functions $\sum_{i=1}^{n} \alpha_i \mathbb{1}_{B_i}$. Since any nonnegative measurable function g is the limit of an increasing sequence of measurable simple functions $(g_n)_{n=1}^{\infty}$, the sequence $((g_n \circ T) \cdot \mathfrak{J}_2)_{n=1}^{\infty}$ is increasing to $(g \circ T) \cdot \mathfrak{J}_2$. Then the monotone convergence theorem (Theorem A.1.35) guarantees that

$$\int_B g\, d\nu = \lim_{n\to\infty} \int_B g_n\, d\nu = \lim_{n\to\infty} \int_{T^{-1}(B)} (g_n \circ T) \cdot \mathfrak{J}_2\, d\mu = \int_{T^{-1}(B)} (g \circ T) \cdot \mathfrak{J}_2\, d\mu$$

for all nonnegative measurable g. As any measurable function g is equal to the difference between its positive and negative parts ($g = g^+ - g^-$), the proof of formula (13.20) is complete. The proof of formula (13.19) is similar. □

Definition 13.3.2. We call $\mathfrak{J}_1^{-1}$ the *Jacobian* of the map T^{-1} with respect to the measures ν and μ. Similarly, $\mathfrak{J}_2$ is called the *Jacobian* of the map T with respect to the measures μ and ν.

Finally, we identify the relation between $\mathfrak{J}_1^{-1}$ and $\mathfrak{J}_2$ when both are defined.

Lemma 13.3.3. *Let $T : X \to Y$ be an invertible Borel-measurable transformation and μ and ν be finite Borel measures on X and Y, respectively. If $\mu \circ T^{-1} \ll \nu$ and $\nu \circ T \ll \mu$, then $\mathfrak{J}_1^{-1} \cdot \mathfrak{J}_2 \equiv 1$ μ-almost everywhere.*

Proof. Let $A \subseteq X$ be a Borel set. Let $g : Y \to \mathbb{R}$ be defined by

$$g := \mathbb{1}_{T(A)} \cdot (\mathfrak{J}_1^{-1} \circ T^{-1}).$$

On one hand, using the change-of-variables formula (13.19), we obtain

$$\int_Y g\, d\nu = \int_Y \mathbb{1}_{T(A)} \cdot (\mathfrak{J}_1^{-1} \circ T^{-1})\, d\nu = \int_{T(A)} (1 \cdot \mathfrak{J}_1^{-1}) \circ T^{-1}\, d\nu = \int_A 1\, d\mu = \mu(A).$$

On the other hand, thanks to the change-of-variables formula (13.20), we have

$$\int_Y g\, d\nu = \int_X (g \circ T) \cdot \mathfrak{J}_2\, d\mu = \int_X (\mathbb{1}_{T(A)} \circ T) \cdot (\mathfrak{J}_1^{-1} \circ T^{-1} \circ T) \cdot \mathfrak{J}_2\, d\mu = \int_X \mathbb{1}_A \cdot \mathfrak{J}_1^{-1} \cdot \mathfrak{J}_2\, d\mu = \int_A \mathfrak{J}_1^{-1} \cdot \mathfrak{J}_2\, d\mu.$$

Hence, $\mu(A) = \int_A \mathfrak{J}_1^{-1} \cdot \mathfrak{J}_2\, d\mu$ for any Borel set $A \subseteq X$. Thus, $\mathfrak{J}_1^{-1} \cdot \mathfrak{J}_2 = 1$ μ-a. e.. □

13.4 Construction of invariant measures from quasi-invariant ones

We now return to our original setting. Let T be an open, distance expanding system of a compact metric space (X, d). Recall from Definition 10.1.1 that a Borel probability measure m on X is quasi-T-invariant if $m \circ T^{-1} \ll m$. Of course, any T-invariant measure is quasi-T-invariant. Though the converse is generally not true, when the map is open and expanding there is a way to construct from a quasi-invariant measure an absolutely continuous and invariant one.

Indeed, suppose that m is a quasi-T-invariant measure. Let $g : X \to \mathbb{R}$ be a nonnegative function in $L^1(m)$. Define the finite Borel measure

$$gm(A) := \int_A g\,dm, \quad \forall A \in \mathcal{B}(X).$$

Observe that $gm \ll m$ and thereby

$$gm \circ T^{-1} \ll m \circ T^{-1} \ll m.$$

This allows us to define an operator $\mathcal{L}_m : L^1(m) \to L^1(m)$ as follows. For a nonnegative function $g \in L^1(m)$ we define, using the Radon–Nikodym theorem (Theorem A.1.51), the function $\mathcal{L}_m(g)$ by

$$\mathcal{L}_m(g) := \frac{d(gm \circ T^{-1})}{dm}.$$

By definition, $\mathcal{L}_m(g) \in L^1(m)$. Moreover,

$$\int_X \mathcal{L}_m(g)\,dm = \int_X \frac{d(gm \circ T^{-1})}{dm}\,dm = gm \circ T^{-1}(X) = gm(X) = \int_X g\,dm.$$

For any $g \in L^1(m)$, define $\mathcal{L}_m(g) := \mathcal{L}_m(g^+) - \mathcal{L}_m(g^-)$. Defining $\mathcal{L}_m$ in this way yields a linear operator on $L^1(m)$ and, using the above integral equality, we get

$$\int_X \mathcal{L}_m(g)\,dm = \int_X g\,dm, \quad \forall g \in L^1(m). \tag{13.21}$$

Furthermore,

$$\int_X |\mathcal{L}_m(g)|\,dm \le \int_X |g|\,dm, \quad \forall g \in L^1(m). \tag{13.22}$$

Therefore, $\mathcal{L}_m : L^1(m) \to L^1(m)$ is a positive bounded linear operator with operator norm $\|\mathcal{L}_m\| \le 1$.

The next theorem is particularly important because it gives a necessary and sufficient condition on g under which the measure gm is T-invariant. More specifically, the theorem states that the measure-theoretic problem of finding a T-invariant measure gm from a quasi-T-invariant measure m reduces to the functional-analytic problem of finding a fixed point for the operator $\mathcal{L}_m$.

Theorem 13.4.1. *Suppose that m is a quasi-T-invariant measure and that $g \in L^1(m)$ is such that $g \geq 0$ and $\int_X g\,dm = 1$. Then the Borel probability measure gm is T-invariant if and only if $\mathcal{L}_m(g) = g$.*

Proof. Since $gm(A) := \int_A g\,dm$, we know that $gm \ll m$. In view of the Radon–Nikodym Theorem (Theorem A.1.51), we have

$$\int_A g\,dm = gm(A) = \int_A \frac{d(gm)}{dm}\,dm, \quad \forall A \in \mathcal{B}(X).$$

Thus, $g = \frac{d(gm)}{dm}$ m-almost everywhere. As $\mathcal{L}_m(g) := \frac{d(gm\circ T^{-1})}{dm}$, we deduce that

$$\mathcal{L}_m(g) = g \quad \Longleftrightarrow \quad \frac{d(gm \circ T^{-1})}{dm} = \frac{d(gm)}{dm} \quad m\text{-a. e.} \quad \Longleftrightarrow \quad gm \circ T^{-1} = gm.$$

That is, $\mathcal{L}_m(g) = g$ if and only if gm is T-invariant. □

Of course, it is desirable to have a formula for $\mathcal{L}_m$. The next proposition will provide one. Before, though, we need to make some observations on a local level. For every $x \in X$, consider the restriction $T_x : T_x^{-1}(B(T(x),\xi)) \to B(T(x),\xi)$. Recall that these restrictions of T are homeomorphisms. If m is a quasi-T-invariant measure, then

$$m \circ T_x^{-1}|_{B(T(x),\xi)} \ll m|_{B(T(x),\xi)}$$

for all $x \in X$, i. e., m is quasi-T_x-invariant for all $x \in X$. We can thus define the Jacobian of m with respect to T_x by

$$\begin{array}{rcccl} \mathfrak{J}_m^{-1} & : & T_x^{-1}(B(T(x),\xi)) & \longrightarrow & [0,\infty) \\ & & z & \longmapsto & \mathfrak{J}_m^{-1}(z) := \mathfrak{J}_1^{-1}(z) = \dfrac{dm \circ T_x^{-1}|_{B(T(x),\xi)}}{dm|_{B(T(x),\xi)}}(T(z)). \end{array}$$

Note that $\mathfrak{J}_m^{-1}(z)$ is independent of x for $z \in T_x^{-1}(B(T(x),\xi))$. By this, we mean that if $z \in T_x^{-1}(B(T(x),\xi)) \cap T_y^{-1}(B(T(y),\xi))$ for some x and y, then it does not matter whether we pick x or y in the definition of the Jacobian. For this reason, we do not adjoin to $\mathfrak{J}_m^{-1}$ a subscript x and we view $\mathfrak{J}_m^{-1}$ as being defined globally on X from this local definition.

We can now give to $\mathcal{L}_m$ an explicit expression.

Proposition 13.4.2. *Let $T : X \to X$ be an open, distance expanding system. Suppose that m is a quasi-T-invariant measure. For every $g \in L^1(m)$, we have*

$$\mathcal{L}_m(g)(z) = \sum_{y\in T^{-1}(z)} g(y)\mathfrak{J}_m^{-1}(y) \quad \textit{for } m\textit{-a. e. } z \in X.$$

Proof. Let $g \in L^1(m)$ be such that $g \geq 0$. Fix any $w \in X$. Let A be a Borel subset of $B(w, \xi)$. Recall that per Lemma 4.2.8, the inverse branches of T have mutually disjoint images on open balls of radius ξ. By the change-of-variables formula (13.19), we obtain

$$\begin{aligned} gm \circ T^{-1}(A) &= gm\Big(\bigcup_{x \in T^{-1}(w)} T_x^{-1}(A) \Big) \\ &= \sum_{x \in T^{-1}(w)} gm(T_x^{-1}(A)) \\ &= \sum_{x \in T^{-1}(w)} \int_{T_x^{-1}(A)} g \, dm \\ &= \sum_{x \in T^{-1}(w)} \int_A (g \circ T_x^{-1}) \cdot (\mathfrak{J}_m^{-1} \circ T_x^{-1}) \, dm \\ &= \int_A \sum_{x \in T^{-1}(w)} g(T_x^{-1}(z)) \mathfrak{J}_m^{-1}(T_x^{-1}(z)) \, dm(z). \end{aligned}$$

As this is true for all Borel sets $A \subseteq B(w, \xi)$, we deduce that

$$\frac{d(gm \circ T^{-1})}{dm}(z) = \sum_{x \in T^{-1}(w)} g(T_x^{-1}(z)) \mathfrak{J}_m^{-1}(T_x^{-1}(z)) = \sum_{y \in T^{-1}(z)} g(y) \mathfrak{J}_m^{-1}(y)$$

for m-a. e. $z \in B(w, \xi)$. Since $w \in X$ was chosen arbitrarily, we conclude that

$$\mathcal{L}_m(g)(z) = \sum_{y \in T^{-1}(z)} g(y) \mathfrak{J}_m^{-1}(y)$$

for m-a. e. $z \in X$. For an arbitrary $g \in L^1(m)$, the result follows from the above and the fact that $\mathcal{L}_m(g) = \mathcal{L}_m(g^+) - \mathcal{L}_m(g^-)$. □

13.5 Transfer operators

Inspired by the expression found for the operator $\mathcal{L}_m$ in Proposition 13.4.2, we introduce a new kind of operators called transfer operators.

Definition 13.5.1. Let $T : X \to X$ be an open, distance expanding system. Let $\varphi \in C(X)$ be a potential and $m \in M(X)$ a Borel probability measure. A positive bounded linear operator $\mathcal{L}_\varphi : L^1(m) \to L^1(m)$ is said to be of *transfer-type* if there exists a constant $C > 0$ such that

$$\mathcal{L}_\varphi(g)(x) = C \sum_{y \in T^{-1}(x)} g(y) e^{\varphi(y)}, \quad \forall x \in X, \ \forall g \in L^1(m). \tag{13.23}$$

The dual operator of $\mathcal{L}_\varphi$ shall be denoted by $\mathcal{L}_\varphi^* : L^1(m)^* \to L^1(m)^*$ and is defined in the usual manner by

$$\mathcal{L}_\varphi^*(F)(g) := F(\mathcal{L}_\varphi(g)) \tag{13.24}$$

for every bounded linear functional $F : L^1(m) \to \mathbb{R}$ and $g \in L^1(m)$. In particular, we have $\mathcal{L}_\varphi^*(m)(g) = m(\mathcal{L}_\varphi(g))$, where m denotes the positive bounded linear functional

$$m(g) = \int_X g\, dm. \tag{13.25}$$

The following proposition reveals that any positive fixed point of the dual $\mathcal{L}_\varphi^*$ of a transfer-type operator $\mathcal{L}_\varphi$ is a quasi-invariant measure.

Proposition 13.5.2. *Let $T : X \to X$ be an open, distance expanding system. If a transfer-type operator $\mathcal{L}_\varphi : L^1(m) \to L^1(m)$ is such that $\mathcal{L}_\varphi^*(m) = m$, then m is quasi-T-invariant. Moreover, $m \circ T \ll m$. Furthermore, the Jacobian of m is*

$$\mathfrak{J}_m^{-1}(z) = Ce^{\varphi(z)} \quad \text{and} \quad \mathfrak{J}_m(z) = C^{-1}e^{-\varphi(z)} \quad \text{for } m\text{-a. e. } z \in X,$$

when $\mathcal{L}_\varphi(g)(x) = C\sum_{y\in T^{-1}(x)} g(y)e^{\varphi(y)}$.

Note: We slightly abuse notation when writing $m \circ T \ll m$. By this, we mean that $m(T(A)) = 0$ whenever A is a Borel subset of X such that $m(A) = 0$. (Observe that $T(A)$ is a Borel set for all Borel sets A since the map T is open.) This is equivalent to $m \circ T_x|_{T_x^{-1}(B(T(x),\xi))} \ll m|_{T_x^{-1}(B(T(x),\xi))}$ for all $x \in X$.

Proof. Let $w \in X$ and A a Borel subset of $B(w, \xi)$. Then

$$\begin{aligned}
m(T^{-1}(A)) = m(\mathbb{1}_{T^{-1}(A)}) &= \mathcal{L}_\varphi^*(m)(\mathbb{1}_{T^{-1}(A)}) \\
&= m(\mathcal{L}_\varphi(\mathbb{1}_{T^{-1}(A)})) \\
&= \int_X \mathcal{L}_\varphi(\mathbb{1}_{T^{-1}(A)})(z)\, dm(z) \\
&= \int_X C \sum_{y\in T^{-1}(z)} \mathbb{1}_{T^{-1}(A)}(y)e^{\varphi(y)}\, dm(z) \\
&= \int_X C \sum_{y\in T^{-1}(z)} \mathbb{1}_A(T(y))e^{\varphi(y)}\, dm(z) \\
&= \int_X \mathbb{1}_A(z) \cdot C \sum_{y\in T^{-1}(z)} e^{\varphi(y)}\, dm(z) \\
&= \int_A \mathcal{L}_\varphi(\mathbb{1})\, dm.
\end{aligned}$$

So, if $m(A) = 0$ then $m \circ T^{-1}(A) = 0$, and it follows that $m \circ T^{-1} \ll m$.

Furthermore, for any Borel set A we have

$$
\begin{aligned}
m(A) = m(\mathbb{1}_A) = \mathcal{L}_\varphi^*(m)(\mathbb{1}_A) \\
&= \int_X \mathcal{L}_\varphi(\mathbb{1}_A)(z)\,dm(z) \\
&= \int_X C \sum_{y\in T^{-1}(z)} \mathbb{1}_A(y)e^{\varphi(y)}\,dm(z) \\
&\geq \int_{T(A)} C \sum_{y\in T^{-1}(z)} \mathbb{1}_A(y)e^{\varphi(y)}\,dm(z) \\
&\geq \frac{1}{(\#T^{-1})e^{2\|\varphi\|_\infty}} \int_{T(A)} C \sum_{y\in T^{-1}(z)} e^{\varphi(y)}\,dm(z) \\
&= \frac{1}{(\#T^{-1})e^{2\|\varphi\|_\infty}} \int_{T(A)} \mathcal{L}_\varphi(\mathbb{1})\,dm,
\end{aligned}
$$

where $\#T^{-1} := \max_{x\in X} \#T^{-1}(x) < \infty$ by the expanding property of T and the compactness of X. Since $\mathcal{L}_\varphi$ is a positive operator, we deduce that $m(T(A)) = 0$ whenever $m(A) = 0$.

Take $x \in X$ and a Borel set $A \subseteq B(T(x), \xi)$. We identify the Jacobians $\mathfrak{J}_m^{-1}(z) := \frac{dm\circ T_x^{-1}|_{B(T(x),\xi)}}{dm|_{B(T(x),\xi)}}(T(z))$ and $\mathfrak{J}_m(z) := \frac{dm\circ T_x|_{T_x^{-1}(B(T(x),\xi))}}{dm|_{T_x^{-1}(B(T(x),\xi))}}(z) = (\mathfrak{J}_m^{-1}(z))^{-1}$ that are induced by the homeomorphism $T_x : T_x^{-1}(B(T(x), \xi)) \to B(T(x), \xi)$. Observe that

$$
\begin{aligned}
m \circ T_x^{-1}(A) = m(\mathbb{1}_{T_x^{-1}(A)}) = \mathcal{L}_\varphi^*(m)(\mathbb{1}_{T_x^{-1}(A)}) \\
&= m(\mathcal{L}_\varphi(\mathbb{1}_{T_x^{-1}(A)})) \\
&= \int_X C \sum_{y\in T^{-1}(z)} \mathbb{1}_{T_x^{-1}(A)}(y)e^{\varphi(y)}\,dm(z) \\
&= \int_A Ce^{\varphi(T_x^{-1}(z))}\,dm(z).
\end{aligned}
$$

So $\frac{dm\circ T_x^{-1}}{dm}(z) = Ce^{\varphi(T_x^{-1}(z))}$ for m-a. e. $z \in B(T(x), \xi)$. Thus, for m-a. e. $w \in T_x^{-1}(B(T(x), \xi))$,

$$\mathfrak{J}_m^{-1}(w) = Ce^{\varphi(T_x^{-1}(T(w)))} = Ce^{\varphi(w)}.$$

As $\{T_x^{-1}(B(T(x), \xi))\}_{x\in X}$ is an open cover of the compact space X, one can extract a finite subcover and conclude that $\mathfrak{J}_m^{-1}(w) = Ce^{\varphi(w)}$ m-a. e. on X. According to Lemma 13.3.3, $\mathfrak{J}_m(w) = C^{-1}e^{-\varphi(w)}$ m-a. e. on X. □

13.6 Nonnecessarily-invariant Gibbs states

In this section, we will establish the existence of at least one Gibbs state for every topologically transitive, open, distance expanding system under a Hölder continuous potential. This Gibbs state will turn out to be ergodic, though perhaps not invariant. In general, Gibbs states are not unique. In the next section, we will pursue the question of existence and uniqueness of an invariant Gibbs state.

We now outline the basic structure of the argument to come in this section.

(a) Start with a restriction of a transfer-type operator $\mathcal{L}_\varphi : C(X) \to C(X)$, where

$$\mathcal{L}_\varphi(g)(x) := \sum_{y \in T^{-1}(x)} g(y) e^{\varphi(y)} \tag{13.26}$$

for all $g \in C(X)$ and $x \in X$. Here, φ is a continuous potential and there is not yet any involvement of a measure.

(b) Define the dual operator $\mathcal{L}_\varphi^* : C(X)^* \to C(X)^*$ by setting

$$\mathcal{L}_\varphi^*(F)(g) := F(\mathcal{L}_\varphi(g)) \tag{13.27}$$

for every $F \in C(X)^*$ and $g \in C(X)$.

(c) Under the additional assumption that T is surjective, find an eigenfunctional $m_\varphi \in C(X)^*$ of the dual $\mathcal{L}_\varphi^* : C(X)^* \to C(X)^*$ such that m_φ is positive and $m_\varphi(\mathbb{1}) = 1$. Let $\Lambda_\varphi \in \mathbb{R}$ be the associated eigenvalue, i. e.,

$$\mathcal{L}_\varphi^*(m_\varphi) = \Lambda_\varphi m_\varphi. \tag{13.28}$$

(d) In view of Riesz representation theorem, $m_\varphi \in M(X)$. Thus, it makes sense to examine the extension $\mathcal{L}_\varphi : L^1(m_\varphi) \to L^1(m_\varphi)$ and its dual operator $\mathcal{L}_\varphi^* : L^1(m_\varphi)^* \to L^1(m_\varphi)^*$. Also, $m_\varphi \in L^1(m_\varphi)^*$. Since m_φ is regular and X is compact, the subspace $C(X)$ is dense in $L^1(m_\varphi)$. Demonstrate that $\mathcal{L}_\varphi^*(m_\varphi) = \Lambda_\varphi m_\varphi$ for the extension, too.

(e) Next, normalize the operators $\mathcal{L}_\varphi$ and $\mathcal{L}_\varphi^*$ by setting $\widehat{\mathcal{L}}_\varphi := \Lambda_\varphi^{-1}\mathcal{L}_\varphi$, and hence $\widehat{\mathcal{L}}_\varphi^* = \Lambda_\varphi^{-1}\mathcal{L}_\varphi^*$.

(f) Then $\widehat{\mathcal{L}}_\varphi^*(m_\varphi) = m_\varphi$, i. e., m_φ is a fixed point of $\widehat{\mathcal{L}}_\varphi^*$. In light of Proposition 13.5.2, deduce that $m_\varphi \circ T^{-1} \ll m_\varphi$ and $m_\varphi \circ T \ll m_\varphi$, and that the Jacobian of m_φ is $\mathfrak{J}_{m_\varphi}^{-1}(z) = \Lambda_\varphi^{-1} e^{\varphi(z)}$. By Proposition 13.4.2, it further follows that $\widehat{\mathcal{L}}_\varphi = \mathcal{L}_{m_\varphi}$.

(g) Investigate the iterates of $\mathcal{L}_\varphi$ and $\widehat{\mathcal{L}}_\varphi$, and their duals. Infer formulas for the measure of backward and forward images of small sets, and establish that m_φ is a $(\Lambda_\varphi e^{-\varphi})$-conformal measure. See Proposition 13.6.9, Corollaries 13.6.10–13.6.11 and Definition 13.6.12.

(h) Show that m_φ is a Gibbs measure. It is at that point that the supplementary assumptions that φ is Hölder continuous and that T is transitive are needed. It further turns out that m_φ is ergodic.

(i) Finally, if T is topologically exact then the Gibbs state m_φ is metrically exact.

Recall that throughout this chapter, $T : X \to X$ is an open, distance expanding system on a compact metric space (X, d) and δ, λ and ξ are constants arising from Definition 4.1.1 and relations (4.29), (4.30) and (4.31); alternatively, see (13.1) and (13.2).

13.6.1 Existence of eigenmeasures for the dual of the transfer operator

Let us begin with the restriction of the transfer-type operator $\mathcal{L}_\varphi : C(X) \to C(X)$ as defined in (13.26).

Lemma 13.6.1. *For every open, distance expanding system $T : X \to X$ and every potential $\varphi \in C(X)$, the positive linear operator $\mathcal{L}_\varphi : C(X) \to C(X)$ is well defined and bounded.*

Proof. First, let us show that $\mathcal{L}_\varphi$ is well defined, i. e., that it maps continuous functions to continuous functions. Let $g \in C(X)$ and $\varepsilon > 0$. By the expanding property of T and the compactness of X, we know that $\#T^{-1} := \max_{x\in X} \#T^{-1}(x) < \infty$. As the functions g and e^φ are uniformly continuous on X, there exists $\zeta > 0$ such that if $d(z, w) < \zeta$ then

$$|g(z) - g(w)| < \frac{\varepsilon}{\#T^{-1} \cdot e^{\sup\varphi}} \quad \text{and} \quad |\varphi(z) - \varphi(w)| < \frac{\varepsilon}{\#T^{-1} \cdot e^{\sup\varphi} \cdot \|g\|_\infty}.$$

Set $\chi = \min\{\zeta, \xi\}$. Let $x, x_0 \in X$ be such that $d(x, x_0) < \chi$. We recycle part of the argument employed in the proof of Lemma 13.1.1. According to Lemmas 4.2.8 and 4.2.9, there is a bijection between the ξ-separated sets $T^{-1}(x)$ and $T^{-1}(x_0)$ defined by

$$\begin{aligned} b : T^{-1}(x) &\longrightarrow T^{-1}(x_0) \\ y &\longmapsto y_0 := b(y) := T_y^{-1}(x_0). \end{aligned}$$

Since the inverse branches of T on any open ball of radius ξ are contractions with ratio λ^{-1}, we deduce that

$$d(y_0, y) = d(T_y^{-1}(x_0), T_y^{-1}(x)) \le \lambda^{-1} d(x_0, x) < \chi.$$

Then

$$\begin{aligned} |\mathcal{L}_\varphi(g)(x) - \mathcal{L}_\varphi(g)(x_0)| &= \left| \sum_{y\in T^{-1}(x)} g(y)e^{\varphi(y)} - \sum_{y\in T^{-1}(x)} g(y_0)e^{\varphi(y_0)} \right| \\ &= \left| \sum_{y\in T^{-1}(x)} \left[(g(y) - g(y_0))e^{\varphi(y)} + g(y_0)(e^{\varphi(y)} - e^{\varphi(y_0)}) \right] \right| \\ &\le e^{\sup\varphi} \sum_{y\in T^{-1}(x)} \left[|g(y) - g(y_0)| + \|g\|_\infty |\varphi(y) - \varphi(y_0)| \right] \\ &< 2\varepsilon. \end{aligned}$$

Thus, $\mathcal{L}_\varphi(g) \in C(X)$, and hence $\mathcal{L}_\varphi$ is well defined. Moreover,

$$\|\mathcal{L}_\varphi(g)\|_\infty \le \sup_{x\in X} \sum_{y\in T^{-1}(x)} |g(y)|e^{\varphi(y)} \le \#T^{-1} \cdot \|g\|_\infty \cdot e^{\sup\varphi}.$$

So $\mathcal{L}_\varphi : C(X) \to C(X)$ is bounded and its norm $\|\mathcal{L}_\varphi\| \le \#T^{-1} \cdot e^{\sup\varphi}$. □

The following theorem establishes the existence of an eigenmeasure m_φ for the dual operator $\mathcal{L}_\varphi^* : C(X)^* \to C(X)^*$.

Theorem 13.6.2. *Let $T : X \to X$ be an open, distance expanding system and $\varphi \in C(X)$ a potential. If T is surjective, then there exist a measure $m_\varphi \in M(X)$ and a constant $\Lambda_\varphi > 0$ such that $\mathcal{L}_\varphi^*(m_\varphi) = \Lambda_\varphi m_\varphi$.*

Proof. For each $m \in M(X)$, set

$$H(m) := \frac{\mathcal{L}_\varphi^*(m)}{\mathcal{L}_\varphi^*(m)(\mathbb{1})}.$$

Because T is surjective, $\mathcal{L}_\varphi^*(m)(\mathbb{1}) := m(\mathcal{L}_\varphi(\mathbb{1})) \ge e^{\inf\varphi} > 0$. Then $H : M(X) \to C(X)^*$ is a positive linear operator with $H(m)(\mathbb{1}) = 1$. So, by the Riesz representation theorem (Theorem A.1.54), $H(m) \in M(X)$. We claim that $H : M(X) \to M(X)$ is a continuous operator in the weak* topology on $M(X)$. As $M(X)$ is metrizable in the weak* topology, it is enough to show sequential continuity, i. e., that if $(m_n)_{n=1}^\infty$ weak* converges to m, then $(H(m_n))_{n=1}^\infty$ weak* converges to $H(m)$. This amounts to establishing that $\mathcal{L}_\varphi^*(m_n)(g) \to \mathcal{L}_\varphi^*(m)(g)$ for all $g \in C(X)$, which holds as $\mathcal{L}_\varphi^*(m_n)(g) := m_n(\mathcal{L}_\varphi g) \to m(\mathcal{L}_\varphi g) =: \mathcal{L}_\varphi^*(m)(g)$.

Since $M(X)$ is a compact convex subset of the locally convex topological vector space $C(X)^*$, by virtue of the Schauder–Tychonov fixed-point theorem (cf. Theorem V.10.5 in Dunford and Schwartz [26]), the continuous map $H : M(X) \to M(X)$ has a fixed point $m_\varphi \in M(X)$. Thus,

$$\frac{\mathcal{L}_\varphi^*(m_\varphi)}{\mathcal{L}_\varphi^*(m_\varphi)(\mathbb{1})} = m_\varphi.$$

Setting $\Lambda_\varphi = \mathcal{L}_\varphi^*(m_\varphi)(\mathbb{1}) > 0$, we see that $\mathcal{L}_\varphi^*(m_\varphi) = \Lambda_\varphi m_\varphi$. □

As pointed out earlier, since $m_\varphi \in M(X)$ it makes sense to examine the extension $\mathcal{L}_\varphi : L^1(m_\varphi) \to L^1(m_\varphi)$ and its dual operator $\mathcal{L}_\varphi^* : L^1(m_\varphi)^* \to L^1(m_\varphi)^*$. First, let us make sure that this extension exists.

Theorem 13.6.3. *For every surjective, open, distance expanding system $T : X \to X$ and every potential $\varphi \in C(X)$, the positive linear operator $\mathcal{L}_\varphi : C(X) \to C(X)$ is uniformly continuous and extends continuously to a positive linear operator $\mathcal{L}_\varphi : L^1(m_\varphi) \to L^1(m_\varphi)$*

defined m_φ-a. e. by formula (13.26) for every $g \in L^1(m_\varphi)$, where m_φ is an eigenmeasure for the dual operator $\mathcal{L}_\varphi^* : C(X)^* \to C(X)^*$.

Proof. According to Theorem 13.6.2, the positive linear operator $\mathcal{L}_\varphi : C(X) \to C(X)$ satisfies $m_\varphi(\mathcal{L}_\varphi(f)) = \Lambda_\varphi m_\varphi(f)$ for all $f \in C(X)$. Let $g \in C(X)$. Using (13.26),

$$\|\mathcal{L}_\varphi(g)\|_1 \le \|\mathcal{L}_\varphi(|g|)\|_1 = m_\varphi(\mathcal{L}_\varphi(|g|)) = \Lambda_\varphi m_\varphi(|g|) = \Lambda_\varphi \|g\|_1.$$

It immediately follows that $\mathcal{L}_\varphi : C(X) \to C(X)$ is uniformly continuous. Since $C(X)$ is dense in $L^1(m_\varphi)$, the operator $\mathcal{L}_\varphi : C(X) \to C(X)$ extends continuously to a positive linear operator $\mathcal{L}_\varphi : L^1(m_\varphi) \to L^1(m_\varphi)$.

Let

$$Z := \left\{ f \in L^1(m_\varphi) \;\middle|\; \mathcal{L}_\varphi(f)(x) = \sum_{y \in T^{-1}(x)} f(y) e^{\varphi(y)} \text{ for } m_\varphi\text{-a. e. } x \in X \right\}.$$

We will show that $Z = L^1(m_\varphi)$ through a series of claims.

Recall that $\mathbb{R}^X$ denotes the set of all real-valued functions defined on X. Let β be an ordinal number such that $\beta > |\mathbb{R}^X|$. Let $\mathcal{B}_0 := C(X)$. Let γ be an ordinal such that $\gamma < \beta$ and suppose that $\mathcal{B}_\alpha$ has been defined for all ordinals $\alpha < \gamma$. Set

$$\mathcal{B}_\gamma := \left\{ f \in \mathbb{R}^X \;\middle|\; f = \lim_{n\to\infty} f_n, \text{ where } \forall n \in \mathbb{N}, f_n \in \mathcal{B}_{\alpha_n} \text{ for some } \alpha_n < \gamma \right\}.$$

It is obvious that $\mathcal{B}_\alpha \subseteq \mathcal{B}_{\alpha'}$ whenever $\alpha \le \alpha'$.

Claim 1.

$$\bigcup_{\gamma<\beta} \mathcal{B}_\gamma = \{f \in \mathbb{R}^X \mid f \text{ Borel function}\}.$$

Proof of Claim 1. It is evident that $\mathcal{B}_0 \subseteq \{f \text{ Borel}\}$. Let $\gamma < \beta$ and suppose that $\mathcal{B}_\alpha \subseteq \{f \text{ Borel}\}$ for all $\alpha < \gamma$. Let $g \in \mathcal{B}_\gamma$. Then $g = \lim_{n\to\infty} g_n$ for some $g_n \in \mathcal{B}_{\alpha_n}$ and $\alpha_n < \gamma$. Thus, g is a pointwise limit of Borel functions, and hence is Borel. So $\mathcal{B}_\gamma \subseteq \{f \text{ Borel}\}$. By transfinite induction, $\bigcup_{\gamma<\beta} \mathcal{B}_\gamma \subseteq \{f \text{ Borel}\}$.

On the other hand, the family $\bigcup_{\gamma<\beta} \mathcal{B}_\gamma$ contains all continuous functions and is closed under taking pointwise limits of sequences of functions. This means that it comprises the smallest family with these properties, namely the Baire functions. According to the Lebesgue–Hausdorff theorem (see Theorem 3.1.36 in Srivastava [120]) every real-valued Borel function defined on a metrizable space is Baire. Therefore, $\bigcup_{\gamma<\beta} \mathcal{B}_\gamma \supseteq \{f \text{ Borel}\}$. ■

Claim 2. *For all $\gamma < \beta$,*

$$g, h \in \mathcal{B}_\gamma \quad \Longrightarrow \quad \min\{g,h\} \in \mathcal{B}_\gamma \textit{ and } \max\{g,h\} \in \mathcal{B}_\gamma.$$

Proof of Claim 2. The reader can easily show that the statement holds for $\gamma = 0$. Let $\gamma < \beta$. Suppose that the statement holds for all $\alpha < \gamma$. Let $g, h \in \mathcal{B}_\gamma$. Then $g = \lim_{n\to\infty} g_n$ for some $g_n \in \mathcal{B}_{\alpha_n}$ and $\alpha_n < \gamma$. Likewise, $h = \lim_{n\to\infty} h_n$ for some $h_n \in \mathcal{B}_{\alpha'_n}$ and $\alpha'_n < \gamma$. For every $n \in \mathbb{N}$, define $\gamma_n := \max\{\alpha_n, \alpha'_n\} < \gamma$. Then $g_n, h_n \in \mathcal{B}_{\gamma_n}$. By the inductive hypothesis, $\min\{g_n, h_n\} \in \mathcal{B}_{\gamma_n}$ and $\max\{g_n, h_n\} \in \mathcal{B}_{\gamma_n}$. Since $\min\{g, h\} = \lim_{n\to\infty} \min\{g_n, h_n\}$, it follows that $\min\{g, h\} \in \mathcal{B}_\gamma$. Likewise, $\max\{g, h\} \in \mathcal{B}_\gamma$. So the statement holds for the arbitrary γ. The claim then ensues from transfinite induction. ■

Now, let $M \geq 0$. Let $\mathcal{B}_0^+(M) := \mathcal{B}_0 \cap \{0 \leq f \leq M\}$. Let $\gamma < \beta$ and suppose that $\mathcal{B}_\alpha^+(M)$ has been defined for all $\alpha < \gamma$. Set

$$\mathcal{B}_\gamma^+(M) := \Big\{f \in \mathbb{R}^X \ \Big|\ f = \lim_{n\to\infty} f_n, \text{ where } \forall n \in \mathbb{N}, f_n \in \mathcal{B}_{\alpha_n}^+(M) \text{ and } \alpha_n < \gamma\Big\}.$$

Claim 3. *For all $\gamma < \beta$, we have*

$$\mathcal{B}_\gamma^+(M) = \mathcal{B}_\gamma \cap \{0 \leq f \leq M\}.$$

Proof of Claim 3. By definition, the statement holds when $\gamma = 0$. Let $\gamma < \beta$ and suppose that the statement holds for all $\alpha < \gamma$.

Let $g \in \mathcal{B}_\gamma^+(M)$. Then $g = \lim_{n\to\infty} g_n$ for some $g_n \in \mathcal{B}_{\alpha_n}^+(M)$ and $\alpha_n < \gamma$. By hypothesis, $\mathcal{B}_{\alpha_n}^+(M) = \mathcal{B}_{\alpha_n} \cap \{0 \leq f \leq M\}$ for all $n \in \mathbb{N}$. Therefore, $g = \lim_{n\to\infty} g_n$ for some $g_n \in \mathcal{B}_{\alpha_n} \cap \{0 \leq f \leq M\}$ and $\alpha_n < \gamma$. So $g \in \mathcal{B}_\gamma \cap \{0 \leq f \leq M\}$.

Conversely, let $h \in \mathcal{B}_\gamma \cap \{0 \leq f \leq M\}$. Since $h \in \mathcal{B}_\gamma$, there are $\alpha_n < \gamma$ and $h_n \in \mathcal{B}_{\alpha_n}$ such that $h = \lim_{n\to\infty} h_n$. Since $0 \leq h \leq M$, observe that $h = \lim_{n\to\infty} h'_n$, where $h'_n := \max\{0, \min\{h_n, M\}\}$. But $h'_n \in \mathcal{B}_{\alpha_n}$ by Claim 2 and $0 \leq h'_n \leq M$. From the inductive hypothesis, we deduce that $h'_n \in \mathcal{B}_{\alpha_n}^+(M)$ for all n. Consequently, $h \in \mathcal{B}_\gamma^+(M)$.

We have thus proved that the statement holds for γ. The claim follows by transfinite induction. ■

Claim 4.

$$\bigcup_{\gamma<\beta} \mathcal{B}_\gamma^+(M) = \{f \text{ Borel function}\} \cap \{0 \leq f \leq M\}.$$

Proof of Claim 4. This is an immediate repercussion of Claims 1 and 3. ■

Let

$$Z^+(M) := Z \cap \{0 \leq f \leq M\}.$$

Claim 5.

$$Z^+(M) = \{f \text{ Borel function}\} \cap \{0 \leq f \leq M\}.$$

Proof of Claim 5. The inclusion $\subseteq$ is clear. In order to prove the opposite one, by virtue of Claim 4, it suffices to show that $Z^+(M) \supseteq \mathcal{B}^+_\gamma(M)$ for every $\gamma < \beta$. This is true for $\gamma = 0$. Let $\gamma < \beta$. Suppose that $Z^+(M) \supseteq \mathcal{B}^+_\alpha(M)$ for all $\alpha < \gamma$. Let $f \in \mathcal{B}^+_\gamma(M)$. Then $f = \lim_{n\to\infty} f_n$ for some $f_n \in \mathcal{B}^+_{\alpha_n}(M)$ and $\alpha_n < \gamma$. Lebesgue's dominated convergence theorem (Theorem A.1.38) then asserts that $(f_n)_{n=1}^\infty$ converges to f not only pointwise but also in $L^1(m_\varphi)$. Since $\mathcal{L}_\varphi : L^1(m_\varphi) \to L^1(m_\varphi)$ is a continuous operator, it follows that $(\mathcal{L}_\varphi(f_n))_{n=1}^\infty$ converges to $\mathcal{L}_\varphi(f)$ in $L^1(m_\varphi)$. By Lemma A.1.47 and Theorem A.1.48, there is a subsequence $(\mathcal{L}_\varphi(f_{n_k}))_{k=1}^\infty$ which converges pointwise m_φ-a. e. to $\mathcal{L}_\varphi(f)$. For m_φ-a. e. $x \in X$ we thereby get that

$$\begin{aligned}
\mathcal{L}_\varphi(f)(x) &= \lim_{k\to\infty} \mathcal{L}_\varphi(f_{n_k})(x) \\
&= \lim_{k\to\infty} \sum_{y\in T^{-1}(x)} f_{n_k}(y)e^{\varphi(y)} \quad \text{as } f_{n_k} \in \mathcal{B}^+_{\alpha_{n_k}}(M) \subseteq Z \\
&= \sum_{y\in T^{-1}(x)} \lim_{k\to\infty} f_{n_k}(y) \cdot e^{\varphi(y)} \\
&= \sum_{y\in T^{-1}(x)} f(y)e^{\varphi(y)}.
\end{aligned} \tag{13.29}$$

That is, $f \in Z$. Since $f \in \mathcal{B}^+_\gamma(M)$, we also know that $0 \le f \le M$. Thus, $f \in Z^+(M)$. So $Z^+(M) \supseteq \mathcal{B}^+_\gamma(M)$ for each $\gamma < \beta$. By transfinite induction, $Z^+(M) \supseteq \bigcup_{\gamma<\beta} \mathcal{B}^+_\gamma(M)$. This completes the demonstration of Claim 5. ∎

Now, let

$$Z^+ := Z \cap \{0 \le f\}.$$

Claim 6.

$$Z^+ = L^1(m_\varphi) \cap \{0 \le f\}.$$

Proof of Claim 6. The inclusion $\subseteq$ is clear. In order to prove the reverse one, first observe that by virtue of Claim 5 we know that

$$Z^+ \supseteq \bigcup_{M\ge 0} Z^+(M) = \{f \text{ nonnegative bounded Borel function}\}.$$

Let $f \in L^1(m_\varphi)$ be nonnegative. By Theorem A.1.17, there exists an increasing sequence $(s_n)_{n=1}^\infty$ of nonnegative simple Borel functions on X such that $\lim_{n\to\infty} s_n = f$. By the relation above, $s_n \in Z^+$ for all $n \in \mathbb{N}$. Moreover, the monotone convergence theorem (Theorem A.1.35) affirms that $(s_n)_{n=1}^\infty$ converges to f not only pointwise but also in $L^1(m_\varphi)$. Since $\mathcal{L}_\varphi : L^1(m_\varphi) \to L^1(m_\varphi)$ is a continuous operator, it follows that $(\mathcal{L}_\varphi(s_n))_{n=1}^\infty$ converges to $\mathcal{L}_\varphi(f)$ in $L^1(m_\varphi)$. By Lemma A.1.47 and Theorem A.1.48, there is a subsequence $(\mathcal{L}_\varphi(s_{n_k}))_{k=1}^\infty$ which converges pointwise m_φ-a. e. to $\mathcal{L}_\varphi(f)$. In fact, the entire sequence $(\mathcal{L}_\varphi(s_n))_{n=1}^\infty$ converges pointwise m_φ-a. e. to $\mathcal{L}_\varphi(f)$ since, like $(s_n)_{n=1}^\infty$, it is increasing. Recall that $s_n \in Z^+$ for all $n \in \mathbb{N}$. The calculation in (13.29) with f_{n_k} replaced by s_n confirms that $f \in Z$. So $Z^+ \supseteq L^1(m_\varphi) \cap \{0 \le f\}$. ■

Claim 7.

$$Z = L^1(m_\varphi).$$

Proof of Claim 7. By definition, $Z \subseteq L^1(m_\varphi)$. So let $g \in L^1(m_\varphi)$. Writing $g = g_+ - g_-$, we have that $g_+, g_- \in Z^+$ according to Claim 6. For m_φ-a. e. $x \in X$ we then obtain that

$$\begin{aligned}
\mathcal{L}_\varphi(g)(x) &= \mathcal{L}_\varphi(g_+)(x) - \mathcal{L}_\varphi(g_-)(x) \quad \text{as } \mathcal{L}_\varphi \text{ is linear} \\
&= \sum_{y\in T^{-1}(x)} g_+(y)e^{\varphi(y)} - \sum_{y\in T^{-1}(x)} g_-(y)e^{\varphi(y)} \quad \text{as } g_+, g_- \in Z \\
&= \sum_{y\in T^{-1}(x)} g(y)e^{\varphi(y)}.
\end{aligned}$$

That is, $g \in Z$. So $Z \supseteq L^1(m_\varphi)$. ■

This completes the proof of Theorem 13.6.3, which confirms that $\mathcal{L}_\varphi : C(X) \to C(X)$ extends continuously to a positive linear operator $\mathcal{L}_\varphi : L^1(m_\varphi) \to L^1(m_\varphi)$ defined m_φ-a. e. by formula (13.26). □

The following result is a consequence of Theorems 13.6.2–13.6.3.

Corollary 13.6.4. *$\mathcal{L}_\varphi^*(m_\varphi) = \Lambda_\varphi m_\varphi$ for the operator $\mathcal{L}_\varphi^* : L^1(m_\varphi)^* \to L^1(m_\varphi)^*$.*

Proof. Let $g \in L^1(m_\varphi)$. Since m_φ is regular and X is compact, the subspace $C(X)$ is dense in $L^1(m_\varphi)$, and hence there exists a sequence $(g_n)_{n=1}^\infty$ in $C(X)$ such that $g_n \to g$ in $L^1(m_\varphi)$ as $n \to \infty$. As $\mathcal{L}_\varphi : L^1(m_\varphi) \to L^1(m_\varphi)$ is a continuous operator according to Theorem 13.6.3 while $m_\varphi \in L^1(m_\varphi)^*$ is a continuous functional, it follows that

$$\begin{aligned}
\mathcal{L}_\varphi^*(m_\varphi)(g) := m_\varphi(\mathcal{L}_\varphi(g)) &= m_\varphi\Big(\mathcal{L}_\varphi\big(\lim_{n\to\infty} g_n\big)\Big) \\
&= m_\varphi\Big(\lim_{n\to\infty} \mathcal{L}_\varphi(g_n)\Big) \quad \text{as } \mathcal{L}_\varphi \text{ is continuous} \\
&= \lim_{n\to\infty} m_\varphi(\mathcal{L}_\varphi(g_n)) \quad \text{as } m_\varphi \text{ is continuous} \\
&= \lim_{n\to\infty} \mathcal{L}_\varphi^*(m_\varphi)(g_n) \\
&= \lim_{n\to\infty} \Lambda_\varphi m_\varphi(g_n) \quad \text{by Theorem 13.6.2} \\
&= \Lambda_\varphi m_\varphi\Big(\lim_{n\to\infty} g_n\Big) \quad \text{as } m_\varphi \text{ is continuous} \\
&= \Lambda_\varphi m_\varphi(g).
\end{aligned}$$

So $\mathcal{L}_\varphi^*(m_\varphi)(g) = \Lambda_\varphi m_\varphi(g)$ for all $g \in L^1(m_\varphi)$. □

Define the normalized operator $\widehat{\mathcal{L}}_\varphi : L^1(m_\varphi) \to L^1(m_\varphi)$ by $\widehat{\mathcal{L}}_\varphi := \Lambda_\varphi^{-1}\mathcal{L}_\varphi$. Its dual $\widehat{\mathcal{L}}_\varphi^* : L^1(m_\varphi)^* \to L^1(m_\varphi)^*$ satisfies $\widehat{\mathcal{L}}_\varphi^* = \Lambda_\varphi^{-1}\mathcal{L}_\varphi^*$. We further have the following result.

Corollary 13.6.5. *The measure m_φ is a fixed point of the normalized transfer-type operator's dual $\widehat{\mathcal{L}}_\varphi^* : L^1(m_\varphi)^* \to L^1(m_\varphi)^*$. Moreover,*
(a) $m_\varphi \circ T^{-1} \ll m_\varphi$ *and* $m_\varphi \circ T \ll m_\varphi$.
(b) *The Jacobian of m_φ is $\mathfrak{J}_{m_\varphi}^{-1}(z) = \Lambda_\varphi^{-1}e^{\varphi(z)}$.*
(c) $\operatorname{supp}(m_\varphi)$ *is completely T-invariant.*

Proof. For every $g \in L^1(m_\varphi)$,

$$\widehat{\mathcal{L}}_\varphi^*(m_\varphi)(g) = m_\varphi(\widehat{\mathcal{L}}_\varphi(g)) = \Lambda_\varphi^{-1}m_\varphi(\mathcal{L}_\varphi(g)) = \Lambda_\varphi^{-1}\mathcal{L}_\varphi^*(m_\varphi)(g) = m_\varphi(g)$$

by Corollary 13.6.4. Thus, $\widehat{\mathcal{L}}_\varphi^*(m_\varphi) = m_\varphi$, i. e., m_φ is a fixed point of $\widehat{\mathcal{L}}_\varphi^*$.

Statements (a) and (b) then follow from Proposition 13.5.2. The proof of (c) is a consequence of (a) and is left to the reader. □

A subsequent implication is the following.

Corollary 13.6.6. $\widehat{\mathcal{L}}_\varphi = \mathcal{L}_{m_\varphi}$.

Proof. Let $g \in L^1(m_\varphi)$. Using Proposition 13.4.2, Corollary 13.6.5(b) and Theorem 13.6.3, we get

$$\mathcal{L}_{m_\varphi}(g)(x) = \sum_{y\in T^{-1}(x)} g(y)\mathfrak{J}_{m_\varphi}^{-1}(y) = \Lambda_\varphi^{-1} \sum_{y\in T^{-1}(x)} g(y)e^{\varphi(y)} = \widehat{\mathcal{L}}_\varphi(g)(x)$$

for m_φ-a. e. $x \in X$. □

We now derive a formula for the iterates of the operator $\mathcal{L}_\varphi$. Let us denote by $\mathcal{L}_{T^n,S_n\varphi}$ the corresponding transfer-type operator for the nth iterate T^n of T under the potential $S_n\varphi$. In particular, $\mathcal{L}_{T,\varphi} = \mathcal{L}_\varphi$.

Proposition 13.6.7. *Let $T : X \to X$ be a surjective, open, distance expanding system and $\varphi \in C(X)$ a potential. Then $\mathcal{L}_\varphi^n = \mathcal{L}_{T^n,S_n\varphi}$ for every $n \in \mathbb{N}$. That is, for each $g \in L^1(m_\varphi)$,*

$$\mathcal{L}_\varphi^n(g)(x) = \sum_{y\in T^{-n}(x)} g(y)e^{S_n\varphi(y)} = \mathcal{L}_{T^n,S_n\varphi}(g)(x) \tag{13.30}$$

for m_φ-a. e. $x \in X$, where m_φ is an eigenmeasure for the dual operator $\mathcal{L}_\varphi^* : C(X)^* \to C(X)^*$.

Proof. Let $g \in L^1(m_\varphi)$. The case $n = 1$ corresponds to the definition of $\mathcal{L}_\varphi$. Thus, there exists $M_1 \in \mathcal{B}(X)$ such that $m_\varphi(M_1) = 1$ and on the entirety of which (13.30) holds with $n = 1$. Suppose now that the proposition holds for some $n \in \mathbb{N}$. Then there exists $M_n \in \mathcal{B}(X)$ such that $m_\varphi(M_n) = 1$ and on which (13.30) holds with g replaced by $\mathcal{L}_\varphi(g)$.

By Corollary 13.6.5, $m_\varphi \circ T^{-n} \ll m_\varphi$. So $m_\varphi(T^{-n}(M_n)) = 1$. Then $m_\varphi(T^{-n}(M_n) \cap M_1) = 1$ and for all $x \in T^{-n}(M_n) \cap M_1$ we obtain that

$$\begin{aligned}
\mathcal{L}_\varphi^{n+1}(g)(x) = \mathcal{L}_\varphi^n(\mathcal{L}_\varphi(g))(x) &= \sum_{y \in T^{-n}(x)} \mathcal{L}_\varphi(g)(y)e^{S_n\varphi(y)} \\
&= \sum_{y \in T^{-n}(x)} \Big[\sum_{z \in T^{-1}(y)} g(z)e^{\varphi(z)} \Big] e^{S_n\varphi(y)} \\
&= \sum_{y \in T^{-n}(x)} \sum_{z \in T^{-1}(y)} g(z)e^{S_n\varphi(y)}e^{\varphi(z)} \\
&= \sum_{y \in T^{-n}(x)} \sum_{z \in T^{-1}(y)} g(z)e^{S_n\varphi(T(z))+\varphi(z)} \\
&= \sum_{z \in T^{-(n+1)}(x)} g(z)e^{S_{n+1}\varphi(z)}.
\end{aligned}$$

This completes the inductive step of the proof. □

A direct consequence for the iterates of $\widehat{\mathcal{L}}_\varphi$ is the following.

Corollary 13.6.8. *$\widehat{\mathcal{L}}_\varphi^n = \Lambda_\varphi^{-n}\mathcal{L}_\varphi^n = \Lambda_\varphi^{-n}\mathcal{L}_{T^n,S_n\varphi} = \widehat{\mathcal{L}}_{T^n,S_n\varphi}$. Furthermore, $\widehat{\mathcal{L}}^*_{T^n,S_n\varphi}(m_\varphi) = m_\varphi$.*

Proof. This follows from Proposition 13.6.7 and Corollary 13.6.5. □

The last part of this corollary implies that we may set $m_{T^n,S_n\varphi} := m_\varphi$ since all that is required of $m_{T^n,S_n\varphi}$ is that it be a fixed point of $\widehat{\mathcal{L}}^*_{T^n,S_n\varphi}$.

13.6.2 Eigenmeasures are conformal measures

We now obtain an integral expression for the measure of backward images of small enough sets.

Proposition 13.6.9. *Let $T : X \to X$ be a surjective, open, distance expanding system and $\varphi \in C(X)$ be a potential. For all $x \in X$, all $n \in \mathbb{N}$ and all Borel sets $A \subseteq B(T^n(x), \xi)$, we have*

$$m_\varphi(T_x^{-n}(A)) = \Lambda_\varphi^{-n} \int_A e^{(S_n\varphi)\circ T_x^{-n}}\, dm_\varphi. \tag{13.31}$$

Recall that ξ is the constant arising from relations (4.29), (4.30) *and* (4.31)*; alternatively, see* (13.2).

Proof. By Corollary 13.6.5 and by the change-of-variables formula (13.19) applied to the homeomorphism $T_x : T_x^{-1}(B(T(x), \xi)) \to B(T(x), \xi)$, we have

$$m_\varphi(T_x^{-1}(A)) = \int_A \mathfrak{J}_{m_\varphi}^{-1} \circ T_x^{-1}\, dm_\varphi = \Lambda_\varphi^{-1} \int_A e^{\varphi\circ T_x^{-1}}\, dm_\varphi.$$

With (1) m_φ replaced by $m_{T^n,S_n\varphi}$; (2) T_x replaced by $T_x^n : T_x^{-n}(B(T^n(x),\xi)) \to B(T^n(x),\xi)$; (3) φ replaced by $S_n\varphi$; and (4) Λ_φ replaced by Λ_φ^n, we deduce that

$$m_\varphi(T_x^{-n}(A)) = m_{T^n,S_n\varphi}(T_x^{-n}(A)) = \Lambda_\varphi^{-n} \int_A e^{(S_n\varphi)\circ T_x^{-n}}\, dm_{T^n,S_n\varphi} = \Lambda_\varphi^{-n} \int_A e^{(S_n\varphi)\circ T_x^{-n}}\, dm_\varphi. \quad \square$$

For the measure of forward images of small enough sets, we infer the following integral expression, whose proof is left to the reader.

Corollary 13.6.10. *For all $x \in X$, all $n \in \mathbb{N}$ and all Borel sets $B \subseteq T_x^{-n}(B(T^n(x),\xi))$, we have*

$$m_\varphi(T^n(B)) = \Lambda_\varphi^n \int_B e^{-S_n\varphi}\, dm_\varphi. \tag{13.32}$$

More generally, we have the following statement.

Corollary 13.6.11. *For all $n \in \mathbb{N}$ and all Borel sets $B \subseteq X$ such that $T^n|_B$ is injective, we have*

$$m_\varphi(T^n(B)) = \Lambda_\varphi^n \int_B e^{-S_n\varphi}\, dm_\varphi. \tag{13.33}$$

Proof. Simply partition the Borel set B into finitely many Borel subsets B_i such that $B_i \subseteq T_{x_i}^{-n}(B(T^n(x_i),\xi))$ for some $x_i \in X$ and use Corollary 13.6.10. $\square$

In Chapters 16, 19, 21 and in several chapters of the third volume, we will see that measures that satisfy a property like (13.33) play an important role. We accordingly introduce a new notion.

Definition 13.6.12. Let $T : X \to X$ be a surjective, open, distance expanding system and $f \in C(X)$. A measure $m \in M(X)$ is said to be *f-conformal* if

$$m(T(B)) = \int_B f\, dm \tag{13.34}$$

for all Borel sets $B \subseteq X$ such that $T|_B$ is injective.

Corollary 13.6.11 affirms that m_φ is $(\Lambda_\varphi e^{-\varphi})$-conformal.

By a classical measure-theoretic argument, relation (13.34) implies that

$$\int_{T(B)} g\, dm = \int_B (g \circ T) f\, dm \tag{13.35}$$

for every g nonnegative or integrable on $T(B)$ and for every Borel set $B \subseteq X$ such that $T|_B$ is injective.

Lemma 13.6.13. *Let $T : X \to X$ be a surjective, open, distance expanding system and $\varphi \in C(X)$. A measure $m \in M(X)$ is $(\Lambda_\varphi e^{-\varphi})$-conformal if and only if m is a fixed point of the dual of the normalized transfer operator $\widehat{\mathcal{L}}_\varphi$, i. e., $\widehat{\mathcal{L}}^*_\varphi(m) = m$.*

Proof. Let m be a $(\Lambda_\varphi e^{-\varphi})$-conformal measure. There exists a finite Borel partition $\{X_i\}_{i=1}^n$ of X such that $T|_{X_i} : X_i \to T(X_i)$ is a homeomorphism for every $1 \le i \le n$. Indeed, recall that T (and all its iterates) has local inverse branches on all open balls of radius ξ (cf. (13.2)). The family $\{T_x^{-1}(B(T(x),\xi))\}_{x\in X}$ forms an open cover of X. The compactness of X implies that there exists a finite subcover $\{T_{x_i}^{-1}(B(T(x_i),\xi))\}_{i=1}^n$. Then define $X_1 := T_{x_1}^{-1}(B(T(x_1),\xi))$ and $X_k := T_{x_k}^{-1}(B(T(x_k),\xi)) \setminus \bigcup_{j<k} X_j$ for all $1 < k \le n$. Let $g \in L^1(m)$. Fix temporarily $1 \le i \le n$. Replacing g by $(ge^\varphi) \circ T|_{X_i}^{-1}$ in (13.35), we obtain that

$$\int_{T(X_i)} (ge^\varphi) \circ T|_{X_i}^{-1}\, dm = \int_{X_i} ((ge^\varphi) \circ T|_{X_i}^{-1} \circ T)(\Lambda_\varphi e^{-\varphi})\, dm = \Lambda_\varphi \int_{X_i} g\, dm.$$

Summing over i yields

$$\begin{aligned}
\Lambda_\varphi \int_X g\, dm &= \Lambda_\varphi \sum_{i=1}^n \int_{X_i} g\, dm \\
&= \sum_{i=1}^n \int_{T(X_i)} (ge^\varphi) \circ T|_{X_i}^{-1}\, dm \\
&= \int_X \sum_{i=1}^n \mathbb{1}_{T(X_i)}(x) \cdot (ge^\varphi)(T|_{X_i}^{-1}(x))\, dm(x) \\
&= \int_X \sum_{y\in T^{-1}(x)} g(y)e^{\varphi(y)}\, dm(x).
\end{aligned}$$

Therefore,

$$m(g) = \int_X g\, dm = \Lambda_\varphi^{-1} \int_X \mathcal{L}_\varphi(g)(x)\, dm(x) = m(\widehat{\mathcal{L}}_\varphi(g)) = \widehat{\mathcal{L}}^*_\varphi(m)(g).$$

Since this holds for all $g \in L^1(m)$, we conclude that $\widehat{\mathcal{L}}^*_\varphi(m) = m$.

The converse has already been proved for m_φ. Let us repeat the argument. Assume that $\widehat{\mathcal{L}}^*_\varphi$ fixes m. Let B be a Borel set such that $T|_B$ is injective. Then

$$\int_B f\, dm = \int_X (\mathbb{1}_B \cdot f)\, d\widehat{\mathcal{L}}^*_\varphi(m) = \Lambda_\varphi^{-1} \int_X \mathcal{L}_\varphi(\mathbb{1}_B \cdot f)\, dm = \Lambda_\varphi^{-1} \int_X \sum_{y\in T^{-1}(x)} \mathbb{1}_B(y) f(y) e^{\varphi(y)}\, dm(x).$$

Letting $f = \Lambda_\varphi e^{-\varphi}$ results in

$$\int_B \Lambda_\varphi e^{-\varphi}\,dm = \int_X \sum_{y\in T^{-1}(x)} \mathbb{1}_B(y)\,dm(x) = m(T(B)).$$

Thus, m is $(\Lambda_\varphi e^{-\varphi})$-conformal. □

13.6.3 Eigenmeasures are Gibbs states for transitive systems

By definition, Gibbs measures are supported on the entire space. So in order for m_φ to have a chance to be a Gibbs state, we need to make sure that it has full support. However, according to Corollary 13.6.5, the support of m_φ is completely T-invariant. So is its complement $X \setminus \mathrm{supp}(m_\varphi)$, which is an open set. Thus, the forward orbit of $X \setminus \mathrm{supp}(m_\varphi)$ is contained in $X \setminus \mathrm{supp}(m_\varphi)$, and hence the only way this forward orbit can cover X is if $X \setminus \mathrm{supp}(m_\varphi) = X$, i. e., $\mathrm{supp}(m_\varphi) = \emptyset$, which is impossible. So the forward orbit of $X \setminus \mathrm{supp}(m_\varphi)$ cannot cover the entire space. If the system T is such that all nonempty open sets have a forward orbit which covers the space, i. e., if T is strongly transitive, then m_φ must be such that $X\setminus\mathrm{supp}(m_\varphi) = \emptyset$, i. e., $\mathrm{supp}(m_\varphi) = X$. In fact, in Lemma 4.2.10 it was observed that the notions of transitivity and (very) strong transitivity coincide for open, distance expanding systems.

Proposition 13.6.14. *If $T : X \to X$ is a topologically transitive, open and distance expanding system, then T is very strongly transitive and* $\mathrm{supp}(m_\varphi) = X$.

Proof. The proof of the very strong transitivity has been given in Lemma 4.2.10. Strong transitivity means that the forward orbit of any nonempty open set covers the space in finite time, i. e., for every nonempty open set $U \subseteq X$ there is $N = N(U) \in \mathbb{N}$ such that $\bigcup_{j=0}^N T^j(U) = X$.

If it turned out that $m_\varphi(U) = 0$ then, since $m_\varphi \circ T \ll m_\varphi$ according to Corollary 13.6.5, we would deduce that $m_\varphi(T^n(U)) = 0$ for all $n \in \mathbb{N}$, and thus $m_\varphi(X) = m(\bigcup_{j=0}^N T^j(U)) = 0$. But $m_\varphi(X) = 1$, so $m_\varphi(U) > 0$. □

Measures with full support confer to balls of any given radius weights that are uniformly bounded away from 0.

Lemma 13.6.15. *If μ is a Borel probability measure on a compact metrizable space X and* $\mathrm{supp}(\mu) = X$, *then*

$$\inf\{\mu(B(x,r)) : x \in X\} > 0$$

for each $r > 0$.

Proof. Fix $r > 0$. Let F be a finite $(r/2)$-spanning set for X. This means that for every point $x \in X$, there exists a point $y \in F$ with $x \in B(y, r/2)$. Then $B(y, r/2) \subseteq B(x, r)$, and hence $\mu(B(x,r)) \ge \min\{\mu(B(z,r/2)) : z \in F\} > 0$. Since x is arbitrary, taking the infimum on the left-hand side gives the desired result. □

The forthcoming proposition is particularly important. It asserts that m_φ is a Gibbs state and identifies Λ_φ as $e^{\mathrm{P}(T,\varphi)}$.

Proposition 13.6.16. *If $T : X \to X$ is a transitive, open, distance expanding system and $\varphi : X \to \mathbb{R}$ is a Hölder continuous potential, then the measure m_φ is a Gibbs state for φ and* $\log \Lambda_\varphi = \mathrm{P}(T,\varphi)$.

Proof. Suppose $\varphi \in H_\alpha(X)$. Let $0 < \xi' \le \xi$. Let also $x \in X$ and $n \in \mathbb{N}$. In light of Proposition 13.6.14, if we set $\mu = m_\varphi$ and $r = \xi'$ in Lemma 13.6.15 we deduce that

$$\kappa(\xi') := \inf_{y \in X} m_\varphi(B(T^n(y), \xi')) > 0.$$

Furthermore, by Proposition 13.6.9 and Lemma 13.1.1 we obtain that

$$\begin{aligned} m_\varphi(T_x^{-n}(B(T^n(x),\xi'))) &= \Lambda_\varphi^{-n} \int_{B(T^n(x),\xi')} e^{S_n\varphi(T_x^{-n}(z))}\, dm_\varphi(z) \\ &\le \Lambda_\varphi^{-n} D_\alpha(\varphi) e^{S_n\varphi(x)} m_\varphi(B(T^n(x),\xi')) \\ &\le D_\alpha(\varphi) e^{S_n\varphi(x) - n\log\Lambda_\varphi}. \end{aligned} \tag{13.36}$$

Similarly,

$$\begin{aligned} m_\varphi(T_x^{-n}(B(T^n(x),\xi'))) &= \Lambda_\varphi^{-n} \int_{B(T^n(x),\xi')} e^{S_n\varphi(T_x^{-n}(z))}\, dm_\varphi(z) \\ &\ge \Lambda_\varphi^{-n} D_\alpha(\varphi)^{-1} e^{S_n\varphi(x)} m_\varphi(B(T^n(x),\xi')) \\ &\ge D_\alpha(\varphi)^{-1} \kappa(\xi') e^{S_n\varphi(x) - n\log\Lambda_\varphi}. \end{aligned} \tag{13.37}$$

Per Remark 13.2.2, m_φ is a Gibbs state for φ with constant $C(\xi') = D_\alpha(\varphi) \cdot (\kappa(\xi'))^{-1}$ (or $C = C(\xi)$ in Definition 13.2.1) and $\log \Lambda_\varphi = \mathrm{P}(T,\varphi)$ by Proposition 13.2.4. □

13.6.4 Ergodicity of the eigenmeasures for transitive systems

We now demonstrate that the Gibbs measure m_φ, whose existence has just been established under the additional assumptions that T is transitive and φ Hölder, is ergodic. To establish this fact, we need the following lemma.

Lemma 13.6.17. *Let $T : X \to X$ be a transitive, open, distance expanding system and $\varphi : X \to \mathbb{R}$ a Hölder continuous potential. Let E be a finite ξ-spanning set for X and let*

$A \in \mathcal{B}(X)$ be such that $m_\varphi(A) > 0$. Then $\forall q \geq 0\ \exists c_q > 0\ \forall \varepsilon > 0\ \exists k_\varepsilon \in \mathbb{N}\ \forall k \geq k_\varepsilon\ \exists z_k \in E$ such that

$$m_\varphi(T^q(B(z_k,\xi) \setminus T^k(A))) \leq c_q \frac{\varepsilon}{m_\varphi(A)}.$$

Proof. Let $q \geq 0$ and $\varepsilon > 0$. Recall that every Borel probability measure on a compact metrizable space X is regular. By the inner regularity of m_φ, it suffices to prove the lemma for all compact sets A. As in the proof of Proposition 13.2.4, for each $k \in \mathbb{N}$ put

$$X_k := \bigcup_{e \in E} X_k(e),$$

where E is a finite ξ-spanning set and

$$X_k(e) := \{x \in T^{-k}(e) \mid A \cap T_x^{-k}(B(T^k(x),\xi)) \neq \emptyset\}.$$

Like in (13.16),

$$A \subseteq \bigcup_{x \in X_k} T_x^{-k}(B(T^k(x),\xi)) \subseteq B(A, 2\lambda^{-k}\xi). \tag{13.38}$$

On one hand, it ensues that

$$\sum_{x \in X_k} m_\varphi(T_x^{-k}(B(T^k(x),\xi))) \geq m_\varphi(A). \tag{13.39}$$

On the other hand, by the outer regularity of m_φ and the compactness of A, there exists $k_\varepsilon \in \mathbb{N}$ so large that for all $k \geq k_\varepsilon$,

$$m_\varphi(B(A, 2\lambda^{-k}\xi) \setminus A) \leq \varepsilon. \tag{13.40}$$

For each $e \in E$, the family $\{T_x^{-k}(B(e,\xi))\}_{x \in T^{-k}(e)}$ consists of preimages under different inverse branches of T^k at e. According to Lemma 4.2.8, these preimages are mutually disjoint. This and relations (13.38) and (13.40) give

$$\begin{aligned} \sum_{x \in X_k} m_\varphi(T_x^{-k}(B(T^k(x),\xi)) \setminus A) &= \sum_{e \in E} \sum_{x \in X_k(e)} m_\varphi(T_x^{-k}(B(T^k(x),\xi)) \setminus A) \\ &= \sum_{e \in E} m_\varphi\Big(\bigcup_{x \in X_k(e)} T_x^{-k}(B(T^k(x),\xi)) \setminus A\Big) \\ &\leq \sum_{e \in E} m_\varphi(B(A, 2\lambda^{-k}\xi) \setminus A) \\ &\leq \#E \cdot \varepsilon. \end{aligned} \tag{13.41}$$

From (13.39) and (13.41), it follows that

$$\frac{\sum_{x\in X_k} m_\varphi(T_x^{-k}(B(T^k(x),\xi))\setminus A)}{\sum_{x\in X_k} m_\varphi(T_x^{-k}(B(T^k(x),\xi)))} \le \frac{\#E\cdot\varepsilon}{m_\varphi(A)}. \tag{13.42}$$

We claim that there exists $y \in X_k$ such that

$$\frac{m_\varphi(T_y^{-k}(B(T^k(y),\xi))\setminus A)}{m_\varphi(T_y^{-k}(B(T^k(y),\xi)))} \le \frac{\#E\cdot\varepsilon}{m_\varphi(A)}. \tag{13.43}$$

Otherwise, i. e., if this ratio was greater than $\frac{\#E\cdot\varepsilon}{m_\varphi(A)}$ for every $y \in X_k$, summing over all y would contradict (13.42). Set $z_k := T^k(y) \in E$. Using Corollary 13.6.10, Proposition 13.6.16 and Lemma 13.1.1, we get

$$\begin{aligned}
\frac{m_\varphi(B(z_k,\xi)\setminus T^k(A))}{m_\varphi(B(z_k,\xi))} &\le \frac{m_\varphi(T^k(T_y^{-k}(B(z_k,\xi))\setminus A))}{m_\varphi(B(z_k,\xi))} \\
&= \frac{1}{m_\varphi(B(z_k,\xi))}\int\limits_{T_y^{-k}(B(z_k,\xi))\setminus A} e^{\mathrm{P}(T,\varphi)k-S_k\varphi}\,dm_\varphi \\
&\le \frac{D_\alpha(\varphi)e^{\mathrm{P}(T,\varphi)k-S_k\varphi(y)}m_\varphi(T_y^{-k}(B(z_k,\xi))\setminus A)}{m_\varphi(T^k(T_y^{-k}(B(z_k,\xi))))} \\
&\le \frac{D_\alpha(\varphi)e^{\mathrm{P}(T,\varphi)k-S_k\varphi(y)}m_\varphi(T_y^{-k}(B(z_k,\xi))\setminus A)}{D_\alpha(\varphi)^{-1}e^{\mathrm{P}(T,\varphi)k-S_k\varphi(y)}m_\varphi(T_y^{-k}(B(z_k,\xi)))} \\
&\le \frac{(D_\alpha(\varphi))^2\cdot\#E\cdot\varepsilon}{m_\varphi(A)}.
\end{aligned} \tag{13.44}$$

Let $0 < \eta \le \xi$ be a Lebesgue number for the open cover $\{T_x^{-q}(B(T^q(x),\xi))\}_{x\in X}$ of the compact set X. Let $\mathcal{A} = \{A_1, A_2, \ldots, A_n\}$ be a finite Borel partition of X with $\operatorname{diam}(\mathcal{A}) < \eta$. Using Corollary 13.6.10 and Proposition 13.6.16 once again, for any Borel subset Y of X we obtain that

$$\begin{aligned}
m_\varphi(T^q(Y)) &= m_\varphi\Big(\bigcup_{i=1}^n T^q(Y\cap A_i)\Big) \\
&\le \sum_{i=1}^n m_\varphi(T^q(Y\cap A_i)) \\
&= \sum_{i=1}^n \int\limits_{Y\cap A_i} e^{\mathrm{P}(T,\varphi)q-S_q\varphi}\,dm_\varphi \\
&\le e^{\mathrm{P}(T,\varphi)q+\|\varphi\|_\infty q}\sum_{i=1}^n m_\varphi(Y\cap A_i) = e^{q(\mathrm{P}(T,\varphi)+\|\varphi\|_\infty)}m_\varphi(Y).
\end{aligned} \tag{13.45}$$

From (13.44) and (13.45), we conclude that

$$\begin{aligned} m_\varphi(T^q(B(z_k,\xi)\setminus T^k(A))) &\le e^{q(\mathrm{P}(T,\varphi)+\|\varphi\|_\infty)} m_\varphi(B(z_k,\xi)\setminus T^k(A)) \\ &\le (D_\alpha(\varphi))^2\cdot \#E\cdot e^{q(\mathrm{P}(T,\varphi)+\|\varphi\|_\infty)}\frac{\varepsilon}{m_\varphi(A)}. \end{aligned}$$

Setting $c_q = (D_\alpha(\varphi))^2\cdot \#E\cdot e^{q(\mathrm{P}(T,\varphi)+\|\varphi\|_\infty)}$, we are done. □

Theorem 13.6.18. *If $T : X \to X$ is a transitive, open, distance expanding system and $\varphi : X \to \mathbb{R}$ is a Hölder continuous potential, then m_φ is ergodic.*

Proof. Let $A \in \mathcal{B}(X)$ be such that $T^{-1}(A) = A$. Then $T^n(A) \subseteq A$ for all $n \in \mathbb{N}$. Suppose further that $m_\varphi(A) > 0$. Let E be a finite ξ-spanning set. By Proposition 13.6.14, there exists $Q \ge 0$ such that

$$\bigcup_{q=0}^{Q} T^q(B(z,\xi)) = X, \quad \forall z \in E.$$

Let $\varepsilon > 0$. Take $c_0,\dots,c_Q$, take k big enough, and $z_k \in E$ satisfying Lemma 13.6.17. Then

$$\begin{aligned} m_\varphi(X\setminus A) &\le m_\varphi\Big(X \setminus \bigcup_{q=0}^{Q} T^{q+k}(A)\Big) \\ &= m_\varphi\Big(\bigcup_{q=0}^{Q} T^q(B(z_k,\xi)) \setminus \bigcup_{q=0}^{Q} T^{q+k}(A)\Big) \\ &\le m_\varphi\Big(\bigcup_{q=0}^{Q} T^q(B(z_k,\xi)) \setminus T^{q+k}(A)\Big) \\ &\le \sum_{q=0}^{Q} m_\varphi\Big(T^q(B(z_k,\xi)\setminus T^k(A))\Big) \\ &\le \frac{\varepsilon}{m_\varphi(A)}\sum_{q=0}^{Q} c_q. \end{aligned}$$

Since this is true for an arbitrary $\varepsilon > 0$, we deduce that $m_\varphi(X\setminus A) = 0$. Hence, m_φ is ergodic. □

13.6.5 Metric exactness of the eigenmeasures for topologically exact systems

Theorem 13.6.19. *If $T : X \to X$ is a topologically exact, open, distance expanding system and $\varphi : X \to \mathbb{R}$ is a Hölder continuous potential, then m_φ is metrically exact.*

Proof. Let $A \in \mathcal{B}(X)$ be such that $m_\varphi(A) > 0$. Let E be a finite ξ-spanning set. The topological exactness of T guarantees that there exists some $q \ge 0$ such that $T^q(B(z,\xi)) = X$

for all $z \in E$. Let $\varepsilon > 0$. Take c_q, k_ε, and for all $k \geq k_\varepsilon$ pick $z_k \in E$ so that Lemma 13.6.17 is satisfied. Then

$$m_\varphi(X \setminus T^{q+k}(A)) = m_\varphi(T^q(B(z_k,\xi)) \setminus T^{q+k}(A)) \leq m_\varphi(T^q(B(z_k,\xi) \setminus T^k(A))) \leq c_q \frac{\varepsilon}{m_\varphi(A)}.$$

Thus, for all $k \geq k_\varepsilon$,

$$m_\varphi(T^{q+k}(A)) \geq 1 - c_q \frac{\varepsilon}{m_\varphi(A)}.$$

So

$$\liminf_{n\to\infty} m_\varphi(T^n(A)) \geq 1 - c_q \frac{\varepsilon}{m_\varphi(A)}.$$

Since this holds for an arbitrary $\varepsilon > 0$, we conclude that

$$\lim_{n\to\infty} m_\varphi(T^n(A)) = 1,$$

i. e., m_φ is metrically exact (cf. Definition 8.4.4 and Proposition 8.4.5). □

13.7 Invariant Gibbs states

In this section, we will show that under the assumptions that T is transitive and φ is Hölder continuous, there exists a unique invariant Gibbs state and that this state is ergodic. Recall that invariant Gibbs states are equilibrium states according to Proposition 13.2.9. We will prove conversely that every equilibrium state is an invariant Gibbs state. More precisely, we will:

(a) Find a fixed point ϱ_φ of $\widehat{\mathcal{L}}_\varphi : L^1(m_\varphi) \to L^1(m_\varphi)$ such that $\varrho_\varphi \geq 0$ and $\int_X \varrho_\varphi \, dm_\varphi = 1$. Thus, $\mathcal{L}_{m_\varphi}(\varrho_\varphi) = \widehat{\mathcal{L}}_\varphi(\varrho_\varphi) = \varrho_\varphi$ and Theorem 13.4.1 affirms that the measure $\mu_\varphi := \varrho_\varphi m_\varphi$ is T-invariant.
(b) Further show that ϱ_φ, which is the Radon–Nikodym derivative $\frac{d\mu_\varphi}{dm_\varphi}$, is a Hölder continuous function uniformly bounded away from 0 and ∞. It follows from Lemma 13.2.3 that μ_φ is a Gibbs measure, like m_φ. It is also ergodic, like m_φ.
(c) Prove the uniqueness of an invariant Gibbs state.
(d) Demonstrate that a measure is an invariant Gibbs state if and only if it is an equilibrium state.

13.7.1 Almost periodicity of normalized transfer operators

We first examine a few more properties of the iterates of the transfer operator $\mathcal{L}_\varphi$ and its normalized counterpart $\widehat{\mathcal{L}}_\varphi$.

Lemma 13.7.1. *If $T : X \to X$ is a topologically transitive, open, distance expanding system and $\varphi \in H_\alpha(X)$, then for all $n \in \mathbb{N}$ and all $x, y \in X$ we have*

$$(\overline{D}_\alpha(\varphi))^{-1} \le \frac{\mathcal{L}_\varphi^n(\mathbb{1})(y)}{\mathcal{L}_\varphi^n(\mathbb{1})(x)} = \frac{\widehat{\mathcal{L}}_\varphi^n(\mathbb{1})(y)}{\widehat{\mathcal{L}}_\varphi^n(\mathbb{1})(x)} \le \overline{D}_\alpha(\varphi).$$

Proof. This follows from Proposition 13.6.7 and Corollary 13.1.2. □

It ensues from this lemma that the functions $\widehat{\mathcal{L}}_\varphi^n(\mathbb{1})$, $n \in \mathbb{N}$, are uniformly bounded away from 0 and ∞.

Proposition 13.7.2. *If $T : X \to X$ is a transitive, open, distance expanding system and $\varphi \in H_\alpha(X)$, then*

$$(\overline{D}_\alpha(\varphi))^{-1} \le \inf \widehat{\mathcal{L}}_\varphi^n(\mathbb{1}) \le \sup \widehat{\mathcal{L}}_\varphi^n(\mathbb{1}) \le \overline{D}_\alpha(\varphi), \quad \forall n \in \mathbb{N}.$$

Proof. Fix $y \in X$. Using Lemma 13.7.1, Corollary 13.6.6 as well as (13.21) with $m = m_\varphi$, we obtain that

$$\begin{aligned}\widehat{\mathcal{L}}_\varphi^n(\mathbb{1})(y) \le \overline{D}_\alpha(\varphi) \int_X \widehat{\mathcal{L}}_\varphi^n(\mathbb{1})(x)\, dm_\varphi(x) &= \overline{D}_\alpha(\varphi) \int_X \mathcal{L}_{m_\varphi}^n(\mathbb{1})(x)\, dm_\varphi(x) \\ &= \overline{D}_\alpha(\varphi) \int_X \mathbb{1}(x)\, dm_\varphi(x) = \overline{D}_\alpha(\varphi).\end{aligned}$$

The lower bound results from a similar argument. □

In particular, this means that the supremum norms of the iterates of the normalized restricted operator $\widehat{\mathcal{L}}_\varphi : C(X) \to C(X)$ are uniformly bounded from above (by $\overline{D}_\alpha(\varphi)$).

Let us recall the concept of almost periodicity of an operator.

Definition 13.7.3. A bounded linear operator $L : E \to E$ on a Banach space E is said to be *almost periodic* if the forward orbit of every point $x \in E$ under L, i. e., the set $\{L^n(x) : n \ge 0\}$, is relatively compact in E.

Theorem 13.7.4. *If $T : X \to X$ is a transitive, open, distance expanding system and $\varphi : X \to \mathbb{R}$ is a Hölder continuous potential, then for every $g \in C(X)$ the family $\{\widehat{\mathcal{L}}_\varphi^n(g)\}_{n=0}^\infty$ is equicontinuous and pointwise bounded in $C(X)$. Hence the operator $\widehat{\mathcal{L}}_\varphi : C(X) \to C(X)$ is almost periodic.*

Proof. Let $g \in C(X)$ and $n \ge 0$. Fix $x, y \in X$ with $d(x,y) < \xi$. Let $\widetilde{\varphi} = \varphi - \mathrm{P}(T,\varphi)$. Then $S_n\widetilde{\varphi} = S_n\varphi - n\mathrm{P}(T,\varphi)$ and one can set $C_\alpha(\widetilde{\varphi}) = C_\alpha(\varphi)$ and $D_\alpha(\widetilde{\varphi}) = D_\alpha(\varphi)$ in Lemma 13.1.1. Observe also that

$$\widehat{\mathcal{L}}_\varphi = e^{-\mathrm{P}(T,\varphi)}\mathcal{L}_\varphi = \mathcal{L}_{\widetilde{\varphi}}, \quad \text{and hence} \quad \widehat{\mathcal{L}}_\varphi^n = e^{-n\mathrm{P}(T,\varphi)}\mathcal{L}_\varphi^n = \mathcal{L}_{\widetilde{\varphi}}^n.$$

Using Proposition 13.6.7, the second part of the proof of Lemma 13.1.1, that lemma itself, the mean value theorem, and Proposition 13.7.2, we deduce that

$$
\begin{aligned}
&|\widehat{\mathcal{L}}_\varphi^n(g)(y) - \widehat{\mathcal{L}}_\varphi^n(g)(x)| \\
&= |\mathcal{L}_{\tilde{\varphi}}^n(g)(y) - \mathcal{L}_{\tilde{\varphi}}^n(g)(x)| \\
&= \Big| \sum_{z\in T^{-n}(y)} e^{S_n\tilde{\varphi}(T_z^{-n}(y))} g(T_z^{-n}(y)) - \sum_{w\in T^{-n}(x)} e^{S_n\tilde{\varphi}(T_w^{-n}(x))} g(T_w^{-n}(x)) \Big| \\
&= \Big| \sum_{w\in T^{-n}(x)} \Big[e^{S_n\tilde{\varphi}(T_w^{-n}(y))} g(T_w^{-n}(y)) - e^{S_n\tilde{\varphi}(T_w^{-n}(x))} g(T_w^{-n}(x)) \Big] \Big| \\
&\le \sum_{w\in T^{-n}(x)} e^{S_n\tilde{\varphi}(T_w^{-n}(y))} |g(T_w^{-n}(y)) - g(T_w^{-n}(x))| \\
&\quad + \sum_{w\in T^{-n}(x)} |g(T_w^{-n}(x))| \Big| e^{S_n\tilde{\varphi}(T_w^{-n}(y))} - e^{S_n\tilde{\varphi}(T_w^{-n}(x))} \Big| \\
&\le \sum_{w\in T^{-n}(y)} e^{S_n\tilde{\varphi}(T_w^{-n}(y))} \max\Big\{ |g(T_z^{-n}(y)) - g(T_z^{-n}(x))| : z \in T^{-n}(x) \Big\} \\
&\quad + \sum_{w\in T^{-n}(x)} |g(T_w^{-n}(x))| D_\alpha(\varphi) e^{S_n\tilde{\varphi}(T_w^{-n}(x))} |S_n\varphi(T_w^{-n}(y)) - S_n\varphi(T_w^{-n}(x))| \\
&\le \widehat{\mathcal{L}}_\varphi^n(\mathbb{1})(y) \max\Big\{ |g(y') - g(x')| : d(y',x') \le \lambda^{-n} d(y,x) \Big\} \\
&\quad + \|g\|_\infty D_\alpha(\varphi) \widehat{\mathcal{L}}_\varphi^n(\mathbb{1})(x) C_\alpha(\varphi)\, d^\alpha(x,y) \\
&\le \overline{D}_\alpha(\varphi) \max_{d(y',x')\le\lambda^{-n}d(y,x)} |g(y') - g(x')| + \|g\|_\infty D_\alpha(\varphi) \overline{D}_\alpha(\varphi) C_\alpha(\varphi)\, d^\alpha(x,y) \\
&= \overline{D}_\alpha(\varphi) \max_{d(y',x')\le\lambda^{-n}d(y,x)} |g(y') - g(x')| + \widetilde{D}_\alpha(\varphi) \|g\|_\infty\, d^\alpha(x,y),
\end{aligned}
\tag{13.46}
$$

where

$$
\widetilde{D}_\alpha(\varphi) := C_\alpha(\varphi) D_\alpha(\varphi) \overline{D}_\alpha(\varphi). \tag{13.47}
$$

As g is uniformly continuous, we conclude that the family $\{\widehat{\mathcal{L}}_\varphi^n(g)\}_{n=0}^\infty$ is equicontinuous. By Proposition 13.7.2, this family is also uniformly bounded since

$$
\|\widehat{\mathcal{L}}_\varphi^n(g)\|_\infty \le \|\widehat{\mathcal{L}}_\varphi^n(\|g\|_\infty \cdot \mathbb{1})\|_\infty = \|g\|_\infty \|\widehat{\mathcal{L}}_\varphi^n(\mathbb{1})\|_\infty \le \overline{D}_\alpha(\varphi) \|g\|_\infty, \quad \forall n \ge 0.
$$

The assumptions of the Arzelà–Ascoli theorem (Theorem A.2.3) being satisfied, we conclude that the family $\{\widehat{\mathcal{L}}_\varphi^n(g)\}_{n=0}^\infty$ is relatively compact. Consequently, the operator $\widehat{\mathcal{L}}_\varphi : C(X) \to C(X)$ is almost periodic. □

13.7.2 Existence, uniqueness and ergodicity of invariant Gibbs states for transitive systems

Lemma 13.7.5. *If $T : X \to X$ is a topologically transitive, open, distance expanding system and $\varphi \in H_\alpha(X)$, then there exists $\varrho_\varphi \in H_\alpha(X)$ with the following properties:*
(a) $\widehat{\mathcal{L}}_\varphi(\varrho_\varphi) = \varrho_\varphi$.
(b) $(\overline{D}_\alpha(\varphi))^{-1} \le \varrho_\varphi \le \overline{D}_\alpha(\varphi)$.
(c) $\int_X \varrho_\varphi \, dm_\varphi = 1$.
(d) $v_\alpha(\varrho_\varphi) \le \max\{\widetilde{D}_\alpha(\varphi), 2\xi^{-\alpha}\overline{D}_\alpha(\varphi)\}$,

where $\widetilde{D}_\alpha(\varphi)$ is defined in (13.47) *and $\overline{D}_\alpha(\varphi)$ in Corollary* 13.1.2.

Proof. It is a general fact, proved via the triangle inequality, that if a family $\{g_n\}_{n=0}^\infty$ is equicontinuous, then so are its Cesàro averages $\{\frac{1}{n}\sum_{j=0}^{n-1} g_j\}_{n=1}^\infty$. Per Theorem 13.7.4, the family $\{\widehat{\mathcal{L}}_\varphi^n(g)\}_{n=0}^\infty$ is equicontinuous for every $g \in C(X)$. In particular, $\{\widehat{\mathcal{L}}_\varphi^n(\mathbb{1})\}_{n=0}^\infty$ is equicontinuous. Consequently, the Cesàro family $\{\frac{1}{n}\sum_{j=0}^{n-1} \widehat{\mathcal{L}}_\varphi^j(\mathbb{1})\}_{n=1}^\infty$ is equicontinuous. By Proposition 13.7.2, we further know that this Cesàro family is uniformly bounded from below by $(\overline{D}_\alpha(\varphi))^{-1}$ and from above by $\overline{D}_\alpha(\varphi)$. By the Arzelà–Ascoli theorem (Theorem A.2.3), there is then a strictly increasing sequence $(n_k)_{k=1}^\infty$ such that the sequence $(\frac{1}{n_k}\sum_{j=0}^{n_k-1} \widehat{\mathcal{L}}_\varphi^j(\mathbb{1}))_{k=1}^\infty$ converges uniformly to a continuous function, denoted by ϱ_φ, which is bounded below by $(\overline{D}_\alpha(\varphi))^{-1}$ and above by $\overline{D}_\alpha(\varphi)$. This establishes (b).

To see that ϱ_φ satisfies part (a), since $\widehat{\mathcal{L}}_\varphi : C(X) \to C(X)$ is a continuous linear operator we obtain that

$$\begin{aligned}
\widehat{\mathcal{L}}_\varphi(\varrho_\varphi) &= \widehat{\mathcal{L}}_\varphi\Big(\lim_{k\to\infty}\frac{1}{n_k}\sum_{j=0}^{n_k-1}\widehat{\mathcal{L}}_\varphi^j(\mathbb{1})\Big) = \lim_{k\to\infty}\frac{1}{n_k}\sum_{j=0}^{n_k-1}\widehat{\mathcal{L}}_\varphi^{j+1}(\mathbb{1}) \\
&= \lim_{k\to\infty}\Big[\frac{1}{n_k}\sum_{j=0}^{n_k-1}\widehat{\mathcal{L}}_\varphi^j(\mathbb{1}) + \frac{1}{n_k}(\widehat{\mathcal{L}}_\varphi^{n_k}(\mathbb{1}) - \mathbb{1})\Big] \\
&= \lim_{k\to\infty}\frac{1}{n_k}\sum_{j=0}^{n_k-1}\widehat{\mathcal{L}}_\varphi^j(\mathbb{1}) \quad \text{using Proposition 13.7.2} \\
&= \varrho_\varphi.
\end{aligned}$$

We now show that ϱ_φ fulfills (c). By Corollary 13.6.5, we have $(\widehat{\mathcal{L}}_\varphi^j)^*(m_\varphi) = m_\varphi$ for each $j \in \mathbb{N}$ and in particular $m_\varphi(\widehat{\mathcal{L}}_\varphi^j(\mathbb{1})) = m_\varphi(\mathbb{1}) = 1$. By Lebesgue's dominated convergence theorem (Theorem A.1.38), we get

$$\begin{aligned}
\int_X \varrho_\varphi \, dm_\varphi = \int_X \lim_{k\to\infty}\frac{1}{n_k}\sum_{j=0}^{n_k-1}\widehat{\mathcal{L}}_\varphi^j(\mathbb{1})\, dm_\varphi &= \lim_{k\to\infty}\frac{1}{n_k}\sum_{j=0}^{n_k-1}\int_X \widehat{\mathcal{L}}_\varphi^j(\mathbb{1})\, dm_\varphi \\
&= \lim_{k\to\infty}\frac{1}{n_k}\sum_{j=0}^{n_k-1} m_\varphi(\widehat{\mathcal{L}}_\varphi^j(\mathbb{1})) = 1.
\end{aligned}$$

Finally, we establish that $\varrho_\varphi \in H_\alpha(X)$ whenever $\varphi \in H_\alpha(X)$. Fix $x, y \in X$. Suppose first that $d(x,y) < \xi$. From (13.46) in the proof of Theorem 13.7.4, we have

$$|\widehat{\mathcal{L}}_\varphi^n(\mathbb{1})(x) - \widehat{\mathcal{L}}_\varphi^n(\mathbb{1})(y)| \le \widetilde{D}_\alpha(\varphi)\, d^\alpha(x,y), \quad \forall n \in \mathbb{N}.$$

This also holds for the Cesàro averages $\frac{1}{n_k}\sum_{j=0}^{n_k-1} \widehat{\mathcal{L}}_\varphi^j(\mathbb{1})$, $k \in \mathbb{N}$:

$$\left|\frac{1}{n_k}\sum_{j=0}^{n_k-1} \widehat{\mathcal{L}}_\varphi^j(\mathbb{1})(x) - \frac{1}{n_k}\sum_{j=0}^{n_k-1} \widehat{\mathcal{L}}_\varphi^j(\mathbb{1})(y)\right| \le \widetilde{D}_\alpha(\varphi)\, d^\alpha(x,y), \quad \forall k \in \mathbb{N}.$$

Passing to the limit $k \to \infty$, we deduce that if $d(x,y) < \xi$ then

$$|\varrho_\varphi(x) - \varrho_\varphi(y)| \le \widetilde{D}_\alpha(\varphi)\, d^\alpha(x,y). \tag{13.48}$$

If $d(x,y) \ge \xi$, then

$$|\varrho_\varphi(x) - \varrho_\varphi(y)| \le 2\|\varrho_\varphi\|_\infty \le \frac{2\|\varrho_\varphi\|_\infty}{\xi^\alpha}\, d^\alpha(x,y). \tag{13.49}$$

By (13.48) and (13.49), we conclude that $\varrho_\varphi \in H_\alpha(X)$.

Note that part (d) is a consequence of (13.48), (13.49) and part (b). □

Remark 13.7.6. Carefully tracing the constants $\overline{D}_\alpha(\varphi)$ and $\widetilde{D}_\alpha(\varphi)$, we observe that there exist two monotone increasing functions $\overline{V}_\alpha, \widetilde{V}_\alpha : [0,\infty) \to (0,\infty)$ such that

$$\overline{D}_\alpha(\varphi) \le \overline{V}_\alpha(\|\varphi\|_\alpha) \quad \text{and} \quad \widetilde{D}_\alpha(\varphi) \le \widetilde{V}_\alpha(\|\varphi\|_\alpha). \tag{13.50}$$

By parts (b) and (d) of Lemma 13.7.5, we conclude that

$$\|\varrho_\varphi\|_\alpha \le V_\alpha(\|\varphi\|_\alpha) := \overline{V}_\alpha(\|\varphi\|_\alpha) + \max\{\widetilde{V}_\alpha(\|\varphi\|_\alpha), 2\xi^{-\alpha}\overline{V}_\alpha(\|\varphi\|_\alpha)\} \tag{13.51}$$

and $V_\alpha : [0,\infty) \to (0,\infty)$ is a monotone increasing function.

We are now in a position to demonstrate the existence of an invariant Gibbs state. We further prove its uniqueness as well as its ergodicity.

Theorem 13.7.7. *If $T : X \to X$ is a topologically transitive, open, distance expanding system and $\varphi : X \to \mathbb{R}$ is a Hölder continuous potential, then the measure $\mu_\varphi := \varrho_\varphi m_\varphi$, i. e., the measure determined by the requirement that*

$$\frac{d\mu_\varphi}{dm_\varphi} = \varrho_\varphi,$$

is the unique T-invariant Gibbs state for φ, is ergodic, and is equivalent to m_φ with Radon–Nikodym derivative ϱ_φ bounded away from 0 and ∞.

Proof. By virtue of its definition and Lemma 13.7.5(c), we know that μ_φ is a Borel probability measure.

By virtue of its definition and Lemma 13.7.5(b), it is clear that μ_φ is equivalent to m_φ with Radon–Nikodym derivative ϱ_φ bounded away from 0 and ∞.

In particular, μ_φ is absolutely continuous with respect to m_φ, and hence the ergodicity of m_φ is transmitted to μ_φ.

From Corollaries 13.6.5–13.6.6, Lemma 13.7.5 and Theorem 13.4.1, we deduce that μ_φ is T-invariant.

Since m_φ is a Gibbs state and since μ_φ is equivalent to m_φ with Radon–Nikodym derivative ϱ_φ bounded away from 0 and ∞, we know by Lemma 13.2.3 that μ_φ is a Gibbs state, too.

Finally, to show uniqueness, suppose that μ is a T-invariant Gibbs state for φ. By Proposition 13.2.4, we know that μ is equivalent to μ_φ. So, like μ_φ, the measure μ is ergodic and T-invariant. But Theorem 8.2.21 then tells us that either $\mu = \mu_\varphi$ or $\mu \perp \mu_\varphi$. This latter case is impossible given that μ is equivalent to μ_φ. Thus, $\mu = \mu_\varphi$. □

This result has several fundamental repercussions. The following three corollaries are rather straightforward.

Corollary 13.7.8. *Let $T : X \to X$ be a topologically transitive, open, distance expanding system and $\varphi : X \to \mathbb{R}$ a Hölder continuous potential. Any nonnegative Borel measurable function satisfying Lemma 13.7.5(a,c) must be m_φ-almost everywhere equal to ϱ_φ. Furthermore, any nonnegative continuous function satisfying Lemma 13.7.5(a,c) must be equal to ϱ_φ.*

Proof. Let ϱ be a nonnegative Borel measurable function satisfying Lemma 13.7.5(a,c), i. e. such that $\widehat{\mathcal{L}}_\varphi(\varrho) = \varrho$ and $\int_X \varrho \, dm_\varphi = 1$. The latter property makes ϱm_φ a Borel probability measure. The former property guarantees that ϱm_φ is T-invariant. Obviously, ϱm_φ is absolutely continuous with respect to m_φ. Therefore the ergodicity of m_φ is passed on to ϱm_φ. This signifies that the measures ϱm_φ and $\varrho_\varphi m_\varphi$ are two ergodic T-invariant Borel probability measures. Theorem 8.2.21 then affirms that either $\varrho m_\varphi = \varrho_\varphi m_\varphi$ or $\varrho m_\varphi \perp \varrho_\varphi m_\varphi$. This latter case being impossible since $\varrho_\varphi > 0$ and $\operatorname{supp}(m_\varphi) = X$, the former case holds. This means that $\varrho = \varrho_\varphi$ m_φ-almost everywhere. □

Since the only invariant Borel probability measure absolutely continuous with respect to an ergodic invariant Borel probability measure is this ergodic measure itself, another simple consequence of Theorem 13.7.7 is the following.

Corollary 13.7.9. *Let $T : X \to X$ be a topologically transitive, open, distance expanding system. If $\varphi : X \to \mathbb{R}$ is a Hölder continuous potential, then μ_φ is the only T-invariant Borel probability measure which is absolutely continuous with respect to m_φ.*

After Corollary 13.6.8, we observed that we may set $m_{T^n,S_n\varphi} := m_\varphi$ since all that is required of $m_{T^n,S_n\varphi}$ is that it be a fixed point of $\widehat{\mathcal{L}}^*_{T^n,S_n\varphi}$. Adopting this, we obtain yet another direct ramification of Theorem 13.7.7.

Corollary 13.7.10. *Let $T : X \to X$ be a topologically transitive, open, distance expanding system and $\varphi : X \to \mathbb{R}$ be a Hölder potential. Then $\mu_{T^n,S_n\varphi} = \mu_\varphi$ and $\varrho_{T^n,S_n\varphi} = \varrho_\varphi$ for every $n \in \mathbb{N}$.*

Proof. We give three different proofs of this fact. In all three cases, we need to observe that $T^n : X \to X$ is a topologically transitive, open, distance expanding system and that $S_n\varphi : X \to \mathbb{R}$ is a Hölder continuous potential.

First, Corollary 13.7.9 (with the pair (T,φ) replaced with $(T^n, S_n\varphi)$) tells us that $\mu_{T^n,S_n\varphi}$ is the only T^n-invariant Borel probability measure which is absolutely continuous with respect to $m_{T^n,S_n\varphi}$. But given that $m_{T^n,S_n\varphi} := m_\varphi$ and that μ_φ is a (in fact, the only) T-invariant Borel probability measure which is absolutely continuous with respect to m_φ, we know that μ_φ is a T^n-invariant Borel probability measure which is absolutely continuous with respect to $m_{T^n,S_n\varphi}$. We deduce that $\mu_{T^n,S_n\varphi} = \mu_\varphi$.

An alternative argument consists in using Exercise 13.11.15(a) and the uniqueness of invariant Gibbs states from Theorem 13.7.7. We leave it to the reader to fill in the blanks.

Finally, we will demonstrate in Proposition 13.7.12 that μ_φ is the unique equilibrium state for φ. A third argument relies on that and on Exercise 13.11.15(b). The details are left to the reader.

In all three cases,

$$\varrho_\varphi = \frac{d\mu_\varphi}{dm_\varphi} = \frac{d\mu_{T^n,S_n\varphi}}{dm_{T^n,S_n\varphi}} = \varrho_{T^n,S_n\varphi}. \qquad \square$$

We now demonstrate that from a measure-theoretic viewpoint, the dynamics of T are essentially the same as those of a subshift of finite type.

Theorem 13.7.11. *Let $T : X \to X$ be a topologically transitive, open, distance expanding system and let $\varphi : X \to \mathbb{R}$ be a Hölder continuous potential. For any Markov partition $\mathcal{R}$ of sufficiently small diameter, the coding map $\pi : (E_A^\infty, \sigma, \mu_{\varphi\circ\pi}) \to (X, T, \mu_{\varphi\circ\pi} \circ \pi^{-1})$ induced by $\mathcal{R}$ is a measure-preserving isomorphism, where $\mu_{\varphi\circ\pi}$ is the unique σ-invariant Gibbs state for the potential $\varphi\circ\pi : E_A^\infty \to \mathbb{R}$. Moreover, the set $\pi^{-1}(X\backslash\bigcup_{n=0}^\infty T^{-n}(\partial\mathcal{R}))$ is of full $\mu_{\varphi\circ\pi}$-measure and the restriction of π to that set is injective. Furthermore, $\mathrm{P}(T,\varphi) = \mathrm{P}(\sigma, \varphi\circ\pi)$.*

Proof. As before, the letters δ, λ and ξ are the constants arising from Definition 4.1.1 and relations (4.29), (4.30) and (4.31); alternatively, see (13.1) and (13.2).

Since T is open and distance expanding, Theorem 4.4.6 asserts that there exists a Markov partition $\mathcal{R}$ for T with $\operatorname{diam}(\mathcal{R}) < \xi$. Write $\mathcal{R} = \{R_1, R_2, \ldots, R_p\}$. Let $E = \{1, 2, \ldots, p\}$, let

$$A_{ij} := \begin{cases} 1 & \text{if } T(\operatorname{Int}(R_i)) \cap \operatorname{Int}(R_j) \neq \emptyset \\ 0 & \text{otherwise.} \end{cases}$$

(cf. (4.46)) and let $\sigma : E_A^\infty \to E_A^\infty$ be the symbolic representation of T induced by $\mathcal{R}$. To shorten notation, write

$$\partial\mathcal{R} := \bigcup_{e\in E} \partial R_e.$$

Let $\pi : E_A^\infty \to X$ be the coding map induced by $\mathcal{R}$ per (4.48), i. e.,

$$\begin{aligned} \pi : (E_A^\infty, \sigma, \mu_{\varphi\circ\pi}) &\longrightarrow \quad (X, T, \mu_{\varphi\circ\pi} \circ \pi^{-1}) \\ \omega \quad &\longmapsto \pi(\omega) := \textstyle\bigcap_{n=0}^\infty T^{-n}(R_{\omega_n}). \end{aligned}$$

According to Theorem 4.5.2, the map π is a Hölder continuous surjection which is injective on $\pi^{-1}(X \setminus \bigcup_{n=0}^\infty T^{-n}(\partial\mathcal{R}))$. Furthermore, every point of the forward T-invariant, dense G_δ-set $X \setminus \bigcup_{n=0}^\infty T^{-n}(\partial\mathcal{R})$ has a unique preimage under π. Finally, the coding map π is a factor map between $\sigma : E_A^\infty \to E_A^\infty$ and $T : X \to X$.

As both $\varphi : X \to \mathbb{R}$ and $\pi : E_A^\infty \to X$ are Hölder continuous, so is their composition $\varphi \circ \pi : E_A^\infty \to \mathbb{R}$. Moreover, since T is topologically transitive, so is σ according to Lemma 4.5.4. We have also observed on several occasions that the map σ is open and distance expanding. By virtue of Theorem 13.7.7, the potential $\varphi \circ \pi$ then admits a unique T-invariant Gibbs state $\mu_{\varphi\circ\pi}$ and this state is ergodic.

Given that $\operatorname{diam}(\mathcal{R}) < \xi$, the map T is a local homeomorphism on a neighborhood of any element of the Markov partition $\mathcal{R}$. According to Lemma 4.5.4, the set $\pi^{-1}(X \setminus \partial\mathcal{R})$ is then backward σ-invariant. By Exercise 8.5.36 and the ergodicity of $\mu_{\varphi\circ\pi}$, we deduce that $\mu_{\varphi\circ\pi}(\pi^{-1}(X \setminus \partial\mathcal{R})) \in \{0, 1\}$. However, since the set $\partial\mathcal{R}$ is closed in X and π is continuous, the set $\pi^{-1}(X \setminus \partial\mathcal{R})$ is open in E_A^∞. As $\mu_{\varphi\circ\pi}$ is a Gibbs measure, it has full support and hence $\mu_{\varphi\circ\pi}(\pi^{-1}(X \setminus \partial\mathcal{R})) > 0$. We conclude that

$$\mu_{\varphi\circ\pi}(\pi^{-1}(X \setminus \partial\mathcal{R})) = 1.$$

By Exercise 8.5.36 again, it follows that

$$\begin{aligned} 1 = \mu_{\varphi\circ\pi}(\pi^{-1}(X \setminus \partial\mathcal{R})) &= \mu_{\varphi\circ\pi}\Big(\bigcap_{n=0}^\infty \sigma^{-n}(\pi^{-1}(X \setminus \partial\mathcal{R}))\Big) \\ = \mu_{\varphi\circ\pi}\Big(\bigcap_{n=0}^\infty (\pi \circ \sigma^n)^{-1}(X \setminus \partial\mathcal{R})\Big) &= \mu_{\varphi\circ\pi}\Big(\bigcap_{n=0}^\infty (T^n \circ \pi)^{-1}(X \setminus \partial\mathcal{R})\Big) \end{aligned}$$

$$= \mu_{\varphi\circ\pi}\Big(\bigcap_{n=0}^{\infty}\pi^{-1}(T^{-n}(X\setminus\partial\mathcal{R}))\Big) = \mu_{\varphi\circ\pi}\Big(\pi^{-1}\Big(\bigcap_{n=0}^{\infty}T^{-n}(X\setminus\partial\mathcal{R})\Big)\Big)$$
$$= \mu_{\varphi\circ\pi}\Big(\pi^{-1}\Big(\bigcap_{n=0}^{\infty}X\setminus T^{-n}(\partial\mathcal{R})\Big)\Big) = \mu_{\varphi\circ\pi}\Big(\pi^{-1}\Big(X\setminus\bigcup_{n=0}^{\infty}T^{-n}(\partial\mathcal{R})\Big)\Big).$$

So $\pi^{-1}(X\setminus\bigcup_{n=0}^{\infty}T^{-n}(\partial\mathcal{R}))$ is of full $\mu_{\varphi\circ\pi}$-measure, and hence the map $\pi : (E_A^\infty, \sigma, \mu_{\varphi\circ\pi}) \to (X, T, \mu_{\varphi\circ\pi}\circ\pi^{-1})$ is a measure-preserving isomorphism.

Turning our attention to the relationship between the pressures $\mathrm{P}(T,\varphi)$ and $\mathrm{P}(\sigma,\varphi\circ\pi)$, one inequality is obvious: since π is a factor map, Proposition 11.1.20 asserts that

$$\mathrm{P}(T,\varphi) \le \mathrm{P}(\sigma,\varphi\circ\pi). \tag{13.52}$$

Regarding the opposite inequality, since the map $\pi : (E_A^\infty, \sigma, \mu_{\varphi\circ\pi}) \to (X, T, \mu_{\varphi\circ\pi}\circ\pi^{-1})$ is a measure-preserving isomorphism, Exercise 9.7.15 affirms that $\mathrm{h}_{\mu_{\varphi\circ\pi}}(\sigma) = \mathrm{h}_{\mu_{\varphi\circ\pi}\circ\pi^{-1}}(T)$. By Proposition 13.2.9, we also know that $\mu_{\varphi\circ\pi}$ is an equilibrium state for $\varphi\circ\pi$ and the variational principle (Theorem 12.1.1) then yields that

$$\mathrm{P}(\sigma,\varphi\circ\pi) = \mathrm{h}_{\mu_{\varphi\circ\pi}}(\sigma) + \int_{E_A^\infty}\varphi\circ\pi\,d\mu_{\varphi\circ\pi} = \mathrm{h}_{\mu_{\varphi\circ\pi}\circ\pi^{-1}}(T) + \int_X \varphi\,d\mu_{\varphi\circ\pi}\circ\pi^{-1} \le \mathrm{P}(T,\varphi). \tag{13.53}$$

By (13.52)–(13.53) the equality follows. □

13.7.3 Invariant Gibbs states and equilibrium states coincide and are unique

According to Proposition 13.2.9, every T-invariant Gibbs state for a potential φ on an open, distance expanding system T is an equilibrium state for φ. If, in addition, T is topologically transitive and φ is Hölder continuous, then Theorem 13.7.7 affirms that φ admits a unique T-invariant Gibbs state and that this state is ergodic. This measure is denoted by μ_φ.

In this section, we show the converse.

Proposition 13.7.12. *Let $T : X \to X$ be a topologically transitive, open, distance expanding system and let $\varphi : X \to \mathbb{R}$ be a Hölder continuous potential. Then φ has a unique equilibrium state and this state is μ_φ, the unique T-invariant Gibbs state for φ. This state is ergodic.*

Rather than proving directly this result in its full generality, we will first establish it in a symbolic setting.

Proposition 13.7.13. *Let E be a finite alphabet and $A : E \to E$ an irreducible matrix. Let $\sigma : E_A^\infty \to E_A^\infty$ be the subshift of finite type generated by A and $\varphi : E_A^\infty \to \mathbb{R}$ be a Hölder continuous potential. Then φ has a unique equilibrium state and this state is μ_φ, the unique σ-invariant Gibbs state for φ. This state is ergodic.*

Let us first convince ourselves that this is really a special case of Proposition 13.7.12. We have observed before that E_A^∞ is a compact space when E is finite and that the shift map σ is open and distance expanding with respect to any one of the usual metrics on E_A^∞ (the distance between two infinite words depends on the length of their initial common word). Per Theorem 3.2.14, the irreducibility of A is equivalent to the transitivity of σ. So Proposition 13.7.13 is indeed a special case of Proposition 13.7.12.

For the sake of completeness, we prove an auxiliary lemma which is a weak version of a well known density lemma from probability theory (see Theorem 35.8 in Billingsley [8]). In its full generality, this density lemma is proved as a consequence of the martingale convergence theorem (Theorem A.1.65) (see also Williams [138]).

Recall that there is a bijection between the set of finite measurable partitions of a measurable space $(X, \mathcal{F})$ and the set of finite sub-σ-algebras of $\mathcal{F}$. Indeed, given a finite measurable partition $\alpha = \{A_1, \dots, A_n\}$ of $(X, \mathcal{F})$ and letting $A_k^{(0)} := A_k$ while $A_k^{(1)} := \emptyset$ for all $1 \le k \le n$, the family $\mathcal{A}(\alpha) = \{\bigcup_{k=1}^n A_k^{(\omega_k)} : \omega \in \{0,1\}^n\}$ of all sets that can be expressed as unions of atoms of α, is a finite sub-σ-algebra of $\mathcal{F}$. Conversely, given a finite sub-σ-algebra $\mathcal{A} = \{B_1, \dots, B_m\}$ of $\mathcal{F}$ and letting $B_k^{(0)} := B_k$ while $B_k^{(1)} := X \backslash B_k$ for all $1 \le k \le m$, the family $\alpha(\mathcal{A}) = \{\bigcap_{k=1}^m B_k^{(\omega_k)} : \omega \in \{0,1\}^m\}$ is a finite measurable partition of $(X, \mathcal{F})$. It is easy to see that $\alpha(\mathcal{A}(\alpha)) = \alpha$ and $\mathcal{A}(\alpha(\mathcal{A})) = \mathcal{A}$. Moreover, given two finite measurable partitions α, β, it is obvious that β is a refinement of α ($\alpha \prec \beta$) if and only if $\mathcal{A}(\alpha) \subseteq \mathcal{A}(\beta)$. Finally, given $x \in X$ recall that $\alpha(x)$ denotes the atom of α that contains the point x.

Lemma 13.7.14. *Let $(X, \mathcal{F})$ be a measurable space and let $(\alpha_n)_{n=1}^\infty$ be a sequence of increasingly finer countable measurable partitions generating $\mathcal{F}$, i. e., $\sigma(\bigcup_{n=1}^\infty \alpha_n) = \mathcal{F}$. If μ and ν are mutually singular probability measures on $(X, \mathcal{F})$, then for every $t > 0$ we have*

$$\lim_{n\to\infty} \nu\Big(\{x \in X : \nu(\alpha_n(x)) \le t\mu(\alpha_n(x))\}\Big) = 0.$$

Proof. Suppose for a contradiction that there exists $t > 0$ such that

$$\Delta := \limsup_{n\to\infty} \nu(X_n) > 0 \quad \text{where} \quad X_n := \{x \in X : \nu(\alpha_n(x)) \le t\mu(\alpha_n(x))\}.$$

Let

$$S_n := \bigcup_{k=n}^\infty X_k \quad \text{and} \quad S := \bigcap_{n=1}^\infty S_n.$$

Clearly $(S_n)_{n=1}^\infty$ forms a descending family of sets. Moreover, $\nu(S_n) \ge \Delta$ for all $n \in \mathbb{N}$ and thus $\nu(S) \ge \Delta$. Since the measures ν and μ are mutually singular, there exists a set $Y \in \mathcal{F}$ such that $\nu(Y) = 1$ and $\mu(Y) = 0$. Setting

$$F := S \cap Y,$$

we then have that

$$\nu(F) = \nu(S) \geq \Delta \quad \text{and} \quad \mu(F) = 0. \tag{13.54}$$

Since $(\alpha_n)_{n=1}^{\infty}$ is a sequence of increasingly finer countable measurable partitions generating $\mathcal{F}$, Exercise 13.11.12 states that there exists a countable family $\mathcal{I} \subseteq \bigcup_{n=1}^{\infty} \alpha_n$ of mutually disjoint sets such that

$$F \subseteq \bigcup_{I \in \mathcal{I}} I \text{ and } \mu\Big(\bigcup_{I \in \mathcal{I}} I\Big) \leq \frac{\Delta}{2t}. \tag{13.55}$$

Then for every $x \in F$ there exists a unique $I(x) \in \mathcal{I}$ such that $x \in I(x)$. Given that for every $I \in \mathcal{I}$ there is $n(I) \in \mathbb{N}$ such that $I \in \alpha_{n(I)}$, we deduce that $I(x) = \alpha_{n(I(x))}(x)$. Let

$$\mathcal{J} := \{\alpha_{n(I(x))}(x) : x \in F\} = \{I(x) : x \in F\} \subseteq \mathcal{I}.$$

As $\mathcal{I}$ is countable, so is $\mathcal{J}$. Similarly, as the elements of $\mathcal{I}$ are mutually disjoint sets, so are the elements of $\mathcal{J}$. Furthermore, observe that

$$F \subseteq \bigcup_{J \in \mathcal{J}} J = \bigcup_{x \in F} \alpha_{n(I(x))}(x) = \bigcup_{x \in F} I(x) \subseteq \bigcup_{I \in \mathcal{I}} I. \tag{13.56}$$

In addition, every x in F is in S and, therefore, lies in $S_{n(I(x))}$. Consequently, $x \in X_{n(I(x))}$. By definition of the sets X_n, it ensues that $\nu(\alpha_{n(I(x))}(x)) \leq t\mu(\alpha_{n(I(x))}(x))$. So $\nu(J) \leq t\mu(J)$ for every $J \in \mathcal{J}$. Using this and (13.54)–(13.56), we deduce that

$$\frac{\Delta}{2t} \geq \mu\Big(\bigcup_{I \in \mathcal{I}} I\Big) \geq \mu\Big(\bigcup_{J \in \mathcal{J}} J\Big) = \sum_{J \in \mathcal{J}} \mu(J) \geq t^{-1} \sum_{J \in \mathcal{J}} \nu(J) = t^{-1}\nu\Big(\bigcup_{J \in \mathcal{J}} J\Big) \geq t^{-1}\nu(F) \geq \frac{\Delta}{t}.$$

This is the contradiction that was sought. □

Using this result, we prove Proposition 13.7.13.

Proof of Proposition 13.7.13. Exercise 12.4.4 asserts that for any topological dynamical system $T : X \to X$ and continuous potential $\varphi : X \to \mathbb{R}$, the set of all equilibrium states for φ is a convex subset of the set $M(T)$ of all T-invariant measures. Furthermore, Theorem 8.2.23 affirms that the ergodic measures $E(T)$ constitute the extreme points of $M(T)$. It ensues that if φ has a unique ergodic equilibrium state, then it admits a unique equilibrium state. So it suffices to demonstrate that any ergodic equilibrium state ν coincides with μ_φ.

Let $T = \sigma : E_A^\infty \to E_A^\infty$ be a subshift of finite type generated by an irreducible matrix A over a finite alphabet E, and let $\varphi : E_A^\infty \to \mathbb{R}$ be a Hölder continuous potential. Pursuant to Theorem 13.7.7, let μ_φ be the unique σ-invariant Gibbs state for φ. Recall that this state is ergodic. Let ν be an ergodic equilibrium state for φ. In particular, ν is σ-invariant. Our goal is to show that $\nu = \mu_\varphi$.

Let $\alpha := \{[e]\}_{e\in E}$ be the partition of E_A^∞ into its one-cylinders. It is easy to see that $\alpha^n := \bigvee_{i=0}^{n-1} \sigma^{-i}(\alpha) = \{[\omega]\}_{\omega\in E_A^n}$, i. e., α^n is the partition of E_A^∞ into its n-cylinders. Recall that α is a generator for σ (see Definition 9.4.19 and Example 9.4.23). By Theorem 9.4.20 and Definition 9.4.10, we know that

$$\mathrm{h}_\nu(\sigma) = \mathrm{h}_\nu(\sigma,\alpha) = \inf_{n\in\mathbb{N}} \frac{1}{n}\mathrm{H}_\nu(\alpha^n).$$

For every $n \in \mathbb{N}$, the variational principle (Theorem 12.1.1) affirms that

$$\begin{aligned}
0 &= n\Big[\mathrm{h}_\nu(\sigma) + \int_{E_A^\infty} (\varphi - \mathrm{P}(\sigma,\varphi))\,d\nu\Big] \\
&\le \mathrm{H}_\nu(\alpha^n) + \int_{E_A^\infty} (S_n\varphi - n\mathrm{P}(\sigma,\varphi))\,d\nu \\
&= -\sum_{A\in\alpha^n} \nu(A)\Big[\log\nu(A) - \frac{1}{\nu(A)}\int_A (S_n\varphi - n\mathrm{P}(\sigma,\varphi))\,d\nu\Big] \\
&\le -\sum_{A\in\alpha^n} \nu(A)[\log\nu(A) - (S_n\varphi(\omega_A) - n\mathrm{P}(\sigma,\varphi))]
\end{aligned}$$

for some $\omega_A \in A$ for each $A \in \alpha^n$. Using the fact that μ_φ is a Gibbs state for φ, we can continue this inequality as follows:

$$\begin{aligned}
0 &\le -\sum_{A\in\alpha^n} \nu(A)\log[\nu(A)\exp(-(S_n\varphi(\omega_A) - n\mathrm{P}(\sigma,\varphi)))] \\
&\le -\sum_{A\in\alpha^n} \nu(A)\log[\nu(A)C^{-1}(\mu_\varphi(A))^{-1}] \\
&= \log C - \sum_{A\in\alpha^n} \nu(A)\log\frac{\nu(A)}{\mu_\varphi(A)},
\end{aligned} \tag{13.57}$$

where the constant C comes from the Gibbs property of μ_φ.

Assume, for a contradiction, that the measure ν is an ergodic equilibrium state distinct from μ_φ. Then, in light of Theorem 8.2.21, the measures ν and μ_φ must be mutually singular. In search of a contradiction it suffices, in view of (13.57), to show that

$$\lim_{n\to\infty} \sum_{A\in\alpha^n} \nu(A)\log\frac{\nu(A)}{\mu_\varphi(A)} = \infty. \tag{13.58}$$

To do this, for every $j \in \mathbb{Z}$ and every $n \in \mathbb{N}$ set

$$F_{n,j} := \Big\{\omega \in E_A^\infty : e^j < \frac{\nu(\alpha^n(\omega))}{\mu_\varphi(\alpha^n(\omega))} \le e^{j+1}\Big\}.$$

Since for any $A \in \alpha^n$ either $A \cap F_{n,j} = \emptyset$ or $A \subseteq F_{n,j}$, we have

$$\nu(F_{n,j}) = \sum_{A \in \alpha^n : A \subseteq F_{n,j}} \nu(A) = \int_{F_{n,j}} \frac{\nu(\alpha^n(\omega))}{\mu_\varphi(\alpha^n(\omega))} \, d\mu_\varphi(\omega) \le e^{j+1}. \tag{13.59}$$

We further obtain that

$$\begin{aligned} \sum_{A \in \alpha^n} \nu(A) \log \frac{\nu(A)}{\mu_\varphi(A)} &= \int_{E_A^\infty} \log \frac{\nu(\alpha^n(\omega))}{\mu_\varphi(\alpha^n(\omega))} \, d\nu(\omega) \\ &= \sum_{j \in \mathbb{Z}} \int_{F_{n,j}} \log \frac{\nu(\alpha^n(\omega))}{\mu_\varphi(\alpha^n(\omega))} \, d\nu(\omega) \\ &\ge \sum_{j \in \mathbb{Z}} j\nu(F_{n,j}). \end{aligned} \tag{13.60}$$

Using (13.59)–(13.60), we deduce for every $k \in \mathbb{N}$ that

$$\begin{aligned} \sum_{A \in \alpha^n} \nu(A) \log \frac{\nu(A)}{\mu_\varphi(A)} &\ge \sum_{j=k}^{\infty} j\nu(F_{n,j}) + \sum_{j=-\infty}^{-1} j\nu(F_{n,j}) \\ &\ge k \sum_{j=k}^{\infty} \nu(F_{n,j}) + \sum_{j=-\infty}^{-1} je^{j+1} \\ &= k\nu\Big(\bigcup_{j=k}^{\infty} F_{n,j}\Big) - \sum_{j=1}^{\infty} je^{-j+1} \\ &= k\nu\left(\left\{\omega \in E_A^\infty : \frac{\nu(\alpha^n(\omega))}{\mu_\varphi(\alpha^n(\omega))} > e^k\right\}\right) - e\sum_{j=1}^{\infty} je^{-j}. \end{aligned}$$

As the measures ν and μ_φ are mutually singular, applying Lemma 13.7.14 with $\alpha_n = \alpha^n$ and $t = e^k$, we infer that

$$\lim_{n\to\infty} \nu\left(\left\{\omega \in E_A^\infty : \frac{\nu(\alpha^n(\omega))}{\mu_\varphi(\alpha^n(\omega))} > e^k\right\}\right) = 1.$$

Therefore,

$$\liminf_{n\to\infty} \sum_{A \in \alpha^n} \nu(A) \log \frac{\nu(A)}{\mu_\varphi(A)} \ge k - e\sum_{j=1}^{\infty} je^{-j}.$$

Since the series $\sum_{j=1}^{\infty} je^{-j}$ is convergent, letting $k \to \infty$ yields (13.58). This contradicts (13.57). Consequently, the only ergodic equilibrium state for φ is the unique σ-invariant Gibbs state μ_φ. This ends the proof in the special case of a topologically transitive subshift of finite type. □

In order to prove the result in its full generality, i. e., Proposition 13.7.12, we will need a measure-lifting lemma. It is kind of, but not quite, a Hahn–Banach theorem, and does not really follow from this latter. We will formulate and prove it in a slightly more general case than we need. In a sense, a more general result can be found in [85], where, with the notation of Theorem 13.7.16 below, neither X nor Y were assumed to be compact but merely Polish spaces; however, X was therein a subshift of finite type over a countable alphabet and all the fibers $\pi^{-1}(y)$, $y \in Y$, were assumed to be compact. The strongest theorem, with the weakest hypotheses, can be found in [20]; however, compactness of all fibers was still assumed. This will not be assumed here.

We start with the following well-known fact from measure theory. We provide its proof since it is easy, short and this fact is not so commonly formulated and proved in textbooks. We do it also for the sake of completeness and convenience for the reader.

Lemma 13.7.15. *Let W and Z be Polish spaces. Let μ be a Borel probability measure on Z, let $\hat{\mu}$ be its completion, and denote by $\widehat{\mathcal{B}}_\mu$ the complete σ-algebra of all $\hat{\mu}$-measurable subsets of Z. Let $f : W \to Z$ be a Borel measurable surjection and let $g : W \to \overline{\mathbb{R}} =: [-\infty, +\infty]$ be a Borel measurable function. Define the functions $g_*, g^* : Z \to \overline{\mathbb{R}}$ respectively by*

$$g_*(z) := \inf\{g(w) \mid w \in f^{-1}(z)\} \quad \text{and} \quad g^*(z) := \sup\{g(w) \mid w \in f^{-1}(z)\}.$$

Then g_*, g^* are measurable with respect to the σ-algebra $\widehat{\mathcal{B}}_\mu$. If, in addition, the map f is locally one-to-one, then g_*, g^* are Borel measurable.

Proof. Replacing g by $-g$, it suffices to prove the lemma for the function $g^* : Z \to \overline{\mathbb{R}}$. Fix $t \in \mathbb{R}$. For any $z \in Z$, we have that $g^*(z) \in (t, \infty]$ if and only if $g(w) \in (t, \infty]$ for some $w \in f^{-1}(z)$. Thus, $(g^*)^{-1}((t,\infty]) = f(g^{-1}((t,\infty]))$. Hence, $(g^*)^{-1}((t,\infty])$ is an analytic set since $g^{-1}((t,\infty])$ is a Borel set, $f : W \to Z$ is a Borel map, and both spaces W and Z are Polish. The first assertion now follows from the fact that all analytic subsets of Z belong to $\widehat{\mathcal{B}}_\mu$. If, in addition, the map $f : W \to Z$ is locally 1-to-1, then the f-images of all Borel subsets of W are Borel in Z, so $f(g^{-1}((t,\infty])) \subseteq Z$ is Borel. □

We now return to measure-theoretic dynamical systems.

Theorem 13.7.16. *Let X and Y be compact metrizable spaces, and let $T : X \to X$ and $S : Y \to Y$ be continuous maps. Assume that there is a continuous surjection $\pi : X \to Y$ making S a topological factor of T, i. e.,*

$$\pi \circ T = S \circ \pi,$$

or, equivalently, that the following diagram commutes:

$$\begin{array}{ccc} X & \xrightarrow{T} & X \\ \pi \downarrow & & \downarrow \pi \\ Y & \xrightarrow[S]{} & Y \end{array}$$

Then the map

$$M(T) \ni \nu \longmapsto \nu \circ \pi^{-1} \in M(S)$$

is surjective. Recall that $M(T)$ (resp., $M(S)$) denotes the set of all T-invariant (resp., S-invariant) Borel probability measures on X (resp., Y).

Proof. Fix $\mu \in M(S)$. Let $\mathcal{B}_b(X)$ and $\mathcal{B}_b(Y)$ be the vector spaces of all bounded Borel measurable real-valued functions defined on X and Y, respectively. Let $\mathcal{B}_b^+(X)$ and $\mathcal{B}_b^+(Y)$ be the corresponding convex cones consisting of all nonnegative functions. Let

$$\widehat{\mathcal{B}}_b(X) := \{g \circ \pi \mid g \in \mathcal{B}_b(Y)\}.$$

Clearly $\widehat{\mathcal{B}}_b(X)$ is a vector subspace of $\mathcal{B}_b(X)$ and, as $\pi : X \to Y$ is a surjection, for each $f \in \widehat{\mathcal{B}}_b(X)$ there exists a unique $g \in \mathcal{B}_b(Y)$ such that $f = g \circ \pi$. Thus, treating (via integration) μ as a linear functional from $\mathcal{B}_b(Y)$ to $\mathbb{R}$, the formula

$$\widehat{\mathcal{B}}_b(X) \ni g \circ \pi \longmapsto \widehat{\mu}(g \circ \pi) := \mu(g) \in \mathbb{R}$$

defines a positive linear functional $\widehat{\mu}$ from $\widehat{\mathcal{B}}_b(X)$ to $\mathbb{R}$. By Lemma 13.7.15 applied to $\pi : X \to Y$, for every $h \in \mathcal{B}_b(X)$ the function $h_* \circ \pi : X \to \mathbb{R}$ belongs to $\widehat{\mathcal{B}}_b(X)$. Also, $h - h_* \circ \pi \geq 0$. Hence,

$$h - h_* \circ \pi \in \mathcal{B}_b^+(X).$$

Riesz's Theorem (Theorem A.2.6) then produces a positive linear functional $\mu^* : \mathcal{B}_b(X) \to \mathbb{R}$ such that

$$\mu^*(f) = \widehat{\mu}(f), \quad \forall h \in \widehat{\mathcal{B}}_b(X).$$

And the restriction of μ^* to the space $C(X)$ of continuous functions on X, is obviously linear and positive.

Claim 1. *If $(g_n)_{n=1}^{\infty}$ is a monotone decreasing sequence of nonnegative functions in $C(X)$ converging pointwise to the constant function 0, then $\lim_{n\to\infty} \mu^*(g_n) = 0$.*

Proof of Claim 1. Clearly, $(g_n^*)_{n=1}^{\infty}$ is a monotone decreasing sequence of nonnegative bounded functions that, by Lemma 13.7.15, belong to $\mathcal{B}(Y)$, and thus to $\mathcal{B}_b^+(Y)$. Fix

$y \in Y$. As the set $\pi^{-1}(y) \subseteq X$ is compact, Dini's theorem (Theorem A.1.43) confirms that the sequence $(g_n|_{\pi^{-1}(y)})_{n=1}^{\infty}$ converges uniformly to zero. Since all these functions are nonnegative, the sequence $(g_n^*)_{n=1}^{\infty}$ converges pointwise to zero. In summary, $(g_n^*)_{n=1}^{\infty}$ is a monotone decreasing sequence of functions in $\mathcal{B}_b^+(Y)$ converging pointwise to zero. Therefore, using the fact that $g_n \le g_n^* \circ \pi$ and the monotone convergence theorem (Theorem A.1.35), we get

$$0 \le \overline{\lim_{n\to\infty}}\, \mu^*(g_n) \le \overline{\lim_{n\to\infty}}\, \mu^*(g_n^* \circ \pi) = \overline{\lim_{n\to\infty}}\, \widehat{\mu}(g_n^* \circ \pi) = \overline{\lim_{n\to\infty}}\, \mu(g_n^*) = 0.$$

So $\lim_{n\to\infty} \mu^*(g_n) = 0$. ■

Having Claim 1, the Daniell–Stone representation theorem (a generalization of the Riesz representation theorem (Theorem A.1.54)) immediately entails the following.

Claim 2. *The positive linear functional $\mu^* : \mathcal{B}_b(X) \to \mathbb{R}$ extends uniquely from $C(X)$ to an element of $M(X)$. We denote this extension by μ^* as well.*

Now, for every $n \in \mathbb{N}$ set

$$\mu_n^* := \frac{1}{n}\sum_{j=0}^{n-1} \mu^* \circ T^{-j}.$$

Since the set $M(X)$ of all Borel probability measures on X is compact with respect to the weak* topology on $C(X)^*$, there exists an increasing sequence $(n_k)_{k=1}^{\infty}$ of positive integers such that the sequence $(\mu_{n_k}^*)_{k=1}^{\infty}$ weak* converges. Denote its limit by $\nu \in M(X)$. A standard argument, employed for instance in the proofs of the Krylov–Bogolyubov theorem (Theorem 8.1.22) and the variational principle (Theorem 12.1.1), shows that $\nu \in M(T)$. By the definitions of $\widehat{\mu}$ and μ^*, we obtain for every $g \in C(X)$ and every $n \in \mathbb{N}$ that

$$\begin{aligned}
\mu_n^* \circ \pi^{-1}(g) = \mu_n^*(g \circ \pi) &= \frac{1}{n}\sum_{j=0}^{n-1} \mu^* \circ T^{-j}(g \circ \pi) = \frac{1}{n}\sum_{j=0}^{n-1} \mu^*(g \circ \pi \circ T^j) \\
&= \frac{1}{n}\sum_{j=0}^{n-1} \mu^*(g \circ S^j \circ \pi) = \frac{1}{n}\sum_{j=0}^{n-1} \widehat{\mu}((g \circ S^j) \circ \pi) \\
&= \frac{1}{n}\sum_{j=0}^{n-1} \mu(g \circ S^j) = \frac{1}{n}\sum_{j=0}^{n-1} \mu(g) = \mu(g).
\end{aligned}$$

So $\mu_n^* \circ \pi^{-1} = \mu$ for all $n \in \mathbb{N}$. Consequently,

$$\nu \circ \pi^{-1} = \lim_{k\to\infty} \mu_{n_k}^* \circ \pi^{-1} = \lim_{k\to\infty} \mu = \mu. \qquad \square$$

Finally, we can prove the main result of this section, Proposition 13.7.12.

Proof of Proposition 13.7.12. Let $T : X \to X$ be a topologically transitive, open, distance expanding system subject to a Hölder continuous potential $\varphi : X \to \mathbb{R}$. By Theorem 13.7.11, given a Markov partition $\mathcal{R} = \{R_1, \dots, R_p\}$ of sufficiently small diameter, we know that the coding map

$$\begin{aligned} \pi : (E_A^\infty, \sigma, \mu_{\varphi\circ\pi}) &\to (X, T, \mu_{\varphi\circ\pi} \circ \pi^{-1}) \\ \{\pi(\omega)\} &= \bigcap_{n=0}^{\infty} T^{-n}(R_{\omega_n}) \end{aligned}$$

is a measure-preserving isomorphism, where $\mu_{\varphi\circ\pi}$ is the unique σ-invariant Gibbs state for the Hölder potential $\varphi \circ \pi : E_A^\infty \to \mathbb{R}$. Furthermore, $\mathrm{P}(T, \varphi) = \mathrm{P}(\sigma, \varphi \circ \pi)$. By Proposition 13.7.13 (the symbolic case already proved), we know that the only equilibrium state for $\varphi \circ \pi$ is $\mu_{\varphi\circ\pi}$.

Let ν be any equilibrium state for φ on X. We will show that $\nu = \mu_{\varphi\circ\pi} \circ \pi^{-1} = \mu_\varphi$. Per Theorem 4.5.2, the coding map $\pi : E_A^\infty \to X$ is a factor map between (E_A^∞, σ) and (X, T). According to Theorem 13.7.16, there then exists a σ-invariant Borel probability measure $\tilde{\nu}$ on E_A^∞ such that

$$\tilde{\nu} \circ \pi^{-1} = \nu.$$

Since π is a factor map, Exercise 9.7.15(j) states that $\mathrm{h}_{\tilde{\nu}}(\sigma) \geq \mathrm{h}_\nu(T)$. Using the variational principle (Theorem 12.1.1), we get

$$\begin{aligned} \mathrm{h}_{\tilde{\nu}}(\sigma) + \int_{E_A^\infty} \varphi \circ \pi \, d\tilde{\nu} = \mathrm{h}_{\tilde{\nu}}(\sigma) + \int_X \varphi \, d\tilde{\nu} \circ \pi^{-1} &= \mathrm{h}_{\tilde{\nu}}(\sigma) + \int_X \varphi \, d\nu \\ &\geq \mathrm{h}_\nu(T) + \int_X \varphi \, d\nu \\ &= \mathrm{P}(T, \varphi) = \mathrm{P}(\sigma, \varphi \circ \pi). \end{aligned}$$

Thus, $\tilde{\nu}$ is an equilibrium state for $\varphi \circ \pi$. From Proposition 13.7.13, we infer that

$$\tilde{\nu} = \mu_{\varphi\circ\pi}.$$

Hence,

$$\nu = \tilde{\nu} \circ \pi^{-1} = \mu_{\varphi\circ\pi} \circ \pi^{-1}.$$

Therefore, φ has a unique equilibrium state, namely $\mu_{\varphi\circ\pi} \circ \pi^{-1}$. Since μ_φ, the unique T-invariant Gibbs state for φ, is an equilibrium state for φ by Theorem 13.7.7 and Proposition 13.2.9, we conclude that $\mu_{\varphi\circ\pi} \circ \pi^{-1} = \mu_\varphi$. □

13.7.4 Hölder continuous potentials with the same Gibbs states

We now provide a complete characterization of when two Hölder continuous potentials share the same invariant Gibbs/equilibrium state. As an immediate consequence of Theorem 13.2.7 and Proposition 13.7.12, we have the following.

Theorem 13.7.17. *If $T : X \to X$ is a topologically transitive, open, distance expanding system and $\varphi, \psi : X \to \mathbb{R}$ are Hölder continuous potentials, then the following statements are equivalent:*

(a) *The invariant Gibbs/equilibrium states of φ and ψ coincide, i. e., $\mu_\varphi = \mu_\psi$.*

(b) *There exists a constant $R_1 \in \mathbb{R}$ such that*

$$S_n\varphi(x) - S_n\psi(x) = R_1 n$$

whenever $n \in \mathbb{N}$ and $x \in X$ is a periodic point of T of period n.

(c) *There exist constants $R_2 \in \mathbb{R}$ and $C \geq 0$ such that*

$$|S_n\varphi(x) - S_n\psi(x) - R_2 n| \leq C, \quad \forall n \in \mathbb{N}, \forall x \in X.$$

(d) *$\varphi - \psi$ is cohomologous to a constant $R_3 \in \mathbb{R}$ in the additive group of bounded functions on X.*

(e) *$\varphi - \psi$ is cohomologous to a constant $R_4 \in \mathbb{R}$ in the additive group of continuous functions on X.*

(f) *$\varphi - \psi$ is cohomologous to a constant $R_5 \in \mathbb{R}$ in the additive group of Hölder continuous functions on X.*

(g) *$\mathcal{G}_\varphi = \mathcal{G}_\psi$.*

(h) *$\mathcal{G}_\varphi \cap \mathcal{G}_\psi \neq \emptyset$.*

In addition, if any of the above statements holds, then all the constants R_1–R_5 are the same and equal to $\mathrm{P}(T, \varphi) - \mathrm{P}(T, \psi)$.

We would like to stress that this theorem shows that a transitive, open, distance expanding system admits an abundance of different Gibbs states. For instance, take a Hölder continuous potential φ. Fix two different periodic orbits. While preserving the Hölder continuity, perturb slightly φ on a neighborhood of one point of the first orbit, a neighborhood so small that it does not intersect the second orbit, and call the resulting potential ψ. Then from the second orbit we would deduce that $R_1 = 0$ in condition (b), whereas from the first orbit we would infer that $R_1 \neq 0$. From this contradiction, we conclude that the perturbed potential must have a different Gibbs state from the one of the original potential. As the reader can foresee, condition (b) is a very powerful tool to show that Gibbs states are different.

13.7.5 Invariant Gibbs states have positive entropy; pressure gap

Definition 13.7.18. Let $T : X \to X$ be a topological dynamical system. A continuous potential $\varphi : X \to \mathbb{R}$ is said to have a *pressure gap* with respect to T if there exists $k \in \mathbb{N}$ such that

$$P(T^k, S_k\varphi) > \sup(S_k\varphi).$$

Our first goal in this section is to show that if T is a transitive, open, distance expanding map and the space X is infinite, then each Hölder continuous potential $\varphi : X \to \mathbb{R}$ has a pressure gap with respect to T. We will deduce from this that $h_{\mu_\varphi}(T) > 0$.

We start with the following well-known fact from measure theory.

Proposition 13.7.19. *Let (X, d) be a compact metric space and μ a Borel probability measure on X. For every $r > 0$, define*

$$S_\mu(r) := \sup\{\mu(B(x, r)) \mid x \in X\}.$$

If the measure μ is atomless, then

$$\lim_{r\to 0} S_\mu(r) = 0.$$

Proof. Suppose that $\lim_{r\to 0} S_\mu(r) > 0$. As the function $(0, \infty) \ni r \mapsto S_\mu(r)$ is monotone increasing, this assumption implies that there exists $\varepsilon > 0$ such that

$$S_\mu(r) > \varepsilon, \quad \forall r > 0.$$

Therefore, for every $r > 0$ there is $x(r) \in X$ such that

$$\mu(B(x(r), r)) > \varepsilon.$$

Given that X is compact, there is a strictly increasing sequence $(n_k)_{k=1}^\infty$ of positive integers such that the sequence $(x(1/n_k))_{k=1}^\infty$ converges. Denote its limit by x. For every $s > 0$, there exists $k \in \mathbb{N}$ so large that

$$d(x, x(1/n_k)) < \frac{s}{2} \quad \text{and} \quad \frac{1}{n_k} < \frac{s}{2}.$$

Then

$$B(x(1/n_k), 1/n_k) \subseteq B(x, s).$$

Thus,

$$\mu(B(x, s)) \geq \mu(B(x(1/n_k), 1/n_k)) > \varepsilon.$$

Hence,

$$\mu(\{x\}) = \lim_{s \to 0} \mu(B(x,s)) \geq \varepsilon > 0.$$

This contradicts the atomless nature of μ. □

Now we can establish the announced existence of pressure gaps.

Theorem 13.7.20. *If $T : X \to X$ is a transitive, open, distance expanding system on an infinite space X, then each Hölder potential $\varphi : X \to \mathbb{R}$ has a pressure gap with respect to T.*

Proof. Since the Gibbs state μ_φ is ergodic, $\operatorname{supp}(\mu_\varphi) = X$ and X is infinite, we infer that μ_φ is atomless. Let $C_\varphi \geq 1$ be the constant C from Definition 13.2.1 ascribed to the potential φ and $P = \mathrm{P}(T, \varphi)$. As

$$\operatorname{diam}(T_x^{-n}(B(T^n(x), \xi))) \leq \xi \lambda^{-n}$$

for all $x \in X$ and all $n \geq 0$, we deduce from Proposition 13.7.19 that there exists $k \in \mathbb{N}$ such that

$$\mu_\varphi(T_x^{-k}(B(T^k(x), \xi))) \leq (eC_\varphi)^{-1}, \quad \forall x \in X.$$

Given any $x \in X$, it follows from this and (13.13) that

$$S_k\varphi(x) - k\mathrm{P}(T,\varphi) \leq \log C_\varphi + \log \mu_\varphi(T_x^{-k}(B(T^k(x), \xi))) \leq \log C_\varphi - \log e - \log C_\varphi = -1.$$

Using this and Theorem 11.1.22 ($\mathrm{P}(T^k, S_k\varphi) = \mathrm{P}(T, \varphi)$), we conclude that

$$S_k\varphi(x) \leq k\mathrm{P}(T,\varphi) - 1 = \mathrm{P}(T^k, S_k\varphi) - 1$$

for any $x \in X$. There is a pressure gap. □

A pressure gap entails strictly positive entropy with respect to the unique equilibrium state μ_φ.

Theorem 13.7.21. *If $T : X \to X$ is a transitive, open, distance expanding map of an infinite space X and $\varphi : X \to \mathbb{R}$ is a Hölder continuous potential, then $\mathrm{h}_{\mu_\varphi}(T) > 0$.*

Proof. We will provide two proofs.

By Theorem 13.7.20, there exists $k \in \mathbb{N}$ such that $\mathrm{P}(T^k, S_k\varphi) > \sup(S_k\varphi)$. Using successively Theorem 9.4.13 ($\mathrm{h}_{\mu_\varphi}(T^k) = k\mathrm{h}_{\mu_\varphi}(T)$), Corollary 13.7.10 and Theorem 12.2.2 (applied to the pair $(T^k, S_k\varphi)$ rather than (T, φ)), we infer that

$$\mathrm{h}_{\mu_\varphi}(T) = \frac{1}{k}\mathrm{h}_{\mu_\varphi}(T^k) = \frac{1}{k}\mathrm{h}_{\mu_{T^k, S_k\varphi}}(T^k) > 0.$$

An alternative argument goes as follows. Per Theorem 11.1.22, we know that $\mathrm{P}(T^k, S_k\varphi) = k\mathrm{P}(T, \varphi)$. Since μ_φ is an equilibrium state, it ensues from the variational principle (Theorem 12.1.1) and the T-invariance of μ that

$$\begin{aligned}
\mathrm{h}_{\mu_\varphi}(T) &= \mathrm{P}(T, \varphi) - \int_X \varphi \, d\mu_\varphi \\
&= \frac{1}{k}\Big[k\mathrm{P}(T, \varphi) - k\int_X \varphi \, d\mu_\varphi\Big] \\
&= \frac{1}{k}\Big[\mathrm{P}(T^k, S_k\varphi) - \int_X S_k\varphi \, d\mu_\varphi\Big] \\
&> \frac{1}{k}\Big[\sup(S_k\varphi) - \int_X S_k\varphi \, d\mu_\varphi\Big] \\
&\geq 0.
\end{aligned}$$

So $\mathrm{h}_{\mu_\varphi}(T) > 0$. □

13.7.6 Absolutely continuous invariant measures for Shub expanding maps

If M is a compact connected smooth manifold, then each Riemannian metric ρ on M induces a unique volume (Lebesgue) measure λ_ρ on M and all such measures are mutually equivalent. Call the corresponding measure class the Lebesgue measure class on M.

As announced at the end of Chapter 6, using the theory of Gibbs states we prove that every $C^{1+\varepsilon}$ ($\varepsilon > 0$) Shub expanding endomorphism admits a unique invariant Borel probability measure absolutely continuous with respect to the Lebesgue measure class. This fact was first proved for C^2 maps in [64]. It is also in that paper that the appropriate transfer operator was for the first time explicitly used in dynamical systems. As it is based on the theory of Gibbs states developed in this chapter, our proof is different but nevertheless presents significant similarities with that of Krzyżewski and Szlenk.

Theorem 13.7.22. *If $T : M \to M$ is a $C^{1+\varepsilon}$ Shub expanding endomorphism on a compact connected smooth manifold M, then there exists a unique T-invariant Borel probability measure μ on M which is absolutely continuous with respect to the Lebesgue measure class on M. In fact, μ is equivalent to the Lebesgue measure class on M and $\mu = \mu_\varphi$, where $\varphi(x) = -\log\det(D_xT)$.*

Proof. Let ρ be a Riemannian metric on M. For every $x \in M$, let

$$D_xT : T_xM \to T_{T(x)}M$$

be the corresponding derivative map and λ_x be the Lebesgue measure on T_xM induced by the Euclidean structure on T_xM which is engendered by the inner product ρ_x. Let $\det(D_xT)$ be the determinant of the map D_xT. It is positive since M is orientable and T is orientation preserving. As T is a $C^{1+\varepsilon}$ endomorphism, the potential

$$M \ni x \longmapsto \varphi(x) = -\log\det(D_xT) \in \mathbb{R}$$

is Hölder continuous with exponent ε. Let $\mathcal{L}_\varphi : H_\varepsilon(M) \to H_\varepsilon(M)$ be the associated transfer operator defined by the formula

$$\mathcal{L}_\varphi(g)(x) = \sum_{y\in T^{-1}(x)} (\det(D_yT))^{-1} g(y).$$

Let ξ be the number coming from the distance expandingness of T. Then for every $x \in M$ and every $n \geq 0$, we have

$$\lambda_\rho(T_x^{-n}(B_\rho(T^n(x),\xi))) = \int_{B_\rho(T^n(x),\xi)} \det(D_zT_x^{-n})\, d\lambda_\rho(z) \asymp (\det(D_xT^n))^{-1} = \exp(S_n\varphi(x)).$$

Thus, λ_ρ is a Gibbs measure for the potential φ, and $\mathrm{P}(T,\varphi) = 0$ according to Proposition 13.2.4. By that same proposition, the ergodic T-invariant Gibbs state μ_φ produced in Theorem 13.7.7 is equivalent to λ_ρ, and the proof of the existence part is complete. Since the only T-invariant Borel probability measure absolutely continuous with respect to an ergodic T-invariant Borel probability measure, is this ergodic measure itself, the uniqueness part is an immediate consequence of the ergodicity of μ_φ. □

13.8 Finer properties of transfer operators and Gibbs states

13.8.1 Iterates of transfer operators

Let $T : X \to X$ be a topologically transitive, open and distance expanding system and $\varphi \in H_\alpha(X)$. Define

$$\widehat{\varphi} := \varphi + \log\varrho_\varphi - \log\varrho_\varphi \circ T,$$

where ϱ_φ comes from Lemma 13.7.5. Note that $\widehat{\varphi} \in C(X)$ may not be Hölder continuous. (Nevertheless, Lemma 13.7.5 affirms that $\varrho_\varphi \in H_\alpha(X)$ and that ϱ_φ is bounded away from 0 and ∞. As log is Lipschitz on any compact subset of $(0,\infty)$, we then have that $\log\varrho_\varphi \in H_\alpha(X)$. If T is additionally Hölder, then $\log\varrho_\varphi \circ T$ is Hölder, and thus $\widehat{\varphi}$ is Hölder. Though this special case is worth mentioning, we will stick to the general case where $\widehat{\varphi}$ may not be Hölder continuous.) In any case, since $\widehat{\varphi}$ is cohomologous to φ in the additive group of bounded (even Hölder continuous) functions, it follows from

Lemma 13.2.6 that $\mathrm{P}(T,\widehat{\varphi}) = \mathrm{P}(T,\varphi)$ and $\mathcal{G}_{\widehat{\varphi}} = \mathcal{G}_\varphi$. In particular, this implies that the system T, under the potential $\widehat{\varphi}$, has a unique invariant Gibbs state, namely μ_φ the unique invariant Gibbs state under the potential φ (see Theorem 13.7.7). A straightforward calculation shows that the operator $\mathcal{L}_{\widehat{\varphi}} : C(X) \to C(X)$ satisfies

$$\mathcal{L}_{\widehat{\varphi}}(g) = \frac{\mathcal{L}_\varphi(\varrho_\varphi g)}{\varrho_\varphi}.$$

Yet another calculation yields that its dual $\mathcal{L}^*_{\widehat{\varphi}} : C(X)^* \to C(X)^*$ is such that $\mathcal{L}^*_{\widehat{\varphi}}(\mu_\varphi) = e^{\mathrm{P}(T,\varphi)}\mu_\varphi$. Therefore, one can set $\Lambda_{\widehat{\varphi}} = e^{\mathrm{P}(T,\varphi)}$ and $m_{\widehat{\varphi}} = \mu_\varphi$ in Theorem 13.6.2 and subsequent applicable results for the potential $\widehat{\varphi}$. Among others, extending and normalizing leads to the positive bounded linear operator $\widehat{\mathcal{L}}_{\widehat{\varphi}} : L^1(\mu_\varphi) \to L^1(\mu_\varphi)$ and its dual $\widehat{\mathcal{L}}^*_{\widehat{\varphi}} : L^1(\mu_\varphi)^* \to L^1(\mu_\varphi)^*$ determined by the formula

$$\widehat{\mathcal{L}}_{\widehat{\varphi}}(g)(x) = \frac{\widehat{\mathcal{L}}_\varphi(\varrho_\varphi g)(x)}{\varrho_\varphi(x)} = \frac{e^{-\mathrm{P}(T,\varphi)}}{\varrho_\varphi(x)} \sum_{y\in T^{-1}(x)} \varrho_\varphi(y)g(y)e^{\varphi(y)} \quad \text{for } \mu_\varphi\text{-a. e. } x \in X,$$

with equality holding everywhere if $g \in C(X)$. Moreover, $\widehat{\mathcal{L}}^*_{\widehat{\varphi}}(\mu_\varphi) = \mu_\varphi$. Then

$$\widehat{\mathcal{L}}_{\widehat{\varphi}} = R_\varphi^{-1} \circ \widehat{\mathcal{L}}_\varphi \circ R_\varphi,$$

where $R_\varphi : L^1(\mu_\varphi) \to L^1(\mu_\varphi)$ is given by

$$R_\varphi(g) := \varrho_\varphi \cdot g.$$

Consequently, for every $n \in \mathbb{N}$,

$$\widehat{\mathcal{L}}^n_{\widehat{\varphi}} = R_\varphi^{-1} \circ \widehat{\mathcal{L}}^n_\varphi \circ R_\varphi \quad \text{and} \quad \widehat{\mathcal{L}}^n_\varphi = R_\varphi \circ \widehat{\mathcal{L}}^n_{\widehat{\varphi}} \circ R_\varphi^{-1}.$$

So

$$\widehat{\mathcal{L}}^n_{\widehat{\varphi}}(g)(x) = \frac{\widehat{\mathcal{L}}^n_\varphi(\varrho_\varphi g)(x)}{\varrho_\varphi(x)} = \frac{e^{-n\mathrm{P}(T,\varphi)}}{\varrho_\varphi(x)} \sum_{y\in T^{-n}(x)} \varrho_\varphi(y)g(y)e^{S_n\varphi(y)} \quad \text{for } \mu_\varphi\text{-a. e. } x \in X, \tag{13.61}$$

with equality holding everywhere if $g \in C(X)$.

Lemma 13.7.1 and Proposition 13.7.2 for $\widehat{\mathcal{L}}_\varphi$ have counterparts for $\widehat{\mathcal{L}}_{\widehat{\varphi}}$. The latter directly follow from the former through relation (13.61) and Lemma 13.7.5.

Lemma 13.8.1. *If $T : X \to X$ is a topologically transitive, open, distance expanding system and $\varphi \in H_\alpha(X)$, then for all $n \in \mathbb{N}$ and all $x, y \in X$ we have*

$$(\overline{D}_\alpha(\varphi))^{-3} \le \frac{\widehat{\mathcal{L}}^n_{\widehat{\varphi}}(\mathbb{1})(y)}{\widehat{\mathcal{L}}^n_{\widehat{\varphi}}(\mathbb{1})(x)} \le (\overline{D}_\alpha(\varphi))^3.$$

Proposition 13.8.2. *If $T : X \to X$ is a transitive, open, distance expanding system and $\varphi \in H_\alpha(X)$, then*

$$(\overline{D}_\alpha(\varphi))^{-3} \le \inf \widehat{\mathcal{L}}^n_{\hat{\varphi}}(\mathbb{1}) \le \sup \widehat{\mathcal{L}}^n_{\hat{\varphi}}(\mathbb{1}) \le (\overline{D}_\alpha(\varphi))^3, \quad \forall n \in \mathbb{N}.$$

Theorem 13.7.4 also has a counterpart. The latter follows from the former, relation (13.61) and Lemma 13.7.5.

Theorem 13.8.3. *If $T : X \to X$ is a transitive, open, distance expanding system and $\varphi : X \to \mathbb{R}$ is a Hölder potential, then for every $g \in C(X)$ the family $\{\widehat{\mathcal{L}}^n_{\hat{\varphi}}(g)\}_{n=0}^{\infty}$ is equicontinuous and pointwise bounded in $C(X)$. Hence the operator $\widehat{\mathcal{L}}_{\hat{\varphi}} : C(X) \to C(X)$ is almost periodic.*

Now we present an analogue of Lemma 13.7.5 and a little more. While ϱ_φ is a fixed point of $\widehat{\mathcal{L}}_\varphi$, the constant function $\mathbb{1}$ is a fixed point of $\widehat{\mathcal{L}}_{\hat{\varphi}}$. Moreover, whereas the measure m_φ is a fixed point for $\widehat{\mathcal{L}}^*_\varphi$, the measure μ_φ is a fixed point for $\widehat{\mathcal{L}}^*_{\hat{\varphi}}$.

Proposition 13.8.4. *If $T : X \to X$ is a transitive, open, distance expanding system and $\varphi : X \to \mathbb{R}$ is a Hölder potential, then the operator $\widehat{\mathcal{L}}_{\hat{\varphi}}$ possesses the following properties:*

(a) $\widehat{\mathcal{L}}_{\hat{\varphi}}(\mathbb{1}) = \mathbb{1}$.
Moreover, $\|\widehat{\mathcal{L}}_{\hat{\varphi}}(f)\|_\infty \le \|f\|_\infty$ for every $f \in C(X)$, i. e. the operator $\widehat{\mathcal{L}}_{\hat{\varphi}} : C(X) \to C(X)$ has for norm $\|\widehat{\mathcal{L}}_{\hat{\varphi}}\|_\infty = 1$.

(b) $\widehat{\mathcal{L}}^*_{\hat{\varphi}}(\mu_\varphi) = \mu_\varphi$, *i. e.* $\int_X \widehat{\mathcal{L}}_{\hat{\varphi}}(g)\, d\mu_\varphi = \int_X g\, d\mu_\varphi$ *for all* $g \in L^1(\mu_\varphi)$.
Also, $\|\widehat{\mathcal{L}}_{\hat{\varphi}}(g)\|_{L^1(\mu_\varphi)} \le \|g\|_{L^1(\mu_\varphi)}$, $\forall g \in L^1(\mu_\varphi)$, i. e. the operator $\widehat{\mathcal{L}}_{\hat{\varphi}} : L^1(\mu_\varphi) \to L^1(\mu_\varphi)$ has for norm $\|\widehat{\mathcal{L}}_{\hat{\varphi}}\|_{L^1(\mu_\varphi)} = 1$.

(c) *For all $g \in C(X)$, the sequence $(\inf(\widehat{\mathcal{L}}^n_{\hat{\varphi}}(g)))_{n=0}^{\infty}$ is increasing while the sequence $(\sup(\widehat{\mathcal{L}}^n_{\hat{\varphi}}(g)))_{n=0}^{\infty}$ is decreasing. In particular,*

$$\inf(g) \le \inf(\widehat{\mathcal{L}}^n_{\hat{\varphi}}(g)) \le \sup(\widehat{\mathcal{L}}^n_{\hat{\varphi}}(g)) \le \sup(g), \quad \forall n \in \mathbb{N}.$$

Proof. Using Lemma 13.7.5(a), we get

$$\widehat{\mathcal{L}}_{\hat{\varphi}}(\mathbb{1}) = R_\varphi^{-1} \circ \widehat{\mathcal{L}}_\varphi \circ R_\varphi(\mathbb{1}) = \frac{1}{\varrho_\varphi} \widehat{\mathcal{L}}_\varphi(\varrho_\varphi) = \mathbb{1}.$$

The positivity and linearity of $\widehat{\mathcal{L}}_{\hat{\varphi}}$ yield $\|\widehat{\mathcal{L}}_{\hat{\varphi}}(f)\|_\infty \le \|f\|_\infty$ for any $f \in C(X)$, so (a) holds.

For part (b), let $g \in L^1(\mu_\varphi)$. Using Corollary 13.6.6 and (13.21), observe that

$$\begin{aligned}\int_X \widehat{\mathcal{L}}_{\hat{\varphi}}(g)\, d\mu_\varphi &= \int_X \frac{1}{\varrho_\varphi} \cdot \widehat{\mathcal{L}}_\varphi \circ R_\varphi(g) \cdot \varrho_\varphi\, dm_\varphi = \int_X \widehat{\mathcal{L}}_\varphi(R_\varphi(g))\, dm_\varphi \\ &= \int_X \mathcal{L}_{m_\varphi}(R_\varphi(g))\, dm_\varphi = \int_X R_\varphi(g)\, dm_\varphi = \int_X g \cdot \varrho_\varphi\, dm_\varphi = \int_X g\, d\mu_\varphi.\end{aligned}$$

Furthermore, using (13.22) we have

$$\|\widehat{\mathcal{L}}_{\widehat{\varphi}}(g)\|_{L^1(\mu_\varphi)} = \int_X |\widehat{\mathcal{L}}_{\widehat{\varphi}}(g)|\, d\mu_\varphi = \int_X \frac{1}{\varrho_\varphi} \cdot |\widehat{\mathcal{L}}_\varphi \circ R_\varphi(g)| \cdot \varrho_\varphi\, dm_\varphi = \int_X |\widehat{\mathcal{L}}_\varphi(R_\varphi(g))|\, dm_\varphi$$
$$= \int_X |\mathcal{L}_{m_\varphi}(R_\varphi(g))|\, dm_\varphi \le \int_X |R_\varphi(g)|\, dm_\varphi = \int_X |g|\varrho_\varphi\, dm_\varphi = \|g\|_{L^1(\mu_\varphi)}.$$

Concerning part (c), as $\widehat{\mathcal{L}}_{\widehat{\varphi}}$ is positive, it is monotone and thus for all $g \in C(X)$,

$$\widehat{\mathcal{L}}_{\widehat{\varphi}}(g)(x) \le \widehat{\mathcal{L}}_{\widehat{\varphi}}(\sup(g)\cdot \mathbb{1})(x) = \sup(g)\cdot \widehat{\mathcal{L}}_{\widehat{\varphi}}(\mathbb{1})(x) = \sup(g)\cdot \mathbb{1}(x) = \sup(g)$$

for all $x \in X$. So $\sup(\widehat{\mathcal{L}}_{\widehat{\varphi}}(g)) \le \sup(g)$. Similarly, $\inf(g) \le \inf(\widehat{\mathcal{L}}_{\widehat{\varphi}}(g))$. Replacing g by $\widehat{\mathcal{L}}^n_{\widehat{\varphi}}(g)$ yields (c). □

In light of Proposition 13.8.4(c), one may ask when the extreme inequalities turn out to be strict. Under the stronger assumption of topological exactness of T, the next lemma provides such conditions on g.

Lemma 13.8.5. *Let $T : X \to X$ be a topologically exact, open, distance expanding system and $\varphi : X \to \mathbb{R}$ a Hölder continuous potential. If $g \in C(X)$ is not identically equal to 0 but $\int_X g\, d\mu = 0$ for some $\mu \in M(X)$ with $\operatorname{supp}(\mu) = X$, then there exists $n \in \mathbb{N}$ with*

$$\inf(g) < \inf(\widehat{\mathcal{L}}^n_{\widehat{\varphi}}(g)) \le \sup(\widehat{\mathcal{L}}^n_{\widehat{\varphi}}(g)) < \sup(g).$$

Proof. Since g is not identically 0, the open set $U_0 := \{x \in X \mid g(x) \neq 0\}$ is not empty. Because $\operatorname{supp}(\mu) = X$, we deduce that $\mu(U_0) > 0$. As $\int g\, d\mu = 0$, there exist $z_1, z_2 \in X$ such that $g(z_1) < 0$ and $g(z_2) > 0$. Since g is a continuous function, there are neighborhoods U_1 and U_2 of z_1 and z_2, respectively, on which $g(y) \le g(z_1)/2 < 0$ for all $y \in U_1$ and $g(y) \ge g(z_2)/2 > 0$ for all $y \in U_2$. The system T being topologically exact, there exists $n \in \mathbb{N}$ such that $T^n(U_1) = X = T^n(U_2)$. Therefore, for all $x \in X$,

$$\begin{aligned}
\widehat{\mathcal{L}}^n_{\widehat{\varphi}}(g)(x) &= e^{-nP(T,\widehat{\varphi})}\Bigg[\sum_{y\in T^{-n}(x)\setminus U_1} g(y)e^{\widehat{\varphi}(y)} + \sum_{y\in T^{-n}(x)\cap U_1} g(y)e^{\widehat{\varphi}(y)}\Bigg]\\
&\le e^{-nP(T,\widehat{\varphi})}\Bigg[\sup(g)\sum_{y\in T^{-n}(x)\setminus U_1} e^{\widehat{\varphi}(y)} + \frac{g(z_1)}{2}\sum_{y\in T^{-n}(x)\cap U_1} e^{\widehat{\varphi}(y)}\Bigg]\\
&= e^{-nP(T,\widehat{\varphi})}\Bigg[\sup(g)\sum_{y\in T^{-n}(x)} e^{\widehat{\varphi}(y)} + \left(\frac{g(z_1)}{2} - \sup(g)\right)\sum_{y\in T^{-n}(x)\cap U_1} e^{\widehat{\varphi}(y)}\Bigg]\\
&= \sup(g)\widehat{\mathcal{L}}^n_{\widehat{\varphi}}(\mathbb{1})(x) + \left(\frac{g(z_1)}{2} - \sup(g)\right)e^{-nP(T,\widehat{\varphi})}\sum_{y\in T^{-n}(x)\cap U_1} e^{\widehat{\varphi}(y)}\\
&\le \sup(g) + \left(\frac{g(z_1)}{2} - \sup(g)\right)e^{-(\|\widehat{\varphi}\|_\infty + nP(T,\widehat{\varphi}))}.
\end{aligned}$$

Taking the supremum over all x on the left-hand side yields $\sup(\widehat{\mathcal{L}}_{\widetilde{\varphi}}^n(g)) < \sup(g)$. The proof that $\inf(\widehat{\mathcal{L}}_{\widetilde{\varphi}}^n(g)) > \inf(g)$ proceeds in an analogous fashion using the set U_2 to split $T^{-n}(x)$. □

From this point on, we consider $C(X)$ and $H_\alpha(X)$ as complex Banach spaces. Nevertheless, the potentials, usually denoted by φ, will remain real valued, which we will indicate by adding $\mathbb{R}$ to the notation of the space; for example, $\varphi \in H_\alpha(X,\mathbb{R})$. The bounded linear operators $\mathcal{L}_\varphi, \widehat{\mathcal{L}}_\varphi, \widehat{\mathcal{L}}_{\widetilde{\varphi}} : C(X) \to C(X)$ naturally extend to complex-valued functions $g \in C(X)$ and it is easy to see that

$$\mathcal{L}_\varphi(g) = \mathcal{L}_\varphi(\mathrm{Re}(g)) + i\mathcal{L}_\varphi(\mathrm{Im}(g)).$$

Consequently, corresponding relations for $\widehat{\mathcal{L}}_\varphi$ and $\widehat{\mathcal{L}}_{\widetilde{\varphi}}$ hold. We have already seen in Theorems 13.7.4 and 13.8.3 that the real operators $\widehat{\mathcal{L}}_\varphi, \widehat{\mathcal{L}}_{\widetilde{\varphi}} : C(X,\mathbb{R}) \to C(X,\mathbb{R})$ are almost periodic and their complex counterparts inherit this property, and many others.

We observed in Proposition 13.8.4 that for every real-valued $g \in C(X)$, the sequence $(\inf(\widehat{\mathcal{L}}_{\widetilde{\varphi}}^n(g)))_{n=0}^\infty$ is increasing while the sequence $(\sup(\widehat{\mathcal{L}}_{\widetilde{\varphi}}^n(g)))_{n=0}^\infty$ is decreasing. It turns out that the sequence $(\widehat{\mathcal{L}}_{\widetilde{\varphi}}^n(g))_{n=0}^\infty$ converges uniformly and we will identify its limit. This holds for any real-valued g, and thus for complex-valued ones, too, by linearity. To accomplish this, we will need the following auxiliary result, which is also interesting on its own.

Proposition 13.8.6. *If $T : X \to X$ is a topologically exact, open, distance expanding system then*

$$\lim_{n\to\infty} \sup_{x,y\in X} \mathrm{dist}(x, T^{-n}(y)) = 0.$$

Proof. Fix $\varepsilon > 0$. As X is compact, there exists a finite $(\varepsilon/2)$-spanning set $F \subseteq X$, i. e.

$$\bigcup_{z\in F} B(z,\varepsilon/2) = X.$$

Since the map $T : X \to X$ is topologically exact, for every $z \in F$ there exists $N_z \geq 0$ such that $T^{N_z}(B(z,\varepsilon/2)) = X$. Setting $N_\varepsilon = \max\{N_z : z \in F\}$ yields

$$T^n(B(z,\varepsilon/2)) = X, \quad \forall z \in F, \forall n \geq N_\varepsilon.$$

Take $x, y \in X$. Then $x \in B(z,\varepsilon/2)$ for some $z \in F$. For every $n \geq N_\varepsilon$, there exists $y_n \in B(z,\varepsilon/2)$ such that $T^n(y_n) = y$. Thus,

$$\mathrm{dist}(x, T^{-n}(y)) \leq d(x,y_n) \leq d(x,z) + d(z,y_n) < \varepsilon/2 + \varepsilon/2 = \varepsilon, \quad \forall n \geq N_\varepsilon.$$ □

Lemma 13.8.7. *If $T : X \to X$ is a topologically exact, open, distance expanding system and $\varphi : X \to \mathbb{R}$ is a Hölder continuous potential, then for every function $g \in C(X)$, the sequence $(\widehat{\mathcal{L}}_{\varphi}^{n}(g))_{n=0}^{\infty}$ converges uniformly in the complex Banach space $C(X)$ and*

$$\lim_{n\to\infty} \widehat{\mathcal{L}}_{\varphi}^{n}(g) = \mu_{\varphi}(g)\mathbb{1}.$$

Proof. Splitting g into its real and imaginary parts, we may assume without loss of generality that g is a real-valued function. Take any subsequence of $(\widehat{\mathcal{L}}_{\varphi}^{n}(g))_{n=0}^{\infty}$. To lighten notation, we will identify this subsequence in the same way as the sequence. By Theorem 13.8.3, there exists a strictly increasing sequence $(n_k)_{k=1}^{\infty}$ such that the sequence $(\widehat{\mathcal{L}}_{\varphi}^{n_k}(g))_{k=1}^{\infty}$ converges uniformly. Denote its limit by $g^* \in C(X)$. By Proposition 13.8.4(c), the sequence $(\sup(\widehat{\mathcal{L}}_{\varphi}^{n}(g^*)))_{n=0}^{\infty}$ is decreasing and

$$\sup(\widehat{\mathcal{L}}_{\varphi}^{n}(g^*)) \le \sup(g^*), \quad \forall n \ge 0. \tag{13.62}$$

But the uniform convergence imposes that $\sup(g^*) = \lim_{k\to\infty} \sup(\widehat{\mathcal{L}}_{\varphi}^{n_k}(g))$. The decreasing nature of the sequence $(\sup(\widehat{\mathcal{L}}_{\varphi}^{n}(g)))_{n=0}^{\infty}$ implies that

$$\inf_{n\ge 0} \sup(\widehat{\mathcal{L}}_{\varphi}^{n}(g)) = \lim_{n\to\infty} \sup(\widehat{\mathcal{L}}_{\varphi}^{n}(g)) = \lim_{k\to\infty} \sup(\widehat{\mathcal{L}}_{\varphi}^{n_k}(g)) = \sup(g^*).$$

Hence,

$$\sup(\widehat{\mathcal{L}}_{\varphi}^{n}(g^*)) \ge \sup(g^*), \quad \forall n \ge 0. \tag{13.63}$$

Inequalities (13.62)–(13.63) give

$$\sup(\widehat{\mathcal{L}}_{\varphi}^{n}(g^*)) = \sup(g^*), \quad \forall n \ge 0.$$

Since X is compact, for every $n \ge 0$ there exists a point $x_n \in X$ such that

$$\sup(\widehat{\mathcal{L}}_{\varphi}^{n}(g^*)) = \widehat{\mathcal{L}}_{\varphi}^{n}(g^*)(x_n).$$

Therefore,

$$\sup(g^*) = \widehat{\mathcal{L}}_{\varphi}^{n}(g^*)(x_n) = \frac{e^{-\mathrm{P}(T,\varphi)n}}{\varrho_{\varphi}(x_n)} \sum_{y\in T^{-n}(x_n)} \varrho_{\varphi}(y) g^*(y) e^{S_n\varphi(y)}.$$

By Proposition 13.8.4(a), the right-hand side of this equation is a convex combination of real numbers smaller than or equal to $\sup(g^*)$. Consequently,

$$g^*(y) = \sup(g^*), \quad \forall y \in T^{-n}(x_n),\ \forall n \ge 0.$$

As the function $g^* : X \to \mathbb{R}$ is continuous, it follows from Proposition 13.8.6 that g^* is constant. From Proposition 13.8.4(b), we then get that

$$g^* = \mu_\varphi(g^*) = \mu_\varphi\Big(\lim_{k\to\infty} \widehat{\mathcal{L}}_{\widehat{\varphi}}^{n_k}(g)\Big) = \lim_{k\to\infty} \mu_\varphi(\widehat{\mathcal{L}}_{\widehat{\varphi}}^{n_k}(g)) = \lim_{k\to\infty} \mu_\varphi(g) = \mu_\varphi(g).$$

So we have proved that every subsequence $(n_j)_{j=1}^\infty$ contains a sub-subsequence $(n_{j_i})_{i=1}^\infty$ such that $(\widehat{\mathcal{L}}_{\widehat{\varphi}}^{n_{j_i}}(g))_{i=1}^\infty$ converges to $\mu_\varphi(g)\mathbb{1}$ in $C(X)$. This means that the sequence $(\widehat{\mathcal{L}}_{\widehat{\varphi}}^{n}(g))_{n=0}^\infty$ converges uniformly. □

Let

$$\mathbb{C}^\perp := \{g \in C(X) \mid \mu_\varphi(g) = 0\}.$$

As a straightforward consequence of Lemma 13.8.7 and Proposition 13.8.4, the space $C(X)$ is the direct sum of two simply expressed, $\widehat{\mathcal{L}}_{\widehat{\varphi}}$-invariant subspaces and the $\widehat{\mathcal{L}}_{\widehat{\varphi}}$-orbits of its points converge as follows.

Theorem 13.8.8. *Let $T : X \to X$ be a topologically exact, open, distance expanding system and let $\varphi : X \to \mathbb{R}$ be a Hölder continuous potential. Then:*
(a) *$\mathbb{C}\mathbb{1}$ and $\mathbb{C}^\perp$ are both closed vector subspaces (so are Banach subspaces) of $C(X)$.*
(b) *$\widehat{\mathcal{L}}_{\widehat{\varphi}}(\mathbb{C}\mathbb{1}) = \mathbb{C}\mathbb{1}$ while $\widehat{\mathcal{L}}_{\widehat{\varphi}}(\mathbb{C}^\perp) \subseteq \mathbb{C}^\perp$.*
(c) *$C(X) = \mathbb{C}\mathbb{1} \bigoplus \mathbb{C}^\perp$.*
(d) *If $g \in C(X)$ then the sequence $(\widehat{\mathcal{L}}_{\widehat{\varphi}}^{n}(g))_{n=0}^\infty$ converges uniformly, i. e., in the Banach space $C(X)$, and*

$$\lim_{n\to\infty} \widehat{\mathcal{L}}_{\widehat{\varphi}}^{n}(g) = \mu_\varphi(g)\mathbb{1}.$$

(e) *In particular, if $g \in \mathbb{C}^\perp$ then*

$$\lim_{n\to\infty} \widehat{\mathcal{L}}_{\widehat{\varphi}}^{n}(g) = 0.$$

Let

$$(\mathbb{C}\varrho_\varphi)^\perp := \{g \in C(X) \mid m_\varphi(g) = 0\}.$$

As a direct repercussion of Theorem 13.8.8 and (13.61), the space $C(X)$ is the direct sum of two simply expressed, $\widehat{\mathcal{L}}_\varphi$-invariant subspaces and the $\widehat{\mathcal{L}}_\varphi$-orbits of its points converge as follows.

Theorem 13.8.9. *Let $T : X \to X$ be a topologically exact, open, distance expanding system and let $\varphi : X \to \mathbb{R}$ be a Hölder continuous potential. Then:*
(a) *$\mathbb{C}\varrho_\varphi$ and $(\mathbb{C}\varrho_\varphi)^\perp$ are closed vector subspaces (so are Banach subspaces) of $C(X)$.*
(b) *$\widehat{\mathcal{L}}_\varphi(\mathbb{C}\varrho_\varphi) = \mathbb{C}\varrho_\varphi$ while $\widehat{\mathcal{L}}_\varphi((\mathbb{C}\varrho_\varphi)^\perp) \subseteq (\mathbb{C}\varrho_\varphi)^\perp$.*
(c) *$C(X) = (\mathbb{C}\varrho_\varphi) \bigoplus (\mathbb{C}\varrho_\varphi)^\perp$.*

(d) *If* $g \in C(X)$ *then the sequence* $(\widehat{\mathcal{L}}_\varphi^n(g))_{n=0}^\infty$ *converges uniformly and*

$$\lim_{n\to\infty} \widehat{\mathcal{L}}_\varphi^n(g) = m_\varphi(g)\varrho_\varphi.$$

(e) *In particular, if* $g \in (\mathbb{C}\varrho_\varphi)^\perp$ *then*

$$\lim_{n\to\infty} \widehat{\mathcal{L}}_\varphi^n(g) = 0.$$

Finally, we establish the uniqueness of the eigenfunction ϱ_φ for the operator $\widehat{\mathcal{L}}_\varphi$ and the uniqueness of the eigenpair (eigenvalue, eigenmeasure) = $(\Lambda_\varphi, m_\varphi)$ for the dual operator $\mathcal{L}_\varphi^*$.

Proposition 13.8.10. *Let* $T : X \to X$ *be a topologically exact, open, distance expanding system and let* $\varphi : X \to \mathbb{R}$ *be a Hölder continuous potential. Then:*
(a) *There exists a unique pair* (Λ, m) *with* $\Lambda > 0$ *and* $m \in M(X)$ *such that* $\mathcal{L}_\varphi^*(m) = \Lambda m$; *namely,* $\Lambda = \Lambda_\varphi = e^{\mathrm{P}(T,\varphi)}$ *and* $m = m_\varphi$.
(b) *There exists a unique* $\varrho \in C(X)$ *such that* $\widehat{\mathcal{L}}_\varphi(\varrho) = \varrho$ *with* $\varrho \geq 0$ *and* $\int_X \varrho \, dm_\varphi = 1$; *namely* $\varrho = \varrho_\varphi$.

Proof. Starting with Theorem 13.6.2, all that was assumed about m_φ and Λ_φ was that m_φ is an eigenmeasure for the dual operator $\mathcal{L}_\varphi^* : C(X)^* \to C(X)^*$ with eigenvalue $\Lambda_\varphi > 0$, i. e., $\mathcal{L}_\varphi^*(m_\varphi) = \Lambda_\varphi m_\varphi$. Thus, all the results from that point on apply to any pair (Λ, m), with $\Lambda \in (0, \infty)$ and $m \in M(X)$ such that $\mathcal{L}_\varphi^*(m) = \Lambda m$. Eventually, we proved in Proposition 13.6.9 that such a measure is conformal and this ultimately helped us establish in Proposition 13.6.16 that such a measure is a Gibbs state for φ. The uniqueness of Λ as $\Lambda_\varphi = e^{\mathrm{P}(T,\varphi)}$ ensues from Proposition 13.2.4. For this, only transitivity of T was needed.

Likewise, under the stronger assumption that T is topologically exact but assuming only the equality $\mathcal{L}_\varphi^*(m) = \Lambda m$, we derived Theorem 13.8.9. Part (d) of that theorem (with $g = \mathbb{1}$) reveals that

$$\varrho = \lim_{n\to\infty} \widehat{\mathcal{L}}_\varphi^n(\mathbb{1}),$$

which is completely independent of m while Λ has already been shown to be unique. So ϱ is unique, i. e., $\varrho = \varrho_\varphi$, and the above is a formula for it. Per Corollary 13.7.8, the function ϱ_φ is the unique fixed point of the operator $\widehat{\mathcal{L}}_\varphi : C(X) \to C(X)$ such that $\varrho_\varphi \geq 0$ and $\int_X \varrho_\varphi \, dm_\varphi = 1$. In fact, ϱ_φ is bounded away from 0 and ∞.

Replacing the above expression for ϱ_φ in part (d) of Theorem 13.8.9 yields

$$m(g) = \lim_{n\to\infty} \frac{\widehat{\mathcal{L}}_\varphi^n(g)(x)}{\widehat{\mathcal{L}}_\varphi^n(\mathbb{1})(x)}, \quad \forall x \in X.$$

This formula establishes the uniqueness of m as m_φ. □

Corollary 13.8.11. *Let $T : X \to X$ be a topologically exact, open, distance expanding, Hölder continuous system and let $\varphi : X \to \mathbb{R}$ be a Hölder continuous potential. Then:*

(a) *There exists a unique pair (Λ, m) with $\Lambda > 0$ and $m \in M(X)$ such that $\mathcal{L}^*_{\widehat{\varphi}}(m) = \Lambda m$; namely, $\Lambda = \Lambda_\varphi = e^{\mathrm{P}(T,\varphi)}$ and $m = m_{\widehat{\varphi}} = \mu_\varphi$.*

(b) *There exists a unique $\varrho \in C(X)$ such that $\widehat{\mathcal{L}}_{\widehat{\varphi}}(\varrho) = \varrho$ with $\varrho \geq 0$ and $\int_X \varrho \, dm_{\widehat{\varphi}} = 1$; namely $\varrho = \varrho_{\widehat{\varphi}} = \mathbb{1}$.*

(c) $\mu_{\widehat{\varphi}} = \mu_\varphi$.

Proof. When we introduced $\widehat{\varphi}$ at the start of Subsection 13.8.1, we observed that $\widehat{\varphi}$ is Hölder continuous when T is. So we can use Proposition 13.8.10 with φ replaced with $\widehat{\varphi}$. We also noted at the beginning of Subsection 13.8.1 that $\mathrm{P}(T, \widehat{\varphi}) = \mathrm{P}(T, \varphi)$ and $\mathcal{L}^*_{\widehat{\varphi}}(\mu_\varphi) = e^{\mathrm{P}(T,\varphi)}\mu_\varphi$. Thus, (a) holds. Furthermore, Proposition 13.8.4 shows that $\varrho_{\widehat{\varphi}} = \mathbb{1}$. Hence, (b) holds. Finally, we noticed at the outset of Subsection 13.8.1 that $\mathcal{G}_{\widehat{\varphi}} = \mathcal{G}_\varphi$. The uniqueness of the T-invariant Gibbs states μ_φ and $\mu_{\widehat{\varphi}}$ implies that $\mu_{\widehat{\varphi}} = \mu_\varphi$. □

In Theorem 13.8.8, we described the uniform convergence of the iterates of a continuous function g under the operator $\widehat{\mathcal{L}}_{\widehat{\varphi}}$. We now establish the $L^1(\mu_\varphi)$-convergence of the iterates of a $L^1(\mu_\varphi)$-function g.

Theorem 13.8.12. *Let $T : X \to X$ be a topologically exact, open, distance expanding system and let $\varphi : X \to \mathbb{R}$ be a Hölder continuous potential. For every $g \in L^1(\mu_\varphi)$,*

$$\lim_{n\to\infty} \|\widehat{\mathcal{L}}^n_{\widehat{\varphi}}(g) - \mu_\varphi(g)\mathbb{1}\|_{L^1(\mu_\varphi)} = 0.$$

Proof. Fix $\varepsilon > 0$. As $C(X)$ is a dense subset of $L^1(\mu_\varphi)$, there exists $f \in C(X)$ such that

$$\|g - f\|_{L^1(\mu_\varphi)} < \varepsilon/2.$$

Then

$$|\mu_\varphi(g) - \mu_\varphi(f)| < \varepsilon/2.$$

It follows from Proposition 13.8.4(b) and Lemma 13.8.7 that

$$\begin{aligned}
&\limsup_{n\to\infty} \|\widehat{\mathcal{L}}^n_{\widehat{\varphi}}(g) - \mu_\varphi(g)\mathbb{1}\|_{L^1(\mu_\varphi)} \\
&\quad = \limsup_{n\to\infty} \|\widehat{\mathcal{L}}^n_{\widehat{\varphi}}(g - f) + (\widehat{\mathcal{L}}^n_{\widehat{\varphi}}(f) - \mu_\varphi(f)\mathbb{1}) + (\mu_\varphi(f) - \mu_\varphi(g))\mathbb{1}\|_{L^1(\mu_\varphi)} \\
&\quad \leq \limsup_{n\to\infty} \|\widehat{\mathcal{L}}^n_{\widehat{\varphi}}(g - f)\|_{L^1(\mu_\varphi)} + \limsup_{n\to\infty} \|\widehat{\mathcal{L}}^n_{\widehat{\varphi}}(f) - \mu_\varphi(f)\mathbb{1}\|_{L^1(\mu_\varphi)} \\
&\qquad + \|(\mu_\varphi(f) - \mu_\varphi(g))\mathbb{1}\|_{L^1(\mu_\varphi)} \\
&\quad \leq \|g - f\|_{L^1(\mu_\varphi)} + \limsup_{n\to\infty} \|\widehat{\mathcal{L}}^n_{\widehat{\varphi}}(f) - \mu_\varphi(f)\mathbb{1}\|_\infty + |\mu_\varphi(f) - \mu_\varphi(g)| \\
&\quad < \varepsilon/2 + 0 + \varepsilon/2 = \varepsilon.
\end{aligned}$$

The arbitrariness of ε means that $\limsup_{n\to\infty} \|\widehat{\mathcal{L}}^n_{\widehat{\varphi}}(g) - \mu_\varphi(g)\mathbb{1}\|_{L^1(\mu_\varphi)} = 0$. □

As a fairly immediate consequence of this theorem, we obtain the following remarkable result.

Theorem 13.8.13. *Let $T : X \to X$ be a topologically exact, open, distance expanding system and let $\varphi : X \to \mathbb{R}$ be a Hölder continuous potential. Then the measure-preserving dynamical system (T, μ_φ) is totally ergodic. That is, all iterates $T^q : X \to X$, $q \in \mathbb{N}$, are ergodic with respect to the measure μ_φ, including T itself.*

Proof. Fix $q \in \mathbb{N}$. Suppose that $A \subseteq X$ is a completely T^q-invariant Borel set, i. e., $T^{-q}(A) = A$. Then $T^{-qn}(A) = A$ for all $n \ge 0$. Equivalently, $\mathbb{1}_A \circ T^{qn} = \mathbb{1}_A$ for all $n \ge 0$. Therefore,

$$\widehat{\mathcal{L}}_{\widehat{\varphi}}^{qn}(\mathbb{1}_A) = \widehat{\mathcal{L}}_{\widehat{\varphi}}^{qn}(\mathbb{1}_A \circ T^{qn}) = \mathbb{1}_A \cdot \widehat{\mathcal{L}}_{\widehat{\varphi}}^{qn}(\mathbb{1}) = \mathbb{1}_A.$$

It follows from Theorem 13.8.12 that $\mathbb{1}_A = \mu_\varphi(A)\mathbb{1}$ μ_φ-almost everywhere. Thus, $\mu_\varphi(A) \in \{0, 1\}$, yielding the ergodicity of T^q. □

In fact, we can easily prove more.

Theorem 13.8.14. *Let $T : X \to X$ be a topologically exact, open, distance expanding system and let $\varphi : X \to \mathbb{R}$ be a Hölder continuous potential. Then the measure-preserving dynamical system (T, μ_φ) is metrically exact, and hence its Rokhlin natural extension is a K-system.*

Proof. Let $\mathcal{B}(X)$ be the Borel σ-algebra of X. Recall that the measure-preserving dynamical system (X, T, μ_φ) is metrically exact if and only if the tail σ-algebra

$$\bigcap_{n=0}^{\infty} T^{-n}(\mathcal{B}(X))$$

consists solely of sets of μ_φ-measure 0 or 1 (cf. Definition 8.4.4 and Proposition 8.4.5). Let $A \in \bigcap_{n=0}^{\infty} T^{-n}(\mathcal{B}(X))$. For every $n \ge 0$ there is a set $A_n \in \mathcal{B}(X)$ such that $A = T^{-n}(A_n)$. Equivalently, $\mathbb{1}_A = \mathbb{1}_{A_n} \circ T^n$. Consequently,

$$\widehat{\mathcal{L}}_{\widehat{\varphi}}^{n}(\mathbb{1}_A) = \widehat{\mathcal{L}}_{\widehat{\varphi}}^{n}(\mathbb{1}_{A_n} \circ T^{n}) = \mathbb{1}_{A_n} \cdot \widehat{\mathcal{L}}_{\widehat{\varphi}}^{n}(\mathbb{1}) = \mathbb{1}_{A_n}.$$

It therefore follows from Theorem 13.8.12 that $\mathbb{1}_{A_n}$ converges in $L^1(\mu_\varphi)$ to the constant function $\mu_\varphi(A)\mathbb{1}$. But since $\mathbb{1}_{A_n}$ takes on only the two values 0 and 1, we conclude that

$$\mu_\varphi(A) \in \{0, 1\}.$$

The metric exactness of (X, T, μ_φ) is thus proved. That its Rokhlin natural extension is a K-system follows directly from Theorem 8.4.6. □

13.8.2 Ionescu-Tulcea and Marinescu inequality and spectral gap

The complex Banach space $(H_\alpha(X), \|\cdot\|_\alpha)$ is a vector subspace of the complex Banach space $(C(X), \|\cdot\|_\infty)$. We shall now derive a Ionescu-Tulcea and Marinescu type of inequality (see condition (c1) at the beginning of Appendix B). We first get such an inequality with respect to $\upsilon_\alpha(\cdot)$; see Section 13.1.

Theorem 13.8.15. *Let $T : X \to X$ be a topologically transitive, open, distance expanding system and $\varphi \in H_\alpha(X, \mathbb{R})$. If $g \in H_\alpha(X)$, then for all $n \in \mathbb{N}$ we have $\widehat{\mathcal{L}}_\varphi^n(g) \in H_\alpha(X)$ and*

$$\upsilon_\alpha(\widehat{\mathcal{L}}_\varphi^n(g)) \le \overline{D}_\alpha(\varphi)\lambda^{-\alpha n}\upsilon_\alpha(g) + D'_\alpha(\varphi)\|g\|_\infty,$$

where $D'_\alpha(\varphi) := \max\{\widetilde{D}_\alpha(\varphi), 2\overline{D}_\alpha(\varphi)/\xi^\alpha\}$ and the constants $\overline{D}_\alpha(\varphi)$ and $\widetilde{D}_\alpha(\varphi)$ respectively arise from Corollary 13.1.2 and (13.47) in the proof of Theorem 13.7.4.

Proof. Fix $x, y \in X$. Suppose first that $d(x, y) < \xi$. From a slight alteration of (13.46) in the proof of Theorem 13.7.4, we obtain

$$\begin{aligned}
|\widehat{\mathcal{L}}_\varphi^n(g)(y) - \widehat{\mathcal{L}}_\varphi^n(g)(x)| &\le \overline{D}_\alpha(\varphi)\max\{|g(T_z^{-n}(y)) - g(T_z^{-n}(x))| : z \in T^{-n}(x)\} \\
&\quad + \widetilde{D}_\alpha(\varphi)\|g\|_\infty d^\alpha(x, y) \\
&\le \overline{D}_\alpha(\varphi)\upsilon_\alpha(g)\max\{d^\alpha(T_z^{-n}(y), T_z^{-n}(x)) : z \in T^{-n}(x)\} \\
&\quad + \widetilde{D}_\alpha(\varphi)\|g\|_\infty d^\alpha(x, y) \\
&\le \overline{D}_\alpha(\varphi)\upsilon_\alpha(g)\lambda^{-\alpha n} d^\alpha(x, y) + \widetilde{D}_\alpha(\varphi)\|g\|_\infty d^\alpha(x, y) \\
&= [\overline{D}_\alpha(\varphi)\lambda^{-\alpha n}\upsilon_\alpha(g) + \widetilde{D}_\alpha(\varphi)\|g\|_\infty]d^\alpha(x, y).
\end{aligned} \tag{13.64}$$

If $d(x, y) \ge \xi$, then using Proposition 13.7.2 we get

$$|\widehat{\mathcal{L}}_\varphi^n(g)(y) - \widehat{\mathcal{L}}_\varphi^n(g)(x)| \le 2\|\widehat{\mathcal{L}}_\varphi^n(g)\|_\infty \le 2\|g\|_\infty\|\widehat{\mathcal{L}}_\varphi^n(\mathbb{1})\|_\infty \le 2\|g\|_\infty\overline{D}_\alpha(\varphi)\frac{d^\alpha(x, y)}{\xi^\alpha}. \tag{13.65}$$

Combining (13.64) and (13.65), we conclude that the expression $\overline{D}_\alpha(\varphi)\lambda^{-\alpha n}\upsilon_\alpha(g) + \max\{\widetilde{D}_\alpha(\varphi), 2\overline{D}_\alpha(\varphi)/\xi^\alpha\}\|g\|_\infty$ is a Hölder constant for $\widehat{\mathcal{L}}_\varphi^n(g)$. □

We now obtain the sought Ionescu-Tulcea and Marinescu inequality.

Corollary 13.8.16. *Let $T : X \to X$ be a topologically transitive, open, distance expanding system and $\varphi \in H_\alpha(X, \mathbb{R})$. There exists $N \in \mathbb{N}$ such that for all $n \ge N$ and all $g \in H_\alpha(X)$, we have*

$$\|\widehat{\mathcal{L}}_\varphi^n(g)\|_\alpha \le \frac{1}{2}\|g\|_\alpha + D''_\alpha(\varphi)\|g\|_\infty,$$

where $D''_\alpha(\varphi) := D'_\alpha(\varphi) + \overline{D}_\alpha(\varphi)$.

Proof. Using Theorem 13.8.15 and Proposition 13.7.2, we obtain for all $n \in \mathbb{N}$ that

$$\begin{aligned}\|\widehat{\mathcal{L}}_\varphi^n(g)\|_\alpha := v_\alpha(\widehat{\mathcal{L}}_\varphi^n(g)) + \|\widehat{\mathcal{L}}_\varphi^n(g)\|_\infty &\le \overline{D}_\alpha(\varphi)\lambda^{-\alpha n} v_\alpha(g) + D'_\alpha(\varphi)\|g\|_\infty + \|\widehat{\mathcal{L}}_\varphi^n(\mathbb{1})\|_\infty \|g\|_\infty \\ &\le \overline{D}_\alpha(\varphi)\lambda^{-\alpha n}\|g\|_\alpha + (D'_\alpha(\varphi) + \overline{D}_\alpha(\varphi))\|g\|_\infty.\end{aligned}$$

Take N so large that $\overline{D}_\alpha(\varphi)\lambda^{-\alpha N} \le 1/2$. □

As a consequence of the extended Ionescu-Tulcea and Marinescu Theorem (see Appendix B), our main result affirms that the normalized transfer operator is quasi-compact. We first recall the definition of this concept.

Definition 13.8.17. Let B be a Banach space. A bounded linear operator $L : B \to B$ is called *quasi-compact* if it exhibits a spectral gap, i. e., if its spectral radius $r(L)$ is larger than its essential spectral radius $r_{\mathrm{ess}}(L)$.

The spectral radius $r(L)$ is the largest modulus of the eigenvalues of the operator L (i. e., the supremum of the modulus of all elements in the spectrum of L). In terms of the operator norm,

$$r(L) = \lim_{n\to\infty} \|L^n\|^{1/n}.$$

The essential spectral radius is

$$r_{\mathrm{ess}}(L) := \lim_{n\to\infty} \Big[\inf_K \|L^n - K\|^{1/n}\Big],$$

where the infimum is taken over all compact operators $K : B \to B$.

Theorem 13.8.18. *If $T : X \to X$ is a topologically exact, open, distance expanding system and $\varphi \in H_\alpha(X, \mathbb{R})$, then:*

(a) *The number 1 is a simple isolated eigenvalue of $\widehat{\mathcal{L}}_\varphi : H_\alpha(X) \to H_\alpha(X)$.*
(b) *The eigenspace of the eigenvalue 1 is generated by the function $\varrho_\varphi \in H_\alpha(X, \mathbb{R})$ arising from Lemma 13.7.5 (see also Theorem 13.7.7), i. e., the eigenspace of 1 is $\mathbb{C}\varrho_\varphi$.*
(c) *For all $n \in \mathbb{N}$,*

$$\widehat{\mathcal{L}}_\varphi^n = L_1 + S^n,$$

where

(d) *$L_1 : H_\alpha(X) \to H_\alpha(X)$ is the projector onto the eigenspace $\mathbb{C}\varrho_\varphi$ given by*

$$L_1(\psi) = \Big[\int_X \psi \, dm_\varphi\Big]\varrho_\varphi; \tag{13.66}$$

(e) *$S : H_\alpha(X) \to H_\alpha(X)$ is simply $S = \widehat{\mathcal{L}}_\varphi - L_1$;*
(f) *$L_1 \circ S = S \circ L_1 = 0$;*

(g) $r(S) < 1$, *where* $r(S)$ *denotes the spectral radius of* S; *and*
(h) $\sigma(\widehat{\mathcal{L}}_\varphi) \setminus \{1\} = \sigma(S) \subseteq \overline{B(0, r(S))}$, *where* $\sigma(R)$ *denotes the spectrum of* R.

Proof. We want to apply the Ionescu-Tulcea and Marinescu theorem (Theorem B.1.1) with $(B, \|\cdot\|) = (H_\alpha(X), \|\cdot\|_\alpha)$, with $(E, |\cdot|) = (C(X), \|\cdot\|_\infty)$, with $T = \widehat{\mathcal{L}}_\varphi$ and with N coming from Corollary 13.8.16. We must check that conditions (a1)–(d1) are satisfied. Condition (b1) follows from Proposition 13.7.2 while (c1) is precisely Corollary 13.8.16 with $n = N$, $r = 1/2$ and $R = D''_\alpha(\varphi)$. By Arzelà–Ascoli's theorem (Theorem A.2.3), every bounded subset of $(H_\alpha(X), \|\cdot\|_\alpha)$ is relatively compact in $(C(X), \|\cdot\|_\infty)$. Since $\widehat{\mathcal{L}}_\varphi : H_\alpha(X) \to H_\alpha(X)$ is a bounded operator, this establishes (d1) per Remark B.1.2. Finally, for (a1), suppose that $(g_n)_{n=1}^\infty \subseteq H_\alpha(X)$ with $\|g_n\|_\alpha \le K$ and $g_n \to g$ uniformly in $C(X)$. We need to verify that $g \in H_\alpha(X)$ and $\|g\|_\alpha \le K$. Let $x, y \in X$. For every $n \in \mathbb{N}$, we have

$$|g_n(x) - g_n(y)| \le v_\alpha(g_n)\, d^\alpha(x, y),$$

where $0 \le v_\alpha(g_n) \le \|g_n\|_\alpha \le K$. Letting $n \to \infty$, we get

$$|g(x) - g(y)| \le \liminf_{n\to\infty} v_\alpha(g_n)\, d^\alpha(x, y).$$

Hence, $v_\alpha(g) \le \liminf_{n\to\infty} v_\alpha(g_n) \le K$. Moreover, since $g_n \to g$ uniformly, we have $\|g_n\|_\infty \to \|g\|_\infty$ as $n \to \infty$. Therefore,

$$\|g\|_\alpha = v_\alpha(g) + \|g\|_\infty \le \liminf_{n\to\infty} v_\alpha(g_n) + \lim_{n\to\infty} \|g_n\|_\infty = \liminf_{n\to\infty} \|g_n\|_\alpha \le K.$$

So (a1) is satisfied. We can thus apply the Ionescu-Tulcea and Marinescu theorem. By (f2) in that theorem, we know that $\sigma(\widehat{\mathcal{L}}_\varphi) \setminus \{\lambda_1, \ldots, \lambda_p\} = \sigma(S) \subseteq \overline{B(0, r(S))} \subseteq \mathbb{C}$, where $\lambda_1, \ldots, \lambda_p \in \mathbb{S}^1$ and $0 < r(S) < 1$.

Claim 1. *The number 1 is a simple isolated eigenvalue for* $\widehat{\mathcal{L}}_\varphi : H_\alpha(X) \to H_\alpha(X)$ *and its corresponding eigenspace is* $\mathbb{C}\varrho_\varphi$.

Proof of Claim 1. We already know from Lemma 13.7.5 that $\varrho_\varphi \in H_\alpha(X, \mathbb{R})$ is an eigenvector for $\widehat{\mathcal{L}}_\varphi$ with eigenvalue 1 and it is clear from (f2) above that the number 1 is an isolated eigenvalue of $\widehat{\mathcal{L}}_\varphi$.

Suppose that $\widehat{\mathcal{L}}_\varphi(g) = 1 \cdot g = g$ for some nonzero function $g \in H_\alpha(X)$. Since g is nonzero, at least one of its real and imaginary part functions $\mathrm{Re}(g)$ and $\mathrm{Im}(g)$ does not identically vanish. Denote by ψ whichever is nonvanishing. Then $\psi \in H_\alpha(X)$ and $\widehat{\mathcal{L}}_\varphi(\psi) = \psi$. It remains to show that ψ is a real multiple of ϱ_φ. By Lemma 13.7.5(b), there exists $M > 0$ so large that $c^{-1} \le \psi + M\varrho_\varphi \le c$ for some constant $c \ge 1$. But $\widehat{\mathcal{L}}_\varphi(\psi + M\varrho_\varphi) = \psi + M\varrho_\varphi$ and, therefore, the function

$$\frac{\psi + M\varrho_\varphi}{\int_X (\psi + M\varrho_\varphi)\, dm_\varphi}$$

satisfies properties (a–c) in Lemma 13.7.5. According to Corollary 13.7.8, this function is nothing else than ϱ_φ. Consequently,

$$\psi + M\varrho_\varphi = \left[\int_X (\psi + M\varrho_\varphi)\, dm_\varphi\right]\varrho_\varphi.$$

Equivalently,

$$\psi = \left[\int_X (\psi + M\varrho_\varphi)\, dm_\varphi - M\right]\varrho_\varphi =: C\varrho_\varphi.$$

Thus, the number 1 is a simple eigenvalue and its corresponding eigenspace is $\mathbb{C}\varrho_\varphi$.

An alternative argument for this latter property is the following.

Subclaim 1*. *If $h : X \to [0, \infty)$ is a continuous function such that $\widehat{\mathcal{L}}_\varphi(h) = h$ and $h(x) = 0$ for some $x \in X$, then $h \equiv 0$.*

Proof of Subclaim 1*. Since h is continuous, relation (13.30) in Proposition 13.6.7 holds everywhere. Then for every $n \in \mathbb{N}$ we have that

$$0 = h(x) = \widehat{\mathcal{L}}_\varphi^n(h)(x) = e^{-n\mathrm{P}(T,\varphi)}\mathcal{L}_\varphi^n(h)(x) = e^{-n\mathrm{P}(T,\varphi)} \sum_{y\in T^{-n}(x)} h(y)e^{S_n\varphi(y)}.$$

Therefore, $h(y) = 0$ for every $y \in \bigcup_{n=0}^\infty T^{-n}(x)$. Since h is continuous, it ensues that $h(y) = 0$ for every $y \in \overline{\bigcup_{n=0}^\infty T^{-n}(x)}$. As T is transitive, open and distance expanding, recall from Proposition 13.6.14 that T is very strongly transitive (alternatively, see Lemma 4.2.10). This implies that T is strongly transitive, i. e. the backward T-orbit of any point is dense in X (see Definition 1.5.14). So h is the constant function 0 and Subclaim 1* follows. ♦

Now assume that $g \in C(X)$ is an eigenfunction associated with the eigenvalue 1 of the operator $\widehat{\mathcal{L}}_\varphi : C(X) \to C(X)$. Given that $\widehat{\mathcal{L}}_\varphi$ preserves the vector subspace of $C(X)$ consisting of real-valued functions, both the real and imaginary parts of g are also eigenfunctions associated with the eigenvalue 1. So we may assume without loss of generality that g is real valued. By Lemma 13.7.5, the function $g/\varrho_\varphi : X \to \mathbb{R}$ is continuous. Since X is compact, this function attains its minimum value; denote this latter by s. Then $g - s\varrho_\varphi \geq 0$ and there exists $x \in X$ such that $(g - s\varrho_\varphi)(x) = 0$. Since $g - s\varrho_\varphi$ is an eigenfunction associated with the eigenvalue 1, too, it follows from Subclaim 1* that $g - s\varrho_\varphi \equiv 0$; equivalently, $g = s\varrho_\varphi$. So, again, the number 1 is a simple eigenvalue and its corresponding eigenspace is $\mathbb{C}\varrho_\varphi$. Note that this alternative argument considers the operator $\widehat{\mathcal{L}}_\varphi : C(X) \to C(X)$, as opposed to $\widehat{\mathcal{L}}_\varphi : H_\alpha(X) \to H_\alpha(X)$, and does not require the use of Corollary 13.7.8.

In any case, Claim 1 is proved. ■

Note that until this point, topological transitivity of T is sufficient. We will need topological exactness to establish the next claim.

Claim 2. *The number* 1 *is the only eigenvalue of* $\widehat{\mathcal{L}}_\varphi : H_\alpha(X) \to H_\alpha(X)$ *having modulus* 1.

Proof of Claim 2. Suppose that $\lambda \neq 1$ is a unitary eigenvalue of $\widehat{\mathcal{L}}_\varphi$. Then there exists a nonzero function $g \in C(X)$ such that $\widehat{\mathcal{L}}_\varphi(g) = \lambda g$. Therefore,

$$\lambda \int_X g\, dm_\varphi = \int_X \widehat{\mathcal{L}}_\varphi(g)\, dm_\varphi = \int_X \mathcal{L}_{m_\varphi}(g)\, dm_\varphi = \int_X g\, dm_\varphi.$$

Since $\lambda \neq 1$, we deduce that $\int g\, dm_\varphi = 0$. By Theorem 13.8.9(d), $\lambda^n g = \widehat{\mathcal{L}}_\varphi^n(g) \to 0$ uniformly in $C(X)$. So g is identically equal to 0. This contradiction finishes the proof of Claim 2. ∎

In particular, this shows that $\sigma(\widehat{\mathcal{L}}_\varphi) \setminus \{1\} = \sigma(S) \subseteq \overline{B(0, r(S))}$.

All that is left to show (the rest follows directly from the Ionescu-Tulcea and Marinescu theorem) is (13.66). From (b), it follows that $L_1(\psi) = a\varrho_\varphi$ for some $a \in \mathbb{C}$. By (c),

$$\widehat{\mathcal{L}}_\varphi^n(\psi) = L_1(\psi) + S^n(\psi) = a\varrho_\varphi + S^n(\psi).$$

Using Lemma 13.7.5(c), we deduce that

$$\begin{aligned} a = \int_X a\varrho_\varphi\, dm_\varphi &= \int_X \widehat{\mathcal{L}}_\varphi^n(\psi)\, dm_\varphi - \int_X S^n(\psi)\, dm_\varphi \\ &= \int_X \mathcal{L}_{m_\varphi}^n(\psi)\, dm_\varphi - \int_X S^n(\psi)\, dm_\varphi \\ &= \int_X \psi\, dm_\varphi - \int_X S^n(\psi)\, dm_\varphi \\ &\to \int_X \psi\, dm_\varphi \quad \text{when } n \to \infty \text{ by (g).} \end{aligned}$$

Hence $a = \int_X \psi\, dm_\varphi$ and (13.66) holds. □

Corollary 13.8.19. *If* $T : X \to X$ *is a topologically exact, open, distance expanding system and* $\varphi \in H_\alpha(X, \mathbb{R})$, *then the normalized transfer operator* $\widehat{\mathcal{L}}_\varphi$ *is quasi-compact.*

Proof. Per Theorem 13.8.18, the spectral radius of $\widehat{\mathcal{L}}_\varphi$ is $r(\widehat{\mathcal{L}}_\varphi) = 1$. Note that each finite-rank operator (i. e., whose range is finite-dimensional) is compact. So the one-dimensional projection operator L_1 onto the eigenspace $\mathbb{C}\rho_\varphi$ is compact. By Theorem 13.8.18 still, we know that $\lim_{n\to\infty} \|\widehat{\mathcal{L}}_\varphi^n - L_1\|^{1/n} = \lim_{n\to\infty} \|S^n\|^{1/n} = r(S) < 1$ and thus $r_{\text{ess}}(\widehat{\mathcal{L}}_\varphi) < 1$. So $r(\widehat{\mathcal{L}}_\varphi) - r_{\text{ess}}(\widehat{\mathcal{L}}_\varphi) > 0$, i. e., $\widehat{\mathcal{L}}_\varphi$ has a spectral gap; in other words, it is quasi-compact.

Note: For some dynamicists, a spectral gap means quasi-compactness of the operator plus simplicity of its leading eigenvalue as well as the absence of any other eigenvalue of maximum modulus (i. e., equal to the operator's spectral radius). Theorem 13.8.18 asserts that the normalized transfer operator has all of these features. □

Distinguishing those parts of the proofs of Theorem 13.8.18 and Corollary 13.8.19 that require topological exactness from those that do not, we get a similar result assuming only topological transitivity.

Theorem 13.8.20. *If $T : X \to X$ is a transitive, open, distance expanding dynamical system and $\varphi \in H_\alpha(X, \mathbb{R})$, then:*

(a) *The number 1 is a simple isolated eigenvalue of $\widehat{\mathcal{L}}_\varphi : H_\alpha(X) \to H_\alpha(X)$.*
(b) *The eigenspace of the eigenvalue 1 is generated by the function $\varrho_\varphi \in H_\alpha(X, \mathbb{R})$ arising from Lemma 13.7.5 (see also Theorem 13.7.7), i. e., the eigenspace of 1 is $\mathbb{C}\varrho_\varphi$.*
(c) *For all $n \in \mathbb{N}$,*

$$\widehat{\mathcal{L}}_\varphi^n = \sum_{j=1}^{p} \lambda_j^n L_j + S^n,$$

where

(d) *$L_1 : H_\alpha(X) \to H_\alpha(X)$ is the projector onto the eigenspace $\mathbb{C}\varrho_\varphi$ given by*

$$L_1(\psi) = \left[\int_X \psi \, dm_\varphi\right] \varrho_\varphi; \tag{13.67}$$

(d′) *For every $j \geq 2$, L_j is a finite-rank projector, meaning that $L_j^2 = L_j$ and $\dim(\mathrm{Range}(L_j)) < \infty$; moreover, λ_j is an eigenvalue of $\widehat{\mathcal{L}}_\varphi$ with modulus 1 and $\mathrm{Ker}(\lambda_j \mathrm{Id} - \widehat{\mathcal{L}}_\varphi) = \mathrm{Range}(L_j)$;*
(e) *$S : H_\alpha(X) \to H_\alpha(X)$ is simply $S = \widehat{\mathcal{L}}_\varphi - \sum_{j=1}^{p} \lambda_j L_j$;*
(f) *$L_j \circ S = S \circ L_j = 0$ for all $j = 1, 2, \ldots, p$ and $L_i \circ L_j = 0$ whenever $i \neq j$;*
(g) *$r(S) < 1$, where $r(S)$ denotes the spectral radius of S;*
(h) *$\sigma(\widehat{\mathcal{L}}_\varphi) \setminus \{\lambda_j : 1 \leq j \leq p\} = \sigma(S) \subseteq \overline{B(0, r(S))}$, where $\sigma(S)$ is the spectrum of S; and*
(i) *$\widehat{\mathcal{L}}_\varphi$ is quasi-compact.*

13.8.3 Continuity of Gibbs states

Theorem 13.8.21. *If $T : X \to X$ is a topologically exact, open, distance expanding system, then for every $\alpha > 0$ and every $R > 0$ the function*

$$\overline{B_{H_\alpha(X)}(0, R)} \ni \varphi \longmapsto \varrho_\varphi \in C(X)$$

is continuous with respect to the topologies of uniform convergence on both the domain and codomain, and the functions

$$\overline{B_{H_\alpha(X)}(0,R)} \ni \varphi \longmapsto m_\varphi, \mu_\varphi \in M(X)$$

are continuous with respect to the topology of uniform convergence on the domain and the weak* *topology on the codomain.*

Proof. We begin with the assertion about the eigenmeasures m_φ. Let $(\psi_n)_{n=1}^\infty$ be a sequence of potentials in $\overline{B_{H_\alpha(X)}(0,R)}$ converging uniformly to some potential $\psi : X \to \mathbb{R}$. This latter belongs to $\overline{B_{H_\alpha(X)}(0,R)}$ as that set is compact by Arzelà–Ascoli's theorem (Theorem A.2.3 in Appendix A). Per Theorem A.1.59 the set $M(X)$ is compact, so let $(m_{\psi_{n_k}})_{k=1}^\infty$ be any convergent subsequence of the sequence $(m_{\psi_n})_{n=1}^\infty$. Denote its limit by m. We obviously have the following.

Claim 1. *The function*

$$C(X) \ni h \longmapsto \mathcal{L}_h \in L(C(X)) \tag{13.68}$$

is continuous.

As an immediate consequence, we get the following.

Claim 2. *The function*

$$C(X) \ni h \longmapsto \mathcal{L}_h^* \in L(C(X)^*) \tag{13.69}$$

is continuous.

According to Exercise 11.5.7, the topological pressure function $\mathrm{P} : C(X) \to \mathbb{R}$ is Lipschitz with constant 1, and thus $\lim_{n\to\infty} \mathrm{P}(T,\psi_n) = \mathrm{P}(T,\psi)$. We then deduce that

$$m = \lim_{k\to\infty} m_{\psi_{n_k}} = \lim_{k\to\infty} e^{-\mathrm{P}(T,\psi_{n_k})} \mathcal{L}_{\psi_{n_k}}^*(m_{\psi_{n_k}}) = e^{-\mathrm{P}(T,\psi)} \mathcal{L}_\psi^*(m).$$

So $\mathcal{L}_\psi^*(m) = e^{\mathrm{P}(T,\psi)} m$, whence $m = m_\psi$ by Proposition 13.8.10(a). The proof of the continuity of $\varphi \mapsto m_\varphi$ is complete.

Moving on to the assertion about the eigenfunctions ϱ_φ, suppose for a contradiction that its continuity claim fails at some potential $\psi \in \overline{B_{H_\alpha(X)}(0,R)}$. This means that there exist $\delta > 0$ and a sequence $(\psi_k)_{k=1}^\infty$ of potentials in $\overline{B_{H_\alpha(X)}(0,R)}$ which converge uniformly to ψ but are such that

$$\|\varrho_{\psi_k} - \varrho_\psi\|_\infty > \delta, \quad \forall k \in \mathbb{N}. \tag{13.70}$$

By Remark 13.7.6, $\|\varrho_{\psi_k}\|_\alpha \le V_\alpha(R)$ for all $k \in \mathbb{N}$. By virtue of Arzelà–Ascoli's theorem (Theorem A.2.3) and by passing to a subsequence if necessary, we may assume that the sequence $(\varrho_{\psi_k})_{k=1}^{\infty}$ converges uniformly to some function $\varrho \in C(X)$. By (13.70),

$$\varrho \neq \varrho_{\psi}. \tag{13.71}$$

On the other hand, using Claim 1 and the continuity of the pressure function, we get

$$\mathcal{L}_{\psi}(\varrho) = e^{\mathrm{P}(T,\psi)}\varrho. \tag{13.72}$$

Furthermore, using the known continuity of $\varphi \mapsto m_{\varphi}$ and Lemma 13.7.5, we have

$$\int_X \varrho \, dm_{\psi} = \lim_{k\to\infty} \int_X \varrho_{\psi_k} \, dm_{\psi_k} = \lim_{k\to\infty} 1 = 1. \tag{13.73}$$

Finally, the nonnegativity of the functions $(\varrho_{\psi_k})_{k=1}^{\infty}$ ensures that

$$\varrho \geq 0. \tag{13.74}$$

According to Proposition 13.8.10, it ensues from relations (13.72)–(13.74) that $\varrho = \varrho_{\psi}$, contrary to (13.71).

The continuity of the invariant Gibbs states μ_{φ} is a consequence of the continuity of the eigenmeasures m_{φ} and of the eigenfunctions ϱ_{φ}. Indeed, if $\psi \in \overline{B_{H_{\alpha}(X)}(0,R)}$ and $(\psi_n)_{n=1}^{\infty}$ is a sequence of functions in $\overline{B_{H_{\alpha}(X)}(0,R)}$ converging uniformly to ψ, then for every function $g \in C(X)$ we obtain that

$$\lim_{n\to\infty} \int_X g \, d\mu_{\psi_n} = \lim_{n\to\infty} \int_X g\varrho_{\psi_n} \, dm_{\psi_n} = \int_X g\varrho_{\psi} \, dm_{\psi} = \int_X g \, d\mu_{\psi}.$$

This means that the sequence $(\mu_{\psi_n})_{n=1}^{\infty}$ converges to μ_{ψ} in the weak* topology on $M(X)$. The proof of Theorem 13.8.21 is thus complete. □

13.9 Stochastic laws

Let $(X, \mathcal{F}, \mu)$ be a probability space and $(Y_n)_{n=0}^{\infty}$ a sequence of identically distributed random variables in $L^2(\mu)$. Define the (m,n)-*correlation* of this sequence to be

$$C_{m,n} := \left| \int_X Y_m \cdot Y_n \, d\mu - \int_X Y_m \, d\mu \cdot \int_X Y_n \, d\mu \right|.$$

If the variables $(Y_n)_{n=0}^{\infty}$ are independent, then $C_{m,n} = 0$ whenever $m \neq n$. The random variables $(Y_n)_{n=0}^{\infty}$ are *mixing* if

$$C_{m,n} \to 0 \text{ as } |m-n| \to \infty.$$

Intuitively, the faster the correlations approach 0, the bigger the mixing.

Now suppose that $T : (X, \mathcal{F}, \mu) \to (X, \mathcal{F}, \mu)$ is a measure-preserving transformation. Suppose also that $f, g \in L^2(\mu)$ and set $Y_n = f \circ T^n$. The sequence $(f \circ T^n)_{n=0}^{\infty}$ is identically distributed because of the T-invariance of μ. Let $m \le n$. Then

$$\begin{aligned} C_{m,n}(f) &:= \left| \int_X (f \circ T^m)(f \circ T^n)\, d\mu - \int_X f \circ T^m\, d\mu \cdot \int_X f \circ T^n\, d\mu \right| \\ &= \left| \int_X (f \cdot (f \circ T^{n-m})) \circ T^m\, d\mu - \int_X f\, d\mu \cdot \int_X f\, d\mu \right| \\ &= \left| \int_X f \cdot (f \circ T^{n-m})\, d\mu - \left(\int_X f\, d\mu \right)^2 \right| = C_{n-m}(f, f), \end{aligned}$$

where (with $E_\mu(g) := \mu(g) := \int_X g\, d\mu$)

$$C_n(f, g) := \left| \int_X g \cdot (f \circ T^n)\, d\mu - \int_X g\, d\mu \cdot \int_X f\, d\mu \right| = \left| E_\mu(g \cdot (f \circ T^n)) - E_\mu(g) \cdot E_\mu(f) \right| \tag{13.75}$$

are the *correlations of f and g*. Observe that

$$\begin{aligned} &\int_X (g - E_\mu(g)) \cdot [(f - E_\mu(f)) \circ T^n]\, d\mu \\ &\quad = \int_X g \cdot (f \circ T^n)\, d\mu - \int_X g \cdot E_\mu(f)\, d\mu - \int_X E_\mu(g) \cdot (f \circ T^n)\, d\mu + \int_X E_\mu(g) \cdot E_\mu(f)\, d\mu \\ &\quad = \int_X g \cdot (f \circ T^n)\, d\mu - E_\mu(f) E_\mu(g) - E_\mu(g) E_\mu(f) + E_\mu(g) E_\mu(f) \\ &\quad = E_\mu(g \cdot (f \circ T^n)) - E_\mu(g) E_\mu(f). \end{aligned}$$

Thus,

$$C_n(f, g) = \left| \int_X (g - E_\mu(g)) \cdot [(f - E_\mu(f)) \circ T^n]\, d\mu \right|. \tag{13.76}$$

Note further that

$$C_n(f, g) = C_n(f - E_\mu(f), g - E_\mu(g)). \tag{13.77}$$

So, when calculating correlations, we may replace f by $f - E_\mu(f)$ and g by $g - E_\mu(g)$. And, theoretically, we may assume without loss of generality that $E_\mu(f) = 0 = E_\mu(g)$.

We shall abbreviate

$$C_n(f) := C_n(f, f). \tag{13.78}$$

Obviously, the correlations $C_n(f,g)$ and $C_n(f)$ depend on the underlying measure μ. When pertinent, we will use the respective notations $C_{\mu,n}(f,g)$ and $C_{\mu,n}(f)$.

13.9.1 Exponential decay of correlations

Let us go back to dynamical systems and potentials. We will demonstrate that the correlations $C_n(f,g)$ decay exponentially under the additional condition that T is Hölder continuous. This entails that $\widehat{\varphi}$ is Hölder continuous, too.

Theorem 13.9.1. *Let $T : X \to X$ be a topologically exact, open, distance expanding, Hölder continuous system and $\varphi \in H_\alpha(X,\mathbb{R})$ a potential. Per Proposition 13.7.12, let μ_φ be the unique T-invariant Gibbs state and equilibrium state for φ. There exist constants $c \geq 0$ and $0 < \zeta < 1$ such that for all $f \in L^1(\mu_\varphi)$ and all $g \in H_\alpha(X)$ we have*

$$C_{\mu_\varphi,n}(f,g) \leq c\zeta^n \|f - E_{\mu_\varphi}(f)\|_{L^1(\mu_\varphi)} \|g - E_{\mu_\varphi}(g)\|_\alpha, \quad \forall n \in \mathbb{N}. \tag{13.79}$$

Proof. Replacing f by $f - E_{\mu_\varphi}(f)$ and g by $g - E_{\mu_\varphi}(g)$, we may assume without loss of generality that $E_{\mu_\varphi}(f) = E_{\mu_\varphi}(g) = 0$. Using (13.76) and Proposition 13.8.4(b), we obtain

$$C_{\mu_\varphi,n}(f,g) = \left|\int_X g \cdot (f \circ T^n)\, d\mu_\varphi\right| = \left|\int_X \widehat{\mathcal{L}}^n_{\widehat{\varphi}}(g \cdot (f \circ T^n))\, d\mu_\varphi\right| = \left|\int_X f \cdot \widehat{\mathcal{L}}^n_{\widehat{\varphi}}(g)\, d\mu_\varphi\right|. \tag{13.80}$$

Applying Theorem 13.8.18 with φ replaced by $\widehat{\varphi}$ and using Corollary 13.8.11 as well as the fact that $E_{\mu_\varphi}(g) = 0$, we infer that $L_1(g) = 0$, and thus $\widehat{\mathcal{L}}^n_{\widehat{\varphi}}(g) = S^n(g)$. Therefore,

$$\|\widehat{\mathcal{L}}^n_{\widehat{\varphi}}(g)\|_\infty = \|S^n(g)\|_\infty \leq \|S^n(g)\|_\alpha \leq c\zeta^n \|g\|_\alpha$$

for some $c \geq 0$ and $r(S) < \zeta < 1$. Inserting this into (13.80), we conclude that

$$C_{\mu_\varphi,n}(f,g) \leq \int_X |f \cdot \widehat{\mathcal{L}}^n_{\widehat{\varphi}}(g)|\, d\mu_\varphi \leq \|\widehat{\mathcal{L}}^n_{\widehat{\varphi}}(g)\|_\infty \int_X |f|\, d\mu_\varphi \leq c\zeta^n \|g\|_\alpha \|f\|_{L^1(\mu_\varphi)}. \qquad \square$$

Example 13.9.2. Let $\sigma : E_A^\infty \to E_A^\infty$ be a subshift of finite type generated by a primitive matrix A on a finite alphabet E. According to Theorems 3.2.16 and 3.2.12 and Example 4.1.3, such a subshift is a topologically exact, open, distance expanding, Hölder system. Let $\varphi \in H_\alpha(E_A^\infty, \mathbb{R})$. Per Proposition 13.7.12, let μ_φ be the unique σ-invariant Gibbs state and equilibrium state for φ. Let $f \in L^1(\mu_\varphi)$ and g be constant on cylinders of length k for some $k \in \mathbb{N}$. Using the metric $d(\omega,\tau) = e^{-|\omega\wedge\tau|}$, it can be observed that

$$\|g - E_{\mu_\varphi}(g)\|_\infty \leq 2\|g\|_\infty \quad \text{and} \quad v_\alpha(g - E_{\mu_\varphi}(g)) = v_\alpha(g) \leq 2\|g\|_\infty e^{-\alpha k}.$$

By Theorem 13.9.1, there exist constants $c \geq 0$ and $0 < \zeta < 1$ such that for all such f and g we have

$$C_{\mu_\varphi,n}(f,g) \leq c\zeta^n(1+e^{-\alpha k})\|g\|_\infty \int_{E_A^\infty} |f - E_{\mu_\varphi}(f)|\, d\mu_\varphi, \quad \forall n \in \mathbb{N}.$$

13.9.2 Asymptotic variance

Lemma 13.9.3. *Let $(X,\mathcal{F},\mu)$ be a probability space. Let $T : (X,\mathcal{F},\mu) \to (X,\mathcal{F},\mu)$ be a measure-preserving transformation and $g \in L^2(\mu)$. If*

$$\sum_{n=1}^{\infty}\left|\int_X g \cdot g \circ T^n\, d\mu\right| < \infty,$$

then

$$\lim_{n\to\infty}\frac{1}{n}\int_X\left(\sum_{k=0}^{n-1} g\circ T^k\right)^2 d\mu = \int_X g^2\, d\mu + 2\sum_{n=1}^{\infty}\int_X g\cdot g\circ T^n\, d\mu.$$

Proof. Almost all of this proof is based on sheer calculations. For every $n \in \mathbb{N}$, using the T-invariance of μ we obtain

$$\begin{aligned}\int_X (S_n g)^2\, d\mu &= \sum_{i,j=0}^{n-1}\int_X g\circ T^i \cdot g\circ T^j\, d\mu \\ &= \sum_{i=0}^{n-1}\int_X g\circ T^i\cdot g\circ T^i\, d\mu + 2\sum_{0\leq i<j\leq n-1}\int_X g\circ T^i\cdot g\circ T^j\, d\mu \\ &= n\int_X g^2\, d\mu + 2\sum_{i=0}^{n-2}\sum_{j=i+1}^{n-1}\int_X g\cdot g\circ T^{j-i}\, d\mu \\ &= n\int_X g^2\, d\mu + 2\sum_{k=1}^{n-1}(n-k)\int_X g\cdot g\circ T^k\, d\mu. \end{aligned} \tag{13.81}$$

Consequently,

$$\frac{1}{n}\int_X (S_n g)^2\, d\mu = \int_X g^2\, d\mu + 2\sum_{k=1}^{n-1}\left(1-\frac{k}{n}\right)\int_X g\cdot g\circ T^k\, d\mu.$$

Hence,

$$\frac{1}{n}\int_X (S_n g)^2\, d\mu - \left(\int_X g^2\, d\mu + 2\sum_{j=1}^{\infty}\int_X g\cdot g\circ T^j\, d\mu\right) = -2\sum_{k=1}^{n-1}\frac{k}{n}\int_X g\cdot g\circ T^k\, d\mu - 2\sum_{j=n}^{\infty}\int_X g\cdot g\circ T^j\, d\mu.$$

Setting

$$\Delta_n := \left| \frac{1}{n} \int_X (S_n g)^2 \, d\mu - \left(\int_X g^2 \, d\mu + 2 \sum_{j=1}^{\infty} \int_X g \cdot g \circ T^j \, d\mu \right) \right|,$$

we thus know that

$$\frac{\Delta_n}{2} \le \sum_{k=1}^{n-1} \frac{k}{n} \left| \int_X g \cdot g \circ T^k \, d\mu \right| + \sum_{j=n}^{\infty} \left| \int_X g \cdot g \circ T^j \, d\mu \right|. \tag{13.82}$$

Now fix any $\varepsilon > 0$. By hypothesis, the series $\sum_{j=1}^{\infty} \left| \int_X g \cdot g \circ T^j \, d\mu \right|$ converges, so there exists $N_1 \in \mathbb{N}$ such that

$$\sum_{j=N_1}^{\infty} \left| \int_X g \cdot g \circ T^j \, d\mu \right| < \frac{\varepsilon}{3}. \tag{13.83}$$

Therefore,

$$\sum_{k=N_1}^{n-1} \frac{k}{n} \left| \int_X g \cdot g \circ T^k \, d\mu \right| \le \sum_{k=N_1}^{n-1} \left| \int_X g \cdot g \circ T^k \, d\mu \right| < \frac{\varepsilon}{3}. \tag{13.84}$$

Finally, observe that there exists $N_2 \ge N_1$ such that

$$\sum_{k=1}^{N_1-1} \frac{k}{n} \left| \int_X g \cdot g \circ T^k \, d\mu \right| < \frac{\varepsilon}{3}, \quad \forall n \ge N_2.$$

Inserting this, (13.84) and (13.83) into (13.82), we get for every $n \ge N_2$ that

$$\frac{\Delta_n}{2} \le \sum_{k=1}^{N_1-1} \frac{k}{n} \left| \int_X g \cdot g \circ T^k \, d\mu \right| + \sum_{k=N_1}^{n-1} \frac{k}{n} \left| \int_X g \cdot g \circ T^k \, d\mu \right| + \sum_{j=n}^{\infty} \left| \int_X g \cdot g \circ T^j \, d\mu \right| < 3\frac{\varepsilon}{3} = \varepsilon.$$

Hence $\lim_{n\to\infty} \Delta_n = 0$. The proof of Lemma 13.9.3 is complete. □

As a direct consequence of this lemma, we have the following. Recall that $C_{\mu,n}(g)$ is defined through (13.75)–(13.78).

Proposition 13.9.4. *Let $T : (X, \mathcal{F}, \mu) \to (X, \mathcal{F}, \mu)$ be a measure-preserving transformation of a probability space $(X, \mathcal{F}, \mu)$ and $g \in L^2(\mu)$. If*

$$\sum_{n=1}^{\infty} C_{\mu,n}(g) = \sum_{n=1}^{\infty} \left| \int_X (g - \mu(g)\mathbb{1}) \cdot (g - \mu(g)\mathbb{1}) \circ T^n \, d\mu \right| < \infty, \tag{13.85}$$

then the limit

$$\sigma^2_{\mu,\infty}(g) := \lim_{n\to\infty} \frac{1}{n} \int_X \left(\sum_{k=0}^{n-1} (g - \mu(g)\mathbb{1}) \circ T^k \right)^2 d\mu \tag{13.86}$$

exists and

$$\sigma^2_{\mu,\infty}(g) = \int_X (g - \mu(g)\mathbb{1})^2 \, d\mu + 2 \sum_{n=1}^{\infty} \int_X (g - \mu(g)\mathbb{1}) \cdot (g - \mu(g)\mathbb{1}) \circ T^n \, d\mu. \tag{13.87}$$

The number $\sigma^2_{\mu,\infty}(g)$ is called the asymptotic variance *of g (with respect to T and μ). It is in general different than the common variance*

$$\sigma^2_{\mu}(g) := \int_X (g - \mu(g)\mathbb{1})^2 \, d\mu.$$

However, if the random variables $g \circ T^n$, $n \geq 0$, are independent with respect to μ, then $\sigma^2_{\mu,\infty}(g) = \sigma^2_{\mu}(g)$.

Our goal in this subsection is to provide several characterizations of the positivity of the asymptotic variance $\sigma^2_{\mu,\infty}(g)$. Essentially all of those equivalencies will be expressed by means of cohomology of functions. Knowing that $\sigma^2_{\mu,\infty}(g) > 0$ is important because it is needed for the central limit theorem and the law of the iterated logarithm, which are the subject of the next two subsections. We start with the following result, which has as a by-product the addition of one item to Theorem 13.7.17 characterizing potentials that have the same Gibbs/equilibrium states.

Theorem 13.9.5. *Let $(X, \mathcal{F}, \mu)$ be a probability space. Let $T : (X, \mathcal{F}, \mu) \to (X, \mathcal{F}, \mu)$ be a measure-preserving transformation and $g \in L^2(\mu)$. If*

$$\sum_{n=0}^{\infty} n C_{\mu,n}(g) < \infty, \tag{13.88}$$

then $\sigma^2_{\mu,\infty}(g)$ exists and the following three conditions are equivalent:

(a) $\sigma^2_{\mu,\infty}(g) = 0$.

(b) $\sup\left\{ \|S_n(g - \mu(g)\mathbb{1})\|_{L^2(\mu)} : n \in \mathbb{N} \right\} < \infty$.

(c) *g is cohomologous to a constant in the Hilbert space $L^2(\mu)$.*

Proof. By Proposition 13.9.4, we know that $\sigma^2_{\mu,\infty}(g)$ exists.

The implication (c)⇒(a) follows immediately from (13.87) after substituting $g = h \circ T - h$, where $h \in L^2(\mu)$.

For all other implications established later in this proof, assume without loss of generality that $\mu(g) = 0$.

Let us prove that (a)⇒(b). Set

$$C_k^*(g) := \int_X g \cdot g \circ T^k \, d\mu, \quad \forall k \geq 0.$$

Using (13.81), we get

$$\begin{aligned}
\int_X (S_n g)^2 \, d\mu &= nC_0^*(g) + 2\sum_{k=1}^{n-1}(n-k)C_k^*(g) \\
&= n\Big(C_0^*(g) + 2\sum_{k=1}^{\infty} C_k^*(g)\Big) - 2n\sum_{k=n}^{\infty} C_k^*(g) - 2\sum_{k=1}^{n-1} kC_k^*(g) \\
&= n\sigma_{\mu,\infty}^2(g) - 2n\sum_{k=n}^{\infty} C_k^*(g) - 2\sum_{k=1}^{n-1} kC_k^*(g) \\
&= -2n\sum_{k=n}^{\infty} C_k^*(g) - 2\sum_{k=1}^{n-1} kC_k^*(g).
\end{aligned}$$

By (13.88), the first summand in this formula converges to 0 as $n \to \infty$ and the second summand is uniformly bounded. We conclude that (b) holds.

Finally, we show that (b)⇒(c). By our hypothesis there exists $M \geq 1$ such that $\|S_n g\|_{L^2(\mu)} \leq M$ for all $n \in \mathbb{N}$. Consequently, the same holds for the Cesàro averages:

$$\left\| \frac{1}{n}\sum_{j=1}^{n} S_j g \right\|_{L^2(\mu)} \leq M, \quad \forall n \in \mathbb{N}.$$

As a Hilbert space, $L^2(\mu)$ is reflexive and hence its closed ball $\overline{B_{L^2(\mu)}(0,M)}$ is weakly compact in $L^2(\mu)$. Thus there is $u \in \overline{B_{L^2(\mu)}(0,M)}$ and an increasing sequence $(n_k)_{k=1}^{\infty}$ of positive integers such that the sequence $(\frac{1}{n_k}\sum_{j=1}^{n_k} S_j g)_{k=1}^{\infty}$ converges weakly to u. Then

$$\lim_{k\to\infty} \frac{1}{n_k}\sum_{j=1}^{n_k} S_j(g \circ T) = u \circ T \quad \text{weakly in } L^2(\mu) \tag{13.89}$$

and

$$\frac{1}{n_k}\sum_{j=1}^{n_k} S_j(g \circ T) = \frac{1}{n_k}\Big(\sum_{j=1}^{n_k} S_{j+1}g - n_k g\Big) = \frac{1}{n_k}\sum_{j=1}^{n_k} S_j g + \frac{1}{n_k}(S_{n_k+1}g - g) - g.$$

Taking the weak limits of both sides of this equation and invoking (13.89) we get $u \circ T = u - g$ or, equivalently, $g = u - u \circ T$. The proof of the implication (b)⇒(c) and of the whole theorem is complete. □

As an immediate consequence of Theorems 13.9.5 and 13.9.1, we get the following.

Corollary 13.9.6. *Let $T : X \to X$ be a topologically exact, open, distance expanding, Hölder continuous system and let $\varphi : X \to \mathbb{R}$ be a Hölder potential. If $g : X \to \mathbb{R}$ is a Hölder function, then $\sigma^2_{\mu_\varphi,\infty}(g)$ exists and the following three conditions are equivalent:*

(a) $\sigma^2_{\mu_\varphi,\infty}(g) = 0$.

(b) $\sup\left\{\|S_n(g - \mu_\varphi(g)\mathbb{1})\|_{L^2(\mu_\varphi)} : n \in \mathbb{N}\right\} < \infty$.

(c) *g is cohomologous to a constant in the Hilbert space $L^2(\mu_\varphi)$.*

Our next goal is to promote the $L^2(\mu_\varphi)$-cohomology in item (c) of Corollary 13.9.6 to cohomology in the class of Hölder continuous functions. Of course, we need a much more special setting and much stronger hypotheses. Though we ultimately aim at carrying out this promotion for transitive, open, distance expanding, Hölder continuous systems, in a first stage we will accomplish this promotion for irreducible/transitive subshifts of finite type and then use the Markov partitions and the symbolic representation from Sections 4.4–4.5, to propagate it to all transitive, open, distance expanding, Hölder continuous maps.

As we will also encounter the issue of characterization of the equation $\sigma^2_{\mu,\infty}(g) = 0$ in Section 18.4 in the context of transitive subshifts of finite type over a countable (possibly infinite) alphabet, we will work in that setting for a while. The reader who is not interested in countable state shifts may assume without losing anything that the countable set E below is finite. The reader familiar with countable state shifts can smoothly read what follows and the reader interested in countable state subshifts but not familiar with them yet, is advised to first read Chapter 17.

Let E be a countable set and $A : E \times E \to \{0, 1\}$ be a finitely irreducible incidence matrix. Let a potential $f : E_A^\infty \to \mathbb{R}$ be summable and Hölder continuous on cylinders. In addition to the usual one-sided subshift $E_A^\infty = E_A^{\mathbb{N}} = E_A^{\{1,2,\dots\}}$, let

$$E_A^{\mathbb{Z}} := \{(\omega_n)_{n\in\mathbb{Z}} \mid A_{\omega_n\omega_{n+1}} = 1, \forall n \in \mathbb{Z}\}$$

be the two-sided subshift generated by the transition matrix A and let

$$E_A^{-\infty} := \{(\omega_n)_{n\le 0} \mid A_{\omega_n\omega_{n+1}} = 1, \forall n \le -1\}.$$

Given $-\infty \le m \le n \le \infty$ and $\omega \in E_A^{\mathbb{Z}}$, let $\omega|_m^n = \omega_m\omega_{m+1}\dots\omega_n$ and let

$$[\omega]_m^n := \{\tau \in E_A^{\mathbb{Z}} \mid \tau_k = \omega_k, \forall m \le k \le n\}.$$

Given $e \in E$, let

$$E_A^{\infty,e} := \{\rho \in E_A^\infty \mid A_{e\rho_1} = 1\} \quad \text{and} \quad E_A^{*,e} := \{\tau \in E_A^* \mid A_{e\tau_1} = 1\}.$$

Let $\alpha_- := \{[\chi]\}_{\chi \in E_A^{-\infty}}$ be the partition of $E_A^{\mathbb{Z}}$ whose atoms are

$$[\chi] = \{\chi\tau : \tau \in E_A^{\infty,\chi_0}\} = \{\gamma \in E_A^{\mathbb{Z}} : \gamma|_{-\infty}^0 = \chi\} = [\chi]_{-\infty}^0,$$

where $\chi \in E_A^{-\infty}$.

For every $k \in \mathbb{Z}$, the map $\pi_k : E_A^{\mathbb{Z}} \to E$ is the projection onto the kth coordinate given by the formula $\pi_k(\omega) := \omega_k$.

Also, for every set $B \subseteq E_A^{\mathbb{Z}}$, the set $B|_1^\infty$ is the projection of B onto E_A^∞.

Finally, let μ_f be the unique one-sided shift-invariant Gibbs state for the potential f, and let $\widetilde{\mu}_f$ be its Rokhlin's natural extension to the two-sided subshift $E_A^{\mathbb{Z}}$. Let us recall that, given finitely many sets $C_1, C_2, \ldots, C_k$ contained in E, the measure $\widetilde{\mu}_f$ is defined on a cylinder

$$C = \pi_{n_1}^{-1}(C_1) \cap \pi_{n_2}^{-1}(C_2) \cap \cdots \cap \pi_{n_k}^{-1}(C_k)$$

with $-\infty < n_1 \le n_2 \le \cdots \le n_k < \infty$, by the formula

$$\widetilde{\mu}_f(C) = \mu_f(\widetilde{\sigma}^{n_1-1}(C)|_1^\infty),$$

where $\widetilde{\sigma} : E_A^{\mathbb{Z}} \to E_A^{\mathbb{Z}}$ is the left-shift map on $E_A^{\mathbb{Z}}$. Let us recall that $\widetilde{\mu}_f$ is two-sided shift-invariant. We shall prove the following.

Lemma 13.9.7. *For every $\omega \in E_A^{\mathbb{Z}}$, there exists a Borel probability measure $\widetilde{\mu}_{f,\alpha_-(\omega)}$ on the atom $\alpha_-(\omega) = [\omega]_{-\infty}^0$ with the following properties:*

(a) *For every Borel set $B \subseteq E_A^{\mathbb{Z}}$ the function*

$$E_A^{\mathbb{Z}} \ni \tau \longmapsto \widetilde{\mu}_{f,\alpha_-(\tau)}(B \cap \alpha_-(\tau)) \in [0,1]$$

is measurable.

(b) *For every Borel set $B \subseteq E_A^{\mathbb{Z}}$,*

$$\widetilde{\mu}_f(B) = \int_{E_A^{\mathbb{Z}}} \widetilde{\mu}_{f,\alpha_-(\tau)}(B \cap \alpha_-(\tau))\, d\widetilde{\mu}_f(\tau).$$

(c) *The measure $\mu_{f,\alpha_-(\omega)}$ defined as the projection of $\widetilde{\mu}_{f,\alpha_-(\omega)}$ on $E_A^{\infty,\omega_0} \subseteq E_A^\infty$ is equivalent to μ_f restricted to E_A^{∞,ω_0}.*

(d) *The Radon–Nikodym derivative $\dfrac{d\mu_{f,\alpha_-(\omega)}}{d\mu_f}$ is uniformly bounded away from 0 and ∞.*

Proof. Let $l_B : \ell_\infty \to \mathbb{R}$ be a Banach limit. Given $\omega \in E_A^{\mathbb{Z}}$ and a Borel set $B \subseteq E_A^{\mathbb{Z}}$, define

$$\widetilde{\mu}_{f,\alpha_-(\omega)}(B) := l_B\left(\left(\frac{\widetilde{\mu}_f(B \cap [\omega]_{-n}^0)}{\widetilde{\mu}_f([\omega]_{-n}^0)}\right)_0^\infty\right). \tag{13.90}$$

Since the functional l_B is linear, as an immediate consequence of this definition, we get the following.

Claim 1. *For every $\omega \in E_A^{\mathbb{Z}}$, the function $B \longmapsto \widetilde{\mu}_{f,\alpha_-(\omega)}(B \cap \alpha_-(\omega)) \in [0,1]$, defined on the Borel σ-algebra of $E_A^{\mathbb{Z}}$, is finitely additive.*

Now we shall show that for every $\tau \in E_A^{*,\omega_0}$,

$$Q^{-3} \le \frac{\widetilde{\mu}_{f,\alpha_-(\omega)}([\omega|_{-\infty}^0 \tau])}{\mu_f([\tau])} \le Q^3, \tag{13.91}$$

where $Q \ge 1$ is a Gibbs constant for μ_f. (Note: To lighten notation, we will use $\mathrm{P}(f)$ in lieu of $\mathrm{P}(\sigma, f)$ in the following calculations.) Let $n \ge 0$. In view of the Gibbs property, we get

$$\begin{aligned}\widetilde{\mu}_f([\omega]_{-n}^0) &= \mu_f(\widetilde{\sigma}^{-(n+1)}([\omega]_{-n}^0)|_1^\infty) \\ &\le Q\exp\Big(\inf(S_{n+1}f|_{\widetilde{\sigma}^{-(n+1)}([\omega]_{-n}^0)|_1^\infty}) - \mathrm{P}(f)(n+1)\Big) \\ &= Q\exp\Big(\inf(S_{n+1}f|_{[\omega_{-n}\dots\omega_0]}) - \mathrm{P}(f)(n+1)\Big).\end{aligned} \tag{13.92}$$

Similarly,

$$\widetilde{\mu}_f([\omega]_{-n}^0) \ge Q^{-1}\exp\Big(\sup(S_{n+1}f|_{[\omega_{-n}\dots\omega_0]}) - \mathrm{P}(f)(n+1)\Big). \tag{13.93}$$

Setting $k = |\tau|$, by the same Gibbs property we obtain that

$$\begin{aligned}&\widetilde{\mu}_f([\omega]_{-n}^0 \cap [\tau]_1^k) = \\ &= \mu_f\Big(\widetilde{\sigma}^{-(n+1)}([\omega]_{-n}^0 \cap [\tau]_1^k)|_1^\infty\Big) \\ &\ge Q^{-1}\exp\Big(\inf(S_{n+1+k}f|_{\widetilde{\sigma}^{-(n+1)}([\omega]_{-n}^0\cap[\tau]_1^k)|_1^\infty}) - \mathrm{P}(f)(n+1+k)\Big) \\ &= Q^{-1}\exp\Big(\inf(S_{n+1+k}f|_{[\omega_{-n}\dots\omega_0\tau_1\dots\tau_k]}) - \mathrm{P}(f)(n+1+k)\Big) \\ &\ge Q^{-1}\exp\Big(\inf(S_{n+1}f|_{[\omega_{-n}\dots\omega_0]}) + \inf(S_kf|_{[\tau_1\dots\tau_k]}) - \mathrm{P}(f)(n+1+k)\Big) \\ &= Q^{-1}\exp\Big(\inf(S_{n+1}f|_{[\omega_{-n}\dots\omega_0]}) - \mathrm{P}(f)(n+1)\Big)\exp\Big(\inf(S_kf|_{[\tau]}) - \mathrm{P}(f)k\Big).\end{aligned} \tag{13.94}$$

Similarly,

$$\widetilde{\mu}_f([\omega]_{-n}^0\cap[\tau]_1^k) \le Q\exp\Big(\sup(S_{n+1}f|_{[\omega_{-n}\dots\omega_0]}) - \mathrm{P}(f)(n+1) + \sup(S_kf|_{[\tau]}) - \mathrm{P}(f)k\Big). \tag{13.95}$$

Using (13.92) and (13.94) and applying the Gibbs property once more, we deduce that

$$\frac{\widetilde{\mu}_f([\omega]_{-n}^0 \cap [\tau]_1^k)}{\widetilde{\mu}_f([\omega]_{-n}^0)} \ge Q^{-2}\exp(\inf(S_kf|_{[\tau]}) - \mathrm{P}(f)k) \ge Q^{-3}\mu_f([\tau]).$$

On the other hand, using (13.93) and (13.95) and applying the Gibbs property once more, we arrive at

$$\frac{\widetilde{\mu}_f([\omega]_{-n}^0 \cap [\tau]_1^k)}{\widetilde{\mu}_f([\omega]_{-n}^0)} \le Q^2 \exp(\sup(S_k f|_{[\tau]}) - \mathrm{P}(f)k) \le Q^3 \mu_f([\tau]).$$

Inserting these two inequalities into (13.90) and using the monotonicity of Banach limits, we conclude that (13.91) holds.

Let $\widehat{\mu}_f$ be the measure on $\alpha_-(\omega) = [\omega]_{-\infty}^0$ given by

$$\widehat{\mu}_f(\omega|_{-\infty}^0 F) := \mu_f(F)$$

for every Borel set $F \subseteq E_A^{\infty,\omega_0}$, where $\omega|_{-\infty}^0 F := \{\omega|_{-\infty}^0 \rho : \rho \in F\}$. Let $\mathcal{B}_\omega^*$ be the algebra of sets on $[\omega]_{-\infty}^0$ generated by the sets of the form $[\omega|_{-\infty}^0 \tau]$, $\tau \in E_A^{*,\omega_0}$. Since each element of this algebra can be represented as a disjoint union of sets of the form $[\omega|_{-\infty}^0 \tau]$, $\tau \in E_A^{*,\omega_0}$, a consequence of (13.91) is that

$$Q^{-3}\widehat{\mu}_f(G) \le \widetilde{\mu}_{f,\alpha_-(\omega)}(G) \le Q^3 \widehat{\mu}_f(G), \quad \forall G \in \mathcal{B}_\omega^*. \tag{13.96}$$

In particular, $\widetilde{\mu}_{f,\alpha_-(\omega)}$ is absolutely continuous with respect to $\widehat{\mu}_f$ on $\mathcal{B}_\omega^*$. In conjunction with Claim 1 and Lemma A.1.31, this gives that $\widetilde{\mu}_{f,\alpha_-(\omega)}$ is a (σ-additive) Borel probability measure on $\alpha_-(\omega)$. Furthermore, items (c,d) of our lemma follow from (13.96).

It immediately ensues from Theorem A.1.68 (martingale convergence theorem for conditional expectations) and Example A.1.66 applied with the function $\varphi = \mathbb{1}_B$ and $\mathcal{A}_n = \{[\omega|^0_{\infty}\tau] : \tau \in E_A^{n,\omega_0}\}$, that

$$\widetilde{\mu}_{f,\alpha_-(\omega)}(B) := \lim_{n\to\infty} \frac{\widetilde{\mu}_f(B \cap [\omega]_{-n}^0)}{\widetilde{\mu}_f([\omega]_{-n}^0)} = E_{\widetilde{\mu}_f}(\mathbb{1}_B|\alpha_-)(\omega) \tag{13.97}$$

for $\widetilde{\mu}_f$-a. e. $\omega \in E_A^{\mathbb{Z}}$ (depending on B). Consequently, item (a) holds. Item (b) directly follows from the definition of conditional expected value. The proof of Lemma 13.9.7 is complete. □

An immediate ramification of this lemma is the following.

Corollary 13.9.8. *For every $\omega \in E_A^{\mathbb{Z}}$,*

$$\operatorname{supp}(\widetilde{\mu}_{f,\alpha_-(\omega)}) = \alpha_-(\omega).$$

We now make a first step toward establishing cohomology in the class of Hölder continuous functions for symbolic systems.

Lemma 13.9.9. *Let E be a countable set and $A : E \times E \to \{0,1\}$ be a finitely irreducible incidence matrix. Let a potential $f : E_A^\infty \to \mathbb{R}$ be summable and Hölder continuous on cylinders. Suppose that $g \in H_\beta(E_A^\infty, \mathbb{R})$ has the following properties:*

(a) $g \in L^2(\mu_f)$;

(b) $\sum_{n=0}^{\infty} nC_{\mu_f,n}(g) < \infty$ (*in particular,* $\sigma^2_{\mu_f,\infty}(g)$ *exists by Proposition* 13.9.4); *and*

(c) $\sigma^2_{\mu_f,\infty}(g) = 0$.

Then there exists $u \in H^b_\beta(E^\infty_A, \mathbb{R})$ *such that*

$$g = \mu_f(g) + u - u \circ \sigma.$$

In particular, the function g turns out to be bounded, so $g \in H^b_\beta(E^\infty_A, \mathbb{R})$.

Proof. Assume without loss of generality that $\mu_f(g) = 0$. It follows from Theorem 13.9.5 that there exists $u \in L^2(\mu_f)$ such that

$$g = u - u \circ \sigma \quad \mu_f\text{-a. e. on } E^\infty_A. \tag{13.98}$$

Our goal is to show that u can be extended to a version which is Hölder continuous of order β on cylinders. We first extend g and u to the two-sided shift space $E^{\mathbb{Z}}_A$ by declaring

$$\tilde{g}(\omega) := g(\omega|_1^\infty) \quad \text{and} \quad \tilde{u}(\omega) := u(\omega|_1^\infty)$$

wherever $u(\omega|_1^\infty)$ is defined and fulfills (13.98). Like g, the function $\tilde{g} : E^{\mathbb{Z}}_A \to \mathbb{R}$ is Hölder continuous of order β on cylinders, while $\tilde{u}$ is measurable (and even belongs to $L^2(\tilde{\mu}_f)$, though we will not need this fact). The cohomological equation (13.98) carries over to $\tilde{g}$ and $\tilde{u}$ on a $\tilde{\mu}_f$-a. e. basis since

$$\tilde{u}(\omega) - \tilde{u} \circ \tilde{\sigma}(\omega) = u(\omega|_1^\infty) - u(\tilde{\sigma}(\omega)|_1^\infty) = u(\omega|_1^\infty) - u(\sigma(\omega|_1^\infty)) = g(\omega|_1^\infty) = \tilde{g}(\omega) \tag{13.99}$$

wherever $u(\omega|_1^\infty)$ is defined and satisfies (13.98).

By virtue of Luzin's theorem (Theorem A.1.49), there exists a compact set $D \subseteq E^{\mathbb{Z}}_A$ such that $\tilde{\mu}_f(D) > 1/2$ and the restricted function $\tilde{u}|_D$ is continuous. Corollary 8.2.14 (Ergodic case of Birkhoff's ergodic theorem), or Corollary 8.2.15, then affirms that there exists a Borel set $B \subseteq E^{\mathbb{Z}}_A$ such that (1) $\tilde{\mu}_f(B) = 1$; (2) for every $\gamma \in B$,

$$\lim_{n\to\infty} \frac{1}{n} \#\{0 \le j < n : \tilde{\sigma}^{-n}(\gamma) \in D\} = \tilde{\mu}_f(D) > 1/2; \tag{13.100}$$

(3) the function $\tilde{u}$ is well-defined on the union $\bigcup_{n\in\mathbb{Z}} \tilde{\sigma}^{-n}(B)$; and (4) relation (13.99) holds on the same set $\bigcup_{n\in\mathbb{Z}} \tilde{\sigma}^{-n}(B)$. By Lemma 13.9.7(b), there is a Borel set $F \subseteq E^{\mathbb{Z}}_A$ such that $\tilde{\mu}_f(F) = 1$ and

$$\tilde{\mu}_{\alpha_-(\omega)}(B \cap \alpha_-(\omega)) = 1, \quad \forall \omega \in F.$$

It then follows from Corollary 13.9.8 that the set $B \cap \alpha_-(\omega)$ is dense in $\alpha_-(\omega)$ for every $\omega \in F$. Now fix an arbitrary element $e \in E$. Since $\tilde{\mu}_f([e]) > 0$, there exists $\omega \in F$ such that $\omega_1 = e$. Let

$$B_e := \{\chi \in E_A^\infty : \chi_1 = e \text{ and } \omega|_{-\infty}^0 \chi \in B\}.$$

Now consider two arbitrary elements $\rho, \tau \in B_e$. Because of (13.100), there exists a strictly increasing sequence $(n_j)_{j=1}^\infty$ of positive integers such that

$$\tilde{\sigma}^{-n_j}(\omega|_{-\infty}^0 \rho),\ \tilde{\sigma}^{-n_j}(\omega|_{-\infty}^0 \tau) \in D, \quad \forall j \in \mathbb{N}.$$

Write $\underline{\omega} := \omega|_{-\infty}^0$. Using (13.99), we get

$$\begin{aligned}
|u(\rho) - u(\tau)| &= |\tilde{u}(\underline{\omega}\rho) - \tilde{u}(\underline{\omega}\tau)| \\
&= \left|\left(\tilde{u}(\tilde{\sigma}^{-n_j}(\underline{\omega}\rho)) - \sum_{k=1}^{n_j} \tilde{g}(\tilde{\sigma}^{-k}(\underline{\omega}\rho))\right) - \left(\tilde{u}(\tilde{\sigma}^{-n_j}(\underline{\omega}\tau)) - \sum_{k=1}^{n_j} \tilde{g}(\tilde{\sigma}^{-k}(\underline{\omega}\tau))\right)\right| \\
&\le |\tilde{u}(\tilde{\sigma}^{-n_j}(\underline{\omega}\rho)) - \tilde{u}(\tilde{\sigma}^{-n_j}(\underline{\omega}\tau))| + \sum_{k=1}^{n_j} |\tilde{g}(\tilde{\sigma}^{-k}(\underline{\omega}\rho)) - \tilde{g}(\tilde{\sigma}^{-k}(\underline{\omega}\tau))|. \qquad (13.101)
\end{aligned}$$

Since $\lim_{j\to\infty} d(\tilde{\sigma}^{-n_j}(\underline{\omega}\rho), \tilde{\sigma}^{-n_j}(\underline{\omega}\tau)) = 0$, since both $\tilde{\sigma}^{-n_j}(\underline{\omega}\rho)$ and $\tilde{\sigma}^{-n_j}(\underline{\omega}\tau)$ belong to D, and since $u|_D$ is uniformly continuous (as D is compact), we deduce that

$$\lim_{j\to\infty} |\tilde{u}(\tilde{\sigma}^{-n_j}(\underline{\omega}\rho)) - \tilde{u}(\tilde{\sigma}^{-n_j}(\underline{\omega}\tau))| = 0. \qquad (13.102)$$

As $\tilde{g}$ is Hölder continuous of order β on cylinders, we obtain that

$$\begin{aligned}
\sum_{k=1}^{n_j} |\tilde{g}(\tilde{\sigma}^{-k}(\underline{\omega}\rho)) - \tilde{g}(\tilde{\sigma}^{-k}(\underline{\omega}\tau))| &\le \sum_{k=1}^{n_j} v_\beta(\tilde{g})[d(\tilde{\sigma}^{-k}(\underline{\omega}\rho), \tilde{\sigma}^{-k}(\underline{\omega}\tau))]^\beta \\
&= \sum_{k=1}^{n_j} v_\beta(\tilde{g}) e^{-\beta k}[d(\underline{\omega}\rho, \underline{\omega}\tau)]^\beta \\
&< \sum_{k=1}^{n_j} v_\beta(\tilde{g}) e^{-\beta k}[d(\rho, \tau)]^\beta \\
&\le \frac{v_\beta(\tilde{g})}{1 - e^{-\beta}}[d(\rho, \tau)]^\beta. \qquad (13.103)
\end{aligned}$$

It follows from (13.101)–(13.103) that

$$|u(\rho) - u(\tau)| \le \frac{v_\beta(\tilde{g})}{1 - e^{-\beta}}[d(\rho, \tau)]^\beta.$$

As B_e is dense in $[e]$, it ensues that u has a Hölder extension from B_e to $[e]$. Since $e \in E$ is arbitrary, we have thus defined an extension $\overline{u}$ of u to E_A^∞ which is Hölder continuous of order β on cylinders. Given that (13.98) holds on a dense subset of E_A^∞, we infer that

$$g = \overline{u} - \overline{u} \circ \sigma \qquad (13.104)$$

holds on this dense set, and consequently on the whole of E_A^∞.

We are left to prove that the function $\overline{u} : E_A^\infty \to \mathbb{R}$ is bounded. Let $\Lambda \subseteq E_A^*$ be a finite set of words witnessing the finite irreducibility of the matrix A. Let

$$|\Lambda| := \max\{|\alpha| : \alpha \in \Lambda\} \quad \text{and} \quad \Lambda_1 := \{\tau_1 : \tau \in \Lambda\}.$$

As $\overline{u}$ and g are Hölder continuous on cylinders, we have that $\sup(|g||_{[e]})$ and $\sup(|\overline{u}||_{[e]})$ are finite for each $e \in E$. So

$$M_g := \max\{\sup(|g||_{[e]}) : e \in \Lambda_1\} < \infty \quad \text{and} \quad M_{\overline{u}} := \max\{\sup(|\overline{u}||_{[e]}) : e \in \Lambda_1\} < \infty.$$

For every $\rho \in E_A^\infty$, choose any $\widehat{\rho} \in \Lambda$ such that $\widehat{\rho}\rho \in E_A^\infty$. Let $k := |\widehat{\rho}| \le |\Lambda|$. It follows from (13.104) iterated k times that

$$S_k g(\widehat{\rho}\rho) = \overline{u}(\widehat{\rho}\rho) - \overline{u}(\sigma^k(\widehat{\rho}\rho)) = \overline{u}(\widehat{\rho}\rho) - \overline{u}(\rho).$$

Therefore,

$$\big|\overline{u}(\rho)\big| \le \big|S_k g(\widehat{\rho}\rho)\big| + \big|\overline{u}(\widehat{\rho}\rho)\big| \le S_k|g|(\widehat{\rho}\rho) + \sup(|\overline{u}||_{[\widehat{\rho}_1]}) \le kM_g + M_{\overline{u}} \le |\Lambda|M_g + M_{\overline{u}}.$$

Hence $\|\overline{u}\|_\infty \le |\Lambda|M_g + M_{\overline{u}} < \infty$ and the proof of Lemma 13.9.9 is complete. □

When the alphabet E is finite, Lemma 13.9.9's statement reduces to the following.

Lemma 13.9.10. *Let E be a finite set and $A : E \times E \to \{0,1\}$ be an irreducible matrix. Let also $f : E_A^\infty \to \mathbb{R}$ be a Hölder continuous potential. If $g \in H_\beta(E_A^\infty, \mathbb{R})$ is such that*

$$\sum_{n=0}^{\infty} nC_{\mu_f,n}(g) < \infty, \tag{13.105}$$

then $\sigma^2_{\mu_f,\infty}(g)$ exists according to Proposition 13.9.4. If furthermore $\sigma^2_{\mu_f,\infty}(g) = 0$, then there exists $u \in H_\beta(E_A^\infty, \mathbb{R})$ such that

$$g = \mu_f(g) + u - u \circ \sigma.$$

A straightforward repercussion of Lemma 13.9.10 and Theorem 13.9.1 is next.

Lemma 13.9.11. *Let E be a finite set and $A : E \times E \to \{0,1\}$ be a primitive incidence matrix. Let also $f : E_A^\infty \to \mathbb{R}$ be a Hölder continuous potential. If $g \in H_\beta(E_A^\infty, \mathbb{R})$, then $\sigma^2_{\mu_f,\infty}(g)$ exists. Furthermore, if $\sigma^2_{\mu_f,\infty}(g) = 0$ then there exists $u \in H_\beta(E_A^\infty, \mathbb{R})$ such that*

$$g = \mu_f(g) + u - u \circ \sigma.$$

The cohomology in the class of Hölder continuous functions established in a finite-state symbolic setting in Lemma 13.9.10, will now be transferred to the more general class of transitive, open, distance expanding, Hölder continuous systems. To accomplish this, we will use the existence of Markov partitions and of a symbolic representation derived in Sections 4.4–4.5.

Lemma 13.9.12. *Let $T : X \to X$ be a topologically transitive, open, distance expanding, Hölder continuous map and let $\varphi : X \to \mathbb{R}$ be a Hölder continuous potential. If $g : X \to \mathbb{R}$ is a Hölder continuous function such that*

$$\sum_{n=0}^{\infty} nC_{\mu_\varphi,n}(g) < \infty, \tag{13.106}$$

then $\sigma^2_{\mu_\varphi,\infty}(g)$ exists according to Proposition 13.9.4. If furthermore $\sigma^2_{\mu_\varphi,\infty}(g) = 0$, then there exists a Hölder continuous function $u : E_A^\infty \to \mathbb{R}$ such that

$$g = \mu_\varphi(g) + u - u \circ \sigma.$$

Proof. By Theorem 4.4.6, let $\mathcal{R} = \{R_e\}_{e\in E}$ be a Markov partition with diameter smaller than an expansive constant for the map $T : X \to X$. Let A be the corresponding irreducible matrix defined in (4.46), and let $\pi : E_A^\infty \to X$ be the coding map in Theorem 4.5.2. By Theorem 13.7.11, this map π is a metric isomorphism of the measure-preserving dynamical systems $(\sigma : E_A^\infty \to E_A^\infty, \mu_{\varphi\circ\pi})$ and $(T : X \to X, \mu_\varphi)$. Since π is Hölder continuous per Theorem 4.5.2, the function $g \circ \pi : E_A^\infty \to \mathbb{R}$ is also Hölder continuous and $\mu_{\varphi\circ\pi}(g \circ \pi) = \mu_\varphi(g)$ and $C_{\mu_{\varphi\circ\pi},n}(g \circ \pi) = C_{\mu_\varphi,n}(g)$ for every $n \in \mathbb{N}$. Therefore, Lemma 13.9.10 applies and asserts that $\sigma^2_{\mu_{\varphi\circ\pi},\infty}(g\circ\pi)$ exists. As $\sigma^2_{\mu_{\varphi\circ\pi},\infty}(g\circ\pi) = \sigma^2_{\mu_\varphi,\infty}(g)$, we know that this latter quantity exists. If it equals 0, then Lemma 13.9.10 produces a Hölder continuous function $\hat{u} : E_A^\infty \to \mathbb{R}$ such that

$$g \circ \pi = \mu_{\varphi\circ\pi}(g \circ \pi) + \hat{u} - \hat{u} \circ \sigma = \mu_\varphi(g) + \hat{u} - \hat{u} \circ \sigma.$$

Assume without loss of generality that $\mu_\varphi(g) = 0$. The above equation reduces to

$$g \circ \pi = \hat{u} - \hat{u} \circ \sigma. \tag{13.107}$$

Let

$$\mathcal{R}_\infty := X \setminus \bigcup_{n=0}^{\infty} T^{-n}\Big(\bigcup_{e\in E} \partial R_e\Big).$$

By Theorem 4.5.2, each point $x \in \mathcal{R}_\infty$ has a unique coding $\pi^{-1}(x) \in E_A^\infty$. So there is a well-defined function $u : \mathcal{R}_\infty \to \mathbb{R}$ given by the formula

$$u(x) := \hat{u}(\pi^{-1}(x)).$$

Fix any $e \in E$. Take $x, y \in \mathcal{R}_\infty \cap R_e$. Let $\omega = \pi^{-1}(x)$ and $\tau = \pi^{-1}(y)$. Then $\omega_1 = \tau_1 = e$, $x = \pi(\omega)$ and $y = \pi(\tau)$. Pick any word $\rho \in E_A^*$ such that $A_{\rho_k e} = 1$, where $k = |\rho|$. Then

$\rho\omega, \rho\tau \in E_A^*$. Using (13.107) and Lemma 13.1.1, we get

$$\begin{aligned}|u(y) - u(x)| &= |\hat{u}(\tau) - \hat{u}(\omega)| \\ &= |(\hat{u}(\rho\tau) - S_k^\sigma(g \circ \pi)(\rho\tau)) - (\hat{u}(\rho\omega) - S_k^\sigma(g \circ \pi)(\rho\omega))| \\ &\leq |\hat{u}(\rho\tau) - \hat{u}(\rho\omega)| + |S_k^\sigma(g \circ \pi)(\rho\omega) - S_k^\sigma(g \circ \pi)(\rho\tau)| \\ &= |\hat{u}(\rho\tau) - \hat{u}(\rho\omega)| + |S_k^T g(\pi(\rho\omega)) - S_k^T g(\pi(\rho\tau))| \\ &\leq |\hat{u}(\rho\tau) - \hat{u}(\rho\omega)| + C_\beta(g) d^\beta(\pi(\omega), \pi(\tau)) \\ &= |\hat{u}(\rho\tau) - \hat{u}(\rho\omega)| + C_\beta(g) d^\beta(x, y).\end{aligned}$$

Taking words ρ with length $k = |\rho|$ increasing to ∞ and noting that $\hat{u} : E_A^\infty \to \mathbb{R}$ is uniformly continuous, we deduce that

$$|u(y) - u(x)| \leq C_\beta(g) d^\beta(y, x). \tag{13.108}$$

This means that u is Hölder continuous on $\mathcal{R}_\infty \cap R_e = \mathcal{R}_\infty \cap \operatorname{Int}(R_e)$.

We would like to show that u is locally Hölder continuous and uniformly continuous on $\mathcal{R}_\infty$ to later extend it to X. Take any $s \in \operatorname{Int}(R_e)$. Let $\eta > 0$ be so small that

$$B(s, 2\eta) \subseteq \operatorname{Int}(R_e). \tag{13.109}$$

By Proposition 13.6.14, the map T is very strongly transitive, i. e., there is $N \geq 0$ such that

$$\bigcup_{k=0}^{N} T^k(B(s, \eta)) = X. \tag{13.110}$$

Let $\lambda > 1$ and $\xi > 0$ come from the definition of $T : X \to X$ being open and distance expanding ((13.1)–(13.2)). Let $\xi' := \min\{\xi, \eta\}$. Fix any two points $w, z \in \mathcal{R}_\infty$ such that $d(w, z) < \xi'$. By (13.110), there exist $z' \in B(s, \eta)$ and $0 \leq k \leq N$ such that $T^k(z') = z$. Let

$$w' := T_{z'}^{-k}(w). \tag{13.111}$$

Since

$$d(w', z') \leq \lambda^{-k} d(w, z) < \xi' \leq \eta \tag{13.112}$$

and since $z' \in B(s, \eta)$, we deduce that $w' \in B(s, 2\eta)$ and it ensues from (13.109) that

$$z', w' \in \operatorname{Int}(R_e). \tag{13.113}$$

As T is a local homeomorphism on a neighbourhood of each element of $\mathcal{R}$, Lemma 4.5.4 states that $T(\bigcup_{e' \in E} \partial R_{e'}) \subseteq \bigcup_{e' \in E} \partial R_{e'}$. As $w, z \in \mathcal{R}_\infty$, we infer that

$$T^n(z'), T^n(w') \in \mathcal{R}_\infty, \quad \forall n \geq 0. \tag{13.114}$$

In consequence, using (13.107) we obtain that

$$S_k^T g(z') = S_k^\sigma(g \circ \pi)(\pi^{-1}(z')) = \hat{u}(\pi^{-1}(z')) - \hat{u}(\sigma^k(\pi^{-1}(z'))) = u(z') - u(z). \tag{13.115}$$

Likewise, $S_k^T g(w') = u(w') - u(w)$.

Moreover, using (13.108) in conjunction with (13.114) (when $n = 0$), we infer that

$$|u(w') - u(z')| \le C_\beta(g) d^\beta(w', z'). \tag{13.116}$$

Furthermore, using $d(w,z) < \xi$ and the definitions of z' and w' (cf. (13.111)), Lemma 13.1.1 yields

$$|S_k^T g(z') - S_k^T g(w')| \le C_\beta(g) d^\beta(w, z). \tag{13.117}$$

From (13.115)–(13.117) and (13.112), it ensues that

$$\begin{aligned} |u(w) - u(z)| &= |(u(w') - u(z')) + (S_k^T g(z') - S_k^T g(w'))| \\ &\le |u(w') - u(z')| + |S_k^T g(z') - S_k^T g(w')| \\ &\le C_\beta(g) d^\beta(w', z') + C_\beta(g) d^\beta(w, z) \le 2C_\beta(g) d^\beta(w, z). \end{aligned} \tag{13.118}$$

This holds for any $z, w \in \mathcal{R}_\infty$ such that $d(z,w) < \xi$. This inequality shows that the function $u : \mathcal{R}_\infty \to \mathbb{R}$ is locally Hölder continuous and also uniformly continuous. The uniform continuity implies that u has a unique continuous extension to $\overline{\mathcal{R}_\infty} = X$. Call it $\overline{u} : X \to \mathbb{R}$. Inequality (13.118) clearly extends to $\overline{u}$. So $\overline{u}$ is locally Hölder continuous, and hence Hölder continuous since X is compact.

Finally, (13.107) and the definition u imply that $g = u - u \circ T$ on $\mathcal{R}_\infty$. As the functions g, $\overline{u}$ and $\overline{u} \circ T$ are continuous, we conclude that $g = \overline{u} - \overline{u} \circ T$ on $\overline{\mathcal{R}_\infty} = X$. □

As an immediate consequence of Lemma 13.9.12 and Theorem 13.9.1, we get the following.

Lemma 13.9.13. *Let $T : X \to X$ be a topologically exact, open, distance expanding, Hölder system and let $\varphi : X \to \mathbb{R}$ be a Hölder potential. If $g : X \to \mathbb{R}$ is a Hölder function, then $\sigma^2_{\mu,\infty}(g)$ exists. Furthermore, if $\sigma^2_{\mu_\varphi,\infty}(g) = 0$ then there exists a Hölder function $u : E_A^\infty \to \mathbb{R}$ such that*

$$g = \mu_\varphi(g) + u - u \circ \sigma.$$

Thanks to Lemma 13.9.13 and Corollary 13.9.6, we can add two more equivalences to Theorem 13.7.17 when the system T is exact and Hölder continuous.

Theorem 13.9.14. *If $T : X \to X$ is a topologically exact, open, distance expanding, Hölder continuous system and $\varphi, \psi : X \to \mathbb{R}$ are Hölder continuous potentials, then the following statements are equivalent:*

(a) *The T-invariant Gibbs/equilibrium states of φ and ψ coincide, i. e., $\mu_\varphi = \mu_\psi$.*
(b) *There exists a constant $R_1 \in \mathbb{R}$ such that*

$$S_n\varphi(x) - S_n\psi(x) = R_1 n$$

whenever $n \in \mathbb{N}$ and $x \in X$ is a periodic point of T of period n.
(c) *There exist constants $R_2 \in \mathbb{R}$ and $C \geq 0$ such that*

$$|S_n\varphi(x) - S_n\psi(x) - R_2 n| \leq C, \quad \forall n \in \mathbb{N},\ \forall x \in X.$$

(d) *$\varphi - \psi$ is cohomologous to a constant $R_3 \in \mathbb{R}$ in the additive group of bounded functions on X.*
(e) *$\varphi - \psi$ is cohomologous to a constant $R_4 \in \mathbb{R}$ in the additive group of continuous functions on X.*
(f) *$\varphi - \psi$ is cohomologous to a constant $R_5 \in \mathbb{R}$ in the additive group of Hölder continuous functions on X.*
(g) $\mathcal{G}_\varphi = \mathcal{G}_\psi$.
(h) $\mathcal{G}_\varphi \cap \mathcal{G}_\psi \neq \emptyset$.
(i) $\sigma^2_{\mu_f,\infty}(\varphi - \psi) = 0$ *for some Hölder continuous potential $f : X \to \mathbb{R}$.*
(j) $\sigma^2_{\mu_f,\infty}(\varphi - \psi) = 0$ *for all Hölder continuous potentials $f : X \to \mathbb{R}$.*

In addition, if any of the above conditions holds, then all the constants R_1–R_5 are the same and equal to $\mathrm{P}(T,\varphi) - \mathrm{P}(T,\psi)$.

13.9.3 Central limit theorem

We now have all the tools to provide a short demonstration of a dynamical variation of the famous central limit theorem, which generally states that the distribution of the difference between (1) the sample average of a sequence of independent and identically distributed random variables with mean μ and variance σ^2 and (2) their expected value is, after adjusting for the size of the sample, approaching the normal (also called bell curve or Gaussian) distribution with the same mean μ and the same variance σ^2 as that of the variables.

Our approach, stemming from the one in [91], is based on the concept of characteristic functions (Fourier transforms). For a short recap on this topic, please see Subsection A.1.10 near the end of Appendix A.

Recall that the normal (Gaussian) distribution $\mathcal{N}_0(\sigma^2)$ with mean zero and variance $\sigma^2 > 0$ is the Borel probability measure on the real line $\mathbb{R}$ whose Radon–Nikodym derivative with respect to the Lebesgue measure is

$$\frac{d}{dt}\mathcal{N}_0(\sigma^2) = \exp\left(\frac{-t^2}{2\sigma^2}\right).$$

Theorem 13.9.15 (Central limit theorem). *Let $T : X \to X$ be a topologically exact, open, distance expanding, Hölder continuous system and $\varphi : X \to \mathbb{R}$ a Hölder continuous potential. If $g : X \to \mathbb{R}$ is Hölder continuous but is not cohomological to a constant in the class of Hölder continuous functions on X, then for every Lebesgue measurable set $B \subseteq \mathbb{R}$ with boundary of Lebesgue measure zero we have that*

$$\lim_{n\to\infty} \mu_\varphi\left(\left\{x \in X : \frac{S_n g(x) - \mu_\varphi(g)n}{\sqrt{n}} \in B\right\}\right) = \mathcal{N}_0(\sigma^2_{\mu_\varphi,\infty}(g))(B)$$
$$= \frac{1}{\sqrt{2\pi\sigma^2_{\mu_\varphi,\infty}(g)}} \int_B \exp\left(\frac{-t^2}{2\sigma^2_{\mu_\varphi,\infty}(g)}\right) dt,$$

where $\sigma^2_{\mu_\varphi,\infty}(g)$ is the asymptotic variance of g, whose existence and positivity are guaranteed by Lemma 13.9.13.

Proof. Let $\varphi, g \in H_\alpha(X, \mathbb{R})$. Replacing φ by $\varphi - \mathrm{P}(T, \varphi)$, we may assume without loss of generality that $\mathrm{P}(T, \varphi) = 0$. In particular, this implies that $\widehat{\mathcal{L}}_\varphi = \mathcal{L}_\varphi$. Replacing g by $g - \mu_\varphi(g)\mathbb{1}$, we may assume without loss of generality that $\mu_\varphi(g) = 0$.

For every $n \in \mathbb{N}$, consider the Hölder continuous function $Y_n : X \to \mathbb{R}$ defined by

$$Y_n(x) = \frac{S_n g(x)}{\sqrt{n}}.$$

As $\mu_\varphi = \varrho_\varphi m_\varphi$, $\mathcal{L}^*_\varphi(m_\varphi) = e^{\mathrm{P}(T,\varphi)} m_\varphi = m_\varphi$ and $\mathcal{L}^n_\varphi(e^{itY_n} \cdot h) = \mathcal{L}^n_{\varphi + \frac{it}{\sqrt{n}} g}(h)$, we get

$$\int_X e^{itY_n}\, d\mu_\varphi = \int_X e^{itY_n} \varrho_\varphi\, dm_\varphi = \int_X \mathcal{L}^n_\varphi(e^{itY_n} \varrho_\varphi)\, dm_\varphi = \int_X \mathcal{L}^n_{\varphi + \frac{it}{\sqrt{n}} g}(\varrho_\varphi)\, dm_\varphi. \tag{13.119}$$

For every $s \in \mathbb{C}$, let $\varphi_s := \varphi + sg$. Thanks to Theorem 13.8.18, we may apply Theorem A.2.5 (Kato–Rellich perturbation theorem) with $L_0 = \mathcal{L}_{\varphi_0} = \mathcal{L}_\varphi$. Set $Q_s = Q_{\mathcal{L}_{\varphi_s}}$, $S_s = S_{\mathcal{L}_{\varphi_s}}$, and $\lambda_s = \alpha(\mathcal{L}_{\varphi_s})$ for all $s \in \mathbb{C}$ in some small enough neighborhood of zero, say $s \in B_\mathbb{C}(0, R)$, determined by that theorem. This application of Kato–Rellich perturbation theorem, in conjunction with (13.119), yields

$$\int_X e^{itY_n}\, d\mu_\varphi = \int_X \mathcal{L}^n_{\varphi + \frac{it}{\sqrt{n}} g}(\varrho_\varphi)\, dm_\varphi = \int_X (\lambda^n_{\frac{it}{\sqrt{n}}} Q_{\frac{it}{\sqrt{n}}}(\varrho_\varphi) + S^n_{\frac{it}{\sqrt{n}}}(\varrho_\varphi))\, dm_\varphi \tag{13.120}$$

for all $n \in \mathbb{N}$ large enough. Given that $\varphi_0 = \varphi$ and $\mu_{\varphi_0}(g) = 0$, this same application of Kato–Rellich perturbation theorem, this time in combination with Taylor's theorem and Theorems 13.10.6–13.10.7, further gives

$$\lambda_s = \exp(\log \lambda_s) = \exp\left(\frac{\sigma^2_{\mu_\varphi,\infty}(g)}{2} s^2 + O(|s|^3)\right), \quad \forall s \in B_\mathbb{C}(0, R). \tag{13.121}$$

Therefore, for all $n \in \mathbb{N}$ large enough,

$$\lambda^n_{\frac{it}{\sqrt{n}}} = \exp\left(-\frac{\sigma^2_{\mu_\varphi,\infty}(g)t^2}{2} + |t|^3 O(n^{-1/2})\right). \tag{13.122}$$

Also by virtue of Theorem 13.8.18 and Theorem A.2.5, we have

$$0 \le \limsup_{n\to\infty} \|S^n_{\frac{it}{\sqrt{n}}}(\varrho_\varphi)\|_\infty \le \limsup_{n\to\infty} \|S^n_{\frac{it}{\sqrt{n}}}(\varrho_\varphi)\|_\alpha = 0,$$

which means that

$$\lim_{n\to\infty} \|S^n_{\frac{it}{\sqrt{n}}}(\varrho_\varphi)\|_\infty = 0. \tag{13.123}$$

Again by means of these same two theorems and (13.122),

$$\lambda^n_{\frac{it}{\sqrt{n}}} Q_{\frac{it}{\sqrt{n}}}(\varrho_\varphi) \underset{\text{uniformly on } X}{\overset{n\to\infty}{\longrightarrow}} \exp\left(-\frac{\sigma^2_{\mu_\varphi,\infty}(g)t^2}{2}\right) Q_0(\varrho_\varphi) = \exp\left(-\frac{\sigma^2_{\mu_\varphi,\infty}(g)t^2}{2}\right)\varrho_\varphi.$$

Using this uniform convergence and (13.120), we conclude that

$$\lim_{n\to\infty} \int_X e^{itY_n}\, d\mu_\varphi = \exp\left(-\frac{\sigma^2_{\mu_\varphi,\infty}(g)t^2}{2}\right) m_\varphi(\varrho_\varphi) = \exp\left(-\frac{\sigma^2_{\mu_\varphi,\infty}(g)t^2}{2}\right).$$

Since the function $\mathbb{R} \ni t \mapsto \exp(-\frac{\sigma^2_{\mu_\varphi,\infty}(g)t^2}{2}) \in \mathbb{R}$ is the characteristic function (Fourier transform) of the Gaussian (normal) distribution $\mathcal{N}_0(\sigma^2_{\mu_\varphi,\infty}(g))$, the proof is completed by invoking Theorem A.1.78. □

Remark 13.9.16. The central limit theorem (Theorem 13.9.15) also follows from a result of Gordin [40] once we know that the series

$$\sum_{n=0}^{\infty} \|\mathcal{L}^n_{\hat{\varphi}}(g)\|^2_{L^2(\mu_\varphi)}$$

converges. And indeed, assuming without loss of generality that $\mu_\varphi(g) = 0$, we get

$$\begin{aligned}\sum_{n=0}^{\infty} \|\mathcal{L}^n_{\hat{\varphi}}(g)\|^2_{L^2(\mu_\varphi)} &= \sum_{n=0}^{\infty} \int_X |\mathcal{L}^n_{\hat{\varphi}}(g)|^2\, d\mu_\varphi \\ &= \sum_{n=0}^{\infty} \int_X |S^n(g)|^2\, d\mu_\varphi \le \sum_{n=0}^{\infty} \int_X (c\zeta^n \|g\|_\alpha)^2\, d\mu_\varphi \\ &= (c\|g\|_\alpha)^2 \sum_{n=0}^{\infty} \zeta^{2n} < \infty,\end{aligned}$$

where, as in the proof of Theorem 13.9.1, the second equality sign and the following inequality sign both result from Theorem 13.8.18 applied with $\widehat{\mathcal{L}}_\varphi$ replaced by $\widehat{\mathcal{L}}_{\widehat{\varphi}}$.

13.9.4 Law of the iterated logarithm

For the sake of completeness, we state a dynamical version of the celebrated law of the iterated logarithm. Its full proof relies on deep results from probability theory and is too involved to be presented in this book. We instead refer the reader to [98], where a complete proof is provided. Given the spectral results derived in this chapter, the law of the iterated logarithm also easily ensues from the seminal paper [41].

Theorem 13.9.17 (Law of the iterated logarithm). *Let $T : X \to X$ be a topologically exact, open, distance expanding, Hölder continuous system and $\varphi : X \to \mathbb{R}$ a Hölder continuous potential. If $g : X \to \mathbb{R}$ is Hölder continuous but is not cohomological to a constant in the class of Hölder continuous functions on X, then*

$$\limsup_{n\to\infty} \frac{S_n g(x) - n\int_X g\, d\mu_\varphi}{\sqrt{n \log\log n}} = \sqrt{2\sigma^2_{\mu_\varphi,\infty}(g)} \quad \text{for } \mu_\varphi\text{-a. e. } x \in X.$$

13.9.5 Metric exactness, K-mixing and weak Bernoulli property

An immediate consequence of Theorems 8.4.6, 13.6.19 and 13.7.7 is the following.

Theorem 13.9.18. *If $T : X \to X$ is a topologically exact, open, distance expanding map and $\varphi : X \to \mathbb{R}$ is a Hölder continuous potential, then the T-invariant Gibbs state μ_φ is metrically exact and the Rokhlin's natural extension $\widetilde{T} : \widetilde{X} \to \widetilde{X}$ is K-mixing.*

In fact, a much stronger result holds: the Rokhlin's natural extension $\widetilde{T}$ is metrically isomorphic to a two-sided Bernoulli shift. In order to establish this, we need to introduce and discuss an important concept commonly referred to as weak Bernoulli.

Definition 13.9.19. Given two finite measurable partitions $\mathcal{A}$ and $\mathcal{B}$ of a probability space $(X, \mathcal{F}, \mu)$ and $\varepsilon \geq 0$, we say that the partition $\mathcal{B}$ is *ε-independent* of $\mathcal{A}$ if there exists a subfamily $\mathcal{A}' \subseteq \mathcal{A}$ such that

$$\mu\Big(\bigcup_{A\in\mathcal{A}'} A\Big) \geq 1-\varepsilon \quad \text{and} \quad \sum_{B\in\mathcal{B}} \left|\frac{\mu(A\cap B)}{\mu(A)} - \mu(B)\right| \leq \varepsilon, \quad \forall A \in \mathcal{A}'.$$

Definition 13.9.20. Given an ergodic measure-preserving endomorphism $T : X \to X$ of a Lebesgue space $(X, \mathcal{F}, \mu)$, a finite measurable partition $\mathcal{A}$ is called *weakly Bernoulli* if for every $\varepsilon > 0$ there exists an integer $k = k(\varepsilon) \geq 0$ such that the partition

$$\mathcal{A}_{m+k}^{n+1} := \bigvee_{j=m+k}^{n} T^{-j}(\mathcal{A})$$

is ε-independent of the partition

$$\mathcal{A}_0^{m+1} := \bigvee_{j=0}^{m} T^{-j}(\mathcal{A})$$

for all integers $m \geq 0$ and $n \geq m + k$. The endomorphism $(T : X \to X, \mu)$ is said to be *weakly Bernoulli* if it has a weakly Bernoulli one-sided generating partition $\mathcal{A}$. "One-sided generating" means that $\sigma(\bigcup_{n=1}^{\infty} \mathcal{A}^n) = \mathcal{F}$.

If $\varepsilon = 0$ and $k = 1$, then all partitions $T^{-j}(\mathcal{A})$, $j \geq 0$, are mutually independent, which means that

$$\mu(A \cap B) = \mu(A)\mu(B)$$

for all $A \in T^{-i}(\mathcal{A})$, all $B \in T^{-j}(\mathcal{A})$ and all $i \neq j$. We then say that the partition $\mathcal{A}$ is *Bernoulli*. If $\mathcal{A}$ is a one-sided generating partition (resp., a two-sided generating partition), then obviously the measure-preserving endomorphism $T : X \to X$ is isomorphic to a one-sided (resp., a two-sided) Bernoulli shift with $\#\mathcal{A}$ symbols. The following famous theorem of Friedman and Ornstein [33] shows the high importance of the "weak Bernoulli" concept.

Theorem 13.9.21. *If an ergodic measure-preserving automorphism $T : X \to X$ of a Lebesgue space $(X, \mathcal{F}, \mu)$ is weakly Bernoulli, then T is isomorphic to a two-sided Bernoulli shift.*

In order to formulate an appropriate "Bernoulli theorem" for noninvertible systems, we bring up the following simple observation.

Observation 13.9.22. If an ergodic measure-preserving endomorphism $T : X \to X$ of a Lebesgue space $(X, \mathcal{F}, \mu)$ has a weakly Bernoulli (one-sided) generating partition $\mathcal{A}$, then $\pi_0^{-1}(\mathcal{A}) = \{\pi_0^{-1}(A) : A \in \mathcal{A}\}$ is a weakly Bernoulli generating partition for the Rokhlin's natural extension $\widetilde{T} : \widetilde{X} \to \widetilde{X}$.

As a direct consequence of this observation, Theorem 8.4.7 and Theorem 13.9.21, we get the following.

Theorem 13.9.23. *If an ergodic measure-preserving endomorphism $T : X \to X$ of a Lebesgue space $(X, \mathcal{F}, \mu)$ is weakly Bernoulli, then its Rokhlin's natural extension is isomorphic to a two-sided Bernoulli shift.*

However, remark that being weakly Bernoulli is usually far from sufficient to conclude that the endomorphism is isomorphic to a one-sided Bernoulli shift. The reason/obstacle is that the endomorphism's Jacobian is invariant under an isomorphism and the Jacobian of each one-sided Bernoulli shift is locally constant.

In order to have a more complete picture, we now formulate another famous theorem of Ornstein [90].

Theorem 13.9.24. *Two Bernoulli shifts (with a finite alphabet) are metrically isomorphic if and only if they have the same entropy.*

Heading toward proving that natural extensions of Gibbs measures of distance expanding maps are isomorphic to two-sided Bernoulli shifts, we shall first show an intermediate (and interesting in itself) version for subshifts of finite type. We start with the following improvement of Theorem 13.9.1 on the exponential decay of correlations in the symbolic case.

Theorem 13.9.25. *Let E be a finite set having at least two elements and $A : E\times E \to \{0,1\}$ be a primitive incidence matrix. Let $\mathcal{A} := \{[e] : e \in E\}$ be the partition of E_A^∞ into cylinders of length one and $\sigma(\mathcal{A})$ be the σ-algebra generated by $\mathcal{A}$. Let $\varphi : E_A^\infty \to \mathbb{R}$ be a Hölder continuous potential. Then there exists a constant $D > 0$ such that if $f \in L^1(\mu_\varphi)$ and $g : E_A^\infty \to \mathbb{R}$ is measurable with respect to the σ-algebra $\sigma(\mathcal{A}^k)$ for some $k \in \mathbb{N}$, then*

$$C_{\mu_\varphi,n}(f,g) \le D\zeta^{n-k}\|f - E_{\mu_\varphi}(f)\|_{L^1(\mu_\varphi)}\|g - E_{\mu_\varphi}(g)\|_{L^1(\mu_\varphi)}, \quad \forall n \ge k,$$

where $\zeta \in (0,1)$ comes from Theorem 13.9.1.

Proof. Let $\varphi \in H_\alpha(E_A^\infty, \mathbb{R})$ and $n \ge k$. Replacing f by $f - E_{\mu_\varphi}(f)$ and g by $g - E_{\mu_\varphi}(g)$, we may assume that $E_{\mu_\varphi}(f) = 0 = E_{\mu_\varphi}(g)$. Observe that $\widehat{\mathcal{L}}_{\widehat{\varphi}}^k(g\cdot(f\circ\sigma^n)) = \widehat{\mathcal{L}}_{\widehat{\varphi}}^k(g)\cdot(f\circ\sigma^{n-k})$. Moreover, $E_{\mu_\varphi}(f\circ\sigma^{n-k}) = 0 = E_{\mu_\varphi}(\widehat{\mathcal{L}}_{\widehat{\varphi}}^k(g))$ since μ_φ is σ-invariant and since $\widehat{\mathcal{L}}_{\widehat{\varphi}}^*(\mu_\varphi) = \mu_\varphi$ by Proposition 13.8.4, respectively. Using all of the above and Theorem 13.9.1, we get

$$\begin{aligned} C_{\mu_\varphi,n}(f,g) &= \left|\int_{E_A^\infty} g\cdot(f\circ\sigma^n)\,d\mu_\varphi\right| = \left|\int_{E_A^\infty} \widehat{\mathcal{L}}_{\widehat{\varphi}}^k(g\cdot(f\circ\sigma^n))\,d\mu_\varphi\right| \\ &= \left|\int_{E_A^\infty} \widehat{\mathcal{L}}_{\widehat{\varphi}}^k(g)\cdot(f\circ\sigma^{n-k})\,d\mu_\varphi\right| \\ &= C_{\mu_\varphi,n-k}(f,\widehat{\mathcal{L}}_{\widehat{\varphi}}^k(g)) \\ &\le c\zeta^{n-k}\|f\|_{L^1(\mu_\varphi)}\|\widehat{\mathcal{L}}_{\widehat{\varphi}}^k(g)\|_\alpha. \end{aligned} \tag{13.124}$$

This last inequality arises from Theorem 13.9.1 and presupposes that $\widehat{\mathcal{L}}_{\widehat{\varphi}}^k(g) \in H_\alpha(E_A^\infty)$, though g may not be in $H_\alpha(E_A^\infty)$. It thus remains to find an appropriate upper bound for $\|\widehat{\mathcal{L}}_{\widehat{\varphi}}^k(g)\|_\alpha$.

Decomposing $g = g_+ - g_-$ into its positive and negative parts, we may assume that $g \geq 0$. Since the function $g : E_A^\infty \to \mathbb{R}$ depends only on the first k coordinates, using Lemma 13.1.1 we obtain for all $\omega, \tau \in E_A^\infty$ with $\omega_1 = \tau_1$ that

$$\begin{aligned}
\mathcal{L}_\varphi^k(g)(\omega) &= \sum_{\gamma \in E_A^k : A_{\gamma_k \omega_1} = 1} g(\gamma\omega) \exp(S_k\varphi(\gamma\omega)) \\
&= \sum_{\gamma \in E_A^k : A_{\gamma_k \tau_1} = 1} g(\gamma\tau) \exp(S_k\varphi(\gamma\omega)) \\
&\leq \exp(C_\alpha(\varphi) d^\alpha(\omega, \tau)) \sum_{\gamma \in E_A^k : A_{\gamma_k \tau_1} = 1} g(\gamma\tau) \exp(S_k\varphi(\gamma\tau)) \\
&= \exp(C_\alpha(\varphi) d^\alpha(\omega, \tau)) \mathcal{L}_\varphi^k(g)(\tau) \leq e^{C_\alpha(\varphi)} \mathcal{L}_\varphi^k(g)(\tau).
\end{aligned} \tag{13.125}$$

Therefore,

$$\sup(\mathcal{L}_\varphi^k(g)|_{[e]}) \leq e^{C_\alpha(\varphi)} \inf(\mathcal{L}_\varphi^k(g)|_{[e]}), \quad \forall e \in E.$$

Using this, Corollary 13.6.5 ($\widehat{\mathcal{L}}_\varphi^*(m_\varphi) = m_\varphi$), the fact that $\mu_\varphi := \varrho_\varphi m_\varphi$ and Lemma 13.7.5, it ensues that

$$\begin{aligned}
\sup(\widehat{\mathcal{L}}_\varphi^k(g)|_{[e]}) &\leq e^{C_\alpha(\varphi)} \inf(\widehat{\mathcal{L}}_\varphi^k(g)|_{[e]}) \leq e^{C_\alpha(\varphi)} \frac{1}{m_\varphi([e])} \int_{[e]} \widehat{\mathcal{L}}_\varphi^k(g)\, dm_\varphi \\
&\leq e^{C_\alpha(\varphi)} \frac{1}{m_\varphi([e])} \int_{E_A^\infty} \widehat{\mathcal{L}}_\varphi^k(g)\, dm_\varphi = e^{C_\alpha(\varphi)} \frac{1}{m_\varphi([e])} \int_{E_A^\infty} g\, dm_\varphi \\
&= e^{C_\alpha(\varphi)} \frac{1}{m_\varphi([e])} \|g\|_{L^1(m_\varphi)} \leq e^{C_\alpha(\varphi)} \frac{1}{m_\varphi([e])} \frac{\|g\|_{L^1(\mu_\varphi)}}{\inf(\varrho_\varphi)} \\
&\leq C_1(\varphi) \|g\|_{L^1(\mu_\varphi)},
\end{aligned}$$

where

$$C_1(\varphi) := \overline{D}_\alpha(\varphi) e^{C_\alpha(\varphi)} \max_{a \in E} \frac{1}{m_\varphi([a])}.$$

Hence,

$$\|\widehat{\mathcal{L}}_\varphi^k(g)\|_\infty = \sup(\widehat{\mathcal{L}}_\varphi^k(g)) \leq C_1(\varphi) \|g\|_{L^1(\mu_\varphi)}. \tag{13.126}$$

Using (13.125)– (13.126), we have for all $\omega, \tau \in E_A^\infty$ with $\omega_1 = \tau_1$ that

$$\begin{aligned}
|\widehat{\mathcal{L}}_\varphi^k(g)(\omega) - \widehat{\mathcal{L}}_\varphi^k(g)(\tau)| &\leq [\exp(C_\alpha(\varphi) d^\alpha(\omega, \tau)) - 1] \max\{\widehat{\mathcal{L}}_\varphi^k(g)(\omega), \widehat{\mathcal{L}}_\varphi^k(g)(\tau)\} \\
&\leq [\exp(C_\alpha(\varphi) d^\alpha(\omega, \tau)) - 1] C_1(\varphi) \|g\|_{L^1(\mu_\varphi)} \\
&\leq C_2(\varphi) C_\alpha(\varphi) d^\alpha(\omega, \tau) \cdot C_1(\varphi) \|g\|_{L^1(\mu_\varphi)},
\end{aligned}$$

where $C_2(\varphi) > 0$ is a constant such that $e^x \leq 1 + C_2(\varphi)x$ for all $x \in [0, C_\alpha(\varphi)]$. Consequently,

$$v_\alpha(\widehat{\mathcal{L}}_\varphi^k(g)) \leq C_2(\varphi)C_\alpha(\varphi) \cdot C_1(\varphi)\|g\|_{L^1(\mu_\varphi)}.$$

Along with (13.126), this gives

$$\|\widehat{\mathcal{L}}_\varphi^k(g)\|_\alpha \leq (1 + C_2(\varphi)C_\alpha(\varphi)) \cdot C_1(\varphi) \cdot \|g\|_{L^1(\mu_\varphi)}.$$

Now replace φ by $\widehat{\varphi}$ (this is possible because $\widehat{\varphi} \in H_\alpha(E_A^\infty, \mathbb{R})$, as the shift map σ is Lipschitz; see the start of Subsection 13.8.1). Since $\mu_{\widehat{\varphi}} = \mu_\varphi$ (see Corollary 13.8.11), we deduce that

$$\|\widehat{\mathcal{L}}_{\widehat{\varphi}}^k(g)\|_\alpha \leq (1 + C_2(\widehat{\varphi})C_\alpha(\widehat{\varphi})) \cdot C_1(\widehat{\varphi}) \cdot \|g\|_{L^1(\mu_\varphi)}.$$

Inserting this into (13.124), we conclude that

$$C_{\mu_\varphi,n}(f,g) \leq [c(1 + C_2(\widehat{\varphi})C_\alpha(\widehat{\varphi})) \cdot C_1(\widehat{\varphi})] \cdot \zeta^{n-k}\|f\|_{L^1(\mu_\varphi)}\|g\|_{L^1(\mu_\varphi)}. \qquad \square$$

Theorem 13.9.26. *Let E be a finite set with $\#E \geq 2$ and let $A : E \times E \to \{0,1\}$ be a primitive incidence matrix. Then the partition $\mathcal{A} := \{[e] : e \in E\}$ of E_A^∞ into cylinders of length one is a weak Bernoulli one-sided generator for the Gibbs state μ_φ for every Hölder continuous potential $\varphi : E_A^\infty \to \mathbb{R}$. In consequence, the Rokhlin's natural extension of the dynamical system $(\sigma : E_A^\infty \to E_A^\infty, \mu_\varphi)$, which is in fact the double-sided shift $(\widetilde{\sigma} : E_A^{\mathbb{Z}} \to E_A^{\mathbb{Z}}, \widetilde{\mu}_\varphi)$, is isomorphic to a two-sided Bernoulli shift.*

Proof. Fix integers $k \in \mathbb{N}$, $\ell \geq 0$ and $N \geq k + \ell$. Fix also $\omega \in E_A^k$ and $\tau \in E_A^{N-(k+\ell)}$. Recall that μ_φ is σ-invariant. Applying Theorem 13.9.25 with $n = k + \ell$, $k = k$, $f = \mathbb{1}_{[\tau]}$ and $g = \mathbb{1}_{[\omega]}$ and using (13.75), we obtain that

$$\begin{aligned}
&\left|\mu_\varphi(\sigma^{-(k+\ell)}([\tau]) \cap [\omega]) - \mu_\varphi(\sigma^{-(k+\ell)}([\tau]))\mu_\varphi([\omega])\right| \\
&\qquad = \left|\mu_\varphi(\mathbb{1}_{[\omega]} \cdot (\mathbb{1}_{[\tau]} \circ \sigma^{k+\ell})) - \mu_\varphi(\mathbb{1}_{[\omega]})\mu_\varphi(\mathbb{1}_{[\tau]})\right| \\
&\qquad = C_{\mu_\varphi,k+\ell}(\mathbb{1}_{[\tau]}, \mathbb{1}_{[\omega]}) \\
&\qquad \leq D\zeta^\ell \|\mathbb{1}_{[\tau]} - \mu_\varphi(\mathbb{1}_{[\tau]})\|_{L^1(\mu_\varphi)} \|\mathbb{1}_{[\omega]} - \mu_\varphi(\mathbb{1}_{[\omega]})\|_{L^1(\mu_\varphi)} \\
&\qquad \leq 4D\zeta^\ell \mu_\varphi([\tau])\mu_\varphi([\omega]).
\end{aligned}$$

Therefore, for every $\omega \in E_A^k$ we get

$$\sum_{\tau \in E_A^{N-(k+\ell)}} \left|\frac{\mu_\varphi(\sigma^{-(k+\ell)}([\tau]) \cap [\omega])}{\mu_\varphi([\omega])} - \mu_\varphi(\sigma^{-(k+\ell)}([\tau]))\right| \leq 4D\zeta^\ell \sum_{\tau \in E_A^{N-(k+\ell)}} \mu_\varphi([\tau]) = 4D\zeta^\ell.$$

As $\zeta \in (0,1)$, given $\varepsilon > 0$ there is $\ell = \ell(\varepsilon) \in \mathbb{N}$ so large that $4D\zeta^{\ell} < \varepsilon$. Hence, the partition $\mathcal{A}_{k+\ell}^{N+1}$ is ε-independent of the partition $\mathcal{A}_0^{k+1}$. Since, moreover, $\mathcal{A}$ is a topological generator for the shift map $\sigma : E_A^{\infty} \to E_A^{\infty}$, the measure-preserving dynamical system $(\sigma : E_A^{\infty} \to E_A^{\infty}, \mu_{\varphi})$ is weakly Bernoulli. Its Rokhlin's natural extension $(\tilde{\sigma} : E_A^{\mathbb{Z}} \to E_A^{\mathbb{Z}}, \tilde{\mu}_{\varphi})$ is isomorphic to a two-sided Bernoulli shift by Theorem 13.9.23. □

A straightforward consequence of this theorem and Theorem 13.7.11 is next.

Theorem 13.9.27. *If $T : X \to X$ is a topologically exact, open, distance expanding system and $\varphi : X \to \mathbb{R}$ is a Hölder continuous potential, then the measure-preserving dynamical system $(T : X \to X, \mu_{\varphi})$ is weakly Bernoulli and its Rokhlin's natural extension is isomorphic to a two-sided Bernoulli shift.*

13.10 Real analyticity of topological pressure

Let $T : X \to X$ be an open, distance expanding system on a compact metric space (X, d). Let $\varepsilon > 0$ be so small that all the inverse branches of T are well defined on any ball $B(x, \varepsilon)$. For instance, take any $\varepsilon \le \xi$, where ξ is the constant arising from relation (13.2) (cf. (4.29)).

Fix $\alpha > 0$. Recall that $C(X)$ and $H_\alpha(X)$ are now considered complex Banach spaces. The supremum norm $\|\cdot\|_\infty$, the Hölder variation v_α and the Hölder norm $\|\cdot\|_\alpha$ are defined exactly as in Section 13.1.

For every point $x \in X$, let $H_{\alpha,x,\varepsilon}(X)$ be the Banach space of all complex-valued Hölder continuous functions with exponent α whose domain is the ball $B(x, \varepsilon)$. The supremum norm $\|\cdot\|_{\infty,x,\varepsilon}$, the Hölder variation $v_{\alpha,x,\varepsilon}$ and the Hölder norm $\|\cdot\|_{\alpha,x,\varepsilon}$ are defined in a similar way as in Section 13.1 by simply replacing X with $B(x, \varepsilon)$.

Let F, G be complex Banach spaces. Let $L(F)$ and $L(F, G)$ respectively denote the Banach spaces of bounded $\mathbb{C}$-linear operators from F to itself and from F to G. For any operator $L \in L(F, G)$, recall that the operator norm of L is defined by

$$\|L\| := \inf\{c \ge 0 : \|L(v)\|_G \le c\|v\|_F,\ \forall v \in F\} = \sup\left\{\frac{\|L(v)\|_G}{\|v\|_F} : v \in F \setminus \{0\}\right\}.$$

In particular, if $L \in L(H_\alpha(X), H_{\alpha,x,\varepsilon}(X))$ then

$$\|L\|_\alpha := \inf\{c \ge 0 : \|L(g)\|_{\alpha,x,\varepsilon} \le c\|g\|_\alpha,\ \forall g \in H_\alpha(X)\}.$$

Given an arbitrary set Λ and a function

$$\Phi : \Lambda \to L(H_\alpha(X)),$$

for every point $x \in X$ define the function

$$\Phi_x : \Lambda \to L(H_\alpha(X), H_{\alpha,x,\varepsilon}(X))$$

by

$$\Phi_x(\lambda)(g) := \Phi(\lambda)(g)|_{B(x,\varepsilon)}.$$

Obviously, if $\Lambda \subseteq \mathbb{C}$ and the function Φ is holomorphic then so is the function Φ_x for all $x \in X$. A natural endeavor then consists in identifying conditions under which the converse holds. This is the motivation behind the following statement.

Lemma 13.10.1. *Let Λ be an open subset of $\mathbb{C}$. If a function $\Phi : \Lambda \to L(H_\alpha(X))$ is such that the function $\Phi_x : \Lambda \to L(H_\alpha(X), H_{\alpha,x,\varepsilon}(X))$ is holomorphic for every $x \in X$ and*

$$\sup\{\|\Phi_x(\lambda)\|_\alpha : x \in X,\ \lambda \in \Lambda\} < \infty,$$

then $\Phi : \Lambda \to L(H_\alpha(X))$ is holomorphic.

Proof. Fix $\lambda^0 \in \Lambda$ and take $r > 0$ so small that the disc $D(\lambda^0, 2r)$ centered at λ^0 and of radius $2r$ is contained in Λ. By hypothesis, for each $x \in X$ the Taylor series expansion of Φ_x around λ_0, i. e.,

$$\Phi_x(\lambda) := \sum_{n=0}^{\infty} a_{x,n}(\lambda - \lambda^0)^n, \quad \lambda \in D(\lambda^0, r),$$

with some $a_{x,n} \in L(H_\alpha(X), H_{\alpha,x,\varepsilon}(X))$, converges uniformly in the Banach space $L(H_\alpha(X), H_{\alpha,x,\varepsilon}(X))$. Set $M = \sup\{\|\Phi_x(\lambda)\|_\alpha : x \in X,\ \lambda \in \Lambda\} < \infty$. It follows from Cauchy's estimates (Theorem A.3.2 in Appendix A) that

$$\|a_{x,n}\|_\alpha \le Mr^{-n}, \quad \forall x \in X, \forall n \ge 0. \tag{13.127}$$

For every $n \ge 0$, define the operator a_n by

$$a_n(g)(x) := a_{x,n}(g)(x), \quad \forall g \in H_\alpha(X),\ \forall x \in X.$$

Let $g \in H_\alpha(X)$. Then

$$\|a_n(g)\|_\infty \le \sup_{x \in X} \|a_{x,n}(g)\|_\alpha \le \sup_{x \in X} \|a_{x,n}\|_\alpha \|g\|_\alpha \le Mr^{-n}\|g\|_\alpha. \tag{13.128}$$

Moreover, if $d(x,y) < \varepsilon$ then for every $w \in B(x,\varepsilon) \cap B(y,\varepsilon)$, we have

$$\begin{aligned}\sum_{n=0}^{\infty} a_{x,n}(g)(w)(\lambda - \lambda^0)^n &= (\Phi_x(\lambda)(g))(w) = (\Phi(\lambda)(g))(w) = (\Phi_y(\lambda)(g))(w) \\ &= \sum_{n=0}^{\infty} a_{y,n}(g)(w)(\lambda - \lambda^0)^n, \quad \forall \lambda \in D(\lambda^0, r).\end{aligned}$$

The uniqueness of the coefficients of Taylor series expansions imposes that

$$a_{x,n}(g)(w) = a_{y,n}(g)(w), \quad \forall n \geq 0. \tag{13.129}$$

Since $x, y \in B(x, \varepsilon) \cap B(y, \varepsilon)$, we get from (13.129) and (13.127) that

$$\begin{aligned}|a_n(g)(y) - a_n(g)(x)| &= |a_{y,n}(g)(y) - a_{x,n}(g)(x)| = |a_{x,n}(g)(y) - a_{x,n}(g)(x)| \\ &\leq \|a_{x,n}(g)\|_{\alpha,x,\varepsilon} d^{\alpha}(x,y) \leq \|a_{x,n}\|_{\alpha} \|g\|_{\alpha} d^{\alpha}(x,y) \\ &\leq Mr^{-n} \|g\|_{\alpha} d^{\alpha}(x,y).\end{aligned}$$

Consequently, $v_{\alpha,x,\varepsilon}(a_n(g)|_{B(x,\varepsilon)}) \leq Mr^{-n}\|g\|_{\alpha}$ for all $x \in X$. Due to the compactness of X, one can extract a finite number of finite chains of open balls of radius ε covering X, chains that are positively separated from one another. There hence exists a constant $C \geq 1$ such that

$$v_{\alpha}(a_n(g)) \leq CMr^{-n}\|g\|_{\alpha}.$$

Combining this with (13.128), we obtain

$$\|a_n(g)\|_{\alpha} \leq (C+1)Mr^{-n}\|g\|_{\alpha}.$$

Thus, $a_n \in L(H_{\alpha}(X))$ and

$$\|a_n\|_{\alpha} \leq (C+1)Mr^{-n}.$$

Therefore, the series

$$\sum_{n=0}^{\infty} a_n(\lambda - \lambda^0)^n, \quad \lambda \in D(\lambda^0, r/2),$$

converges absolutely uniformly in the Banach space $L(H_{\alpha}(X))$ with

$$\left\| \sum_{n=0}^{\infty} a_n(\lambda - \lambda^0)^n \right\|_{\alpha} \leq 2(C+1)M, \quad \forall \lambda \in D(\lambda^0, r/2).$$

The function $D(\lambda^0, r/2) \ni \lambda \longmapsto \sum_{n=0}^{\infty} a_n(\lambda - \lambda^0)^n \in L(H_{\alpha}(X))$ is thereby holomorphic.

Finally, for every $g \in H_{\alpha}(X)$ and every $x \in X$,

$$\begin{aligned}\left(\sum_{n=0}^{\infty} a_n(\lambda - \lambda^0)^n\right)(g)(x) &= \sum_{n=0}^{\infty} a_n(g)(x)(\lambda - \lambda^0)^n = \sum_{n=0}^{\infty} a_{x,n}(g)(x)(\lambda - \lambda^0)^n \\ &= \left(\sum_{n=0}^{\infty} a_{x,n}(\lambda - \lambda^0)^n\right)(g)(x) = \Phi_x(\lambda)(g)(x) \\ &= \Phi(\lambda)(g)(x).\end{aligned}$$

So

$$\Phi(\lambda) = \sum_{n=0}^{\infty} a_n(\lambda - \lambda^0)^n, \quad \forall \lambda \in D(\lambda^0, r/2)$$

and Φ is thus holomorphic on $D(\lambda^0, r/2)$. Since holomorphicity is a local property, the result ensues. □

The main technical result of this section concerns the analytic dependence of the transfer operator $\mathcal{L}_{\varphi_\lambda}$ on the parameter λ.

Theorem 13.10.2. *Let $T : X \to X$ be an open, distance expanding system. Suppose that Λ is an open subset of $\mathbb{C}^k$ for some $k \in \mathbb{N}$. If $\varphi_\lambda \in H_\alpha(X)$ for every $\lambda \in \Lambda$, if $\sup\{\|\varphi_\lambda\|_\alpha : \lambda \in \Lambda\} < \infty$ and if the function $\Lambda \ni \lambda \longmapsto \varphi_\lambda(x) \in \mathbb{C}$ is holomorphic for every $x \in X$, then the map*

$$\Lambda \ni \lambda \longmapsto \mathcal{L}_{\varphi_\lambda} \in L(H_\alpha(X)),$$

is holomorphic, where $\mathcal{L}_{\varphi_\lambda}$ is the natural complexification of the real transfer operator $\mathcal{L}_\varphi$.

Proof. By Hartogs' theorem (Theorem A.4.1 in Appendix A), we may assume without loss of generality that $k = 1$, i. e., $\Lambda \subseteq \mathbb{C}$. Let $H := \sup\{\|\varphi_\lambda\|_\alpha : \lambda \in \Lambda\} < \infty$. For all $\lambda \in \Lambda$, all $x \in X$ and all $y \in T^{-1}(x)$, we know that

$$\|\exp(\varphi_\lambda \circ T_y^{-1})\|_{\infty,x,\varepsilon} \le e^H, \tag{13.130}$$

where T_y^{-1} is the branch of T^{-1} on $B(x,\varepsilon)$ mapping x to y. Fix $\lambda^0 \in \Lambda$ and take a radius $r > 0$ so small that $D(\lambda^0, 2r) \subseteq \Lambda$. Let also $x \in X$ and $y \in T^{-1}(x)$. By hypothesis, the function $D(\lambda^0, 2r) \ni \lambda \longmapsto \exp(\varphi_\lambda \circ T_y^{-1}(w))$ is holomorphic for every $w \in B(x,\varepsilon)$. Consider its Taylor series expansion

$$\exp(\varphi_\lambda \circ T_y^{-1}(w)) = \sum_{n=0}^{\infty} a_{y,n}(w)(\lambda - \lambda^0)^n, \quad \forall \lambda \in D(\lambda^0, r).$$

In view of Cauchy's estimates (Theorem A.3.2) and (13.130), we get

$$|a_{y,n}(w)| \le e^H r^{-n}, \quad \forall n \ge 0, \ \forall w \in B(x,\varepsilon). \tag{13.131}$$

Moreover, as the local inverse branches of T are contractions, we know that $d(T_y^{-1}(w), T_y^{-1}(v)) \le d(w,v)$ for all $w, v \in B(x,\varepsilon)$. Cauchy's estimates and the mean value inequality then yield

$$
\begin{aligned}
|a_{y,n}(w) - a_{y,n}(v)| &\le r^{-n} \sup_{\lambda\in\Lambda} |\exp(\varphi_\lambda \circ T_y^{-1}(w)) - \exp(\varphi_\lambda \circ T_y^{-1}(v))| \\
&\le r^{-n} \sup_{\lambda\in\Lambda} e^{\max \mathrm{Re}(\varphi_\lambda)} |\varphi_\lambda \circ T_y^{-1}(w) - \varphi_\lambda \circ T_y^{-1}(v)| \\
&\le r^{-n} \sup_{\lambda\in\Lambda} e^{\|\varphi_\lambda\|_\alpha} v_\alpha(\varphi_\lambda) d^\alpha(T_y^{-1}(w), T_y^{-1}(v)) \\
&\le e^H H r^{-n} d^\alpha(w, v).
\end{aligned}
\tag{13.132}
$$

Take an arbitrary $g \in H_\alpha(X)$. In view of relations (13.131)–(13.132), we obtain that

$$
\begin{aligned}
&|a_{y,n}(w)g(T_y^{-1}(w)) - a_{y,n}(v)g(T_y^{-1}(v))| \\
&\quad \le |(a_{y,n}(w) - a_{y,n}(v))g(T_y^{-1}(w)) + a_{y,n}(v)(g(T_y^{-1}(w)) - g(T_y^{-1}(v)))| \\
&\quad \le |a_{y,n}(w) - a_{y,n}(v)| \cdot \|g\|_\infty + |a_{y,n}(v)| \cdot v_\alpha(g) d^\alpha(T_y^{-1}(w), T_y^{-1}(v)) \\
&\quad \le e^H H r^{-n} d^\alpha(w, v) \cdot \|g\|_\alpha + e^H r^{-n} \cdot \|g\|_\alpha d^\alpha(w, v) \\
&\quad = (H + 1)e^H r^{-n} \|g\|_\alpha d^\alpha(w, v).
\end{aligned}
$$

Combining this and (13.131), we deduce that the formula

$$
N_{y,n}(g)(w) := a_{y,n}(w) \cdot g(T_y^{-1}(w))
$$

engenders a bounded linear operator $N_{y,n} : H_\alpha(X) \longrightarrow H_{\alpha,x,\varepsilon}(X)$ such that $v_{\alpha,x,\varepsilon}(N_{y,n}(g)) \le (H + 1)e^H r^{-n} \|g\|_\alpha$ and $\|N_{y,n}(g)\|_{\infty,x,\varepsilon} \le e^H r^{-n} \|g\|_\alpha$. So $\|N_{y,n}(g)\|_{\alpha,x,\varepsilon} \le (H + 2)e^H r^{-n} \|g\|_\alpha$, and hence

$$
\|N_{y,n}\|_\alpha \le (H + 2)e^H r^{-n}.
$$

Consequently, the function

$$
D(\lambda^0, r/2) \ni \lambda \longmapsto N_{y,n}(\lambda - \lambda^0)^n
$$

is holomorphic and

$$
\|N_{y,n}(\lambda - \lambda^0)^n\|_\alpha \le 2^{-n}(H + 2)e^H, \quad \forall \lambda \in D(\lambda^0, r/2).
$$

Thus, the series

$$
A_{\lambda,y} := \sum_{n=0}^{\infty} N_{y,n}(\lambda - \lambda^0)^n, \quad \lambda \in D(\lambda^0, r/2),
$$

converges absolutely uniformly in the Banach space $L(H_\alpha(X), H_{\alpha,x,\varepsilon}(X))$, with norm $\|A_{\lambda,y}\|_\alpha \le 2(H+2)e^H$, and the function

$$D(\lambda^0, r/2) \ni \lambda \longmapsto A_{\lambda,y} \in L(H_\alpha(X), H_{\alpha,x,\varepsilon}(X))$$

is holomorphic. Hence,

$$\mathcal{L}_{\lambda,x} := \sum_{y \in T^{-1}(x)} A_{\lambda,y} \in L(H_\alpha(X), H_{\alpha,x,\varepsilon}(X)), \tag{13.133}$$

with $\|\mathcal{L}_{\lambda,x}\|_\alpha \le 2\#T^{-1}(H+2)e^H$, and the function $D(\lambda^0, r/2) \ni \lambda \longmapsto \mathcal{L}_{\lambda,x}$ is holomorphic. Note that the above estimate on $\|\mathcal{L}_{\lambda,x}\|_\alpha$ is independent of λ and λ_0, and thus holds for all $\lambda \in \Lambda$. It is also independent of x. Moreover, analyticity is a local property. A straightforward calculation establishes that $\mathcal{L}_{\lambda,x}(g)(x) = \mathcal{L}_{\varphi_\lambda}(g)(x)$ for all $g \in H_\alpha(X)$, all $x \in X$ and all $\lambda \in \Lambda$. Invoking Lemma 13.10.1 concludes the proof. □

A function from a complex vector space to a complex Banach space is called holomorphic if its restriction to any complex finite-dimensional affine subspace is holomorphic; see [26, Definition VI.10.5]. Theorem 13.10.2 yields the holomorphicity of

$$H_\alpha(X) \ni \varphi \longmapsto \mathcal{L}_\varphi \in L(H_\alpha(X)) \tag{13.134}$$

mentioned in the introduction to this section. It further yields real analyticity after restricting this function to the real Banach space $H_\alpha(X, \mathbb{R})$.

From Theorem 13.10.2, we can infer the real analyticity of the pressure function for a "nicely" parametrized family of Hölder continuous potentials.

Theorem 13.10.3. *Let $T : X \to X$ be a topologically transitive, open, distance expanding system. Suppose that Λ is an open subset of $\mathbb{C}^k$ for some $k \in \mathbb{N}$. Assume that $\varphi_\lambda \in H_\alpha(X)$ for every $\lambda \in \Lambda$, that $\sup\{\|\varphi_\lambda\|_\alpha : \lambda \in \Lambda\} < \infty$, and that the function $\Lambda \ni \lambda \longmapsto \varphi_\lambda(x) \in \mathbb{C}$ is holomorphic for every $x \in X$. If M is a real-analytic submanifold of Λ such that*

$$\varphi_\lambda(X) \subseteq \mathbb{R}, \quad \forall \lambda \in M,$$

then the pressure function

$$M \ni \lambda \longmapsto \mathrm{P}(T, \varphi_\lambda) \in \mathbb{R}$$

is real analytic.

Proof. Fix $\lambda^0 \in M$. Since $\Lambda \subseteq \mathbb{C}^k$ is open, there exists $R > 0$ such that $B_{\mathbb{C}^k}(\lambda^0, R) \subseteq \Lambda$. By Theorem 13.8.20, the number $\exp(\mathrm{P}(T, \varphi_\lambda))$ is a simple isolated eigenvalue of the operator $\mathcal{L}_{\varphi_\lambda} \in L(H_\alpha(X))$ for all $\lambda \in M \cap B_{\mathbb{C}^k}(\lambda^0, R)$. It follows from Theorem 13.10.2 and the Kato–Rellich theorem for perturbations of linear operators (Theorem A.2.5) that there exist $r \in (0, R]$ and a holomorphic function

$$\gamma : B_{\mathbb{C}^k}(\lambda^0, r) \to \mathbb{C}$$

such that $\gamma(\lambda)$ is an eigenvalue of the operator $\mathcal{L}_{\varphi_\lambda}$ for all $\lambda \in B_{\mathbb{C}^k}(\lambda^0, r)$ and

$$\gamma(\lambda) = \exp(\mathrm{P}(T, \varphi_\lambda)), \quad \forall \lambda \in M \cap B_{\mathbb{C}^k}(\lambda^0, r).$$

Consequently, the function $M \cap B_{\mathbb{C}^k}(\lambda^0, r) \ni \lambda \longmapsto \mathrm{P}(T, \varphi_\lambda) \in \mathbb{R}$ is real analytic. Since real analyticity is a local property, the result holds. □

As a special case of this theorem, we get the real analyticity of the pressure function for the parametrized family $\{t\varphi + \psi\}_{t \in \mathbb{R}}$ for any Hölder functions φ, ψ.

Corollary 13.10.4. *Let $T : X \to X$ be a topologically transitive, open, distance expanding system. If $\varphi, \psi : X \to \mathbb{R}$ are Hölder continuous functions, then the pressure function*

$$\mathbb{R} \ni t \longmapsto \mathrm{P}(T, t\varphi + \psi) \in \mathbb{R}$$

is real analytic.

Proof. Let α be a Hölder exponent common to φ and ψ. Fix $t \in \mathbb{R}$. Let $\Lambda = B_{\mathbb{C}}(t, 1)$ and $\varphi_\lambda = \lambda\varphi + \psi$ for every $\lambda \in \Lambda$. Obviously, $\varphi_\lambda \in H_\alpha(X)$ for every $\lambda \in \Lambda$ and

$$\begin{aligned} \sup\{\|\varphi_\lambda\|_\alpha : \lambda \in \Lambda\} &= \sup\{\|\lambda\varphi + \psi\|_\alpha : \lambda \in \Lambda\} \le \sup\{|\lambda| \cdot \|\varphi\|_\alpha + \|\psi\|_\alpha : \lambda \in \Lambda\} \\ &= \sup\{|\lambda| : \lambda \in B_{\mathbb{C}}(t, 1)\}\|\varphi\|_\alpha + \|\psi\|_\alpha \\ &= \max\{-(t-1), t+1\}\|\varphi\|_\alpha + \|\psi\|_\alpha < \infty. \end{aligned}$$

Furthermore, the function $\Lambda \ni \lambda \longmapsto \varphi_\lambda(x) \in \mathbb{C}$ is clearly holomorphic for all $x \in X$. Moreover, $\varphi_\lambda(X) \subseteq \mathbb{R}$ for all $\lambda \in (t-1, t+1) = \Lambda \cap \mathbb{R} = M$. So Theorem 13.10.3 applies, giving the desired outcome. □

We will need another real analyticity result similar to Theorem 13.10.3.

Theorem 13.10.5. *Let $T : X \to X$ be a topologically transitive, open, distance expanding dynamical system. Suppose that Λ is an open subset of $\mathbb{C}^k$ for some $k \in \mathbb{N}$. Assume also that $\varphi_\lambda \in H_\alpha(X)$ for every $\lambda \in \Lambda$, that $\sup\{\|\varphi_\lambda\|_\alpha : \lambda \in \Lambda\} < \infty$, and that the function $\Lambda \ni \lambda \longmapsto \varphi_\lambda(x) \in \mathbb{C}$ is holomorphic for every $x \in X$. If M is a real-analytic submanifold of Λ such that*

$$\varphi_\lambda(X) \subseteq \mathbb{R}, \quad \forall \lambda \in M,$$

then the function

$$M \ni \lambda \longmapsto \varrho_{\varphi_\lambda}(x) \in \mathbb{R}$$

is real analytic for every $x \in X$, where $\varrho_{\varphi_\lambda} \in H_\alpha(X)$ arises from Lemma 13.7.5.

Proof. Like the proof of Theorem 13.10.3, the current theorem follows immediately from Theorem 13.8.20, Theorem 13.10.2 and Kato–Rellich theorem for perturbations of linear operators (Theorem A.2.5), applied with the vector $v_0 = \mathbb{1}$. □

It follows from Corollary 13.10.4 that both the first and second derivatives $\mathrm{P}'(t\varphi+\psi)$ and $\mathrm{P}''(t\varphi + \psi)$ exist. We will now express them in familiar dynamical terms.

Theorem 13.10.6. *Let $T : X \to X$ be a topologically transitive, open, distance expanding dynamical system. If $\varphi, \psi : X \to \mathbb{R}$ are Hölder continuous functions, then for every $s \in \mathbb{R}$,*

$$\frac{d}{dt}\mathrm{P}(T, t\varphi + \psi)\Big|_{t=s} = \int_X \varphi \, d\mu_{s\varphi+\psi}, \tag{13.135}$$

where $\mu_{s\varphi+\psi}$ is the unique T-invariant Gibbs state for $s\varphi + \psi$ (see Theorem 13.7.7).

Proof. With the notation and terminology of Theorems 13.10.2–13.10.3 and Corollary 13.10.4, it follows from these results and the proof of Theorem 13.10.2 (particularly formula (13.133) and the fact that $\mathcal{L}_{\lambda,x}(g)(x) = \mathcal{L}_{\varphi_\lambda}(g)(x))$ that the derivative of the operator $\mathcal{L}_\lambda = \mathcal{L}_{\varphi_\lambda}$, $\lambda \in \Lambda = B_{\mathbb{C}}(s, 1)$, is given by the formula

$$\frac{d}{d\lambda}\mathcal{L}_\lambda(g)(x) = \sum_{y \in T^{-1}(x)} g(y)\frac{d}{d\lambda}e^{\varphi_\lambda(y)} - \sum_{y \in T^{-1}(x)} g(y)\varphi(y)e^{\varphi_\lambda(y)} = \mathcal{L}_\lambda(g\varphi)(x).$$

So

$$\frac{d}{d\lambda}\mathcal{L}_\lambda(g) = \mathcal{L}_\lambda(g\varphi). \tag{13.136}$$

For every $t \in \mathbb{R}$, set $\mu_t := \mu_{t\varphi+\psi}$ and $m_t := m_{t\varphi+\psi}$. Then

$$\mathrm{P}(T, t\varphi + \psi) = \mathrm{P}(T, t(\varphi - \mu_s(\varphi)) + \psi + t\mu_s(\varphi)) = \mathrm{P}(T, t(\varphi - \mu_s(\varphi)) + \psi) + t\mu_s(\varphi).$$

Therefore,

$$\frac{d}{dt}\mathrm{P}(T, t\varphi + \psi)\Big|_{t=s} = \frac{d}{dt}\mathrm{P}(T, t(\varphi - \mu_s(\varphi)) + \psi)\Big|_{t=s} + \mu_s(\varphi). \tag{13.137}$$

Furthermore, as $s(\varphi - \mu_s(\varphi)) + \psi - (s\varphi + \psi) = -s\mu_s(\varphi)$, Theorem 13.7.17 asserts that

$$\mu_{s\varphi+\psi} = \mu_{s(\varphi-\mu_s(\varphi))+\psi}. \tag{13.138}$$

Replacing φ by $\varphi - \mu_s(\varphi)$, it ensues from (13.137)–(13.138) that we may assume that

$$\mu_s(\varphi) = 0. \tag{13.139}$$

Proving (13.135) then reduces to showing that

$$\frac{d}{dt}\mathrm{P}(T, t\varphi + \psi)\Big|_{t=s} = 0. \tag{13.140}$$

Moreover, since $\mathrm{P}(T, t\varphi + \psi - \mathrm{P}(T, s\varphi + \psi)) = \mathrm{P}(T, t\varphi + \psi) - \mathrm{P}(T, s\varphi + \psi)$, we get

$$\frac{d}{dt}\mathrm{P}(T, t\varphi + \psi)\Big|_{t=s} = \frac{d}{dt}\mathrm{P}(T, t\varphi + \psi - \mathrm{P}(T, s\varphi + \psi))\Big|_{t=s}. \tag{13.141}$$

Also, $\mu_{s\varphi+\psi} = \mu_{s\varphi+\psi-\mathrm{P}(T,s\varphi+\psi)}$ by Theorem 13.7.17, whence

$$\mu_{s\varphi+\psi}(\varphi) = \mu_{s\varphi+\psi-\mathrm{P}(T,s\varphi+\psi)}(\varphi). \tag{13.142}$$

By (13.141)–(13.142), we may replace ψ by $\psi - \mathrm{P}(T, s\varphi + \psi)$ and assume without losing generality that

$$\mathrm{P}(T, s\varphi + \psi) = 0. \tag{13.143}$$

Set $\mathrm{P}(t) := \mathrm{P}(T, t\varphi + \psi)$ and $\varrho_t := \varrho_{t\varphi+\psi}$, where $\varrho_{t\varphi+\psi}$ arises from Lemma 13.7.5. Take $t \in (s-1, s+1)$. By Theorem 13.10.5, the function $(s-1, s+1) \ni t \longmapsto \varrho_t(x) \in \mathbb{R}$ is real analytic for every $x \in X$. Differentiating with respect to t the equation

$$\mathcal{L}_t(\varrho_t) = e^{\mathrm{P}(t)}\varrho_t$$

with the help of (13.136), we obtain that

$$\mathcal{L}_t(\varrho_t\varphi) + \mathcal{L}_t(\varrho_t') = e^{\mathrm{P}(t)}(\mathrm{P}'(t)\varrho_t + \varrho_t'). \tag{13.144}$$

Taking $t = s$ and using (13.143), we deduce that

$$\mathcal{L}_s(\varrho_s\varphi) + \mathcal{L}_s(\varrho_s') = \mathrm{P}'(s)\varrho_s + \varrho_s'.$$

Integrating this equality with respect to the measure m_s, we have

$$m_s(\mathcal{L}_s(\varrho_s\varphi)) + m_s(\mathcal{L}_s(\varrho_s')) = \mathrm{P}'(s)m_s(\varrho_s) + m_s(\varrho_s').$$

Using Corollary 13.6.4, this equation boils down to

$$m_s(\varrho_s\varphi) + m_s(\varrho_s') = \mathrm{P}'(s)m_s(\varrho_s) + m_s(\varrho_s').$$

By virtue of Theorem 13.7.7 and Lemma 13.7.5, this reduces to

$$\mu_s(\varphi) + m_s(\varrho_s') = \mathrm{P}'(s) + m_s(\varrho_s').$$

Invoking (13.139), we conclude that

$$\mathrm{P}'(s) = \mu_s(\varphi) = 0,$$

meaning that (13.140) holds, as sought. □

An alternative proof, analogous to that of Theorems 16.4.10 and 20.2.4 and based on the variational principle (Theorem 12.1.1), is left to the reader as an exercise.

We now provide a formula for the second derivative of the topological pressure function for those systems that are additionally exact and Hölder continuous.

Theorem 13.10.7. *Let $T : X \to X$ be a topologically exact, open, distance expanding, Hölder continuous system. If $\varphi, \psi : X \to \mathbb{R}$ are Hölder continuous functions, then for every $s \in \mathbb{R}$,*

$$\frac{d^2}{dt^2}\mathrm{P}(T, t\varphi + \psi)\Big|_{t=s} = \sigma^2_{\mu_{s\varphi+\psi,\infty}}(\varphi),$$

where $\sigma^2_{\mu_{s\varphi+\psi,\infty}}(\varphi)$ is the asymptotic variance of φ, is defined in (13.86) *and satisfies* (13.87) *in Proposition* 13.9.4.

Proof. Adapting the proof of Theorem 13.10.6 we may again assume without loss of generality that both (13.139) and (13.143) hold. Equation (13.144) holds, too. Differentiating it we get, with the use of (13.136), that

$$\mathcal{L}_t(\varrho_t\varphi^2) + 2\mathcal{L}_t(\varrho_t'\varphi) + \mathcal{L}_t(\varrho_t'') = e^{\mathrm{P}(t)}\mathrm{P}'(t)(\mathrm{P}'(t)\varrho_t + \varrho_t') + e^{\mathrm{P}(t)}(\mathrm{P}''(t)\varrho_t + \mathrm{P}'(t)\varrho_t' + \varrho_t'').$$

Setting $t = s$ and using (13.140), we deduce that

$$\mathcal{L}_s(\varrho_s\varphi^2) + 2\mathcal{L}_s(\varrho_s'\varphi) + \mathcal{L}_s(\varrho_s'') = \mathrm{P}''(s)\varrho_s + \varrho_s''.$$

Integrating this equality with respect to the measure m_s, we obtain

$$\mu_s(\varphi^2) + 2m_s(\varrho_s'\varphi) + m_s(\varrho_s'') = \mathrm{P}''(s) + m_s(\varrho_s'').$$

Equivalently,

$$\mathrm{P}''(s) = \mu_s(\varphi^2) + 2m_s(\varrho_s'\varphi). \tag{13.145}$$

Now, fix any $n \in \mathbb{N}$. The iterate $T^n : X \to X$ is also a topologically exact, open, distance expanding, Hölder continuous dynamical system and $S_n\varphi, S_n\psi : X \to \mathbb{R}$ are

Hölder continuous potentials. So is the potential $tS_n\varphi + S_n\psi = S_n(t\varphi + \psi)$. According to Theorem 11.1.22,

$$\mathrm{P}(T^n, tS_n\varphi + S_n\psi) = n\mathrm{P}(T, t\varphi + \psi).$$

Moreover, per Corollaries 13.6.8 and 13.7.10, we know that

$$m_{T^n,sS_n\varphi+S_n\psi} = m_{T,s\varphi+\psi}, \quad \mu_{T^n,sS_n\varphi+S_n\psi} = \mu_{T,s\varphi+\psi} \quad \text{and} \quad \varrho_{T^n,sS_n\varphi+S_n\psi} = \varrho_{T,s\varphi+\psi}.$$

Applying (13.145) to the iterate T^n and the potential $sS_n\varphi + S_n\psi$, we deduce that

$$n\mathrm{P}''(s) = \mu_s((S_n\varphi)^2) + 2m_s(\varrho_s' S_n\varphi).$$

Equivalently,

$$\mathrm{P}''(s) = \frac{1}{n}\mu_s((S_n\varphi)^2) + 2m_s\Big(\varrho_s'\frac{1}{n}S_n\varphi\Big). \tag{13.146}$$

The sequence $(\varrho_s'\frac{1}{n}S_n\varphi)_{n=1}^{\infty}$ is uniformly bounded by $\|\varrho_s'\|_\infty \cdot \|\varphi\|_\infty$, and by Birkhoff's Ergodic theorem (Theorem 8.2.11) and (13.139) it converges μ_s-a. e. to the zero function. But since the measures m_s and μ_s are equivalent, that sequence converges m_s-a. e. to the zero function as well. Therefore, it follows from Lebesgue's dominated convergence theorem (Theorem A.1.38) that

$$\lim_{n\to\infty} 2m_s\Big(\varrho_s'\frac{1}{n}S_n\varphi\Big) = 0.$$

Thus, (13.146) yields

$$\mathrm{P}''(s) = \lim_{n\to\infty}\frac{1}{n}\mu_s((S_n\varphi)^2) = \sigma^2_{\mu_{s\varphi+\psi,\infty}}(\varphi).$$

This is the sought result. Note that the last equality follows from the definition of $\sigma^2_{\mu_{s\varphi+\psi,\infty}}(\varphi)$ (cf. (13.86)) and the reduction of the proof to the case where $\mu_{s\varphi+\psi}(\varphi) = 0$ per (13.139). Proposition 13.9.4 applies according to Theorem 13.9.1. □

13.11 Exercises

Exercise 13.11.1. Let (X, d) be a metric space. A function $\varphi : X \to \mathbb{R}$ is said to be locally Hölder continuous with exponent $\alpha > 0$ if for every $z \in X$ there exist constants $c_z \geq 0$ and $\delta_z > 0$ such that

$$|\varphi(x) - \varphi(y)| \leq c_z[d(x, y)]^\alpha, \quad \forall x, y \in B(z, \delta_z).$$

Show that local Hölder continuity and (global) Hölder continuity are equivalent notions when the space X is compact.

Exercise 13.11.2. Let (X, d) be a compact metric space and $\alpha > 0$. Show that the space $(H_\alpha(X), \|\cdot\|_\alpha)$ is a Banach space. Show further that $(H_\alpha(X), \|\cdot\|_\alpha)$ is a vector subspace of the Banach space $(C(X), \|\cdot\|_\infty)$. Finally, show that $(H_\alpha(X), \|\cdot\|_\alpha)$ is not a Banach subspace of $(C(X), \|\cdot\|_\infty)$.

Exercise 13.11.3. Let (X, d) be a compact metric space. If $\varphi \in H_\alpha(X)$ and $\psi \in H_\beta(X)$, then show that $\varphi, \psi \in H_{\min\{\alpha,\beta\}}(X)$.

Exercise 13.11.4. Let μ be a Gibbs state for a potential φ and for some constant P. Suppose that a Borel probability measure ν is equivalent to μ and is such that $0 < \operatorname{ess\,inf}(\frac{d\mu}{d\nu}) \leq \operatorname{ess\,sup}(\frac{d\mu}{d\nu}) < \infty$. Show that ν is also a Gibbs state for the same potential φ and for the same constant P.

Exercise 13.11.5. Let $T : X \to X$ be a dynamical system with at least one periodic point. Show that if a function $g : X \to \mathbb{R}$ is cohomologous to a constant c, then c is unique.

Exercise 13.11.6. Prove the first change-of-variables formula in Proposition 13.3.1.

Exercise 13.11.7. Prove relation (13.22).

Exercise 13.11.8. Show that $\mathfrak{J}_m^{-1}(z)$ is independent of x for $z \in T_x^{-1}(B(T(x), \xi))$.

Exercise 13.11.9. Consider the transfer-type operator $\mathcal{L}_\varphi : L^1(m) \to L^1(m)$ introduced in Definition 13.5.1 and its dual $\mathcal{L}_\varphi^* : L^1(m)^* \to L^1(m)^*$.
(a) Show that $\mathcal{L}_\varphi$ is positive, bounded and linear.
(b) Prove that $\mathcal{L}_\varphi(C(X)) \subseteq C(X)$.
(c) Show that the dual operator $\mathcal{L}_\varphi^*$ is well defined, positive, bounded and linear.
(d) Prove that the functional $m(g) = \int_X g\, dm$ is positive, bounded and linear.

Exercise 13.11.10. Prove part (c) of Corollary 13.6.5.

Exercise 13.11.11. Prove Corollary 13.6.10.

Exercise 13.11.12. Let $(X, \mathcal{F})$ be a measurable space and let $(\alpha_n)_{n=1}^\infty$ be a sequence of increasingly finer countable measurable partitions generating $\mathcal{F}$, i. e., $\sigma(\cup_{n=1}^\infty \alpha_n) = \mathcal{F}$. Let also μ be a probability measure on $(X, \mathcal{F})$.
(a) Let $A, B \in \bigcup_{n\in\mathbb{N}} \alpha_n$. Prove that $A \cap B = A$ or $A \cap B = B$ or $A \cap B = \emptyset$.
(b) Without loss of generality, assume that $\emptyset \in \alpha_1$. Let α^+ be the family of all sets $Z \subseteq X$ that can be expressed as a countable union of sets in $\bigcup_{n\in\mathbb{N}} \alpha_n$, i. e., for which there exists a countable set $\mathcal{I}^+ \subseteq \bigcup_{n\in\mathbb{N}} \alpha_n$ such that $Z = \bigcup_{I\in\mathcal{I}^+} I$. Show that any such set Z can be expressed as a countable union of disjoint sets in $\bigcup_{n\in\mathbb{N}} \alpha_n$. Show also that α^+ is closed under countable unions and under finite intersections.
(c) Let

$$\alpha^- := \{X \setminus Z \mid Z \in \alpha^+\}.$$

Deduce that α^- is closed under countable intersections and under finite unions.

(d) Let

$$\widehat{\alpha} := \left\{ W \in \mathcal{F} \,\middle|\, \begin{array}{l} \forall \varepsilon > 0 \; \exists F_\varepsilon \in \alpha^- \; \exists G_\varepsilon \in \alpha^+ \text{ such that} \\ F_\varepsilon \subseteq W \subseteq G_\varepsilon \text{ and } \mu(G_\varepsilon \setminus F_\varepsilon) < \varepsilon \end{array} \right\}.$$

Observe that

$$\bigcup_{n \in \mathbb{N}} \alpha_n \subseteq \alpha^+ \cap \alpha^- \subseteq \widehat{\alpha}.$$

(e) Demonstrate that $\widehat{\alpha}$ is a σ-algebra.

(f) Infer that $\widehat{\alpha} = \mathcal{F}$.

Exercise 13.11.13. Let E be a finite set having at least two elements and $P : E \to (0,1)$ a probability vector. Let μ_P be the Bernoulli measure on the shift space $E^{\mathbb{N}}$, as defined in Examples 8.1.14 and 8.2.32. Show that μ_P is the unique shift-invariant Gibbs state and the unique equilibrium state for the potential $\varphi_P : E^{\mathbb{N}} \to \mathbb{R}$ given by the formula

$$\varphi_P(\omega) = \log P_{\omega_1}.$$

More generally, let $A : E \times E \to (0,1)$ be a stochastic matrix and μ_A be the Markov chain measure on the shift space $E^{\mathbb{N}}$, as defined in Exercise 8.5.48. Show that μ_A is the unique shift-invariant Gibbs state and the unique equilibrium state for the potential $\varphi_A : E^{\mathbb{N}} \to \mathbb{R}$ given by the formula

$$\varphi_A(\omega) = \log A_{\omega_1 \omega_2}.$$

Exercise 13.11.14. Establish that the equilibrium states that you identified in Exercise 12.4.2 are Gibbs states for the potential φ dealt with therein. Compare this exercise with Exercise 13.11.13.

Exercise 13.11.15. Let $T : X \to X$ be an open, distance expanding system and $\varphi : X \to \mathbb{R}$ a continuous potential. Let also μ be a Borel probability measure on X. Fix any $k \in \mathbb{N}$. Prove the following statements:

(a) If μ is a Gibbs state for the potential φ on the system T, then μ is a Gibbs state for the potential $S_k\varphi$ on the system T^k.

(b) If μ is an equilibrium state for the potential φ on the system T, then μ is an equilibrium state for the potential $S_k\varphi$ on the system T^k.

Exercise 13.11.16. Inspiring yourself from Lemma 13.9.3, show the following generalization of Proposition 13.9.4. Suppose that $T : (X, \mathcal{F}, \mu) \to (X, \mathcal{F}, \mu)$ is a measure-preserving transformation of a probability space $(X, \mathcal{F}, \mu)$ and that $f, g \in L^2(\mu)$. Show that if

$$\sum_{n=1}^{\infty} C_{\mu,n}(f,g) < \infty \quad \text{and} \quad \sum_{n=1}^{\infty} C_{\mu,n}(g,f) < \infty, \tag{13.147}$$

then the limit

$$\sigma^2_{\mu,\infty}(f,g) := \lim_{n\to\infty} \frac{1}{n} \int_X \left[\sum_{k=0}^{n-1} (g - \mu(g)\mathbb{1}) \circ T^k\right] \cdot \left[\sum_{j=0}^{n-1} (f - \mu(f)\mathbb{1}) \circ T^j\right] d\mu$$

exists and

$$\begin{aligned}\sigma^2_{\mu,\infty}(f,g) = \int_X (f - \mu(f)\mathbb{1})(g - \mu(g)\mathbb{1})\, d\mu + \sum_{n=1}^{\infty} \int_X (f - \mu(f)\mathbb{1}) \cdot [(g - \mu(g)\mathbb{1}) \circ T^n]\, d\mu \\ + \sum_{n=1}^{\infty} \int_X (g - \mu(g)\mathbb{1}) \cdot [(f - \mu(f)\mathbb{1}) \circ T^n]\, d\mu.\end{aligned} \tag{13.148}$$

The number $\sigma^2_{\mu,\infty}(f,g)$ is called the *asymptotic covariance* of f and g (with respect to T and μ). It is in general different than the usual *covariance*

$$\sigma^2_{\mu}(f,g) := \int_X (f - \mu(f)\mathbb{1})(g - \mu(g)\mathbb{1})\, d\mu.$$

However, if the random variables $f \circ T^n$, $g \circ T^n$, $n \geq 0$, are independent with respect to μ, then $\sigma^2_{\mu,\infty}(f,g) = \sigma^2_{\mu}(f,g)$. Moreover, $\sigma^2_{\mu,\infty}(g,g) = \sigma^2_{\mu,\infty}(g)$ is simply the asymptotic variance.

14 Lasota–Yorke maps

In this chapter, we study Lasota–Yorke maps. Roughly speaking, Lasota–Yorke maps are piecewise monotone, piecewise differentiable expanding maps of a compact interval. They have finitely many pieces and the reciprocal of the absolute value of their derivative is of bounded variation. These maps appeared for the first time in the literature in Andrzej Lasota and James Yorke's 1973 paper [65]. After defining them in Section 14.1, we describe the form of their transfer (Perron–Frobenius) operator in Section 14.2 and use this operator in Section 14.3 to establish the existence of an invariant measure absolutely continuous with respect to the Lebesgue measure. Moreover, we demonstrate in Section 14.4 the exponential decay of correlations with respect to that invariant measure, under two additional assumptions:

(1) the transfer operator has 1 as its only eigenvalue of modulus one, and
(2) the eigenspace of the eigenvalue 1 is one-dimensional.

Lasota–Yorke maps have attracted dynamicists' attention ever since they were introduced. After Lasota and Yorke's original paper, Gerhard Keller and Franz Hofbauer's article [54] substantially contributed to their rise in popularity. Marek Rychlik [114] successfully extended the class of Lasota–Yorke maps to the ones having infinitely many pieces of monotonicity. A book exposition of Lasota–Yorke maps is provided in [13].

Finally, it is worthwhile to point out that it is in Lasota and Yorke's groundbreaking paper [65] that a Ionescu-Tulcea and Marinescu inequality was for the first time used in dynamical systems.

14.1 Definition

Let $I = [a, b] \subseteq \mathbb{R}$ be a nondegenerate compact interval endowed with the normalized Lebesgue measure λ, i. e., $\lambda(I) = 1$. In the sequel, a partition $\mathcal{Q}$ of I shall be a finite set $\{x_0, x_1, \ldots, x_n\}$ such that $x_{i-1} < x_i$ for all $1 \le i \le n$, with $a = x_0$ and $x_n = b$. The set of compact subintervals determined by $\mathcal{Q}$ will be denoted by $\overline{\mathcal{Q}}$, i. e., $\overline{\mathcal{Q}} = \{J_1, J_2, \ldots, J_n\}$, where $J_i := [x_{i-1}, x_i]$. The restriction of T to a subinterval $J \in \overline{\mathcal{Q}}$ will be denoted by $T_J := T|_J : J \to T(J)$.

Definition 14.1.1. A map $T : I \to I$ is called *Lasota–Yorke* if:

(a) T is piecewise expanding, meaning that there exists a partition $\mathcal{P}$ of I and a constant $\alpha > 1$ such that T_J is differentiable for every $J \in \overline{\mathcal{P}}$ and $|T_J'(x)| \ge \alpha$ for all $x \in J$ and all $J \in \overline{\mathcal{P}}$. (The derivative at an endpoint of a subinterval $J \in \overline{\mathcal{P}}$ is the one-sided derivative defined by approaching that endpoint from within J.)
(b) $1/|T_J'|$ is a function of bounded variation for every $J \in \overline{\mathcal{P}}$. (For more information on functions of bounded variation, see Exercises 14.5.8–14.5.17.)

https://doi.org/10.1515/9783110702699-014

14.2 Transfer operator

Let $(X, \mathcal{B}(X))$ be a Borel measurable space and let $T : X \to X$ be a Borel measurable transformation. Recall from Sections 10.1 and 8.1 that a Borel probability measure m on X is quasi-T-invariant if $m \circ T^{-1} \ll m$ and T-invariant if $m \circ T^{-1} = m$. Clearly, any T-invariant measure is quasi-T-invariant. Though the converse is generally not true, we have learned in Section 13.4 a way of constructing an invariant measure from a quasi-invariant one. Suppose that m is a quasi-T-invariant measure. Let $g : X \to \mathbb{R}$ be any m-integrable nonnegative function. Define the finite Borel measure $gm(A) := \int_A g\,dm$ for every $A \in \mathcal{B}(X)$. Observe that $gm \ll m$ and thereby $gm \circ T^{-1} \ll m \circ T^{-1} \ll m$. This allows us to define a transfer operator $\mathcal{L}_m : L^1(m) \to L^1(m)$ as follows. Using the Radon–Nikodym theorem (Theorem A.1.51), for every nonnegative $g \in L^1(m)$ define the function $\mathcal{L}_m g$ (or, more formally, $\mathcal{L}_m(g)$) by

$$\mathcal{L}_m g := \frac{d(gm \circ T^{-1})}{dm}.$$

The following equality characterizes this transfer operator:

$$\int_A \mathcal{L}_m g\,dm = \int_{T^{-1}(A)} g\,dm, \quad \forall A \in \mathcal{B}(X). \tag{14.1}$$

For any $g \in L^1(m)$, set $\mathcal{L}_m g := \mathcal{L}_m(g^+) - \mathcal{L}_m(g^-)$. Defining $\mathcal{L}_m : L^1(m) \to L^1(m)$ in this way makes the transfer operator a positive bounded linear operator that conforms to (14.1) for all $g \in L^1(m)$ and whose norm satisfies $\|\mathcal{L}_m\|_1 \le 1$. A counterpart of Theorem 13.4.1 for Lasota-Yorke maps provides a characterization of the nonnegative fixed points of $\mathcal{L}_m$. Indeed, it affirms that if m is a quasi-T-invariant measure and if $g \in L^1(m)$ is such that $g \ge 0$ and $\int_X g\,dm = 1$, then $\mathcal{L}_m g = g$ if and only if $gm \circ T^{-1} = gm$.

Clearly, the Lebesgue measure λ is quasi-T-invariant for any Lasota–Yorke map T since $T_J : J \to T(J)$ is invertible and satisfies $|(T_J^{-1})'(x)| = |T_J'(T_J^{-1}(x))|^{-1} \le \alpha^{-1} < 1$ for all $x \in T(J)$. The transfer operator $\mathcal{L}_\lambda$ associated to a Lasota–Yorke map has the following form.

Theorem 14.2.1. *If $T : I \to I$ is a Lasota–Yorke map, then for all $g \in L^1(\lambda)$ we have*

$$\mathcal{L}_\lambda g(x) = \sum_{J \in \overline{\mathcal{P}}} g(T_J^{-1}(x))|(T_J^{-1})'(x)|\mathbb{1}_{T(J)}(x) = \sum_{y \in T^{-1}(x)} \frac{g(y)}{|T'(y)|} \quad \textit{for } \lambda\textit{-a.e. } x \in I.$$

Proof. Let $B \subseteq I$ be a Borel set. Since λ is atomless and T is bounded-to-one, the change-of-variables formula (13.19) applied to each invertible map T_J yields

$$\begin{aligned}\int_B \mathcal{L}_\lambda g\,d\lambda &= \int_{T^{-1}(B)} g\,d\lambda = \int_{\bigcup_{J \in \overline{\mathcal{P}}} T^{-1}(B) \cap J} g\,d\lambda \\ &= \sum_{J \in \overline{\mathcal{P}}} \int_{T^{-1}(B) \cap J} g\,d\lambda = \sum_{J \in \overline{\mathcal{P}}} \int_{T_J^{-1}(B \cap T(J))} g\,d\lambda\end{aligned}$$

$$= \sum_{J \in \overline{\mathcal{P}}} \int_{B \cap T(J)} g(T_J^{-1}(x))|(T_J^{-1})'(x)|\, d\lambda(x)$$
$$= \sum_{J \in \overline{\mathcal{P}}} \int_B g(T_J^{-1}(x))|(T_J^{-1})'(x)|\mathbb{1}_{T(J)}(x)\, d\lambda(x)$$
$$= \int_B \sum_{J \in \overline{\mathcal{P}}} g(T_J^{-1}(x))|T_J'(T_J^{-1}(x))|^{-1}\mathbb{1}_{T(J)}(x)\, d\lambda(x)$$
$$= \int_B \sum_{y \in T^{-1}(x)} g(y)|T'(y)|^{-1}\, d\lambda(x).$$

Since the Borel set B is arbitrary, the result follows immediately. □

The iterates of the transfer operator have the following form.

Corollary 14.2.2. *If $T : I \to I$ is a Lasota–Yorke map, then for every $n \in \mathbb{N}$ and all $g \in L^1(\lambda)$ we have*

$$\mathcal{L}_\lambda^n g(x) = \sum_{J \in \overline{\mathcal{P}}^n} g(T_J^{-n}(x))|(T_J^{-n})'(x)|\mathbb{1}_{T^n(J)}(x) = \sum_{y \in T^{-n}(x)} \frac{g(y)}{|(T^n)'(y)|} \quad \text{for } \lambda\text{-a.e. } x \in I.$$

Proof. The proof by induction is left to the reader as an exercise. □

The advantage of the first expression for $\mathcal{L}_\lambda^n g(x)$ is that it is defined for all $x \in I$, whereas the second expression generally is not (as $(T^n)'$ is not defined at the endpoints of the subintervals in $\overline{\mathcal{P}}^n$, except at a and b). For this reason, we will henceforth define $\mathcal{L}_\lambda^n g(x)$ as the first expression.

We now show that $\mathcal{L}_\lambda g$ is of bounded variation whenever g is.

Lemma 14.2.3. *Let $T : I \to I$ be a Lasota–Yorke map and $\delta = \min_{J \in \overline{\mathcal{P}}} \lambda(J)$. Then for every $g \in \mathrm{BV}(I)$, we have that*

$$V_I(\mathcal{L}_\lambda g) \le A V_I(g) + B\|g\|_1,$$

where $A = \frac{3}{\alpha} + \max_{J \in \overline{\mathcal{P}}} V_J(|T_J'|^{-1})$ and $B = \frac{2}{\alpha\delta} + \frac{1}{\delta} \max_{J \in \overline{\mathcal{P}}} V_J(|T_J'|^{-1})$. Recall that α arises from Definition 14.1.1 *while* $\mathrm{BV}(I)$ *and V_I are defined in Exercise* 14.5.8.

Proof. Let $\mathcal{Q}$ be an arbitrary partition of I, generally unrelated to the partition $\mathcal{P}$ defining T as a Lasota–Yorke map. For each $K \in \overline{\mathcal{Q}}$, write $K = [a_K, b_K]$. Using the triangle inequality, we get

$$\sum_{K \in \overline{\mathcal{Q}}} |\mathcal{L}_\lambda g(b_K) - \mathcal{L}_\lambda g(a_K)| = \sum_{K \in \overline{\mathcal{Q}}} \Bigg| \sum_{J \in \overline{\mathcal{P}}} g(T_J^{-1}(b_K))|(T_J^{-1})'(b_K)|\mathbb{1}_{T(J)}(b_K)$$
$$- \sum_{J \in \overline{\mathcal{P}}} g(T_J^{-1}(a_K))|(T_J^{-1})'(a_K)|\mathbb{1}_{T(J)}(a_K)\Bigg|$$

$$\leq \sum_{K\in\overline{\mathcal{Q}}}\sum_{J\in\overline{\mathcal{P}}}\Big|g(T_J^{-1}(b_K))|(T_J^{-1})'(b_K)|\mathbb{1}_{T(J)}(b_K)$$
$$- g(T_J^{-1}(a_K))|(T_J^{-1})'(a_K)|\mathbb{1}_{T(J)}(a_K)\Big|.$$

We divide the sum on the right-hand side into three subsums:

$\sum_1$ = sum over the pairs of intervals for which $\mathbb{1}_{T(J)}(b_K) = \mathbb{1}_{T(J)}(a_K) = 1$.

$\sum_2$ = sum over the pairs for which $\mathbb{1}_{T(J)}(b_K) = 1$ and $\mathbb{1}_{T(J)}(a_K) = 0$.

$\sum_3$ = sum over the pairs for which $\mathbb{1}_{T(J)}(b_K) = 0$ and $\mathbb{1}_{T(J)}(a_K) = 1$.

Then

$$\begin{aligned}
\sum_1 &= \sum_{\substack{(K,J)\in\overline{\mathcal{Q}}\times\overline{\mathcal{P}}:\\ a_K,b_K\in T(J)}} \Big|g(T_J^{-1}(b_K))|(T_J^{-1})'(b_K)| - g(T_J^{-1}(a_K))|(T_J^{-1})'(a_K)|\Big| \\
&\leq \sum_{\substack{(J,K)\in\overline{\mathcal{P}}\times\overline{\mathcal{Q}}:\\ a_K,b_K\in T(J)}} |g(T_J^{-1}(b_K))|\Big||(T_J^{-1})'(b_K)| - |(T_J^{-1})'(a_K)|\Big| \\
&\quad + \sum_{\substack{(J,K)\in\overline{\mathcal{P}}\times\overline{\mathcal{Q}}:\\ a_K,b_K\in T(J)}} |(T_J^{-1})'(a_K)|\Big|g(T_J^{-1}(b_K)) - g(T_J^{-1}(a_K))\Big| \\
&= \sum_{J\in\overline{\mathcal{P}}}\sum_{\substack{K\in\overline{\mathcal{Q}}:\\ a_K,b_K\in T(J)}} |g(T_J^{-1}(b_K))|\Big||T_J'(T_J^{-1}(b_K))|^{-1} - |T_J'(T_J^{-1}(a_K))|^{-1}\Big| \\
&\quad + \sum_{J\in\overline{\mathcal{P}}}\sum_{\substack{K\in\overline{\mathcal{Q}}:\\ a_K,b_K\in T(J)}} |T_J'(T_J^{-1}(a_K))|^{-1}\Big|g(T_J^{-1}(b_K)) - g(T_J^{-1}(a_K))\Big| \\
&\leq \sum_{J\in\overline{\mathcal{P}}} \sup_{x\in J}|g(x)|\cdot V_J(|T_J'|^{-1}) + \sum_{J\in\overline{\mathcal{P}}}\max_{x\in J}|T_J'(x)|^{-1}\cdot V_J(g).
\end{aligned}$$

Using Exercises 14.5.15 and 14.5.8(d) successively, we obtain

$$\begin{aligned}
\sum_1 &\leq \max_{J\in\overline{\mathcal{P}}} V_J(|T_J'|^{-1})\cdot\sum_{J\in\overline{\mathcal{P}}}\Big(V_J(|g|) + \frac{1}{\lambda(J)}\int_J |g|\,d\lambda\Big) + \max_{J\in\overline{\mathcal{P}}}\max_{x\in J}|T_J'(x)|^{-1}\cdot\sum_{J\in\overline{\mathcal{P}}} V_J(g) \\
&\leq \max_{J\in\overline{\mathcal{P}}} V_J(|T_J'|^{-1})\Big(V_I(g) + \frac{1}{\delta}\int_I |g|\,d\lambda\Big) + \frac{1}{\alpha}V_I(g).
\end{aligned}$$

We consider $\sum_2$ and $\sum_3$ together. Since $\mathbb{1}_{T(J)}(b_K) = 1$ and $\mathbb{1}_{T(J)}(a_K) = 0$ if and only if $b_K \in T(J)$ and $a_K \notin T(J)$, for each $J \in \overline{\mathcal{P}}$ there is at most one $K(J) \in \overline{\mathcal{Q}}$ for which these two conditions are satisfied. Similarly, there exists at most one $K'(J)$ such that

$a_{K'} \in T(J)$ and $b_{K'} \notin T(J)$. Thus,

$$\sum_2 + \sum_3 \le \sum_{J\in\overline{\mathcal{P}}} \Big(|g(T_J^{-1}(b_{K(J)}))| \cdot |(T_J^{-1})'(b_{K(J)})| + |g(T_J^{-1}(a_{K'(J)}))| \cdot |(T_J^{-1})'(a_{K'(J)})| \Big)$$
$$\le \frac{1}{\alpha} \sum_{J\in\overline{\mathcal{P}}} \Big(|g(T_J^{-1}(b_{K(J)}))| + |g(T_J^{-1}(a_{K'(J)}))| \Big).$$

For each $J \in \overline{\mathcal{P}}$, choose $y_J \in J$ such that

$$|g(y_J)| \le \frac{1}{\lambda(J)} \int_J |g|\, d\lambda.$$

Then

$$\sum_2 + \sum_3 \le \frac{1}{\alpha} \sum_{J\in\overline{\mathcal{P}}} \Big(2|g(y_J)| + |g(T_J^{-1}(b_{K(J)})) - g(y_J)| + |g(T_J^{-1}(a_{K'(J)})) - g(y_J)| \Big)$$
$$\le \frac{2}{\alpha} \sum_{J\in\overline{\mathcal{P}}} \Big(\frac{1}{\lambda(J)} \int_J |g|\, d\lambda + V_J(g) \Big) \le \frac{2}{\alpha} \Big[\frac{1}{\delta} \int_I |g|\, d\lambda + V_I(g) \Big].$$

Hence,

$$\sum_1 + \sum_2 + \sum_3 \le \Big(\frac{3}{\alpha} + \max_{J\in\overline{\mathcal{P}}} V_J(|T_J'|^{-1}) \Big) V_I(g) + \Big(\frac{2}{\alpha\delta} + \frac{1}{\delta} \max_{J\in\overline{\mathcal{P}}} V_J(|T_J'|^{-1}) \Big) \int_I |g|\, d\lambda.$$

Since the partition $\mathcal{Q}$ was chosen arbitrarily, we conclude that

$$V_I(\mathcal{L}_\lambda g) \le \Big(\frac{3}{\alpha} + \max_{J\in\overline{\mathcal{P}}} V_J(|T_J'|^{-1}) \Big) V_I(g) + \Big(\frac{2}{\alpha\delta} + \frac{1}{\delta} \max_{J\in\overline{\mathcal{P}}} V_J(|T_J'|^{-1}) \Big) \|g\|_1. \qquad \square$$

The set $\mathrm{BV}(I)$ of all functions of bounded variation on I is a vector space in itself (see Exercise 14.5.12). Functions of bounded variation are also Lebesgue integrable (see Exercise 14.5.14). By identifying functions that are λ-almost everywhere equal on I, the set $\mathrm{BV}(I)$ can be seen as a vector subspace of $L^1(\lambda)$. Under a norm $\|\cdot\|_{\mathrm{BV}} := \|\cdot\|_{\mathrm{BV}}^* + \|\cdot\|_1$, this set becomes a Banach space on its own. The definition and basic properties of that norm are given in Exercise 14.5.17. It is then desirable to derive a form of Ionescu-Tulcea and Marinescu inequality (see condition (c1) at the beginning of Appendix B). We make a first attempt now.

Corollary 14.2.4. *Let $T : I \to I$ be a Lasota–Yorke map. For every $g \in \mathrm{BV}(I)$,*

$$\|\mathcal{L}_\lambda g\|_{\mathrm{BV}} \le A\|g\|_{\mathrm{BV}} + (B+1)\|g\|_1,$$

where A and B are defined in Lemma 14.2.3.

Proof. Let $g \in \mathrm{BV}(I)$. Recall that $\|\mathcal{L}_\lambda\|_1 \leq 1$. It is also easy to see that if $h \in [g]_1$, then $\mathcal{L}_\lambda h \in [\mathcal{L}_\lambda g]_1$. Using Lemma 14.2.3, we obtain that

$$\begin{aligned}
\|\mathcal{L}_\lambda g\|_{\mathrm{BV}} &= \|\mathcal{L}_\lambda g\|_{\mathrm{BV}}^* + \|\mathcal{L}_\lambda g\|_1 \\
&= \inf_{H\in[\mathcal{L}_\lambda g]_1} V_I(H) + \|\mathcal{L}_\lambda g\|_1 \\
&\leq \inf_{h\in[g]_1\cap \mathrm{BV}(I)} V_I(\mathcal{L}_\lambda h) + \|\mathcal{L}_\lambda\|_1\|g\|_1 \\
&\leq \inf_{h\in[g]_1\cap \mathrm{BV}(I)} (AV_I(h) + B\|h\|_1) + \|g\|_1 \\
&= A \inf_{h\in[g]_1} V_I(h) + B\|g\|_1 + \|g\|_1 \\
&\leq A\|g\|_{\mathrm{BV}} + (B+1)\|g\|_1.
\end{aligned}$$

□

We would like $A < 1$. For this, we look at the iterates of T and of $\overline{\mathcal{P}}$.

Lemma 14.2.5. *Let T be a Lasota–Yorke map. For every $n \in \mathbb{N}$, we have*

$$W_n := \max_{J\in\overline{\mathcal{P}}^n} V_J(|(T_J^n)'|^{-1}) \leq \frac{n}{\alpha^{n-1}} W_1.$$

Proof. We proceed by induction on n. The case $n = 1$ is trivial. Now suppose that the inequality holds for some $n \in \mathbb{N}$. Let $J \in \overline{\mathcal{P}}^{n+1} = \overline{\mathcal{P}}^n \vee T^{-n}(\overline{\mathcal{P}})$ and let $\mathcal{Q}$ be a partition of J. Then

$$\begin{aligned}
&\sum_{K\in\overline{\mathcal{Q}}} \Big||(T_J^{n+1})'(b_K)|^{-1} - |(T_J^{n+1})'(a_K)|^{-1}\Big| \\
&\quad= \sum_{K\in\overline{\mathcal{Q}}} \Big||(T_J^n)'(T(b_K))|^{-1}\cdot|T_J'(b_K)|^{-1} - |(T_J^n)'(T(a_K))|^{-1}\cdot|T_J'(a_K)|^{-1}\Big| \\
&\quad\leq \sum_{K\in\overline{\mathcal{Q}}} \Big||(T_J^n)'(T(b_K))|^{-1} - |(T_J^n)'(T(a_K))|^{-1}\Big|\cdot|T_J'(b_K)|^{-1} \\
&\qquad+ \sum_{K\in\overline{\mathcal{Q}}} |(T_J^n)'(T(a_K))|^{-1}\Big||T_J'(b_K)|^{-1} - |T_J'(a_K)|^{-1}\Big| \\
&\quad\leq \frac{1}{\alpha}W_n + \frac{1}{\alpha^n}W_1 \leq \frac{1}{\alpha}\cdot\frac{n}{\alpha^{n-1}}W_1 + \frac{1}{\alpha^n}W_1 = \frac{n+1}{\alpha^n}W_1.
\end{aligned}$$

Since the partition $\mathcal{Q}$ of J is arbitrary, we deduce that $V_J(|(T_J^{n+1})'|^{-1}) \leq \frac{n+1}{\alpha^n}W_1$. As J was any subinterval in $\overline{\mathcal{P}}^{n+1}$, we conclude that $W_{n+1} \leq \frac{n+1}{\alpha^n}W_1$. □

As a consequence of Corollary 14.2.4 and Lemma 14.2.5, we get the following Ionescu-Tulcea and Marinescu inequality.

Proposition 14.2.6. *There exists $N \in \mathbb{N}$ and $C > 0$ such that*

$$\|\mathcal{L}_\lambda^N g\|_{\mathrm{BV}} \leq \frac{1}{2}\|g\|_{\mathrm{BV}} + C\|g\|_1, \quad \forall g \in \mathrm{BV}(I).$$

14.3 Existence of absolutely continuous invariant probability measures

We can now prove the main result about Lasota–Yorke maps.

Theorem 14.3.1. *Every Lasota–Yorke map $T : I \to I$ admits a T-invariant Borel probability measure μ which is absolutely continuous with respect to the Lebesgue measure λ on I. More precisely, $\mu = \rho\lambda$ for some $\rho \in \mathrm{BV}(I)$ such that $\rho \geq 0$, $\int_I \rho\, d\lambda = 1$ and $\mathcal{L}_\lambda \rho = \rho$.*

Proof. From Lemma B.1.6 in the proof of the Ionescu-Tulcea Marinescu theorem (see Appendix B), there exists a constant $M \geq 0$ such that $\|\mathcal{L}_\lambda^n \mathbb{1}\|_{\mathrm{BV}} \leq M$ for all $n \geq 0$. Consequently,

$$\left\| \frac{1}{n} \sum_{j=0}^{n-1} \mathcal{L}_\lambda^j \mathbb{1} \right\|_{\mathrm{BV}} \leq M, \quad \forall n \in \mathbb{N}.$$

So the sequence $(\frac{1}{n} \sum_{j=0}^{n-1} \mathcal{L}_\lambda^j \mathbb{1})_{n=1}^\infty$ contains a subsequence $(\frac{1}{n_k} \sum_{j=0}^{n_k-1} \mathcal{L}_\lambda^j \mathbb{1})_{k=1}^\infty$ converging to, say, ρ in $L^1(\lambda)$ since all bounded subsets of $\mathrm{BV}(I)$ are relatively compact in $L^1(\lambda)$. In fact, $\rho \in \mathrm{BV}(I)$ per Exercise 14.5.17(f). Furthermore, $\rho \geq 0$ and

$$\int_I \rho\, d\lambda = \lim_{k\to\infty} \frac{1}{n_k} \sum_{j=0}^{n_k-1} \int_I \mathcal{L}_\lambda^j \mathbb{1}\, d\lambda = \lim_{k\to\infty} \frac{1}{n_k} \sum_{j=0}^{n_k-1} \int_I \mathbb{1}\, d\lambda = 1.$$

Moreover,

$$\begin{aligned}\left\| \mathcal{L}_\lambda \left(\frac{1}{n_k} \sum_{j=0}^{n_k-1} \mathcal{L}_\lambda^j \mathbb{1} \right) - \frac{1}{n_k} \sum_{j=0}^{n_k-1} \mathcal{L}_\lambda^j \mathbb{1} \right\|_1 &= \left\| \frac{1}{n_k} \sum_{j=0}^{n_k-1} \mathcal{L}_\lambda^{j+1} \mathbb{1} - \frac{1}{n_k} \sum_{j=0}^{n_k-1} \mathcal{L}_\lambda^j \mathbb{1} \right\|_1 \\ &= \left\| \frac{1}{n_k} (\mathcal{L}_\lambda^{n_k} \mathbb{1} - \mathbb{1}) \right\|_1 \leq \frac{2}{n_k} \to 0 \quad \text{as } k \to \infty.\end{aligned}$$

Therefore, $\|\mathcal{L}_\lambda \rho - \rho\|_1 = 0$, that is, $\mathcal{L}_\lambda \rho = \rho$ in $L^1(\lambda)$. As ρ satisfies $\rho \geq 0$, $\int_I \rho\, d\lambda = 1$ and $\mathcal{L}_\lambda \rho = \rho$, and as λ is quasi-T-invariant, Theorem 13.4.1 affirms that $\rho\lambda$ is T-invariant. Obviously, $\rho\lambda$ is absolutely continuous with respect to λ. □

14.4 Exponential decay of correlations

By the Ionescu-Tulcea and Marinescu theorem (Theorem B.1.1 in Appendix B), there exist a finite number of distinct eigenvalues of $\mathcal{L}_\lambda : \mathrm{BV}(I) \to \mathrm{BV}(I)$ denoted by $\lambda_1, \ldots, \lambda_p \in \mathbb{S}^1 := \{z \in \mathbb{C} : |z| = 1\}$ and operators $L_1, \ldots, L_p, S \in C(\mathrm{BV}(I), L^1(\lambda))$ such that

$$\mathcal{L}_\lambda^n = \sum_{i=1}^p \lambda_i^n L_i + S^n$$

for all $n \in \mathbb{N}$, with $L_i^2 = L_i$, $L_i \circ L_j = 0$ whenever $i \neq j$, $L_i \circ S = S \circ L_i = 0$ and $r(S) < 1$.

Under two additional assumptions, we shall prove the exponential decay of correlations with respect to the absolutely continuous T-invariant measure $\mu = \rho\lambda$ uncovered in Theorem 14.3.1.

Theorem 14.4.1. *Let $T : I \to I$ be a Lasota–Yorke map and $\mu = \rho\lambda$ be the T-invariant Borel probability measure from Theorem 14.3.1. Assume further that:*
(1) $\{\Lambda_i\}_{i=1}^{p} = \{1\}$; *and*
(2) $\dim(L_1(L^1(\lambda))) = 1$.

Then there exist constants $c \geq 0$ and $r(S) < r < 1$ such that for all $f \in L^\infty(\lambda)$ and all $g \in L^1(\lambda)$ we have

$$C_n(f,g) \leq cr^n \|f\|_\infty \|g\rho\|_{\mathrm{BV}}, \quad \forall n \in \mathbb{N}.$$

Proof. The two additional assumptions ensure that

$$\mathcal{L}_\lambda^n = L_1 + S^n$$

for all $n \in \mathbb{N}$, and

$$\lim_{n\to\infty} \mathcal{L}_\lambda^n g = L_1 g = C(g)\rho$$

for some $C(g) \in \mathbb{R}$ and all $g \in L^1(\lambda)$. Therefore,

$$C(g) = \int_I C(g)\rho \, d\lambda = \lim_{n\to\infty} \int_I \mathcal{L}_\lambda^n g \, d\lambda = \lim_{n\to\infty} \int_I g \, d\lambda = \int_I g \, d\lambda.$$

So $L_1 g = (\int g \, d\lambda)\rho$.

Now, let $f \in L^\infty(\lambda)$ and $g \in L^1(\lambda)$. Thanks to (13.77), we may assume without loss of generality that $E_\mu(f) = E_\mu(g) = 0$ by replacing f with $f - E_\mu(f)$ and g with $g - E_\mu(g)$. Then $C(g\rho) = \int g\rho \, d\lambda = \int g \, d\mu = 0$. Hence, $L_1(g\rho) = 0$ and $\mathcal{L}_\lambda^n(g\rho) = S^n(g\rho)$. By (13.76),

$$\begin{aligned}
C_n(f,g) = \left| \int_I g \cdot (f \circ T^n) \, d\mu \right| &= \left| \int_I (g\rho) \cdot (f \circ T^n) \, d\lambda \right| \\
&= \left| \int_I \mathcal{L}_\lambda^n((g\rho) \cdot (f \circ T^n)) \, d\lambda \right| \\
&= \left| \int_I f \cdot \mathcal{L}_\lambda^n(g\rho) \, d\lambda \right| \leq \|f\|_\infty \int_I |\mathcal{L}_\lambda^n(g\rho)| \, d\lambda \\
&\leq \|f\|_\infty \|\mathcal{L}_\lambda^n(g\rho)\|_{\mathrm{BV}} = \|f\|_\infty \|S^n(g\rho)\|_{\mathrm{BV}} \\
&\leq \|f\|_\infty \|S^n\|_{\mathrm{BV}} \|(g\rho)\|_{\mathrm{BV}} \leq \|f\|_\infty \cdot cr^n \|g\rho\|_{\mathrm{BV}}
\end{aligned}$$

for some constants $c \geq 0$ and $r(S) < r < 1$. That is, the correlations $C_n(f, g)$ decay exponentially (to 0 as $n \to \infty$). □

14.5 Exercises

Exercise 14.5.1. Show that the Lebesgue measure λ is quasi-T-invariant for any Lasota–Yorke map T, i. e. $\lambda \circ T^{-1} \ll \lambda$.

Exercise 14.5.2. Is the composition of any two Lasota–Yorke maps a Lasota–Yorke map?

Exercise 14.5.3. Establish that under the hypotheses of Theorem 14.3.1 there exists at least one ergodic measure μ claimed therein.

Exercise 14.5.4. Prove that under the hypotheses of Theorem 14.3.1 there exists at most finitely many ergodic measures μ claimed therein.

Exercise 14.5.5. Show that under the hypotheses of Theorem 14.3.1, if the map T is additionally topologically exact then the hypotheses of Theorem 14.4.1 are satisfied.

Exercise 14.5.6. Prove that under the hypotheses of Theorem 14.4.1 there exists exactly one measure μ as claimed in Theorem 14.3.1. Show that this measure is ergodic and even more, that the Rokhlin's natural extension of the measure-preserving dynamical system (T, μ) is K–mixing.

Exercise 14.5.7. Deduce in the same way as in Chapter 13 that under the hypotheses of Theorem 14.4.1 the central limit theorem and the law of the iterated logarithm hold.

The remainder of the exercises in this chapter pertain to functions of bounded variation. For all of them, let $I = [a, b]$ be a nondegenerate compact interval in $\mathbb{R}$.

Exercise 14.5.8. Let $f : I \to \mathbb{R}$. Given a partition $\mathcal{P} = \{x_i\}_{i=0}^n$ of I, the *variation* of f over the interval I with respect to the partition $\mathcal{P}$ is defined as

$$V_I(f, \mathcal{P}) = \sum_{i=1}^{n} |f(x_i) - f(x_{i-1})|.$$

The variation of f over the interval I is then defined by

$$V_I(f) = \sup\{V_I(f, \mathcal{P}) : \mathcal{P} \text{ is a partition of } I\}.$$

The function f is said to be *of bounded variation* on I if $V_I(f) < \infty$. We shall denote the set of functions of bounded variation on I by $\mathrm{BV}(I)$.

(a) Let $\mathcal{P}$, $\mathcal{Q}$ be two partitions of I such that $\mathcal{Q}$ is a refinement of $\mathcal{P}$, i. e., $\mathcal{P} \subseteq \mathcal{Q}$. Prove that $V_I(f, \mathcal{P}) \leq V_I(f, \mathcal{Q})$.

(b) Let $\mathcal{P}$ be a partition of I and suppose that $f, g : I \to \mathbb{R}$ are functions that differ only at one point $x \in I$. Show that $|V_I(f, \mathcal{P}) - V_I(g, \mathcal{P})| \leq 2|f(x) - g(x)|$.
(c) Deduce that if $f, g : I \to \mathbb{R}$ differ at at most finitely many points, then $f \in \mathrm{BV}(I)$ if and only if $g \in \mathrm{BV}(I)$.
(d) Show that $V_I(|f|) \leq V_I(f)$. Hence, $f \in \mathrm{BV}(I) \Longrightarrow |f| \in \mathrm{BV}(I)$.
(e) Find a function f such that $|f| \in \mathrm{BV}(I)$ but $f \notin \mathrm{BV}(I)$.
(f) Prove that if f is continuous on I, then $|f| \in \mathrm{BV}(I) \Longleftrightarrow f \in \mathrm{BV}(I)$.

Exercise 14.5.9. Let $a < c < b$. Show that

$$V_{[a,b]}(f) = V_{[a,c]}(f) + V_{[c,b]}(f).$$

Exercise 14.5.10.
(a) Prove that $V_I(f) \geq |f(b) - f(a)|$, and that the equality holds if and only if f is monotone on I. In particular, this implies that any monotone function on I is of bounded variation on I.
(b) Show that $|f(x)| \leq |f(a)| + V_I(f)$ for all $x \in I$. In particular, this means that any function of bounded variation on I is bounded on I.
(c) Identify bounded functions which are not of bounded variation.

Exercise 14.5.11.
(a) Prove that any function $f \in C^1(I)$ is of bounded variation on I. That is, $C^1(I) \subseteq \mathrm{BV}(I)$.
(b) Show that there exist functions of bounded variation which are not continuously differentiable. Thus, $C^1(I) \subsetneq \mathrm{BV}(I)$.
(c) Find continuous functions which are not of bounded variation. So $C(I) \not\subset \mathrm{BV}(I)$.

Exercise 14.5.12. The aim of this exercise is to establish that $\mathrm{BV}(I)$ is a vector space and that the variation of functions is a seminorm on that space.
(a) Show that $V_I(f \pm g) \leq V_I(f) + V_I(g)$. Deduce that $f, g \in \mathrm{BV}(I) \Longrightarrow f \pm g \in \mathrm{BV}(I)$.
(b) Prove that $V_I(cf) = |c| V_I(f)$. Deduce that $f \in \mathrm{BV}(I) \Longrightarrow cf \in \mathrm{BV}(I)$ for all $c \in \mathbb{R}$.
(c) Show that $V_I(f) \geq 0$, and that the equality prevails if and only if f is a constant function.

Exercise 14.5.13. The goal of this exercise is to study the operations of multiplication and division on functions of bounded variation.
(a) Prove that $V_I(fg) \leq M(V_I(f) + V_I(g))$, where $M = \max\{\sup |f|, \sup |g|\}$. Deduce that $f, g \in \mathrm{BV}(I) \Longrightarrow fg \in \mathrm{BV}(I)$.
(b) Let $f \in \mathrm{BV}(I)$ be such that $|f| \geq \varepsilon$ for some $\varepsilon > 0$. Prove that $V_I(1/f) \leq (1/\varepsilon^2) V_I(f)$. Deduce that $1/f \in \mathrm{BV}(I)$ for such a function f.
(c) Show that $f \in \mathrm{BV}(I) \not\Longrightarrow 1/f \in \mathrm{BV}(I)$ in general.

Exercise 14.5.14. Exercise 14.5.10(a) affirms that any nondecreasing (resp., nonincreasing) function on I is of bounded variation on I. Moreover, Exercise 14.5.12(a) states that the sum and difference of two functions of bounded variation on I are also of bounded variation. The aim of this exercise is to prove a converse statement, namely, that any function of bounded variation on I is the difference of two nondecreasing functions on I. This fact was uncovered by Jordan. We further deduce that a function of bounded variation has at most countably many points of discontinuity and thus is Riemann integrable.

Let $f : I \to \mathbb{R}$. The total variation of f is the function $F : I \to [0, \infty]$ defined by $F(a) = 0$ and

$$F(x) = V_{[a,x]}(f).$$

(a) Prove that F is a nondecreasing function.
(b) Show that $|f(y) - f(x)| \le F(y) - F(x)$ for all $a \le x \le y \le b$.
(c) Deduce that $F - f$ is a nondecreasing function.
(d) Deduce further that $V_I(f) \le V_I(F)$.
(e) Deduce from (a) and (c) that $f \in \mathrm{BV}(I)$ if and only if there exist nondecreasing functions $g, h : I \to \mathbb{R}$ such that $f = g - h$.
(f) Deduce from (e) that any $f \in \mathrm{BV}(I)$ has a left-hand limit at each $x \in (a, b]$ and a right-hand limit at each $x \in [a, b)$. (This can also be proved by contradiction.)
(g) Deduce from (f) that any $f \in \mathrm{BV}(I)$ has at most countably many points of discontinuity.
(h) Conclude that any $f \in \mathrm{BV}(I)$ is Riemann integrable and hence Lebesgue integrable, i. e., $\mathrm{BV}(I) \subseteq L^1(\lambda)$.

Exercise 14.5.15. Let $f \in \mathrm{BV}(I)$. Recall that the average of f over I is defined by

$$\mathrm{Avg}_I(f) = \frac{1}{\lambda(I)} \int_I f \, d\lambda.$$

(This average is well defined according to Exercise 14.5.14(h).) Prove that

$$\mathrm{Avg}_I(f) - V_I(f) \le \inf_I f \le \sup_I f \le \mathrm{Avg}_I(f) + V_I(f).$$

Exercise 14.5.16. The objective of this exercise is to show that pointwise convergence, and even uniform convergence, are insufficient on their own to ensure the transmission of the bounded variation property from a sequence of functions to its limit. Under an additional condition, though, the bounded variation property is preserved.

(a) Construct a sequence $(f_n)_{n=1}^{\infty}$ in $\mathrm{BV}(I)$ which converges pointwise to a function $f \notin \mathrm{BV}(I)$.

(b) Build a sequence $(f_n)_{n=1}^{\infty}$ in $\mathrm{BV}(I)$ which converges uniformly to a function $f \notin \mathrm{BV}(I)$.
(c) Let $(f_n)_{n=1}^{\infty}$ be a sequence in $\mathrm{BV}(I)$ which converges pointwise to a function f. Show that if $\sup_{n\in\mathbb{N}} V_I(f_n) < \infty$, then $f \in \mathrm{BV}(I)$. More specifically, prove that $V_I(f) = \lim_{n\to\infty} V_I(f_n)$.

Exercise 14.5.17. The objective of this exercise is to study the properties of a norm that makes $\mathrm{BV}(I)$ a Banach space. Given that any two functions that are λ-almost everywhere equal are identified in $L^1(\lambda)$, they must be identified in $\mathrm{BV}(I)$ as well and we cannot simply use the variation of functions $V_I(\cdot)$ as a seminorm (cf. Exercise 14.5.12). Rather, we need to take the infimum of the variations over all functions in the same $L^1(\lambda)$-class. Let

$$\|f\|_{\mathrm{BV}}^{*} = \inf_{g\in[f]_1} V_I(g) \quad \text{and} \quad \|f\|_{\mathrm{BV}} = \|f\|_{\mathrm{BV}}^{*} + \|f\|_1,$$

where $[f]_1$ denotes the equivalence class of f in $L^1(\lambda)$ and $\|f\|_1 = \int_I |f|\, d\lambda$ is its L^1-norm.
(a) Prove that the set $\mathrm{BV}(I)$ resulting from the aforementioned identification is a real vector subspace of $L^1(\lambda)$.
(b) Show that the function $\|\cdot\|_{\mathrm{BV}}^{*} : \mathrm{BV}(I) \to \mathbb{R}$ defines a seminorm on that vector space. Demonstrate also that $\|f\|_{\mathrm{BV}}^{*} = 0$ if and only if f is constant λ-almost everywhere.
(c) Prove that the function $\|\cdot\|_{\mathrm{BV}} : \mathrm{BV}(I) \to \mathbb{R}$ defines a norm on that vector space.
(d) Demonstrate that $\mathrm{BV}(I)$ is a Banach space, i. e., a complete normed vector space, when endowed with the norm $\|\cdot\|_{\mathrm{BV}}$. One way to do this is the following. Let $(f_n)_{n=1}^{\infty}$ be a Cauchy sequence in $(\mathrm{BV}(I), \|\cdot\|_{\mathrm{BV}})$.
   1. Deduce that $(f_n)_{n=1}^{\infty}$ is a Cauchy sequence in $(\mathrm{BV}(I), \|\cdot\|_{\mathrm{BV}}^{*})$ and in $L^1(\lambda)$.
   2. Deduce that $(f_n)_{n=1}^{\infty}$ converges to some function f in $L^1(\lambda)$.
   3. Prove that $(f_n)_{n=1}^{\infty}$ admits a subsequence $(g_m)_{m=1}^{\infty}$ that converges λ-almost everywhere to f.
   4. Show that the subsequence $(g_m)_{m=1}^{\infty}$ can be chosen so that $\|g_{m+1}-g_m\|_{\mathrm{BV}}^{*} < 1/2^m$ for all $m \in \mathbb{N}$.
   5. Establish the existence of a sequence $(g_m^*)_{m=1}^{\infty}$ such that $g_m^* \in [g_m]_1$ and $V_I(g_{m+1}^* - g_m^*) < 1/2^m$ for all $m \in \mathbb{N}$.
   6. Deduce that the sequence $(g_m^*)_{m=1}^{\infty}$ converges pointwise to a function $f^* \in [f]_1$.
   7. Using Exercise 14.5.16(c), deduce that f^* has bounded variation.
   8. Show that $V_I(g_m^* - f^*) \to 0$ as $m \to \infty$.
   9. Deduce that $\|g_m - f\|_{\mathrm{BV}}^{*} \to 0$ as $m \to \infty$.
   10. Deduce that $\|f_n - f\|_{\mathrm{BV}}^{*} \to 0$ as $n \to \infty$.
   11. Conclude that $(f_n)_{n=1}^{\infty}$ converges to f in $(\mathrm{BV}(I), \|\cdot\|_{\mathrm{BV}})$.
(e) Show that the Banach space $(\mathrm{BV}(I), \|\cdot\|_{\mathrm{BV}})$ is not a Banach subspace of the Banach space $(L^1(\lambda), \|\cdot\|_1)$.

(f) Let $(f_n)_{n=1}^{\infty}$ be a sequence in $\mathrm{BV}(I)$ such that

$$\lim_{n\to\infty} \|f_n - f\|_1 = 0 \text{ for some } f \in L^1(\lambda) \quad \text{and} \quad \sup_{n\in\mathbb{N}} \|f_n\|_{\mathrm{BV}} \le K \text{ for some } K \ge 0.$$

Show that

$$f \in \mathrm{BV}(I) \quad \text{and} \quad \|f\|_{\mathrm{BV}} \le K.$$

Note: This demonstrates that condition (a1) in the Ionescu-Tulcea and Marinescu theorem (Theorem B.1.1 at the beginning of Appendix B) is fulfilled.

15 Fractal measures and dimensions

In a sense, this chapter is exceptional as it does not directly involve any dynamics: only fractal geometry. However, the concepts and results it contains will be frequently used in most of the forthcoming chapters.

In this chapter, we shall discuss different ways of evaluating the size of sets, especially fractal sets. Unlike classical geometrical objects such as lines, squares or cubes which have integer Euclidean and topological dimensions, fractal sets cannot, because of their irregularities at all scales, be accurately described and distinguished from one another by these conventional dimensions. We need less common but more befitting geometric measures, such as the Hausdorff and packing measures, as well as the dimensions they induce, to examine the local and global structures of these sets at arbitrarily small scales. It is important to emphasize that unlike the topological dimension, the Hausdorff and packing dimensions are not topological invariants, meaning that they are generally not preserved by homeomorphisms. They are however almost trivially preserved by isometries and, much more significantly, by bi-Lipschitz maps. In particular, these dimensions constitute useful tools for discerning metric spaces which, despite being homeomorphic, are not bi-Lipschitz equivalent.

To be more specific, in this chapter we introduce several basic geometric concepts on metric spaces. These are: Hausdorff measures and the Hausdorff dimension; packing measures and packing dimensions; box-counting dimensions; and dimensions of measures. In Sections 15.2 and 15.3, we establish the fundamental properties of these measures and dimensions. While Hausdorff measures and the Hausdorff dimension were introduced in 1919 by Felix Hausdorff [47], it took many more decades for packing measures and packing dimensions to be defined. This was done in stages in [126, 130, 131]. There are now plenty of books on these concepts; we refer the reader to [28–30, 72, 98], among others. Also interesting (not only for historical reasons) is the classical book [101] by C. A. Rogers, which originally appeared in 1970. The 1998 edition is particularly appealing thanks to Kenneth Falconer's foreword.

In Section 15.5, we consider volume lemmas and their geometrical consequences. Their main purpose is to tell when a Hausdorff or a packing measure is null/positive and finite/infinite. We refer to them as Frostman converse theorems. Somewhat strangely, these lemmas/theorems are frequently called mass redistribution principle in the fractal geometry literature, as if to indicate that such measures always appear in the context of an iterative construction. At the end of that section, we formulate Frostman's (direct) lemma and compare it with the Frostman converse theorems. As already mentioned, the advantage of the latter is that they provide tools to calculate, or at least to estimate, both Hausdorff and packing measures and dimensions.

In Section 15.6, we deal specifically with dimensions of measures. We approach them by means of pointwise dimensions and show how to calculate them by proving results frequently referred to as volume lemmas.

https://doi.org/10.1515/9783110702699-015

The exposition given in this chapter is brief but self-contained, and comprises those results that are relevant for many dynamical applications.

We start with outer measures, a classical concept in measure theory. (Basic measure-theoretic concepts and results can be found at the beginning of Appendix A.)

15.1 Outer measures

While the family $\mathcal{P}(X)$ of all subsets of a set X is a σ-algebra on X, some measures are not and cannot be defined on $\mathcal{P}(X)$. For example, Vitali (see 5.7, The Vitali Monsters on p. 120 of [51]) constructed a subset of $\mathbb{R}$ which is not measurable under the Lebesgue measure. That is, the Lebesgue measure cannot be defined on all subsets of $\mathbb{R}$. However, some measures arise from so-called "outer measures". Outer measures are generally not measures themselves but they are defined on $\mathcal{P}(X)$ and when restricted to an appropriate sub-σ-algebra of $\mathcal{P}(X)$, they constitute a measure.

Definition 15.1.1. Let X be a set. An *outer measure* on X is a function $\mu : \mathcal{P}(X) \to [0,\infty]$ satisfying the following three conditions:
(a) $\mu(\emptyset) = 0$.
(b) If $A \subseteq B \subseteq X$, then $\mu(A) \le \mu(B)$.
(c) $\mu(\bigcup_{i=1}^{\infty} A_i) \le \sum_{i=1}^{\infty} \mu(A_i)$ for any sequence $(A_i)_{i=1}^{\infty}$ in $\mathcal{P}(X)$.

The great Greek mathematician Carathéodory discovered that there is a natural σ-algebra that can be associated to every outer measure. Given an outer measure μ, he defined a μ-measurable set to be a set which splits every other set without losing any measure in the process.

Definition 15.1.2. Let $\mu : \mathcal{P}(X) \to [0,\infty]$ be an outer measure. A set $A \subseteq X$ is called *μ-measurable* if

$$\mu(B \cap A) + \mu(B \setminus A) = \mu(B), \quad \forall B \subseteq X.$$

Remark 15.1.3. Due to property (c) of an outer measure (see Definition 15.1.1), the above condition is equivalent to

$$\mu(B \cap A) + \mu(B \setminus A) \le \mu(B), \quad \forall B \subseteq X.$$

Theorem 15.1.4. *Let $\mu : \mathcal{P}(X) \to [0,\infty]$ be an outer measure and let $\mathcal{F}_\mu$ denote the family of all μ-measurable sets. Then $\mathcal{F}_\mu$ is a σ-algebra and $\mu|_{\mathcal{F}_\mu}$ is a measure.*

Proof. The proof is left to the reader as an exercise. □

When the underlying set X has a metric structure, one can define a concept of metric outer measure as follows.

Definition 15.1.5. Let (X, d) be a metric space. An outer measure μ on X is said to be a *metric outer measure* if

$$d(A, B) > 0 \Longrightarrow \mu(A \cup B) = \mu(A) + \mu(B),$$

where $d(A, B) = \inf\{d(x, y) : x \in A, y \in B\}$.

In other words, a metric outer measure is an outer measure that behaves like a measure for any pair of sets that lie at a positive distance from each other. Henceforth, let μ be a metric outer measure on a metric space (X, d). Our first goal is to show that the Borel σ-algebra is contained in $\mathcal{F}_\mu$, i. e., to show that all Borel sets are μ-measurable. The following result, which will help us achieve our goal, is well known for any measure and any ascending sequence of measurable sets. Here, we prove it for any metric outer measure and any ascending sequence of (not necessarily measurable) sets whose boundaries stand at a positive distance from each other.

Lemma 15.1.6. *Let μ be a metric outer measure on a metric space (X, d). Let $(A_n)_{n=1}^\infty$ be an ascending sequence of subsets of X and set $A = \bigcup_{n=1}^\infty A_n$. If $d(A_n, A \setminus A_{n+1}) > 0$ for all $n \in \mathbb{N}$, then*

$$\mu(A) = \lim_{n\to\infty} \mu(A_n).$$

Proof. Since the sequence $(\mu(A_n))_{n=1}^\infty$ is nondecreasing, it converges (perhaps to infinity). Also, $A_n \subseteq A$ implies that $\mu(A_n) \le \mu(A)$ for all $n \in \mathbb{N}$. Thus, $\lim_{n\to\infty} \mu(A_n) \le \mu(A)$. It remains to show the opposite inequality, namely, that

$$M := \lim_{n\to\infty} \mu(A_n) \ge \mu(A).$$

This is trivial if $M = \infty$. So assume that $M < \infty$. Define the sequence $(B_n)_{n=1}^\infty$ in the following way. Let $B_1 := A_1$, and $B_n := A_n \setminus A_{n-1}$ for each $n > 1$. The sequence $(B_n)_{n=1}^\infty$ consists of pairwise disjoint sets. If $n, m \in \mathbb{N}$ are such that $n \ge m + 2$, then $B_m \subseteq A_m$ and, because the sequence $(A_n)_{n=1}^\infty$ satisfies $A_n \subseteq A_{n+1}$ for all $n \in \mathbb{N}$, we have that

$$B_n \subseteq A \setminus A_{n-1} \subseteq A \setminus A_{m+1}.$$

As $d(A_m, A \setminus A_{m+1}) > 0$ for all $m \in \mathbb{N}$, it follows that for all $n \ge m + 2$,

$$d(B_m, B_n) \ge d(A_m, A \setminus A_{m+1}) > 0.$$

Thus, for each $k \in \mathbb{N}$ we have

$$\sum_{i=1}^{k} \mu(B_{2i-1}) = \mu\Big(\bigcup_{i=1}^{k} B_{2i-1}\Big) \le \mu(A_{2k-1}) \le M \quad \text{and similarly} \quad \sum_{i=1}^{k} \mu(B_{2i}) \le \mu(A_{2k}) \le M.$$

Hence, $\sum_{i=1}^{\infty} \mu(B_i) \leq 2M < \infty$. Then, for all $n \in \mathbb{N}$ we obtain that

$$\mu(A) = \mu\left(\bigcup_{k=n}^{\infty} A_k\right) = \mu\left(A_n \cup \bigcup_{k=n+1}^{\infty} B_k\right) \leq \mu(A_n) + \sum_{k=n+1}^{\infty} \mu(B_k) \leq \lim_{j\to\infty} \mu(A_j) + \sum_{k=n+1}^{\infty} \mu(B_k).$$

Since the series $\sum_{k=n+1}^{\infty} \mu(B_k)$ converges, letting $n \to \infty$ yields $\mu(A) \leq \lim_{j\to\infty} \mu(A_j)$. □

Using this result, we can now prove that every Borel set is μ-measurable whenever μ is a metric outer measure.

Theorem 15.1.7. *If $\mu : \mathcal{P}(X) \to [0, \infty]$ is a metric outer measure on a metric space (X, d), then $\mathcal{B}(X) \subseteq \mathcal{F}_\mu$, where $\mathcal{B}(X)$ denotes the Borel σ-algebra on X, and the restriction $\mu : \mathcal{B}(X) \to [0, \infty]$ is a measure.*

Proof. Theorem 15.1.4 affirms that $\mathcal{F}_\mu$ is a σ-algebra and that $\mu|_{\mathcal{F}_\mu}$ is a measure. Therefore, it is enough to show that all the closed sets are in $\mathcal{F}_\mu$, since these generate the Borel sets. Let A be a closed subset of X and B an arbitrary subset of X. It suffices to show that $\mu(B \cap A) + \mu(B \setminus A) \leq \mu(B)$ (cf. Remark 15.1.3).

For each $n \in \mathbb{N}$, define

$$B_n := \{x \in B \setminus A : d(x, A) \geq 1/n\}.$$

Then $d(A, B_n) \geq 1/n$. Moreover, $(B_n)_{n=1}^{\infty}$ is an ascending sequence of sets such that $B \setminus A = \bigcup_{n=1}^{\infty} B_n$. Let $x \in (B\setminus A) \setminus B_{n+1}$ and $y \in B_n$. Then $d(x, A) < 1/(n+1)$. Thus, there exists $z \in A$ such that $d(x, z) < 1/(n + 1)$. Therefore,

$$d(y, x) \geq d(y, z) - d(z, x) \geq \frac{1}{n} - \frac{1}{n+1} = \frac{1}{n(n+1)},$$

and hence $d(B_n, (B\setminus A) \setminus B_{n+1}) \geq 1/[n(n + 1)] > 0$. Consequently, Lemma 15.1.6 asserts that $\mu(B \setminus A) = \lim_{n\to\infty} \mu(B_n)$. It follows that

$$\begin{aligned}
\mu(B \cap A) + \mu(B \setminus A) &= \mu(B \cap A) + \lim_{n\to\infty} \mu(B_n) \\
&= \lim_{n\to\infty} (\mu(B \cap A) + \mu(B_n)) \\
&= \lim_{n\to\infty} \mu((B \cap A) \cup B_n) \\
&\leq \lim_{n\to\infty} \mu((B \cap A) \cup B) \\
&= \lim_{n\to\infty} \mu(B) = \mu(B).
\end{aligned}$$

The third equality above is a consequence of the fact that μ is a metric outer measure and $d(B \cap A, B_n) \geq d(A, B_n) > 0$ for all n. □

15.2 Geometric (Hausdorff and packing) outer measures and dimensions

In this section, we introduce two families of outer measures which are in some sense defined in a geometrical way. The first one is the family of Hausdorff measures while the second is the family of packing measures. We show how they give rise to Borel measures, also referred to as Hausdorff and packing measures, and we prove some of their fundamental properties and relations between them.

15.2.1 Gauge functions

Definition 15.2.1. A function $\varphi : [0, \infty) \to [0, \infty)$ is called a *gauge function* provided that:
(a) φ is nondecreasing.
(b) $\varphi(0) = 0$.
(c) $\varphi((0, \infty)) \subseteq (0, \infty)$.
(d) $0 = \varphi(0) = \lim_{t \to 0} \varphi(t)$, i. e., φ is continuous at 0.

Conditions (b) and (c) immediately imply that $\varphi^{-1}(0) = \{0\}$.

Example 15.2.2. The foremost family of such functions is $\varphi(t) = t^{\alpha}$, $\alpha > 0$. Later in the chapter, we shall see why these functions are especially important.

A family of gauges for which it is somewhat easier to calculate the Hausdorff and packing measures is the following.

Definition 15.2.3. A gauge function $\varphi : [0, \infty) \to [0, \infty)$ is called *evenly varying* if there exists a function $c_{\varphi} : [0, \infty) \to [1, \infty)$ such that for all $a > 0$ and for all sufficiently small $t > 0$,

$$(c_{\varphi}(a))^{-1} \varphi(t) \leq \varphi(at) \leq c_{\varphi}(a)\varphi(t).$$

Example 15.2.4. For each $\alpha > 0$, the gauge function $\varphi_{\alpha}(t) = t^{\alpha}$ is evenly varying. To see this, note that $\varphi_{\alpha}(at) = a^{\alpha} t^{\alpha}$. So it suffices to let $c_{\varphi}(a) := \max\{a^{-\alpha}, a^{\alpha}\}$.

15.2.2 Hausdorff measures

Given a set $A \subseteq X$ and $\delta > 0$, we endeavor to confer a "measure" to A by covering it with the "best" δ-covers possible. How well a particular covering covers a given set shall be assessed by the values that the gauge function assigns to the diameter of each set in the covering. The lower the sum of these values, the better the covering is considered to be (since the gauge function is nondecreasing). This is the justification for the name given to the gauge functions—they allow us to gauge the quality of a cover. Note that there may not actually be a best cover, in this sense. This explains the use of the infimum, rather than a minimum, in the following definition.

Definition 15.2.5. Let (X,d) be a metric space and $\varphi : [0,\infty) \to [0,\infty)$ be a gauge function. Let also $\delta > 0$ and $A \subseteq X$. A countable cover $\{C_i\}_{i=1}^{\infty}$ of A such that $\sup\{\text{diam}(C_i) : i \in \mathbb{N}\} \leq \delta$ is said to be a *δ-cover* of A. Define

$$H_\varphi^\delta(A) := \inf\left\{\sum_{i=1}^{\infty} \varphi(\text{diam}(C_i)) : \{C_i\}_{i=1}^{\infty} \text{ is a } \delta\text{-cover of } A\right\}.$$

Lemma 15.2.6. $H_\varphi^\delta : \mathcal{P}(X) \to [0,\infty]$ *is an outer measure.*

Proof. We must check that the three conditions in Definition 15.1.1 are satisfied. First, it is clear that $H_\varphi^\delta(\emptyset) = 0$. For the second condition, let $A \subseteq B \subseteq X$. Then $H_\varphi^\delta(A) \leq H_\varphi^\delta(B)$ since any δ-cover of B is also a δ-cover of A.

It remains to show that $H_\varphi^\delta(\bigcup_{n=1}^{\infty} A_n) \leq \sum_{n=1}^{\infty} H_\varphi^\delta(A_n)$ for every sequence $(A_n)_{n=1}^{\infty}$ in $\mathcal{P}(X)$. This is certainly true if $\sum_{n=1}^{\infty} H_\varphi^\delta(A_n) = \infty$. So assume that $\sum_{n=1}^{\infty} H_\varphi^\delta(A_n) < \infty$. Let $\varepsilon > 0$ and for each $n \in \mathbb{N}$, let $\{C_{n,i}\}_{i=1}^{\infty}$ be a δ-cover of A_n such that

$$\sum_{i=1}^{\infty} \varphi(\text{diam}(C_{n,i})) \leq H_\varphi^\delta(A_n) + \frac{\varepsilon}{2^n}.$$

Then the countable family $\{C_{n,i}\}_{n,i=1}^{\infty}$ is a δ-cover of the set $\bigcup_{n=1}^{\infty} A_n$. As φ is a nonnegative function, we have

$$H_\varphi^\delta\left(\bigcup_{n=1}^{\infty} A_n\right) \leq \sum_{n=1}^{\infty}\sum_{i=1}^{\infty} \varphi(\text{diam}(C_{n,i})) \leq \sum_{n=1}^{\infty}\left[H_\varphi^\delta(A_n) + \frac{\varepsilon}{2^n}\right] = \sum_{n=1}^{\infty} H_\varphi^\delta(A_n) + \varepsilon.$$

Since this holds for any $\varepsilon > 0$, letting $\varepsilon \to 0$ yields the desired result. □

In general, H_φ^δ is not a metric outer measure. However, as δ decreases, the family of δ-covers shrinks. Thus, $H_\varphi^\delta(A)$ increases. Passing to the limit as δ tends to zero, we obtain a metric outer measure.

Definition 15.2.7. Let (X,d) be a metric space and φ be a gauge function. For every $A \subseteq X$, define

$$H_\varphi(A) := \lim_{\delta\to 0} H_\varphi^\delta(A) = \sup_{\delta>0} H_\varphi^\delta(A).$$

Lemma 15.2.8. $H_\varphi : \mathcal{P}(X) \to [0,\infty]$ *is a metric outer measure and its restriction* $H_\varphi : \mathcal{B}(X) \to [0,\infty]$ *is a measure.*

Proof. Clearly, $H_\varphi(\emptyset) = 0$, and $H_\varphi(A) \le H_\varphi(B)$ for all $A \subseteq B$, by the corresponding properties of H^δ_φ, $\delta > 0$. To prove countable subadditivity, note first that $H_\varphi(A) \ge H^\delta_\varphi(A)$ for all $\delta > 0$ and all $A \subseteq X$. Let $A = \bigcup_{n=1}^\infty A_n$ and let $\delta > 0$ be arbitrary. Then

$$H^\delta_\varphi\Big(\bigcup_{n=1}^\infty A_n\Big) \le \sum_{n=1}^\infty H^\delta_\varphi(A_n) \le \sum_{n=1}^\infty H_\varphi(A_n).$$

Letting $\delta \to 0$ yields

$$H_\varphi\Big(\bigcup_{n=1}^\infty A_n\Big) \le \sum_{n=1}^\infty H_\varphi(A_n).$$

So H_φ is an outer measure. It remains to show that H_φ is a metric outer measure, i. e., to prove that $H_\varphi(A\cup B) = H_\varphi(A)+H_\varphi(B)$ for all $A, B \subseteq X$ such that $d(A, B) > 0$. That $H_\varphi(A \cup B) \le H_\varphi(A) + H_\varphi(B)$ comes from the subadditivity of H_φ. For the opposite inequality, let $\delta < d(A, B)/2$ and choose a δ-cover $\{C_i\}_{i=1}^\infty$ of $A \cup B$. Without loss of generality, we may assume that each C_i is contained in $A \cup B$. Also, by the choice of δ, if $C_i \cap A \ne \emptyset$ then $C_i \cap B = \emptyset$, while $C_i \cap B \ne \emptyset$ implies $C_i \cap A = \emptyset$. So we can split the cover as follows:

$$C_A = \{i \in \mathbb{N} : C_i \cap A \ne \emptyset\}$$
$$C_B = \{i \in \mathbb{N} : C_i \cap B \ne \emptyset\}.$$

Then the subfamily $\{C_i\}_{i\in C_A}$ covers A whereas the subfamily $\{C_i\}_{i\in C_B}$ covers B. Therefore,

$$\sum_{i=1}^\infty \varphi(\operatorname{diam}(C_i)) = \sum_{i\in C_A} \varphi(\operatorname{diam}(C_i)) + \sum_{i\in C_B} \varphi(\operatorname{diam}(C_i)) \ge H^\delta_\varphi(A) + H^\delta_\varphi(B).$$

Taking the infimum over all δ-covers of $A \cup B$ on the left-hand side, we deduce that

$$H^\delta_\varphi(A \cup B) \ge H^\delta_\varphi(A) + H^\delta_\varphi(B).$$

Letting $\delta \to 0$ yields $H_\varphi(A\cup B) \ge H_\varphi(A)+H_\varphi(B)$. So H_φ is a metric outer measure. Then Theorem 15.1.7 states that $H_\varphi|_{\mathcal{B}(X)}$ is a measure. □

Definition 15.2.9. The metric outer measure H_φ is called the *Hausdorff outer measure* induced by the gauge function φ.

Remark 15.2.10.
(a) If two gauge functions φ and ψ are such that $\varphi|_{[0,\eta)} = \psi|_{[0,\eta)}$ for some $\eta > 0$, then $H_\varphi = H_\psi$.
(b) H_φ has no atoms, i. e., $H_\varphi(\{x\}) = 0$ for every $x \in X$. This is because $H^\delta_\varphi(\{x\}) \le \varphi(\operatorname{diam}(\{x\})) = \varphi(0) = 0$ for each $\delta > 0$.

(c) If $\{C_i\}_{i=1}^{\infty}$ is a δ-cover of a set $A \subseteq X$, then so is $\{C_i \cap A\}_{i=1}^{\infty}$. Moreover,

$$\sum_{i=1}^{\infty} \varphi(\operatorname{diam}(C_i \cap A)) \le \sum_{i=1}^{\infty} \varphi(\operatorname{diam}(C_i)).$$

This means that the structure of the space outside of A is irrelevant, i. e., if d, ρ are two metrics on X with the property that $d|_{A\times A} = \rho|_{A\times A}$, then the quantities $H_\varphi(A)$ with respect to d and ρ are equal. In other terms, the Hausdorff measure $H_\varphi(A)$ is an intrinsic property of the metric subspace $(A, d|_{A\times A})$.

We shall soon see that the case when $\varphi(t) := \varphi_\alpha(t) = t^\alpha$ for some $\alpha > 0$ is of particular interest and for this reason we abridge the notation $H_{\varphi_\alpha}(A) = H_{t^\alpha}(A)$ to $H_\alpha(A)$. The outer measure H_α also deserves a special name.

Definition 15.2.11. The metric outer measure H_α is called the *α-dimensional Hausdorff measure.*

It is sometimes more convenient to consider covers consisting of balls instead of arbitrary sets.

Definition 15.2.12. Let (X, d) be a metric space, and let $A \subseteq X$ and $r > 0$. A countable family of open balls $\{B(x_i, r_i)\}_{i=1}^{\infty}$ is said to be an *r-ball cover of A* if it is a cover of A such that the centers x_i belong to A and

$$\sup\{r_i : i \in \mathbb{N}\} \le r.$$

Let

$$H_\varphi^{B,r}(A) := \inf\left\{\sum_{i=1}^{\infty} \varphi(r_i) : \{B(x_i, r_i)\}_{i=1}^{\infty} \text{ is an } r\text{-ball cover of } A\right\}.$$

Following the same limit procedure as in the definition of H_φ, define

$$H_\varphi^{B}(A) := \lim_{r\to 0} H_\varphi^{B,r}(A) = \sup_{r>0} H_\varphi^{B,r}(A).$$

Remark 15.2.13. Ball covers can also be defined using closed balls rather than open ones. The avid reader may check that the most fundamental forthcoming results also hold in that case.

In contrast with H_φ, the set function H_φ^B is not necessarily an outer measure. For example, if $A \subseteq B$, a ball cover of B need not be a ball cover of A, since some centers that are in B may not lie in A. Nevertheless, the set functions H_φ^B and H_φ are closely related to one another when φ is evenly varying.

Lemma 15.2.14. *Let φ be an evenly varying gauge function. For all $A \subseteq X$, we have*

$$H_\varphi^B(A) \le c_\varphi(2)H_\varphi(A) \le (c_\varphi(2))^2 H_\varphi^B(A),$$

where $c_\varphi(2)$ is the constant from Definition 15.2.3.

Proof. Let $\{B(x_i, r_i)\}_{i=1}^\infty$ be an r-ball cover of A. Since every r-ball cover of A is a $(2r)$-cover of A, we get that

$$H_\varphi^{2r}(A) \le \sum_{i=1}^\infty \varphi(\operatorname{diam}(B(x_i, r_i))) \le \sum_{i=1}^\infty \varphi(2r_i) \le c_\varphi(2) \sum_{i=1}^\infty \varphi(r_i).$$

As the r-ball cover of A was chosen arbitrarily, we conclude that $H_\varphi^{2r}(A) \le c_\varphi(2)H_\varphi^{B,r}(A)$. Letting $r \to 0$ yields $H_\varphi(A) \le c_\varphi(2)H_\varphi^B(A)$.

For the other inequality, let $\delta > 0$ and let $\{C_i\}_{i=1}^\infty$ be a δ-cover of A. We may assume that $C_i \cap A \ne \emptyset$ for all i. Let $x_i \in C_i \cap A$ and let r_i be the diameter of C_i. Then $\{B(x_i, 2r_i)\}_{i=1}^\infty$ is a (2δ)-ball cover of A since $C_i \subseteq B(x_i, 2r_i)$ for each $i \in \mathbb{N}$. Thus,

$$H_\varphi^{B,2\delta}(A) \le \sum_{i=1}^\infty \varphi(2r_i) \le c_\varphi(2) \sum_{i=1}^\infty \varphi(\operatorname{diam}(C_i)).$$

As $\{C_i\}_{i=1}^\infty$ is an arbitrary δ-cover of A, we deduce that $H_\varphi^{B,2\delta}(A) \le c_\varphi(2)H_\varphi^\delta(A)$. Letting $\delta \to 0$ gives $H_\varphi^B(A) \le c_\varphi(2)H_\varphi(A)$. □

Remark 15.2.15. As noted above, $H_\varphi^B(A)$ need not be an outer measure. However, as long as we are only interested in whether $H_\varphi(A)$ is positive and finite (properties which determine the Hausdorff dimension of a set; see Subsection 15.3.1), we can consider either $H_\varphi^B(A)$ or $H_\varphi(A)$.

15.2.3 Packing measures

We now aim to introduce the packing measure that any gauge function induces. Let us first define the notion of packing.

Definition 15.2.16. Let (X, d) be a metric space and $\varphi : [0, \infty) \to [0, \infty)$ be a gauge function. A countable collection $\{(x_i, r_i)\}_{i=1}^\infty$ in $X \times (0, \infty)$ is said to be a *packing* of a set $A \subseteq X$ if

$$x_i \in A, \quad \forall i \in \mathbb{N} \quad \text{and} \quad d(x_i, x_j) \ge r_i + r_j, \quad \forall i, j \in \mathbb{N} \text{ with } i \ne j.$$

Moreover, a packing $\{(x_i, r_i)\}_{i=1}^\infty$ is called an *r-packing* if

$$\sup\{r_i : i \in \mathbb{N}\} \le r.$$

Observe that if $\{(x_i, r_i)\}_{i=1}^{\infty}$ is a packing of A, then the balls $\{B(x_i, r_i)\}_{i=1}^{\infty}$ are mutually disjoint. In this sense, a packing can be thought of as a counterpart to a covering. Whereas the Hausdorff measure of a set is determined by the way the set can be covered, the packing measure is defined by the way the set can be packed. The Hausdorff measure is obtained by covering the set as well as possible at every scale, which means finding "minimal" coverings with respect to the gauge function at every scale. So it is natural to define the packing measure by packing the set the "best" way (i. e., as "much" as) possible at every scale. This justifies the use of the supremum, rather than the infimum, in the following definition.

Definition 15.2.17. Let (X, d) be a metric space and $\varphi : [0, \infty) \to [0, \infty)$ be a gauge function. For every $r > 0$ and $A \subseteq X$, define

$$P_{\varphi}^{*r}(A) := \sup\left\{\sum_{i=1}^{\infty} \varphi(r_i) : \{(x_i, r_i)\}_{i=1}^{\infty} \text{ is an } r\text{-packing of } A\right\}.$$

In general, P_{φ}^{*r} is not an outer measure. Nevertheless, as r decreases, the family of r-packings shrinks. Thus, $P_{\varphi}^{*r}(A)$ decreases. Passing to the limit as r tends to zero, we obtain the following quantity.

Definition 15.2.18. Let (X, d) be a metric space and φ be a gauge function. For every $A \subseteq X$, define

$$P_{\varphi}^{*}(A) := \lim_{r\to 0} P_{\varphi}^{*r}(A) = \inf_{r>0} P_{\varphi}^{*r}(A).$$

It is natural to ask whether P_{φ}^{*} is an outer measure. Obviously, $P_{\varphi}^{*}(\emptyset) = 0$. If $A \subseteq B$, then each r-packing of A is an r-packing of B. Therefore, $P_{\varphi}^{*r}(A) \le P_{\varphi}^{*r}(B)$. Letting $r \to 0$ yields $P_{\varphi}^{*}(A) \le P_{\varphi}^{*}(B)$.

However, countable subadditivity may fail to hold. For example, if $X = [0, 1]$ and $Q = \mathbb{Q} \cap [0, 1]$, then $P_{\varphi}^{*}(\{x\}) = 0$ and so $\sum_{x\in Q} P_{\varphi}^{*}(\{x\}) = 0$ for any gauge function φ. On the other hand, take $\varphi(t) = t$. For all $n \in \mathbb{N}$, the family $\{(\frac{i}{2^n}, \frac{1}{2^{n+1}})\}_{i=0}^{2^n}$ is a $\frac{1}{2^{n+1}}$-packing of Q. Thus,

$$P_{\varphi}^{*2^{-(n+1)}}(Q) \ge \sum_{i=0}^{2^n} \frac{1}{2^{n+1}} = \frac{2^n + 1}{2^{n+1}}.$$

Hence, $P_{\varphi}^{*}(Q) \ge 1/2$ and P_{φ}^{*} is not countably subadditive in this case.

Though countable subadditivity does not hold in general, perhaps finite subadditivity does. This latter can be reduced to two sets, say A and B. Pick any r-packing $\{(z_k, r_k)\}_{k=1}^{\infty}$ of $A \cup B$. Observe that the couples $\{(z_k, r_k) : z_k \in A\}$ and $\{(z_k, r_k) : z_k \in B\}$ form r-packings of A and B, respectively. Consequently,

$$\sum_{k=1}^{\infty} \varphi(r_k) \le \sum_{z_k\in A} \varphi(r_k) + \sum_{z_k\in B} \varphi(r_k) \le P_{\varphi}^{*r}(A) + P_{\varphi}^{*r}(B).$$

Since the r-packing $\{(z_k, r_k)\}_{k=1}^{\infty}$ of $A \cup B$ was arbitrarily chosen, we infer that

$$P_{\varphi}^{*r}(A \cup B) \le P_{\varphi}^{*r}(A) + P_{\varphi}^{*r}(B), \quad \forall r > 0. \tag{15.1}$$

Letting $r \to 0$ shows that finite subadditivity prevails:

$$P_{\varphi}^{*}(A \cup B) \le P_{\varphi}^{*}(A) + P_{\varphi}^{*}(B).$$

Let us push our investigation one step further. Suppose that $d(A, B) > 0$. We want to establish that $P_{\varphi}^{*r}(A \cup B) = P_{\varphi}^{*r}(A) + P_{\varphi}^{*r}(B)$ for every $0 < r < d(A, B)/2$. It suffices to show that $P_{\varphi}^{*r}(A \cup B) \ge P_{\varphi}^{*r}(A) + P_{\varphi}^{*r}(B)$, as the opposite inequality is simply (15.1). Let $\{(x_i, r_i)\}_{i=1}^{\infty}$ be an r-packing of A and let $\{(y_j, s_j)\}_{j=1}^{\infty}$ be an r-packing of B. Then $\{(x_i, r_i)\}_{i=1}^{\infty} \cup \{(y_j, s_j)\}_{j=1}^{\infty}$ is an r-packing of $A \cup B$ since $d(x_i, y_j) \ge d(A, B) > 2r \ge r_i + s_j$ for all i, j. Therefore,

$$P_{\varphi}^{*r}(A \cup B) \ge \sum_{i=1}^{\infty} \varphi(r_i) + \sum_{j=1}^{\infty} \varphi(s_j).$$

Upon taking the supremum over all r-packings of A and B, it follows that

$$P_{\varphi}^{*r}(A \cup B) \ge P_{\varphi}^{*r}(A) + P_{\varphi}^{*r}(B).$$

In summary,

$$P_{\varphi}^{*r}(A \cup B) = P_{\varphi}^{*r}(A) + P_{\varphi}^{*r}(B), \quad \forall 0 < r < d(A, B)/2.$$

Letting $r \to 0$ yields

$$P_{\varphi}^{*}(A \cup B) = P_{\varphi}^{*}(A) + P_{\varphi}^{*}(B) \quad \text{when} \quad d(A, B) > 0. \tag{15.2}$$

In retrospect, only countable subadditivity is missing in order for P_{φ}^{*} to be a metric outer measure. To obtain one, we have to go through one further step.

Definition 15.2.19. For each $A \subseteq X$, let

$$\begin{aligned} P_{\varphi}(A) &:= \inf\left\{\sum_{i=1}^{\infty} P_{\varphi}^{*}(C_i) : \{C_i\}_{i=1}^{\infty} \text{ is a countable partition of } A\right\} \\ &= \inf\left\{\sum_{i=1}^{\infty} P_{\varphi}^{*}(C_i) : \{C_i\}_{i=1}^{\infty} \text{ is a countable cover of } A\right\}. \end{aligned}$$

Remark 15.2.20. Clearly, $P_{\varphi}(A) \le P_{\varphi}^{*}(A)$ for all $A \subseteq X$. The reader also ought to convince themself that countable partitions and countable covers can be used interchangeably in the definition of $P_{\varphi}(A)$. We can thus use partitions or covers, whichever are more suitable to the situation at hand.

Lemma 15.2.21. *$P_\varphi : \mathcal{P}(X) \to [0,\infty]$ is a metric outer measure and its restriction $P_\varphi : \mathcal{B}(X) \to [0,\infty]$ is a measure.*

Proof. For the first two conditions of an outer measure, notice that $P_\varphi(\emptyset) = 0$ since $P^*_\varphi(\emptyset) = 0$ while the monotonicity of P_φ follows directly from that of P^*_φ. It remains to establish countable subadditivity. To that end, let $A = \bigcup_{n=1}^\infty A_n$. Let $\varepsilon > 0$ and for each $n \in \mathbb{N}$, choose a countable cover $\{C_{n,i}\}_{i=1}^\infty$ of A_n such that

$$\sum_{i=1}^\infty P^*_\varphi(C_{n,i}) \le P_\varphi(A_n) + \frac{\varepsilon}{2^n}.$$

Then $\{C_{n,i}\}_{n,i=1}^\infty$ is a countable cover of A and thus

$$P_\varphi(A) \le \sum_{n=1}^\infty \sum_{i=1}^\infty P^*_\varphi(C_{n,i}) \le \sum_{n=1}^\infty \left[P_\varphi(A_n) + \frac{\varepsilon}{2^n}\right] = \sum_{n=1}^\infty P_\varphi(A_n) + \varepsilon.$$

Letting $\varepsilon \to 0$ leads to the inequality $P_\varphi(A) \le \sum_{n=1}^\infty P_\varphi(A_n)$. So P_φ is an outer measure.

It remains to show that P_φ is a metric one. Suppose that $d(A,B) > 0$. We want to show that $P_\varphi(A \cup B) \ge P_\varphi(A) + P_\varphi(B)$ (the opposite inequality follows from the subadditivity of P_φ). Let $\{C_i\}_{i=1}^\infty$ be any countable cover of $A \cup B$. Then $\{C_i \cap A\}_{i=1}^\infty$ covers A and $\{C_i \cap B\}_{i=1}^\infty$ covers B. Moreover, $d(C_i \cap A, C_i \cap B) \ge d(A,B) > 0$ for all i. Hence $P^*_\varphi((C_i \cap A) \cup (C_i \cap B)) = P^*_\varphi(C_i \cap A) + P^*_\varphi(C_i \cap B)$ for all i by (15.2). Thus,

$$\begin{aligned}\sum_{i=1}^\infty P^*_\varphi(C_i) &\ge \sum_{i=1}^\infty P^*_\varphi(C_i \cap (A \cup B)) \\ &= \sum_{i=1}^\infty P^*_\varphi((C_i \cap A) \cup (C_i \cap B)) \\ &= \sum_{i=1}^\infty P^*_\varphi(C_i \cap A) + \sum_{i=1}^\infty P^*_\varphi(C_i \cap B) \\ &\ge P_\varphi(A) + P_\varphi(B).\end{aligned}$$

Taking the infimum over all countable covers of $A \cup B$ gives

$$P_\varphi(A \cup B) \ge P_\varphi(A) + P_\varphi(B).$$

So P_φ is a metric outer measure. Theorem 15.1.7 states that $P_\varphi|_{\mathcal{B}(X)}$ is a measure. □

Definition 15.2.22. The metric outer measure P_φ is called the *packing measure* induced by the gauge function φ.

We shall soon see that the case $\varphi(t) := \varphi_\alpha(t) = t^\alpha$, with $\alpha > 0$, is of particular importance and for this reason we shorten the notations $P^*_{t^\alpha}(A)$ and $P_{t^\alpha}(A)$ to P^*_α and P_α, respectively. The outer measure P_α accordingly deserves a special name.

Definition 15.2.23. The metric outer measure P_α is called the *α-dimensional packing measure.*

15.2.4 Packing versus Hausdorff measures

There is a general relationship between the Hausdorff and packing measures induced by an evenly varying gauge function.

Proposition 15.2.24. *Let $A \subseteq X$. If φ is an evenly varying gauge function, then*

$$H_\varphi(A) \leq (c_\varphi(2))^2 P_\varphi(A).$$

Proof. Let $r > 0$. First, we will show that $H_\varphi^{B,2r}(A) \leq c_\varphi(2) P_\varphi^{*r}(A)$. Suppose that there is no finite, maximal (in the sense of cardinality) r-packing of A of the form $\{(x_i, r)\}_{i=1}^\infty$. Then for every $k \in \mathbb{N}$ there is a packing $\{(x_i, r)\}_{i=1}^k$ of A. Hence,

$$P_\varphi^{*r}(A) \geq \sum_{i=1}^k \varphi(r) = k\varphi(r) \to \infty \quad \text{as } k \to \infty.$$

Thus, $P_\varphi^{*r}(A) = \infty$ and we are done in this case.

Otherwise, there exists a finite maximal r-packing $\{(x_i, r)\}_{i=1}^\ell$ of A. Then $\{B(x_i, 2r)\}_{i=1}^\ell$ is a $(2r)$-ball cover of A. Therefore,

$$H_\varphi^{B,2r}(A) \leq \sum_{i=1}^\ell \varphi(2r) \leq c_\varphi(2) \sum_{i=1}^\ell \varphi(r) \leq c_\varphi(2) P_\varphi^{*r}(A).$$

Upon taking the limit as r tends to zero, we obtain that

$$H_\varphi^B(A) \leq c_\varphi(2) P_\varphi^*(A).$$

Now, if $A \subseteq \bigcup_{i=1}^\infty C_i$ then using the above and Lemma 15.2.14 we get

$$H_\varphi(A) \leq \sum_{i=1}^\infty H_\varphi(C_i) \leq c_\varphi(2) \sum_{i=1}^\infty H_\varphi^B(C_i) \leq (c_\varphi(2))^2 \sum_{i=1}^\infty P_\varphi^*(C_i).$$

Taking the infimum over all countable covers of A, we finally deduce that

$$H_\varphi(A) \leq (c_\varphi(2))^2 P_\varphi(A). \qquad \square$$

Though we do not provide a proof of the following fact, this latter will prove quite useful in Chapter 16 (in particular, see Definition 16.2.1 and Theorem 16.3.2; for a definition of conformality, see Section 16.1).

Lemma 15.2.25. *Let $n \in \mathbb{N}$. If U is an open subset of $\mathbb{R}^n$ and $f : U \to \mathbb{R}^n$ is a conformal C^1 local diffeomorphism, then for every Borel set $B \subseteq U$ it turns out that*

$$H_\alpha(f(B)) = \int_B |f'(x)|^\alpha \, dH_\alpha(x) \quad \text{and} \quad P_\alpha(f(B)) = \int_B |f'(x)|^\alpha \, dP_\alpha(x).$$

15.3 Dimensions of sets

In this section, we shall see that Hausdorff and packing measures that are induced by some parameterized families of gauge functions change drastically at a certain parameter. This special parameter, which depends on the set $A \subseteq X$ (and the metric d on X) under consideration, is called the Hausdorff (resp., packing) dimension of the set A. The changing behavior of these measures can be described as follows. At any parameter smaller than the dimension of the set, the measure of the set is infinite (as the gauge function is then too fine to estimate the size of the set), while at any parameter larger than the dimension of the set, the measure of the set is zero (as the gauge function is then too coarse to perceive the structure of the set). At the critical/transition parameter, the Hausdorff (resp., packing) measure of the set can be zero, positive and finite, or infinite; anything is possible.

15.3.1 Hausdorff dimension

Let us begin by proving a straightforward lemma that will allow us to formulate the definition of the Hausdorff dimension of a set. As we will shortly see, the Hausdorff dimension has the advantage of being defined for any set. We restrict ourselves to the case $\varphi_\alpha(t) = t^\alpha$.

Lemma 15.3.1. *Let $A \subseteq X$. If $0 < \alpha_1 \leq \alpha_2$, then $H^\delta_{\alpha_2}(A) \leq \delta^{\alpha_2-\alpha_1} H^\delta_{\alpha_1}(A)$ for all $\delta > 0$.*

Proof. Let $\delta > 0$ and $\{C_i\}_{i=1}^\infty$ be a δ-cover of A. Then

$$\sum_{i=1}^\infty (\operatorname{diam}(C_i))^{\alpha_2} = \sum_{i=1}^\infty (\operatorname{diam}(C_i))^{\alpha_2-\alpha_1} (\operatorname{diam}(C_i))^{\alpha_1} \leq \delta^{\alpha_2-\alpha_1} \sum_{i=1}^\infty (\operatorname{diam}(C_i))^{\alpha_1}.$$

Taking the infimum over all δ-covers of A gives $H^\delta_{\alpha_2}(A) \leq \delta^{\alpha_2-\alpha_1} H^\delta_{\alpha_1}(A)$. □

Corollary 15.3.2. *Let $A \subseteq X$. If $H_{\alpha_1}(A) < \infty$ for some $\alpha_1 > 0$, then $H_{\alpha_2}(A) = 0$ for all $\alpha_2 > \alpha_1$.*

Proof. Assume that $H_{\alpha_1}(A) < \infty$ for some $\alpha_1 \geq 0$ and let $\alpha_2 > \alpha_1$. By Lemma 15.3.1,

$$H_{\alpha_2}(A) = \lim_{\delta\to 0} H^\delta_{\alpha_2}(A) \leq \lim_{\delta\to 0} \delta^{\alpha_2-\alpha_1} \cdot \lim_{\delta\to 0} H^\delta_{\alpha_1}(A) = 0 \cdot H_{\alpha_1}(A) = 0.$$ □

Hence, there is a unique parameter $\alpha \in [0,\infty]$ such that $H_\beta(A) = \infty$ for all $\beta < \alpha$ and $H_\beta(A) = 0$ for all $\beta > \alpha$. This unique α is called the *Hausdorff dimension* of the set A. The value of $H_\alpha(A)$ may be 0, positive and finite, or ∞. Figure 15.1 shows a graph of the function $\alpha \mapsto H_\alpha(A)$.

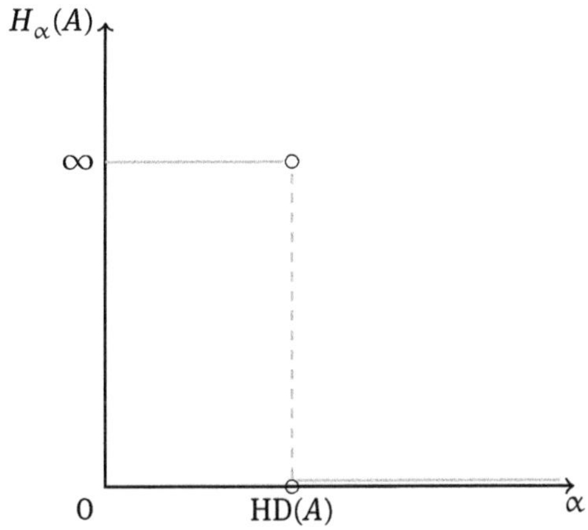

Figure 15.1: As the parameter α crosses the Hausdorff dimension HD(A), the α-dimensional Hausdorff measure $H_\alpha(A)$ collapses from ∞ to 0.

Definition 15.3.3. Let $A \subseteq X$. The *Hausdorff dimension* of A is defined to be

$$\mathrm{HD}(A) := \sup\{\alpha > 0 : H_\alpha(A) = \infty\} = \inf\{\alpha > 0 : H_\alpha(A) = 0\}.$$

We now examine the basic properties of the Hausdorff dimension. First of all, the Hausdorff dimension is an increasing set function.

Theorem 15.3.4. *If $A \subseteq B$, then* $\mathrm{HD}(A) \le \mathrm{HD}(B)$.

Proof. If $\mathrm{HD}(B) = \infty$, we are done. So suppose that $\mathrm{HD}(B) < \infty$. Let $\mathrm{HD}(B) < \alpha$. Then $H_\alpha(B) = 0$, and thus $H_\alpha(A) = 0$ since $A \subseteq B$. Hence, $\mathrm{HD}(A) \le \alpha$. Since this is true for all $\mathrm{HD}(B) < \alpha$, we conclude that $\mathrm{HD}(A) \le \mathrm{HD}(B)$. □

We shall now show that the Hausdorff dimension function is *countably stable* (also called *σ-stable*).

Theorem 15.3.5. *If $\{A_n\}_{n=1}^\infty$ is a countable family of subsets of X, then*

$$\mathrm{HD}\Big(\bigcup_{n=1}^{\infty} A_n\Big) = \sup_{n\in\mathbb{N}} \mathrm{HD}(A_n).$$

Proof. By Theorem 15.3.4, it is clear that $\mathrm{HD}(A_n) \le \mathrm{HD}(\cup_{m=1}^\infty A_m)$ for every $n \in \mathbb{N}$. Taking the supremum over all $n \in \mathbb{N}$, we obtain the inequality $\sup_{n\in\mathbb{N}} \mathrm{HD}(A_n) \le \mathrm{HD}(\cup_{n=1}^\infty A_n)$.

For the opposite inequality, let $s := \sup_{n\in\mathbb{N}} \mathrm{HD}(A_n)$. If $s = \infty$, we are done. So assume that $s < t < \infty$. Then $\mathrm{HD}(A_n) \le s < t$ for all $n \in \mathbb{N}$, and hence $H_t(A_n) = 0$ for all n. Recall that H_t is an outer measure, so is countably subadditive. Therefore $H_t(\bigcup_{n=1}^{\infty} A_n) \le \sum_{n=1}^{\infty} H_t(A_n) = 0$. Thus, $\mathrm{HD}(\bigcup_{n=1}^{\infty} A_n) \le t$ for every $t > s$. That is, $\mathrm{HD}(\bigcup_{n=1}^{\infty} A_n) \le s$. □

An immediate consequence of this result is that any countable set has Hausdorff dimension zero.

Proposition 15.3.6. *If $A \subseteq X$ is countable, then* $\mathrm{HD}(A) = 0$.

Proof. By Theorem 15.3.5, it suffices to show that $\mathrm{HD}(\{x\}) = 0$ for every $x \in X$. We already know from Remark 15.2.10(b) that each H_α is atomless, i. e., $H_\alpha(\{x\}) = 0$ for every $x \in X$ and every $\alpha > 0$. Thus, $\mathrm{HD}(\{x\}) = 0$ for all x. □

15.3.2 Packing dimensions

We can also define packing* and packing dimensions in a similar natural way, from the definitions of the quantity P_φ^* and the metric outer measure P_φ. We again restrict ourselves to the case $\varphi_\alpha(t) = t^\alpha$.

Lemma 15.3.7. *Let $A \subseteq X$. If $0 < \alpha_1 \le \alpha_2$, then for all $r > 0$,*

$$P_{\alpha_2}^{*r}(A) \le r^{\alpha_2-\alpha_1} P_{\alpha_1}^{*r}(A).$$

Proof. The proof goes along the same lines as that of Lemma 15.3.1 and is left to the reader as an exercise. □

Corollary 15.3.8. *Let $A \subseteq X$. If $P_{\alpha_1}^*(A) < \infty$ for some α_1, then $P_{\alpha_2}^*(A) = 0$ for all $\alpha_2 > \alpha_1$.*

Proof. The proof follows the same lines as that of Corollary 15.3.2. □

Thus, the graph of the function $\alpha \mapsto P_\alpha^*(A)$ has the same general shape as that of $\alpha \mapsto H_\alpha(A)$, as there exists a unique parameter $\alpha^* \in [0,\infty]$ such that $P_\beta^*(A) = \infty$ for all $\beta < \alpha^*$ and $P_\beta^*(A) = 0$ for all $\beta > \alpha^*$. This unique α^* is called the *packing* dimension* of A. The value of $P_\alpha^*(A)$ may be 0, positive and finite, or ∞.

Definition 15.3.9. Let $A \subseteq X$. The *packing* dimension* of A is

$$\mathrm{PD}^*(A) := \sup\{\alpha > 0 : P_\alpha^*(A) = \infty\} = \inf\{\alpha > 0 : P_\alpha^*(A) = 0\}.$$

By taking the infimum over all countable covers of the set A, we further obtain some information about the packing measures.

Corollary 15.3.10. *Let $A \subseteq X$. If $P_{\alpha_1}(A) < \infty$ for some α_1, then $P_{\alpha_2}(A) = 0$ for all $\alpha_2 > \alpha_1$.*

Proof. This ensues from Definition 15.2.19 and Corollary 15.3.8. □

Consequently, the graph of the function $\alpha \mapsto P_\alpha(A)$ resembles those of $\alpha \mapsto P^*_\alpha(A)$ and $\alpha \mapsto H_\alpha(A)$, as there exists a unique $\alpha \geq 0$ such that $P_\beta(A) = \infty$ for all $\beta < \alpha$ and $P_\beta(A) = 0$ for all $\beta > \alpha$. This unique α is called the *packing dimension* of A. The value of $P_\alpha(A)$ may be 0, positive and finite, or ∞.

Definition 15.3.11. Let $A \subseteq X$. The *packing dimension* of A is

$$\mathrm{PD}(A) := \sup\{\alpha > 0 : P_\alpha(A) = \infty\} = \inf\{\alpha > 0 : P_\alpha(A) = 0\}.$$

Like the Hausdorff dimension, the packing dimension respects the set-inclusion order and is countably stable.

Theorem 15.3.12. *If* $A \subseteq B \subseteq X$, *then* $\mathrm{PD}^*(A) \leq \mathrm{PD}^*(B)$ *and* $\mathrm{PD}(A) \leq \mathrm{PD}(B)$. *Moreover, if* $\{A_n\}_{n=1}^{\infty}$ *is a countable family of subsets of* X, *then*

$$\mathrm{PD}\Big(\bigcup_{n=1}^{\infty} A_n\Big) = \sup_{n\in\mathbb{N}} \mathrm{PD}(A_n).$$

Proof. A proof, analog to those of Theorems 15.3.4–15.3.5, is left as an exercise. □

The following is a direct consequence of this theorem and is proved like its counterpart for the Hausdorff dimension, Proposition 15.3.6.

Proposition 15.3.13. *If* $A \subseteq X$ *is countable, then* $\mathrm{PD}(A) = 0$.

15.3.3 Packing versus Hausdorff dimensions

We now describe a relationship between the Hausdorff, packing* and packing dimensions of any given set.

Proposition 15.3.14. *For any* $A \subseteq X$, *we have that* $\mathrm{HD}(A) \leq \mathrm{PD}(A) \leq \mathrm{PD}^*(A)$.

Proof. By Proposition 15.2.24 and the fact that $\varphi_\alpha(t) = t^\alpha$ is evenly varying, we get

$$H_\alpha(A) \leq 2^{2\alpha} P_\alpha(A).$$

Thus, $P_\alpha(A) = 0$ implies that $H_\alpha(A) = 0$ and, therefore, $\mathrm{HD}(A) \leq \mathrm{PD}(A)$.

On the other hand, since $\{A\}$ is a cover of A, it is clear that $P_\alpha(A) \leq P^*_\alpha(A)$ for every α, and hence $\mathrm{PD}(A) \leq \mathrm{PD}^*(A)$. □

15.3.4 Box-counting dimensions

We now examine a slightly different type of dimension, the box-counting dimension. This dimension is not defined via an outer measure. It behaves somewhat awkwardly:

it is not σ-stable, and a set and its closure have the same dimension. However, its definition is substantially simpler than those of the Hausdorff and packing dimensions; it is frequently easier to calculate or to estimate; it often agrees with the Hausdorff and packing dimensions; and it is widely used in physics.

Definition 15.3.15. Let $r > 0$, and let $A \subseteq X$ be a bounded set. Define $N(A, r)$ to be the minimum cardinality of an r-ball cover of A. Then the *upper and lower box-counting* (or, more simply, *box*) dimensions of A are respectively defined to be

$$\overline{\mathrm{BD}}(A) := \limsup_{r\to 0} \frac{\log N(A,r)}{-\log r} \quad \text{and} \quad \underline{\mathrm{BD}}(A) := \liminf_{r\to 0} \frac{\log N(A,r)}{-\log r}.$$

If these two quantities are equal, their common value is called the *box-counting* (or simply *box*) dimension of A and is denoted by $\mathrm{BD}(A)$.

The box-counting dimensions do not share all the congenial properties of the Hausdorff dimension. In particular, they are not σ-stable. To see this, observe that $\mathrm{BD}(\mathbb{Q} \cap [0,1]) = 1 \neq 0 = \sup\{\mathrm{BD}(\{q\}) : q \in \mathbb{Q} \cap [0,1]\}$.

15.3.5 Alternate definitions of box dimensions

The terminology "box-counting" comes from the fact that in Euclidean spaces we may use boxes from a lattice, rather than balls, to cover the set under scrutiny. Indeed, let $n \in \mathbb{N}$, $X = \mathbb{R}^n$ and let $\mathcal{L}(r)$ be any lattice in $\mathbb{R}^n$ consisting of closed n-dimensional cubes (also called boxes) with edges of length r. For any $A \subseteq X$, define

$$L(A, r) := \#\{C \in \mathcal{L}(r) : C \cap A \neq \emptyset\}.$$

Proposition 15.3.16. *If A is a bounded subset of $\mathbb{R}^n$, then*

$$\overline{\mathrm{BD}}(A) = \limsup_{r\to 0} \frac{\log L(A,r)}{-\log r} \quad \textit{and} \quad \underline{\mathrm{BD}}(A) = \liminf_{r\to 0} \frac{\log L(A,r)}{-\log r}.$$

Proof. Without loss of generality, let $0 < r < 1$. Select points $x_i \in A$ so that

$$A \subseteq \bigcup_{i=1}^{N(A,r)} B(x_i, r).$$

Momentarily fix $1 \le i \le N(A,r)$. If $C \in \mathcal{L}(r)$ is such that $d(C, x_i) < r$, where d denotes the standard Euclidean metric on $\mathbb{R}^n$, one immediately verifies that $C \subseteq B(x_i, (1+\sqrt{n})r)$. Thus, for any $1 \le i \le N(A,r)$, we have that

$$\#\{C \in \mathcal{L}(r) : d(C, x_i) < r\} \le \frac{\lambda_n(B(x_i,(1+\sqrt{n})r))}{\lambda_n(\text{cube of side } r)} \le \frac{c_n[(1+\sqrt{n})r]^n}{r^n} = c_n(1+\sqrt{n})^n,$$

where λ_n denotes the Lebesgue measure on $\mathbb{R}^n$ and c_n denotes the volume of the unit ball in $\mathbb{R}^n$. Since every $C \in L(A,r)$ admits at least one $1 \leq i \leq N(A,r)$ such that $d(C,x_i) < r$, we deduce that $L(A,r) \leq N(A,r)c_n(1+\sqrt{n})^n$. Therefore,

$$\log L(A,r) \leq \log[c_n(1+\sqrt{n})^n] + \log N(A,r).$$

Hence,

$$\frac{\log L(A,r)}{-\log r} \leq \frac{\log[c_n(1+\sqrt{n})^n]}{-\log r} + \frac{\log N(A,r)}{-\log r}.$$

So

$$\limsup_{r\to 0} \frac{\log L(A,r)}{-\log r} \leq \overline{\mathrm{BD}}(A) \quad \text{and} \quad \liminf_{r\to 0} \frac{\log L(A,r)}{-\log r} \leq \underline{\mathrm{BD}}(A).$$

For the opposite inequalities, let again $0 < r < 1$ and for each $C \in L(A,r)$ choose $x_C \in C \cap A$. Then $C \cap A \subseteq B(x_C, 2r\sqrt{n})$. Thus, the family of balls $\{B(x_C, 2r\sqrt{n}) : C \in L(A,r)\}$ covers A. Therefore, $N(A, 2r\sqrt{n}) \leq L(A,r)$. It then follows that

$$\overline{\mathrm{BD}}(A) \leq \limsup_{r\to 0} \frac{\log L(A,r)}{-\log r} \quad \text{and} \quad \underline{\mathrm{BD}}(A) \leq \liminf_{r\to 0} \frac{\log L(A,r)}{-\log r}. \qquad \square$$

15.3.6 Hausdorff versus packing versus box dimensions

We now present general relationships between the Hausdorff, packing and box dimensions. For this, we return to our general setting. Let (X,d) be a metric space, let $A \subseteq X$ and $r > 0$. Define $P(A,r)$ to be the supremum of the cardinalities of all r-packings of A of the form $\{(x_i,r)\}_{i=1}^k$, i. e.,

$$P(A,r) := \sup\{k : \{(x_i,r)\}_{i=1}^k \text{ is an } r\text{-packing}\}. \tag{15.3}$$

Lemma 15.3.17. *If $A \subseteq X$ and $r > 0$, then $N(A,2r) \leq P(A,r) \leq N(A,r)$.*

Proof. The first inequality certainly holds if $P(A,r) = \infty$. So assume this is not the case and let $\{(x_i,r)\}_{i=1}^k$ be an r-packing of A with maximal cardinality $k = P(A,r) < \infty$. Then $\{B(x_i,2r)\}_{i=1}^k$ is a $(2r)$-cover of A. Consequently, $N(A,2r) \leq k = P(A,r)$.

For the second inequality, there is nothing to prove if $N(A,r) = \infty$. So, again, let $\{(x_i,r)\}_{i=1}^k$ be a finite r-packing of A. Let also $\{B(y_j,r)\}_{j=1}^{\ell}$ be a finite r-ball cover of A. Then, for each $1 \leq i \leq k$ there exists $1 \leq j(i) \leq \ell$ such that $x_i \in B(y_{j(i)}, r)$. We will show that $k \leq \ell$. It is enough to show that the function $i \mapsto j(i)$ is injective. But for each $1 \leq j \leq \ell$, the cardinality of the set $\{x_i\}_{i=1}^k \cap B(y_j,r)$ is at most 1; otherwise, $\{(x_i,r)\}_{i=1}^k$ would not be an r-packing. Thus, $i \mapsto j(i)$ is injective, as required. Hence, $k \leq l$ and $P(A,r) \leq N(A,r)$. $\square$

These inequalities have the following immediate implications.

Corollary 15.3.18. *If $A \subseteq X$, then*

$$\overline{\mathrm{BD}}(A) = \limsup_{r\to 0} \frac{\log P(A,r)}{-\log r} \quad \textit{and} \quad \underline{\mathrm{BD}}(A) = \liminf_{r\to 0} \frac{\log P(A,r)}{-\log r}.$$

Proof. This follows directly from Lemma 15.3.17. □

A surprising consequence of this is that the packing* and upper box dimensions coincide, as we show in the next lemma.

Lemma 15.3.19. *For every set $A \subseteq X$, we have* $\mathrm{PD}^*(A) = \overline{\mathrm{BD}}(A)$.

Proof. Let $t < \overline{\mathrm{BD}}(A)$. We will show that $\mathrm{PD}^*(A) \geq t$ for all such t, which will ensure that $\mathrm{PD}^*(A) \geq \overline{\mathrm{BD}}(A)$. Indeed, Corollary 15.3.18 implies that there is a sequence $(r_n)_{n=1}^{\infty}$, with $0 < r_n < 1$, such that $\lim_{n\to\infty} r_n = 0$ and

$$\frac{\log P(A, r_n)}{-\log r_n} > t.$$

Then

$$P_t^{*r_n}(A) = \sup\left\{\sum_{i=1}^{\infty} s_i^t : \{(x_i, s_i)\}_{i=1}^{\infty} \text{ is an } r_n\text{-packing of } A\right\} \geq r_n^t P(A, r_n) > 1.$$

Letting $n \to \infty$, we infer that $P_t^*(A) \geq 1$. So $\mathrm{PD}^*(A) \geq t$, as sought.

For the converse inequality, let $s < t < \mathrm{PD}^*(A)$. We must show that $\overline{\mathrm{BD}}(A) \geq s$. Since $t < \mathrm{PD}^*(A)$, we have that $P_t^*(A) = \infty$. Thus, for each $n \in \mathbb{N}$ large enough there is a (2^{-n})-packing $\{(x_{n,i}, r_{n,i})\}_{i=1}^{k(n)}$ of A such that $\sum_{i=1}^{k(n)} r_{n,i}^t > 1$.

Fix $n \in \mathbb{N}$ large enough. The idea now is to split the (2^{-n})-packing into subpackings with nearly the same radius in order to get closer to the definition of the box-counting dimension. For each $m \geq n$, let

$$I_{n,m} := \{1 \leq i \leq k(n) : 2^{-(m+1)} < r_{n,i} \leq 2^{-m}\}.$$

Observe that $\{(x_{n,i}, r_{n,i})\}_{i\in I_{n,m}}$ is a $2^{-(m+1)}$-packing of A. We would like to estimate the cardinality of $I_{n,m}$. First, note that

$$1 < \sum_{i=1}^{k(n)} r_{n,i}^t = \sum_{m=n}^{\infty} \sum_{i\in I_{n,m}} r_{n,i}^t \leq \sum_{m=n}^{\infty} (\#I_{n,m}) 2^{-mt}.$$

Now, if it turned out that $\#I_{n,m} < 2^{ms}(1 - 2^{s-t})$ for all $m \geq n$, then we would have

$$1 < \sum_{m=n}^{\infty} (\#I_{n,m}) 2^{-mt} < (1 - 2^{s-t}) \sum_{m=1}^{\infty} 2^{m(s-t)} = 1,$$

which is impossible. Thus, for all $n \in \mathbb{N}$, there exists $m(n) \geq n$ with $\#I_{n,m(n)} \geq 2^{m(n)s}(1 - 2^{s-t})$. Obviously, $m(n) \to \infty$ as $n \to \infty$. Hence,

$$P(A, 2^{-(m(n)+1)}) \geq \#I_{n,m(n)} \geq 2^{m(n)s}(1 - 2^{s-t}).$$

Therefore, for all $n \in \mathbb{N}$ we deduce that

$$\frac{\log P(A, 2^{-(m(n)+1)})}{-\log 2^{-(m(n)+1)}} \geq \frac{m(n)s \log 2 + \log(1 - 2^{s-t})}{(m(n) + 1)\log 2}.$$

By Corollary 15.3.18, we conclude that

$$\overline{\mathrm{BD}}(A) = \limsup_{r\to 0} \frac{\log P(A, r)}{-\log r} \geq \limsup_{n\to\infty} \frac{\log P(A, 2^{-(m(n)+1)})}{-\log 2^{-(m(n)+1)}} \geq s. \qquad \square$$

We then have the following general relationships between the various dimensions introduced so far.

Theorem 15.3.20. *If $A \subseteq X$, then*

$$\mathrm{HD}(A) \leq \min\{\mathrm{PD}(A), \underline{\mathrm{BD}}(A)\} \leq \max\{\mathrm{PD}(A), \underline{\mathrm{BD}}(A)\} \leq \overline{\mathrm{BD}}(A) = \mathrm{PD}^*(A).$$

Proof. That $\overline{\mathrm{BD}}(A) = \mathrm{PD}^*(A)$ is just the statement of Lemma 15.3.19. The second inequality is obvious. The last inequality follows from Proposition 15.3.14 and the trivial fact that $\underline{\mathrm{BD}}(A) \leq \overline{\mathrm{BD}}(A)$.

Concerning the first inequality, since $\mathrm{HD}(A) \leq \mathrm{PD}(A)$ by Proposition 15.3.14, all that remains to be shown is that $\mathrm{HD}(A) \leq \underline{\mathrm{BD}}(A)$. If $\mathrm{HD}(A) = 0$ then the inequality readily holds. So suppose that $\mathrm{HD}(A) > 0$ and let $0 < \alpha < \mathrm{HD}(A)$. Then

$$\lim_{\delta\to 0} H_\alpha^{2\delta}(A) = H_\alpha(A) = \infty.$$

Since A can be covered by $N(A, \delta)$ balls of radii at most δ, it ensues that $H_\alpha^{2\delta}(A) \leq N(A, \delta)(2\delta)^\alpha$. Thus, for all sufficiently small $\delta > 0$,

$$\log N(A, \delta) + \alpha \log(2\delta) \geq \log H_\alpha^{2\delta}(A) > 0.$$

Therefore,

$$\alpha \leq \liminf_{\delta\to 0} \frac{\log N(A, \delta)}{-\log(2\delta)} = \liminf_{\delta\to 0} \frac{\log N(A, \delta)}{-\log \delta} = \underline{\mathrm{BD}}(A).$$

Since this holds for all $\alpha < \mathrm{HD}(A)$, we deduce that $\mathrm{HD}(A) \leq \underline{\mathrm{BD}}(A)$. $\square$

15.3.7 Hausdorff, packing and box dimensions under (bi-)Lipschitz mappings

We now investigate the behavior of the Hausdorff and packing measures and dimensions, as well as the box-counting dimensions, under Lipschitz and bi-Lipschitz maps. It is obvious that isometries preserve all these quantities. However, it is fairly easy to see that, unlike the topological dimension of a set, the Hausdorff dimension of a set is not invariant under homeomorphisms. We will show that bi-Lipschitz maps preserve dimensions, and preserve "so much" of Hausdorff and packing measures (namely, their nullness and finiteness/infiniteness) that it would be well justified to call them (iso)morphisms for the category of metric spaces with respect to fractal measures and dimensions.

We first look at the behavior of the Hausdorff measures and dimension.

Theorem 15.3.21. *If (X, d_X) and (Y, d_Y) are metric spaces and $f : X \to Y$ is a Lipschitz map with Lipschitz constant $L > 0$, then:*
(a) $H_\alpha(f(X)) \le L^\alpha H_\alpha(X)$ *for every* $\alpha \ge 0$.
(a′) *If* $H_\alpha(X) < \infty$, *then* $H_\alpha(f(X)) < \infty$.
(b) $\mathrm{HD}(f(X)) \le \mathrm{HD}(X)$.

If f is bi-Lipschitz, meaning that f is bijective and both f and f^{-1} are Lipschitz, then
(c) $H_\alpha(Y) \le L^\alpha H_\alpha(X) \le L^\alpha L_{-1}^\alpha H_\alpha(Y)$ *for every* $\alpha \ge 0$, *where* L_{-1} *is a Lipschitz constant for* f^{-1}.
(c′) $H_\alpha(Y) > 0$ *if and only if* $H_\alpha(X) > 0$ *and* $H_\alpha(Y) < \infty$ *if and only if* $H_\alpha(X) < \infty$.
(d) $\mathrm{HD}(Y) = \mathrm{HD}(X)$.

Proof. Let $\alpha \ge 0$. Fix $\delta > 0$ and let $\{A_n\}_{n=1}^\infty$ be a $(L^{-1}\delta)$-cover of X. Then $\{f(A_n)\}_{n=1}^\infty$ is a δ-cover of $f(X)$ and

$$\sum_{n=1}^{\infty}(\mathrm{diam}(f(A_n)))^\alpha \le L^\alpha \sum_{n=1}^{\infty}(\mathrm{diam}(A_n))^\alpha.$$

Thus,

$$H_\alpha^\delta(f(X)) \le L^\alpha H_\alpha^{L^{-1}\delta}(X).$$

Letting $\delta \searrow 0$, part (a) ensues. The other parts are direct consequences of (a) and are left as an exercise. □

The behavior exhibited by the packing measures and dimension is identical.

Theorem 15.3.22. *If (X, d_X) and (Y, d_Y) are metric spaces and $f : X \to Y$ is a Lipschitz map with Lipschitz constant $L > 0$, then:*
(a) $P_\alpha(f(X)) \le L^\alpha P_\alpha(X)$ *for every* $\alpha \ge 0$.
(a′) *If* $P_\alpha(X) < \infty$, *then* $P_\alpha(f(X)) < \infty$.
(b) $\mathrm{PD}(f(X)) \le \mathrm{PD}(X)$.

If f is bi-Lipschitz, then

(c) $P_\alpha(Y) \le L^\alpha P_\alpha(X) \le L^\alpha L_{-1}^\alpha P_\alpha(Y)$ *for every* $\alpha \ge 0$, *where* L_{-1} *is a Lipschitz constant for* f^{-1}.

(c′) $P_\alpha(Y) > 0$ *if and only if* $P_\alpha(X) > 0$ *and* $P_\alpha(Y) < \infty$ *if and only if* $P_\alpha(X) < \infty$.

(d) $\mathrm{PD}(Y) = \mathrm{PD}(X)$.

Proof. Let $A \subseteq X$. Fix $r > 0$ and let $\{(y_j, r_j)\}_{j=1}^\infty$ be an r-packing of $f(A)$. For every $j \in \mathbb{N}$, pick an element $x_j \in f^{-1}(y_j) \cap A$. If $k \ne \ell$, then

$$d_X(x_k, x_\ell) \ge L^{-1} r_k + L^{-1} r_\ell$$

as, otherwise, we would have

$$r_k + r_\ell \le d_Y(y_k, y_\ell) = d_Y(f(x_k), f(x_\ell)) \le L d_X(x_k, x_\ell) < L(L^{-1} r_k + L^{-1} r_\ell) = r_k + r_\ell,$$

which is impossible. Hence, $\{(x_j, L^{-1} r_j)\}_{j=1}^\infty$ is an $(L^{-1} r)$-packing of A. Thus,

$$\sum_{j=1}^\infty r_j^\alpha = L^\alpha \sum_{j=1}^\infty (L^{-1} r_j)^\alpha \le L^\alpha P_\alpha^{*L^{-1} r}(A).$$

Taking the supremum over all r-packings of $f(A)$, it follows that

$$P_\alpha^{*r}(f(A)) \le L^\alpha P_\alpha^{*L^{-1} r}(A).$$

So

$$P_\alpha^*(f(A)) = \inf_{r>0} P_\alpha^{*r}(f(A)) \le L^\alpha \inf_{r>0} P_\alpha^{*L^{-1} r}(A) = L^\alpha P_\alpha^*(A).$$

Therefore, if $\{A_k\}_{k=1}^\infty$ is a countable cover of X then $\{f(A_k)\}_{k=1}^\infty$ is a countable cover of $f(X)$, and consequently,

$$P_\alpha(f(X)) \le \sum_{k=1}^\infty P_\alpha^*(f(A_k)) \le L^\alpha \sum_{k=1}^\infty P_\alpha^*(A_k).$$

Taking the infimum over all covers of X, we deduce that $P_\alpha(f(X)) \le L^\alpha P_\alpha(X)$. Part (a) is established. The other parts are repercussions of (a) and are left as an exercise. □

The following is a fairly immediate consequence of Theorems 15.3.21–15.3.22.

Theorem 15.3.23. *If (X, d_X) is a separable metric space, (Y, d_Y) is a metric space and $f : X \to Y$ is a locally Lipschitz map on some co-countable subset of X, then*

$$\mathrm{HD}(f(X)) \le \mathrm{HD}(X) \quad \textit{and} \quad \mathrm{PD}(f(X)) \le \mathrm{PD}(X).$$

Proof. By hypothesis, there exists a countable set $F \subseteq X$ such that for every $x \in X\backslash F$ there is $r(x) > 0$ for which $f|_{B(x,r(x))}$ is Lipschitz continuous. Since X is separable, so is $X\backslash F$. Therefore, there is a countable set $E \subseteq X\backslash F$ such that

$$\bigcup_{x\in E} B(x, r(x)) \supseteq X\backslash F.$$

Hence, $\{f(B(x,r(x))) : x \in E\}$ is a countable cover of $f(X\backslash F)$. It follows from Proposition 15.3.6 and Theorems 15.3.5 and 15.3.21(b) that

$$\begin{aligned}\mathrm{HD}(f(X)) &= \max\{\mathrm{HD}(f(F)), \mathrm{HD}(f(X\backslash F))\} = \max\{0, \mathrm{HD}(f(X\backslash F))\}\\ &= \mathrm{HD}(f(X\backslash F)) \le \sup\{\mathrm{HD}(f(B(x,r(x)))) : x \in E\}\\ &\le \sup\{\mathrm{HD}(B(x,r(x))) : x \in E\} \le \mathrm{HD}(X).\end{aligned}$$

The inequality $\mathrm{PD}(f(X)) \le \mathrm{PD}(X)$ is proved in the same way. □

Finally, let us glance at the behavior of the box-counting dimensions.

Theorem 15.3.24. *If (X, d_X) and (Y, d_Y) are metric spaces and $f : X \to Y$ is a Lipschitz map with Lipschitz constant $L > 0$, then:*
(a) $\underline{\mathrm{BD}}(f(X)) \le \underline{\mathrm{BD}}(X)$.
(b) $\overline{\mathrm{BD}}(f(X)) \le \overline{\mathrm{BD}}(X)$.
(c) *If* $\mathrm{BD}(X)$ *exists, then* $\underline{\mathrm{BD}}(f(X)) \le \overline{\mathrm{BD}}(f(X)) \le \mathrm{BD}(X)$.

If f is bi-Lipschitz, then
(d) $\underline{\mathrm{BD}}(Y) = \underline{\mathrm{BD}}(X)$.
(e) $\overline{\mathrm{BD}}(Y) = \overline{\mathrm{BD}}(X)$.
(f) *If* $\mathrm{BD}(X)$ *exists, then so does* $\mathrm{BD}(Y)$ *and* $\mathrm{BD}(Y) = \mathrm{BD}(X)$.

Proof. Fix $r > 0$. Since $f(B(x, L^{-1}r)) \subseteq B(f(x), r)$ for all $x \in X$, it is obvious that

$$N(f(X), r) \le N(X, L^{-1}r).$$

Therefore,

$$\begin{aligned}\underline{\mathrm{BD}}(f(X)) &= \liminf_{r\to 0}\frac{\log N(f(X),r)}{-\log r} \le \liminf_{r\to 0}\frac{\log N(X,L^{-1}r)}{-\log r} = \liminf_{s\to 0}\frac{\log N(X,s)}{-\log(Ls)}\\ &= \liminf_{s\to 0}\frac{\log N(X,s)}{-\log L - \log s} = \liminf_{s\to 0}\frac{\log N(X,s)}{-\log s(1+\frac{\log L}{\log s})} = \liminf_{s\to 0}\frac{\log N(X,s)}{-\log s}\\ &= \underline{\mathrm{BD}}(X).\end{aligned}$$

Thus, part (a) holds. Part (b) is proved in the same way. The other parts follow immediately from (a) and (b). □

15.4 A digression into geometric measure theory

In order to estimate measures and dimensions of sets, as well as dimensions of measures, we will need covering theorems. In geometric measure theory, there are essentially three fundamental covering theorems: Vitaly, $5r$ (or $4r$), and Besicovitch. We will look at the last two.

In what follows, we use the notation $tB(x,r) := B(x,tr)$. Furthermore, for every ball $B := B(x,r)$, we let $r(B) := r$ and $c(B) := x$.

Theorem 15.4.1 ($5r$-covering theorem). *Let (X,d) be a metric space and $\mathcal{B}$ a family of balls in X such that $\sup\{r(B) : B \in \mathcal{B}\} < \infty$. Then there exists a subfamily $\mathcal{B}' \subseteq \mathcal{B}$ consisting of mutually disjoint balls such that for every $B \in \mathcal{B}$ there is $B' \in \mathcal{B}'$ with $B \cap B' \neq \emptyset$ and $r(B) \leq 2r(B')$. In particular, this implies that*

$$\bigcup_{B\in\mathcal{B}} B \subseteq \bigcup_{B'\in\mathcal{B}'} 5B'.$$

In addition, if X is separable and $r(B) > 0$ for all but countably many $B \in \mathcal{B}$, then $\mathcal{B}'$ is countable.

For the proof, recall the following version of Zorn's lemma: If every chain in a nonempty partially ordered set has an upper bound, then the partially ordered set has a maximal element.

Proof. Let Ω denote the partially ordered (by inclusion) set consisting of all disjointed subfamilies ω of $\mathcal{B}$ with the following property: If a ball $B \in \mathcal{B}$ intersects some ball from ω, then there is $B' \in \omega$ with $B \cap B' \neq \emptyset$ and $r(B) \leq 2r(B')$.

First, note that $\Omega \neq \emptyset$ because the one-ball family $\omega = \{B_1\}$ is in Ω whenever $B_1 \in \mathcal{B}$ is such that $r(B_1) \geq \frac{1}{2}\sup\{r(B) : B \in \mathcal{B}\}$.

Second, if C is a chain in Ω then it is easy to see that

$$\omega_C := \bigcup_{\omega\in C} \omega$$

belongs to Ω. Thus every chain in Ω has an upper bound and Ω has a maximal element, denoted by $\mathcal{B}'$, according to Zorn's lemma. By construction, $\mathcal{B}'$ is disjointed.

Third, if there were a ball $B \in \mathcal{B}$ that did not meet any ball from $\mathcal{B}'$, then one could pick a ball $B_0 \in \mathcal{B}$ such that the radius of B_0 is at least half of the radius of any other ball that does not meet the balls from $\mathcal{B}'$. Then, if a ball $B \in \mathcal{B}$ met a ball from $\mathcal{B}'' := \mathcal{B}' \cup \{B_0\}$, by construction it would meet at least one ball from $\mathcal{B}''$ whose radius is at least half that of B, showing that $\mathcal{B}''$ belongs to Ω. But this would contradict the maximality of $\mathcal{B}'$. Consequently, every ball $B \in \mathcal{B}$ meets a ball $B' \in \mathcal{B}'$ such that $r(B) \leq 2r(B')$. The triangle inequality shows that $B \subseteq 5B'$.

The last part of the theorem follows immediately from the fact that any family of mutually disjoint open subsets of a separable space is countable. □

Remark 15.4.2.

(a) The proof of this theorem comes from Heinonen [48]. It applies to more general coverings than those by balls (see 2.8 in Federer [32]).

(b) Zorn's lemma is equivalent to the axiom of choice; see, e. g., Chapter 1 in [52] or [51].

(c) If there exists a finite Borel measure μ on X such that $\mu(B) > 0$ for all $B \in \mathcal{B}$, then $\mathcal{B}'$ is countable.

(d) Simple examples show that the assumption of uniformly bounded radii, i. e., $\sup\{r(B) : B \in \mathcal{B}\} < \infty$, is necessary.

(e) As a matter of fact, $5r$ in Theorem 15.4.1 can be replaced by $4r$. This improvement has virtually no consequences whatsoever, but its proof is much longer and much more involved as it relies on transfinite induction, and thus requires knowledge of ordinal numbers. Such a proof can be found in [63] but stems from the one in [72], where it is assumed that all closed balls are compact and an ordinary induction suffices.

In the following section, we shall derive several geometrical consequences of the $5r$-covering theorem.

Later, we will also need the following theorem, a proof of which can be found in [63]. That reference also contains the proofs of all other basic classical covering theorems used in geometric measure theory.

Theorem 15.4.3 (Besicovitch covering theorem). *Let $n \in \mathbb{N}$. There exists a constant $c(n) \in \mathbb{N}$ such that the following holds: If A is a bounded subset of $\mathbb{R}^n$ then for any function $r : A \to (0,\infty)$ there exists a countable subset $\{x_k\}_{k=1}^{\infty}$ of A such that the collection*

$$\mathcal{B}(A,r) := \{B(x_k, r(x_k)) : k \in \mathbb{N}\}$$

covers A and can be decomposed into $c(n)$ packings of A.

15.5 Volume lemmas—Frostman converse theorems

In this section, we identify conditions under which some sets have a positive or finite Hausdorff (resp., packing) measure. In general, this is much harder than showing that this measure is zero or infinite. But this will be instrumental in getting upper and lower bounds for the Hausdorff (resp., packing) dimension of those sets. Moreover, this will help us in estimating the dimension of some measures in Section 15.6. Throughout, we adopt the convention that $\frac{1}{\infty} := 0$ and $\frac{1}{0} := \infty$.

Theorem 15.5.1 (Frostman converse theorem for generalized Hausdorff measures). *Let $\varphi : [0,\infty) \to [0,\infty)$ be a continuous, evenly varying gauge function. Let μ be a Borel probability measure on a metric space X, and $A \subseteq X$.*

(a) *If there exists a constant* $c \in (0, \infty]$ *such that for all but countably many* $x \in A$,

$$\limsup_{r \to 0} \frac{\mu(B(x,r))}{\varphi(r)} \geq c, \tag{15.4}$$

then $H_\varphi(E) \leq c^{-1} c_\varphi(10)\mu(E)$ *for every Borel set* $E \subseteq A$. *In particular,* $H_\varphi(E) < \infty$, *and* $H_\varphi(E) = 0$ *if* $c = \infty$ *or* $\mu(E) = 0$.

(b) *If there exists a constant* $c \in [0, \infty)$ *such that for all* $x \in A$,

$$\limsup_{r \to 0} \frac{\mu(B(x,r))}{\varphi(r)} \leq c, \tag{15.5}$$

then $H_\varphi(E) \geq (cc_\varphi(2))^{-1}\mu(E)$ *for every Borel set* $E \subseteq A$. *In particular,* $H_\varphi(E) > 0$ *whenever* $\mu(E) > 0$.

Proof. (a) We shall prove this statement when $c < \infty$. The case $c = \infty$ is left as an exercise. Let E be a Borel set contained in A. Since H_φ is an outer measure that equals 0 on any countable set, we may assume that E does not contain any point of the exceptional countable set at which (15.4) is not satisfied. Fix $\varepsilon > 0$. By the regularity of μ, there exists an open set $G \supseteq E$ such that $\mu(G) \leq \mu(E) + \varepsilon$. Fix $R > 0$. By (15.4), for each $x \in E$ there exists $0 < r(x) < R$ such that $B(x, r(x)) \subseteq G$ and

$$\frac{\mu(B(x,r(x)))}{\varphi(r(x))} \geq c - \varepsilon \quad \text{or, equivalently,} \quad \varphi(r(x)) \leq \frac{\mu(B(x,r(x)))}{c - \varepsilon}. \tag{15.6}$$

In particular, this implies that the family $\{B(x, r(x))\}_{x \in E}$ consists of balls of positive μ-measure. According to the $5r$-covering theorem (Theorem 15.4.1) and Remark 15.4.2(c), there exists a countable subfamily of mutually disjoint balls $\{B(x_k, r(x_k))\}_{k=1}^{\infty}$ such that

$$E \subseteq \bigcup_{x \in E} B(x, r(x)) \subseteq \bigcup_{k=1}^{\infty} B(x_k, 5r(x_k))$$

for appropriately chosen $x_k \in E$. For each $k \in \mathbb{N}$, we obviously have that

$$\operatorname{diam}(B(x_k, 5r(x_k))) \leq 10r(x_k) \leq 10R.$$

Using this and (15.6), we obtain that

$$\begin{aligned} H_\varphi^{10R}(E) &\leq \sum_{k=1}^{\infty} \varphi(\operatorname{diam}(B(x_k, 5r(x_k)))) \\ &\leq \sum_{k=1}^{\infty} c_\varphi(10)\varphi(r(x_k)) \end{aligned}$$

$$\begin{aligned}
&\le c_\varphi(10)\sum_{k=1}^{\infty}\frac{1}{c-\varepsilon}\mu(B(x_k,r(x_k)))\\
&=\frac{c_\varphi(10)}{c-\varepsilon}\mu\Big(\bigcup_{k=1}^{\infty}B(x_k,r(x_k))\Big)\\
&\le\frac{c_\varphi(10)}{c-\varepsilon}\mu(G)\\
&\le\frac{c_\varphi(10)}{c-\varepsilon}(\mu(E)+\varepsilon).
\end{aligned}$$

Letting successively $R \to 0$ and $\varepsilon \to 0$ yields

$$H_\varphi(E) \le \frac{c_\varphi(10)}{c}\mu(E).$$

(b) For every $r > 0$, the regularity of μ implies that the function from X to $\mathbb{R}$ given by $x \mapsto \mu(B(x,r))/\varphi(r)$ is lower semicontinuous, and hence Borel measurable. Since the supremum of a countable family of measurable functions is a measurable function, it follows that for every $k \in \mathbb{N}$ the function

$$x \longmapsto \psi_k(x) := \sup\Big\{\frac{\mu(B(x,r))}{\varphi(r)} : r \in \mathbb{Q}\cap\Big(0,\frac{1}{k}\Big]\Big\}$$

is Borel measurable. Observe also that $(\psi_k)_{k=1}^{\infty}$ forms a decreasing sequence of functions. Let $s > c$. Let $A_0 = \emptyset$. For each $k \in \mathbb{N}$, let $A_k := \psi_k^{-1}((0,s])$. Since $(\psi_k)_{k=1}^{\infty}$ is decreasing, the sequence $(A_k)_{k=1}^{\infty}$ is ascending. Moreover, since $s > c$, it follows from (15.5) that $\bigcup_{k=1}^{\infty} A_k \supseteq A$. Furthermore, since $(0,s]$ is a Borel subset of $\mathbb{R}$, the measurability of ψ_k ensures that A_k is a Borel subset of X. Fix $k \in \mathbb{N}$. Choose some arbitrary $r \in (0,\frac{1}{k}]$ and pick a sequence of rational numbers $(r_j)_{j=1}^{\infty}$ such that r_j increases to r. Since μ is a measure and φ is continuous, we obtain for all $x \in A_k$ that

$$\frac{\mu(B(x,r))}{\varphi(r)} = \lim_{j\to\infty}\frac{\mu(B(x,r_j))}{\varphi(r_j)} \le \psi_k(x) \le s.$$

So, if $x \in A_k$ then

$$\sup\Big\{\frac{\mu(B(x,r))}{\varphi(r)} : r \in \Big(0,\frac{1}{k}\Big]\Big\} \le s. \tag{15.7}$$

Now, fix any set $F \subseteq A_k$, any $r \in (0,\frac{1}{k}]$, and let $\{B(x_i,t_i)\}_{i=1}^{\infty}$ be an r-ball cover of F. By (15.7),

$$\sum_{i=1}^{\infty}\varphi(t_i) \ge s^{-1}\sum_{i=1}^{\infty}\mu(B(x_i,t_i)) \ge s^{-1}\mu\Big(\bigcup_{i=1}^{\infty}B(x_i,t_i)\Big) \ge s^{-1}\mu(F).$$

Taking the infimum over all r-ball covers yields $H_\varphi^{B,r}(F) \geq s^{-1}\mu(F)$. As this holds for all $0 < r \leq 1/k$, we deduce that $H_\varphi^B(F) \geq s^{-1}\mu(F)$. By Lemma 15.2.14, we obtain that

$$H_\varphi(F) \geq \left(sc_\varphi(2)\right)^{-1}\mu(F), \quad \forall F \subseteq A_k, \ \forall k \in \mathbb{N}.$$

To complete the proof, consider the sequence $(B_k)_{k=1}^\infty$, where $B_k = (A_k \backslash A_{k-1}) \cap A$. Given that the sequence $(A_k)_{k=0}^\infty$ is ascending and $\bigcup_{k=1}^\infty A_k \supseteq A$, the sequence $(B_k)_{k=1}^\infty$ consists of mutually disjoint sets such that $B_k \subseteq A_k$ for all k and $\bigcup_{k=1}^\infty B_k = \bigcup_{k=1}^\infty A_k \cap A = A$. Hence, if E is a Borel subset of A then

$$H_\varphi(E) = H_\varphi\left(E \cap \bigcup_{k=1}^\infty B_k\right) = \sum_{k=1}^\infty H_\varphi(E \cap B_k) \geq \left(sc_\varphi(2)\right)^{-1}\sum_{k=1}^\infty \mu(E \cap B_k) = \left(sc_\varphi(2)\right)^{-1}\mu(E).$$

Letting $s \searrow c$ finishes the proof. □

Theorem 15.5.2 (Frostman converse theorem for generalized packing measures). *Let φ be a continuous, evenly varying gauge function. Let μ be a Borel probability measure on a metric space X, and $A \subseteq X$.*

(a) *If there exists a constant $c \in (0, \infty]$ such that for all but countably many $x \in A$,*

$$\liminf_{r \to 0} \frac{\mu(B(x,r))}{\varphi(r)} \geq c, \tag{15.8}$$

then $P_\varphi(E) \leq c^{-1}\mu(E)$ for each Borel set $E \subseteq A$. In particular, $P_\varphi(E) < \infty$, and $P_\varphi(E) = 0$ if $c = \infty$ or $\mu(E) = 0$.

(b) *If X is separable and there exists a constant $c \in [0, \infty)$ such that for all $x \in A$,*

$$\liminf_{r \to 0} \frac{\mu(B(x,r))}{\varphi(r)} \leq c, \tag{15.9}$$

then $P_\varphi(E) \geq (cc_\varphi(10))^{-1}\mu(E)$ for every Borel set $E \subseteq A$. In particular, $P_\varphi(E) > 0$ whenever $\mu(E) > 0$.

Proof. (a) Since P_φ is an outer measure that equals 0 on any countable set, we may assume that A does not contain any point of the exceptional countable set at which (15.8) is not satisfied. The sequence of functions $(\psi_k)_{k=1}^\infty$, where

$$x \longmapsto \psi_k(x) := \inf\left\{\frac{\mu(B(x,r))}{\varphi(r)} : r \in \mathbb{Q} \cap \left(0, \frac{1}{k}\right]\right\}$$

forms an increasing sequence of Borel measurable functions. Let $0 < s < c$. For each $k \in \mathbb{N}$, let $A_k := \psi_k^{-1}([s, \infty))$. As $(\psi_k)_{k=1}^\infty$ is increasing, so is the sequence $(A_k)_{k=1}^\infty$. Moreover, since $s < c$, it follows from (15.8) that $\bigcup_{k=1}^\infty A_k \supseteq A$. Furthermore, since $[s, \infty)$ is a Borel subset of $\mathbb{R}$, the measurability of ψ_k ensures that A_k is a Borel subset of X. Fix $k \in \mathbb{N}$. Choose some arbitrary $r \in (0, \frac{1}{k}]$ and pick a sequence of rational numbers $(r_j)_{j=1}^\infty$

such that r_j increases to r. Since μ is a measure and φ is continuous, we obtain for all $x \in A_k$ that

$$\frac{\mu(B(x,r))}{\varphi(r)} = \lim_{j\to\infty} \frac{\mu(B(x,r_j))}{\varphi(r_j)} \geq \psi_k(x) \geq s.$$

So, if $x \in A_k$ then $\inf\{\frac{\mu(B(x,r))}{\varphi(r)} : r \in (0, \frac{1}{k}]\} \geq s$. Now, fix any set $F \subseteq A_k$, any $r \in (0, \frac{1}{k}]$, and let $\{(x_i, t_i)\}_{i=1}^{\infty}$ be an r-packing of F. Then the balls $\{B(x_i, t_i)\}_{i=1}^{\infty}$ are mutually disjoint, and hence

$$\sum_{i=1}^{\infty} \varphi(t_i) \leq s^{-1} \sum_{i=1}^{\infty} \mu(B(x_i, t_i)) = s^{-1}\mu\Big(\bigcup_{i=1}^{\infty} B(x_i, t_i)\Big) \leq s^{-1}\mu(B(F, r)),$$

where $B(F, r) = \{x \in X : d(x, F) < r\}$ is the open r-neighborhood of F. Taking the supremum over all r-packings yields $P_{\varphi}^{*r}(F) \leq s^{-1}\mu(B(F, r))$. Thus, $P_{\varphi}(F) \leq s^{-1}\mu(B(F, r))$ for all $r \in (0, 1/k]$ and each subset $F \subseteq A_k$. Consequently, $P_{\varphi}(F) \leq s^{-1}\mu(\cap_{n\in\mathbb{N}} B(F, 1/n)) = s^{-1}\mu(\overline{F})$ for all $F \subseteq A_k$. In particular, if C is a closed subset of E then

$$P_{\varphi}(C \cap A_k) \leq s^{-1}\mu(\overline{C \cap A_k}) \leq s^{-1}\mu(C) \leq s^{-1}\mu(E).$$

As this holds for all $k \in \mathbb{N}$ and since $C \subseteq E \subseteq A \subseteq \cup_{k=1}^{\infty} A_k$ and the sequence $(A_k)_{k=1}^{\infty}$ is ascending, we deduce that $P_{\varphi}(C) \leq s^{-1}\mu(E)$. Upon taking the supremum over all closed sets C contained in E, the regularity of μ confirms that $P_{\varphi}(E) \leq s^{-1}\mu(E)$. Letting $s \nearrow c$ completes the proof.

(b) Let $\varepsilon > 0$. Choose an arbitrary subset $F \subseteq A$. Define a descending sequence $(G_n)_{n=0}^{\infty}$ of open sets containing F, as follows. Let $G_0 = X$. By (15.9), for every $x \in A$ there exists $0 < r_1(x) < 1$ such that

$$\frac{\mu(B(x, r_1(x)))}{\varphi(r_1(x))} \leq c + \varepsilon.$$

Take the family of balls $\{B(x, \frac{1}{5}r_1(x))\}_{x\in F}$. According to the $5r$-covering theorem (Theorem 15.4.1), there is a countable set $F_1 \subseteq F$ such that the subfamily $\{B(x, \frac{1}{5}r_1(x))\}_{x\in F_1}$ consists of mutually disjoint balls satisfying

$$F \subseteq \bigcup_{x\in F} B\Big(x, \frac{1}{5}r_1(x)\Big) \subseteq \bigcup_{x\in F_1} B(x, r_1(x)).$$

Set $G_1 := \bigcup_{x\in F_1} B(x, r_1(x))$. For the inductive step, suppose that G_n has been defined for some $n \in \mathbb{N}$. By (15.9), for every $x \in A$ there exists some $0 < r_{n+1}(x) < \frac{1}{n+1}$ such that $B(x, r_{n+1}(x)) \subseteq G_n$ and

$$\frac{\mu(B(x, r_{n+1}(x)))}{\varphi(r_{n+1}(x))} \leq c + \varepsilon. \tag{15.10}$$

Consider the family of balls $\{B(x, \frac{1}{5}r_{n+1}(x))\}_{x\in F}$. According to the $5r$-covering theorem (Theorem 15.4.1), there exists a countable set $F_{n+1} \subseteq F$ such that the subfamily $\{B(x, \frac{1}{5}r_{n+1}(x))\}_{x\in F_{n+1}}$ consists of mutually disjoint balls satisfying

$$F \subseteq \bigcup_{x\in F} B\Big(x, \frac{1}{5}r_{n+1}(x)\Big) \subseteq \bigcup_{x\in F_{n+1}} B(x, r_{n+1}(x)).$$

Let $G_{n+1} := \bigcup_{x\in F_{n+1}} B(x, r_{n+1}(x))$. It is clear that G_{n+1} is an open set and $F \subseteq G_{n+1} \subseteq G_n$. Moreover, for all pairs $x, y \in F_{n+1} \subseteq F$ we know that

$$d(x,y) \geq \frac{1}{5}\max\{r_{n+1}(x), r_{n+1}(y)\} \geq \frac{1}{10}(r_{n+1}(x) + r_{n+1}(y)).$$

Therefore, the collection $\{(x, \frac{1}{10}r_{n+1}(x))\}_{x\in F_{n+1}}$ forms a $(\frac{1}{n+1})$-packing of F. Using (15.10), it follows that

$$\begin{aligned}
P_\varphi^{*\frac{1}{n+1}}(F) &\geq \sum_{x\in F_{n+1}} \varphi\Big(\frac{1}{10}r_{n+1}(x)\Big) \\
&\geq (c_\varphi(10))^{-1} \sum_{x\in F_{n+1}} \varphi(r_{n+1}(x)) \\
&\geq (c_\varphi(10))^{-1} \sum_{x\in F_{n+1}} \frac{\mu(B(x, r_{n+1}(x)))}{C+\varepsilon} \\
&= \frac{(c_\varphi(10))^{-1}}{C+\varepsilon}\mu\Big(\bigcup_{x\in F_{n+1}} B(x, r_{n+1}(x))\Big) \\
&= \frac{(c_\varphi(10))^{1}}{C+\varepsilon}\mu(G_{n+1}).
\end{aligned}$$

Letting $n \to \infty$, it ensues that

$$P_\varphi^*(F) \geq \frac{(c_\psi(10))^{-1}}{C+\varepsilon} \lim_{n\to\infty} \mu(G_n) = \frac{(c_\varphi(10))^{-1}}{C+\varepsilon}\mu(G_F),$$

where $G_F := \bigcap_{n=1}^\infty G_n$ is a G_δ-set, and thereby a Borel set, containing F. Consequently, for every Borel set $E \subseteq A$ we have

$$\begin{aligned}
P_\varphi(E) &= \inf\Big\{\sum_{i=1}^\infty P_\varphi^*(C_i) : \{C_i\}_{i=1}^\infty \text{ is a partition of } E\Big\} \\
&\geq \frac{(c_\varphi(10))^{-1}}{C+\varepsilon} \inf\Big\{\sum_{i=1}^\infty \mu(G_{C_i}) : \{C_i\}_{i=1}^\infty \text{ is a partition of } E\Big\} \\
&\geq \frac{(c_\varphi(10))^{-1}}{C+\varepsilon} \inf\Big\{\mu\Big(\bigcup_{i=1}^\infty G_{C_i}\Big) : \{C_i\}_{i=1}^\infty \text{ is a partition of } E\Big\}
\end{aligned}$$

$$\geq \frac{(c_\varphi(10))^{-1}}{c+\varepsilon} \inf\left\{\mu\left(\bigcup_{i=1}^{\infty} C_i\right) : \{C_i\}_{i=1}^{\infty} \text{ is a partition of } E\right\}$$
$$= \frac{(c_\varphi(10))^{-1}}{c+\varepsilon}\mu(E).$$

Since this holds for all $\varepsilon > 0$, we deduce that $P_\varphi(E) \geq (cc_\varphi(10))^{-1}\mu(E)$. □

Replacing $\varphi(r)$ by r^t, H_φ by H_t, and P_φ by P_t in Theorems 15.5.1 and 15.5.2 respectively, we immediately get the following two results.

Theorem 15.5.3 (Frostman converse theorem for Hausdorff measures). *Fix $t > 0$. Let μ be a Borel probability measure on a metric space X, and $A \subseteq X$.*

(a) *If there exists a constant $c \in (0, \infty]$ such that*

$$\limsup_{r\to 0} \frac{\mu(B(x,r))}{r^t} \geq c$$

for all but countably many points $x \in A$, then the Hausdorff measure H_t satisfies

$$H_t(E) \leq c^{-1}10^t\mu(E)$$

for every Borel set $E \subseteq A$. In particular,

$$H_t(E) < \infty \quad \text{and} \quad (H_t(E) = 0 \text{ if } c = \infty \text{ or } \mu(E) = 0).$$

(b) *If there exists a constant $c \in [0, \infty)$ such that*

$$\limsup_{r\to 0} \frac{\mu(B(x,r))}{r^t} \leq c$$

for all $x \in A$, then

$$H_t(E) \geq (c2^t)^{-1}\mu(E)$$

for every Borel set $E \subseteq A$. In particular,

$$\mu(E) > 0 \Longrightarrow H_t(E) > 0.$$

Theorem 15.5.4 (Frostman converse theorem for packing measures). *Fix $t > 0$. Let μ be a Borel probability measure on a metric space X, and $A \subseteq X$.*

(a) *If there exists a constant $c \in (0, \infty]$ such that*

$$\liminf_{r\to 0} \frac{\mu(B(x,r))}{r^t} \geq c$$

for all but countably many $x \in A$, then the packing measure P_t satisfies

$$P_t(E) \le c^{-1}\mu(E)$$

for every Borel set $E \subseteq A$. In particular,

$$P_t(E) < \infty \quad and \quad (P_t(E) = 0 \text{ if } c = \infty \text{ or } \mu(E) = 0).$$

(b) *If X is separable and there exists a constant $c \in [0, \infty)$ such that*

$$\liminf_{r \to 0} \frac{\mu(B(x,r))}{r^t} \le c$$

for all $x \in A$, then

$$P_t(E) \ge (c10^t)^{-1}\mu(E)$$

for every Borel set $E \subseteq A$. In particular,

$$\mu(E) > 0 \Longrightarrow P_t(E) > 0.$$

In the literature, Frostman converse theorems are sometimes referred to as mass redistribution principles (especially for Hausdorff measures) or volume lemmas.

In the opposite direction to Frostman converse theorems, there is the following well-known fact (for a proof of this result, see Theorems 8.8 and 8.17 in [72]).

Theorem 15.5.5 (Frostman direct lemma). *If X is a Borel subset of a Euclidean space $\mathbb{R}^n$ or an arbitrary compact metric space and if $H_t(X) > 0$ for some $t > 0$, then there exists a Borel probability measure μ on X such that*

$$\mu(B(x,r)) \le r^t$$

for every point $x \in X$ and all radii $r > 0$.

This is an interesting theorem, although Frostman converse theorems are more convenient to estimate and calculate Hausdorff and packing measures and dimensions.

15.6 Dimensions of measures

The dimension of a measure is defined as the infimum of the dimensions of the sets of full measure.

Definition 15.6.1. Let μ be a Borel probability measure on a metric space (X, d).

(a) The *Hausdorff dimension of μ* is

$$\mathrm{HD}(\mu) := \inf\{\mathrm{HD}(Y) : \mu(Y) = 1\} = \inf\{\mathrm{HD}(Y) : \mu(X \setminus Y) = 0\}.$$

(b) The *packing dimension of* μ is

$$\mathrm{PD}(\mu) := \inf\{\mathrm{PD}(Y) : \mu(Y) = 1\} = \inf\{\mathrm{PD}(Y) : \mu(X \setminus Y) = 0\}.$$

We now introduce the well-studied and extremely useful concept of pointwise dimensions of a measure. These will be featured in Chapters 16 and 21 where multifractal analysis will be discussed.

Definition 15.6.2. Let μ be a Borel probability measure on a metric space X. The *lower* and *upper pointwise dimensions* of μ at a point $x \in X$ are respectively defined by

$$\underline{d}_\mu(x) := \liminf_{r\to 0} \frac{\log \mu(B(x,r))}{\log r} \quad \text{and} \quad \overline{d}_\mu(x) := \limsup_{r\to 0} \frac{\log \mu(B(x,r))}{\log r}.$$

If $\underline{d}_\mu(x) = \overline{d}_\mu(x)$, this common value is called the *pointwise dimension* of μ at x and is denoted by $d_\mu(x)$.

As a direct consequence of Theorem 15.5.3, we obtain an upper (resp., lower) bound for the Hausdorff dimension of sets whose lower pointwise dimensions have common upper (resp., lower) bounds.

Theorem 15.6.3. *Let μ be a Borel probability measure on a metric space X and $A \subseteq X$ be a Borel set.*

(a) *If $\mu(A) > 0$ and there exists $\theta_1 \geq 0$ such that for all $x \in A$ we have*

$$\underline{d}_\mu(x) \geq \theta_1, \tag{15.11}$$

then $\mathrm{HD}(A) \geq \theta_1$.

(b) *If there exists $\theta_2 \geq 0$ such that for all $x \in A$ we have*

$$\underline{d}_\mu(x) \leq \theta_2, \tag{15.12}$$

then $\mathrm{HD}(A) \leq \theta_2$.

Proof. (a) If $\theta_1 = 0$, the statement is trivial. So assume $\theta_1 > 0$ and pick an arbitrary $0 \leq \theta < \theta_1$. By (15.11), for every $x \in A$ there exists $0 < R < 1$ such that for all $0 < r < R$,

$$\frac{\log \mu(B(x,r))}{\log r} \geq \theta \quad \text{or, equivalently,} \quad \mu(B(x,r)) \leq r^\theta.$$

Therefore, $\limsup_{r\to 0} \frac{\mu(B(x,r))}{r^\theta} \leq 1$. Applying Theorem 15.5.3(b), we obtain that

$$H_\theta(A) \geq 2^{-\theta}\mu(A) > 0.$$

Thus, $\mathrm{HD}(A) \geq \theta$. Letting $\theta \nearrow \theta_1$, we conclude that $\mathrm{HD}(A) \geq \theta_1$.

(b) Fix $\theta > \theta_2$. By (15.12), for each $x \in A$ there is a sequence $(r_n)_{n=1}^{\infty}$ with $0 < r_n < 1$ and $r_n \to 0$ such that

$$\frac{\log \mu(B(x, r_n))}{\log r_n} \leq \theta \quad \text{or, equivalently,} \quad \mu(B(x, r_n)) \geq r_n^{\theta}.$$

Hence, $\limsup_{n\to\infty} \frac{\mu(B(x,r_n))}{r_n^{\theta}} \geq 1$. By Theorem 15.5.3(a), we deduce that

$$H_\theta(A) \leq 10^\theta \mu(A) < \infty.$$

Thus, $\mathrm{HD}(A) \leq \theta$. Letting $\theta \searrow \theta_2$, we conclude that $\mathrm{HD}(A) \leq \theta_2$. □

As a corollary to this theorem, we can also estimate the Hausdorff dimension of certain measures.

Corollary 15.6.4. *Let μ be a Borel probability measure on a metric space X.*
(a) *If there exist $\theta_1 \geq 0$ and a Borel set $A \subseteq X$ such that $\mu(A) > 0$ and*

$$\underline{d}_\mu(x) \geq \theta_1, \quad \forall x \in A, \tag{15.13}$$

then $\mathrm{HD}(\mu) \geq \theta_1$.
(b) *If there exists $\theta_2 \geq 0$ such that*

$$\underline{d}_\mu(x) \leq \theta_2, \quad \textit{for } \mu\textit{-a.e. } x \in X, \tag{15.14}$$

then $\mathrm{HD}(\mu) \leq \theta_2$.

Proof. (a) Let $Y \subseteq X$ be a Borel set such that $\mu(Y) = 1$. Then $Y \cap A$ is a Borel set such that $\mu(Y \cap A) = \mu(A) > 0$ and (15.13) is satisfied at each point of that set. By Theorem 15.6.3(a), we deduce that $\mathrm{HD}(Y) \geq \mathrm{HD}(Y \cap A) \geq \theta_1$. The arbitrary choice of Y implies that $\mathrm{HD}(\mu) \geq \theta_1$.

(b) Let $\widehat{X}$ be the Borel set of all points $x \in X$ at which (15.14) holds. By hypothesis, $\mu(\widehat{X}) = 1$. Moreover, $\mathrm{HD}(\widehat{X}) \leq \theta_2$ by Theorem 15.6.3(b). Hence, $\mathrm{HD}(\mu) \leq \mathrm{HD}(\widehat{X}) \leq \theta_2$. □

Our next aim is to obtain a characterization of the Hausdorff dimension of any measure. Before embarking upon this, though, let us recall the definition of the essential infimum and supremum of a measurable function.

Definition 15.6.5. Let μ be a Borel probability measure on a metric space X and $f : X \to \mathbb{R}$ be a Borel measurable function. The μ-*essential infimum and supremum of* f are defined by

$$\operatorname{ess\,inf}(f) := \sup_{\mu(N)=0} \inf_{x\in X\setminus N} f(x) \quad \text{and} \quad \operatorname{ess\,sup}(f) := \inf_{\mu(N)=0} \sup_{x\in X\setminus N} f(x).$$

The following characterizations of the essential infimum and supremum, which we state here without proof, are also often used in practice.

Lemma 15.6.6. *Let μ be a Borel probability measure on a metric space X and $f : X \to \mathbb{R}$ be a Borel measurable function. Then:*

(a) $\operatorname{ess\,inf}(f) = \theta_1$*, where θ_1 is the unique real number such that*

$$\mu(\{x \in X : f(x) < \theta_1\}) = 0 \quad \textit{and} \quad \mu(\{x \in X : f(x) < \theta\}) > 0, \quad \forall \theta > \theta_1.$$

(b) $\operatorname{ess\,sup}(f) = \theta_2$*, where θ_2 is the unique real number such that*

$$\mu(\{x \in X : f(x) > \theta_2\}) = 0 \quad \textit{and} \quad \mu(\{x \in X : f(x) > \theta\}) > 0, \quad \forall \theta < \theta_2.$$

We can now formulate a characterization of the Hausdorff dimension of any measure.

Corollary 15.6.7. *If μ is a Borel probability measure on a metric space X, then*

$$\mathrm{HD}(\mu) = \operatorname{ess\,sup}(\underline{d}_\mu).$$

In particular, if there exists $\theta \geq 0$ such that $\underline{d}_\mu(x) = \theta$ for μ-a. e. $x \in X$, then $\mathrm{HD}(\mu) = \theta$.

Proof. Let $\theta_2 := \operatorname{ess\,sup}(\underline{d}_\mu)$. By Lemma 15.6.6(b), it immediately follows that $\mu(\{x \in X : \underline{d}_\mu(x) > \theta_2\}) = 0$ or, equivalently, $\mu(\{x : \underline{d}_\mu(x) \leq \theta_2\}) = 1$. By Corollary 15.6.4(b), we deduce that $\mathrm{HD}(\mu) \leq \theta_2$.

On the other hand, fix $Y \subseteq X$ with $\mu(Y) = 1$. Let $\theta < \theta_2$. By Lemma 15.6.6(b), we have that $\mu(\{x : \underline{d}_\mu(x) > \theta\}) > 0$, and consequently $\mu(Y \cap \{x : \underline{d}_\mu(x) > \theta\}) > 0$. It follows from Theorem 15.6.3(a) that $\mathrm{HD}(Y \cap \{x : \underline{d}_\mu(x) > \theta\}) \geq \theta$. So $\mathrm{HD}(Y) \geq \theta$ for all $\theta < \theta_2$. In other words, $\mathrm{HD}(Y) \geq \theta_2$. As this holds for all $Y \subseteq X$ such that $\mu(Y) = 1$, we infer that $\mathrm{HD}(\mu) \geq \theta_2$. □

Let us now present results analogous to Theorem 15.6.3 as well as Corollaries 15.6.4 and 15.6.7 for packing measures and dimensions.

Theorem 15.6.8. *Let μ be a Borel probability measure on a metric space X and $A \subseteq X$ be a Borel set.*

(a) *If X is separable, $\mu(A) > 0$, and there exists $\theta_1 \geq 0$ such that*

$$\overline{d}_\mu(x) \geq \theta_1, \quad \forall x \in A,$$

then $\mathrm{PD}(A) \geq \theta_1$.

(b) *If there exists $\theta_2 \geq 0$ such that*

$$\overline{d}_\mu(x) \leq \theta_2, \quad \forall x \in A,$$

then $\mathrm{PD}(A) \leq \theta_2$.

Proof. Similar to that of Theorem 15.6.3 and hence is left as an exercise. □

Corollary 15.6.9. *Let μ be a Borel probability measure on a metric space X.*
(a) *If X is separable and there exist $\theta_1 \geq 0$ and a Borel set $A \subseteq X$ such that $\mu(A) > 0$ and*

$$\overline{d}_\mu(x) \geq \theta_1, \quad \forall x \in A,$$

then $\mathrm{PD}(\mu) \geq \theta_1$.
(b) *If there exists $\theta_2 \geq 0$ such that*

$$\overline{d}_\mu(x) \leq \theta_2, \quad \text{for } \mu\text{-a.e. } x \in X,$$

then $\mathrm{PD}(\mu) \leq \theta_2$.

Proof. Similar to that of Corollary 15.6.4 and hence is left as an exercise. □

Corollary 15.6.10. *If μ is a Borel probability measure on a separable metric space X, then*

$$\mathrm{PD}(\mu) = \operatorname{ess\,sup}(\overline{d}_\mu).$$

In particular, if there exists $\theta \geq 0$ such that $\overline{d}_\mu(x) = \theta$ for μ-a.e. $x \in X$, then $\mathrm{PD}(\mu) = \theta$.

Proof. Similar to that of Corollary 15.6.7 and hence is left as an exercise. □

As an immediate consequence of this corollary and Corollary 15.6.7, we get the following result.

Corollary 15.6.11. *Let μ be a Borel probability measure on a separable metric space X. If $d_\mu(x)$ exists for μ-a.e. $x \in X$ and takes on the same value d for μ-a.e. $x \in X$, then*

$$\mathrm{HD}(\mu) = \mathrm{PD}(\mu) = d.$$

Any measure satisfying the hypotheses of this corollary is said to be dimensionally exact.

Measures which confer to open balls weights that are comparable to their radii, as long as those radii are sufficiently small, are called geometric measures.

Definition 15.6.12. A Borel probability measure μ on a metric space X is said to be *geometric or Ahlfors* if there exist constants $t > 0$ and $c \geq 1$ such that

$$c^{-1} \leq \frac{\mu(B(x,r))}{r^t} \leq c, \quad \forall x \in X, \ \forall 0 < r \leq 1. \tag{15.15}$$

We call t the *exponent* of the geometric measure μ. Note that t is unique.

The following observation is immediate.

Lemma 15.6.13. *Any geometric measure μ is dimensionally exact. More precisely, $d_\mu(x) = t$ for every $x \in X$, where t is the exponent of μ.*

We end this section and this chapter with fundamental properties of geometric measures.

Theorem 15.6.14. *Suppose that X is a separable metric space and μ is a geometric measure on X with exponent t. Then the following statements hold:*

(a) *The measures μ, H_t and P_t are mutually equivalent. Furthermore,*

$$0 < \inf \frac{dH_t}{d\mu} \le \sup \frac{dH_t}{d\mu} < \infty \quad \textit{and} \quad 0 < \inf \frac{dP_t}{d\mu} \le \sup \frac{dP_t}{d\mu} < \infty.$$

In particular, $0 < H_t(X), P_t(X) < \infty$.

(b)

$$t = \mathrm{HD}(X) = \mathrm{PD}(X) = \mathrm{BD}(X).$$

Proof. Statement (a) follows from Theorems 15.5.3–15.5.4 with $A = X$. Consequently, $t = \mathrm{HD}(X) = \mathrm{PD}(X)$. In view of Theorem 15.3.20, we need only show that $\overline{\mathrm{BD}}(X) \le t$. To this end, let $\{(x_i, r)\}_{i=1}^k$ be a packing of X, where $0 < r \le 1$ and $k < \infty$. Then the balls $\{B(x_i, r)\}_{i=1}^k$ are mutually disjoint and by (15.15),

$$kr^t = \sum_{i=1}^k r^t \le c \sum_{i=1}^k \mu(B(x_i, r)) = c\mu\Big(\bigcup_{i=1}^k B(x_i, r)\Big) \le c\mu(X) = c.$$

Hence, $k \le cr^{-t}$, which implies that $P(X, r) \le cr^{-t}$ (cf. (15.3)). Thus, by Corollary 15.3.18,

$$\overline{\mathrm{BD}}(X) = \limsup_{r \to 0} \frac{\log P(X, r)}{-\log r} \le \limsup_{r \to 0} \frac{\log c - t \log r}{-\log r} = t. \qquad \square$$

15.7 Exercises

Exercise 15.7.1. Prove Remark 15.2.20.

Exercise 15.7.2. Prove Lemma 15.2.25 in the case of a similarity map.

Exercise 15.7.3. Prove Theorem 15.3.12.

Exercise 15.7.4. In contrast with Theorem 15.3.12, show that the packing* dimension is not countably stable.

Exercise 15.7.5. Show that the Hausdorff dimension of a set is generally not invariant under homeomorphisms.

Exercise 15.7.6. Let X be a metric space and μ be a Borel probability measure on X. For every fixed $r > 0$, show that the function from X to $\mathbb{R}$ given by $x \mapsto \mu(B(x,r))$ is lower semicontinuous.

Exercise 15.7.7. Prove Theorem 15.5.1(a) when $c = \infty$.

Exercise 15.7.8. Prove Lemma 15.6.6.

Exercise 15.7.9. Prove Theorem 15.6.8.

Exercise 15.7.10. Prove Corollary 15.6.9.

Exercise 15.7.11. Prove Corollary 15.6.10.

Exercise 15.7.12. Show that the upper bound 1 on r in Definition 15.6.12 can be replaced by any $r^* > 0$.

Exercise 15.7.13. Show that for every $t \in (0,\infty)$ there exists $n \in \mathbb{N}$ and a Borel set $B_t \subseteq \mathbb{R}^n$ such that $\mathrm{HD}(B_t) = t$.

Exercise 15.7.14. Show that for every $t \in (0,\infty)$ there exists $n \in \mathbb{N}$ and a Borel set $B_t \subseteq \mathbb{R}^n$ such that $\mathrm{PD}(B_t) = t$.

Exercise 15.7.15. Does there exist for every $t \in (0,\infty)$ some $n \in \mathbb{N}$ and a Borel set $B_t \subseteq \mathbb{R}^n$ such that $\mathrm{HD}(B_t) \cdot \mathrm{PD}(B_t) = t$?

Exercise 15.7.16. Given $n \in \mathbb{N}$, prove that the Lebesgue measure λ_n, the Hausdorff measure H_n and the packing measure P_n are nonzero constant multiples of each other.

Exercise 15.7.17. Show that for every $n \in \mathbb{N}$ there is a set $E \subseteq \mathbb{R}^n$ such that $\mathrm{HD}(E) = n$ but $\lambda_n(E) = 0$.

Exercise 15.7.18. Given $n \in \mathbb{N}$, show that every Lebesgue nonmeasurable set $F \subseteq \mathbb{R}^n$ has a Hausdorff dimension equal to n.

Exercise 15.7.19. For every $p > 0$, calculate the box-counting dimension of the set $\{n^p : n \in \mathbb{N}\} \subseteq \mathbb{R}$.

Exercise 15.7.20. Give an example of a set $G \subseteq \mathbb{R}$ whose dimensions $\mathrm{HD}(G)$, $\underline{\mathrm{BD}}(G)$, $\overline{\mathrm{BD}}(G)$ and $\mathrm{PD}(G)$, are all distinct.

Exercise 15.7.21. Let $f : [0,1] \to \mathbb{R}$ be a Hölder continuous function with exponent $\alpha \in [0,1]$. Show that the upper box-counting dimension of the graph of f is bounded above by $2 - \alpha$.

Exercise 15.7.22. Show that there exists a continuous function $f : [0,1] \to \mathbb{R}$ whose graph has Hausdorff dimension equal to 2. Can it have a positive 2–dimensional Lebesgue measure?

16 Conformal expanding repellers

In this chapter, we return to the class of expanding repellers that were introduced in Chapter 4. As these maps are open and distance expanding, the theory of Gibbs states developed in Chapter 13 applies to them. However, this class is a little too big to achieve many interesting results about the fractal structure of the limit set these repellers act on. So in this chapter we will impose the extra condition that the repellers be conformal.

In Section 16.1, we briefly review the notion of conformality. In Section 16.2, we define the subclass of conformal expanding repellers, provide some examples and derive some consequences of results presented in Chapters 4 and 13, in particular, bounded distortion properties.

Beginning with Section 16.3, we deal with the fractal geometry of the limit set of conformal expanding repellers. We consider all fractal dimensions (Hausdorff, packing and box-counting dimensions) of the limit set, as well as the dimensions of invariant measures on that limit set. To do this, we heavily rely on results and techniques from Chapter 15.

In Section 16.3, we prove the celebrated Bowen's formula, according to which the unique zero of the pressure function for a particular, geometric potential is equal to the Hausdorff dimension of the limit set of the repeller. The history of Bowen's formula began with Bowen's breakthrough work on quasi-Fuchsian groups [12]. Since then, it has attracted the attention of many researchers and has been adapted, in various forms, to many contexts of conformal dynamics. Conformal expanding repellers are just one such instance. Conformal graph directed Markov systems are another one (see Chapter 19, and especially Section 19.6). Bowen's formula is a generalization of Hutchinson's formula [56], a well-known result in fractal geometry.

The fine observer will note that Bowen's formula is entirely expressed in dynamical terms, more precisely in thermodynamic formalism terms. Using transfer (i. e., Perron–Frobenius) operators, this enables us to study in Section 16.4 the behavior of the Hausdorff dimension of the limit sets of complex repellers under perturbations. This technically-involved study also requires the use of powerful methods from functional analysis, more specifically the perturbation theory of bounded linear operators. It further necessitates the use of quasiconformal mappings, holomorphic motion and the celebrated λ-lemma, whose complete proof is provided in Theorem 16.4.4. Ultimately, we prove in Theorem 16.4.11 that the Hausdorff dimension of the limit set of repellers varies real analytically under analytic perturbations of the repellers.

In Section 16.5, we develop a formula expressing the Hausdorff dimension of an ergodic invariant Borel probability measure as the ratio of its measure-theoretic entropy and its Lyapunov exponent. For (partly justified) historical reasons, this formula is frequently referred to as a volume lemma. Its first forms can be traced back to the works of Eggleston [27] and Billingsley [7]. From a dynamical viewpoint, a certain breakthrough was achieved by Lai-Sang Young [139]. Since then, a multitude of papers appeared.

https://doi.org/10.1515/9783110702699-016

Some of the first few are [69, 71, 97]. The early papers [12] and [99] shed some light on the nature of dimensions of measures.

Finally, in Section 16.6 we perform a multifractal analysis of the Hausdorff dimension of the level sets of Gibbs states for any Hölder continuous potential. This is a fertile field of applications of the theory of conformal expanding repellers and more largely of this book. In addition to its intrinsic interest and importance, the multifractal analysis conducted in this chapter bears witness to the strength and sophistication of the tools and methods developed in this book.

To our knowledge, rigorous mathematical research on multifractal analysis began in 1989 with David Rand's work on cookie-cutter Cantor sets [100]. In its full generality, nearly as in Section 16.6, a multifractal analysis was carried out by Pesin and Weiss in the paper [96]. Other valuable resources are the books [95] and [98]. Our exposition stems from this latter.

By now, there are legions of books and papers on multifractal analysis of Hausdorff dimension of level sets of various measures and other functions, both from a mathematical perspective and from a more informal, but rich in ideas and stimulating, physical perspective. A quite general approach to the subject can be found in Lars Olsen's papers [87, 88]. We regret that, due to a sheer lack of space, we did not do a multifractal analysis of the Hausdorff dimension of level sets of Birkhoff's averages of Hölder continuous functions. This is an important, quickly developing research field and we encourage the reader to familiarize themself with it.

16.1 Conformal maps

Let $d \in \mathbb{N}$. The usual inner product of two vectors $v, w \in \mathbb{R}^d$ will be denoted by $\langle v, w\rangle := \sum_{j=1}^{d} v_j w_j$. The norm $\|v\| \geq 0$ of a vector v is defined by $\|v\|^2 := \langle v, v\rangle = \sum_{j=1}^{d} v_j^2$. The angle $0 \leq \angle(v, w) \leq \pi$ between two vectors v, w satisfies

$$\cos \angle(v, w) = \frac{\langle v, w\rangle}{\|v\| \cdot \|w\|}.$$

By definition, a *conformal map* is an angle-preserving transformation. To be precise, let $\{e_j\}_{j=1}^{d}$ be the standard basis of $\mathbb{R}^d$. Let $I \subseteq \mathbb{R}$ be an interval. Let $\gamma, \Gamma : I \to \mathbb{R}^d$ be two differentiable curves that meet at a point $P = \gamma(t_0) = \Gamma(t_0)$. If the curve $\gamma : I \to \mathbb{R}^d$, given by $\gamma(t) = (\gamma_1(t), \ldots, \gamma_d(t)) = \sum_{j=1}^{d} \gamma_j(t) e_j$, is such that $\gamma_i'(t_0) \neq 0$ for some $1 \leq i \leq d$, then γ has a tangent vector at $\gamma(t_0)$, namely $\gamma'(t_0) = \sum_{j=1}^{d} \gamma_j'(t_0) e_j$. Likewise for Γ. By definition, the angle between γ and Γ at P is the angle between their tangent vectors at P, i. e.,

$$\angle_P(\gamma, \Gamma) := \angle(\gamma'(t_0), \Gamma'(t_0)).$$

Let U be a nonempty open subset of $\mathbb{R}^d$. A transformation $T : U \to \mathbb{R}^d$ is said to be *conformal at a point* $P \in U$ if it preserves the angle between any two curves at P, i. e.,

$$\angle_{T(P)}(T(\gamma), T(\Gamma)) = \angle_P(\gamma, \Gamma)$$

for all differentiable curves with a tangent vector at P. This map is said to be *conformal* if it is conformal at each point of its domain.

Analytically, the conformality of a differentiable map can be characterized in terms of its derivative. For instance, a C^1 local diffeomorphism $T : U \to \mathbb{R}^d$ is conformal if and only if for all vectors $v, w \in \mathbb{R}^d$,

$$\langle T'(x)v, T'(x)w \rangle = e^{\sigma(x)} \langle v, w \rangle$$

for some real-valued function $\sigma : U \to \mathbb{R}$ (see Theorem 3.8 in Blair [10]). In slightly different terms, a C^1 local diffeomorphism $T : U \to \mathbb{R}^d$ is conformal if and only if at every point $x \in U$ its derivative $T'(x) : \mathbb{R}^d \to \mathbb{R}^d$ is a similarity without a translation component, i. e., of the form rA, where $r > 0$ and A is an orthogonal matrix.

Recall that the operator norm of $T'(x)$ is

$$|T'(x)| := \sup\{\|T'(x)v\| : \|v\| \leq 1\}.$$

A conformal C^1 local diffeomorphism $T : U \to \mathbb{R}^d$ has, among others, the following properties:

$$\|T'(x)v\| = |T'(x)| \cdot \|v\|, \quad \forall v \in \mathbb{R}^d, \forall x \in U \tag{16.1}$$

and

$$\langle T'(x)v, T'(x)w \rangle = |T'(x)|^2 \langle v, w \rangle, \quad \forall v, w \in \mathbb{R}^d, \forall x \in U. \tag{16.2}$$

Moreover, if $S : V \to \mathbb{R}^d$ and $T : U \to \mathbb{R}^d$ are two conformal C^1 local diffeomorphisms such that $T(U) \subseteq V$, then

$$|(S \circ T)'(x)| = |S'(T(x))| \cdot |T'(x)|, \quad \forall x \in U. \tag{16.3}$$

16.2 Conformal expanding repellers

Now that we have described conformal maps, let us define the family of conformal expanding repellers. The larger class of expanding repellers was introduced in Definition 4.1.4.

Definition 16.2.1. Let U be an nonempty open subset of $\mathbb{R}^d$ and $T : U \to \mathbb{R}^d$ a map. Also let X be a nonempty compact subset of U. The triple (X, U, T) is said to be a *conformal expanding repeller* provided that it satisfies the following conditions:

(a) $T \in C^{1+\epsilon}(U)$, i. e., $T \in C^1(U)$ and the map $U \ni x \mapsto T'(x) \in L(\mathbb{R}^d)$ is Hölder continuous with exponent ϵ, where $L(\mathbb{R}^d)$ is the d^2-dimensional Banach space of linear operators from $\mathbb{R}^d$ to $\mathbb{R}^d$.
(b) T is conformal.
(c) $T(X) = X$.
(d) There exists $\lambda > 1$ such that $|T'(x)| \geq \lambda$ for all $x \in X$.
(e) $\bigcap_{n=0}^{\infty} T^{-n}(U) = X$.
(f) The restriction $T|_X : X \to X$ is topologically transitive.

The set X is sometimes called the *limit set* of the repeller T.

In other words, T is an expanding repeller satisfying the additional conditions (a), (b) and (f). Condition (b) in Definition 4.1.4 reduces to condition (d) by (16.1).

In Section 4.1.2, we constructed conformal expanding repellers called hyperbolic Cantor sets. These sets were built within a compact set X_0, from which a neighborhood U, and thereafter the limit set X, were constructed. We now present an alternative construction method. We begin with an open, bounded, convex set B, construct U and then X.

Example 16.2.2. Let $2 \leq k \in \mathbb{N}$. For each $1 \leq i \leq k$, let $F_i : \mathbb{R}^d \to \mathbb{R}^d$ be a contracting similarity map, i. e., a map of the form $F_i(x) = r_i A_i(x) + b_i$, where $0 < r_i < 1$, A_i is an orthogonal matrix and $b_i \in \mathbb{R}^d$. The maps F_i, $1 \leq i \leq k$, are often called *generators*. Bundled together, they form a so-called *iterated function system* (*IFS*).

Let B be an open, bounded, convex subset of $\mathbb{R}^d$. By choosing the r_i's small enough and adjusting the b_i's appropriately, the images $F_i(B)$, $1 \leq i \leq k$, can be made to be:
(1) mutually disjoint, i. e., $F_i(B) \cap F_j(B) = \emptyset$ whenever $i \neq j$; this is frequently called the *Open Set Condition* (abbreviated to *OSC*); and
(2) such that $F_i(\overline{B}) \subseteq B$. (Note that $F_i(\overline{B}) = \overline{F_i(B)}$.)

Together, these two conditions are commonly called the *Strong Separation Condition* (*SSC*) in the theory of IFSs. Nevertheless, a lot of the theory of IFSs can be developed without condition (2), i. e., by only requiring the OSC. The reader will encounter that situation in much greater generality in Chapter 19: condition (2) will not be assumed, the generators will be conformal (not merely affine), the alphabet will be countable (not necessarily finite) and, lastly, the system will be graph directed and Markov (so some transitions will be forbidden, unlike in IFSs where all transitions are allowed).

Let $U := \bigcup_{i=1}^{k} F_i(B) \subseteq B$ and define the map $f : U \to \mathbb{R}^d$ by setting

$$f(x) := F_i^{-1}(x) \quad \text{if } x \in F_i(B).$$

By virtue of property (1), the map f is well defined. Furthermore, $f'(x) = r_i^{-1} A_i^{-1}(x)$ for each $x \in F_i(B)$. Thus, f satisfies conditions (a) and (b) in the definition of a conformal expanding repeller. Defining $X := \bigcap_{n=0}^{\infty} f^{-n}(U)$, condition (e) is automatically satisfied. Setting $\lambda = (\max\{r_i : 1 \le i \le k\})^{-1} > 1$, condition (d) holds on $U \supseteq X$.

Regarding condition (c), observe that $f^{-1}(U) \subseteq U$ by definition. Therefore, $f^{-(n+1)}(U) \subseteq f^{-n}(U)$ for all $n \ge 0$, and hence $f^{-1}(X) \subseteq X$. The opposite inclusion being obvious from the definition of X, we deduce that $f^{-1}(X) = X$. It immediately follows that $f(X) \subseteq X$. To prove that $f(X) \supseteq X$, let $x \in X$, fix any $1 \le i \le k$, and let $y = F_i(x)$. Since $x \in X \subseteq U \subseteq B$, we have that $y \in F_i(B) \subseteq U$. Moreover, $f(y) = f(F_i(x)) = F_i^{-1}(F_i(x)) = x \in X$ and so $f^n(y) = f^{n-1}(x) \in X$ for all $n \in \mathbb{N}$. Hence, $f^n(y) \in U$ for all $n \ge 0$, i. e., $y \in \bigcap_{n=0}^{\infty} f^{-n}(U) = X$. Consequently, $x = f(y) \in f(X)$. So $X \subseteq f(X)$. In summary, $f(X) = X$, i. e., condition (c) is fulfilled.

We also need to show that X is nonempty and compact. For any set $C \subseteq B$, observe that $f^{-1}(C) = \bigcup_{i=1}^{k} F_i(C)$. In particular, $f^{-1}(B) = U$. Therefore,

$$f^{-(n+1)}(C) \subseteq f^{-(n+1)}(B) = f^{-n}(U), \quad \forall n \ge 0. \tag{16.4}$$

So

$$\bigcap_{n=0}^{\infty} f^{-n}(C) \subseteq \bigcap_{n=0}^{\infty} f^{-n}(U) = X. \tag{16.5}$$

Let $K := \bigcup_{i=1}^{k} F_i(\overline{B})$. By the definitions of U and K and property (2),

$$U \subseteq K \subseteq B. \tag{16.6}$$

Hence, by (16.6) and (16.4),

$$f^{-1}(K) \subseteq f^{-1}(B) = U \subseteq K. \tag{16.7}$$

So $(f^{-n}(K))_{n=0}^{\infty}$ is a descending sequence. Moreover, by (16.6) and (16.5) with $C = K$, observe that

$$X := \bigcap_{n=0}^{\infty} f^{-n}(U) \subseteq \bigcap_{n=0}^{\infty} f^{-n}(K) \subseteq X.$$

That is, $X = \bigcap_{n=0}^{\infty} f^{-n}(K)$.

Since $\overline{B}$ is bounded and closed, it is compact according to the Heine–Borel theorem. So is each $F_i(\overline{B})$, as the image of a compact set under a continuous map. Thus, as a finite union of compact sets, K is compact. Every $f^{-n}(K)$ is compact since it is a bounded preimage of a closed set under a continuous map. Hence, X is the intersection of a descending sequence of nonempty compact sets, and is consequently nonempty and compact.

Finally, we prove that $f|_X : X \to X$ is topologically exact. We will use symbolic dynamics notation. Let $E = \{1, \ldots, k\}$ be the alphabet. For every $\omega \in E^* := \bigcup_{n=0}^{\infty} E^n$, we have that $f^{|\omega|}(x) = F_\omega^{-1}(x)$ for all $x \in F_\omega(B)$, where $F_\omega = F_{\omega_1} \circ \cdots \circ F_{\omega_{|\omega|}}$. Notice that $F_\omega(B)$ is open in $\mathbb{R}^d$ and $\operatorname{diam}(F_\omega(B)) \le \lambda^{-|\omega|}\operatorname{diam}(B)$ for all $\omega \in E^*$ since B is convex. Moreover, $f^{|\omega|}(F_\omega(B)) = B \supseteq X$. As $f(X) = X = f^{-1}(X)$, we deduce that $f^{|\omega|}(F_\omega(B) \cap X) = X$. Let V be a nonempty open subset of X. That is, $V = W \cap X$ for some nonempty open set $W \subseteq B$. Since $X = \bigcap_{n=0}^{\infty} f^{-n}(U) = \bigcap_{n=0}^{\infty} \bigcup_{\omega \in E^n} F_\omega(B)$ and $\operatorname{diam}(F_\omega(B)) \le \lambda^{-|\omega|}\operatorname{diam}(B)$ for all $\omega \in E^*$, there exists $\tau \in E^*$ such that $F_\tau(B) \subseteq W$. Then $F_\tau(B) \cap X \subseteq V$. Therefore, $X = f^{|\tau|}(F_\tau(B) \cap X) \subseteq f^{|\tau|}(V) \subseteq f^{|\tau|}(X) = X$. Hence, $f^{|\tau|}(V) = X$. So $f : X \to X$ is topologically exact and thus transitive. Condition (f) is established.

Remark 16.2.3.

(a) Note that any conformal expanding repeller built from an open, bounded, convex set B in Example 16.2.2 results in the same hyperbolic Cantor set as the one constructed by setting $X_0 = \bigcup_{i=1}^{k} F_i(\overline{B})$ in Section 4.1.2.
Conversely, any conformal expanding repeller built from a compact set X_0 in Section 4.1.2 leads to the same hyperbolic Cantor set constructed by setting $B = B(X_0, \varepsilon)$ in Example 16.2.2.
These two construction methods are equivalent.

(b) In Example 16.2.2, property (1) ensures that the map f is well defined. Property (2) guarantees that X is nonempty and compact. See Exercises 16.7.1–16.7.3.

Example 16.2.4. Fix $0 < \delta < 1/8$ and let $U := (-\delta, (1/4) + \delta) \cup ((1/2) - \delta, 1 + \delta) \subseteq \mathbb{R}$. Define the map $T : U \to \mathbb{R}$ by setting

$$T(x) := \begin{cases} 4x & \text{if } x \in (-\delta, \frac{1}{4} + \delta) \\ \frac{1}{x} - 1 & \text{if } x \in (\frac{1}{2} - \delta, 1 + \delta). \end{cases}$$

It is clear that $T \in C^{1+\epsilon}(U)$ for any $\epsilon > 0$. Furthermore, T is conformal since it is a differentiable map on $U \subseteq \mathbb{R}$. Thus, T satisfies conditions (a) and (b) in the definition of a conformal expanding repeller. As in Example 16.2.2, define $X := \bigcap_{n=0}^{\infty} T^{-n}(U)$. So condition (e) is satisfied. Note that $|T'(x)| \le 1$ for all $x \in [1, 1+\delta) \subseteq U$. However, $X \subseteq T^{-1}(U) \subseteq (-\frac{\delta}{4}, \frac{1}{\frac{3}{2}-\delta})$. Then condition (d) holds with $\lambda = (11/8)^2$. We leave the verification of the remaining conditions as an exercise to the reader.

According to Theorem 4.1.5, conformal expanding repellers are distance expanding maps on X. They are also open maps; in fact, they are local diffeomorphisms. Thus, all the theory developed in Chapter 13 for Gibbs states of open, distance expanding maps, applies to them. Let us now investigate some consequences of this theory.

Normally, one would only be working on X, and thus with open balls within X (i. e., the projection onto X of open balls in $\mathbb{R}^d$). However, the use of the mean value theorem in the forthcoming Lemma 16.2.9 requires that Corollary 16.2.8, and accordingly Lemma 16.2.6, apply to sufficiently small open balls in $\mathbb{R}^d$ centered on points of X (and correspondingly the preimages of those balls). The distinction is somewhat subtle but it is for this sole reason that we proved in Theorem 4.1.5 that expanding repellers are actually distance expanding on $B(X,\kappa) = \{y \in \mathbb{R}^d : d(y,X) < \kappa\}$, a κ-neighborhood of X. Accordingly, all balls are from this point on open balls in $B(X,\kappa)$ centered on points of X (as opposed to balls in X).

To be more precise, the proof of Theorem 4.1.5 shows that expanding repellers are distance expanding with constants $\delta = \kappa/2$ and λ replaced by $\frac{\lambda+1}{2}$ in Definition 4.1.1; alternatively, see (13.1). To lighten notation, we will still denote $\frac{\lambda+1}{2}$ by λ. Let ξ be the constant arising in relations (4.29), (4.30) and (4.31); alternatively, see (13.2).

Let $\varphi(x) := -\log|T'(x)|$ for all $x \in U$, where $|T'(x)|$ is the operator norm of the derivative of T at x. This potential function plays a special role for conformal expanding repellers. First, note that φ is locally Hölder continuous with exponent ϵ since $T \in C^{1+\epsilon}(U)$. Moreover, because of the conformality of T, the ergodic sums of this potential enjoy the following crucial property.

Lemma 16.2.5. *Let (X,U,T) be a conformal expanding repeller and let $\varphi(x) := -\log|T'(x)|$ for each $x \in U$. For all $n \geq 0$ and all $x \in \bigcap_{j=0}^{n-1} T^{-j}(U)$, we have that*

$$S_n\varphi(x) = \sum_{j=0}^{n-1} -\log\bigl|T'(T^j(x))\bigr| = -\log\bigl|(T^n)'(x)\bigr|.$$

Proof. The first equality is simply the definition of the ergodic sum. The second equality, and really the crucial one, is a direct consequence of the chain rule and the conformality of T (see (16.3)). □

One key property of conformal expanding repellers is that the distortion their iterates create near one point is comparable to that created near any other point.

Lemma 16.2.6 (Distortion lemma). *Let (X,U,T) be a conformal expanding repeller. There exists a constant $K \geq 1$ such that for all $n \geq 0$, all $z \in X$ and all $x,y \in T_z^{-n}(B(T^n(z),\xi))$, we have that*

$$K^{-1} \leq \frac{\bigl|(T^n)'(x)\bigr|}{\bigl|(T^n)'(y)\bigr|} \leq K.$$

Proof. Let $v,w \in B(T^n(z),\xi)$ be such that $x = T_z^{-n}(v)$ and $y = T_z^{-n}(w)$. Let also $\varphi(x) := -\log|T'(x)|$. It follows from Lemmas 16.2.5 and 13.1.1 that

$$\begin{aligned}\Bigl|\log\bigl|(T^n)'(x)\bigr| - \log\bigl|(T^n)'(y)\bigr|\Bigr| &= |S_n\varphi(x) - S_n\varphi(y)| \\ &= \bigl|S_n\varphi(T_z^{-n}(v)) - S_n\varphi(T_z^{-n}(w))\bigr| \\ &\leq C_\epsilon(\varphi)(\operatorname{diam}(X))^\epsilon.\end{aligned}$$

Rearranging and taking the exponential yields the result. □

Remark 16.2.7. Lemma 13.1.1 was derived for balls in X. In the case of expanding repellers, using the full strength of Theorem 4.1.5, the reader ought to convince themself that Lemma 13.1.1 holds true for balls in $B(X,\kappa)$ centred on points of X.

Corollary 16.2.8 (Bounded distortion property (BDP)). *Let (X,U,T) be a conformal expanding repeller. There exists a constant $K \geq 1$ such that for all $n \geq 0$, all $z \in X$ and all $x,y \in B(T^n(z),\xi)$, we have*

$$K^{-1} \leq \frac{|(T_z^{-n})'(x)|}{|(T_z^{-n})'(y)|} \leq K.$$

Proof. Since $(T^n)'(T_z^{-n}(x)) \circ (T_z^{-n})'(x) = \mathrm{Id}_{\mathbb{R}^d}$ for all $x \in B(T^n(z),\xi)$, the conformality of T implies that $|(T^n)'(T_z^{-n}(x))| \cdot |(T_z^{-n})'(x)| = 1$. So $|(T_z^{-n})'(x)| = |(T^n)'(T_z^{-n}(x))|^{-1}$. The result ensues from Lemma 16.2.6. □

Thanks to the bounded distortion, the preimages of balls are distorted uniformly, in the sense that the nth preimages of balls contain, and are contained in, balls of radii that are comparable to the reciprocal of the norm of the derivative of the nth iterate, and this uniformly in n. This is the object of the next lemma.

Lemma 16.2.9. *Let (X,U,T) be a conformal expanding repeller. There exists a constant $K \geq 1$ such that for all $n \geq 0$ and all $z \in X$, we have*

$$B(z, K^{-1}|(T^n)'(z)|^{-1}\xi) \subseteq T_z^{-n}(B(T^n(z),\xi)) \subseteq B(z, K|(T^n)'(z)|^{-1}\xi).$$

Proof. For any $u,v \in \mathbb{R}^d$, let $[u,v] \subseteq \mathbb{R}^d$ denote the line segment joining u and v.

For the first inclusion, since $T_z^{-n}(B(T^n(z),\xi))$ is an open set, let $R > 0$ be the maximal radius such that $B(z,R) \subseteq T_z^{-n}(B(T^n(z),\xi))$. Then $\partial B(z,R) \cap \partial T_z^{-n}(B(T^n(z),\xi)) \neq \emptyset$. The continuity of T also ensures that

$$T^n(\overline{B(z,R)}) \subseteq \overline{B(T^n(z),\xi)}.$$

So

$$T^n(\overline{B(z,R)}) \cap \partial B(T^n(z),\xi) \neq \emptyset.$$

Therefore, for all $\varepsilon > 0$ there exists $x \in B(z,R)$ such that $\|T^n(x) - T^n(z)\| \geq \xi - \varepsilon$. Thus, by the mean value theorem and Lemma 16.2.6,

$$\begin{aligned}\xi - \varepsilon &\leq \|T^n(x) - T^n(z)\| \\ &\leq \sup_{w \in [x,z]} |(T^n)'(w)| \cdot \|x - z\| \\ &\leq K|(T^n)'(z)|R.\end{aligned}$$

Letting $\varepsilon \to 0$, this implies that

$$K^{-1}\left|(T^n)'(z)\right|^{-1}\xi \le R.$$

Hence, $B(z, K^{-1}|(T^n)'(z)|^{-1}\xi) \subseteq B(z, R) \subseteq T_z^{-n}(B(T^n(z), \xi))$.

For the second inclusion, let $x \in B(T^n(z), \xi)$. By the mean value theorem and Corollary 16.2.8, we have

$$\begin{aligned}
\|T_z^{-n}(x) - z\| &= \|T_z^{-n}(x) - T_z^{-n}(T^n(z))\| \\
&\le \sup_{y \in [x, T^n(z)]} \left|(T_z^{-n})'(y)\right| \cdot \|x - T^n(z)\| \\
&\le K\left|(T_z^{-n})'(T^n(z))\right|\xi \\
&= K\left|(T^n)'(z)\right|^{-1}\xi.
\end{aligned}$$

Thus, $T_z^{-n}(B(T^n(z), \xi)) \subseteq B(z, K|(T^n)'(z)|^{-1}\xi)$. □

Let $x \in X$ and $0 < r < \xi$. Set $k = k(x, r) \ge 0$ to be the largest integer such that

$$B(x, r) \subseteq T_x^{-k}(B(T^k(x), \xi)) \tag{16.8}$$

and set $\ell = \ell(x, r) \ge 0$ to be the smallest integer for which

$$T_x^{-\ell}(B(T^\ell(x), \xi)) \subseteq B(x, r). \tag{16.9}$$

Indeed, the sequence of sets $(T_x^{-n}(B(T^n(x), \xi)))_{n=0}^{\infty}$ is descending and the diameters of the sets $T_x^{-n}(B(T^n(x), \xi))$ decrease to 0 as n increases to ∞, because T is distance expanding, and hence its inverse branches are contracting by a factor $0 < \lambda^{-1} < 1$ for each iteration of T. Therefore, $k = k(x, r)$ and $\ell = \ell(x, r)$ are well defined. Observe that $T_x^{-\ell}(B(T^\ell(x), \xi)) \subseteq T_x^{-k}(B(T^k(x), \xi))$ and thereby $\ell \ge k$.

The next result reveals that r is comparable to both $|(T^k)'(x)|^{-1}$ and $|(T^\ell)'(x)|^{-1}$.

Corollary 16.2.10. *Let (X, U, T) be a conformal expanding repeller. There exists a constant $\widetilde{K} \ge 1$ such that*

$$\widetilde{K}^{-1} \le \frac{r}{|(T^k)'(x)|^{-1}} \le \widetilde{K} \quad \text{and} \quad \widetilde{K}^{-1} \le \frac{r}{|(T^\ell)'(x)|^{-1}} \le \widetilde{K}, \quad \forall x \in X,\ \forall 0 < r < \xi,$$

where $k = k(x, r)$ and $\ell = \ell(x, r)$ were defined just before this corollary.

Proof. By the choice of k and Lemma 16.2.9, it follows that

$$B(x, r) \subseteq T_x^{-k}(B(T^k(x), \xi)) \subseteq B(x, K|(T^k)'(x)|^{-1}\xi)$$

while

$$B(x,r) \not\subseteq T_x^{-(k+1)}(B(T^{k+1}(x),\xi)) \supseteq B(x,K^{-1}|(T^{k+1})'(x)|^{-1}\xi).$$

From these two relations, we deduce that

$$K^{-1}\left|(T^{k+1})'(x)\right|^{-1}\xi < r \leq K\left|(T^k)'(x)\right|^{-1}\xi.$$

Using the chain rule, we conclude that

$$K^{-1}\|T'\|_{\infty}^{-1}\xi < \frac{r}{\left|(T^k)'(x)\right|^{-1}} \leq K\xi.$$

Note that these lower and upper estimates are independent of x and r. We leave to the reader the task of writing a similar argument for $\ell = \ell(x,r)$. □

It ensues from the preceding corollary that $|(T^k)'(x)|$ and $|(T^\ell)'(x)|$ are comparable. Given that T is expanding, this means that the difference between k and ℓ must be uniformly bounded in x and r.

Corollary 16.2.11. *Let (X,U,T) be a conformal expanding repeller. There exists a constant $\overline{K} \geq 0$ such that*

$$0 \leq \ell(x,r) - k(x,r) \leq \overline{K}, \quad \forall x \in X,\ \forall 0 < r < \xi,$$

where $k(x,r)$ and $\ell(x,r)$ were defined just before Corollary 16.2.10.

Proof. Let $x \in X$ and $0 < r < \xi$. Let $k = k(x,r)$ and $\ell = \ell(x,r)$. Just before Corollary 16.2.10, we pointed out that $\ell \geq k$. From Corollary 16.2.10, we infer that

$$\widetilde{K}^{-2} \leq \frac{|(T^\ell)'(x)|}{|(T^k)'(x)|} \leq \widetilde{K}^2.$$

Using the chain rule and the conformality of T, we deduce that

$$\widetilde{K}^{-2} \leq \left|(T^{\ell-k})'(T^k(x))\right| \leq \widetilde{K}^2.$$

But $\inf_{z\in X}|(T^{\ell-k})'(z)| \geq \lambda^{\ell-k}$. Consequently, $\lambda^{\ell-k} \leq \widetilde{K}^2$, that is,

$$\ell - k \leq \frac{2\log\widetilde{K}}{\log\lambda}.$$ □

From the previous results, we can easily deduce the following fact about Gibbs measures for continuous potentials.

Lemma 16.2.12. *Let (X,U,T) be a conformal expanding repeller. Let $\psi : X \to \mathbb{R}$ be a continuous potential such that $\mathrm{P}(T,\psi) = 0$, and suppose that m is a Gibbs state for ψ.*

Then there exists a constant $\widehat{K} \geq 1$ such that for all $x \in X$ and $0 < r < \xi$ we have

$$\frac{\log \widehat{K} + S_k\psi(x)}{-\log \widehat{K} - \log|(T^k)'(x)|} \leq \frac{\log m(B(x,r))}{\log r} \leq \frac{-\log \widehat{K} + S_k\psi(x)}{\log \widehat{K} - \log|(T^k)'(x)|},$$

where $k = k(x,r)$ was defined just before Corollary 16.2.10.

Proof. Let $x \in X$ and $0 < r < \xi$. Let also $k = k(x,r)$ and $\ell = \ell(x,r)$ be as defined before Corollary 16.2.10. By the Gibbs property of m (cf. Definition 13.2.1), there exists a constant $C \geq 1$ such that

$$C^{-1}e^{S_\ell\psi(x)} \leq m(T_x^{-\ell}(B(T^\ell(x),\xi))) \leq m(B(x,r)) \leq m(T_x^{-k}(B(T^k(x),\xi))) \leq Ce^{S_k\psi(x)}.$$

Using Corollary 16.2.11, we obtain that

$$S_\ell\psi(x) = S_k\psi(x) + S_{\ell-k}\psi(T^k(x)) \geq S_k\psi(x) - (\ell-k)\|\psi\|_\infty \geq S_k\psi(x) - \overline{K}\|\psi\|_\infty.$$

Therefore,

$$C^{-1}e^{S_k\psi(x)-\overline{K}\|\psi\|_\infty} \leq m(B(x,r)) \leq Ce^{S_k\psi(x)+\overline{K}\|\psi\|_\infty}.$$

So

$$-\log C + S_k\psi(x) - \overline{K}\|\psi\|_\infty \leq \log m(B(x,r)) \leq \log C + S_k\psi(x) + \overline{K}\|\psi\|_\infty. \tag{16.10}$$

On the other hand, by means of Corollary 16.2.10 we have

$$\widetilde{K}^{-1}|(T^k)'(x)|^{-1} \leq r \leq \widetilde{K}|(T^k)'(x)|^{-1}.$$

Consequently,

$$-\log \widetilde{K} - \log|(T^k)'(x)| \leq \log r \leq \log \widetilde{K} - \log|(T^k)'(x)|. \tag{16.11}$$

Define $\widehat{K}$ by the relation

$$\log \widehat{K} = \max\{\log C + \overline{K}\|\psi\|_\infty, \log \widetilde{K}\}.$$

Note that $\widehat{K} \geq \widetilde{K} \geq 1$ and that relations (16.10) and (16.11) hold with $\log \widehat{K}$ replacing $\log C + \overline{K}\|\psi\|_\infty$ and $\log \widetilde{K}$, respectively.

Combining the resulting relations yields the estimates for the ratio $\frac{\log m(B(x,r))}{\log r}$. The details are left to the reader as an exercise. □

Remark 16.2.13. In Lemma 16.2.12, the integer $k = k(x,r)$ can be replaced by $\ell = \ell(x,r)$.

16.3 Bowen's formula

Bowen's formula links the pressure function generated by a parametrized family of geometric potentials to the Hausdorff dimension of the limit set X.

For every $t \in \mathbb{R}$, let $\varphi_t : X \to X$ be the potential defined by $\varphi_t = t\varphi$, where $\varphi(x) := -\log|T'(x)|$. Like φ, each φ_t is Hölder continuous, and hence continuous. Then the parametrized family of potentials $\{\varphi_t\}_{t\in\mathbb{R}}$ generates the pressure function $t \mapsto \mathrm{P}(t)$, where $\mathrm{P}(t) := \mathrm{P}(T, \varphi_t)$ was defined in Chapter 11. The next proposition gives some basic properties of this function.

Proposition 16.3.1. *For every conformal expanding repeller (X, U, T), the pressure function $\mathbb{R} \ni t \mapsto \mathrm{P}(t) := \mathrm{P}(T, \varphi_t) = \mathrm{P}(T, -t\log|T'|) \in \mathbb{R}$ possesses the following properties:*
(a) *It is Lipschitz with Lipschitz constant $L := \log\|T'\|_\infty$, where $\|T'\|_\infty := \sup_{x\in X}|T'(x)|$.*
(b) *It is strictly decreasing.*
(c) $\lim_{t\to-\infty}\mathrm{P}(t) = \infty$ *whereas* $\lim_{t\to\infty}\mathrm{P}(t) = -\infty$.
(d) $\mathrm{P}(0) = \mathrm{h_{top}}(T) > 0$, *where $\mathrm{h_{top}}(T)$ is the topological entropy of $T : X \to X$.*
(e) *There exists a unique $h \in \mathbb{R}$ such that $\mathrm{P}(h) = 0$. In fact, $h > 0$.*
(f) *The pressure function $\mathbb{R} \ni t \mapsto \mathrm{P}(t) \in \mathbb{R}$ is convex.*

Proof. (a) By Corollary 12.1.11, we obtain that

$$\big|\mathrm{P}(t) - \mathrm{P}(s)\big| \le \|\varphi_t - \varphi_s\|_\infty = \big\|\log|T'|\big\|_\infty \cdot |s-t|.$$

That is, the function $t \mapsto \mathrm{P}(t)$ is Lipschitz continuous, with Lipschitz constant $L := \|\log|T'|\|_\infty = \log\|T'\|_\infty$.

(b, c) By Proposition 13.7.12, there exists a unique equilibrium state μ_t for the Hölder continuous potential φ_t. Suppose that $s \le t$. Then, by the variational principle (Theorem 12.1.1),

$$\mathrm{P}(t) = \mathrm{h}_{\mu_t}(T) + \int_X t\varphi\,d\mu_t = \mathrm{h}_{\mu_t}(T) + \int_X s\varphi\,d\mu_t + (s-t)\int_X \log|T'|\,d\mu_t \le \mathrm{P}(s) + (s-t)\log\lambda,$$

where λ comes from condition (d) of the definition of a conformal expanding repeller (cf. Definition 16.2.1). So

$$\lim_{t\to\infty}\mathrm{P}(t) \le \mathrm{P}(s) + \log\lambda\cdot\lim_{t\to\infty}(s-t) = -\infty$$

while

$$\lim_{s\to-\infty}\mathrm{P}(s) \ge \mathrm{P}(t) + \log\lambda\cdot\lim_{s\to-\infty}(t-s) = \infty.$$

Moreover, if $s < t$, then $\mathrm{P}(t) \le \mathrm{P}(s) + (s-t)\log\lambda < \mathrm{P}(s)$.

(d) By Remark 11.1.19, we know that $\mathrm{P}(0) = \mathrm{h}_{\mathrm{top}}(T)$. Moreover, $\mathrm{h}_{\mathrm{top}}(T) > 0$ according to Exercise 7.6.12, which states that the topological entropy of any transitive, open, expansive dynamical system is positive.

(e) This immediately follows from properties (a)–(d).

(f) This statement follows from the more general fact that the function $C(X) \ni \psi \mapsto \mathrm{P}(T,\psi)$ is convex for any dynamical system $T : X \to X$. This more general case can easily be deduced from the variational principle (Theorem 12.1.1). A direct proof, i. e., from the definition of the pressure function, also exists. Indeed, let $\psi,\chi \in C(X)$ and $p \in [0,1]$. Fix momentarily $n \in \mathbb{N}$. For any finite subset E of X, it follows from Hölder's inequality that

$$\sum_{x\in E} \exp(pS_n\psi(x) + (1-p)S_n\chi(x)) \le \Big(\sum_{x\in E} \exp(S_n\psi(x))\Big)^p \Big(\sum_{x\in E} \exp(S_n\chi(x))\Big)^{1-p}.$$

Given any $\varepsilon > 0$, this inequality holds for all (n,ε)-separated sets. Passing to the supremum over all such sets on both sides yields

$$P_n(T, p\psi + (1-p)\chi, \varepsilon) \le P_n(T,\psi,\varepsilon)^p P_n(T,\chi,\varepsilon)^{1-p}.$$

Therefore,

$$\begin{aligned}\overline{P}(T, p\psi + (1-p)\chi, \varepsilon) &:= \limsup_{n\to\infty} \frac{1}{n} \log P_n(T, p\psi + (1-p)\chi, \varepsilon)\\ &\le p \limsup_{n\to\infty} \frac{1}{n} \log P_n(T,\psi,\varepsilon) + (1-p) \limsup_{n\to\infty} \frac{1}{n} \log P_n(T,\chi,\varepsilon)\\ &= p\overline{P}(T,\psi,\varepsilon) + (1-p)\overline{P}(T,\chi,\varepsilon).\end{aligned}$$

Consequently,

$$\begin{aligned}\mathrm{P}(T, p\psi + (1-p)\chi) &:= \lim_{\varepsilon\to 0} \overline{P}(T, p\psi + (1-p)\chi, \varepsilon)\\ &\le p \lim_{\varepsilon\to 0} \overline{P}(T,\psi,\varepsilon) + (1-p) \lim_{\varepsilon\to 0} \overline{P}(T,\chi,\varepsilon)\\ &= p\mathrm{P}(T,\psi) + (1-p)\mathrm{P}(T,\chi).\end{aligned}$$

□

The next theorem is the main result of this chapter. Before stating it, let us recall that geometric measures were introduced in Definition 15.6.12.

Theorem 16.3.2 (Bowen's formula). *Let (X, U, T) be a conformal expanding repeller. Per Proposition 16.3.1, let h be the unique zero of the pressure function $t \mapsto \mathrm{P}(t)$. Let $\varphi_h : X \to \mathbb{R}$ be the corresponding Hölder continuous potential $\varphi_h(x) = -h\log|T'(x)|$. By virtue of Theorem 13.6.2 and Proposition 13.7.12, let m_h be the eigenmeasure of the transfer operator's dual $\mathcal{L}^*_{\varphi_h} : C(X)^* \to C(X)^*$ and μ_h be the unique equilibrium state and unique T-invariant Gibbs state for the potential φ_h. Then m_h and μ_h are geometric* *measures with exponent h. In particular,*

$$\mathrm{HD}(X) = \mathrm{PD}(X) = \mathrm{BD}(X) = h. \tag{16.12}$$

Moreover,

$$m_h = \frac{1}{H_h(X)} H_h = \frac{1}{P_h(X)} P_h. \tag{16.13}$$

In particular, $H_h(X)$ and $P_h(X)$ are positive and finite.

Proof. Recall that Proposition 13.6.16 asserts that m_h is a Gibbs state for φ_h. Moreover, m_h and μ_h are equivalent according to Theorem 13.7.7. Therefore, it suffices to prove the statement for one of these measures. We will use m_h.

Let $x \in X$ and $0 < r < \xi$. Let also $k = k(x,r)$ be as defined just before Corollary 16.2.10. First, it follows from Lemma 16.2.5 that $S_k\varphi_h(x) = -h\log|(T^k)'(x)|$. Lemma 16.2.12 applies with $\psi = \varphi_h$ since $\mathrm{P}(T,\psi) = \mathrm{P}(T,\varphi_h) = \mathrm{P}(h) = 0$. Using $m = m_h$, there exists a constant $\widehat{K} \geq 1$ such that

$$\frac{\log\widehat{K} - h\log|(T^k)'(x)|}{-\log\widehat{K} - \log|(T^k)'(x)|} \leq \frac{\log m_h(B(x,r))}{\log r} \leq \frac{-\log\widehat{K} - h\log|(T^k)'(x)|}{\log\widehat{K} - \log|(T^k)'(x)|}.$$

Equivalently,

$$\frac{\frac{\log\widehat{K}}{\log|(T^k)'(x)|} - h}{-\frac{\log\widehat{K}}{\log|(T^k)'(x)|} - 1} \leq \frac{\log m_h(B(x,r))}{\log r} \leq \frac{-\frac{\log\widehat{K}}{\log|(T^k)'(x)|} - h}{\frac{\log\widehat{K}}{\log|(T^k)'(x)|} - 1},$$

Letting $r \searrow 0$ forces $k \nearrow \infty$, and hence $|(T^k)'(x)| \nearrow \infty$, so

$$\frac{0-h}{-0-1} \leq \lim_{r\to 0} \frac{\log m_h(B(x,r))}{\log r} \leq \frac{-0-h}{0-1},$$

i. e., $\lim_{r\to 0} \frac{\log m_h(B(x,r))}{\log r} = h$. Thus, m_h is a geometric measure with exponent h. The measure μ_h also has this property since μ_h and m_h are equivalent, as stated by Theorem 13.7.7. Then (16.12) immediately follows from Theorem 15.6.14. Finally, according to Lemma 15.2.25, the measures $H_h|_X$ and $P_h|_X$ are $(\Lambda_{\varphi_h}e^{-\varphi_h})$-conformal since $\Lambda_{\varphi_h} = \exp(\mathrm{P}(T,\varphi_h)) = 1$ (cf. Definition 13.6.12). So, like m_h, both $H_h|_X$ and $P_h|_X$ are fixed points of the dual operator $\mathcal{L}^*_{\varphi_h}$ per Lemma 13.6.13. Then (16.13) can be inferred from Proposition 13.8.10(a). □

Example 16.3.3. Consider the middle-third Cantor set C (cf. Example 4.1.6). Recall that $U = (-\frac{1}{9},\frac{4}{9}) \cup (\frac{5}{9},\frac{10}{9})$ and $T : U \to \mathbb{R}$ is defined by

$$T(x) = \begin{cases} 3x & \text{if } x \in (-\frac{1}{9},\frac{4}{9}) \\ 3x-2 & \text{if } x \in (\frac{5}{9},\frac{10}{9}). \end{cases}$$

By setting $B = (-\frac{1}{3}, \frac{4}{3})$, it follows from Example 16.2.2 that $(C, U, T : U \to \mathbb{R})$ is a conformal expanding repeller. It is easy to see that the map T is topologically conjugate to the shift map $\sigma : \{0,2\}^\infty \to \{0,2\}^\infty$ via the conjugacy map $h : \{0,2\}^\infty \to C$ defined by $h(\omega) = \sum_{j=1}^{\infty} \frac{\omega_j}{3^j}$. Thus, $\mathrm{P}(0) = \mathrm{h}_{\mathrm{top}}(T) = \mathrm{h}_{\mathrm{top}}(\sigma) = \log 2$ by Proposition 7.2.18 and Example 7.2.26. By Proposition 11.3.1, we get

$$\mathrm{P}(t) := \mathrm{P}(T, -t\log|T'|) = \mathrm{P}(T, -t\log 3) = \mathrm{P}(0) - t\log 3 = \log 2 - t\log 3.$$

So the unique t for which $\mathrm{P}(t) = 0$ is $t = \frac{\log 2}{\log 3}$. Therefore, $\mathrm{HD}(C) = \mathrm{PD}(C) = \mathrm{BD}(C) = \frac{\log 2}{\log 3}$ by Bowen's formula (Theorem 16.3.2).

16.3.1 Special case of Hutchinson's formula

In the previous example, the calculation of the pressure function was simplified by the fact that the derivative of T was constant. Using Bowen's formula, we now derive Hutchinson's formula for iterated function systems (IFSs) consisting of finitely many contracting similarities satisfying the strong separation condition (SSC; see Example 16.2.2). The solution to this formula is simultaneously the Hausdorff, packing and box dimensions of the limit set X. Among others, this demonstrates that Bowen's formula is a far-reaching extension of Hutchinson's formula.

Note that Hutchinson's formula holds under the sole open set condition (OSC). We will deduce it from Bowen's formula in an even more general context in Section 19.6.3.

Theorem 16.3.4. *Let $\{F_i(x) := r_iA_i(x) + b_i\}_{i=1}^k$ be an IFS consisting of contracting similarities satisfying the strong separation condition (SSC). The limit set X of this IFS satisfies $\mathrm{HD}(X) = \mathrm{PD}(X) = \mathrm{BD}(X) = h > 0$, where h is the unique number such that*

$$\sum_{i=1}^{k} r_i^h = 1.$$

This equation is called Hutchinson's formula.

Proof. Let (X, U, f) be the conformal expanding repeller engendered by this IFS, as outlined in Example 16.2.2. For every $1 \le i \le k$ and $x_i \in F_i(B)$, observe that

$$\varphi_t(x_i) = -t\log|f'(x_i)| = -t\log|(F_i^{-1})'(x_i)| = -t\log|F_i'(F_i^{-1}(x_i))|^{-1} = \log r_i^t.$$

Thus, the function φ_t is piecewise constant on $U := \bigcup_{i=1}^k F_i(B) \supseteq X$.

According to (4.25) in Section 4.1.2, the conformal expanding repeller (U, X, f) is topologically conjugate to the symbolic system (E^∞, σ), where $E = \{1, \dots, k\}$, via a coding map $\pi : E^\infty \to X$, where $\{\pi(\omega)\} = \bigcap_{n=1}^\infty \varphi_{\omega|_n}(X_0)$ with $X_0 = \bigcup_{i=1}^k F_i(\overline{B})$. More precisely, $\pi \circ \sigma = f \circ \pi$. Corollary 11.1.21 and Example 11.4.1 then jointly affirm that the pressure function is

$$\mathrm{P}(t) = \mathrm{P}(f,\varphi_t) = \mathrm{P}(\sigma,\varphi_t\circ\pi) = \log\sum_{e\in E}\exp(\widetilde{\varphi_t\circ\pi}(e)) = \log\sum_{i=1}^{k}\exp(\varphi_t(x_i)) = \log\sum_{i=1}^{k} r_i^t.$$

By Bowen's formula (Theorem 16.3.2), we know that $\mathrm{HD}(X) = \mathrm{PD}(X) = \mathrm{BD}(X) = h$, where h is the unique zero of the pressure function $t \mapsto \mathrm{P}(t)$. Thus, $\mathrm{P}(t) = 0$ when $\sum_{i=1}^{k} r_i^t = 1$, and Hutchinson's formula holds.

Alternatively, since $f : X \to X$ is expansive, Theorem 11.1.26 guarantees that $\mathrm{P}(f,\varphi_t) = \mathrm{P}(f,\varphi_t,\mathcal{U})$ for any finite open partition $\mathcal{U}$ of X with $\mathrm{diam}(\mathcal{U}) \le \delta$, where δ is an expansive constant for f. Let $\mathcal{U} = \{F_i(B)\cap X : 1 \le i \le k\}$. By the OSC and since $X \subseteq U := \bigcup_{i=1}^{k} F_i(B)$, it is obvious that $\mathcal{U}$ is a finite open partition of X. However, it is unclear whether $\mathrm{diam}(\mathcal{U}) \le \delta$. Nevertheless, observe that $\mathcal{U}^n = \{F_\omega(B)\cap X : \omega \in E^n\}$ and $\mathrm{diam}(\mathcal{U}^n) \le \lambda^{-(n-1)}\mathrm{diam}(\mathcal{U})$ for all $n \in \mathbb{N}$. Thus there exists $N \in \mathbb{N}$ such that $\mathrm{diam}(\mathcal{U}^N) \le \delta$. Lemma 11.1.15 affirms that $\mathrm{P}(f,\varphi_t,\mathcal{U}^N) = \mathrm{P}(f,\varphi_t,\mathcal{U})$. So it suffices to compute $\mathrm{P}(f,\varphi_t,\mathcal{U})$. By Definition 11.1.11, the topological pressure of the potential φ_t with respect to $\mathcal{U}$ is

$$\mathrm{P}(f,\varphi_t,\mathcal{U}) := \lim_{n\to\infty}\frac{1}{n}\log Z_n(\varphi_t,\mathcal{U}),$$

where, by Definition 11.1.2, the nth partition function of $\mathcal{U}$ is

$$Z_n(\varphi_t,\mathcal{U}) = \inf\Big\{\sum_{V\in\mathcal{V}}\exp(\overline{S}_n\varphi_t(V)) : \mathcal{V} \text{ is a subcover of } \mathcal{U}^n\Big\}.$$

As $\mathcal{U}$ is a partition of X, so is $\mathcal{U}^n$, and thus $Z_n(\varphi_t,\mathcal{U})$ reduces to

$$Z_n(\varphi_t,\mathcal{U}) = \sum_{V\in\mathcal{U}^n}\exp(\overline{S}_n\varphi_t(V)) = \sum_{\omega\in E^n}\exp(\overline{S}_n\varphi_t(F_\omega(B)\cap X)).$$

Recall that $\varphi_t(x) = \log r_i^t$ for all $x \in F_i(B)$ and all $1 \le i \le k$. Then $S_{|\omega|}\varphi_t(x) = \log r_\omega^t$ for all $x \in F_\omega(B)$ and all $\omega \in E^*$, where $r_\omega := \prod_{j=1}^{|\omega|} r_{\omega_j}$. Therefore, $\overline{S}_{|\omega|}\varphi_t(F_\omega(B)\cap X) = \log r_\omega^t$ for all $\omega \in E^*$. So

$$Z_n(\varphi_t,\mathcal{U}) = \sum_{\omega\in E^n} r_\omega^t = \Big(\sum_{i=1}^{k} r_i^t\Big)^n.$$

Consequently,

$$\mathrm{P}(t) = \mathrm{P}(f,\varphi_t) = \mathrm{P}(f,\varphi_t,\mathcal{U}) = \lim_{n\to\infty}\frac{1}{n}\log\Big(\sum_{i=1}^{k} r_i^t\Big)^n = \log\sum_{i=1}^{k} r_i^t.$$

Thus, $\mathrm{P}(t) = 0$ when $\sum_{i=1}^{k} r_i^t = 1$, and Hutchinson's formula holds. □

Remark 16.3.5. Hutchinson's formula does not generally hold if the OSC is not satisfied. See Exercise 16.7.4.

16.4 Real-analytic dependence of Hausdorff dimension of repellers in $\mathbb{C}$

Consider a conformal expanding repeller (X, T, U) in $\mathbb{C}$ (i. e. $U \subseteq \mathbb{C}$). Denote by J (rather than X) the associated compact invariant repelling set

$$J := \bigcap_{n=0}^{\infty} T^{-n}(U),$$

i. e., the limit set of T. (The letter J is reminiscent of the Julia set for complex-valued functions of a complex variable.)

We are primarily interested in analytic perturbations of T, defined as follows.

Definition 16.4.1. Let (J, T, U) be a conformal expanding repeller in $\mathbb{C}$. Let also Λ be an open subset of $\mathbb{C}$. A family of conformal maps $\{T_\lambda : U \to \mathbb{C}\}_{\lambda \in \Lambda}$ is called an *holomorphic* (or *analytic*) *perturbation* of T if the function

$$\Lambda \times U \ni (\lambda, z) \longmapsto T_\lambda(z) \in \mathbb{C}$$

is analytic and $T = T_{\lambda_0}$ for some $\lambda_0 \in \Lambda$.

Our goal is to show that on some neighborhood of λ_0, the triples $(J_\lambda, T_\lambda, \widehat{U})$ with yet-to-be-identified sets J_λ but a common neighborhood $\widehat{U}$ of those sets J_λ, constitute conformal expanding repellers, and that the maps $T_\lambda|_{J_\lambda}$ and $T|_J$ are topologically conjugate (and more). For this, we need the powerful tools of holomorphic motion and quasisymmetric (or quasiconformal) homeomorphisms.

Definition 16.4.2. Let (Y, d_Y) and (Z, d_Z) be metric spaces. A map $f : Y \to Z$ is *M-quasisymmetric* for some $M > 0$ if

$$d_Y(y_1, y_2) \le d_Y(y_1, y_3) \quad \Longrightarrow \quad d_Z(f(y_1), f(y_2)) \le M d_Z(f(y_1), f(y_3)).$$

The number M is called a *quasisymmetric constant* for f. Clearly, any $M' > M$ is also a quasisymmetric constant for f. Moreover, if f is not constant then the infimum of all quasisymmetric constants for f is positive and is also a quasisymmetric constant for f. We will call it *the* quasisymmetric constant for f.

When Y is an open subset of a Euclidean space of dimension at least 2, the term *quasiconformal* is usually preferred. In that case, if f is in addition a homeomorphism, several equivalent characterizations of quasiconformality are known and frequently used (for more information, see [1, 3, 14, 35, 55, 66, 67]).

We now define the concept of holomorphic motion, one of the key notions in holomorphic dynamics.

Denote the open unit disk in $\mathbb{C}$ by $\mathbb{D} := \{z : |z| < 1\}$ and the Riemann sphere by $\widehat{\mathbb{C}}$.

Definition 16.4.3. Let A be a subset of $\widehat{\mathbb{C}}$. A family of maps

$$\{\tau_\lambda : A \to \widehat{\mathbb{C}}\}_{\lambda\in\mathbb{D}}$$

is called a *holomorphic motion of* A if the following conditions are satisfied:
(a) τ_0 is the identity map;
(b) For every $\lambda \in \mathbb{D}$, the map $\tau_\lambda : A \to \widehat{\mathbb{C}}$ is injective; and
(c) For every $z \in A$, the map $\mathbb{D} \ni \lambda \mapsto \tau_\lambda(z) \in \widehat{\mathbb{C}}$ is holomorphic.

The result which makes this notion so amazingly useful is the following "lemma" due to Mañé, Sad and Sullivan [70].

Theorem 16.4.4 (λ-lemma). *If $\{\tau_\lambda : A \to \widehat{\mathbb{C}}\}_{\lambda\in\mathbb{D}}$ is a holomorphic motion of a set $A \subseteq \widehat{\mathbb{C}}$, then for every $\lambda \in \mathbb{D}$ the map $\tau_\lambda : A \to \widehat{\mathbb{C}}$ has an injective quasisymmetric extension*

$$\tau_\lambda : \overline{A} \to \widehat{\mathbb{C}}.$$

Furthermore:
(a) *For every $z \in \overline{A}$, the map $\mathbb{D} \ni \lambda \mapsto \tau_\lambda(z) \in \widehat{\mathbb{C}}$ is holomorphic.*
(b) *The map $\mathbb{D} \times \overline{A} \ni (\lambda, z) \mapsto \tau_\lambda(z) \in \widehat{\mathbb{C}}$ is continuous.*
(c) *For every compact set $K \subseteq \mathbb{D}$, there is $0 < M(K) < \infty$ such that the maps τ_λ, $\lambda \in K$, are all $M(K)$-quasisymmetric. That is, the quasisymmetric constants of the maps τ_λ, $\lambda \in K$, are uniformly bounded above.*
(d) *The quasisymmetric constants of the maps τ_λ converge uniformly to 1 as $\lambda \to 0$.*

Proof. If the set A is finite, the theorem is trivially satisfied. So we may assume that A is infinite. The proof is based on the following well-known fact from hyperbolic geometry: any holomorphic map from $\mathbb{D}$ to the thrice-punctured sphere $\widehat{\mathbb{C}} \setminus \{0, 1, \infty\}$ does not increase hyperbolic distances on $\mathbb{D}$ and $\widehat{\mathbb{C}} \setminus \{0, 1, \infty\}$. Its proof is an easy consequence of Schwarz lemma (Lemma A.3.5 in Appendix A) and the fact that the universal cover of the thrice-punctured sphere is conformally the open unit disk $\mathbb{D}$. Fix three points of A and normalize the maps $\tau_\lambda : A \to \widehat{\mathbb{C}}$, $\lambda \in \mathbb{D}$, so that the images of these three points under each τ_λ are 0, 1 and ∞. This can be done by composing each map τ_λ with some appropriate Möbius transformation ($\widehat{\mathbb{C}} \ni z \mapsto \frac{az+b}{cz+d} \in \widehat{\mathbb{C}}$). We do not lose any generality in this way since Möbius transformations are uniformly Lipschitz, and hence uniformly continuous, with respect to the spherical metric on $\widehat{\mathbb{C}}$.

For any three points $x, y, z \in A$, disjoint from those previously fixed, consider the following four holomorphic functions from $\mathbb{D}$ to $\widehat{\mathbb{C}}$:

$$x(\lambda) := \tau_\lambda(x), \quad y(\lambda) := \tau_\lambda(y), \quad z(\lambda) := \tau_\lambda(z), \quad w(\lambda) := \frac{y(\lambda) - x(\lambda)}{y(\lambda)}.$$

By design, these four functions avoid the points 0, 1 and ∞. Fix any numbers

$$0 < m < 1 < M < \infty.$$

Let

$$A(m,M) := \{\xi \in \mathbb{C} : m \le |\xi| \le M\}$$

be the closed annulus centered at 0 with inner radius m and outer radius M. Fix $R, \varepsilon > 0$. Denote the hyperbolic distance on $\widehat{\mathbb{C}} \setminus \{0,1,\infty\}$ by $\mathrm{dist}^*_{\mathrm{hyp}}$. Note that

$$\Delta(M,R) := \sup\Big\{|z| : z \in \overline{B}^*_{\mathrm{hyp}}(A(0,M),R)\Big\} < \infty.$$

There exists $\delta > 0$ so small that

$$\Big[a,b \in \widehat{\mathbb{C}} \setminus \{0,1,\infty\},\ |a| < \frac{\delta}{m} \text{ and } \mathrm{dist}^*_{\mathrm{hyp}}(a,b) \le R\Big] \Longrightarrow |b| < \frac{\varepsilon}{\Delta(M,R)}.$$

Let $\mathrm{dist}^0_{\mathrm{hyp}}$ be the hyperbolic distance on the unit disk $\mathbb{D}$. Suppose that $y \in A \cap A(m,M)$, $|x-y| < \delta$ and $\lambda \in \overline{B}^0_{\mathrm{hyp}}(0,R) := \{z \in \mathbb{D} : \mathrm{dist}^0_{\mathrm{hyp}}(z,0) \le R\}$. Then $|w(0)| < \delta/m$. Moreover, as holomorphic maps do not increase hyperbolic distances (this boils down to Schwarz lemma for holomorphic maps of the unit disk), we know that $\mathrm{dist}^*_{\mathrm{hyp}}(w(\lambda), w(0)) \le R$. Hence, $|w(\lambda)| < \varepsilon/\Delta(M,R)$. On the other hand, like for w we have $\mathrm{dist}^*_{\mathrm{hyp}}(y(\lambda), y(0)) \le R$ and $y(0) = y \in A(m,M)$. Thus $|y(\lambda)| \le \Delta(M,R)$. So

$$\big|\tau_\lambda(x) - \tau_\lambda(y)\big| = \big|x(\lambda) - y(\lambda)\big| = \big|w(\lambda)\big| \cdot \big|y(\lambda)\big| < \varepsilon.$$

Thus we have proved that the family

$$\{\tau_\lambda : A \cap A(m,M) \to \widehat{\mathbb{C}}\}_{\lambda \in \overline{B}^0_{\mathrm{hyp}}(0,R)}$$

is uniformly equicontinuous for every $R > 0$. Letting $R \to \infty$, we deduce that each map

$$\tau_\lambda : A \cap A(m,M) \to \widehat{\mathbb{C}}, \quad \lambda \in \mathbb{D},$$

is uniformly continuous.

Since the annulus $A(m,M)$ contains 1, by exchanging the roles of 0, 1, and ∞ (i. e., by considering the annuli $A'(m,M) := \{\xi \in \mathbb{C} : m \le |\xi - 1| \le M\}$ and $A''(m,M) := \{\xi \in \widehat{\mathbb{C}} : m \le d_{\mathrm{sph}}(\xi,\infty) \le M\}$, where d_{sph} is the spherical metric on $\widehat{\mathbb{C}}$), we know that finitely many of such annuli cover $\widehat{\mathbb{C}}$. In consequence, each map τ_λ, $\lambda \in \mathbb{D}$, has a continuous extension to $\overline{A}$. Furthermore, the family

$$\{\tau_\lambda : \overline{A} \to \widehat{\mathbb{C}}\}_{\lambda \in \overline{B}^0_{\mathrm{hyp}}(0,R)}$$

is uniformly equicontinuous for each $R > 0$.

Now we shall prove that all maps $\tau_\lambda : \overline{A} \to \widehat{\mathbb{C}}$, $\lambda \in \mathbb{D}$, are injective. Fix again $0 < m < 1 < M < \infty$ and $R, \Delta > 0$. There exists $\delta > 0$ so small that

$$\left[a, b \in \widehat{\mathbb{C}} \setminus \{0, 1, \infty\},\ |a| > \frac{\delta}{M} \text{ and } \operatorname{dist}^*_{\mathrm{hyp}}(a, b) \le R\right] \Longrightarrow |b| > \frac{\Delta}{m}.$$

Assume that $y \in A \cap A(m, M)$, $|x - y| > \delta$ and $\lambda \in \overline{B}^0_{\mathrm{hyp}}(0, R)$. Then $|w(0)| < \delta/M$ and $\operatorname{dist}^*_{\mathrm{hyp}}(w(\lambda), w(0)) \le R$. Hence, $|w(\lambda)| > \Delta/m$. Therefore,

$$|\tau_\lambda(x) - \tau_\lambda(y)| = |x(\lambda) - y(\lambda)| > \Delta.$$

Now, letting first $m \to 0$ and $M \to \infty$, and then exchanging the roles of 0, 1 and ∞, we deduce that all maps $\tau_\lambda : \overline{A} \to \widehat{\mathbb{C}}$, $\lambda \in \overline{B}^0_{\mathrm{hyp}}(0, R)$, are injective. Letting $R \to \infty$ leads to the conclusion that the maps $\tau_\lambda : \overline{A} \to \widehat{\mathbb{C}}$, $\lambda \in \mathbb{D}$, are injective.

In order to prove that the maps $\tau_\lambda : \overline{A} \to \widehat{\mathbb{C}}$, $\lambda \in \mathbb{D}$, are quasisymmetric, and to prove parts (c) and (d), consider the function $g : \mathbb{D} \to \widehat{\mathbb{C}}$ given by the formula

$$g(\lambda) := \frac{x(\lambda) - y(\lambda)}{x(\lambda) - z(\lambda)}.$$

This function omits 0, 1 and ∞. Fix $R > 0$. Assume further that x, y, z are chosen such that $|g(0)| = 1$. Then

$$\operatorname{dist}^*_{\mathrm{hyp}}(g(\lambda), \partial\mathbb{D}) \le R, \quad \forall \lambda \in \overline{B}^0_{\mathrm{hyp}}(0, R). \tag{16.14}$$

Therefore,

$$0 < \inf\left\{|g(\lambda)| : \lambda \in \overline{B}^0_{\mathrm{hyp}}(0, R)\right\} \le \sup\left\{|g(\lambda)| : \lambda \in \overline{B}^0_{\mathrm{hyp}}(0, R)\right\} < \infty,$$

yielding that all maps $\tau_\lambda : \overline{A} \to \widehat{\mathbb{C}}$, $\lambda \in \overline{B}^0_{\mathrm{hyp}}(0, R)$, are quasisymmetric with their quasisymmetric constants bounded above by

$$M(R) := \frac{\sup\{|g(\lambda)| : \lambda \in \overline{B}^0_{\mathrm{hyp}}(0, R)\}}{\inf\{|g(\lambda)| : \lambda \in \overline{B}^0_{\mathrm{hyp}}(0, R)\}} < \infty. \tag{16.15}$$

So, item (c) is proved. In addition, by letting $R \to \infty$, we can infer that the maps $\tau_\lambda : \overline{A} \to \widehat{\mathbb{C}}$, $\lambda \in \mathbb{D}$, are quasisymmetric. On the other hand, we deduce from (16.14) that

$$\lim_{R \to 0} M(R) = 1,$$

finishing the proof of item (d).

Finally, pick any $x \in \overline{A}$. There exists a sequence $(x_n)_{n=1}^{\infty}$ of points in A such that $\lim_{n\to\infty} x_n = x$. Then

$$\tau_\lambda(x) = \lim_{n\to\infty} \tau_\lambda(x_n), \quad \forall \lambda \in \mathbb{D}.$$

This, along with the first part of the proof, implies that the map

$$\mathbb{D} \ni \lambda \mapsto \tau_\lambda(x) \in \widehat{\mathbb{C}}$$

is a limit of holomorphic maps $\mathbb{D} \ni \lambda \mapsto \tau_\lambda(x_n) \in \widehat{\mathbb{C}}$, $n \in \mathbb{N}$, converging uniformly on all compact subsets of $\mathbb{D}$. Thus, it is holomorphic, proving item (a). Due to the already proved equicontinuity, the map in item (b) is then continuous. □

Remark 16.4.5. Note that the hypothesis that the domain of λ's is an open subset of the complex plane $\mathbb{C}$ is substantial. If, for instance, the motion is only for $\lambda \in \mathbb{R}$, then the lemma does not hold. Indeed, consider the holomorphic motion of $A = \mathbb{C} \setminus \mathbb{R}$ such that the lower half-plane shifts horizontally in one direction ($\tau_\lambda(z) = z - \lambda$, $\lambda \in \mathbb{R}$) while the upper half-plane moves in the opposite direction ($\tau_\lambda(z) = z + \lambda$, $\lambda \in \mathbb{R}$). Clearly, this motion cannot be extended injectively nor continuously to $\overline{A} = \widehat{\mathbb{C}}$.

Remark 16.4.6. The maps $\tau_\lambda : \overline{A} \to \widehat{\mathbb{C}}$ of the holomorphic motion in Theorem 16.4.4 are Hölder continuous. Moreover, for every compact set $K \subseteq \mathbb{D}$ these maps have a common Hölder exponent $\alpha(K)$ and a uniformly bounded above Hölder norm in $H_{\alpha(K)}(\overline{A})$ and the Hölder exponent converges to 1 when $\lambda \to 0$. This follows from Slodkowski's theorem (Theorem A.3.18 or [119]), which asserts that the motion $\tau_\lambda : \overline{A} \to \widehat{\mathbb{C}}$, $\lambda \in \mathbb{D}$, extends from $\overline{A}$ to the whole Riemann sphere with the same quasisymmetric constants (see also [3]). Having this, we invoke the classical fact that each M-quasisymmetric (M-quasiconformal) homeomorphism is Hölder continuous with exponent $1/M$ and the corresponding Hölder norm uniformly bounded above (see [1, 67] for instance).

We now study the analytic perturbations of a conformal expanding repeller. The family of conjugating homeomorphisms we are after is provided by the next result.

Theorem 16.4.7. *If $\{T_\lambda : U \to \mathbb{C}\}_{\lambda\in\Lambda}$ is an analytic perturbation of a conformal expanding repeller $(J_{\lambda_0}, T_{\lambda_0}, U)$ in $\mathbb{C}$, then there exist a neighborhood $\widehat{\Lambda}$ of λ_0 and an open neighborhood $\widehat{U} \subseteq U$ of J_{λ_0} such that for every $\lambda \in \widehat{\Lambda}$ there are a conformal expanding repeller $(J_\lambda, T_\lambda, \widehat{U})$ and a homeomorphism $\tau_\lambda : J_{\lambda_0} \to J_\lambda$ with the following properties:*

(a) *For every fixed $z \in J_{\lambda_0}$, the map*

$$\widehat{\Lambda} \ni \lambda \longmapsto \tau_\lambda(z)$$

is holomorphic.

(b) *The map τ_{λ_0} is the identity map, and for every $\lambda \in \widehat{\Lambda}$ the map τ_λ, i. e.*

$$J_{\lambda_0} \ni z \longmapsto \tau_\lambda(z) \in J_\lambda,$$

is quasiconformal and Hölder continuous with exponent converging to 1 *uniformly as* $\lambda \to \lambda_0$.

(c) *The homeomorphism* τ_λ *conjugates the maps* T_{λ_0} *and* T_λ, *i. e.,*

$$T_\lambda \circ \tau_\lambda = \tau_\lambda \circ T_{\lambda_0}|_{J_{\lambda_0}}.$$

(d) *The map* $\widehat{\Lambda} \times J_{\lambda_0} \ni (\lambda, z) \longmapsto \tau_\lambda(z) \in \widehat{U}$ *is continuous.*

Proof. As this proof is long, it will be divided into parts and claims.

The neighborhoods $\widehat{U} \subseteq U$ and $\widehat{\Lambda} \subseteq \Lambda$. The following are necessary technicalities to define conveniently the neighborhoods $\widehat{U} \subseteq U$ and $\widehat{\Lambda} \subseteq \Lambda$.

First, observe that the triple $(J_{\lambda_0}, T_{\lambda_0}, V)$ is a conformal expanding repeller for any open set V such that $J_{\lambda_0} \subseteq V \subseteq U$. Second, in Theorem 4.1.5, we proved that any expanding repeller such as $(J_{\lambda_0}, T_{\lambda_0}, U)$, is distance expanding not only on J_{λ_0} but also on one of its neighborhoods, say $B(J_{\lambda_0}, \kappa)$ for some $\kappa > 0$. We will reuse part of that argument to show that this property carries over to all λ in a neighborhood of λ_0. By shrinking U or κ if necessary, we may assume that $U = B(J_{\lambda_0}, \kappa)$. To avoid any confusion, the expansion constant λ from Definition 16.2.1(d) will be replaced by L. Then

$$|T'_{\lambda_0}(z)| \geq \frac{1+L}{2}, \quad \forall z \in U \tag{16.16}$$

and

$$|T_{\lambda_0}(z) - T_{\lambda_0}(w)| \geq \frac{1+L}{2}|z-w|, \quad \forall z, w \in U \text{ with } |z-w| < \kappa. \tag{16.17}$$

As the set J_{λ_0} is compact and (16.16) holds, the analyticity of the perturbation $(\lambda, z) \mapsto T_\lambda(z)$ ensures, by downsizing κ (correspondingly U) and Λ if necessary, that there exist two constants $\Gamma \geq \gamma > 1$ such that

$$\gamma \leq |T'_\lambda(z)| \leq \Gamma, \quad \forall z \in U, \ \forall \lambda \in \Lambda. \tag{16.18}$$

Using (16.17) and after shrinking κ and Λ even further if necessary, the converse to Hurwitz's theorem (Theorem A.3.15) asserts that the map $T_\lambda : B(z, \kappa) \to T_\lambda(B(z, \kappa))$ is univalent for every $z \in J_{\lambda_0}$ and every $\lambda \in \Lambda$. Denote its inverse by

$$(T_\lambda)_z^{-1} : T_\lambda(B(z, \kappa)) \to B(z, \kappa). \tag{16.19}$$

By the Koebe quarter theorem (Theorem A.3.7) and (16.18), we know that

$$T_\lambda(B(z, \kappa)) \supseteq B\Big(T_\lambda(z), \frac{1}{4}\kappa|T'_\lambda(z)|\Big) \supseteq B\Big(T_\lambda(z), \frac{1}{4}\kappa\gamma\Big), \quad \forall z \in J_{\lambda_0}, \ \forall \lambda \in \Lambda.$$

Alternatively, the proof of Lemma 4.2.2 can be generalized to yield $q > 0$ such that

$$B(T_\lambda(z), q) \subseteq T_\lambda(B(z,\kappa)), \quad \forall z \in J_{\lambda_0}, \ \forall \lambda \in \Lambda. \tag{16.20}$$

By reducing Λ appropriately, the continuity of $(\lambda, z) \mapsto T_\lambda(z)$ then guarantees that there exists $0 < \eta \le \kappa$ such that

$$T_\lambda(B(z,\eta)) \subseteq B(T_\lambda(z), q), \quad \forall z \in J_{\lambda_0}, \ \forall \lambda \in \Lambda. \tag{16.21}$$

Combining (16.19)–(16.21), we observe that

$$B(z,\eta) \subseteq (T_\lambda)_z^{-1}(B(T_\lambda(z), q)) \subseteq B(z,\kappa), \quad \forall z \in J_{\lambda_0}, \ \forall \lambda \in \Lambda. \tag{16.22}$$

Let $\lambda \in \Lambda$ and let $z_1, z_2 \in B(J_{\lambda_0}, \eta/2)$ be such that $|z_1 - z_2| < \eta/2$. Then there exists $z \in J_{\lambda_0}$ such that $z_1, z_2 \in B(z,\eta)$. Consequently, $T_\lambda(z_1), T_\lambda(z_2) \in B(T_\lambda(z), q)$ according to (16.21). Let $S = [T_\lambda(z_1), T_\lambda(z_2)]$ be the line segment joining $T_\lambda(z_1)$ and $T_\lambda(z_2)$. Due to the convexity of discs in $\mathbb{C}$, we know that $S \subseteq B(T_\lambda(z), q)$. By (16.22), the curve $(T_\lambda)_z^{-1}(S)$ joins the points $(T_\lambda)_z^{-1}(T_\lambda(z_1)) = z_1$ and $(T_\lambda)_z^{-1}(T_\lambda(z_2)) = z_2$ from within $B(z,\kappa)$. Hence,

$$|z_1 - z_2| \le \ell((T_\lambda)_z^{-1}(S)) = \int_S |((T_\lambda)_z^{-1})'(w)|\, dw, \tag{16.23}$$

where $\ell(C)$ stands for the length of the curve C. By (16.18),

$$1 = |T_\lambda'((T_\lambda)_z^{-1}(w)) \cdot ((T_\lambda)_z^{-1})'(w)| \ge \gamma |((T_\lambda)_z^{-1})'(w)|, \quad \forall w \in S.$$

Consequently, (16.23) yields that

$$|z_1 - z_2| \le \gamma^{-1}\ell(S) = \gamma^{-1}|T_\lambda(z_1) - T_\lambda(z_2)|.$$

Letting $\kappa = \eta/2$, we infer that

$$|T_\lambda(z) - T_\lambda(w)| \ge \gamma|z - w|, \quad \forall \lambda \in \Lambda, \quad \forall z, w \in U \text{ with } |z \ \ w| < \kappa. \tag{16.24}$$

Moreover, the shadowing property of expanding maps (Corollary 4.3.6) states that there is $0 < \alpha < \kappa/3$ such that every infinite α-pseudo-orbit in J_{λ_0} is $(\kappa/3)$-shadowed by a unique point in J_{λ_0}.

Finally, thanks to the continuity of the perturbation, we may shrink Λ and select $0 < \hat{\kappa} < \alpha/(3\Gamma)$ such that

$$|T_\lambda(z) - T_{\lambda_0}(z)| < \frac{\alpha}{3}, \quad \forall z \in B(J_{\lambda_0}, \hat{\kappa}), \ \forall \lambda \in \Lambda. \tag{16.25}$$

We set $\widehat{U} := B(J_{\lambda_0}, \hat{\kappa})$. The neighborhood $\hat{\Lambda}$ will be identified later. All these technicalities will be used at the end of the proof to establish that $(J_\lambda, T_\lambda, \widehat{U})$ are conformal expanding repellers. In the meantime, we can simply work with the sets U and Λ.

The map $(\lambda, z) \mapsto \tau_\lambda(z)$**.** By definition, the conformal expanding repeller $T_{\lambda_0} : J_{\lambda_0} \to J_{\lambda_0}$ is a transitive, open and distance expanding map. By Corollary 4.3.10, the set of periodic points of T_{λ_0} is dense in J_{λ_0}. In particular, there exists at least one periodic orbit P in J_{λ_0}. Denote its prime period by p. As J_{λ_0} is a compact subset of the open set U, there exists $\delta > 0$ such that $\overline{B(J_{\lambda_0}, 2\delta)} \subseteq U$. Given the continuity of the map T_{λ_0} and the openness of $B(J_{\lambda_0}, \delta)$, the set $\bigcap_{k=0}^{p} T_{\lambda_0}^{-k}(B(J_{\lambda_0}, \delta))$ is open. As $T_{\lambda_0}(J_{\lambda_0}) = J_{\lambda_0}$, we know that $J_{\lambda_0} \subseteq \bigcap_{k=0}^{p} T_{\lambda_0}^{-k}(B(J_{\lambda_0}, \delta))$. Since J_{λ_0} is compact, there is $0 < \delta' < \delta$ such that $B(J_{\lambda_0}, \delta') \subseteq \bigcap_{k=0}^{p} T_{\lambda_0}^{-k}(B(J_{\lambda_0}, \delta))$. Therefore

$$\bigcup_{k=0}^{p} T_{\lambda_0}^{k}(B(J_{\lambda_0}, \delta')) \subseteq B(J_{\lambda_0}, \delta) \subseteq \overline{B(J_{\lambda_0}, 2\delta)} \subseteq U.$$

As $\{T_\lambda\}_{\lambda \in \Lambda}$ is an analytic perturbation of T_{λ_0}, the continuity of $(\lambda, z) \mapsto T_\lambda(z)$ signifies that there exists $R' > 0$ such that

$$\bigcup_{k=0}^{p} T_{\lambda}^{k}(B(J_{\lambda_0}, \delta')) \subseteq B(J_{\lambda_0}, 2\delta), \quad \forall \lambda \in B(\lambda_0, R').$$

Fix $\xi \in P$ and consider the function $F : B(J_{\lambda_0}, \delta') \times B(\lambda_0, R') \to \mathbb{C}$ defined by

$$F(z, \lambda) := T_\lambda^p(z) - z.$$

Then F is well defined and (separately) holomorphic with respect to each of the variables z and λ and, therefore, is (jointly) holomorphic by virtue of Hartogs' theorem (Theorem A.4.1 in Appendix A). As T_{λ_0} is a conformal expanding repeller, the periodic point ξ is repelling, i. e.,

$$\left| \frac{\partial T_\lambda^p}{\partial z} \bigg|_{(\xi, \lambda_0)} \right| > 1.$$

Hence,

$$\frac{\partial F}{\partial z} \bigg|_{(\xi, \lambda_0)} = \frac{\partial T_\lambda^p}{\partial z} \bigg|_{(\xi, \lambda_0)} - 1 \neq 0.$$

So the implicit function theorem (Theorem A.3.16) applies and yields the existence of two numbers $\delta_\xi \in (0, \delta']$ and $R_\xi \in (0, R']$ as well as a unique holomorphic function

$$B(\lambda_0, R_\xi) \ni \lambda \longmapsto \tau_\lambda(\xi) \in B(\xi, \delta_\xi)$$

such that

$$\tau_{\lambda_0}(\xi) = \xi \tag{16.26}$$

and for every $\lambda \in B(\lambda_0, R_\xi)$, the number $\tau_\lambda(\xi)$ is the only point $z \in B(\xi, \delta_\xi)$ for which

$$F(z, \lambda) = 0. \tag{16.27}$$

In particular,

$$T_\lambda^p(\tau_\lambda(\xi)) = \tau_\lambda(\xi), \quad \forall \lambda \in B(\lambda_0, R_\xi). \tag{16.28}$$

Doing this for each $\xi \in P$, we set

$$R := \min\{R_\xi : \xi \in P\} \in (0, R']. \tag{16.29}$$

Now fix an arbitrary point

$$\zeta \in \bigcup_{n=1}^{\infty} T_{\lambda_0}^{-n}(P) \setminus P \subseteq J_{\lambda_0}. \tag{16.30}$$

This last inclusion ensues from the fact that $T_{\lambda_0}^{-1}(J_{\lambda_0}) = J_{\lambda_0}$ per Definition 16.2.1(c,e). Let $n(\zeta)$ be the smallest $n \in \mathbb{N}$ such that

$$T_{\lambda_0}^n(\zeta) \in P. \tag{16.31}$$

Denote

$$\xi := T_{\lambda_0}^{n(\zeta)}(\zeta). \tag{16.32}$$

Since $\zeta \in J_{\lambda_0} = T_{\lambda_0}(J_{\lambda_0})$ and since $T_{\lambda_0} : U \to \mathbb{C}$ is continuous, there is $\delta'_\zeta > 0$ such that

$$T_{\lambda_0}^k(B(\zeta, \delta'_\zeta)) \subseteq B(J_{\lambda_0}, \delta), \quad \forall k = 0, 1, \ldots, n(\zeta).$$

Because $\{T_\lambda\}_{\lambda \in \Lambda}$ is an analytic perturbation of T_{λ_0}, there exists $R'_\zeta \in (0, R]$ such that

$$T_\lambda^k(B(\zeta, \delta'_\zeta)) \subseteq B(J_{\lambda_0}, 2\delta), \quad \forall \lambda \in B(\lambda_0, R'_\zeta), \ \forall k = 0, 1, \ldots, n(\zeta). \tag{16.33}$$

Hence, the function $G : B(\zeta, \delta'_\zeta) \times B(\lambda_0, R'_\zeta) \to \mathbb{C}$ given by

$$G(z, \lambda) := T_\lambda^{n(\zeta)}(z) - \tau_\lambda(\xi) = T_\lambda^{n(\zeta)}(z) - \tau_\lambda(T_{\lambda_0}^{n(\zeta)}(\zeta))$$

is well defined, holomorphic and such that

$$\left.\frac{\partial G}{\partial z}\right|_{(\zeta, \lambda_0)} = \left.\frac{\partial T_\lambda^{n(\zeta)}}{\partial z}\right|_{(\zeta, \lambda_0)} \neq 0.$$

So the implicit function theorem (Theorem A.3.16) applies, yielding the existence of two numbers $\delta_\zeta \in (0, \delta'_\zeta]$ and $R_\zeta \in (0, R'_\zeta]$ as well as a unique holomorphic function

$$B(\lambda_0, R_\zeta) \ni \lambda \longmapsto \tau_\lambda(\zeta) \in B(\zeta, \delta_\zeta)$$

such that

$$\tau_{\lambda_0}(\zeta) = \zeta \tag{16.34}$$

and for every $\lambda \in B(\lambda_0, R_\zeta)$, the number $\tau_\lambda(\zeta)$ is the only point $z \in B(\zeta, \delta_\zeta)$ such that

$$G(z, \lambda) = 0. \tag{16.35}$$

In particular,

$$T_\lambda^{n(\zeta)}(\tau_\lambda(\zeta)) = \tau_\lambda(\xi) = \tau_\lambda(T_{\lambda_0}^{n(\zeta)}(\zeta)), \quad \forall \lambda \in B(\lambda_0, R_\zeta). \tag{16.36}$$

We shall prove the following assertion.

Claim 1. *The function $B(\lambda_0, R_\zeta) \ni \lambda \longmapsto \tau_\lambda(\zeta) \in B(\zeta, \delta_\zeta)$ extends holomorphically to $B(\lambda_0, R)$ and is uniquely determined by the relation*

$$T_\lambda^{n(\zeta)}(\tau_\lambda(\zeta)) = \tau_\lambda(\xi) = \tau_\lambda(T_{\lambda_0}^{n(\zeta)}(\zeta)), \quad \forall \lambda \in B(\lambda_0, R).$$

Proof of Claim 1. Write the holomorphic function $B(\lambda_0, R_\zeta) \ni \lambda \longmapsto \tau_\lambda(\zeta) \in B(\zeta, \delta_\zeta)$ as a power series around λ_0, i. e.,

$$\tau_\lambda(\zeta) = \sum_{n=0}^{\infty} a_n(\zeta)(\lambda - \lambda_0)^n, \quad \forall \lambda \in B(\lambda_0, R_\zeta).$$

Let $\rho \geq R_\zeta > 0$ be the minimum of R and the radius of convergence of this series. Set

$$\rho' := \sup\Big\{t \in [0, \rho] : T_\lambda^{n(\zeta)}(\tau_\lambda(\zeta)) \in B(J_{\lambda_0}, \delta'), \ \forall \lambda \in B(\lambda_0, t)\Big\}.$$

We will show that

$$\rho' = \rho. \tag{16.37}$$

Then, by the continuity of the functions $(\lambda, z) \longmapsto T_\lambda^{n(\zeta)}(z)$, $\lambda \longmapsto \tau_\lambda(\zeta)$ and $\lambda \longmapsto \tau_\lambda(\xi)$, we will deduce that

$$T_\lambda^{n(\zeta)}(\tau_\lambda(\zeta)) = \tau_\lambda(\xi), \quad \forall \lambda \in B(\lambda_0, \rho). \tag{16.38}$$

Seeking a contradiction, suppose that $\rho' < \rho$. By (16.36) and the continuity of the functions $(\lambda, z) \longmapsto T_\lambda^{n(\zeta)}(z)$, $\lambda \longmapsto \tau_\lambda(\zeta)$ and $\lambda \longmapsto \tau_\lambda(\xi)$, we would then infer that

$$T_\lambda^{n(\zeta)}(\tau_\lambda(\zeta)) = \tau_\lambda(\xi), \quad \forall \lambda \in \overline{B(\lambda_0, \rho')}. \tag{16.39}$$

As $\tau_\lambda(\xi) \in B(\xi, \delta_\xi) \subseteq B(\xi, \delta')$ for all $\lambda \in B(\lambda_0, R) \supseteq \overline{B(\lambda_0, \rho')}$, we would deduce that

$$T_\lambda^{n(\zeta)}(\tau_\lambda(\zeta)) \in B(J_{\lambda_0}, \delta'), \quad \forall \lambda \in \overline{B(\lambda_0, \rho')}.$$

As the image of the compact set $\overline{B(\lambda_0, \rho')}$ under the composition of the continuous functions $(\lambda, z) \longmapsto T_\lambda^{n(\zeta)}(z)$ and $\lambda \longmapsto \tau_\lambda(\zeta)$, the set $\{T_\lambda^{n(\zeta)}(\tau_\lambda(\zeta)) : \lambda \in \overline{B(\lambda_0, \rho')}\}$ is compact. There would then exist $r \in (\rho', \rho]$ such that

$$T_\lambda^{n(\zeta)}(\tau_\lambda(\zeta)) \in B(J_{\lambda_0}, \delta'), \quad \forall \lambda \in B(\lambda_0, r),$$

contrary to the definition of ρ'. Thus, (16.37) is established.

Now we shall prove that

$$\rho = R.$$

Again seeking a contradiction, suppose that $\rho < R$. Fix any $\gamma \in \mathbb{C}$ with $|\gamma - \lambda_0| = \rho$. Let $(\gamma_k)_{k=1}^\infty$ be any sequence in $B(\lambda_0, \rho)$ converging to γ and such that the sequence $(\tau_{\gamma_k}(\zeta))_{k=1}^\infty$ also converges. Denote this latter limit by $\widehat{\zeta}$. By definition of $\lambda \mapsto \tau_\lambda(\zeta)$, we know that $\widehat{\zeta} \in \overline{B(\zeta, \delta_\zeta)} \subseteq \overline{B(\zeta, \delta'_\zeta)}$. It then follows from (16.33), (16.38), the convergence of $(\tau_{\gamma_k}(\zeta))_{k=1}^\infty$ to $\widehat{\zeta}$ and the continuity of the maps $(\lambda, z) \longmapsto T_\lambda^{n(\zeta)}(z)$ and $\lambda \longmapsto \tau_\lambda(\xi)$, that

$$T_\gamma^{n(\zeta)}(\widehat{\zeta}) = \lim_{k\to\infty} T_{\gamma_k}^{n(\zeta)}(\tau_{\gamma_k}(\zeta)) = \lim_{k\to\infty} \tau_{\gamma_k}(\xi) = \tau_\gamma(\xi).$$

So we can apply the implicit function theorem (Theorem A.3.16) in exactly the same way as before to get two numbers $\widehat{\delta} > 0$ and $\widehat{R} > 0$, and a unique holomorphic function

$$B(\gamma, \widehat{R}) \ni \lambda \longmapsto \widehat{\tau}_\lambda(\zeta) \in B(\widehat{\zeta}, \widehat{\delta})$$

such that

$$\widehat{\tau}_\gamma(\zeta) = \widehat{\zeta} \tag{16.40}$$

and for every $\lambda \in B(\gamma, \widehat{R})$, the number $\widehat{\tau}_\lambda(\zeta)$ is the only point $z \in B(\widehat{\zeta}, \widehat{\delta})$ such that

$$T_\lambda^{n(\zeta)}(\widehat{\tau}_\lambda(\zeta)) = \tau_\lambda(\xi). \tag{16.41}$$

But then the uniqueness part of the implicit function theorem affirms that

$$\widehat{\tau}_\lambda(\zeta) = \tau_\lambda(\zeta), \quad \forall \lambda \in B(\lambda_0, \rho) \cap B(\gamma, \widehat{R}).$$

We deduce that the holomorphic function

$$B(\lambda_0,\rho) \ni \lambda \longmapsto \tau_\lambda(\zeta) \in \mathbb{C}$$

extends holomorphically to some larger ball $B(\lambda_0,\rho'')$ for some $\rho'' > \rho$. This contradicts the definition of ρ, and finishes the proof of Claim 1. ■

We now establish the following fact.

Claim 2. *The maps τ_λ satisfy the following commutativity property:*

$$T_\lambda \circ \tau_\lambda(z) = \tau_\lambda \circ T_{\lambda_0}(z), \quad \forall \lambda \in B(\lambda_0,R),\ \forall z \in \bigcup_{n=0}^{\infty} T_{\lambda_0}^{-n}(P). \tag{16.42}$$

Proof of Claim 2. Consider first $z = \zeta \in \bigcup_{n=0}^{\infty} T_{\lambda_0}^{-n}(P) \setminus P$. Then $n(\zeta) \geq 1$ and $n(T_{\lambda_0}(\zeta)) = n(\zeta) - 1 \geq 0$. By Claim 1, we have

$$T_\lambda^{n(T_{\lambda_0}(\zeta))}(T_\lambda \circ \tau_\lambda(\zeta)) = T_\lambda^{n(\zeta)}(\tau_\lambda(\zeta)) = \tau_\lambda(T_{\lambda_0}^{n(\zeta)}(\zeta)) = \tau_\lambda(T_{\lambda_0}^{n(T_{\lambda_0}(\zeta))}(T_{\lambda_0}(\zeta))), \quad \forall \lambda \in B(\lambda_0,R).$$

Using Claim 1 once again but this time with $T_{\lambda_0}(\zeta)$ instead of ζ, we get

$$T_\lambda^{n(T_{\lambda_0}(\zeta))}(\tau_\lambda \circ T_{\lambda_0}(\zeta)) = \tau_\lambda(T_{\lambda_0}^{n(T_{\lambda_0}(\zeta))}(T_{\lambda_0}(\zeta))), \quad \forall \lambda \in B(\lambda_0,R).$$

In addition, since the function $B(\lambda_0,R) \ni \lambda \longmapsto T_\lambda \circ \tau_\lambda(\zeta) \in \mathbb{C}$ is continuous, there exists $\widehat{R}_\zeta \in (0,R]$ such that

$$T_\lambda \circ \tau_\lambda(\zeta) \in B(T_{\lambda_0} \circ \tau_{\lambda_0}(\zeta), \delta_{T_{\lambda_0} \circ \tau_{\lambda_0}(\zeta)}) = B(T_{\lambda_0}(\zeta), \delta_{T_{\lambda_0}(\zeta)}), \quad \forall \lambda \in B(\lambda_0, \widehat{R}_\zeta).$$

It thereby follows from the uniqueness part of Claim 1 for $T_{\lambda_0}(\zeta)$ that

$$T_\lambda \circ \tau_\lambda(\zeta) = \tau_\lambda \circ T_{\lambda_0}(\zeta), \quad \forall \lambda \in B(\lambda_0, \widehat{R}_\zeta).$$

Since both functions

$$B(\lambda_0,R) \ni \lambda \longmapsto T_\lambda \circ \tau_\lambda(\zeta) \in \mathbb{C} \quad \text{and} \quad B(\lambda_0,R) \ni \lambda \longmapsto \tau_\lambda \circ T_{\lambda_0}(\zeta) \in \mathbb{C}$$

are holomorphic, we conclude that

$$T_\lambda \circ \tau_\lambda(\zeta) = \tau_\lambda \circ T_{\lambda_0}(\zeta), \quad \forall \lambda \in B(\lambda_0,R),\ \forall \zeta \in \bigcup_{n=0}^{\infty} T_{\lambda_0}^{-n}(P) \setminus P. \tag{16.43}$$

Now, suppose that $z = \xi \in P$. Then both points ξ and $T_{\lambda_0}(\xi)$ are periodic with prime period p under the map T_{λ_0}. Hence,

$$T_\lambda^p(\tau_\lambda(T_{\lambda_0}(\xi))) = \tau_\lambda(T_{\lambda_0}(\xi)) \quad \text{and} \quad T_\lambda^p(T_\lambda(\tau_\lambda(\xi))) = T_\lambda(T_\lambda^p(\tau_\lambda(\xi))) = T_\lambda(\tau_\lambda(\xi)).$$

So, in a similar way to the previous case, the uniqueness part of the implicit function theorem (see (16.26)–(16.28) with $T_{\lambda_0}(\xi)$ in lieu of ξ) and the continuity of the function $B(\lambda_0, R) \ni \lambda \mapsto T_\lambda \circ \tau_\lambda(z) \in \mathbb{C}$ lead to the conclusion that $T_\lambda \circ \tau_\lambda(\xi) = \tau_\lambda \circ T_{\lambda_0}(\xi)$. Along with (16.43), this completes the proof of Claim 2. ∎

Since both sets P and $T_{\lambda_0}^{-1}(P)$ are finite and for every $z \in \bigcup_{n=0}^\infty T_{\lambda_0}^{-n}(P)$ the function $B(\lambda_0, R) \ni \lambda \mapsto \tau_\lambda(z)$ is continuous, there exists $\widehat{R} \in (0, R]$ such that

$$\max_{w \in T_{\lambda_0}^{-1}(P)} \sup_{\lambda \in \overline{B(\lambda_0, \widehat{R})}} |\tau_\lambda(w) - \tau_{\lambda_0}(w)| < \frac{1}{2} \min\left\{|\xi - \zeta| : \xi \in P,\ \zeta \in T_{\lambda_0}^{-1}(P) \setminus P\right\}. \tag{16.44}$$

Claim 3. *For every $\lambda \in B(\lambda_0, \widehat{R})$, the function*

$$\bigcup_{n=0}^{\infty} T_{\lambda_0}^{-n}(P) \ni z \longmapsto \tau_\lambda(z) \in \mathbb{C}$$

is injective.

Proof of Claim 3. Suppose that

$$\tau_\gamma(z) = \tau_\gamma(w) \tag{16.45}$$

for some $\gamma \in B(\lambda_0, \widehat{R})$ and $z, w \in \bigcup_{n=0}^\infty T_{\lambda_0}^{-n}(P)$ with $z \neq w$. We will consider several cases.

Assume first that $z, w \in P$, i. e.,

$$n(z) = n(w) = 0.$$

Since $\tau_{\lambda_0}(z) = z \neq w = \tau_{\lambda_0}(w)$, we may assume that

$$\tau_\lambda(z) \neq \tau_\lambda(w), \quad \forall \lambda \in B(\lambda_0, |\gamma - \lambda_0|). \tag{16.46}$$

Let $y = \tau_\gamma(z) = \tau_\gamma(w)$. By Claim 2, $T_\lambda^p(\tau_\lambda(z)) = \tau_\lambda(T_{\lambda_0}^p(z)) = \tau_\lambda(z)$ and $T_\lambda^p(\tau_\lambda(w)) = \tau_\lambda(w)$ for every $\lambda \in B(\lambda_0, R)$. In particular, $T_\gamma^p(y) = y$. It follows from the uniqueness part of the implicit function theorem that $\tau_\lambda(z) = \tau_\lambda(w)$ for every λ belonging simultaneously to $B(\lambda_0, R)$ and some sufficiently small neighborhood of γ. This, however, contradicts (16.46) and we are done in this case.

Now assume that $z \in P$ and $w \in \bigcup_{n=0}^\infty T_{\lambda_0}^{-n}(P) \setminus P$, i. e.,

$$n(z) = 0 \quad \text{and} \quad n(w) \geq 1.$$

Because we are seeking a contradiction, we may assume without loss of generality that z, w are such that $n(w)$ is the least $n \in \mathbb{N}$ for which (16.45) holds. By Claim 2,

$$\tau_\gamma(T_{\lambda_0}(w)) = T_\gamma(\tau_\gamma(w)) = T_\gamma(\tau_\gamma(z)) = \tau_\gamma(T_{\lambda_0}(z)). \tag{16.47}$$

But $T_{\lambda_0}(z) \in P$ while $T_{\lambda_0}(w) \in \bigcup_{n=0}^{\infty} T_{\lambda_0}^{-n}(P)$. So (16.47) states that the pair $T_{\lambda_0}(z), T_{\lambda_0}(w)$ satisfies (16.45). Since $n(T_{\lambda_0}(w)) = n(w) - 1$, we surmise from the minimality of $n(w)$ that $n(T_{\lambda_0}(w)) = 0$, i. e.,

$$n(w) = 1. \tag{16.48}$$

Hence,

$$w \in T_{\lambda_0}^{-1}(P) \setminus P.$$

Since $z \in P \subseteq T_{\lambda_0}^{-1}(P)$, it follows from (16.44) that

$$\left|\tau_\gamma(w) - \tau_\gamma(z)\right| \geq |z-w| - \left|\tau_\gamma(z) - z\right| - \left|\tau_\gamma(w) - w\right| > |z-w| - 2 \cdot \frac{1}{2}|z-w| = 0.$$

This contradicts (16.45) and we are done in this case, too.

Finally, assume that

$$\min\{n(z), n(w)\} \geq 1.$$

We may assume without loss of generality that $n(z) \leq n(w)$. By Claim 2,

$$\tau_\gamma \circ T_{\lambda_0}^{n(z)}(z) = T_\gamma^{n(z)} \circ \tau_\gamma(z) = T_\gamma^{n(z)} \circ \tau_\gamma(w) = \tau_\gamma \circ T_{\lambda_0}^{n(z)}(w).$$

But $n(T_{\lambda_0}^{n(z)}(z)) = 0$, whence it follows from the previous case that

$$T_{\lambda_0}^{n(z)}(z) = T_{\lambda_0}^{n(z)}(w).$$

Since $T_{\lambda_0}^{n(z)}(z) \in P$, this implies that $n(z) = n(w)$. As $\tau_\gamma(z) = \tau_\gamma(w)$, it ensues from Claim 1 that there exists $\delta_\gamma > 0$ such that both functions $B(\gamma, \delta_\gamma) \ni \lambda \longmapsto \tau_\lambda(z), \tau_\lambda(w)$ are well defined and $\tau_\lambda(z) = \tau_\lambda(w)$ for all $\lambda \in B(\gamma, \delta_\gamma)$. Since these two functions are analytic, we conclude that $\tau_\lambda(z) = \tau_\lambda(w)$ for all $\lambda \in B(\lambda_0, \widehat{R})$. In particular, $z = \tau_{\lambda_0}(z) = \tau_{\lambda_0}(w) = w$. The proof of Claim 3 is complete. ■

We have thus proved that the family of maps

$$\left\{\tau_\lambda : \bigcup_{n=0}^{\infty} T_{\lambda_0}^{-n}(P) \to \mathbb{C}\right\}_{\lambda \in B(\lambda_0, \widehat{R})}$$

is a (shifted and rescaled) holomorphic motion in the sense of Definition 16.4.3. As $T_{\lambda_0} : J_{\lambda_0} \to J_{\lambda_0}$ is transitive, open and distance expanding, it is very strongly transitive by Lemma 4.2.10 and Definition 1.5.14 affirms that the backward orbit of any point in J_{λ_0} is dense in J_{λ_0}. So $J_{\lambda_0} = \overline{\bigcup_{n=0}^{\infty} T_{\lambda_0}^{-n}(P)}$. It follows from the λ-lemma (Theorem 16.4.4) that all the maps τ_λ, $\lambda \in B(\lambda_0, \widehat{R})$, have quasisymmetric extensions to J_{λ_0}. Furthermore, all of them are injective and enjoy properties (a)–(d) of Theorem 16.4.4. We keep for them the notation

$$\tau_\lambda : J_{\lambda_0} \to \mathbb{C}.$$

Property (a) of the current theorem follows immediately from property (a) of Theorem 16.4.4. As the maps $\tau_\lambda : J_{\lambda_0} \to \mathbb{C}$, $\lambda \in B(\lambda_0, \widehat{R})$, are continuous per item (b) of Theorem 16.4.4, property (c) of the current theorem is a direct consequence of (16.42), the density of $\bigcup_{n=0}^{\infty} T_{\lambda_0}^{-n}(P)$ in J_{λ_0}, and the continuity of the maps τ_λ. Property (b) follows from item (d) of Theorem 16.4.4 and Remark 16.4.6. Property (d) comes directly from item (b) of the λ-lemma.

Finally, for every $\lambda \in \widehat{\Lambda} := B(\lambda_0, \widehat{R})$ define

$$J_\lambda := \tau_\lambda(J_{\lambda_0}).$$

As the image of a compact set under a continuous map, this set is compact.

$(J_\lambda, T_\lambda, \widehat{U})$ is a conformal expanding repeller. Recall that $U = B(J_{\lambda_0}, \kappa)$ for some κ small enough that (16.18) and (16.24) hold. Then a number $0 < \alpha < \kappa/3$ is selected so that every infinite α-pseudo-orbit in J_{λ_0} is $(\kappa/3)$-shadowed by a unique point in J_{λ_0}. Finally, $\widehat{U} = B(J_{\lambda_0}, \widehat{\kappa})$, where $0 < \widehat{\kappa} < \alpha/(3\Gamma)$ is chosen so that (16.25) is valid.

Given that $T_\lambda : U \to \mathbb{C}$ is holomorphic, it is $C^{1+\epsilon}(U)$ and conformal on $U \supseteq \widehat{U}$. This establishes properties (a) and (b) of Definition 16.2.1.

Since $\tau_\lambda : J_{\lambda_0} \to J_\lambda$ is a continuous bijection from a compact set to a Hausdorff space, it is a homeomorphism. Because $T_{\lambda_0}(J_{\lambda_0}) = J_{\lambda_0}$, using property (c) proved above, it is easy to see that $T_\lambda(J_\lambda) = J_\lambda$. So property (c) of Definition 16.2.1 holds. This also implies that τ_λ is a topological conjugacy between $T_{\lambda_0} : J_{\lambda_0} \to J_{\lambda_0}$ and $T_\lambda : J_\lambda \to J_\lambda$.

The continuity of the function $(\lambda, z) \mapsto \tau_\lambda(z)$ and (16.24) mean that $T_\lambda : J_\lambda \to J_\lambda$ is expanding, i. e., property (d) of Definition 16.2.1 is satisfied, provided λ is close enough to λ_0. Observe that the expansion is uniform in λ.

Given that $T_\lambda(J_\lambda) \subseteq J_\lambda$, it immediately follows that $\bigcap_{n=0}^{\infty} T_\lambda^{-n}(\widehat{U}) \supseteq J_\lambda$. For the opposite inclusion, let $z_0 \in \bigcap_{n=0}^{\infty} T_\lambda^{-n}(\widehat{U})$. This means that $T_\lambda^n(z_0) \in \widehat{U} := B(J_{\lambda_0}, \widehat{\kappa})$ for all $n \geq 0$. That is, for every $n \geq 0$ there exists $z_n \in J_{\lambda_0}$ such that

$$T_\lambda^n(z_0) \in B(z_n, \widehat{\kappa}), \quad \text{and thus} \quad |T_\lambda^n(z_0) - z_n| < \widehat{\kappa} < \frac{\alpha}{3\Gamma} < \frac{\kappa}{9\Gamma}. \tag{16.49}$$

It ensues from this, (16.18) and (16.25) that

$$\begin{aligned} |z_{n+1} - T_{\lambda_0}(z_n)| &\leq |z_{n+1} - T_\lambda^{n+1}(z_0)| + |T_\lambda^{n+1}(z_0) - T_\lambda(z_n)| + |T_\lambda(z_n) - T_{\lambda_0}(z_n)| \\ &< \frac{\alpha}{3\Gamma} + \Gamma |T_\lambda^n(z_0) - z_n| + \frac{\alpha}{3} < 3 \cdot \frac{\alpha}{3} = \alpha. \end{aligned}$$

So the sequence $(z_n)_{n=0}^{\infty}$ is an α-pseudo-orbit for $T_{\lambda_0} : J_{\lambda_0} \to J_{\lambda_0}$. By definition of α, there exists a (unique) point $w \in J_{\lambda_0}$ which $(\kappa/3)$-shadows this pseudo-orbit, i. e.,

$$|T_{\lambda_0}^n(w) - z_n| < \frac{\kappa}{3}, \quad \forall n \geq 0. \tag{16.50}$$

Shrinking Λ if necessary, the continuity of the function $(\lambda, z) \mapsto \tau_\lambda(z)$ guarantees that

$$|\tau_\lambda(z) - \tau_{\lambda_0}(z)| < \widehat{\kappa}, \quad \forall z \in J_{\lambda_0}, \ \forall \lambda \in \Lambda. \tag{16.51}$$

It follows from (16.49)–(16.51) that

$$\begin{aligned}
|T_\lambda^n(\tau_\lambda(w)) - T_\lambda^n(z_0)| &= |\tau_\lambda(T_{\lambda_0}^n(w)) - T_\lambda^n(z_0)| \\
&\leq |\tau_\lambda(T_{\lambda_0}^n(w)) - \tau_{\lambda_0}(T_{\lambda_0}^n(w))| + |T_{\lambda_0}^n(w) - z_n| + |z_n - T_\lambda^n(z_0)| \\
&< \widehat{\kappa} + \frac{\kappa}{3} + \widehat{\kappa} = 2\widehat{\kappa} + \frac{\kappa}{3} < \kappa
\end{aligned}$$

for all $n \geq 0$. Moreover, $T_\lambda^n(\tau_\lambda(w))$ and $T_\lambda^n(z_0)$ lie both in U. Indeed, $T_\lambda^n(z_0) \in \widehat{U} \subseteq U$ while $|T_\lambda^n(\tau_\lambda(w)) - z_n| \leq |T_\lambda^n(\tau_\lambda(w)) - T_\lambda^n(z_0)| + |T_\lambda^n(z_0) - z_n| < 2\widehat{\kappa} + \frac{\kappa}{3} + \widehat{\kappa} < \kappa$ and $z_n \in J_{\lambda_0}$. Hence, $T_\lambda^n(\tau_\lambda(w)) \in B(J_{\lambda_0}, \kappa) = U$. Per (16.24), we then know that

$$|T_\lambda^{n+1}(\tau_\lambda(w)) - T_\lambda^{n+1}(z_0)| \geq \gamma |T_\lambda^n(\tau_\lambda(w)) - T_\lambda^n(z_0)|, \quad \forall n \geq 0.$$

By recurrence,

$$|\tau_\lambda(w) - z_0| \leq \gamma^{-n} |T_\lambda^n(\tau_\lambda(w)) - T_\lambda^n(z_0)| \leq \kappa \gamma^{-n}, \quad \forall n \geq 0.$$

Thus, $|\tau_\lambda(w) - z_0| = 0$. So $z_0 = \tau_\lambda(w) \in \tau_\lambda(J_{\lambda_0}) = J_\lambda$. This shows that $\bigcap_{n=0}^{\infty} T_\lambda^{-n}(\widehat{U}) \subseteq J_\lambda$. Therefore, property (e) of Definition 16.2.1 carries over to T_λ.

Finally, recall that transitivity is a topological invariant (see Remark 1.5.7). Since τ_λ is a topological conjugacy between $T_{\lambda_0} : J_{\lambda_0} \to J_{\lambda_0}$ and $T_\lambda : J_\lambda \to J_\lambda$ and since $T_{\lambda_0} : J_{\lambda_0} \to J_{\lambda_0}$ is transitive, it follows that $T_\lambda : J_\lambda \to J_\lambda$ is transitive, too. So property (f) of Definition 16.2.1 holds. □

We need one more technical result, which is a well-known fact in complex analysis. The proof we provide is taken from Lemma 9.1 in [84].

Fix an integer $d \in \mathbb{N}$. Embed $\mathbb{C}^d$ into $\mathbb{C}^{2d}$ by the formula

$$z = (z_1, z_2, \ldots, z_d) \longmapsto \tilde{z} := (\operatorname{Re} z_1, \operatorname{Im} z_1, \operatorname{Re} z_2, \operatorname{Im} z_2, \ldots, \operatorname{Re} z_d, \operatorname{Im} z_d) \in \mathbb{R}^{2d} \subseteq \mathbb{C}^{2d}.$$

For every $z \in \mathbb{C}^d$ and every $r > 0$, denote by $D_{\mathbb{C}^d}(z, r) := \prod_{k=1}^d B_{\mathbb{C}}(z_k, r)$ the d-dimensional polydisk in $\mathbb{C}^d$ centered at z and of "radius" r (not to be confused with the open ball $B_{\mathbb{C}^d}(z, r)$ in $\mathbb{C}^d$ centered at z and of radius r). Moreover, $|z|^2 := \sum_{k=1}^d |z_k|^2$.

Lemma 16.4.8. *For every $M \geq 0$, every $R > 0$, every $\lambda^0 \in \mathbb{C}^d$, and every analytic function $\psi : D_{\mathbb{C}^d}(\lambda^0, R) \to \mathbb{C}$ whose modulus is bounded above by M, there exists a unique analytic function*

$$\widetilde{\psi} : D_{\mathbb{C}^{2d}}(\widetilde{\lambda^0}, R/4) \to \mathbb{C}$$

whose modulus is bounded above by $4^d M$ and whose restriction to the polydisk $D_{\mathbb{C}^d}(\lambda^0, R/4) \equiv \widetilde{D_{\mathbb{C}^d}(\lambda^0, R/4)} = D_{\mathbb{C}^{2d}}(\widetilde{\lambda^0}, R/4) \cap \mathbb{R}^{2d}$ coincides with $\operatorname{Re}\psi$, the real part of ψ.

Proof. Denote by $\mathbb{N}_0$ the set of all nonnegative integers. Write the analytic function $\psi : D_{\mathbb{C}^d}(\lambda^0, R) \to \mathbb{C}$ in the form of its Taylor series expansion

$$\psi(\lambda_1, \lambda_2, \ldots, \lambda_d) = \sum_{\alpha \in \mathbb{N}_0^d} a_\alpha (\lambda_1 - \lambda_1^0)^{\alpha_1} (\lambda_2 - \lambda_2^0)^{\alpha_2} \cdots (\lambda_d - \lambda_d^0)^{\alpha_d}.$$

By Cauchy's estimates (Theorem A.3.2 in Appendix A),

$$|a_\alpha| \leq \frac{M}{R^{\|\alpha\|_1}}, \quad \forall \alpha \in \mathbb{N}_0^d, \tag{16.52}$$

where $\|\alpha\|_1 := \sum_{j=1}^d |\alpha_j| = \sum_{j=1}^d \alpha_j$. Then

$$\begin{aligned}
\operatorname{Re}\psi(\lambda_1, \lambda_2, \ldots, \lambda_d) &= \sum_{\alpha \in \mathbb{N}_0^d} \operatorname{Re}\Bigg[a_\alpha \Bigg(\sum_{p=0}^{\alpha_1} \binom{\alpha_1}{p} (\operatorname{Re}\lambda_1 - \operatorname{Re}\lambda_1^0)^p (\operatorname{Im}\lambda_1 - \operatorname{Im}\lambda_1^0)^{\alpha_1 - p} i^{\alpha_1 - p} \Bigg) \\
&\quad \cdot \Bigg(\sum_{p=0}^{\alpha_2} \binom{\alpha_2}{p} (\operatorname{Re}\lambda_2 - \operatorname{Re}\lambda_2^0)^p (\operatorname{Im}\lambda_2 - \operatorname{Im}\lambda_2^0)^{\alpha_2 - p} i^{\alpha_2 - p} \Bigg) \cdots \\
&\quad \cdot \Bigg(\sum_{p=0}^{\alpha_d} \binom{\alpha_d}{p} (\operatorname{Re}\lambda_d - \operatorname{Re}\lambda_d^0)^p (\operatorname{Im}\lambda_d - \operatorname{Im}\lambda_d^0)^{\alpha_d - p} i^{\alpha_d - p} \Bigg) \Bigg] \\
&= \sum_{\beta \in \mathbb{N}_0^{2d}} \operatorname{Re}\Bigg[a_{\widehat{\beta}} \prod_{j=1}^d \binom{\beta_j^{(1)} + \beta_j^{(2)}}{\beta_j^{(1)}} i^{\beta_j^{(2)}} (\operatorname{Re}\lambda_j - \operatorname{Re}\lambda_j^0)^{\beta_j^{(1)}} (\operatorname{Im}\lambda_j - \operatorname{Im}\lambda_j^0)^{\beta_j^{(2)}} \Bigg] \\
&= \sum_{\beta \in \mathbb{N}_0^{2d}} \operatorname{Re}\Bigg[a_{\widehat{\beta}} \prod_{j=1}^d \binom{\beta_j^{(1)} + \beta_j^{(2)}}{\beta_j^{(1)}} i^{\beta_j^{(2)}} \Bigg] (\operatorname{Re}\lambda_j - \operatorname{Re}\lambda_j^0)^{\beta_j^{(1)}} (\operatorname{Im}\lambda_j - \operatorname{Im}\lambda_j^0)^{\beta_j^{(2)}},
\end{aligned}$$

where $\beta \in \mathbb{N}_0^{2d}$ is written in the form $(\beta_1^{(1)}, \beta_1^{(2)}, \beta_2^{(1)}, \beta_2^{(2)}, \ldots, \beta_d^{(1)}, \beta_d^{(2)})$ and $\widehat{\beta} = (\beta_1^{(1)} + \beta_1^{(2)}, \beta_2^{(1)} + \beta_2^{(2)}, \ldots, \beta_d^{(1)} + \beta_d^{(2)}) \in \mathbb{N}_0^d$. Note that $\|\beta\|_1 = \|\widehat{\beta}\|_1$. Set

$$c_\beta = \operatorname{Re}\Bigg[a_{\widehat{\beta}} \prod_{j=1}^d \binom{\beta_j^{(1)} + \beta_j^{(2)}}{\beta_j^{(1)}} i^{\beta_j^{(2)}} \Bigg].$$

Using (16.52), we get

$$|c_\beta| \le |a_{\widehat{\beta}}| \prod_{j=1}^{d} \binom{\beta_j^{(1)} + \beta_j^{(2)}}{\beta_j^{(1)}} \le MR^{-\|\widehat{\beta}\|_1} \prod_{j=1}^{d} 2^{\beta_j^{(1)}+\beta_j^{(2)}} = MR^{-\|\beta\|_1} 2^{\|\beta\|_1}.$$

Thus, the formula

$$\widetilde{\psi}(z_1, z_2, \dots, z_{2d-1}, z_{2d}) = \sum_{\beta \in \mathbb{N}_0^{2d}} c_\beta \prod_{j=1}^{d} (z_{2j-1} - \operatorname{Re} \lambda_j^0)^{\beta_j^{(1)}} (z_{2j} - \operatorname{Im} \lambda_j^0)^{\beta_j^{(2)}}$$

defines an analytic function on $D_{\mathbb{C}^{2d}}(\widetilde{\lambda^0}, R/4)$ and

$$|\widetilde{\psi}(z_1, z_2, \dots, z_{2d-1}, z_{2d})| \le 4^d M.$$

Obviously, $\widetilde{\psi}|_{D_{\mathbb{C}^d}(\lambda_0, R/4)} = \operatorname{Re} \psi|_{D_{\mathbb{C}^d}(\lambda_0, R/4)}$, and we are done with the existence part of our lemma. The uniqueness part follows from the fact that the set where any two holomorphic functions are equal is a complex manifold and the smallest complex manifold containing $\mathbb{R}^{2d}$ is $\mathbb{C}^{2d}$. □

We can now prove the real analyticity of the pressure function.

Theorem 16.4.9. *If $\{T_\lambda : U \to \mathbb{C}\}_{\lambda \in \Lambda}$ is an analytic perturbation of a conformal expanding repeller $(J_{\lambda_0}, T_{\lambda_0}, U)$ in $\mathbb{C}$, then the pressure function*

$$\widehat{\Lambda} \times \mathbb{R} \ni (\lambda, t) \longmapsto \mathrm{P}(\lambda, t) := \mathrm{P}(T_\lambda, -t \log|T_\lambda'|) = \mathrm{P}(T_{\lambda_0}, -t \log|T_\lambda'| \circ \tau_\lambda) \in \mathbb{R} \tag{16.53}$$

is real analytic, where $\widehat{\Lambda}$ and τ_λ arise from Theorem 16.4.7.

Proof. By virtue of Theorem 16.4.7, it suffices to prove real analyticity on some open neighborhood in $\widehat{\Lambda} \times \mathbb{R}$ of every point (λ_0, s), $s \in \mathbb{R}$. By Theorem 16.4.7(a), for every $z \in J_{\lambda_0}$ the function

$$\widehat{\Lambda} \ni \lambda \longmapsto \Psi_z(\lambda) := \log|T_\lambda'(\tau_\lambda(z))| - \log|T_{\lambda_0}'(z)|$$

is harmonic with $\Psi_z(\lambda_0) = 0$ (per Theorem 16.4.7(b)). Fix $r > 0$ such that $B(\lambda_0, 8r) \subseteq \widehat{\Lambda}$. The continuity of $(\lambda, z) \mapsto \tau_\lambda(z)$ (per Theorem 16.4.7(d)) combined with the continuity of $(\lambda, z) \mapsto T_\lambda'(z)$ (as $\{T_\lambda\}_{\lambda \in \Lambda}$ is an analytic perturbation of T_{λ_0}) implies that

$$M := \sup\{|\Psi_z(\lambda)| : (z, \lambda) \in J_{\lambda_0} \times B(\lambda_0, 4r)\} < \infty.$$

By Lemma 16.4.8, each function Ψ_z has an holomorphic extension $\widetilde{\Psi}_z : B_{\mathbb{C}^2}(\widetilde{\lambda_0}, r) \to \mathbb{C}$ such that

$$\widetilde{M} := \sup\{|\widetilde{\Psi}_z(\widetilde{\lambda})| : (z, \widetilde{\lambda}) \in J_{\lambda_0} \times B_{\mathbb{C}^2}(\widetilde{\lambda_0}, r)\} \le 4M < \infty.$$

Since all the functions $\tau_\lambda, \lambda \in B_{\mathbb{C}}(\lambda_0, 4r)$, are Hölder continuous with a common Hölder exponent, say α, and common Hölder norm for the exponent α (see Remark 16.4.6), an application of Cauchy's estimates (Theorem A.3.2) affirms that for all $\widetilde{\lambda} \in B_{\mathbb{C}^2}(\widetilde{\lambda_0}, r)$ the function $z \mapsto \widetilde{\Psi}_z(\widetilde{\lambda})$ is Hölder continuous with exponent α and the corresponding Hölder norms are uniformly bounded by, say, $\widetilde{M}_\alpha$.

Given any bounded open set $V \subseteq \mathbb{C}$, consider the two-complex-parameter family of potentials $\varphi_{\widetilde{\lambda},w} : J_{\lambda_0} \to \mathbb{C}$ defined by

$$\varphi_{\widetilde{\lambda},w}(z) := -w\Big[\widetilde{\Psi}_z(\widetilde{\lambda}) + \log|T'_{\lambda_0}(z)|\Big], \quad (\widetilde{\lambda}, w) \in B_{\mathbb{C}^2}(\widetilde{\lambda_0}, r) \times V.$$

Recall that $T_{\lambda_0} \in C^{1+\epsilon}(U)$ for some $\epsilon > 0$ by Definition 16.2.1(a) of a repeller. Then T'_{λ_0}, $|T'_{\lambda_0}|$ and $\log|T'_{\lambda_0}|$ are Hölder continuous with exponent ϵ. Let $L := \|\log|T'_{\lambda_0}|\|_\epsilon$. Replacing α and ϵ by $\min\{\alpha, \epsilon\}$ and adjusting the constants $\widetilde{M}_\alpha$ and L accordingly, the above family satisfies

$$\sup\Big\{\|\varphi_{\widetilde{\lambda},w}\|_\alpha : (\widetilde{\lambda}, w) \in B_{\mathbb{C}^2}(\widetilde{\lambda_0}, r) \times V\Big\} \leq |V|(\widetilde{M}_\alpha + L) < \infty,$$

where $|V| := \sup\{|w| : w \in V\}$. Moreover, for every $z \in J_{\lambda_0}$ the function $B_{\mathbb{C}^2}(\widetilde{\lambda_0}, r) \times V \ni (\widetilde{\lambda}, w) \longmapsto \varphi_{\widetilde{\lambda},w}(z) \in \mathbb{C}$ is jointly holomorphic since it is separately holomorphic in each variable $\widetilde{\lambda}$ and w. So the hypotheses of Theorem 13.10.2 are fulfilled and the function

$$B_{\mathbb{C}^2}(\widetilde{\lambda_0}, r) \times V \ni (\widetilde{\lambda}, t) \longmapsto \mathcal{L}_{\varphi_{\widetilde{\lambda},t}} \in L(H_\alpha(J_{\lambda_0}))$$

is holomorphic. For all $(\lambda, t) \in B_{\mathbb{C}}(\lambda_0, r) \times (V \cap \mathbb{R})$, observe that

$$\varphi_{\widetilde{\lambda},t}(z) := -t\Big[\widetilde{\Psi}_z(\widetilde{\lambda}) + \log|T'_{\lambda_0}(z)|\Big] = -t\Big[\Psi_z(\lambda) + \log|T'_{\lambda_0}(z)|\Big] = -t\log|T'_\lambda \circ \tau_\lambda(z)| \in \mathbb{R}.$$

Thus all the hypotheses of Theorem 13.10.3 are satisfied, too, and hence the function

$$B_{\mathbb{C}}(\lambda_0, r) \times (V \cap \mathbb{R}) \ni (\lambda, t) \longmapsto \mathrm{P}(T_{\lambda_0}, -t\log|T'_\lambda| \circ \tau_\lambda) \in \mathbb{R}$$

is real analytic. Given the arbitrariness of $\lambda_0 \in \widehat{\Lambda}$ and of $V \subseteq \mathbb{C}$ as a bounded open set, the localness of real analyticity allows us to infer that

$$\widehat{\Lambda} \times \mathbb{R} \ni (\lambda, t) \longmapsto \mathrm{P}(T_{\lambda_0}, -t\log|T'_\lambda| \circ \tau_\lambda) \in \mathbb{R}$$

is real analytic.

By Theorem 16.4.7(c) and Corollary 11.1.21, it turns out that

$$\mathrm{P}(\lambda, t) := \mathrm{P}(T_\lambda, -t\log|T'_\lambda|) = \mathrm{P}(T_{\lambda_0}, -t\log|T'_\lambda| \circ \tau_\lambda)$$

and the result follows. □

We now identify the partial derivative with respect to t of the pressure function $\widehat{\Lambda} \times \mathbb{R} \ni (\lambda, t) \longmapsto \mathrm{P}(\lambda, t)$.

Theorem 16.4.10. *If $\{T_\lambda : U \to \mathbb{C}\}_{\lambda\in\Lambda}$ is an analytic perturbation of a conformal expanding repeller $(J_{\lambda_0}, T_{\lambda_0}, U)$ in $\mathbb{C}$, then*

$$\frac{\partial}{\partial t}\mathrm{P}(\lambda, t) = -\int_{J_{\lambda_0}} \log|T_\lambda'| \circ \tau_\lambda \, d\mu_{\lambda,t}, \quad \forall (\lambda, t) \in \widehat{\Lambda} \times \mathbb{R},$$

where $\mu_{\lambda,t}$ is the unique equilibrium state and unique T_{λ_0}-invariant Gibbs state for the potential $\varphi_{\lambda,t} = -t \log|T_\lambda'| \circ \tau_\lambda$.

Proof. By virtue of Theorems 16.4.7 and 16.4.9 as well as Proposition 13.7.12, for every $(\lambda, t) \in \widehat{\Lambda} \times \mathbb{R}$ the potential $\varphi_{\lambda,t} = -t\log|T_\lambda'| \circ \tau_\lambda$ admits a unique equilibrium state and unique T_{λ_0}-invariant Gibbs state $\mu_{\lambda,t}$. Let $s \in \mathbb{R}$. By the variational principle (Theorem 12.1.1), we deduce that

$$\begin{aligned}
\mathrm{P}(\lambda, s) &= \mathrm{P}(T_{\lambda_0}, -s\log|T_\lambda'| \circ \tau_\lambda) \\
&\geq \mathrm{h}_{\mu_{\lambda,t}}(T_{\lambda_0}) + \int_{J_{\lambda_0}} -s\log|T_\lambda'| \circ \tau_\lambda \, d\mu_{\lambda,t} \\
&= \mathrm{h}_{\mu_{\lambda,t}}(T_{\lambda_0}) + \int_{J_{\lambda_0}} -t\log|T_\lambda'| \circ \tau_\lambda \, d\mu_{\lambda,t} + (s-t)\int_{J_{\lambda_0}} -\log|T_\lambda'| \circ \tau_\lambda \, d\mu_{\lambda,t} \\
&= \mathrm{P}(T_{\lambda_0}, -t\log|T_\lambda'| \circ \tau_\lambda) + (s-t)\int_{J_{\lambda_0}} -\log|T_\lambda'| \circ \tau_\lambda \, d\mu_{\lambda,t} \\
&= \mathrm{P}(\lambda, t) + (s-t)\int_{J_{\lambda_0}} -\log|T_\lambda'| \circ \tau_\lambda \, d\mu_{\lambda,t}.
\end{aligned}$$

On one hand,

$$\frac{\partial}{\partial t}\mathrm{P}(\lambda, t) = \lim_{s\to t^+} \frac{\mathrm{P}(\lambda, s) - \mathrm{P}(\lambda, t)}{s-t} \geq -\int_{J_{\lambda_0}} \log|T_\lambda'| \circ \tau_\lambda \, d\mu_{\lambda,t}.$$

On the other hand,

$$\frac{\partial}{\partial t}\mathrm{P}(\lambda, t) = \lim_{s\to t^-} \frac{\mathrm{P}(\lambda, s) - \mathrm{P}(\lambda, t)}{s-t} \leq -\int_{J_{\lambda_0}} \log|T_\lambda'| \circ \tau_\lambda \, d\mu_{\lambda,t}.$$

In conclusion, $\frac{\partial}{\partial t}\mathrm{P}(\lambda, t) = -\int_{J_{\lambda_0}} \log|T_\lambda'| \circ \tau_\lambda \, d\mu_{\lambda,t}$. □

Finally, we demonstrate the real analyticity of the Hausdorff dimension function.

Theorem 16.4.11. *If $\{T_\lambda : U \to \mathbb{C}\}_{\lambda\in\Lambda}$ is an analytic perturbation of a conformal expanding repeller $(J_{\lambda_0}, T_{\lambda_0}, U)$ in $\mathbb{C}$, then the Hausdorff dimension function*

$$\widehat{\Lambda} \ni \lambda \longmapsto \mathrm{HD}(J_\lambda) \in (0,2)$$

is real analytic.

Proof. Denote $h_\lambda := \mathrm{HD}(J_\lambda)$. By Bowen's formula (Theorem 16.3.2) and formula (16.53), we know that

$$\mathrm{P}(T_{\lambda_0}, -h_\lambda \log|T'_\lambda| \circ \tau_\lambda) = 0. \tag{16.54}$$

By Theorem 16.4.10, we further have that

$$\frac{\partial}{\partial t}\mathrm{P}(T_{\lambda_0}, -t \log|T'_\lambda| \circ \tau_\lambda) = -\int_{J_{\lambda_0}} \log|T'_\lambda| \circ \tau_\lambda \, d\mu_{\lambda,t} < 0.$$

It follows from the implicit function theorem, applied to the equation

$$\mathrm{P}(T_{\lambda_0}, -t \log|T'_\lambda| \circ \tau_\lambda) = 0,$$

that h_λ is the unique real-analytic function satisfying equation (16.54) on some sufficiently small neighborhood of λ_0. As real analyticity is a local property, the proof is complete. □

Remark 16.4.12. Let $h_\lambda := \mathrm{HD}(J_\lambda)$. One can wonder whether a result analogous to Theorem 16.4.11 holds on a deeper level, i. e., whether the function

$$\widehat{\Lambda} \ni \lambda \longmapsto H_{h_\lambda}(J_\lambda) \tag{16.55}$$

is real analytic. This issue is very subtle. It is easy to see that the function

$$\widehat{\Lambda} \ni \lambda \longmapsto m_{h_\lambda} \in M(\overline{U})$$

is continuous, as m_{h_λ}, the h_λ-conformal measure for the map $T_\lambda : J_\lambda \to J_\lambda$, is also the normalized h_λ-dimensional Hausdorff measure on J_λ. Indeed, one can just notice that a weak* limit of conformal measures is a conformal measure and apply the uniqueness of conformal measures.

The much more involved continuity of the function in (16.55) was established in [128]. In fact, that continuity was proved therein in a more general setting. This work stemmed from the seminal paper of Olsen [89]. It was furthermore shown in [128] that the function in (16.55) is Hölder continuous. However, the question of its real analyticity remains open; even Lipschitz continuity is open. Nevertheless, piecewise real analyticity was proved in [128] in the context of repellers contained in the real line.

16.5 Dimensions of measures, Lyapunov exponents and measure-theoretic entropy

To describe the dimensions of invariant measures, we now introduce the concept of Lyapunov exponent.

Definition 16.5.1. Let (X, U, T) be a conformal expanding repeller. For any measure $\mu \in M(T)$, i. e., any T-invariant Borel probability measure μ, the quantity

$$\chi_\mu(T) := \int_X \log|T'|\,d\mu$$

is called the (*characteristic*) *Lyapunov exponent* of μ.

The Lyapunov exponent of any measure μ is positive since $\chi_\mu(T) \geq \log\lambda > 0$ per Definition 16.2.1(d).

Conformal expanding repellers exhibit a beautiful property: each of their ergodic invariant Borel probability measures has the same pointwise dimension almost everywhere, and this dimension is the ratio of the measure-theoretic entropy and the Lyapunov exponent of that measure.

Theorem 16.5.2 (Volume lemma for ergodic invariant measures). *Let (X, U, T) be a conformal expanding repeller and μ an ergodic T-invariant Borel probability measure. Then for μ-a. e. $x \in X$,*

$$d_\mu(x) = \frac{\mathrm{h}_\mu(T)}{\chi_\mu(T)} = \mathrm{HD}(\mu) = \mathrm{PD}(\mu),$$

where $d_\mu(x)$ is the pointwise dimension of μ at x and $\mathrm{h}_\mu(T)$ is the measure-theoretic entropy of T with respect to μ.

Moreover, as a geometric measure, the unique equilibrium state and unique T-invariant Gibbs state $\mu_{\mathrm{HD}(X)}$ associated with the Hölder continuous potential $\varphi_{\mathrm{HD}(X)}(x) := -\mathrm{HD}(X)\log|T'(x)|$, has the property that for <u>all</u> $x \in X$,

$$\begin{aligned} d_{\mu_{\mathrm{HD}(X)}}(x) &= \frac{\mathrm{h}_{\mu_{\mathrm{HD}(X)}}(T)}{\chi_{\mu_{\mathrm{HD}(X)}}(T)} = \frac{-\mu_{\mathrm{HD}(X)}(\varphi_{\mathrm{HD}(X)})}{\chi_{\mu_{\mathrm{HD}(X)}}(T)} \\ &= \mathrm{HD}(X) = \mathrm{PD}(X) = \mathrm{BD}(X) = \mathrm{HD}(\mu_{\mathrm{HD}(X)}) = \mathrm{PD}(\mu_{\mathrm{HD}(X)}). \end{aligned}$$

Proof. Let $x \in X$. Let also ξ, $0 < r < \xi$, $k = k(x, r)$, and $\ell = \ell(x, r)$ be as in the proof of Bowen's formula (Theorem 16.3.2). For every $n \geq 0$, we then have that

$$B_{n+1}(x, \xi) = T_x^{-n}(B(T^n(x), \xi)), \tag{16.56}$$

where $B_{n+1}(x, \xi) := \bigcap_{j=0}^{n} T^{-j}(B(T^j(x), \xi))$ denotes the dynamical $(n+1, \xi)$-ball at x (cf. Section 7.3). Indeed, the inclusion $T_x^{-n}(B(T^n(x), \xi)) \subseteq B_{n+1}(x, \xi)$ is immediate while

$B_{n+1}(x,\xi) \subseteq T_x^{-n}(B(T^n(x),\xi))$ follows by induction. By definition (cf. (16.8)–(16.9)),

$$T_x^{-\ell(x,r)}(B(T^{\ell(x,r)}(x),\xi)) \subseteq B(x,r) \subseteq T_x^{-k(x,r)}(B(T^{k(x,r)}(x),\xi)) \tag{16.57}$$

and

$$\lim_{r\to 0} k(x,r) = \lim_{r\to 0} \ell(x,r) = \infty. \tag{16.58}$$

From this point on, we use the lighter notation k for $k(x,r)$ and ℓ for $\ell(x,r)$. Combining (16.56)–(16.57) yields

$$B_{\ell+1}(x,\xi) \subseteq B(x,r) \subseteq B_{k+1}(x,\xi).$$

So

$$\mu(B_{\ell+1}(x,\xi)) \le \mu(B(x,r)) \le \mu(B_{k+1}(x,\xi)).$$

Taking logarithms on both sides, multiplying by -1 and dividing by $-\log r$, we get

$$\frac{-\log\mu(B_{k+1}(x,\xi))}{k+1}\cdot\frac{k+1}{-\log r} \le \frac{-\log\mu(B(x,r))}{-\log r} \le \frac{-\log\mu(B_{\ell+1}(x,\xi))}{\ell+1}\cdot\frac{\ell+1}{-\log r}. \tag{16.59}$$

By Corollary 16.2.10, we know that there is $\widetilde{K} \ge 1$ such that

$$-\log\widetilde{K} + \log|(T^\ell)'(x)| \le -\log r \le \log\widetilde{K} + \log|(T^k)'(x)|.$$

Inserting this into (16.59), we deduce that

$$\begin{aligned}\frac{\log\mu(B(x,r))}{\log r} &\le \frac{-\log\mu(B_{\ell+1}(x,\xi))}{\ell+1}\cdot\frac{\ell+1}{-\log\widetilde{K}+\log|(T^\ell)'(x)|}\\ &\le \frac{-\log\mu(B_{\ell+1}(x,\xi))}{\ell+1}\cdot\frac{1}{\frac{-\log\widetilde{K}}{\ell+1}+\frac{\log|(T^\ell)'(x)|}{\ell+1}}\end{aligned} \tag{16.60}$$

and

$$\begin{aligned}\frac{\log\mu(B(x,r))}{\log r} &\ge \frac{-\log\mu(B_{k+1}(x,\xi))}{k+1}\cdot\frac{k+1}{\log\widetilde{K}+\log|(T^k)'(x)|}\\ &\ge \frac{-\log\mu(B_{k+1}(x,\xi))}{k+1}\cdot\frac{1}{\frac{\log\widetilde{K}}{k+1}+\frac{\log|(T^k)'(x)|}{k+1}}.\end{aligned} \tag{16.61}$$

Since μ is ergodic, we can apply the ergodic case of Birkhoff's ergodic theorem (Corollary 8.2.14) to extract a Borel set $X_1 \subseteq X$ such that $\mu(X_1) = 1$ and

$$\lim_{n\to\infty}\frac{1}{n}\log|(T^n)'(x)| = \lim_{n\to\infty}\frac{1}{n}\sum_{i=0}^{n-1}\log|T'(T^i(x))| = \int_X \log|T'|\,d\mu = \chi_\mu(T), \quad \forall x \in X_1. \tag{16.62}$$

Being distance expanding, the map $T : X \to X$ is expansive. It follows from Theorem 9.6.4 (Brin–Katok local entropy formula for expansive maps) that there exists a Borel set $X_2 \subseteq X$ such that $\mu(X_2) = 1$ and

$$\lim_{n\to\infty} \frac{-\log \mu(B_n(x,\xi))}{n} = \mathrm{h}_\mu(T), \quad \forall x \in X_2.$$

Inserting this and (16.62) into (16.60)–(16.61) and invoking (16.58), we get for every $x \in X_1 \cap X_2$ that

$$\limsup_{r\to 0} \frac{\log \mu(B(x,r))}{\log r} \le \frac{\mathrm{h}_\mu(T)}{\chi_\mu(T)} \le \liminf_{r\to 0} \frac{\log \mu(B(x,r))}{\log r}.$$

Hence,

$$d_\mu(x) := \lim_{r\to 0} \frac{\log \mu(B(x,r))}{\log r} = \frac{\mathrm{h}_\mu(T)}{\chi_\mu(T)}.$$

Since $\mu(X_1 \cap X_2) = 1$, the proof of the first equality in the first part of Theorem 16.5.2 is complete. The second and third equalities in the first part are a direct application of Corollary 15.6.11.

The second part of the statement follows directly from Bowen's formula (Theorem 16.3.2), the first part of the current theorem, as well as the characteristic of $\mu_{\mathrm{HD}(X)}$ of being an equilibrium state, for then

$$\mathrm{h}_{\mu_{\mathrm{HD}(X)}}(T) = \mathrm{P}(T, \varphi_{\mathrm{HD}(X)}) - \int_X \varphi_{\mathrm{HD}(X)}\, d\mu_{\mathrm{HD}(X)} = 0 - \mu_{\mathrm{HD}(X)}(\varphi_{\mathrm{HD}(X)}). \qquad \square$$

Remark 16.5.3. Recall that the almost everywhere existence and constantness of the pointwise dimension of a measure is usually referred to as the *exact dimensionality* of the measure (cf. Corollary 15.6.11).

There is a more general version of the volume lemma for nonnecessarily ergodic but T-invariant probability measures, though it is much more complicated to derive. We will need several intermediate results.

Lemma 16.5.4. *Every increasing function $k : I \to \mathbb{R}$ defined on a compact interval $I \subseteq \mathbb{R}$ is Lipschitz continuous at Lebesgue almost every point in I. In other words, for every $\varepsilon > 0$ there exist $L > 0$ and a set $J \subseteq I$ such that* $\mathrm{Leb}(I \setminus J) < \varepsilon$ *and at each $r \in J$ the function k is Lipschitz continuous with Lipschitz constant L.*

Proof. Suppose, on the contrary, that the set

$$B_\infty := \left\{ x \in I : \sup_{\substack{y\in I\\ y\neq x}} \frac{|k(x)-k(y)|}{|x-y|} = \infty \right\}$$

has positive Lebesgue measure. Write $I = [a, b]$. Because the Lebesgue measure is (inner) regular and every singleton is a null set, there is a compact set $B \subseteq B_\infty \cap (a, b)$ of positive measure. For every $x \in B$, choose $x' \in (a, b) \setminus \{x\}$ such that

$$\frac{|k(x) - k(x')|}{|x - x'|} > 2\frac{k(b) - k(a)}{\mathrm{Leb}(B)}. \tag{16.63}$$

Replace each pair x, x' by a pair y, y' such that $(y, y') \supseteq [\min\{x, x'\}, \max\{x, x'\}]$ and such that y, y' is so close to x, x' that (16.63) still holds for y, y'. Note that $y < y'$. We shall use for y, y' the old notation x, x', so $x < x'$.

From the family of intervals (x, x') choose a finite cover of the compact set B. It is possible to find a subcover $\mathcal{I}$ of B which consists of two subfamilies $\mathcal{I}^1$ and $\mathcal{I}^2$, each made up of pairwise disjoint intervals. Indeed, start with $I_1 = (x_1, x_1') \in \mathcal{I}$ with minimal possible $x = x_1$ and maximal in $\mathcal{I}$ in the sense of inclusion. Having found $I_1 = (x_1, x_1'), \ldots, I_n = (x_n, x_n')$, pick $I_{n+1} = (x_{n+1}, x_{n+1}')$ as follows. Consider the family

$$\mathcal{I}_{n+1} := \left\{(x, x') \in \mathcal{I} : x \in \bigcup_{i=1}^{n} I_i \text{ and } x' > \max_{1 \le i \le n} x_i'\right\}.$$

If $\mathcal{I}_{n+1} \neq \emptyset$, set $I_{n+1} \in \mathcal{I}_{n+1}$ so that $x_{n+1}' = \max\{x' : (x, x') \in \mathcal{I}_{n+1}\}$. If $\mathcal{I}_{n+1} = \emptyset$, set $I_{n+1} \in \mathcal{I}$ so that x_{n+1} is minimal satisfying $x_{n+1} \ge \max_{1 \le i \le n} x_i'$ and maximal in $\mathcal{I}$.

In this construction, the intervals I_n with even n's are pairwise disjoint, as $I_{n+2} \notin \mathcal{I}_{n+1}$. The same is true for odd n's. We define $\mathcal{I}^1$ (resp., $\mathcal{I}^2$) as the family of (x_n, x_n') for odd (resp., even) n's.

In view of the pairwise disjointness of the intervals in the families $\mathcal{I}^1$ and $\mathcal{I}^2$, the monotonicity of k and (16.63), we obtain that

$$k(b) - k(a) \ge \sum_{n \text{ odd}} (k(x_n') - k(x_n)) > 2\frac{k(b) - k(a)}{\mathrm{Leb}(B)} \sum_{n \text{ odd}} (x_n' - x_n)$$

and a corresponding inequality for even n's. Given that $\mathcal{I} = \mathcal{I}^1 \cup \mathcal{I}^2$ covers B, we deduce that

$$2(k(b) - k(a)) > 2\frac{k(b) - k(a)}{\mathrm{Leb}(B)} \sum_{n \in \mathbb{N}} (x_n' - x_n) > 2\frac{k(b) - k(a)}{\mathrm{Leb}(B)} \mathrm{Leb}(B) = 2(k(b) - k(a)),$$

which is a contradiction. □

Corollary 16.5.5. *For every Borel probability measure ν on a compact metric space (X, d) and for every $r > 0$, there exists a finite partition $\mathcal{P} = \{P_t\}_{t=1}^M$ of X into Borel sets of positive ν-measure with* $\mathrm{diam}(\mathcal{P}) < r$ *and for which there is $C > 0$ such that*

$$\nu(\partial_{\mathcal{P},a}) \le Ca, \quad \forall a > 0, \tag{16.64}$$

where $\partial_{\mathcal{P},a} = \bigcap_{t=1}^M (\bigcup_{s \neq t} B(P_s, a))$.

Proof. Let $\{x_1,\dots,x_N\}$ be a finite $(r/4)$-spanning set in X. Fix $\varepsilon \in (0,r/4N)$. For each function $t \mapsto k_i(t) := \nu(B(x_i,t))$, $t \in I = [r/4,r/2]$, apply Lemma 16.5.4 and find appropriate L_i and J_i for all $1 \le i \le N$. Let $L = \max\{L_i : 1 \le i \le N\}$ and let $J = \bigcap_{i=1}^N J_i$. The set J has positive Lebesgue measure by the choice of ε. So we can choose a point $r_0 \in J$ with $r/4 < r_0 < r/2$. For all $0 < a < a_0 := \min\{r_0 - r/4, r/2 - r_0\}$ and for all $1 \le i \le N$, we have $\nu(B(x_i,r_0+a) \setminus B(x_i,r_0-a)) \le 2La$. Setting

$$\Delta(a) = \bigcup_{i=1}^N (B(x_i,r_0+a) \setminus B(x_i,r_0-a)),$$

we get $\nu(\Delta(a)) \le 2LNa$. Define $\mathcal{P} = \{\bigcap_{i=1}^N B^{\kappa(i)}(x_i,r_0)\}$ as a family over all functions $\kappa : \{1,\dots,N\} \to \{+,-\}$, where $B^+(x_i,r_0) := B(x_i,r_0)$ and $B^-(x_i,r_0) := X \setminus B(x_i,r_0)$, but excluding functions κ that yield sets of measure zero. After removing from X a set of measure zero, the partition $\mathcal{P}$ covers X. As $r_0 \ge r/4$, the balls $\{B(x_i,r_0)\}_{i=1}^N$ cover X. Hence, for each nonempty $P_t \in \mathcal{P}$, at least one value of κ is $+$. Thus, $\operatorname{diam}(P_t) \le 2r_0 < r$.

Now note that $\partial_{\mathcal{P},a} \subseteq \Delta(a)$. Indeed, let $x \in \partial_{\mathcal{P},a}$. Since $\mathcal{P}$ covers X there exists t_0 such that $x \in P_{t_0}$, so $x \notin P_t$ for all $t \ne t_0$. However, as $x \in \bigcup_{t \ne t_0} B(P_t,a)$, there exists $t_1 \ne t_0$ such that $\operatorname{dist}(x,P_{t_1}) < a$. Let $B = B(x_i,r_0)$ be such that $P_{t_0} \subseteq B^+$ and $P_{t_1} \subseteq B^-$, or vice versa. In the case where $x \in P_{t_0} \subseteq B^+$, by the triangle inequality $d(x,x_i) > r_0 - a$, and since $d(x,x_i) < r_0$, we get $x \in \Delta(a)$. In the case where $x \in P_{t_1} \subseteq B^-$, we have $x \in B(x_i,r_0+a) \setminus B(x_i,r_0) \subseteq \Delta(a)$.

We conclude that $\nu(\partial_{\mathcal{P},a}) \le \nu(\Delta(a)) \le 2LNa$ for all $a < a_0$. For all $a \ge a_0$, it suffices to take $C \ge 1/a_0$. Set $C = \max\{2LN, 1/a_0\}$. □

Remark 16.5.6. If X is embedded in a compact manifold Y, then we can view ν as a measure on Y; we find a partition $\mathcal{P}$ of Y, and then $\partial_{\mathcal{P},a} = B(\bigcup_{t=1}^M \partial P_t, a)$, provided $M \ge 2$. This justifies the notation $\partial_{\mathcal{P},a}$.

Corollary 16.5.7. *Let ν be a Borel probability measure on a compact metric space (X,d), and let $f : X \to X$ be a measure-preserving transformation. For every $r > 0$, the partition $\mathcal{P}$ constructed in Corollary 16.5.5 is such that for every $\delta > 0$ and ν-a. e. $x \in X$ there exists $N(x) \in \mathbb{N}$ for which*

$$B(f^n(x), e^{-n\delta}) \subseteq \mathcal{P}(f^n(x)), \quad \forall n \ge N(x). \tag{16.65}$$

Proof. Fix $\delta > 0$. By Corollary 16.5.5,

$$\sum_{n=0}^{\infty} \nu(\partial_{\mathcal{P},\exp(-n\delta)}) \le \sum_{n=0}^{\infty} C \exp(-n\delta) < \infty.$$

By the f-invariance of ν, we deduce that

$$\sum_{n=0}^{\infty} \nu(f^{-n}(\partial_{\mathcal{P},\exp(-n\delta)})) < \infty.$$

Applying the Borel–Cantelli lemma (Lemma A.1.20) to $(f^{-n}(\partial_{\mathcal{P},\exp(-n\delta)}))_{n=1}^{\infty}$, we conclude that for ν-a. e. $x \in X$ there exists $N(x) \in \mathbb{N}$ such that $x \notin f^{-n}(\partial_{\mathcal{P},\exp(-n\delta)})$ for every $n \geq N(x)$. So $f^n(x) \notin \partial_{\mathcal{P},\exp(-n\delta)}$ for all $n \geq N(x)$. Hence, by definition of $\partial_{\mathcal{P},\exp(-n\delta)}$, if $f^n(x) \in P_t \in \mathcal{P}$ then $f^n(x) \notin \bigcup_{s\neq t} B(P_s, \exp(-n\delta))$. Thus, $B(f^n(x), \exp(-n\delta)) \subseteq P_t = \mathcal{P}(f^n(x))$. □

Finally, we can establish a generalization of the volume lemma (Theorem 16.5.2). For this, we will need a slight refinement of the definition of Hausdorff dimension of a measure (cf. Definition 15.6.1).

Definition 16.5.8. Let μ be a Borel probability measure on a metric space (X, d). Define

$$\mathrm{HD}_\star(\mu) := \inf\{\mathrm{HD}(Y) \mid \mu(Y) > 0\}.$$

Observe that $\mathrm{HD}_\star(\mu) \leq \mathrm{HD}(\mu)$.

Theorem 16.5.9.

$$\mathrm{HD}_\star(\mu) = \operatorname{ess\,inf}(\underline{d}_\mu) \quad \textit{and} \quad \mathrm{HD}(\mu) = \operatorname{ess\,sup}(\underline{d}_\mu).$$

Proof. The statement about $\mathrm{HD}(\mu)$ is just Corollary 15.6.7.

Let $\theta_1 = \operatorname{ess\,inf}(\underline{d}_\mu)$. Applying Lemma 15.6.6(a), we have $\mu(\{x : \underline{d}_\mu(x) \geq \theta_1\}) = 1$. For every Y with $\mu(Y) > 0$, there then is $Y' \subseteq Y$ with $\mu(Y') = \mu(Y)$ such that $\underline{d}_\mu(x) \geq \theta_1$ for every $x \in Y'$. Hence, $\mathrm{HD}(Y) \geq \mathrm{HD}(Y') \geq \theta_1$ by Theorem 15.6.3(a), and thereby $\mathrm{HD}_\star(\mu) \geq \theta_1$.

On the other hand, by Lemma 15.6.6(a) we know that $\mu(\{x : \underline{d}_\mu(x) < \theta\}) > 0$ for every $\theta > \theta_1$, and thus $\mathrm{HD}(\{x : \underline{d}_\mu(x) < \theta\}) \leq \theta$ by Theorem 15.6.3(b). Therefore, $\mathrm{HD}_\star(\mu) \leq \theta$. Letting $\theta \searrow \theta_1$ we get $\mathrm{HD}_\star(\mu) \leq \theta_1$.

We conclude from the above two paragraphs that $\mathrm{HD}_\star(\mu) = \theta_1$. □

The following is a generalization of Theorem 16.5.2.

Theorem 16.5.10 (Volume lemma for arbitrary invariant measures). *Let (X, U, T) be a conformal expanding repeller and μ a T-invariant Borel probability measure on X. Then*

$$\mathrm{HD}_\star(\mu) \leq \frac{\mathrm{h}_\mu(T)}{\chi_\mu(T)} \leq \mathrm{HD}(\mu). \tag{16.66}$$

In addition, if μ is ergodic then

$$d_\mu(x) = \frac{\mathrm{h}_\mu(T)}{\chi_\mu(T)} = \mathrm{HD}(\mu) = \mathrm{PD}(\mu) \quad \textit{for } \mu\textit{-a. e. } x \in X. \tag{16.67}$$

Proof. Let $r = \xi$ and $\mathcal{P}$ be the corresponding partition constructed in Corollary 16.5.5. As we saw in Lemma 13.2.8,

$$\mathcal{P}^{n+1}(x) \subseteq T_x^{-n}(B(T^n(x), \xi)), \quad \forall x \in X, \ \forall n \geq 0. \tag{16.68}$$

We now work toward obtaining some sort of opposite inclusion. Consider an arbitrary $\delta > 0$, and $x \in X$ so that Corollary 16.5.7 applies for all $n \geq N(x)$. For every $N(x) \leq i \leq n$, define $k(i) = [i\frac{\delta}{\log\lambda} + \frac{\log\xi}{\log\lambda}] + 1$, where $\lambda > 1$ is an expanding constant for T and $[\cdot]$ is the integer part function. Then $\exp(-i\delta) \geq \xi\,\lambda^{-k(i)}$. Therefore,

$$T_{T^i(x)}^{-k(i)}(B(T^{i+k(i)}(x), \xi)) \subseteq B(T^i(x), \exp(-i\delta)).$$

Using Corollary 16.5.7 with i in place of n, we deduce that

$$T_x^{-(i+k(i))}(B(T^{i+k(i)}(x), \xi)) \subseteq T_x^{-i}(\mathcal{P}(T^i(x))), \quad \forall N(x) \leq i \leq n.$$

From this inclusion over all $N(x) \leq i \leq n$, we conclude that

$$T_x^{-(n+k(n))}(B(T^{n+k(n)}(x), \xi)) \subseteq \mathcal{P}_{N(x)}^{n+1}(x). \tag{16.69}$$

Notice that for μ-a. e. $x \in X$ there is $a > 0$ such that $B(x, a) \subseteq \mathcal{P}^{N(x)}(x)$, by definition of $\partial_{\mathcal{P},a}$ in Corollary 16.5.5. It follows from this and (16.69) that for all $n \geq N(x)$ large enough (depending on x),

$$T_x^{-(n+k(n))}(B(T^{n+k(n)}(x), \xi)) \subseteq \mathcal{P}^{n+1}(x). \tag{16.70}$$

It ensues from (16.70), and (16.68) with $n + k(n)$ in lieu of n, that

$$\begin{aligned}
\lim_{n\to\infty} -\frac{1}{n}\log\mu(\mathcal{P}^n(x)) &\leq \liminf_{n\to\infty} \frac{-\log\mu(T_x^{-(n+k(n))}(B(T^{n+k(n)}(x), \xi)))}{n} \\
&\leq \limsup_{n\to\infty} \frac{-\log\mu(T_x^{-(n+k(n))}(B(T^{n+k(n)}(x), \xi)))}{n} \\
&\leq \lim_{n\to\infty} -\frac{1}{n}\log\mu(\mathcal{P}^{n+k(n)+1}(x)).
\end{aligned}$$

The limits on the far left- and far right-hand sides of these inequalities exist for μ-a. e. $x \in X$ by the Shannon–McMillan–Breiman theorem (Theorem 9.5.4), and their ratio is equal to $\lim_{n\to\infty}\frac{n}{n+k(n)} = 1/(1 + \frac{\delta}{\log\lambda})$. Letting $\delta \to 0$, we obtain the existence of the limit for the middle terms and the equality

$$\mathrm{h}_\mu(T, \mathcal{P}, x) := \lim_{n\to\infty} -\frac{1}{n}\log\mu(\mathcal{P}^n(x)) = \lim_{n\to\infty} \frac{-\log\mu(T_x^{-n}(B(T^n(x), \xi)))}{n}. \tag{16.71}$$

In view of Birkhoff's ergodic theorem (Theorem 8.2.11), the limit

$$\chi_\mu(T, x) := \lim_{n\to\infty} \frac{1}{n}\log|(T^n)'(x)| \tag{16.72}$$

exists for μ-a. e. $x \in X$.

Let $K \geq 1$ be the bounded distortion constant from Lemma 16.2.9. Fix $r \in (0, K^{-1}\xi)$, and let $k = k(x, r) \geq 0$ be the largest $k \in \mathbb{N}$ such that

$$|(T^k)'(x)|rK \leq \xi. \tag{16.73}$$

By this and Lemma 16.2.9, we have that

$$T_x^{-k}(B(T^k(x), \xi)) \supseteq B(x, K^{-1}|(T^k)'(x)|^{-1}\xi) \supseteq B(x, r). \tag{16.74}$$

Take $N \in \mathbb{N}$ such that $\lambda^{N-1} \geq K^2$. We then obtain that

$$|(T^{k+N})'(x)| = |(T^{N-1})'(T^{k+1}(x))| \cdot |(T^{k+1})'(x)| \geq \lambda^{N-1} \cdot (rK)^{-1}\xi \geq r^{-1}K\xi. \tag{16.75}$$

Hence, by Lemma 16.2.9,

$$T_x^{-(k+N)}(B(T^{k+N}(x), \xi)) \subseteq B(x, K|(T^{k+N})'(x)|^{-1}\xi) \subseteq B(x, r). \tag{16.76}$$

It follows from (16.73)–(16.76) that

$$\frac{\log \mu(T_x^{-k}(B(T^k(x), \xi)))}{\log(\xi K) - \log|(T^{k+N})'(x)|} \leq \frac{\log \mu(B(x, r))}{\log r} \leq \frac{\log \mu(T_x^{-(k+N)}(B(T^{k+N}(x), \xi)))}{\log(\xi K^{-1}) - \log|(T^k)'(x)|}. \tag{16.77}$$

Given that N is independent of k and x and that $\lambda^N \leq |(T^N)'(y)| \leq \|T'\|_\infty^N$ for all $y \in X$, the terms $\log|(T^k)'(x)|$ and $\log|(T^{k+N})'(x)|$ are comparable with a constant of comparability independent of k and x. Passing to the limit $r \to 0$ (so $k \to \infty$) and using (16.71)–(16.72), we deduce that

$$\lim_{r\to 0} \frac{\log \mu(B(x, r))}{\log r} = \frac{\mathrm{h}_\mu(T, \mathcal{P}, x)}{\chi_\mu(T, x)}.$$

By the Shannon–McMillan–Breiman theorem and Birkhoff's ergodic theorem,

$$\frac{\int_X \mathrm{h}_\mu(T, \mathcal{P}, x)\, d\mu(x)}{\int_X \chi_\mu(T, x)\, d\mu(x)} = \frac{\mathrm{h}_\mu(T, \mathcal{P})}{\chi_\mu(T)} = \frac{\mathrm{h}_\mu(T)}{\chi_\mu(T)}.$$

The latter equality holds since T is expansive and $\mathrm{diam}(\mathcal{P})$ is less than an expansive constant for T. There thus exists a set of positive measure where $\frac{\mathrm{h}_\mu(T,\mathcal{P},x)}{\chi_\mu(T,x)} \leq \frac{\mathrm{h}_\mu(T)}{\chi_\mu(T)}$, and a set of positive measure where the opposite inequality holds. Therefore,

$$\lim_{r\to 0} \frac{\log \mu(B(x, r))}{\log r} \leq \frac{\mathrm{h}_\mu(T)}{\chi_\mu(T)}$$

on a set of positive measure, and the opposite inequality holds on a positive measure set as well. The inequalities (16.66) follow from this and Theorem 16.5.9. In the ergodic case, $\mathrm{h}_\mu(T, \mathcal{P}, x) = \mathrm{h}_\mu(T)$ and $\chi_\mu(T, x) = \chi_\mu(T)$ for μ-a. e. $x \in X$. So (16.67) holds. □

16.6 Multifractal analysis of Gibbs states

In Section 16.3, we derived Bowen's formula for conformal expanding repellers. This formula states that the Hausdorff (as well as the packing and box) dimension of the limit set of such a repeller is equal to the unique zero of the repeller's pressure function.

The Hausdorff dimension of a fractal set gives us an idea of the global structure of that set. More precisely, it describes the complexity there is in covering the entire set at infinitesimal scales.

Nevertheless, two fractal sets with similar global structures may have substantially different local structures. The pointwise or local dimension of a Borel measure contributes to capturing the local structure of a set, and potentially helps in distinguishing that set from another one which may be globally similar.

Multifractal analysis consists in:

(1) determining the range of local dimensions that a set admits; and
(2) for each value in that range, determining the size of the subset of all points of the set where the local dimension is equal to that value.

The resulting map is called *multifractal spectrum*, as it describes the relative sizes of a multitude of fractal subsets which form a splitting of the original fractal set.

Multifractal formalism origins from physics and mathematics (among others, see [34, 42, 45, 68]). In this latter paper, strong hints of parallels between multifractal theory and the theory of statistical physics were suggested. Some of the first rigorous mathematical results on multifractals can be found in [15, 100]. Since then, many papers have been written on this subject. For instance, see [87, 88, 94–96]. In particular, Pesin [95] developed a general framework in which multifractal formalism can be derived.

Formally, let ν be a Borel probability measure on a metric space X. Recall that ν has *pointwise or local dimension* α at a point $x \in X$ if (cf. Definition 15.6.2)

$$d_\nu(x) := \lim_{r\to 0} \frac{\log \nu(B(x,r))}{\log r} = \alpha.$$

For each number $0 \leq \alpha \leq \infty$, let

$$X_\nu(\alpha) := \{x \in X \mid d_\nu(x) = \alpha\}$$

be the subset of all points $x \in X$ where the measure ν has local dimension α.

The domain of the pointwise dimension function d_ν, namely the set

$$X_\nu := \bigcup_{0\leq\alpha\leq\infty} X_\nu(\alpha),$$

is called the *regular part* of X.

Its complement

$$X_\nu^c := X \setminus X_\nu = \{x \in X \mid d_\nu(x) \nexists\}$$

is called the *singular part* of X.

The decomposition (i. e., the partition) of the set X as

$$X = \bigcup_{0 \le \alpha \le \infty} X_\nu(\alpha) \cup X_\nu^c$$

is called the *multifractal decomposition of X with respect to the pointwise dimension.*

Let

$$F_\nu(\alpha) := \mathrm{HD}(X_\nu(\alpha))$$

be the Hausdorff dimension of the fractal subset $X_\nu(\alpha)$. The function

$$\alpha \longmapsto F_\nu(\alpha),$$

whose domain is $\{0 \le \alpha \le \infty : X_\nu(\alpha) \neq \emptyset\}$, is called (*fine Hausdorff*) *multifractal spectrum* of the measure ν. Alternatively, it is called *spectrum for pointwise dimensions* or *pointwise dimension spectrum.*

If $X \subseteq \mathbb{R}^d$, then for every $q \in \mathbb{R} \setminus \{1\}$, let

$$R_q(\nu) := \frac{1}{q-1} \lim_{r \to 0} \frac{\log \sum_{i=1}^{N(r)} \nu(B_i)^q}{\log r}, \tag{16.78}$$

where $N(r)$ is the total number of boxes B_i of the form

$$B_i = \{(x_1, \ldots, x_d) \in \mathbb{R}^d : rk_j \le x_j \le r(k_j + 1),\ \forall j = 1, \ldots, d\}$$

for integers $k_j = k_j(i)$ such that $\nu(B_i) > 0$. The function R_q, whose domain consists of all q for which the limit (16.78) exists, is called *Rényi dimension spectrum.*

It is easy to check that it is equal to the *Hentschel–Procaccia spectrum* (often abbreviated HP-spectrum), which is defined by

$$\mathrm{HP}_q(\nu) := \frac{1}{q-1} \lim_{r \to 0} \frac{\log \inf_{\mathcal{G}_r} \sum_{B(x_i, r) \in \mathcal{G}_r} \nu(B(x_i, r))^q}{\log r},$$

where the infimum is taken over all countable covers $\mathcal{G}_r$ of the topological support of ν by balls of radius r centered at $x_i \in X$, or

$$\mathrm{HP}_q(\nu) := \frac{1}{q-1} \lim_{r \to 0} \frac{\log \int_X \nu(B(x, r))^{q-1}\, d\nu(x)}{\log r}$$

provided those limits exist.

For $q = 1$, we define the *information dimension* $I(\nu)$ in two steps. First, set

$$\mathrm{H}_\nu(r) := \inf_{\mathcal{F}_r}\Big(-\sum_{B\in\mathcal{F}_r} \nu(B)\log\nu(B)\Big),$$

where the infimum is taken over all partitions $\mathcal{F}_r$ of a set of full ν-measure into Borel sets B of diameter at most r. Second, define

$$I(\nu) := \lim_{r\to 0}\frac{\mathrm{H}_\nu(r)}{-\log r} \tag{16.79}$$

provided this limit exists.

Note that for Rényi and HP dimensions it does not make any difference whether we consider covers of the topological support (the smallest closed set of full measure) of a measure or any set of full measure, since all balls have the same radius r, so we can always choose locally finite (number independent of r) subcovers. These are "box type" dimension quantities.

To avoid any confusion with the temperature function that we will introduce later and that will be denoted by T, the conformal repellers will be denoted from this point on by (X, U, f) rather than (X, U, T).

Returning to the pointwise dimension spectrum, a priori there is no reason for the function F_ν to behave nicely. Its domain of definition obviously depends on ν. If ν is a geometric measure (see Definition 15.6.12), then the domain of F_ν is a singleton, namely, the exponent t of the geometric measure. Its range is the Hausdorff dimension of the underlying space X. For instance, we know by the volume lemma (Theorem 16.5.2) that the domain and the range of F_ν are both $\{\mathrm{HD}(X)\}$ for the geometric measure $\nu = \mu_{-\mathrm{HD}(X)\log|f'|}$ associated to a conformal expanding repeller (X, U, f). If ν is an ergodic f-invariant measure, then the volume lemma states that $d_\nu(x) = \alpha_0$ for ν-a. e. $x \in X$, where $\alpha_0 = \mathrm{HD}(\nu)$. In particular, we know that the domain of F_ν is nonempty. However, for any $\alpha \neq \alpha_0$ the set $X_\nu(\alpha)$ is not visible to the measure ν since $\nu(X_\nu(\alpha)) = 0$. Whereas the function $\mathrm{HP}_q(\nu)$ can be determined by statistical properties of ν-typical (a. e.) trajectories, the function $F_\nu(\alpha)$ seems intractable. Nevertheless, if $\nu = \mu_\varphi$ is the unique f-invariant Gibbs state and unique equilibrium state associated with a Hölder continuous potential φ on a conformal expanding repeller (X, U, f), then miraculously the dimension spectra $\alpha \mapsto F_{\mu_\varphi}(\alpha)$ and $q \mapsto \mathrm{HP}_q(\mu_\varphi)$ happen to be real-analytic functions and $-F_{\mu_\varphi}(-p)$ and $\mathrm{HP}_q(\mu_\varphi)$ are mutual Legendre transforms.

Accordingly, let $\varphi : X \to \mathbb{R}$ be a Hölder continuous potential. For $q, t \in \mathbb{R}$, consider the two-parameter family of auxiliary Hölder continuous potentials $\varphi_{q,t} : X \to \mathbb{R}$ defined by

$$\varphi_{q,t} := q\varphi - t\log|f'|.$$

This family naturally generates a two-variable pressure function

$$\mathbb{R}^2 \ni (q,t) \longmapsto \mathrm{P}(q,t) := \mathrm{P}(f,\varphi_{q,t}) = \mathrm{P}(f, q\varphi - t\log|f'|) \in \mathbb{R}.$$

Lemma 16.6.1. *Let (X,U,f) be a conformal expanding repeller and let $\varphi : X \to \mathbb{R}$ be a Hölder continuous potential. For every $q \in \mathbb{R}$, the one-variable pressure function $\mathbb{R} \ni t \longmapsto \mathrm{P}(q,t) \in \mathbb{R}$ possesses the following properties:*

(a) *It is Lipschitz with Lipschitz constant $L := \log\|f'\|_\infty$, where $\|f'\|_\infty := \sup_{x\in X}|f'(x)|$.*
(b) *It is strictly decreasing.*
(c) $\lim_{t\to-\infty}\mathrm{P}(q,t) = \infty$ *whereas* $\lim_{t\to\infty}\mathrm{P}(q,t) = -\infty$.
(d) $\mathrm{P}(q,0) = \mathrm{P}(f,q\varphi)$.
(e) *There is a unique $T(q) \in \mathbb{R}$ such that $\mathrm{P}(q,T(q)) = 0$.*
The resulting function $\mathbb{R} \ni q \longmapsto T(q) \in \mathbb{R}$ is called the temperature function.

Moreover,

(f) *The two-variable pressure function $(q,t) \longmapsto \mathrm{P}(q,t)$ is convex.*

Proof. The proof is left to the reader. It goes along almost the same lines as that of the properties of the pressure function when $q = 0$ in Proposition 16.3.1. □

Per Proposition 13.7.12, let μ_φ be the unique f-invariant Gibbs state and unique equilibrium state for the potential φ. The unique f-invariant Gibbs state and unique equilibrium state for the potential $\varphi_{q,t}$, normally denoted by $\mu_{\varphi_{q,t}}$, will be commonly abbreviated to $\mu_{q,t}$. The unique f-invariant Gibbs state and unique equilibrium state for the potential $\varphi_{q,T(q)}$, normally denoted by $\mu_{\varphi_{q,T(q)}}$ or $\mu_{q,T(q)}$, will be further abbreviated to μ_q. The relation between the measures $\{\mu_q\}_{q\in\mathbb{R}}$ and the measure μ_φ is explained in the following result.

Lemma 16.6.2. *Let (X,U,f) be a conformal expanding repeller and let $\varphi : X \to \mathbb{R}$ be a Hölder continuous potential such that $\mathrm{P}(f,\varphi) = 0$. For every $q \in \mathbb{R}$ there exists $C_q \geq 1$ such that*

$$C_q^{-1} \leq \frac{\mu_q(B(x,r))}{r^{T(q)}\mu_\varphi(B(x,r))^q} \leq C_q, \quad \forall x \in X,\ \forall 0 < r < \xi. \tag{16.80}$$

Proof. Let $k = k(x,r)$ be defined as just before Corollary 16.2.10. By (16.10) in Lemma 16.2.12 (with m equal to μ_φ and μ_q successively) and by Corollary 16.2.10, the ratios

$$\frac{\mu_\varphi(B(x,r))}{\exp(S_k\varphi(x))}, \quad \frac{\mu_q(B(x,r))}{|(f^k)'(x)|^{-T(q)}\exp(qS_k\varphi(x))}, \quad \text{and} \quad \frac{r}{|(f^k)'(x)|^{-1}}$$

are comparable, i. e., they are all bounded from below and above by positive constants independent of x and r. This yields (16.80). □

We now derive a generalization of the volume lemma (Theorem 16.5.2) for the pointwise dimension of the measure μ_φ.

Lemma 16.6.3. *Let (X, U, f) be a conformal expanding repeller and let $\varphi : X \to \mathbb{R}$ be a Hölder continuous potential such that $\mathrm{P}(f, \varphi) = 0$. For any ergodic f-invariant Borel probability measure ν on X, we have*

$$d_{\mu_\varphi}(x) = \frac{-\nu(\varphi)}{\chi_\nu(f)} = \mathrm{HD}(\mu_\varphi) = \mathrm{PD}(\mu_\varphi) \quad \textit{for } \nu\textit{-a. e. } x \in X.$$

Proof. Using Lemma 16.2.12 and the ergodic case of Birkhoff's ergodic theorem (Corollary 8.2.14), we get

$$\begin{aligned} d_{\mu_\varphi}(x) &= \lim_{r\to 0} \frac{\log \mu_\varphi(B(x,r))}{\log r} = \lim_{k\to\infty} \frac{S_k\varphi(x)}{-\log|(f^k)'(x)|} \\ &= \frac{\lim_{k\to\infty} \frac{1}{k} S_k\varphi(x)}{-\lim_{k\to\infty} \frac{1}{k}\log|(f^k)'(x)|} = \frac{\int_X \varphi\, d\nu}{-\int_X \log|f'|\, d\nu} = \frac{\nu(\varphi)}{-\chi_\nu(f)} \end{aligned}$$

for ν-a. e. $x \in X$. The proof of the first equality is complete. The second and third equalities are a direct application of Corollary 15.6.11. □

We can further conclude from Lemma 16.6.3 that the singular part $X^c_{\mu_\varphi}$ of X has zero ν-measure with respect to every ergodic invariant measure ν, including μ_φ. Yet the set $X^c_{\mu_\varphi}$ is usually geometrically big (see Exercise 16.7.15).

On the Legendre transform

Let $k = k(q) : I \to \overline{\mathbb{R}} := \mathbb{R} \cup \{-\infty, \infty\}$ be a continuous convex function on an interval $I = [\alpha_1(k), \alpha_2(k)]$ with $-\infty \le \alpha_1(k) \le \alpha_2(k) \le \infty$, excluding the degenerate cases where I is the singleton $\{-\infty\}$ or $\{\infty\}$. So I is either (1) a point in $\mathbb{R}$; or (2) a closed real interval; or (3) a closed real half-line jointly with one of $-\infty$ or ∞; or (4) $\overline{\mathbb{R}}$. We also assume that k is finite on (α_1, α_2).

The *Legendre transform* of k is the function g of a new variable p defined by

$$g(p) = \sup_{q\in I}\{pq - k(q)\}.$$

Its domain is the closure in $\overline{\mathbb{R}}$ of the set of points $p \in \mathbb{R}$ where $g(p)$ is finite, with g extended continuously to the boundary.

It can be easily proved that the domain of g is also either (1) a real point; or (2) a closed real interval; or (3) a closed real semiline with one of $-\infty$ or ∞; or (4) $\overline{\mathbb{R}}$. More precisely, the domain is $[\alpha_1(g), \alpha_2(g)]$, where $\alpha_1(g) = -\infty$ if $\alpha_1(k)$ is finite, or $\alpha_1(g) = \lim_{x\to-\infty} k'(x)$ if $\alpha_1(k) = -\infty$. The derivative here is a one-sided derivative; it does not matter whether it is left or right. A similar description of $\alpha_2(g)$ is obtained by replacing $-\infty$ with ∞.

It is further easy to show that g is a continuous convex function and that the Legendre transform is involutive. We then say that the functions k and g form a *Legendre transform pair.*

Proposition 16.6.4. *If two convex functions k and g form a Legendre transform pair then $g(k'(q)) = qk'(q) - k(q)$, where $k'(q)$ is any number between the left- and right-hand side derivatives of k at q, defined as $-\infty$ at $q = \alpha_1(k)$ if $\alpha_1(k)$ is finite and defined as ∞ at $q = \alpha_2(k)$ if $\alpha_2(k)$ is finite.*

Note that if k is C^2 with $k'' > 0$ and therefore is strictly convex, then $g'' > 0$ at all points $k'(q)$ for $\alpha_1(k) < q < \alpha_2(k)$ and thus g is strictly convex on $[k'(\alpha_1(k)), k'(\alpha_2(k))]$. Outside this interval, g is affine on the rest of its domain. If the domain of k is a singleton, then g is affine on $\mathbb{R}$ and vice versa.

We now formulate the main theorem of this section.

Theorem 16.6.5. *Let (X, U, f) be an infinite conformal expanding repeller (infinite means $\#X = \infty$). Let $\varphi : X \to \mathbb{R}$ be a Hölder continuous potential such that $\mathrm{P}(f, \varphi) = 0$.*

(a) *For μ_φ-a. e. $x \in X$,*

$$d_{\mu_\varphi}(x) = \frac{-\mu_\varphi(\varphi)}{\chi_{\mu_\varphi}(f)} = \frac{\mathrm{h}_{\mu_\varphi}(f)}{\chi_{\mu_\varphi}(f)} = \mathrm{HD}(\mu_\varphi) = \mathrm{PD}(\mu_\varphi).$$

(b) *The temperature function $\mathbb{R} \ni q \longmapsto T(q) \in \mathbb{R}$ is real analytic and satisfies*

$$T(0) = \mathrm{HD}(X), \quad T(1) = 0, \quad T'(q) = \frac{\mu_q(\varphi)}{\chi_{\mu_q}(f)} \le \frac{-\mathrm{h}_{\mu_q}(f)}{\chi_{\mu_q}(f)} < 0, \quad \text{and} \quad T''(q) \ge 0,$$

where μ_q is the f-invariant Gibbs measure for the potential $\varphi_{q,T(q)}$.

(c) *For all $q \in \mathbb{R}$, we have $\mu_q(X_{\mu_\varphi}(-T'(q))) = 1$. Moreover, $\mathrm{HD}(\mu_q) = \mathrm{HD}(X_{\mu_\varphi}(-T'(q)))$.*

(d) *For every $q \in \mathbb{R}$, we have $F_{\mu_\varphi}(-T'(q)) = T(q) - qT'(q)$, i. e., $p \longmapsto -F_{\mu_\varphi}(-p)$ is the Legendre transform of $T(q)$.*

In particular, F_{μ_φ} is continuous at the boundary points $-T'(\pm\infty)$ of its domain, as the Legendre transform is. Furthermore, the sets $X_{\mu_\varphi}(\alpha)$ are empty for all $\alpha \notin [-T'(\infty), -T'(-\infty)]$, so these α's do not lie in the domain of F_{μ_φ}, as they do not belong to the domain of the Legendre transform. (This emptyness property is called the completeness of the F_{μ_φ}-spectrum.)

If $\mu_\varphi = \mu_{-\mathrm{HD}(X)\log|f'|}$ (equivalently, if φ is cohomologous to $-\mathrm{HD}(X)\log|f'|$ in the class of bounded functions on X), then $T(q)$ is affine and the domain of $F_{\mu_\varphi}(\alpha)$ is one point, namely $\mathrm{HD}(X) = -T'(q)$ for all $q \in \mathbb{R}$.

For this sole paragraph, suppose that the repeller (X, U, f) is topologically exact. If $\mu_\varphi \ne \mu_{-\mathrm{HD}(X)\log|f'|}$, then $T''(q) > 0$ and $F''_{\mu_\varphi}(\alpha) < 0$, i. e., the functions $T(q)$ and $F_{\mu_\varphi}(\alpha)$ are respectively strictly convex on $\mathbb{R}$ and strictly concave on $[-T'(\infty), -T'(-\infty)]$, which is a bounded interval in $\mathbb{R}^+ = \{\alpha \in \mathbb{R} : \alpha > 0\}$.

(e) *For every* $q \neq 1$, *the HP and Rényi spectra exist (i. e., limits in their definitions exist) and* $\frac{T(q)}{1-q} = \mathrm{HP}_q(\mu_\varphi) = R_q(\mu_\varphi)$. *For* $q = 1$, *the information dimension* $I(\mu_\varphi)$ *exists and*

$$\lim_{q\to 1,\, q\neq 1} \frac{T(q)}{1-q} = -T'(1) = \mathrm{HD}(\mu_\varphi) = \mathrm{PD}(\mu_\varphi) = I(\mu_\varphi).$$

For outlines of the graphs of $T(q)$ and $F_{\mu_\varphi}(\alpha)$, see Figures 16.1 and 16.2. See also Exercise 16.7.17. Compare with [95, Figures 17(a,b) on pp. 219–220].

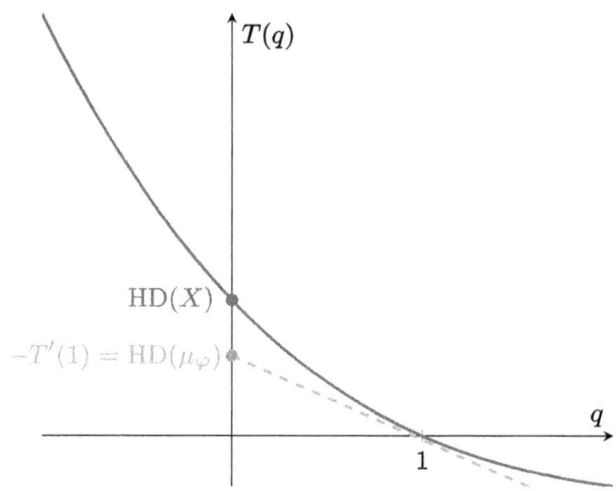

Figure 16.1: Graph of T.

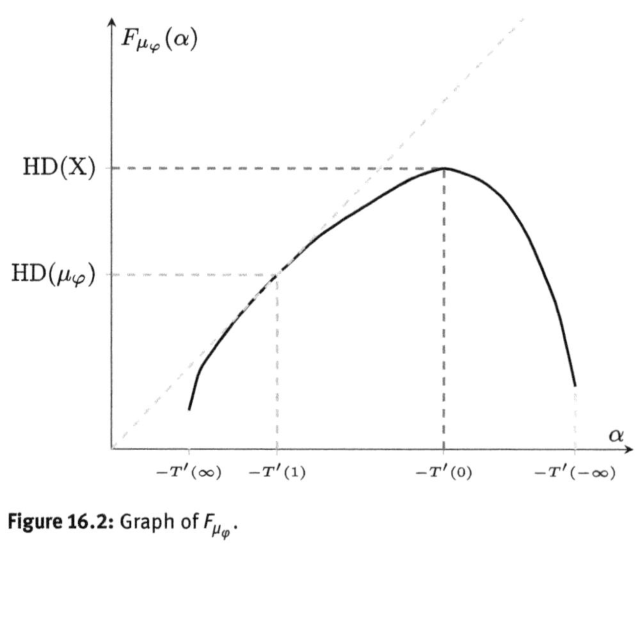

Figure 16.2: Graph of F_{μ_φ}.

Proof. **(a)** As $\mathrm{P}(f,\varphi)=0$, the first equality in part (a) is an immediate consequence of the volume lemma (or of Lemma 16.6.3 with $\nu=\mu_\varphi$) and the fact that μ_φ is an equilibrium state. The second and third equalities follow from Corollary 15.6.11.

(b) The equality $T(0)=\mathrm{HD}(X)$ ensues from Bowen's formula (Theorem 16.3.2) while $T(1)=0$ follows from the equality $\mathrm{P}(f,\varphi)=0$.

Concerning the real analyticity of $q\mapsto T(q)$, let β be a Hölder exponent for the function φ. The function $\mathbb{C}^2\ni(q,t)\mapsto\varphi_{q,t}=q\varphi-t\log|f'|\in H_\beta(X)$ is affine. By Theorem 13.10.3, the pressure function $\mathbb{R}^2\ni(q,t)\mapsto \mathrm{P}(q,t):=\mathrm{P}(f,\varphi_{q,t})\in\mathbb{R}$ is real analytic. Thus, the real analyticity of $q\mapsto T(q)$ will follow from the implicit function theorem once we verify the nondegeneracy assumption. For this, Theorem 13.10.6 (with $T=f$, $\varphi=-\log|f'|$ and $\psi=\varphi$) confirms that for every $(q_0,t_0)\in\mathbb{R}^2$,

$$\frac{\partial \mathrm{P}(q,t)}{\partial t}\bigg|_{(q_0,t_0)}=-\int_X\log|f'|\,d\mu_{q_0,t_0}=-\chi_{\mu_{q_0,t_0}}(f)<0, \tag{16.81}$$

where μ_{q_0,t_0} is the f-invariant Gibbs state for the function φ_{q_0,t_0}, and the real analyticity of $q\mapsto T(q)$ ensues from the implicit function theorem.

Differentiating the equality $\mathrm{P}(q,t)=0$ with respect to q, we obtain

$$\frac{\partial \mathrm{P}(q,t)}{\partial q}\bigg|_{(q,T(q))}+\frac{\partial \mathrm{P}(q,t)}{\partial t}\bigg|_{(q,T(q))}\cdot T'(q)=0. \tag{16.82}$$

Therefore, we get the standard formula

$$T'(q)=-\frac{\partial \mathrm{P}(q,t)}{\partial q}\bigg|_{(q,T(q))}\bigg/\frac{\partial \mathrm{P}(q,t)}{\partial t}\bigg|_{(q,T(q))}.$$

Using Theorem 13.10.6, we infer expressions for the numerator and denominator of $T'(q)$:

$$\frac{\partial \mathrm{P}(q,t)}{\partial q}\bigg|_{(q,T(q))}=\int_X\varphi\,d\mu_q\quad\text{and}\quad\frac{\partial \mathrm{P}(q,t)}{\partial t}\bigg|_{(q,T(q))}=-\int_X\log|f'|\,d\mu_q. \tag{16.83}$$

So

$$T'(q)=\frac{\mu_q(\varphi)}{\chi_{\mu_q}(f)}. \tag{16.84}$$

As $\mathrm{P}(f,\varphi)=0$ and μ_q is f-invariant (i. e., $\mu_q\in M(f)$), the variational principle (Theorem 12.1.1) implies that

$$0=\mathrm{P}(f,\varphi)=\sup_{\mu\in M(f)}\left\{\mathrm{h}_\mu(f)+\int_X\varphi\,d\mu\right\}\geq \mathrm{h}_{\mu_q}(f)+\int_X\varphi\,d\mu_q.$$

Hence,

$$T'(q) \le \frac{-\mathrm{h}_{\mu_q}(f)}{\chi_{\mu_q}(f)} < 0.$$

The strict inequality holds since $\chi_{\mu_q}(f) \ge \log\lambda > 0$ and $\mathrm{h}_{\mu_q}(f) > 0$ by Theorem 13.7.21.

Finally, the inequality $T''(q) \ge 0$ follows from the convexity of $\mathrm{P}(q,t)$ (see Lemma 16.6.1). Indeed, the fact that the part of $\mathbb{R}^3$ above the graph of $\mathrm{P}(q,t)$ is convex implies that its intersection with the plane (q,t) is also convex. Since $\frac{\partial \mathrm{P}(q,t)}{\partial t}|_{(q_0,t_0)} = -\chi_{\mu_{q_0,t_0}}(f) < 0$, this is the part of the plane above the graph of T. Hence, T is a convex function. This completes the proof of (b).

We now establish the **last two paragraphs in part (d).**

Theorem 13.7.17 asserts that $\mu_\varphi = \mu_{-\mathrm{HD}(X)\log|f'|}$ if and only if $\varphi + \mathrm{HD}(X)\log|f'|$ is cohomologous to a constant R in the class of bounded functions on X. By Bowen's formula (Theorem 16.3.2), Lemma 13.2.6 and the hypothesis $\mathrm{P}(f,\varphi) = 0$, this constant R is

$$R = \mathrm{P}(f, -\mathrm{HD}(X)\log|f'|) + R = \mathrm{P}(f,\varphi) = 0.$$

In other words, $\mu_\varphi = \mu_{-\mathrm{HD}(X)\log|f'|}$ if and only if φ is cohomologous to $-\mathrm{HD}(X)\log|f'|$ via a bounded coboundary $\upsilon : X \to \mathbb{R}$. Using this and Lemma 13.2.6 once again, for every $q \in \mathbb{R}$ we get

$$\begin{aligned}
\mathrm{P}(q,T(q)) = 0 = \mathrm{P}(f,\varphi) &= \mathrm{P}(f, -\mathrm{HD}(X)\log|f'|) \\
&= \mathrm{P}(f, -\mathrm{HD}(X)\log|f'| + q\upsilon - q\upsilon\circ f) \\
&= \mathrm{P}\big(f, q(-\mathrm{HD}(X)\log|f'| + \upsilon - \upsilon\circ f) + (q-1)\mathrm{HD}(X)\log|f'|\big) \\
&= \mathrm{P}(f, q\varphi + (q-1)\mathrm{HD}(X)\log|f'|) = \mathrm{P}(f, q\varphi - (1-q)\mathrm{HD}(X)\log|f'|) \\
&= \mathrm{P}(q,(1-q)\mathrm{HD}(X)).
\end{aligned}$$

By the uniqueness of $T(q)$, we infer that $T(q) = (1-q)\mathrm{HD}(X)$. This proves the second last paragraph in (d).

In part (b)'s proof of convexity of T, we avoided an explicit expression of T''. However, to discuss strict convexity it is necessary to compute T''. Differentiating the formula

$$0 = \frac{d\mathrm{P}}{dq}(q,T(q)) = \frac{\partial \mathrm{P}}{\partial q}(q,T(q)) + \frac{\partial \mathrm{P}}{\partial t}(q,T(q)) \cdot T'(q)$$

with respect to q and using (16.81), we obtain that

$$T''(q) = \frac{\frac{\partial^2 \mathrm{P}}{\partial q^2}(q,T(q)) + 2T'(q)\frac{\partial^2 \mathrm{P}}{\partial q \partial t}(q,T(q)) + (T'(q))^2 \frac{\partial^2 \mathrm{P}}{\partial t^2}(q,T(q))}{\chi_{\mu_q}(f)}. \tag{16.85}$$

If the repeller f is topologically exact, then it follows from Theorem 13.10.7 that

$$\frac{\partial^2 \mathrm{P}}{\partial q^2}(q, T(q)) = \sigma^2_{\mu_q,\infty}(\varphi) \quad \text{and} \quad \frac{\partial^2 \mathrm{P}}{\partial t^2}(q, T(q)) = \sigma^2_{\mu_q,\infty}(-\log|f'|). \tag{16.86}$$

From Exercise 13.11.16 and Theorem 13.9.1, one can deduce the analog of Theorem 13.10.7 for the mixed second derivative

$$\frac{\partial^2 \mathrm{P}}{\partial q \partial t}(q, T(q)) = \sigma^2_{\mu_q,\infty}(-\log|f'|, \varphi). \tag{16.87}$$

To shorten notation, let $F := -\log|f'|$. So $\mu_q(F) = -\chi_{\mu_q}(f)$. Using (13.87), (13.148) and (16.84), we can write

$$\begin{aligned}
&(T'(q))^2 \frac{\partial^2 \mathrm{P}}{\partial t^2}(q, T(q)) + 2T'(q)\frac{\partial^2 \mathrm{P}}{\partial q \partial t}(q, T(q)) + \frac{\partial^2 \mathrm{P}}{\partial q^2}(q, T(q)) \\
&= (T'(q))^2\Big[\mu_q((F + \chi_{\mu_q}(f)\mathbb{1})^2) + 2\sum_{k=1}^{\infty}\big[\mu_q(F \cdot F \circ T^k) - (\chi_{\mu_q}(f))^2\big]\Big] \\
&\quad + 2T'(q)\mu_q\big((F + \chi_{\mu_q}(f)\mathbb{1})(\varphi - \mu_q(\varphi)\mathbb{1})\big) \\
&\quad + 2T'(q)\sum_{k=1}^{\infty}\big(\mu_q(F \cdot \varphi \circ T^k) + \chi_{\mu_q}(f)\mu_q(\varphi)\big) \\
&\quad + 2T'(q)\sum_{k=1}^{\infty}\big(\mu_q(F \circ T^k \cdot \varphi) + \chi_{\mu_q}(f)\mu_q(\varphi)\big) \\
&\quad + \Big[\mu_q((\varphi - \mu_q(\varphi)\mathbb{1})^2) + 2\sum_{k=1}^{\infty}\big[\mu_q(\varphi \cdot \varphi \circ T^k) - (\mu_q(\varphi))^2\big]\Big] \\
&= \mu_q\Big(\big[T'(q)(F + \chi_{\mu_q}(f)\mathbb{1}) + (\varphi - \mu_q(\varphi)\mathbb{1})\big]^2\Big) \\
&\quad + 2(T'(q))^2\sum_{k=1}^{\infty}\big[\mu_q(F \cdot F \circ T^k) - (\chi_{\mu_q}(f))^2\big] \\
&\quad + 2T'(q)\sum_{k=1}^{\infty}\big(\mu_q(F \cdot \varphi \circ T^k) + \chi_{\mu_q}(f)\mu_q(\varphi)\big) \\
&\quad + 2T'(q)\sum_{k=1}^{\infty}\big(\mu_q(F \circ T^k \cdot \varphi) + \chi_{\mu_q}(f)\mu_q(\varphi)\big) \\
&\quad + 2\sum_{k=1}^{\infty}\big[\mu_q(\varphi \cdot \varphi \circ T^k) - (\mu_q(\varphi))^2\big] \\
&= \mu_q\Big(\big[(T'(q)F + \varphi) - \mu_q(T'(q)F + \varphi)\mathbb{1}\big]^2\Big) \\
&\quad + 2\sum_{k=1}^{\infty}\mu_q\big(T'(q)F \cdot (T'(q)F \circ T^k + \varphi \circ T^k)\big)
\end{aligned}$$

$$
\begin{aligned}
&+2\sum_{k=1}^{\infty}\mu_q\Big(\varphi\cdot\big(T'(q)F\circ T^k+\varphi\circ T^k\big)\Big)-2\sum_{k=1}^{\infty}\Big(T'(q)\chi_{\mu_q}(f)-\mu_q(\varphi)\Big)^2\\
&=\mu_q\Big(\Big[(T'(q)F+\varphi)-\mu_q(T'(q)F+\varphi)\mathbb{1}\Big]^2\Big)\\
&\quad+2\sum_{k=1}^{\infty}\mu_q\Big((T'(q)F+\varphi)\cdot(T'(q)F+\varphi)\circ T^k\Big)-2\sum_{k=1}^{\infty}(0)^2\\
&=\sigma^2_{\mu_q,\infty}(T'(q)F+\varphi)\geq 0,
\end{aligned}
\tag{16.88}
$$

where the second last equality relies on (16.84) while the last equality follows from (13.87) for the Hölder function $T'(q)F+\varphi = -T'(q)\log|f'|+\varphi$. Substituting (16.88) into (16.85), we get

$$
T''(q)=\frac{\sigma^2_{\mu_q,\infty}(-T'(q)\log|f'|+\varphi)}{\chi_{\mu_q}(f)}\geq 0. \tag{16.89}
$$

In particular, this provides another proof of (b), albeit under the additional assumption that the repeller is topologically exact. By Theorem 13.9.14, we know that $\sigma^2_{\mu_q,\infty}(-T'(q)\log|f'|+\varphi) = 0$ if and only if the function $-T'(q)\log|f'|+\varphi$ is cohomologous to a constant R in the class of Hölder continuous functions. It then follows from (16.84) that $R = \int R\,d\mu_q = \int(\varphi - T'(q)\log|f'|)\,d\mu_q = 0$. Therefore, $T'(q)\log|f'|$ is cohomologous to φ, and hence $\mathrm{P}(f, T'(q)\log|f'|) = \mathrm{P}(f,\varphi) = 0$ by Lemma 13.2.6 and the hypothesis $\mathrm{P}(f,\varphi) = 0$. Thus, $T'(q) = -\mathrm{HD}(X)$ by Bowen's formula (Theorem 16.3.2). Consequently, φ is cohomologous to the function $-\mathrm{HD}(X)\log|f'|$. This means that $\mu_\varphi = \mu_{-\mathrm{HD}(X)\log|f'|}$. In view of (16.89), we conclude that if $\mu_\varphi \neq \mu_{-\mathrm{HD}(X)\log|f'|}$ then $T''(q) > 0$ for all $q\in\mathbb{R}$. This proves the last paragraph in (d), which is the only portion of the result where topological exactness is needed.

(c) By Lemma 16.6.3 (applied with $\nu = \mu_q$) and part (b), there exists a set $X_q \subseteq X$ of full μ_q-measure such that

$$
d_{\mu_\varphi}(x)=\frac{-\mu_q(\varphi)}{\chi_{\mu_q}(f)}=-T'(q),\quad \forall x\in X_q.
$$

Hence, $X_q \subseteq X_{\mu_\varphi}(-T'(q))$ and thereby $\mu_q(X_{\mu_\varphi}(-T'(q))) = 1$. Let $x\in X$ and $0<r<\xi$. By Lemma 16.6.2,

$$
\Big|\log\mu_q(B(x,r))-T(q)\log r-q\log\mu_\varphi(B(x,r))\Big|<\log C_q.
$$

Therefore,

$$
\lim_{r\to 0}\left|\frac{\log\mu_q(B(x,r))}{\log r}-T(q)-q\frac{\log\mu_\varphi(B(x,r))}{\log r}\right|=0. \tag{16.90}
$$

Using this, observe that for every $x \in X_{\mu_\varphi}(-T'(q))$,

$$d_{\mu_q}(x) = \lim_{r\to 0} \frac{\log \mu_q(B(x,r))}{\log r} = T(q) + q \lim_{r\to 0} \frac{\log \mu_\varphi(B(x,r))}{\log r} = T(q) + q d_{\mu_\varphi}(x) = T(q) - qT'(q).$$

In particular, Corollary 15.6.11 tells us that

$$d_{\mu_q}(x) = T(q) - qT'(q) = \mathrm{HD}(\mu_q), \quad \forall x \in X_{\mu_\varphi}(-T'(q)) \supseteq X_q. \tag{16.91}$$

Although X_q can be much smaller than $X_{\mu_\varphi}(-T'(q))$, surprisingly they have the same Hausdorff dimension. Indeed, the measure μ_q restricted to either X_q or $X_{\mu_\varphi}(-T'(q))$ satisfies the hypotheses of Theorem 15.6.3 with $\theta_1 = \theta_2 = T(q) - qT'(q)$. Therefore,

$$\mathrm{HD}(X_q) = \mathrm{HD}(X_{\mu_\varphi}(-T'(q))) = T(q) - qT'(q) \tag{16.92}$$

and thus

$$F_{\mu_\varphi}(-T'(q)) = T(q) - qT'(q). \tag{16.93}$$

Relations (16.91)–(16.92) establish (c) while (16.93) is the first part of (d).

Remarks

(i) Replacing the pointwise dimension d_ν by the lower pointwise dimension $\underline{d}_\nu$ and replacing accordingly the sets $X_\nu(\alpha) := \{x \in X \mid d_\nu(x) = \alpha\}$ by the larger sets $\underline{X}_\nu(\alpha) := \{x \in X \mid \underline{d}_\nu(x) = \alpha\}$ in the definition of $F_\nu(\alpha)$, we obtain the same Hausdorff dimension spectra, again by virtue of Theorem 15.6.3. This means that the F_ν-spectrum is the same; in particular, it is given by the same Legendre transform formula when $\nu = \mu_\varphi$. There is no singular part.

(ii) Notice that (16.92) means that $\mathrm{HD}(X_{\mu_\varphi}(-T'(q)))$ is the vertical intercept of the tangent line to the graph of T at the point $(q, T(q))$.

(iii) We only used the f-invariance of μ_φ when estimating $\mathrm{HD}(X_{\mu_\varphi}(-T'(q)))$ from below (through Birkhoff's ergodic theorem). In the estimate from above, we solely used the Gibbs property of μ_φ. In a more general setting, it is sufficient that μ_φ be conformal.

In the next steps of the proof, the following will be useful.

Claim (Variational principle for $q \mapsto T(q)$). *For every ergodic f-invariant Borel probability measure ν on X, consider the linear equation in the two variables q, t*

$$\mathrm{h}_\nu(f) + \int_X \varphi_{q,t}\, d\nu = 0 \quad \textit{or equivalently} \quad t = t_\nu(q) = \frac{\mathrm{h}_\nu(f)}{\chi_\nu(f)} + q\frac{\nu(\varphi)}{\chi_\nu(f)}. \tag{16.94}$$

Then for every $q \in \mathbb{R}$ it turns out that

$$T(q) = \sup_{\nu} t_\nu(q) = t_{\mu_q}(q),$$

where the supremum is taken over all ergodic f-invariant Borel probability measures ν.

Proof of Claim. Fix $q \in \mathbb{R}$. Recall from Lemma 16.6.1 that $t \mapsto \mathrm{P}(q,t)$ is strictly decreasing, and that $\mathrm{P}(q, T(q)) = 0$ by definition. Take any $t > T(q)$. Then $\mathrm{P}(q,t) < 0$. By the variational principle (Theorem 12.1.1), we know that $\mathrm{h}_\nu(f) + \int \varphi_{q,t}\, d\nu \le \mathrm{P}(q,t) < 0$. So the solution $t_\nu(q)$ to the equation $\mathrm{h}_\nu(f) + \int \varphi_{q,t}\, d\nu = 0$ cannot be any $t > T(q)$, i. e., $t_\nu(q) \le T(q)$. Consequently,

$$\sup_{\nu} t_\nu(q) \le T(q). \tag{16.95}$$

Recall further that μ_q is the unique equilibrium state for $\varphi_{q,T(q)}$. Thus, $\mathrm{h}_{\mu_q}(f) + \int \varphi_{q,T(q)}\, d\mu_q = \mathrm{P}(q, T(q)) = 0$. So the solution $t_{\mu_q}(q)$ to the equation $\mathrm{h}_{\mu_q}(f) + \int \varphi_{q,t}\, d\mu_q = 0$ is precisely $T(q)$. Therefore,

$$T(q) = t_{\mu_q}(q) \le \sup_{\nu} t_\nu(q). \tag{16.96}$$

The claim follows from (16.95)–(16.96). ∎

(d) We resume the proof of Theorem 16.6.5 by showing the missing part of (d). We have already proved that (cf. (16.91)–(16.93))

$$F_{\mu_\varphi}(-T'(q)) = T(q) - qT'(q) = \mathrm{HD}(X_{\mu_\varphi}(-T'(q))) = \mathrm{HD}(X_q) = \mathrm{HD}(\mu_q).$$

Note that $[-T'(\infty), -T'(-\infty)] \subseteq \mathbb{R}^+ \cup \{0,\infty\}$ since $T'(q) < 0$ for all $q \in \mathbb{R}$ according to part (b). Notice further that

$$-T'(-\infty) := -\lim_{q\to-\infty} T'(q) = \lim_{q\to-\infty} \frac{-\mu_q(\varphi)}{\chi_{\mu_q}(f)} \le \frac{\|\varphi\|_\infty}{\inf \log|f'|} < \infty$$

and

$$-T'(\infty) := \lim_{q\to\infty} -T'(q) = \lim_{q\to\infty} \frac{-\mu_q(\varphi)}{\chi_{\mu_q}(f)}.$$

We would like to show that this last limit is strictly positive. As $-T'(q) > 0$ for all $q \in \mathbb{R}$, it suffices to show that $\frac{-\mu_q(\varphi)}{\chi_{\mu_q}(f)}$ is bounded away from 0 as $q \to \infty$. By (b), we know that $T(1) = 0$ and T is strictly decreasing. Choose $q_0 > 1$. Then $T(q_0) < 0$. By the claim (variational principle for T), we have $t_{\mu_q}(q_0) \le T(q_0)$. Since $t_{\mu_q}(0) = \frac{\mathrm{h}_{\mu_q}(f)}{\chi_{\mu_q}(f)} \ge 0$, we get

$$-q_0\frac{\mu_q(\varphi)}{\chi_{\mu_q}(f)} = t_{\mu_q}(0) - t_{\mu_q}(q_0) \ge 0 - T(q_0) = |T(q_0)|.$$

Hence, $\frac{-\mu_q(\varphi)}{\chi_{\mu_q}(f)} \ge |T(q_0)|/q_0 > 0$ for all $q \in \mathbb{R}$.

To end the proof of (d), we need to **(d.1)** derive the formula for F_{μ_φ} at $-T'(\pm\infty)$ when T is not affine and **(d.2)** prove that $X_{\mu_\varphi}(\alpha) = \emptyset$ for each $\alpha \notin [-T'(\infty), -T'(-\infty)]$. First, note the following.

(d*) For any ergodic f-invariant Borel probability measure ν on X, there exists $q \in \mathbb{R} \cup \{\pm\infty\}$ such that

$$\frac{\nu(\varphi)}{\chi_\nu(f)} = \frac{\mu_q(\varphi)}{\chi_{\mu_q}(f)}, \tag{16.97}$$

with $\lim_{q\to\pm\infty}$ in the $\pm\infty$ cases.

Indeed, by the claim (variational principle for T) the graphs of the functions $t_\nu(q)$ and $T(q)$ do not intersect transversally; they can only be tangent when/if they intersect. Hence, the graph of the linear function $t_\nu(q)$ coincides with the tangent to the graph of T at some point $(q_0, T(q_0))$, or with one of the asymptotes of T at $-\infty$ or ∞. Now (16.97) follows from (16.94) for $\nu = \mu_{q_0}$, since the graph of $t_{\mu_{q_0}}$ is tangent to the graph of T at $(q_0, T(q_0))$. (Note that the previous sentence proves the formula $T'(q) = \frac{\mu_q(\varphi)}{\chi_{\mu_q}(f)}$ in a different way than in (b), namely via the variational principle for T.)

(d.2) Proof that $X_{\mu_\varphi}(\alpha) = \emptyset$ for $\alpha \notin [-T'(\infty), -T'(-\infty)]$. Suppose there exists $x \in X$ with $\alpha := d_{\mu_\varphi}(x) \notin [-T'(\infty), -T'(-\infty)]$. Consider any sequence of integers $n_k \to \infty$ and real numbers b_1, b_2 such that (cf. Lemma 16.2.12)

$$\lim_{k\to\infty}\frac{1}{n_k}S_{n_k}\varphi(x) = b_1, \quad \lim_{k\to\infty} -\frac{1}{n_k}\log|(f^{n_k})'(x)| = b_2, \quad \text{and} \quad b_1/b_2 = \alpha.$$

Let τ be any weak*-limit of the sequence of measures

$$\tau_{n_k} := \frac{1}{n_k}\sum_{j=0}^{n_k-1}\delta_{f^j(x)},$$

where $\delta_{f^j(x)}$ is the Dirac measure supported at $f^j(x)$. Straightforward calculations show that

$$\tau(\varphi) := \int_X \varphi\, d\tau = b_1 \quad \text{and} \quad -\chi_\tau(f) := -\int_X \log|f'|\, d\tau = b_2.$$

Therefore, $\frac{\tau(\varphi)}{-\chi_\tau(f)} = b_1/b_2 = \alpha.$

Due to Choquet's representation theorem or the ergodic decomposition theorem (Theorem 8.2.26), we can assume that τ is ergodic. Indeed, τ is an "average" of ergodic measures. So among the ergodic measures ν comprised in that "average", there are ν_1 such that $\frac{\nu_1(\varphi)}{-\chi_{\nu_1}(f)} \le \alpha$ and ν_2 such that $\frac{\nu_2(\varphi)}{-\chi_{\nu_2}(f)} \ge \alpha$. If $\alpha < -T'(\infty)$, take ν_1 as the ergodic τ; if $\alpha > -T'(-\infty)$, pick ν_2 instead. For the ergodic τ found in this way, the limit α can be different than for the original τ, but it will not belong to $[-T'(\infty), -T'(-\infty)]$ and we shall use the same symbol α to denote it. Applying Lemma 16.6.3, we get

$$\alpha = d_{\mu_\varphi}(x) = \frac{\tau(\varphi)}{-\chi_\tau(f)} \quad \text{for } \tau\text{-a. e. } x \in X.$$

But by (16.97) there then exists $q \in \mathbb{R}$ such that $\frac{\mu_q(\varphi)}{-\chi_{\mu_q}(f)} = \frac{\tau(\varphi)}{-\chi_\tau(f)} = \alpha$, whence $\alpha \in [-T'(\infty), -T'(-\infty)]$. This contradiction finishes the proof of (d.2).

Remark. In fact, we proved that for all $x \in X$ any limit number of the quotients $\log \mu_\varphi(B(x,r))/\log r$ as $r \to 0$ lies in $[-T'(\infty), -T'(-\infty)]$, a statement stronger than $d_{\mu_\varphi}(x) \in [-T'(\infty), -T'(-\infty)]$ for all x in the regular part of X.

(d.1) On $F_{\mu_\varphi}(-T'(\pm\infty))$. Consider any weak*-limit τ of a subsequence of measures μ_q as $q \to \infty$ (a similar argument holds when $q \to -\infty$). We shall try to proceed with τ just like we did with μ_q, though we will meet some difficulties. In particular, we do not know whether τ is ergodic and choosing an ergodic one from its ergodic decomposition might result in the loss of the convergence $\mu_q \to \tau$. Nevertheless, using Birkhoff's ergodic theorem (Theorem 8.2.11) and proceeding as in the proof of Lemma 16.6.3, we get

$$\begin{aligned}\frac{\int_X \lim_{n\to\infty} \frac{1}{n} S_n\varphi(x)\, d\tau(x)}{-\int_X \lim_{n\to\infty} \frac{1}{n} \log |(f^n)'(x)|\, d\tau(x)} &= \frac{\int \varphi\, d\tau}{-\int \log |f'|\, d\tau} = \lim_{q\to\infty} \frac{\int \varphi\, d\mu_q}{-\int \log |f'|\, d\mu_q} \\ &= \lim_{q\to\infty} (-T'(q)) = -T'(\infty),\end{aligned}$$

with the convergence over a subsequence of q's. Since we know by (d.2) that

$$d_{\mu_\varphi}(x) = \frac{\lim_{n\to\infty} \frac{1}{n} S_n\varphi(x)}{-\lim_{n\to\infty} \frac{1}{n} \log |(f^n)'(x)|} \ge -T'(\infty),$$

there must be a set X_τ of full τ-measure over which $d_{\mu_\varphi}(x) = -T'(\infty)$ for all $x \in X_\tau$. We conclude that $X_\tau \subseteq X_{\mu_\varphi}(-T'(\infty))$ and hence $\mathrm{HD}(X_{\mu_\varphi}(-T'(\infty))) \ge \mathrm{HD}(X_\tau) \ge \mathrm{HD}(\tau)$.

Now we use the volume lemma with the measure τ. There is no reason for τ to be Gibbs or ergodic, so we rely on the version of the volume lemma in Theorem 16.5.10.

Moreover, thanks to the expansiveness of the repeller, we can use the upper semicontinuity of the measure-theoretic entropy function $\nu \mapsto \mathrm{h}_\nu(f)$ at τ (see Theorem 12.2.5). We then obtain

$$\mathrm{HD}(X_{\mu_\varphi}(-T'(\infty))) \geq \mathrm{HD}(\tau) \geq \frac{\mathrm{h}_\tau(f)}{\chi_\tau(f)} \geq \limsup_{q\to\infty} \frac{\mathrm{h}_{\mu_q}(f)}{\chi_{\mu_q}(f)} = \lim_{q\to\infty}(T(q)-qT'(q)) = F_{\mu_\varphi}(-T'(\infty)).$$

It only remains to estimate $\mathrm{HD}(X_{\mu_\varphi}(-T'(\infty)))$ from above. By (16.90), for every $q \in \mathbb{R}$ and $x \in X_{\mu_\varphi}(-T'(\infty))$ we have that

$$\begin{aligned} d_{\mu_q}(x) &:= \lim_{r\to 0} \frac{\log \mu_q(B(x,r))}{\log r} = T(q) + q \lim_{r\to 0} \frac{\log \mu_\varphi(B(x,r))}{\log r} = T(q) + q d_{\mu_\varphi}(x) \\ &= T(q) - qT'(\infty) \leq T(q) - qT'(q). \end{aligned}$$

Thus, $\mathrm{HD}(X_{\mu_\varphi}(-T'(\infty))) \leq T(q) - qT'(q)$. Letting $q \to \infty$, we get $\mathrm{HD}(X_{\mu_\varphi}(-T'(\infty))) \leq F_{\mu_\varphi}(-T'(\infty))$.

(e) HP and Rényi spectra. Recall that μ_φ and μ_q both have X for topological support, since these measures are Gibbs states and thus do not vanish on open subsets of X (for instance, see Proposition 13.6.14 and recall that μ_φ (resp., μ_q) is equivalent to m_φ (resp., m_q)). For every finite or countable cover $\mathcal{B}_r$ of X by balls of radius r and of multiplicity at most C, we have

$$1 \leq \sum_{B\in\mathcal{B}_r} \mu_q(B) \leq C.$$

By Lemma 16.6.2,

$$(C_q C)^{-1} \leq r^{T(q)} \sum_{B\in\mathcal{B}_r} (\mu_\varphi(B))^q \leq C_q C.$$

Assume $q \neq 1$. Taking logarithms and dividing by $(1-q)\log r$ yields (e) for $q \neq 1$.

Information dimension. For $q = 1$, we have $\lim_{q\to 1, q\neq 1} \frac{T(q)}{1-q} = -T'(1)$ by definition of the derivative. It is equal to $\mathrm{HD}(\mu_\varphi) = \mathrm{PD}(\mu_\varphi)$ by (a) and (b), and equal to $I(\mu_\varphi)$ by Exercise 16.7.19. This is (e) when $q = 1$. □

16.7 Exercises

Exercise 16.7.1. Prove that a C^1 local diffeomorphism $T : U \to \mathbb{R}^d$ is conformal if and only if at every point $x \in U$ its derivative $T'(x) : \mathbb{R}^d \to \mathbb{R}^d$ is a similarity map without a translation component, i. e., is of the form rA, where $r > 0$ and A is an orthogonal matrix.

Exercise 16.7.2. By providing a specific example, show that if condition (2) is not satisfied in Example 16.2.2, then the limit set X may be empty.

Exercise 16.7.3. By providing an example, show that if condition (2) is not satisfied in Example 16.2.2, then the limit set X may not be compact.

Exercise 16.7.4. Construct an IFS that satisfies condition (2) but not condition (1) in the SSC (cf. Example 16.2.2) and for which Hutchinson's formula does not hold.

Exercise 16.7.5. By applying Hutchinson's formula to the conformal expanding repeller from Example 16.3.3, show that the Hausdorff dimension of the middle-third Cantor set is $\log 2/\log 3$.

Exercise 16.7.6. Prove Remark 16.2.13.

Exercise 16.7.7. Fix $d \in \mathbb{N}$. Let $\mathcal{K}_d$ be the space of nonempty compact subsets of the unit cube $[0,1]^d$. In some contrast to Theorem 16.4.11, show that the function

$$\mathcal{K}_d \ni K \longmapsto \mathrm{HD}(K) \in [0,d]$$

has no continuity point.

Exercise 16.7.8. Let $T : X \to X$ be a conformal expanding repeller and $\varphi : X \to \mathbb{R}$ a Hölder continuous potential. Prove that the following three conditions are equivalent:

(a) $\mathrm{HD}(\mu_\varphi) = \mathrm{HD}(X)$.

(b) $\mu_\varphi = \mu_{-\mathrm{HD}(X)\log|T'|}$.

(c) The potentials φ and $-\mathrm{HD}(X)\log|T'|$ are cohomologous modulo a constant in the class of continuous (equivalently, Hölder continuous) functions from X to $\mathbb{R}$.

Exercise 16.7.9. Strengthening Exercise 16.7.8, show that if $T : X \to X$ is a conformal expanding repeller and $\mu \in M(T)$, then $\mathrm{HD}(\mu) = \mathrm{HD}(X)$ if and only if $\mu = \mu_{-\mathrm{HD}(X)\log|T'|}$.

Exercise 16.7.10. Prove that if $T : X \to X$ is a conformal expanding repeller, then the function $H_\alpha(X,\mathbb{R}) \ni \varphi \longmapsto \mathrm{HD}(\mu_\varphi)$ is continuous.

Exercise 16.7.11. Show that if $T : X \to X$ is a conformal expanding repeller and $\varphi : X \to \mathbb{R}$ is a Hölder potential, then the function $\mathbb{R} \ni t \longmapsto \mathrm{HD}(\mu_{t\varphi})$ is real analytic.

Exercise 16.7.12. Under the hypotheses of Theorem 16.4.11, suppose that $\varphi : U \to \mathbb{R}$ is a Hölder continuous function. Prove that the map

$$\widehat{\Lambda} \ni \lambda \longmapsto \mathrm{HD}(\mu_{\varphi|_{J_\lambda}})$$

is real analytic.

Exercise 16.7.13. Prove the equality of the Rényi and Hentschel–Procaccia spectra.

Exercise 16.7.14. Show that Lemma 16.6.3 is a generalization of the volume lemma (Theorem 16.5.2) for the pointwise dimension of the measure μ_φ.

Exercise 16.7.15. Show that if φ is not cohomologous to $-\mathrm{HD}(X)\log|T'|$ then the singular part $X^c_{\mu_\varphi}$ of X is nonempty. Moreover, $\mathrm{HD}(X^c_{\mu_\varphi}) = \mathrm{HD}(X)$.

Hint: Using the shadowing lemma (Corollary 4.3.6), find trajectories that have blocks close to blocks of trajectories typical for $\mu_{-\mathrm{HD}(X)\log|T'|}$ of length N interchanging with blocks close to blocks typical for μ_φ of length εN, for N arbitrarily large and $\varepsilon > 0$ arbitrarily small.

Exercise 16.7.16. Prove Proposition 16.6.4 about Legendre transform pairs and the remarks preceding and following it.

Exercise 16.7.17. For $\alpha = -T'(1)$, show that $F_{\mu_\varphi}(\alpha) = \alpha$, $F'_{\mu_\varphi}(\alpha) = 1$, and that $F'_{\mu_\varphi}(-T'(\pm\infty)) = \pm\infty$. See Figure 16.2.

Exercise 16.7.18. Prove (16.87).

Exercise 16.7.19. Define the *lower and upper information dimensions* $\underline{I}(\nu)$ and $\overline{I}(\nu)$ by replacing in the definition of $I(\nu)$ (see (16.79)) the limit by the liminf and limsup, respectively. Prove that $\mathrm{HD}_*(\nu) \le \underline{I}(\nu) \le \overline{I}(\nu) \le \mathrm{PD}_*(\nu)$.

Exercise 16.7.20. Theorem 16.6.5 applies to infinite conformal expanding repellers. Write a corresponding theorem for finite repellers.

17 Countable state thermodynamic formalism

In Chapter 3, we discussed symbolic dynamics over a finite alphabet. In Chapters 11 and 12, we developed thermodynamic formalism for topological dynamical systems, i. e., for self-maps of a compact metrizable phase space, subject to (Hölder) continuous potentials. In Chapter 13, we described more specific formalism for open, distance expanding maps. In this chapter, we present thermodynamic formalism for countable-alphabet subshifts of finite type E_A^∞ generated by finitely irreducible transition matrices $A : E \times E \to \{0, 1\}$ on a countable alphabet E. If the alphabet E is finite, then the symbolic space E_A^∞ is compact, and thus all the knowledge acquired in Chapters 11–13 about pressure, variational principle, equilibrium states, Gibbs states and transfer operators, applies to the shift map $\sigma : E_A^\infty \to E_A^\infty$. If the alphabet E is infinite and the incidence matrix A is finitely irreducible, then the phase space E_A^∞ is not compact and essentially none of the results proved in the compact case holds directly: thermodynamic formalism in the case of a countably infinite alphabet must be developed anew, although its exposition can be guided and motivated by the compact case to a certain extent. The compact case may also provide valuable hints for some proofs in the non-compact case but there are many important issues specific to the infinite-alphabet case that have no analogue in the finite-alphabet case and must be dealt with entirely on their own.

Thermodynamic formalism for countable state subshifts of finite type originated in the papers [43, 44]. However, the incidence matrix A was therein assumed to be such that the resulting symbolic space E_A^∞ is locally compact. This permitted the performance of a one-point (Alexandroff) compactification of the space E_A^∞ and the use of the methods developed for compact dynamical systems. Finite irreducibility of A, which is the underlying assumption in this and the next chapter, rules out local compactness of E_A^∞ and appropriate thermodynamic formalism, which will be unveiled in those two chapters, must be developed independently. Our exposition stems from [77] and [81]. It expands these two sources, in particular improving some results and/or their proofs, or providing alternative arguments.

The primary goal of this chapter is to introduce the appropriate concept of topological pressure and to prove the existence and uniqueness of equilibrium states and invariant Gibbs states for potentials that are summable and Hölder continuous on cylinders. The definitions of pressure, equilibrium state and Gibbs state reduce to those in [11] in the case of a finite alphabet. These definitions allow us to outline a rich thermodynamic formalism that serves well in the geometric applications detailed in Chapters 19–21 about countable-alphabet conformal graph directed Markov systems and their special subclass of conformal iterated function systems.

In Section 17.1, we give the basic definitions of finite irreducibility and finite primitivity. Unless otherwise stated, the results mentioned in this paragraph assume that the subshift is finitely irreducible. In Section 17.2, we define the pressure. In order to speak of equilibrium states, one needs a variational principle for the pressure and we

https://doi.org/10.1515/9783110702699-017

prove three such principles in Section 17.3. In Section 17.4, we study Gibbs states and prove the existence of a unique invariant Gibbs state. This state is ergodic, and even totally ergodic when the subshift is finitely primitive. In Section 17.5, we identify a simple condition under which the unique invariant Gibbs state is an equilibrium state. Moreover, we show that equilibrium states are unique. Consequently, we deduce that invariant Gibbs states and equilibrium states exist, are unique, and coincide. In Section 17.6, we examine the transfer operator (also called Perron–Frobenius operator) and we demonstrate that its dual admits at most one eigenmeasure, which turns out to be a (nonnecessarily invariant) Gibbs state. In Section 17.7, the existence of such an eigenmeasure is established. So the dual of the transfer operator has a unique eigenmeasure. Its eigenvalue is the exponential of the pressure. The aforementioned facts are summarized in Corollary 17.7.5. Finally, in Section 17.8 we confirm the presence of a pressure gap and, consequently, the positivity of the measure-theoretic entropy of the subshift with respect to the unique equilibrium state.

There are two more aspects we would like to emphasize. First, unlike the case of a finite alphabet, in the realm of countable-alphabet subshifts of finite type there are two disjoint classes of Hölder continuous functions that are involved with the transfer operator: those that are summable (and thus are necessarily unbounded from below) and those that are bounded. The former class serves as the source of potentials for Gibbs and equilibrium states while the latter is the Banach space on which the transfer operator acts. Secondly, the method for proving the existence of an eigenmeasure for the transfer operator is to truncate the subshift to larger and larger finite subsets of the alphabet. The transfer operators associated with the finite-subalphabet subshifts each have an eigenmeasure according to Chapter 13. The resulting sequence of Borel probability measures is tight. Prohorov's theorem then affirms that it has a weak* limit point. This limit point is the eigenmeasure of the transfer operator.

17.1 Finitely irreducible subshifts

Let E be a countable alphabet and let $\sigma : E^\infty \to E^\infty$ be the shift map, i. e., the map cutting off the first coordinate from any word. It is given by the formula

$$\sigma\big((\omega_n)_{n=1}^\infty\big) := (\omega_{n+1})_{n=1}^\infty.$$

We let $E^0 := \{\epsilon\}$, where ϵ is the empty word, and the set of finite words

$$E^* := \bigcup_{n=0}^{\infty} E^n.$$

For every $\omega \in E^*$, we denote by $|\omega|$ the only integer $n \geq 0$ such that $\omega \in E^n$. We call $|\omega|$ the length of ω. If $\omega \in E^\infty$ and $n \in \mathbb{N}$, we set

$$\omega|_n := \omega_1\omega_2\cdots\omega_n \in E^n.$$

Given $\omega, \tau \in E^\infty$, we define $\omega \wedge \tau \in E^\infty \cup E^*$ to be the longest initial block common to ω and τ. Two words $\omega, \tau \in E^\infty \cup E^*$ are said to be *comparable* if one of them is the extension of the other, i. e., if $|\omega \wedge \tau| = \min\{|\omega|, |\tau|\}$. Otherwise, they are *incomparable*.

For each $0 < s < 1$, we define a metric d_s on E^∞ by setting

$$d_s(\omega, \tau) := s^{|\omega \wedge \tau|}.$$

These metrics are all topologically equivalent, and so induce the same topology and hence the same Borel sets on E^∞. A function is uniformly continuous with respect to one of these metrics if and only if it is uniformly continuous with respect to all. If no metric is specifically mentioned, we take it to be

$$d(\omega, \tau) := d_{e^{-1}}(\omega, \tau) = e^{-|\omega \wedge \tau|}.$$

A matrix $A : E \times E \to \{0, 1\}$ will act as an incidence matrix, i. e., it will dictate which letter(s) may follow any given letter. By definition, letter $e_2 \in E$ can follow letter $e_1 \in E$ precisely when $A_{e_1 e_2} = 1$. Accordingly, we set

$$E_A^\infty := \{\omega \in E^\infty \mid A_{\omega_j \omega_{j+1}} = 1, \forall j \in \mathbb{N}\}.$$

The elements of E_A^∞ are called *infinite A-admissible words*. For each $n \geq 0$, let

$$E_A^n := \{\omega \in E^n \mid A_{\omega_j \omega_{j+1}} = 1, \forall 1 \leq j < n\} \quad \text{and set} \quad E_A^* := \bigcup_{n=0}^{\infty} E_A^n.$$

The elements of E_A^n are called *A-admissible n-letter words* or more simply *n-words*, while the elements of E_A^* are called *finite A-admissible words*. For every $\omega \in E_A^*$, let

$$[\omega] = \{\tau \in E_A^\infty : \tau|_{|\omega|} = \omega\}.$$

These sets are called *initial A-cylinders*. They are both the open and the closed balls in E_A^∞, and thus they form a countable base for the topology on E_A^∞. Let also

$$E_A^{\infty,\omega} := \{\gamma \in E_A^\infty \mid \omega\gamma \in E_A^\infty\} = \bigcup_{e \in E:\, A_{\omega_{|\omega|} e} = 1} [e]$$

be the set of all infinite A-admissible suffixes of ω, whereas

$$E_A^{*,\omega} := \{\gamma \in E_A^* \mid \omega\gamma \in E_A^*\} = \{\gamma \in E_A^* \mid A_{\omega_{|\omega|}\gamma_1} = 1\}$$

will denote the set of all finite A-admissible suffixes of ω and

$$E_A^{n,\omega} := \{\gamma \in E_A^n \mid \omega\gamma \in E_A^*\} = \{\gamma \in E_A^n \mid A_{\omega_{|\omega|}\gamma_1} = 1\}$$

will be the set of all A-admissible n-letter suffixes of ω.

The following fact is obvious.

Proposition 17.1.1. *E_A^∞ is a closed invariant subset of E^∞, and hence $\sigma : E_A^\infty \to E_A^\infty$ is a subshift of the full E-shift $\sigma : E^\infty \to E^\infty$.*

The subshifts we will restrict our attention to are generated by a finitely irreducible matrix. They are a generalization of irreducible subshifts over finite alphabets.

Definition 17.1.2. The matrix $A : E \times E \to \{0,1\}$ is said to be *finitely irreducible* if there is a finite set $\Lambda \subseteq E_A^*$ such that for all $e_1, e_2 \in E$ there exists a word $\omega = \omega(e_1, e_2) \in \Lambda$ such that $e_1\omega e_2 \in E_A^*$.

It will sometimes be helpful to replace the original set Λ by a finite set of words of a specific minimal length.

Lemma 17.1.3. *Suppose that a matrix $A : E \times E \to \{0,1\}$ is finitely irreducible and let $\Lambda \subseteq E_A^*$ be a finite set which witnesses the finite irreducibility of A. For any $L \in \mathbb{N}$, there exists a finite set $\Lambda_L \subseteq E_A^*$ which witnesses the finite irreducibility of A and whose words have length at least L.*

Proof. Let $L \in \mathbb{N}$. Recall that ϵ is the empty word, the only word of length 0. If $\Lambda = \{\epsilon\}$, then all transitions are allowed and one can thus set $\Lambda = \{e^L\}$ for any $e \in E$. Assume now that $\Lambda \supseteq \{\epsilon, \eta\}$ for some $\eta \in E_A^* \setminus \{\epsilon\}$. Consider

$$\widetilde{\Lambda} = \{\alpha\eta\omega \in E_A^* : \alpha, \omega \in \Lambda\}.$$

Then $\#\widetilde{\Lambda} \le (\#\Lambda)^2 < \infty$. Let $e_1, e_2 \in E$. As Λ witnesses the finite irreducibility of A, there exists $\alpha, \omega \in \Lambda$ such that $e_1\alpha\eta\omega e_2 \in E_A^*$. Moreover, $|\alpha\eta\omega| = |\alpha| + |\eta| + |\omega| \ge |\eta| \ge 1$. Since $e_1, e_2 \in E$ are arbitrary, it follows that $\widetilde{\Lambda}$ is a finite set of words of length at least 1 witnessing the finite irreducibility of A. Take

$$\Lambda' = (\widetilde{\Lambda})^L \cap E_A^*.$$

Clearly, $\#\Lambda' \le (\#\widetilde{\Lambda})^L < \infty$ and words in Λ' have length at least L. It remains to show that this set witnesses the finite irreducibility of A. Let $e_1, e_2 \in E$. There exists $\alpha_1 \in \widetilde{\Lambda}$ such that $e_1\alpha_1 \in E_A^*$. In turn, there is $\alpha_2 \in \widetilde{\Lambda}$ such that $\alpha_1\alpha_2 \in E_A^*$. Continuing in this way, one can find $\alpha_{k+1} \in \widetilde{\Lambda}$ such that $\alpha_k\alpha_{k+1} \in E_A^*$, and this for every $1 \le k < L-1$. Once α_{L-1} has been found, there is $\alpha_L \in \widetilde{\Lambda}$ such that $\alpha_{L-1}\alpha_L e_2 \in E_A^*$. Then the word $\alpha' = \alpha_1\alpha_2\ldots\alpha_L \in \Lambda'$ is such that $e_1\alpha' e_2 \in E_A^*$. Since $e_1, e_2 \in E$ are arbitrary, the set Λ' witnesses the finite irreducibility of A. □

17.1.1 Finitely primitive subshifts

At times, we will restrict our attention to subshifts that are generated by a finitely primitive matrix. This is a stronger requirement than finite irreducibility.

Definition 17.1.4. The matrix $A : E \times E \to \{0, 1\}$ is said to be *finitely primitive* if there is a finite set $\Omega \subseteq E_A^*$ of words of equal length such that for all $e_1, e_2 \in E$ there is a word $\omega = \omega(e_1, e_2) \in \Omega$ for which $e_1 \omega e_2 \in E_A^*$.

17.2 Topological pressure

We will concentrate our attention on a family of potentials that are natural generalizations of Hölder continuous potentials on compact spaces.

17.2.1 Potentials: acceptability and Hölder continuity on cylinders

Let us recall the notion of Birkhoff (or ergodic) sum, this time in the context of symbolic systems. Given a potential $f : E_A^\infty \to \mathbb{R}$, the nth Birkhoff sum of f at a word $\tau \in E_A^\infty$ is

$$S_n f(\tau) = \sum_{j=0}^{n-1} f(\sigma^j(\tau)).$$

This is the sum of the values of potential f at the first n points in the forward orbit of τ under the shift map $\sigma : E_A^\infty \to E_A^\infty$.

For every $G \subseteq E_A^\infty$, let

$$\overline{S}_n f(G) := \sup_{\tau \in G} S_n f(\tau).$$

We shall be interested in one special class of potentials.

Definition 17.2.1. A potential $f : E_A^\infty \to \mathbb{R}$ is said to be *acceptable* provided that it is uniformly continuous and

$$\operatorname{osc}(f) := \operatorname{osc}(f, \mathcal{U}_E) = \sup_{e \in E} [\sup(f|_{[e]}) - \inf(f|_{[e]})] < \infty.$$

That is, a potential is acceptable if it is uniformly continuous and has finite oscillation over the partition $\mathcal{U}_E := \{[e] : e \in E\}$ of E_A^∞ into its initial 1-cylinders. Note that an acceptable potential need not be bounded when E is infinite.

A particularly interesting subclass of acceptable potentials consists of those which are Hölder continuous on cylinders.

Definition 17.2.2. A function $f : E_A^\infty \to \mathbb{R}$ is said to be *Hölder continuous on cylinders* if there exist constants $\beta > 0$ and $c \geq 0$ for which

$$|\rho \wedge \zeta| \geq 1 \Longrightarrow |f(\rho) - f(\zeta)| \leq c[d(\rho, \zeta)]^\beta.$$

The constant β is called an *Hölder exponent* of f and the least constant c with this property will be denoted by $v_\beta(f)$.

Note that a function is Hölder continuous on cylinders with respect to the metric $d(\rho,\zeta) = e^{-|\rho\wedge\zeta|}$ if and only if it is Hölder continuous on cylinders with respect to all metrics $d_s(\rho,\zeta) = s^{|\rho\wedge\zeta|}$, $0 < s < 1$. Of course, the Hölder exponent β depends on the metric d_s.

Functions that are Hölder continuous on cylinders are obviously bounded on every cylinder but they need not be (globally) bounded when E is infinite. In fact, Hölder continuity on cylinders and (global) boundedness are, together, equivalent to global Hölder continuity.

The ergodic sums of functions that are Hölder continuous on cylinders have bounded variation on cylinders (cf. Lemma 13.1.1). This flexibility is crucial when E is infinite.

Lemma 17.2.3 (Bounded variation principle for ergodic sums). *If a potential $f : E_A^\infty \to \mathbb{R}$ is Hölder continuous on cylinders with exponent β, then for all $n \in \mathbb{N}$, all $\omega \in E_A^n$, and all $\rho,\gamma \in E_A^{\infty,\omega}$, we have*

$$|S_nf(\omega\rho) - S_nf(\omega\gamma)| \le \frac{v_\beta(f)}{e^\beta - 1}[d(\rho,\gamma)]^\beta = \frac{v_\beta(f)}{e^\beta - 1}e^{-\beta|\rho\wedge\gamma|}.$$

In particular, given $\xi,\zeta \in E_A^\infty$,

$$|\xi\wedge\zeta| \ge n \Longrightarrow |S_nf(\xi) - S_nf(\zeta)| \le \frac{v_\beta(f)}{e^\beta - 1}.$$

Consequently, for any $\tau\omega \in E_A^*$,

$$\overline{S}_{|\tau|}f([\tau]) + \overline{S}_{|\omega|}f([\omega]) - \frac{v_\beta(f)}{e^\beta - 1} \le \overline{S}_{|\tau\omega|}f([\tau\omega]) \le \overline{S}_{|\tau|}f([\tau]) + \overline{S}_{|\omega|}f([\omega]).$$

Proof. Let $n \in \mathbb{N}$, $\omega \in E_A^n$ and $\rho,\gamma \in E_A^{\infty,\omega}$. Then

$$\begin{aligned}|S_nf(\omega\rho) - S_nf(\omega\gamma)| &\le \sum_{j=0}^{n-1}|f(\sigma^j(\omega\rho)) - f(\sigma^j(\omega\gamma))| \\ &\le \sum_{j=0}^{n-1} v_\beta(f)[d(\sigma^j(\omega\rho),\sigma^j(\omega\gamma))]^\beta \\ &= v_\beta(f)\sum_{j=0}^{n-1} e^{-\beta(n-j)}[d(\rho,\gamma)]^\beta \\ &\le \frac{v_\beta(f)}{e^\beta - 1}[d(\rho,\gamma)]^\beta \le \frac{v_\beta(f)}{e^\beta - 1}.\end{aligned}$$

For any $\tau\omega \in E_A^*$, this implies that

$$\begin{aligned}
\overline{S}_{|\tau\omega|}f([\tau\omega]) &= \sup_{\xi\in[\tau\omega]} S_{|\tau|+|\omega|}f(\xi) = \sup_{\xi\in[\tau\omega]} [S_{|\tau|}f(\xi) + S_{|\omega|}f(\sigma^{|\tau|}(\xi))] \\
&\geq \inf_{\xi\in[\tau\omega]} S_{|\tau|}f(\xi) + \sup_{\xi\in[\tau\omega]} S_{|\omega|}f(\sigma^{|\tau|}(\xi)) \\
&\geq \inf_{\zeta\in[\tau]} S_{|\tau|}f(\zeta) + \sup_{\chi\in[\omega]} S_{|\omega|}f(\chi) \\
&\geq \sup_{\zeta\in[\tau]} S_{|\tau|}f(\zeta) - \frac{v_\beta(f)}{e^\beta - 1} + \sup_{\chi\in[\omega]} S_{|\omega|}f(\chi) \\
&= \overline{S}_{|\tau|}f([\tau]) + \overline{S}_{|\omega|}f([\omega]) - \frac{v_\beta(f)}{e^\beta - 1}.
\end{aligned}$$

On the other hand,

$$\begin{aligned}
\overline{S}_{|\tau\omega|}f([\tau\omega]) &= \sup_{\xi\in[\tau\omega]} [S_{|\tau|}f(\xi) + S_{|\omega|}f(\sigma^{|\tau|}(\xi))] \\
&\leq \sup_{\xi\in[\tau\omega]} S_{|\tau|}f(\xi) + \sup_{\xi\in[\tau\omega]} S_{|\omega|}f(\sigma^{|\tau|}(\xi)) \\
&\leq \sup_{\zeta\in[\tau]} S_{|\tau|}f(\zeta) + \sup_{\chi\in[\omega]} S_{|\omega|}f(\chi) \\
&= \overline{S}_{|\tau|}f([\tau]) + \overline{S}_{|\omega|}f([\omega]).
\end{aligned}$$

□

17.2.2 Partition functions

Subalphabets will later play an important role since some properties of infinite systems can be derived from the corresponding properties of their finite subsystems. Given a subalphabet $F \subseteq E$, let

$$F^\infty = \{\omega \in E^\infty : \omega_i \in F, \forall i \in \mathbb{N}\} \quad \text{and} \quad F_A^\infty = F^\infty \cap E_A^\infty.$$

Similarly, let

$$F_A^n = F^n \cap E_A^n \quad \text{and} \quad F_A^* = F^* \cap E_A^*.$$

Then $\sigma_F := \sigma|_{F_A^\infty} : F_A^\infty \to F_A^\infty$ is a subshift of $\sigma : E_A^\infty \to E_A^\infty$. For any $\omega \in F_A^*$, let $[\omega]_F = [\omega] \cap F_A^\infty$. Notice that $\sigma_E = \sigma$ and $[\omega]_E = [\omega]$.

Note also that if $A : E \times E \to \{0,1\}$ is finitely irreducible and $\Lambda \subseteq E_A^*$ is a finite set witnessing the finite irreducibility of A, then $A : F \times F \to \{0,1\}$ is finitely irreducible when $\Lambda_E \subseteq F$, where $\Lambda_E = \{\omega_k : \omega \in \Lambda \text{ and } 1 \leq k \leq |\omega|\}$.

If $\mathcal{U}_F = \{[e]_F : e \in F\}$ is the open partition (and hence an open cover) of F_A^∞ into its initial 1-cylinders, then $\mathcal{U}_F^n := \bigvee_{j=0}^{n-1} \sigma_F^{-j}(\mathcal{U}_F) = \{[\omega]_F : \omega \in F_A^n\}$ is the partition of F_A^∞ into its initial n-cylinders. Definition 11.1.2 states that the nth partition function of $\mathcal{U}_F$ (or to lighten terminology and notation, the nth partition function of F) with respect to a potential f is

$$Z_n(f,F) := Z_n(f,\mathcal{U}_F) = \sum_{\omega\in F_A^n} \exp(\overline{S}_nf([\omega]_F)).$$

Remark 17.2.4.

(a) By convention, $\overline{S}_{|\omega|}f([\omega]_F) = -\infty$ when $[\omega]_F = \emptyset$, and $e^{-\infty} = 0$. This is justified by the expectation that $Z_n(f,F)$ be the number of elements in the cover $\mathcal{U}_F^n$ when the potential $f \equiv 0$.

(b) $0 \le Z_n(f,F) \le \infty$ for all $n \in \mathbb{N}$.

(c) If $\#F < \infty$, then $Z_n(f,F) < \infty$ for all $n \in \mathbb{N}$.

(d) $\exists n \in \mathbb{N}, Z_n(f,F) = 0 \iff Z_n(f,F) = 0, \forall n \in \mathbb{N} \implies F_A^\infty = \emptyset$.

(e) If $F = E$, we suppress the subscript and write simply $Z_n(f)$ for $Z_n(f,E)$. That is,

$$Z_n(f) = \sum_{\omega\in E_A^n} \exp(\overline{S}_nf([\omega])). \tag{17.1}$$

We now address the question of bounded sub- and supermultiplicativity of the partition functions. A sequence $(a_n)_{n=1}^\infty$ of positive real numbers is said to be *boundedly submultiplicative* with constant $B > 0$ if

$$a_{m+n} \le Ba_ma_n, \quad \forall m,n \in \mathbb{N}.$$

If $0 < B \le 1$, then the sequence is *submultiplicative*. Similarly, a sequence $(a_n)_{n=1}^\infty$ of positive real numbers is *boundedly supermultiplicative* with constant $C > 0$ if

$$C^{-1}a_ma_n \le a_{m+n}, \quad \forall m,n \in \mathbb{N}.$$

If $0 < C \le 1$, then the sequence is *supermultiplicative*.

Lemma 17.2.5. *For any $F \subseteq E$, the sequence $(Z_n(f,F))_{n=1}^\infty$ is submultiplicative. Moreover, if $f : E_A^\infty \to \mathbb{R}$ is acceptable and A is finitely irreducible, then there exists a constant $R > 0$ such that $Z_{n+1}(f) \ge RZ_n(f)$ for all $n \in \mathbb{N}$. In particular,*

$$Z_1(f) = \infty \iff \exists n \in \mathbb{N}, \quad Z_n(f) = \infty \iff Z_n(f) = \infty, \quad \forall n \in \mathbb{N}.$$

If, in addition, f is Hölder continuous on cylinders, then the sequence $(Z_n(f))_{n=1}^\infty$ is boundedly supermultiplicative.

Proof. Let us first prove the submultiplicativity. We must show that $Z_{m+n}(f,F) \le Z_m(f,F)Z_n(f,F)$ for all $m,n \in \mathbb{N}$. And indeed,

$$\begin{aligned} Z_{m+n}(f,F) &= \sum_{\omega\in F_A^{m+n}} \exp(\overline{S}_{m+n}f([\omega]_F)) \\ &\le \sum_{\omega\in F_A^{m+n}} \exp(\overline{S}_mf([\omega]_F) + \overline{S}_nf([\sigma^m(\omega)]_F)) \end{aligned}$$

$$
\begin{aligned}
&\le \sum_{\chi\in F_A^m}\sum_{\rho\in F_A^n} \exp(\overline{S}_m f([\chi]_F) + \overline{S}_n f([\rho]_F))\\
&= \sum_{\chi\in F_A^m} \exp(\overline{S}_m f([\chi]_F)) \sum_{\rho\in F_A^n} \exp(\overline{S}_n f([\rho]_F))\\
&= Z_m(f,F)Z_n(f,F).
\end{aligned}
$$

Now, suppose that A is finitely irreducible, and that $\inf_{\zeta\in[e]} f(\zeta) > -\infty$ for all $e \in E$ (this is the case when f is acceptable). Let $\Lambda \subseteq E_A^*$ be a finite set which witnesses the finite irreducibility of A. Let also $\Lambda_E := \{\omega_k : \omega \in \Lambda \text{ and } 1 \le k \le |\omega|\}$. Set $m := \min\{\inf f|_{[e]} : e \in \Lambda_E\}$ and observe that $m > -\infty$. For every $\omega \in E_A^*$, let $e_\omega \in \Lambda_E$ be such that $e_\omega\omega \in E_A^*$. Then

$$
\begin{aligned}
Z_{n+1}(f) &= \sum_{\tau\in E_A^{n+1}} \exp(\overline{S}_{n+1} f([\tau]))\\
&\ge \sum_{\omega\in E_A^n} \exp(\overline{S}_{n+1} f([e_\omega\omega]))\\
&= \sum_{\omega\in E_A^n} \exp\Big(\sup_{\rho\in[e_\omega\omega]} (f(\rho) + S_n f(\sigma(\rho)))\Big)\\
&\ge \sum_{\omega\in E_A^n} \exp\Big(\inf_{\zeta\in[e_\omega\omega]} f(\zeta) + \sup_{\rho\in[e_\omega\omega]} S_n f(\sigma(\rho))\Big)\\
&\ge \sum_{\omega\in E_A^n} \exp\Big(\inf_{\zeta\in[e_\omega]} f(\zeta)\Big) \exp\Big(\sup_{\chi\in[\omega]} S_n f(\chi)\Big)\\
&\ge e^m \sum_{\omega\in E_A^n} \exp(S_n f([\omega]))\\
&= e^m Z_n(f).
\end{aligned}
$$

Together with submultiplicativity, this implies that $Z_1(f)e^{(n-1)m} \le Z_n(f) \le (Z_1(f))^n$ for all $n \in \mathbb{N}$. In particular, $Z_1(f) = \infty$ if and only if $Z_n(f) = \infty$ for some $n \in \mathbb{N}$ if and only if $Z_n(f) = \infty$ for all $n \in \mathbb{N}$.

Finally, we show that the partition functions of the system are boundedly supermultiplicative when the potential f is additionally Hölder continuous on cylinders. We must thus show that there exists $0 < Q < \infty$ such that

$$
Z_m(f)Z_n(f) \le QZ_{m+n}(f), \quad \forall m,n \in \mathbb{N}.
$$

Since $Z_1(f) = \infty$ if and only if $Z_n(f) = \infty$ for all $n \in \mathbb{N}$, bounded supermultiplicativity is trivial if $Z_1(f) = \infty$. So assume that $Z_1(f) < \infty$. Let $\Lambda \subseteq E_A^*$ be a finite set which makes the matrix A finitely irreducible. Set $|\Lambda| := \max\{|\omega| : \omega \in \Lambda\}$ and $M := \min\{\overline{S}_{|\omega|} f([\omega]) : \omega \in \Lambda\}$, and notice that $-\infty < M < \infty$ according to Lemma 17.2.3. For every couple $(e_1, e_2) \in E^2$, let $\omega(e_1, e_2) \in \Lambda$ be such that $e_1\omega(e_1, e_2)e_2 \in E_A^*$. More generally, for every couple $(\tau,\chi) \in (E_A^*\backslash\{\epsilon\})^2$ define $\omega(\tau,\chi) := \omega(\tau_{|\tau|},\chi_1)$. Fix $m, n \in \mathbb{N}$. The map

$$\begin{aligned} l : E_A^m \times E_A^n &\longrightarrow \bigcup_{k=0}^{|\Lambda|} E_A^{m+n+k} \\ (\tau,\chi) &\longmapsto \tau\omega(\tau,\chi)\chi \end{aligned}$$

is clearly injective. Using the principle of bounded variation (Lemma 17.2.3), we obtain

$$\begin{aligned} Z_m(f)Z_n(f) &= \sum_{\tau\in E_A^m} \exp(\overline{S}_m f([\tau])) \sum_{\chi\in E_A^n} \exp(\overline{S}_n f([\chi])) \\ &\le \frac{1}{e^M} \sum_{\tau\in E_A^m}\sum_{\chi\in E_A^n} \exp\Big(\overline{S}_m f([\tau]) + \overline{S}_{|\omega(\tau,\chi)|} f([\omega(\tau,\chi)]) + \overline{S}_n f([\chi])\Big) \\ &\le \frac{e^{2\frac{v_\beta(f)}{e^\beta-1}}}{e^M} \sum_{\tau\in E_A^m}\sum_{\chi\in E_A^n} \exp(\overline{S}_{|l(\tau,\chi)|} f([l(\tau,\chi)])) \\ &\le e^{2\frac{v_\beta(f)}{e^\beta-1}-M} \sum_{k=0}^{|\Lambda|} \sum_{\omega\in E_A^{m+n+k}} \exp(\overline{S}_{m+n+k} f([\omega])) \\ &= e^{2\frac{v_\beta(f)}{e^\beta-1}-M} \sum_{k=0}^{|\Lambda|} Z_{m+n+k}(f) \\ &\le e^{2\frac{v_\beta(f)}{e^\beta-1}-M} \sum_{k=0}^{|\Lambda|} Z_k(f)\cdot Z_{m+n}(f) = QZ_{m+n}(f), \end{aligned}$$

where $Q = e^{2\frac{v_\beta(f)}{e^\beta-1}-M} \sum_{k=0}^{|\Lambda|} Z_k(f) \le e^{2\frac{v_\beta(f)}{e^\beta-1}-M}(|\Lambda|+1)\max\{1,(Z_1(f))^{|\Lambda|}\}$. □

We now aim at proving that every partition function of the system coincides with the supremum of the corresponding partition functions of its finite subsystems for acceptable potentials.

Lemma 17.2.6. *If $f : E_A^\infty \to \mathbb{R}$ is acceptable and A is finitely irreducible, then*

$$Z_n(f) = \sup\{Z_n(f,F) : F \subseteq E \text{ and } \#F < \infty\}$$

for every $n \in \mathbb{N}$.

Proof. Fix $n \in \mathbb{N}$. The inequality $Z_n(f) \ge \sup\{Z_n(f,F)\}$ is obvious. To prove the converse, suppose first that $Z_n(f) < \infty$. Let $\varepsilon > 0$. There exists a finite set $\Omega = \Omega(\varepsilon) \subseteq E_A^n$ such that

$$Z_n(f) \le e^\varepsilon \sum_{\omega\in\Omega} \exp(\overline{S}_n f([\omega])). \tag{17.2}$$

Since f is acceptable, $\overline{S}_n f([\omega]) < \infty$ for every $\omega \in E_A^n$. So for every $\omega \in \Omega$, there is $\rho(\omega) \in [\omega]$ such that

$$|S_n f(\rho(\omega)) - \overline{S}_n f([\omega])| \le \varepsilon.$$

It follows from (17.2) that

$$Z_n(f) \le e^{2\varepsilon} \sum_{\omega\in\Omega} \exp(S_n f(\rho(\omega))). \tag{17.3}$$

By the uniform continuity of f, there exists $\ell = \ell(\varepsilon) \in \mathbb{N}$ such that

$$|\chi \wedge \zeta| \ge \ell \Longrightarrow |f(\chi) - f(\zeta)| \le \frac{\varepsilon}{n}.$$

Let $\Lambda \subseteq E_A^*$ be a finite set which makes the matrix A finitely irreducible. Let

$$F = F(\varepsilon) = \{(\rho(\omega))_j : \omega \in \Omega \quad \text{and} \quad 1 \le j \le \ell + n\} \cup \Lambda_E,$$

where $\Lambda_E := \{\tau_k : \tau \in \Lambda \text{ and } 1 \le k \le |\tau|\}$. Clearly, $\#F < \infty$ and $F_A^\infty \ne \emptyset$. In fact, $A : F \times F \to \{0,1\}$ is (finitely) irreducible, so $[\chi]_F \ne \emptyset$ for all $\chi \in F_A^* \backslash \{\epsilon\}$. Note also that $F \supseteq \Omega_E$, where $\Omega_E := \{\omega_k : \omega \in \Omega \text{ and } 1 \le k \le |\omega|\}$. Therefore, $F_A^n \supseteq \Omega$. For each $\omega \in \Omega$, let

$$\widetilde{\rho}(\omega) \in [(\rho(\omega))_1 \cdots (\rho(\omega))_{\ell+n}]_F \subseteq [(\rho(\omega))_1 \cdots (\rho(\omega))_n]_F = [\omega]_F. \tag{17.4}$$

Then $|\rho(\omega) \wedge \widetilde{\rho}(\omega)| \ge \ell + n$, and thus $|\sigma^j(\rho(\omega)) \wedge \sigma^j(\widetilde{\rho}(\omega))| \ge \ell + n - j \ge \ell$ for all $0 \le j \le n$. Consequently,

$$|S_n f(\rho(\omega)) - S_n f(\widetilde{\rho}(\omega))| \le \sum_{j=0}^{n-1} |f(\sigma^j(\rho(\omega))) - f(\sigma^j(\widetilde{\rho}(\omega)))| \le \sum_{j=0}^{n-1} \frac{\varepsilon}{n} = \varepsilon.$$

It follows from this, (17.3) and (17.4) that

$$Z_n(f) \le e^{3\varepsilon} \sum_{\omega\in\Omega} \exp(S_n f(\widetilde{\rho}(\omega))) \le e^{3\varepsilon} \sum_{\omega\in F_A^n} \exp(\overline{S}_n f([\omega]_F)) = e^{3\varepsilon} Z_n(f,F).$$

Therefore $Z_n(f) \le e^{3\varepsilon} \sup\{Z_n(f,F)\}$. Since $\varepsilon > 0$ was chosen arbitrarily, we deduce that $Z_n(f) \le \sup\{Z_n(f,F)\}$.

A similar proof applies to the case $Z_n(f) = \infty$. □

17.2.3 The pressure function

Theorem 11.1.26 suggests defining the topological pressure $\mathrm{P}_F(f)$ of $\sigma_F : F_A^\infty \to F_A^\infty$ when subject to the potential f as the topological pressure of σ_F with respect to the partition $\mathcal{U}_F = \{[e]_F : e \in F\}$, i. e., $\mathrm{P}(\sigma_F, f, \mathcal{U}_F)$ as in Definition 11.1.11. This would amount to defining $\mathrm{P}_F(f)$ as the asymptotic exponential growth rate of the sequence $(Z_n(f,F))_{n=1}^{\infty}$, i. e., as the limit of the sequence $(\frac{1}{n}\log Z_n(f,F))_{n=1}^{\infty}$.

But despite being submultiplicative, the partition functions $(Z_n(f,F))_{n=1}^{\infty}$ may be infinite and Lemma 3.2.17 may not apply. That lemma affirms that for any submultiplicative sequence $(a_n)_{n=1}^{\infty}$ of positive real numbers, the sequence $(\frac{1}{n}\log a_n)_{n=1}^{\infty}$ converges and

$$\lim_{n\to\infty}\frac{1}{n}\log a_n = \inf_{n\in\mathbb{N}}\frac{1}{n}\log a_n. \tag{17.5}$$

However, this result does not generally apply to submultiplicative sequences of extended real numbers (e. g., consider $\infty, 1, \infty, 1, \ldots$). We need to be more cautious.

Definition 17.2.7. The *topological pressure* $\mathrm{P}_F(f)$ *of the shift map* $\sigma_F : F_A^{\infty} \to F_A^{\infty}$ under a potential f is defined to be

$$\mathrm{P}_F(f) := \limsup_{n\to\infty}\frac{1}{n}\log Z_n(f,F).$$

If $F = E$, we suppress the subscript and simply write $\mathrm{P}(f)$ for $\mathrm{P}_E(f)$.

We now prove that the finite or infinite nature of the pressure of a symbolic system is determined by the nature of any of its partition functions.

Theorem 17.2.8. *If $f : E_A^{\infty} \to \mathbb{R}$ is acceptable and A is finitely irreducible, then the following statements hold:*

(a) $\mathrm{P}(f) < \infty \iff Z_1(f) < \infty \iff Z_n(f) < \infty$ *for some* $n \in \mathbb{N} \iff Z_n(f) < \infty, \forall n \in \mathbb{N}$.
(b) $\mathrm{P}(f) = \infty \iff Z_1(f) = \infty \iff Z_n(f) = \infty$ *for some* $n \in \mathbb{N} \iff Z_n(f) = \infty, \forall n \in \mathbb{N}$.
(c) $\mathrm{P}(f) = \lim_{n\to\infty}\frac{1}{n}\log Z_n(f) = \inf_{n\in\mathbb{N}}\frac{1}{n}\log Z_n(f)$.

Proof. Lemma 17.2.5 asserts that $Z_1(f) = \infty$ if and only if $Z_n(f) = \infty$ for some $n \in \mathbb{N}$ if and only if $Z_n(f) = \infty$ for all $n \in \mathbb{N}$. So if $Z_1(f) = \infty$, then

$$\mathrm{P}(f) = \lim_{n\to\infty}\frac{1}{n}\log Z_n(f) = \inf_{n\in\mathbb{N}}\frac{1}{n}\log Z_n(f) = \infty.$$

On the other hand, $Z_1(f) < \infty$ if and only if $Z_n(f) < \infty$ for some $n \in \mathbb{N}$ if and only if $Z_n(f) < \infty$ for all $n \in \mathbb{N}$. So if $Z_1(f) < \infty$, since the $Z_n(f)$, $n \in \mathbb{N}$, are submultiplicative, relation (17.5) applies and gives that

$$\mathrm{P}(f) = \lim_{n\to\infty}\frac{1}{n}\log Z_n(f) = \inf_{n\in\mathbb{N}}\frac{1}{n}\log Z_n(f) < \infty.$$ □

We now demonstrate that, as for the partition functions, the pressure of the system is the supremum of the pressures of all its finite subsystems.

Theorem 17.2.9. *If $f : E_A^\infty \to \mathbb{R}$ is acceptable and A is finitely irreducible, then*

$$\mathrm{P}(f) = \sup\{\mathrm{P}_F(f) : F \subseteq E \text{ and } \#F < \infty\}.$$

Proof. The inequality $\mathrm{P}(f) \geq \sup\{\mathrm{P}_F(f)\}$ is obvious. To prove the converse, suppose first that $\mathrm{P}(f) < \infty$. Fix $\varepsilon > 0$. By the uniform continuity of f, there exists $\ell = \ell(\varepsilon) \in \mathbb{N}$ such that

$$|\rho \wedge \zeta| \geq \ell \Longrightarrow |f(\rho) - f(\zeta)| \leq \varepsilon.$$

Let $M = (\ell - 1)\mathrm{osc}(f)$, and $k \geq \ell$ be such that $k\varepsilon \geq M$. For any $\omega \in E_A^k$ and $\rho, \zeta \in [\omega]$, we have that $|\rho \wedge \zeta| \geq |\omega| = k$, and thus $|\sigma^j(\rho) \wedge \sigma^j(\zeta)| \geq k - j \geq \ell$ whenever $0 \leq j \leq k - \ell$. Therefore,

$$\begin{aligned}
|S_kf(\rho) - S_kf(\zeta)| &\leq \sum_{j=0}^{k-1} |f(\sigma^j(\rho)) - f(\sigma^j(\zeta))| \\
&= \sum_{j=0}^{k-\ell} |f(\sigma^j(\rho)) - f(\sigma^j(\zeta))| + \sum_{j=k-\ell+1}^{k-1} |f(\sigma^j(\rho)) - f(\sigma^j(\zeta))| \\
&\leq (k - \ell + 1)\varepsilon + (\ell - 1)\mathrm{osc}(f) \leq k\varepsilon + M \leq 2k\varepsilon.
\end{aligned}$$

Thus,

$$0 \leq \overline{S}_kf([\omega]) - \underline{S}_kf([\omega]) \leq 2k\varepsilon, \tag{17.6}$$

where $\overline{S}_kf([\omega]) = \sup_{\zeta \in [\omega]} S_kf(\zeta)$ and $\underline{S}_kf([\omega]) = \inf_{\zeta \in [\omega]} S_kf(\zeta)$.

Let $\Lambda \subseteq E_A^*$ be a finite set which witnesses the finite irreducibility of A. Let $|\Lambda| := \max\{|\omega| : \omega \in \Lambda\} < \infty$. By Theorem 17.2.8, we know that $\frac{1}{k} \log Z_k(f) \geq \mathrm{P}(f)$. Choosing k large enough, we may suppose that $\frac{1}{k+|\Lambda|} \log Z_k(f) > \mathrm{P}(f) - \varepsilon$. According to Lemma 17.2.6, there then exists a finite set $F \subseteq E$ such that

$$\frac{1}{k + |\Lambda|} \log Z_k(f, F) \geq \mathrm{P}(f) - \varepsilon. \tag{17.7}$$

Enlarging F if necessary, we may assume that $F \supseteq \Lambda_E$, where $\Lambda_E := \{\omega_i : \omega \in \Lambda \text{ and } 1 \leq i \leq |\omega|\}$. Then $F_A^* \supseteq \Lambda$ and for each $\tau = (\tau^{(1)}, \tau^{(2)}, \ldots, \tau^{(n)}) \in (F_A^k)^n$ (not necessarily in F_A^{kn}), there are words $\alpha^{(1)}, \alpha^{(2)}, \ldots, \alpha^{(n-1)} \in \Lambda$ such that

$$\overline{\tau} := \tau^{(1)}\alpha^{(1)}\tau^{(2)}\alpha^{(2)} \cdots \tau^{(n-1)}\alpha^{(n-1)}\tau^{(n)} \in F_A^*.$$

Note that $kn \leq |\overline{\tau}| \leq kn + |\Lambda|(n-1)$ and that the mapping $\tau \mapsto \overline{\tau}$ is at most u^{n-1}-to-1, where u is the number of different lengths among the words composing Λ. Indeed, suppose that

$$\tau^{(1)}\alpha^{(1)}\tau^{(2)}\alpha^{(2)} \cdots \tau^{(n-1)}\alpha^{(n-1)}\tau^{(n)} = \omega^{(1)}\beta^{(1)}\omega^{(2)}\beta^{(2)} \cdots \omega^{(n-1)}\beta^{(n-1)}\omega^{(n)} \in F_A^*,$$

where $\tau^{(i)}, \omega^{(i)} \in F_A^k$ for all $1 \leq i \leq n$ and $\alpha^{(j)}, \beta^{(j)} \in \Lambda$ for all $1 \leq j < n$. Comparing initial subwords, it is clear that $\tau^{(1)} = \omega^{(1)}$ since $|\tau^{(1)}| = |\omega^{(1)}| = k$. If $|\alpha^{(1)}| = |\beta^{(1)}|$, then $\alpha^{(1)} = \beta^{(1)}$. It follows that $\tau^{(2)} = \omega^{(2)}$. If additionally $|\alpha^{(2)}| = |\beta^{(2)}|$, then $\alpha^{(2)} = \beta^{(2)}$. It follows that $\tau^{(3)} = \omega^{(3)}$. More generally, if $|\alpha^{(j)}| = |\beta^{(j)}|$ for all j, then $\tau^{(i)} = \omega^{(i)}$ for all i and $\alpha^{(j)} = \beta^{(j)}$ for all j. This means that for each $(n-1)$-tuple of lengths for the $\alpha^{(j)}$'s, the mapping $\tau \mapsto \overline{\tau}$ has at most one preimage. Since there are u different potential lengths for each $\alpha^{(j)}$, there are at most u^{n-1} different $(n-1)$-tuples of lengths, and thus the mapping $\tau \mapsto \overline{\tau}$ is at most u^{n-1}-to-1.

Setting $-\infty < m := \min\{\underline{S}_{|\omega|}f([\omega]) : \omega \in \Lambda\} < \infty$, for every $n \in \mathbb{N}$ we obtain that

$$
\begin{aligned}
u^{n-1} \sum_{j=kn}^{kn+|\Lambda|(n-1)} Z_j(f,F) &= u^{n-1} \sum_{j=kn}^{kn+|\Lambda|(n-1)} \sum_{\omega \in E_A^j} \exp(\overline{S}_{|\omega|}f([\omega]_F)) \\
&\geq \sum_{\tau \in (F_A^k)^n} \exp(\overline{S}_{|\overline{\tau}|}f([\overline{\tau}]_F)) \\
&\geq \sum_{\tau \in (F_A^k)^n} \exp(\underline{S}_{|\overline{\tau}|}f([\overline{\tau}]_F)) \\
&\geq \sum_{\tau \in (F_A^k)^n} \exp\left(\sum_{i=1}^{n} \underline{S}_k f([\tau^{(i)}]_F) + m(n-1)\right) \\
&\geq e^{m(n-1)} \sum_{\tau \in (F_A^k)^n} \exp \sum_{i=1}^{n} (\overline{S}_k f([\tau^{(i)}]_F) - 2k\varepsilon) \quad \text{by (17.6)} \\
&= e^{m(n-1)-2k\varepsilon n} \sum_{\tau^{(1)} \in F_A^k} \cdots \sum_{\tau^{(n)} \in F_A^k} \prod_{i=1}^{n} \exp(\overline{S}_k f([\tau^{(i)}]_F)) \\
&= e^{m(n-1)-2k\varepsilon n} \sum_{\tau^{(1)} \in F_A^k} \exp(\overline{S}_k f([\tau^{(1)}]_F)) \cdot \\
&\qquad \cdots \cdot \sum_{\tau^{(n)} \in F_A^k} \exp(\overline{S}_k f([\tau^{(n)}]_F)) \\
&= e^{-m} e^{n(m-2k\varepsilon)} \left(\sum_{\tau' \in F_A^k} \exp(\overline{S}_k f([\tau']_F)) \right)^n \\
&= e^{-m} e^{n(m-2k\varepsilon)} (Z_k(f,F))^n.
\end{aligned}
$$

Hence, there exists $kn \leq j_n \leq kn + |\Lambda|(n-1) \leq (k+|\Lambda|)n$ such that

$$
Z_{j_n}(f,F) \geq \frac{1}{|\Lambda|(n-1)+1} u e^{-m} e^{n(m-2k\varepsilon-\log u)} (Z_k(f,F))^n.
$$

Using (17.7), we obtain that

$$\mathrm{P}_F(f) = \lim_{n\to\infty} \frac{1}{j_n} \log Z_{j_n}(f,F) \geq -\frac{|m| + 2k\varepsilon + \log u}{k} + \mathrm{P}(f) - \varepsilon \geq \mathrm{P}(f) - 5\varepsilon$$

as long as k is chosen large enough. Letting $\varepsilon \searrow 0$, the theorem follows in the case $\mathrm{P}(f) < \infty$. The case $\mathrm{P}(f) = \infty$ can be treated similarly. □

17.3 Variational principles and equilibrium states

The reader may remember the variational principle (Theorem 12.1.1). This principle applies to all dynamical systems living on compact spaces but not necessarily to those which subsist on noncompact spaces. We shall obtain three variational principles that rule in the current symbolic setting.

Definition 17.3.1. A Borel probability measure μ on E_A^∞ is said to be *finitely supported* provided that there exists a finite set $F \subseteq E$ such that $\mathrm{supp}(\mu) \subseteq F_A^\infty$.

Let $f : E_A^\infty \to \mathbb{R}$ be a continuous potential. The variational principle (Theorem 12.1.1), which applies to finitely supported measures since F_A^∞ is compact whenever F is finite, tells us that

$$\mathrm{P}_F(f) = \sup_\mu \left\{ \mathrm{h}_\mu(\sigma) + \int_{E_A^\infty} f \, d\mu \right\} \tag{17.8}$$

for every finite set $F \subseteq E$, where the supremum is taken over the set of all σ-invariant Borel probability measures μ on E_A^∞ that are supported on F_A^∞. In fact, according to Corollary 12.1.10, this supremum can be restricted to the subset of those measures that are ergodic. Applying Theorem 17.2.9, we readily obtain the following principle for acceptable potentials (cf. Definition 17.2.1).

Theorem 17.3.2 (1st variational principle). *If $f : E_A^\infty \to \mathbb{R}$ is an acceptable potential and A is a finitely irreducible matrix, then*

$$\mathrm{P}(f) = \sup_\mu \left\{ \mathrm{h}_\mu(\sigma) + \int_{E_A^\infty} f \, d\mu \right\},$$

where the supremum is taken over the set of all finitely supported σ-invariant Borel probability measures μ on E_A^∞. In fact, the supremum can be restricted to the subset of those measures that are ergodic.

Now, let $\mathcal{U}_E := \{[e] : e \in E\}$ be the partition of E_A^∞ into its initial cylinders of length 1. Then $\mathcal{U}_E^n = \bigvee_{j=0}^{n-1} \sigma^{-j}(\mathcal{U}_E)$ is the partition of E_A^∞ into its initial cylinders of length n.

Theorem 17.3.3 (2nd variational principle). *If $f : E_A^\infty \to \mathbb{R}$ is an acceptable potential and A is a finitely irreducible matrix, then*

$$\mathrm{P}(f) \ge \sup_{\mu}\Big\{\mathrm{h}_\mu(\sigma) + \int_{E_A^\infty} f\,d\mu\Big\},$$

where the supremum is taken over all σ-invariant Borel probability measures μ on E_A^∞ such that $\int f\,d\mu > -\infty$. In addition, if $\mathrm{P}(f) < \infty$ then $\mathrm{H}_\mu(\mathcal{U}_E^n) < \infty$ for all $n \in \mathbb{N}$ and all such measures μ.

Proof. If $\mathrm{P}(f) = \infty$, there is nothing to prove. So suppose that $\mathrm{P}(f) < \infty$. By Theorem 17.2.8, this means that $Z_n(f) < \infty$ for every $n \in \mathbb{N}$. Also, we have

$$\sum_{\omega\in E_A^n} \mu([\omega])\overline{S}_nf([\omega]) \ge \int_{E_A^\infty} S_nf\,d\mu = n\int_{E_A^\infty} f\,d\mu > -\infty.$$

Therefore, using the concavity of the function $k(x) = -x\log x$, we obtain that

$$\begin{aligned}
\mathrm{H}_\mu(\mathcal{U}_E^n) + \int_{E_A^\infty} S_nf\,d\mu &\le \sum_{\omega\in E_A^n} \mu([\omega])(\overline{S}_nf([\omega]) - \log\mu([\omega]))\\
&= Z_n(f)\sum_{\omega\in E_A^n} Z_n(f)^{-1}e^{\overline{S}_nf([\omega])}k(\mu([\omega])e^{-\overline{S}_nf([\omega])})\\
&\le Z_n(f)k\Big(\sum_{\omega\in E_A^n} Z_n(f)^{-1}e^{\overline{S}_nf([\omega])}\mu([\omega])e^{-\overline{S}_nf([\omega])}\Big)\\
&= Z_n(f)k(Z_n(f)^{-1})\\
&= \log Z_n(f).
\end{aligned}$$

Hence, $\mathrm{H}_\mu(\mathcal{U}_E^n) \le \log Z_n(f) - n\int_{E_A^\infty} f\,d\mu < \infty$ for every $n \in \mathbb{N}$. Since the partition $\mathcal{U}_E$ is a generator (cf. Definition 9.4.19), upon dividing both sides by n, passing to the limit as $n \to \infty$ and using a straightforward generalization of Corollary 9.4.18 to countable measurable partitions, we conclude that

$$\mathrm{h}_\mu(\sigma) + \int_{E_A^\infty} f\,d\mu = \lim_{n\to\infty}\frac{1}{n}\Big(\mathrm{H}_\mu(\mathcal{U}_E^n) + \int_{E_A^\infty} S_nf\,d\mu\Big) \le \lim_{n\to\infty}\frac{1}{n}\log Z_n(f) = \mathrm{P}(f). \qquad \square$$

Observe that if f is an acceptable potential and μ is a finitely supported Borel probability measure, then $\int_{E_A^\infty} f\,d\mu > -\infty$. Hence, as a direct consequence of Theorems 17.3.2–17.3.3, we deduce the following.

Theorem 17.3.4 (3rd variational principle). *If $f : E_A^\infty \to \mathbb{R}$ is an acceptable potential and A is a finitely irreducible matrix, then*

$$\mathrm{P}(f) = \sup_{\mu}\left\{\mathrm{h}_\mu(\sigma) + \int_{E_A^\infty} f\,d\mu\right\},$$

where the supremum is taken over the set of all σ-invariant Borel probability measures μ on E_A^∞ such that $\int_{E_A^\infty} f\,d\mu > -\infty$. In fact, the supremum can be restricted to the subset of those measures that are ergodic.

This latter result is the analog of the usual variational principle (Theorem 12.1.1). In light of this, we make the following definition (cf. Definition 12.2.1 and the sentence that follows it).

Definition 17.3.5. Let $f : E_A^\infty \to \mathbb{R}$ be an acceptable potential and A a finitely irreducible matrix. A σ-invariant Borel probability measure μ on E_A^∞ is said to be an *equilibrium state* for f if

$$\int_{E_A^\infty} f\,d\mu > -\infty \quad \text{and} \quad \mathrm{P}(f) = \mathrm{h}_\mu(\sigma) + \int_{E_A^\infty} f\,d\mu.$$

17.4 Gibbs states

Let $f : E_A^\infty \to \mathbb{R}$ be an acceptable potential. For a symbolic system, a Borel probability measure m on E_A^∞ is called a Gibbs state for f (cf. Definition 13.2.1) if there exist a number $P \subset \mathbb{R}$ and a constant $C \geq 1$ such that for every $\omega \in E_A^*$ and every $\rho \in [\omega]$ the following holds:

$$C^{-1} \leq \frac{m([\omega])}{\exp(S_{|\omega|}f(\rho) - P|\omega|)} \leq C. \tag{17.9}$$

This means that the m-measure of any cylinder $[\omega]$ is comparable with the exponential of the difference between:

(1) the sum of the values of f at the first $|\omega|$ iterates of any word in the cylinder, and
(2) P times the length $|\omega|$ of the cylinder.

If additionally m is σ-invariant, then m is called an invariant Gibbs state.

Remark 17.4.1. Since $\mathrm{osc}(f) < \infty$, the ergodic sum $S_{|\omega|}f(\rho)$ in (17.9) can be replaced by $\overline{S}_{|\omega|}f([\omega]) := \sup(S_{|\omega|}f|_{[\omega]})$ or by $\underline{S}_{|\omega|}f([\omega]) := \inf(S_{|\omega|}f|_{[\omega]})$.

We start with the following observations, which are analog to Proposition 13.2.4.

Proposition 17.4.2. *Let $f : E_A^\infty \to \mathbb{R}$ be an acceptable potential.*
(a) *For every Gibbs state, $P = \mathrm{P}(f)$.*
(b) *If m is a Gibbs state for the potential f, then a Borel probability measure μ is a Gibbs state for f if and only if m and μ are boundedly equivalent, i. e., have Radon–Nikodym derivatives bounded away from zero and infinity.*

Proof. (a) Fix $n \in \mathbb{N}$. Recall that $\sum_{\omega\in E_A^n} m([\omega]) = 1$. Replacing $S_{|\omega|}f(\rho)$ by $\overline{S}_{|\omega|}f([\omega])$ in (17.9) and taking the sum over all words $\omega \in E_A^n$, we get

$$C^{-1}e^{-Pn}\sum_{\omega\in E_A^n}\exp(\overline{S}_nf([\omega])) \le 1 \le Ce^{-Pn}\sum_{\omega\in E_A^n}\exp(\overline{S}_nf([\omega])).$$

Applying the natural logarithm to all three terms, dividing by n and taking the limsup as $n \to \infty$, we obtain $-P + \mathrm{P}(f) \le 0 \le -P + \mathrm{P}(f)$, which means that $P = \mathrm{P}(f)$.

(b) Suppose m and μ are Gibbs states for a same potential f. Part (a) and the Gibbs property guarantee the existence of a constant $D \ge 1$ such that

$$D^{-1} \le \frac{m([\omega])}{\mu([\omega])} \le D, \quad \forall \omega \in E_A^*.$$

Since the initial cylinders are the open balls of E_A^∞ and since both measures are regular, it ensues that $D^{-1}\mu(B) \le m(B) \le D\mu(B)$ for all Borel sets $B \subseteq E_A^\infty$. That is, m and μ are boundedly equivalent and $D^{-1} \le \frac{dm}{d\mu} \le D$.

The proof of the converse implication consists in walking back the argument just given. □

As an immediate consequence of (17.9) and Remark 17.4.1, we get the following.

Proposition 17.4.3. *Any uniformly continuous function $f : E_A^\infty \to \mathbb{R}$ that has a Gibbs state is acceptable.*

Throughout the remainder of this chapter, our main considerations will pertain to the existence, the uniqueness and the properties of Gibbs states. Proposition 17.4.3 reveals that we must then restrict ourselves to potentials which are at least acceptable. The next major result in this direction is due to Sarig [115]. It asserts that if an invariant Gibbs state exists and the incidence matrix is irreducible, then it must be finitely irreducible. To establish this fact, we first need an auxiliary lemma.

Lemma 17.4.4. *Suppose that the incidence matrix $A : E \times E \to \{0, 1\}$ is irreducible and m is a shift-invariant Gibbs state for some acceptable potential $f : E_A^\infty \to \mathbb{R}$. Then there is a constant $K > 0$ such that for every $e \in E$,*

$$\min\Big\{\sum_{a\in E:\,A_{ea}=1}\exp(\sup(f|_{[a]})),\ \sum_{a\in E:\,A_{ae}=1}\exp(\sup(f|_{[a]}))\Big\} \ge K.$$

Proof. Fix any $a, b \in E$ such that $ab \in E_A^*$. Since m is a Gibbs state for f, we have by virtue of (17.9) and Proposition 17.4.2 that

$$C^{-1}e^{-2\mathrm{P}(f)}e^{\inf(f|_{[a]})+\inf(f|_{[b]})} \le m([ab]) \le Ce^{-2\mathrm{P}(f)}e^{\sup(f|_{[a]})+\sup(f|_{[b]})}. \tag{17.10}$$

Therefore,

$$m([a]) = \sum_{c\in E:\,A_{ac}=1} m([ac]) \le Ce^{-2\mathrm{P}(f)}e^{\sup(f|_{[a]})} \sum_{c\in E:\,A_{ac}=1} \exp(\sup(f|_{[c]})). \tag{17.11}$$

On the other hand, the Gibbs property of m yields that

$$m([a]) \ge C^{-1}e^{-\mathrm{P}(f)} \exp(\sup(f|_{[a]})). \tag{17.12}$$

It ensues from (17.12) and (17.11) that

$$C^{-2}e^{\mathrm{P}(f)} \le \sum_{c\in E:\,A_{ac}=1} \exp(\sup(f|_{[c]})). \tag{17.13}$$

This is half of the inequality sought. Similarly, as m is shift-invariant, we know that

$$m([a]) = m(\sigma^{-1}([a])) = \sum_{c\in E:\,A_{ca}=1} m([ca]).$$

Using this, (17.12) and (17.10), we deduce that

$$C^{-1}e^{-\mathrm{P}(f)} \exp(\sup(f|_{[a]})) \le Ce^{-2\mathrm{P}(f)} \sum_{c\in E:\,A_{ca}=1} \exp(\sup(f|_{[c]})) \exp(\sup(f|_{[a]})).$$

Consequently,

$$C^{-2}e^{\mathrm{P}(f)} \le \sum_{c\in E:\,A_{ca}=1} \exp(\sup(f|_{[c]})).$$

Along with (17.13), this completes the proof. □

Theorem 17.4.5. *Assume that an incidence matrix $A : E \times E \to \{0, 1\}$ is irreducible. If an acceptable potential $f : E_A^\infty \to \mathbb{R}$ has an invariant Gibbs state, then the incidence matrix A is finitely irreducible.*

Proof. Since A is irreducible, it suffices to find a finite set of letters $F \subseteq E$ such that for every letter $e \in E$ there exist $a, b \in F$ such that

$$A_{ea} = A_{be} = 1.$$

Indeed, as a set Λ yielding the finite irreducibility of A we can then take

$$\{i\omega^{i,j}j : i, j \in F\}$$

where the $\omega^{i,j}$ are words such that $i\omega^{i,j}j \in E_A^*$. Their existence is guaranteed by the irreducibility of A.

Without loss of generality, we may assume that $E = \mathbb{N}$. Since f has a Gibbs state, we have by virtue of (17.9) and Proposition 17.4.2(a) that

$$\sum_{n\in\mathbb{N}} \exp(\sup(f|_{[n]})) \le Ce^{\mathrm{P}(f)} < \infty.$$

So there exists some $q \in \mathbb{N}$ such that

$$\sum_{j>q} \exp(\sup(f|_{[j]})) < K, \tag{17.14}$$

where K is the constant in Lemma 17.4.4. Let

$$F := \{1, 2, \ldots, q\}.$$

It follows from (17.14) and Lemma 17.4.4 that every letter from E is followed by some letter in F and that every letter from E is preceded by some letter in F. □

As irreducibility is the weakest form of topological mixing, from this point on we are fully justified to restrict our attention to finitely irreducible matrices. This is thus what we will do henceforth.

We now address the question of uniqueness and ergodicity of invariant Gibbs states.

Theorem 17.4.6. *If an acceptable potential $f : E_A^\infty \to \mathbb{R}$ has a Gibbs state and if the incidence matrix A is finitely irreducible, then f has a unique invariant Gibbs state μ_f, and this latter is ergodic. Furthermore, if A is finitely primitive then μ_f is totally ergodic, that is, all the iterates of the shift map are ergodic with respect to μ_f, including the shift map σ itself.*

Proof. Given $\omega \in E_A^*$ and $n \in \mathbb{N}$, let $E_A^n(\omega) = \{\tau \in E_A^n : A_{\tau_n\omega_1} = 1\}$ be the set of A-admissible n-letter prefixes of ω. Let m be a Gibbs state for f. Using (17.9), Remark 17.4.1 and Proposition 17.4.2(a), we get

$$\begin{aligned} m(\sigma^{-n}([\omega])) &= \sum_{\tau\in E_A^n(\omega)} m([\tau\omega]) \le \sum_{\tau\in E_A^n(\omega)} C\exp(\overline{S}_{n+|\omega|}f([\tau\omega]) - (n+|\omega|)\mathrm{P}(f)) \\ &\le \sum_{\tau\in E_A^n(\omega)} C\exp(\overline{S}_n f([\tau]) - n\mathrm{P}(f))\exp(\overline{S}_{|\omega|}f([\omega]) - |\omega|\mathrm{P}(f)) \\ &\le \sum_{\tau\in E_A^n(\omega)} C\cdot Cm([\tau])\cdot Cm([\omega]) \le C^3 m([\omega]). \end{aligned}$$

This inequality actually holds for any union of cylinders, since every union of cylinders can be expressed as a union of mutually disjoint cylinders. Furthermore, as cylinders form an algebra generating the Borel σ-algebra on E_A^∞, it ensues from Lemma A.1.32 in Appendix A that this inequality extends to all Borel sets, i. e.,

$$m(\sigma^{-n}(B)) \le C^3 m(B), \quad \forall B \in \mathcal{B}(E_A^\infty), \quad \forall n \in \mathbb{N}. \tag{17.15}$$

On the other hand, let $\Lambda \subseteq E_A^*$ be a finite set of words witnessing the finite irreducibility of the matrix A and let $|\Lambda|$ be the maximal length of these words. Since f is acceptable,

$$\ell := \min\{\underline{S}_{|\alpha|} f([\alpha]) - |\alpha| \mathrm{P}(f) : \alpha \in \Lambda\} > -\infty.$$

For each $\tau, \omega \in E_A^*$, let $\alpha = \alpha(\tau, \omega) \in \Lambda$ be such that $\tau\alpha\omega \in E_A^*$. For all $\omega \in E_A^*$ and all $n \in \mathbb{N}$, we then have that

$$\begin{aligned}
\sum_{j=n}^{n+|\Lambda|} m(\sigma^{-j}([\omega])) &= \sum_{j=n}^{n+|\Lambda|} \sum_{\beta \in E_A^j(\omega)} m([\beta\omega]) \ge \sum_{\tau \in E_A^n} m([\tau\alpha(\tau,\omega)\omega]) \\
&\ge \sum_{\tau \in E_A^n} C^{-1} \exp\Big(\underline{S}_{|\tau|+|\alpha|+|\omega|} f([\tau\alpha\omega]) - (|\tau| + |\alpha| + |\omega|)\mathrm{P}(f)\Big) \\
&\ge C^{-1} \sum_{\tau \in E_A^n} \exp(\underline{S}_n f([\tau]) - n\mathrm{P}(f)) \cdot \exp(\underline{S}_{|\alpha|} f([\alpha]) - |\alpha|\mathrm{P}(f)) \\
&\qquad \cdot \exp(\underline{S}_{|\omega|} f([\omega]) - |\omega|\mathrm{P}(f)) \\
&\ge C^{-1} e^{\ell} \exp(\underline{S}_{|\omega|} f([\omega]) - |\omega|\mathrm{P}(f)) \sum_{\tau \in E_A^n} \exp(\underline{S}_n f([\tau]) - n\mathrm{P}(f)) \\
&\ge C^{-1} e^{\ell} C^{-1} m([\omega]) \sum_{\tau \in E_A^n} C^{-1} m([\tau]) = C^{-3} e^{\ell} m([\omega]).
\end{aligned}$$

By the same argument as above, this inequality extends to all Borel sets, i. e.,

$$\sum_{j=n}^{n+|\Lambda|} m(\sigma^{-j}(B)) \ge C^{-3} e^{\ell} m(B), \quad \forall B \in \mathcal{B}(E_A^\infty), \quad \forall n \in \mathbb{N}. \tag{17.16}$$

Let L be a Banach limit defined on the Banach space of all bounded sequences of real numbers. It is not difficult to check that the formula

$$\mu(B) = L\bigg(\bigg(\frac{1}{|\Lambda| + 1} \sum_{j=n}^{n+|\Lambda|} m(\sigma^{-j}(B)) \bigg)_{n=0}^{\infty} \bigg)$$

defines a finite, nonzero, nonnegative, σ-invariant, finitely additive function on the Borel subsets of E_A^∞. In fact, $\mu(E_A^\infty) = L((1)_{n=0}^\infty) = 1$. It further follows from (17.15) and (17.16) that

$$\frac{C^{-3}e^{\ell}}{|\Lambda|+1}m(B) \le \mu(B) \le C^3 m(B), \quad \forall B \in \mathcal{B}(E_A^\infty). \tag{17.17}$$

This shows that μ is equivalent to m. So Lemma A.1.31 asserts that μ is a countably additive function and hence is a σ-invariant Borel probability measure. Moreover, by Proposition 17.4.2, the measure μ is then, like m, a Gibbs state, albeit with a different constant C_μ.

Let us prove the ergodicity of μ. Let $\omega \in E_A^*$. Going along almost the same lines as before (replacing m by μ), for each $\tau \in E_A^*$ we find that

$$\sum_{j=|\omega|}^{|\omega|+|\Lambda|} \mu(\sigma^{-j}([\tau]) \cap [\omega]) \ge \mu([\omega\alpha(\omega,\tau)\tau]) \ge C_\mu^{-3}e^{\ell_\mu}\mu([\tau])\mu([\omega]). \tag{17.18}$$

Take an arbitrary Borel set $B \subseteq E_A^\infty$. Fix $\varepsilon > 0$. Since the initial cylinders $\{[\tau] : \tau \in E_A^*\}$ are the open balls in E_A^∞ and since μ is regular, for every $\omega \in E_A^*$ there is a set $Z = Z(\omega) \subseteq E_A^*$ consisting of mutually incomparable words such that $B \subseteq \bigcup_{\tau\in Z}[\tau]$ and

$$\mu\Big(\bigcup_{\tau\in Z}[\tau] \setminus B\Big) \le \frac{\varepsilon}{|\Lambda|+1}.$$

It follows from this and the σ-invariance of μ that

$$\sum_{\tau\in Z} \mu(\sigma^{-j}([\tau]) \cap [\omega]) \le \mu(\sigma^{-j}(B) \cap [\omega]) + \frac{\varepsilon}{|\Lambda|+1}, \quad \forall |\omega| \le j \le |\omega| + |\Lambda|.$$

Using also (17.18), we then infer that

$$\begin{aligned}
\sum_{j=|\omega|}^{|\omega|+|\Lambda|} \mu(\sigma^{-j}(B) \cap [\omega]) + \varepsilon &\ge \sum_{j=|\omega|}^{|\omega|+|\Lambda|} \sum_{\tau\in Z} \mu(\sigma^{-j}([\tau]) \cap [\omega]) \\
&\ge \sum_{\tau\in Z} C_\mu^{-3}e^{\ell_\mu}\mu([\tau])\mu([\omega]) \\
&\ge C_\mu^{-3}e^{\ell_\mu}\mu(B)\mu([\omega]).
\end{aligned}$$

Letting $\varepsilon \to 0$, we obtain

$$\sum_{j=|\omega|}^{|\omega|+|\Lambda|} \mu(\sigma^{-j}(B) \cap [\omega]) \ge C_\mu^{-3}e^{\ell_\mu}\mu(B)\mu([\omega]).$$

From this inequality, we deduce that

$$\sum_{j=|\omega|}^{|\omega|+|\Lambda|} \mu(\sigma^{-j}(E_A^\infty \setminus B) \cap [\omega]) \le [|\Lambda| + 1 - C_\mu^{-3}e^{\ell_\mu}\mu(B)]\mu([\omega]).$$

Replacing $E_A^\infty \setminus B$ by B, we thus have

$$\sum_{j=|\omega|}^{|\omega|+|\Lambda|} \mu(\sigma^{-j}(B) \cap [\omega]) \le [|\Lambda| + 1 - C_\mu^{-3} e^{\ell_\mu}(1 - \mu(B))]\mu([\omega]) \tag{17.19}$$

for every Borel set $B \subseteq E_A^\infty$ and every $\omega \in E_A^*$.

In order to conclude the proof of the ergodicity of σ with respect to μ, suppose that $\sigma^{-1}(B) = B$ and that $0 < \mu(B) < 1$ for some $B \in \mathcal{B}(E_A^\infty)$. Let $\zeta = 1 - C_\mu^{-3} e^{\ell_\mu}(1-\mu(B))/(|\Lambda|+1)$. It follows from (17.19) and $\mu(B) < 1$ that $0 < \zeta < 1$ and that for every $\omega \in E_A^*$ there exists $|\omega| \le j_\omega \le |\omega| + |\Lambda|$ such that $\mu(\sigma^{-j_\omega}(B) \cap [\omega]) \le \zeta\mu([\omega])$. The σ-invariance of B implies that $\mu(B \cap [\omega]) \le \zeta\mu([\omega])$ for every $\omega \in E_A^*$. Pick $1 < \eta < 1/\zeta$ and choose a set $R \subseteq E_A^*$ consisting of mutually incomparable words such that $B \subseteq \bigcup_{\omega \in R} [\omega]$ and $\sum_{\omega \in R} \mu([\omega]) \le \eta\mu(B)$. Then

$$\mu(B) = \sum_{\omega \in R} \mu(B \cap [\omega]) \le \sum_{\omega \in R} \zeta\mu([\omega]) \le \zeta\eta\mu(B) < \mu(B).$$

This contradiction finishes the proof of the existence of an ergodic invariant Gibbs state.

According to Theorem 8.2.21, any two distinct ergodic invariant measures must be mutually singular. The uniqueness of an invariant Gibbs state readily follows from the ergodicity of any invariant Gibbs state and from Proposition 17.4.2(b).

Finally, let us prove the complete ergodicity of μ when A is finitely primitive. In essence, we repeat the argument just given. Let Λ be a finite set of words of equal length which witnesses the finite primitivity of A. Fix $r \in \mathbb{N}$. Let $\omega \in E_A^*$. For each $\tau \in E_A^*$, we find the following improvement of (17.18):

$$\begin{aligned}\mu(\sigma^{-(|\omega|+|\Lambda|r)}([\tau]) \cap [\omega]) &\ge \sum_{\alpha \in \Lambda^r \cap E_A^{|\Lambda|r} :\, A_{\omega_{|\omega|}\alpha_1} = A_{\alpha_{|\Lambda|r}\tau_1} = 1} \mu([\omega\alpha\tau]) \\ &\ge C_\mu^{-3} e^{r\ell_\mu} \mu([\tau])\mu([\omega]).\end{aligned} \tag{17.20}$$

Take an arbitrary Borel set $B \subseteq E_A^\infty$. Fix $\varepsilon > 0$. For every $\omega \in E_A^*$, there exists a set $Z \subseteq E_A^*$ consisting of mutually incomparable words such that $B \subseteq \bigcup_{\tau \in Z} [\tau]$ and

$$\sum_{\tau \in Z} \mu(\sigma^{-(|\omega|+|\Lambda|r)}([\tau]) \cap [\omega]) \le \mu(\sigma^{-(|\omega|+|\Lambda|r)}(B) \cap [\omega]) + \varepsilon.$$

Using (17.20), we then get

$$\mu(\sigma^{-(|\omega|+|\Lambda|r)}(B) \cap [\omega]) + \varepsilon \ge \sum_{\tau \in Z} C_\mu^{-3} e^{r\ell_\mu} \mu([\tau])\mu([\omega]) \ge C_\mu^{-3} e^{r\ell_\mu} \mu(B)\mu([\omega]).$$

Letting $\varepsilon \to 0$, we obtain

$$\mu(\sigma^{-(|\omega|+|\Lambda|r)}(B) \cap [\omega]) \ge C(r)\mu(B)\mu([\omega]),$$

where $C(r) = C_\mu^{-3} e^{r\ell_\mu}$. By setting $B = E_A^\infty$, it follows from this last inequality that $C(r) \le 1$. From this same inequality, we also find that

$$\mu(\sigma^{-(|\omega|+|\Lambda|r)}(E_A^\infty \setminus B) \cap [\omega]) \le (1 - C(r)\mu(B))\mu([\omega]).$$

Thus, for every $\omega \in E_A^*$ we have

$$\mu(\sigma^{-(|\omega|+|\Lambda|r)}(B) \cap [\omega]) \le (1 - C(r)(1 - \mu(B)))\mu([\omega]). \tag{17.21}$$

In order to conclude the proof of the complete ergodicity of σ with respect to μ, suppose that $\sigma^{-r}(B) = B$ and $0 < \mu(B) < 1$. Set $\zeta = 1 - C(r)(1 - \mu(B))$. It follows from (17.21) and $\mu(B) < 1$ that $0 < \zeta < 1$ and for every $\omega \in \bigcup_{n=1}^\infty E_A^{nr}$ we have $\mu(B \cap [\omega]) = \mu(\sigma^{-(|\omega|+|\Lambda|r)}(B) \cap [\omega]) \le \zeta\, \mu([\omega])$. Take $1 < \eta < 1/\zeta$ and choose a set $R \subseteq \bigcup_{n=1}^\infty E_A^{nr}$ consisting of mutually incomparable words such that $B \subseteq \bigcup_{\omega \in R}[\omega]$ and $\sum_{\omega \in R} \mu([\omega]) \le \eta\mu(B)$. Then

$$\mu(B) = \sum_{\omega \in R} \mu(B \cap [\omega]) \le \sum_{\omega \in R} \zeta\, \mu([\omega]) \le \zeta\eta\mu(B) < \mu(B).$$

This contradiction completes the proof of the complete ergodicity of μ. □

We shall now characterize the condition $\int f \, d\mu_f > -\infty$, where μ_f is the unique invariant Gibbs state (if it exists) for an acceptable potential f. Recall that this condition is part of the definition of an equilibrium state.

Lemma 17.4.7. *Suppose that the incidence matrix A is finitely irreducible and that an acceptable potential $f : E_A^\infty \to \mathbb{R}$ has a Gibbs state. Denote by μ_f its unique invariant Gibbs state per Theorem 17.4.6. Then the following three conditions are equivalent:*
(a) $\int_{E_A^\infty} f \, d\mu_f > -\infty$.
(b) $\sum_{e\in E} \inf(-f|_{[e]}) \exp(\inf f|_{[e]}) < \infty$.
(c) $\mathrm{H}_{\mu_f}(\mathcal{U}_E) < \infty$, *where $\mathcal{U}_E = \{[e] : e \in E\}$ is the partition of E_A^∞ into its initial cylinders of length* 1.

Proof. [(a)⇒(b)] Suppose that $\int_{E_A^\infty} f \, d\mu_f > -\infty$, i. e. $\sum_{e\in E} \int_{[e]} (-f) \, d\mu_f < \infty$. As μ_f satisfies the Gibbs property (17.9) (see also Remark 17.4.1), we have

$$\begin{aligned}\infty > \sum_{e\in E} \inf(-f|_{[e]})\mu_f([e]) &\ge C_{\mu_f}^{-1} \sum_{e\in E} \inf(-f|_{[e]}) \exp(\inf(f|_{[e]}) - \mathrm{P}(f)) \\ &= C_{\mu_f}^{-1} e^{-\mathrm{P}(f)} \sum_{e\in E} \inf(-f|_{[e]}) \exp(\inf(f|_{[e]})).\end{aligned}$$

[(b)⇒(c)] Assume that $\sum_{e\in E} \inf(-f|_{[e]}) \exp(\inf(f|_{[e]})) < \infty$. Then

$$\begin{aligned}\mathrm{H}_{\mu_f}(\mathcal{U}_E) = \sum_{e\in E} -\mu_f([e]) \log \mu_f([e]) &\le \sum_{e\in E} -\mu_f([e])(\inf(f|_{[e]}) - \mathrm{P}(f) - \log C_{\mu_f}) \\ &= \mathrm{P}(f) + \log C_{\mu_f} + \sum_{e\in E} -\mu_f([e]) \inf(f|_{[e]}).\end{aligned}$$

Thus, it remains to show that $\sum_{e\in E} -\mu_f([e])\inf(f|_{[e]}) < \infty$. And indeed,

$$\begin{aligned}\sum_{e\in E} -\mu_f([e])\inf(f|_{[e]}) &= \sum_{e\in E} \mu_f([e])\sup(-f|_{[e]})\\ &\le \sum_{e\in E} \mu_f([e])(\inf(-f|_{[e]}) + \operatorname{osc}(f))\\ &= \operatorname{osc}(f) + \sum_{e\in E} \mu_f([e])\inf(-f|_{[e]}).\end{aligned}$$

Since $\operatorname{osc}(f) < \infty$, we need to show that $\sum_{e\in E} \mu_f([e])\inf(-f|_{[e]}) < \infty$. Without loss of generality, we may assume that $E = \mathbb{N}$. As μ_f is a probability measure, $\lim_{e\to\infty} \mu_f([e]) = 0$. Then it follows from the Gibbs property that $\lim_{e\to\infty}(\sup f|_{[e]} - \mathrm{P}(f)) = -\infty$. Since $\mathrm{P}(f) < \infty$, we deduce that $\lim_{e\to\infty} \sup f|_{[e]} = -\infty$. It is thus possible to reorder the set E so that there is $k \in E$ such that $\sup f|_{[e]} \ge 0$ for all $e < k$ while $\sup f|_{[e]} < 0$ for all $e \ge k$. Hence $\inf(-f|_{[e]}) \le 0$ for all $e < k$ whereas $\inf(-f|_{[e]}) > 0$ for all $e \ge k$. Using the Gibbs property again, we then get

$$\begin{aligned}\sum_{e\in E} \mu_f([e])\inf(-f|_{[e]}) &\le \sum_{e\ge k} C_{\mu_f} \exp(\inf(f|_{[e]}) - \mathrm{P}(f))\inf(-f|_{[e]})\\ &= C_{\mu_f} e^{-\mathrm{P}(f)} \sum_{e\ge k} \exp(\inf(f|_{[e]}))\inf(-f|_{[e]}),\end{aligned}$$

and this latter sum is finite by assumption. Therefore, $\mathrm{H}_{\mu_f}(\mathcal{U}_E) < \infty$.

[(c)⇒(a)] Suppose that $\mathrm{H}_{\mu_f}(\mathcal{U}_E) < \infty$. Then

$$\begin{aligned}\infty > \mathrm{H}_{\mu_f}(\mathcal{U}_E) &= \sum_{e\in E} -\mu_f([e])\log\mu_f([e])\\ &\ge \sum_{e\in E} -\mu_f([e])(\inf(f|_{[e]}) - \mathrm{P}(f) + \log C_{\mu_f})\\ &= \mathrm{P}(f) - \log C_{\mu_f} + \sum_{e\in E} -\mu_f([e])\inf(f|_{[e]}).\end{aligned}$$

Hence, $\sum_{e\in E} -\mu_f([e])\inf(f|_{[e]}) < \infty$, i. e. $\sum_{e\in E} \mu_f([e])\inf(f|_{[e]}) > -\infty$. Therefore,

$$\int_{E_A^\infty} f\,d\mu_f = \sum_{e\in E} \int_{[e]} f\,d\mu_f \ge \sum_{e\in E} \inf(f|_{[e]})\mu_f([e]) > -\infty. \qquad \square$$

Remark 17.4.8. Observe that the shift-invariance of μ_f was not used anywhere in above proof and the previous lemma actually holds for any Gibbs state.

17.5 Gibbs states versus equilibrium states

The next theorem shows that the condition $\int f\,d\mu_f > -\infty$ is sufficient for the unique invariant Gibbs state μ_f to be the unique equilibrium state for any acceptable potential f.

Theorem 17.5.1. *Suppose that the incidence matrix A is finitely irreducible. Suppose also that $f : E_A^\infty \to \mathbb{R}$ is an acceptable potential which admits a Gibbs state and that $\int_{E_A^\infty} f\,d\mu_f > -\infty$, where μ_f is the unique invariant Gibbs state for the potential f per Theorem 17.4.6. Then μ_f is the unique equilibrium state for f.*

Proof. In order to show that μ_f is an equilibrium state for the potential f, consider the partition $\mathcal{U}_E = \{[e] : e \in E\}$ of E_A^∞ into its initial cylinders of length 1. By Lemma 17.4.7, $\mathrm{H}_{\mu_f}(\mathcal{U}_E) < \infty$. Using the ergodic case of Shannon–McMillan–Breiman's theorem (Corollary 9.5.5), the Gibbs property (17.9) and the ergodic case of Birkhoff's theorem (Corollary 8.2.14), we get for μ_f-a. e. $\rho \in E_A^\infty$,

$$\begin{aligned}
\mathrm{h}_{\mu_f}(\sigma) \geq \mathrm{h}_{\mu_f}(\sigma, \mathcal{U}_E) &= \lim_{n\to\infty} -\frac{1}{n}\log \mu_f([\rho|_n]) \\
&\geq \lim_{n\to\infty} -\frac{1}{n}(\log C_{\mu_f} + S_n f(\rho) - n\mathrm{P}(f)) \\
&= -\lim_{n\to\infty}\frac{1}{n}S_n f(\rho) + \mathrm{P}(f) = -\int_{E_A^\infty} f\,d\mu_f + \mathrm{P}(f).
\end{aligned}$$

In conjunction with the 2nd variational principle (Theorem 17.3.3), this implies that μ_f is an equilibrium state for the potential f.

To prove the uniqueness of the equilibrium state, suppose that ν is an equilibrium state for the potential $f : E_A^\infty \to \mathbb{R}$ and that $\nu \neq \mu_f$. Applying the ergodic decomposition theorem (Theorem 8.2.26), we may assume that ν is ergodic. Using the Gibbs property (17.9), we then have that for every $n \in \mathbb{N}$,

$$\begin{aligned}
0 = n\Big(\mathrm{h}_\nu(\sigma) + \int_{E_A^\infty} (f - \mathrm{P}(f))\,d\nu\Big) &\leq \mathrm{H}_\nu(\mathcal{U}_E^n) + \int_{E_A^\infty} (S_n f - n\mathrm{P}(f))\,d\nu \\
&= -\sum_{\omega\in E_A^n} \nu([\omega])\Big(\log \nu([\omega]) - \frac{1}{\nu([\omega])}\int_{[\omega]} (S_n f - n\mathrm{P}(f))\,d\nu\Big) \\
&\leq -\sum_{\omega\in E_A^n} \nu([\omega])\big(\log \nu([\omega]) - (\overline{S}_n f([\omega]) - n\mathrm{P}(f))\big) \\
&= -\sum_{\omega\in E_A^n} \nu([\omega])\log\big[\nu([\omega])\exp(n\mathrm{P}(f) - \overline{S}_n f([\omega]))\big] \\
&\leq -\sum_{\omega\in E_A^n} \nu([\omega])\log\big[\nu([\omega])(C_{\mu_f}\mu_f([\omega]))^{-1}\big] \\
&= \log C_{\mu_f} - \sum_{\omega\in E_A^n} \nu([\omega])\log\frac{\nu([\omega])}{\mu_f([\omega])}.
\end{aligned}$$

Therefore, in order to complete the proof by contradiction, it suffices to show that

$$\lim_{n\to\infty} -\sum_{\omega\in E_A^n} \nu([\omega])\log\frac{\nu([\omega])}{\mu_f([\omega])} = -\infty.$$

Since both measures ν and μ_f are ergodic and $\nu \neq \mu_f$, the measures ν and μ_f must be mutually singular. According to Lemma 13.7.14,

$$\lim_{n\to\infty} \nu\left(\left\{\omega \in E_A^\infty : \frac{\nu([\omega|_n])}{\mu_f([\omega|_n])} \le t\right\}\right) = 0$$

for every $t > 0$. For every $n \in \mathbb{N}$ and every $j \in \mathbb{Z}$, set

$$F_{n,j} = \left\{\omega \in E_A^\infty : e^{-j} \le \frac{\nu([\omega|_n])}{\mu_f([\omega|_n])} < e^{-j+1}\right\}.$$

Then $F_{n,j}$ is a union of cylinders of length n and

$$\nu(F_{n,j}) = \int_{F_{n,j}} \frac{\nu([\omega|_n])}{\mu_f([\omega|_n])}\, d\mu_f(\omega) \le e^{-j+1}.$$

Notice that

$$\begin{aligned}
-\sum_{\omega\in E_A^n} \nu([\omega]) \log \frac{\nu([\omega])}{\mu_f([\omega])} &= -\int_{E_A^\infty} \log \frac{\nu([\omega|_n])}{\mu_f([\omega|_n])}\, d\nu(\omega) \\
&= -\sum_{j\in\mathbb{Z}} \int_{F_{n,j}} \log \frac{\nu([\omega|_n])}{\mu_f([\omega|_n])}\, d\nu(\omega) \le \sum_{j\in\mathbb{Z}} j\nu(F_{n,j}).
\end{aligned}$$

Thus, for each $k \in -\mathbb{N}$ we obtain that

$$\begin{aligned}
-\sum_{\omega\in E_A^n} \nu([\omega]) \log \frac{\nu([\omega])}{\mu_f([\omega])} &\le k \sum_{j\le k} \nu(F_{n,j}) + \sum_{j\ge 1} je^{-j+1} \\
&= k\nu\left(\left\{\omega \in E_A^\infty : e^{-k} \le \frac{\nu([\omega|_n])}{\mu_f([\omega|_n])}\right\}\right) + \sum_{j\ge 1} je^{-j+1}.
\end{aligned}$$

Hence, for each $k \in -\mathbb{N}$ we get that

$$\limsup_{n\to\infty} -\sum_{\omega\in E_A^n} \nu([\omega]) \log \frac{\nu([\omega])}{\mu_f([\omega])} \le k + \sum_{j\ge 1} je^{-j+1}.$$

Letting $k \to -\infty$ completes the proof. □

17.6 Transfer operator

We will now describe basic properties of the transfer operator. This operator is ultimately used to prove the existence of Gibbs states (see Theorem 17.7.4 and Corollary 17.7.5). It is also used to demonstrate several stochastic laws that Gibbs states obey to and the real analyticity of the topological pressure, though the discussion of these issues is deferred to Chapters 18 and 20, respectively.

Let $C_b(E_A^\infty)$ denote the space of all bounded continuous functions on E_A^∞ equipped with the supremum norm. Given a potential $f : E_A^\infty \to \mathbb{R}$ and a function $g \in C_b(E_A^\infty)$, define the transfer operator $\mathcal{L}_f$ by

$$\mathcal{L}_f(g)(\rho) := \sum_{e\in E:\, A_{e\rho_1}=1} g(e\rho)\exp(f(e\rho)). \tag{17.22}$$

In our generally noncompact symbolic setting, this definition is in line with the one given in (13.26). To make sure that the transfer operator preserves the space $C_b(E_A^\infty)$, we will assume not only that the potential f is acceptable but also summable.

Definition 17.6.1. A potential $f : E_A^\infty \to \mathbb{R}$ is said to be *summable* if

$$\sum_{e\in E} \exp(\sup f|_{[e]}) < \infty.$$

Note that if f has a Gibbs state, then f is summable.

Lemma 17.6.2. *If $f : E_A^\infty \to \mathbb{R}$ is a summable and acceptable potential, then the linear operator $\mathcal{L}_f : C_b(E_A^\infty) \to C_b(E_A^\infty)$ is well-defined and is bounded with $\|\mathcal{L}_f\|_\infty \le \sum_{e\in E} \exp(\sup f|_{[e]})$.*

Proof. The argument is somewhat similar to that of Lemma 13.6.1. □

It is easy to see that the nth iterate of this operator is given by

$$\mathcal{L}_f^n(g)(\rho) = \sum_{\tau\in E_A^n:\, A_{\tau_n\rho_1}=1} g(\tau\rho)\exp(S_nf(\tau\rho)). \tag{17.23}$$

The dual operator $\mathcal{L}_f^* : C_b(E_A^\infty)^* \to C_b(E_A^\infty)^*$ has the following form:

$$\mathcal{L}_f^*(\mu)(g) := \mu(\mathcal{L}_f(g)) := \int_{E_A^\infty} \mathcal{L}_f(g)\,d\mu. \tag{17.24}$$

Throughout the rest of this subsection, we assume that m is a Borel probability eigenmeasure of $\mathcal{L}_f^*$. The corresponding eigenvalue is denoted by λ. Since $\mathcal{L}_f$ is a positive operator, $\lambda \ge 0$. Then, for every $n \in \mathbb{N}$,

$$(\mathcal{L}_f^*)^n(m) = \lambda^n m.$$

The integral version of this equality is

$$\int_{E_A^\infty} \sum_{\tau\in E_A^n:\, A_{\tau_n\rho_1}=1} g(\tau\rho)e^{S_nf(\tau\rho)}\,dm(\rho) = \lambda^n \int_{E_A^\infty} g\,dm, \quad \forall g \in C_b(E_A^\infty). \tag{17.25}$$

In fact, formulas (17.22) to (17.25) extend to the space of all bounded Borel-measurable functions on E_A^∞. In particular, taking $\omega \in E_A^n$, a Borel set $B \subseteq E_A^{\infty,\omega}$ and $g = \mathbb{1}_{\omega B}$, where

$E_A^{\infty,\omega} := \{\gamma \in E_A^\infty \mid \omega\gamma \in E_A^\infty\}$ and $\omega B := \{\omega\rho : \rho \in B\}$, we obtain from (17.25) that

$$m(\omega B) = \lambda^{-n} \int_B e^{S_n f(\omega\rho)}\, dm(\rho). \tag{17.26}$$

Remark 17.6.3. If (17.26) holds, then by representing a Borel set $B \subseteq E_A^\infty$ as the union $\bigcup_{\omega \in E_A^n}(\omega B_\omega)$, where $B_\omega = \{\chi \in E_A^{\infty,\omega} : \omega\chi \in B\}$, a straightforward calculation based on (17.26) demonstrates that (17.25) is satisfied for the characteristic function $\mathbb{1}_B$. It follows from standard approximation arguments that (17.25) is then satisfied for all m-integrable functions. Thus, m is an eigenmeasure of $\mathcal{L}_f^*$ if and only if formula (17.26) holds with $n = 1$.

We now show that any eigenmeasure of the transfer operator's dual is a Gibbs state under the condition that the potential is summable and Hölder continuous on cylinders. Note that if Hölder continuity held globally, then the potential would be bounded and hence would not be summable. In turn, this would imply that such a potential would not have a Gibbs state.

Theorem 17.6.4. *If the incidence matrix A is finitely irreducible and if a potential f is summable and Hölder continuous on cylinders, then any eigenmeasure m of the dual operator $\mathcal{L}_f^*$ is a Gibbs state for f. In addition, its eigenvalue is $\lambda = e^{\mathrm{P}(f)}$. Furthermore,*

$$C_f := e^{\frac{v_\beta(f)}{e^\beta - 1}} \max\Big\{1, \#\Lambda \exp\Big(2|\Lambda| \cdot |\mathrm{P}(f)| - 2\min_{\alpha\in\Lambda} \underline{S}_{|\alpha|} f([\alpha])\Big)\Big\}$$

is a Gibbs constant for m, where $\#\Lambda$ denotes the cardinality of Λ while $|\Lambda| = \max\{|\alpha| : \alpha \in \Lambda\}$.

Proof. Let $\omega \in E_A^* \setminus \{\epsilon\}$ and $n = |\omega| \in \mathbb{N}$. Since f is summable and Hölder continuous on cylinders, relation (17.26) and Lemma 17.2.3 imply that for every $\rho \in [\omega]$,

$$\begin{aligned} m([\omega]) &= m(\omega E_A^{\infty,\omega}) \\ &= \lambda^{-n} \int_{E_A^{\infty,\omega}} e^{S_n f(\omega\gamma)}\, dm(\gamma) \\ &\le \lambda^{-n} \cdot e^{S_n f(\rho) + \frac{v_\beta(f)}{e^\beta - 1}}\, m(E_A^{\infty,\omega}) \\ &\le e^{\frac{v_\beta(f)}{e^\beta - 1}} \cdot \exp(S_n f(\rho) - n \log\lambda) \\ &\le C_f \exp(S_n f(\rho) - n\log\lambda). \end{aligned} \tag{17.27}$$

On the other hand, let $\Lambda \subseteq E_A^*$ be a nonempty finite set which witnesses the finite irreducibility of A. By the definition of Λ,

$$\bigcup_{\alpha\in\Lambda} E_A^{\infty,\alpha} = E_A^\infty.$$

Hence, there exists $\alpha_0 \in \Lambda$ such that $m(E_A^{\infty,\alpha_0}) \geq 1/\#\Lambda$. Then there exists $\xi \in \Lambda$ such that $\omega\xi\alpha_0 \in E_A^*$. Let $\rho \in [\omega]$ and write $n = |\omega|$, $p = |\xi|$ and $q = |\alpha_0|$. Using (17.26) and Lemma 17.2.3 once again, we obtain that

$$\begin{aligned}
m([\omega]) \geq m(\omega\xi\alpha_0 E_A^{\infty,\alpha_0}) &= \lambda^{-(n+p+q)} \int_{E_A^{\infty,\alpha_0}} e^{S_{n+p+q}f(\omega\xi\alpha_0\gamma)}dm(\gamma)\\
&= \lambda^{-(n+p+q)} \int_{E_A^{\infty,\alpha_0}} e^{S_nf(\omega\xi\alpha_0\gamma)}e^{S_pf(\xi\alpha_0\gamma)}e^{S_qf(\alpha_0\gamma)}dm(\gamma)\\
&\geq \lambda^{-(p+q)} \exp\Big(2\min_{\alpha\in\Lambda} \underline{S}_{|\alpha|}f([\alpha])\Big) \int_{E_A^{\infty,\alpha_0}} e^{S_nf(\omega\xi\alpha_0\gamma)}dm(\gamma)\cdot\lambda^{-n}\\
&\geq \min\{1,\lambda^{-2|\Lambda|}\} \exp\Big(2\min_{\alpha\in\Lambda} \underline{S}_{|\alpha|}f([\alpha])\Big)\cdot e^{S_nf(\rho)-\frac{v_\beta(f)}{e^\beta-1}} m(E_A^{\infty,\alpha_0})\cdot\lambda^{-n}\\
&\geq e^{-\frac{v_\beta(f)}{e^\beta-1}} \min\{1,\lambda^{-2|\Lambda|}\} \exp\Big(2\min_{\alpha\in\Lambda} \underline{S}_{|\alpha|}f([\alpha])\Big)\cdot(1/\#\Lambda)\\
&\quad\cdot\exp(S_nf(\rho)-n\log\lambda).
\end{aligned} \tag{17.28}$$

Combining (17.27) and (17.28), we conclude that m is a Gibbs state for f. The equality $\lambda = e^{\mathrm{P}(f)}$ follows immediately from Proposition 17.4.2(a). We can infer from this and (17.27)–(17.28) that C_f is a Gibbs constant for m. □

Let us now introduce the normalized transfer operator

$$\widehat{\mathcal{L}}_f = e^{-\mathrm{P}(f)}\mathcal{L}_f.$$

Theorem 17.6.5. *If the incidence matrix A is finitely irreducible and if a potential f is summable and Hölder continuous on cylinders, then the dual operator $\widehat{\mathcal{L}}_f^*$ fixes at most one Borel probability measure.*

Proof. Suppose that m and m_1 are fixed points under $\widehat{\mathcal{L}}_f^*$. In view of Theorem 17.6.4 and Proposition 17.4.2(b), the measures m and m_1 are boundedly equivalent. Consider the Radon–Nikodym derivative $\rho = dm_1/dm$. Temporarily fix $\omega \in E_A^*$ with $n = |\omega| \geq 2$. We deduce from (17.26) and Theorem 17.6.4 that

$$\begin{aligned}
m([\omega]) &= \int_{E_A^{\infty,\omega}} \exp(S_nf(\omega\gamma) - n\mathrm{P}(f))dm(\gamma)\\
&= \int_{E_A^{\infty,\omega}} \exp(S_{n-1}f(\sigma(\omega\gamma)) - (n-1)\mathrm{P}(f))\exp(f(\omega\gamma)-\mathrm{P}(f))dm(\gamma)\\
&= \int_{E_A^{\infty,\sigma(\omega)}} \exp(S_{n-1}f(\sigma(\omega\gamma)) - (n-1)\mathrm{P}(f))\exp(f(\omega\gamma)-\mathrm{P}(f))dm(\gamma).
\end{aligned}$$

Thus,

$$\inf(\exp(f|_{[\omega]} - \mathrm{P}(f)))m([\sigma(\omega)]) \le m([\omega]) \le \sup(\exp(f|_{[\omega]} - \mathrm{P}(f)))m([\sigma(\omega)]).$$

Since $f : E_A^\infty \to \mathbb{R}$ is continuous, we conclude that for every $\omega \in E_A^\infty$,

$$\lim_{n\to\infty} \frac{m([\omega|_n])}{m([\sigma(\omega)|_{n-1}])} = \exp(f(\omega) - \mathrm{P}(f)). \tag{17.29}$$

Obviously, the same formula holds with m replaced by m_1. But according to Theorem 17.4.6 and Proposition 17.4.2(b), there is a σ-invariant Borel probability measure which is equivalent to m. Using this and Exercise 9.7.14, we infer that there exists a set of points $\omega \in E_A^\infty$ with full m-measure on which

$$\rho(\omega) = \lim_{n\to\infty} \frac{m_1([\omega|_n])}{m([\omega|_n])} \quad \text{and} \quad \rho(\sigma(\omega)) = \lim_{n\to\infty} \frac{m_1([\sigma(\omega)|_n])}{m([\sigma(\omega)|_n])}.$$

From this, (17.29) and its version for m_1, we then obtain

$$\begin{aligned}\rho(\omega) &= \lim_{n\to\infty}\left[\frac{m_1([\omega|_n])}{m_1([\sigma(\omega)|_{n-1}])}\cdot\frac{m_1([\sigma(\omega)|_{n-1}])}{m([\sigma(\omega)|_{n-1}])}\cdot\frac{m([\sigma(\omega)|_{n-1}])}{m([\omega|_n])}\right]\\ &= \exp(f(\omega) - \mathrm{P}(f))\cdot\rho(\sigma(\omega))\cdot\left[\exp(f(\omega) - \mathrm{P}(f))\right]^{-1} = \rho(\sigma(\omega))\end{aligned}$$

on a set of full m-measure. But according to Theorem 17.4.6, the shift map σ is ergodic with respect to a σ-invariant Borel probability measure equivalent to m. By a short argument relying on Theorem 8.2.18, the function ρ is m-almost everywhere constant. So there exists a real number $\rho \ge 0$ such that $\rho(\omega) = \rho$ for m-a.e. $\omega \in E_A^\infty$. Therefore,

$$1 = m(E_A^\infty) = \int_{E_A^\infty} \rho(\omega)dm(\omega) = \rho m(E_A^\infty) = \rho.$$

In consequence, $m_1 = m$. □

17.7 Existence and uniqueness of eigenmeasures of the dual transfer operator, of Gibbs states and of equilibrium states

So far the character of this chapter has been somewhat conditional. To remedy this, we prove in this section the existence of eigenmeasures of the dual transfer operator. We begin with the finite case, which is due to Bowen [11].

Lemma 17.7.1. *If the alphabet E is finite and the incidence matrix A is irreducible, then for every continuous potential $f : E_A^\infty \to \mathbb{R}$ the dual operator $\mathcal{L}_f^*$ admits a Borel probability eigenmeasure m_f.*

Proof. This is a special case of Theorem 13.6.2. □

Remark 17.7.2. In Lemma 17.7.1, it suffices that the matrix contain at least one 1 in every row and every column. Irreducibility will ensure that the eigenmeasure is a Gibbs state per Theorem 17.6.4.

The following simple fact about irreducible matrices will be needed to generalize Lemma 17.7.1 to the infinite case.

Lemma 17.7.3. *If $A : E \to E$ is an irreducible matrix, then there exists an ascending sequence $(F_n)_{n=1}^{\infty}$ of finite subsets of E whose union is E and such that $A|_{F_n \times F_n}$ is an irreducible matrix.*

Lemma 17.7.1 can be generalized to a countably infinite alphabet as follows.

Theorem 17.7.4. *If the incidence matrix A is finitely irreducible and if a potential $f : E_A^{\infty} \to \mathbb{R}$ is summable and Hölder continuous on cylinders, then the dual operator $\mathcal{L}_f^*$ has a Borel probability eigenmeasure m_f.*

Proof. Without loss of generality, we may assume that $E = \mathbb{N}$. Since the incidence matrix is irreducible, Lemma 17.7.3 affirms that we can reorder the set $\mathbb{N}$ so that there exists a strictly increasing sequence $(\ell_n)_{n=1}^{\infty}$ in $\mathbb{N}$ such that the matrix $A|_{\{1,\dots,\ell_n\}\times\{1,\dots,\ell_n\}}$ is irreducible for every $n \in \mathbb{N}$. Let

$$E_{\ell_n}^{\infty} := E_A^{\infty} \cap \{1,\dots,\ell_n\}^{\infty} = \{1,\dots,\ell_n\}_A^{\infty}.$$

Consider the potential $f_n := f|_{E_{\ell_n}^{\infty}}$, and the corresponding transfer operator $\mathcal{L}_{f_n} : C(E_{\ell_n}^{\infty}) \to C(E_{\ell_n}^{\infty})$ and its dual $\mathcal{L}_{f_n}^* : C(E_{\ell_n}^{\infty})^* \to C(E_{\ell_n}^{\infty})^*$. As f is continuous (being Hölder on cylinders), Lemma 17.7.1 guarantees that for every $n \in \mathbb{N}$ there is a Borel probability eigenmeasure $m_n : \mathcal{B}(E_{\ell_n}^{\infty}) \to [0,1]$ of $\mathcal{L}_{f_n}^*$.

Our first goal is to show that the sequence $(m_n)_{n=1}^{\infty}$ is tight when the m_n's are treated as Borel probability measures on E_A^{∞}. That is, for every $\varepsilon > 0$ we must find a compact set $K_\varepsilon \subseteq E_A^{\infty}$ such that $m_n(K_\varepsilon) \geq 1-\varepsilon$ for all $n \in \mathbb{N}$. For every $k \in \mathbb{N}$, let $\pi_k : E_A^{\infty} \to E$ be the projection onto the kth coordinate, i. e., $\pi_k(\omega) = \omega_k$. Let also $\mathrm{P}_n = \mathrm{P}(\sigma|_{E_{\ell_n}^{\infty}}, f|_{E_{\ell_n}^{\infty}})$. Obviously, $\mathrm{P}_{n+1} \geq \mathrm{P}_n$ for all $n \in \mathbb{N}$. By Theorem 17.6.4, the eigenvalue of $\mathcal{L}_{f_n}^*$ corresponding to the eigenmeasure m_n is e^{P_n}. This measure is also a Gibbs state for the potential $f|_{E_{\ell_n}^{\infty}}$ applied to the finite-state subsystem $\sigma|_{E_{\ell_n}^{\infty}}$. We thus obtain for every $n \in \mathbb{N}$, every $k \in \mathbb{N}$, and every $e \in E = \mathbb{N}$ that

$$\begin{aligned} m_n(\pi_k^{-1}(e)) &= \sum_{\omega\in E_{\ell_n}^k:\,\omega_k=e} m_n([\omega]) \leq \sum_{\omega\in E_{\ell_n}^k:\,\omega_k=e} \exp(\overline{S}_k f([\omega]) - \mathrm{P}_n k) \\ &\leq \exp(-\mathrm{P}_n k) \sum_{\omega\in E_{\ell_n}^k:\,\omega_k=e} \exp(\overline{S}_{k-1} f([\omega]) + \sup f|_{[e]}) \\ &\leq \exp(-\mathrm{P}_1 k)\Big(\sum_{i\in E} \exp(\sup f|_{[i]})\Big)^{k-1} \exp(\sup f|_{[e]}). \end{aligned}$$

Therefore,

$$m_n(\pi_k^{-1}([e+1,\infty))) \le e^{-\mathrm{P}_1 k}\Big(\sum_{i\in E} e^{\sup f|_{[i]}}\Big)^{k-1} \sum_{j>e} e^{\sup f|_{[j]}}.$$

Fix $\varepsilon > 0$. Since f is summable, for every $k \in \mathbb{N}$ choose $n_k \in \mathbb{N}$ such that

$$e^{-\mathrm{P}_1 k}\Big(\sum_{i\in E} e^{\sup f|_{[i]}}\Big)^{k-1} \sum_{j>n_k} e^{\sup f|_{[j]}} \le \frac{\varepsilon}{2^k}.$$

Then $m_n(\pi_k^{-1}([n_k+1,\infty))) \le \varepsilon/2^k$ for every $n \in \mathbb{N}$ and every $k \in \mathbb{N}$. Hence,

$$m_n\Big(E_A^\infty \cap \prod_{k=1}^\infty [1,n_k]\Big) \ge 1 - \sum_{k=1}^\infty m_n(\pi_k^{-1}([n_k+1,\infty))) \ge 1 - \sum_{k=1}^\infty \frac{\varepsilon}{2^k} = 1-\varepsilon$$

for every $n \in \mathbb{N}$. As $E_A^\infty \cap \prod_{k=1}^\infty [1,n_k]$ is a compact subset of E_A^∞, the tightness of the sequence $(m_n)_{n=1}^\infty$ is proved. Prohorov's theorem (see Chapter 1, Section 5 in Billingsley [9]) then affirms that $(m_n)_{n=1}^\infty$ has a weak* limit point m.

Naturally, m is our candidate as eigenmeasure of $\mathcal{L}_f^*$. To simplify calculations, we will use the normalized operators $\widehat{\mathcal{L}}_f = e^{-\mathrm{P}(f)}\mathcal{L}_f$ and $\widehat{\mathcal{L}}_{f_n} = e^{-\mathrm{P}_n}\mathcal{L}_{f_n}$, and their respective duals. Fix $g \in C_b(E_A^\infty)$. Let $g_n = g|_{E_{e_n}^\infty}$. By the triangle inequality,

$$\begin{aligned}|\widehat{\mathcal{L}}_f^*(m)(g) - m(g)| \le{}& |\widehat{\mathcal{L}}_f^*(m)(g) - \widehat{\mathcal{L}}_f^*(m_n)(g)| + |\widehat{\mathcal{L}}_f^*(m_n)(g) - \widehat{\mathcal{L}}_{f_n}^*(m_n)(g)| \\ &+ |\widehat{\mathcal{L}}_{f_n}^*(m_n)(g) - m_n(g_n)| + |m_n(g_n) - m_n(g)| \\ &+ |m_n(g) - m(g)|. \end{aligned} \tag{17.30}$$

Observe that

$$\begin{aligned}\widehat{\mathcal{L}}_{f_n}^*(m_n)(g) &= \int_{E_A^\infty} g\, d\widehat{\mathcal{L}}_{f_n}^*(m_n) = \int_{E_{e_n}^\infty} g\, d\widehat{\mathcal{L}}_{f_n}^*(m_n) \\ &= \int_{E_{e_n}^\infty} g_n\, d\widehat{\mathcal{L}}_{f_n}^*(m_n) = \widehat{\mathcal{L}}_{f_n}^*(m_n)(g_n) = m_n(g_n)\end{aligned} \tag{17.31}$$

and

$$m_n(g_n) - m_n(g) = \int_{E_{e_n}^\infty} (g_n - g)\,dm_n = \int_{E_{e_n}^\infty} 0\, dm_n = 0. \tag{17.32}$$

Thus, the right-hand side of (17.30) reduces to the first, second and fifth terms. Let $\varepsilon > 0$. Recall that $(m_n)_{n=1}^\infty$ weak* converges to m, that f is summable, and that

$\lim_{n\to\infty} \mathrm{P}_n = \mathrm{P}(f)$ by Theorem 17.2.9. Take $n \in \mathbb{N}$ so large that

$$|m(g) - m_n(g)| \le \frac{\varepsilon}{3}, \tag{17.33}$$

$$|m(\widehat{\mathcal{L}}_f(g)) - m_n(\widehat{\mathcal{L}}_f(g))| \le \frac{\varepsilon}{3}, \tag{17.34}$$

$$\|g\|_\infty e^{-\mathrm{P}(f)} \sum_{i>\ell_n} \exp(\sup f|_{[i]}) \le \frac{\varepsilon}{6}, \tag{17.35}$$

$$\sum_{i\in E} \|g\|_\infty \exp(\sup f|_{[i]} - \mathrm{P}_1) \cdot (\mathrm{P}(f) - \mathrm{P}_n) \le \frac{\varepsilon}{6}. \tag{17.36}$$

Relations (17.34) and (17.33) state that the first and last summands on the right-hand side of (17.30) are each bounded above by $\varepsilon/3$. Let us look at the second summand. First, from (17.31) notice that

$$\begin{aligned}
\widehat{\mathcal{L}}_{f_n}^*(m_n)(g) = \widehat{\mathcal{L}}_{f_n}^*(m_n)(g_n) &= m_n(\widehat{\mathcal{L}}_{f_n}(g_n)) \\
&= \int_{E_{\ell_n}^\infty} \sum_{i\le\ell_n:\, A_{i\omega_1}=1} g_n(i\omega)\exp(f_n(i\omega) - \mathrm{P}_n)\,dm_n(\omega) \\
&= \int_{E_A^\infty} \sum_{i\le\ell_n:\, A_{i\omega_1}=1} g(i\omega)\exp(f(i\omega) - \mathrm{P}_n)\,dm_n(\omega).
\end{aligned}$$

Using the mean value theorem, (17.36) and (17.35), we then obtain

$$\begin{aligned}
|\widehat{\mathcal{L}}_f^*(m_n)(g) - \widehat{\mathcal{L}}_{f_n}^*(m_n)(g)| &= \Bigg| \int_{E_A^\infty} \sum_{i\le\ell_n:\, A_{i\omega_1}=1} g(i\omega)\Big(\exp(f(i\omega) - \mathrm{P}(f)) \\
&\qquad\qquad - \exp(f(i\omega) - \mathrm{P}_n)\Big)dm_n(\omega) \\
&\quad + \int_{E_A^\infty} \sum_{i>\ell_n:\, A_{i\omega_1}=1} g(i\omega)\exp(f(i\omega) - \mathrm{P}(f))\,dm_n(\omega)\Bigg| \\
&\le \sum_{i\le\ell_n} \|g\|_\infty \exp(\sup f|_{[i]} - \mathrm{P}_1)\cdot(\mathrm{P}(f) - \mathrm{P}_n) \\
&\quad + \sum_{i>\ell_n} \|g\|_\infty \exp(\sup f|_{[i]} - \mathrm{P}(f)) \\
&\le \frac{\varepsilon}{6} + \frac{\varepsilon}{6} = \frac{\varepsilon}{3}.
\end{aligned}$$

So the second summand on the right-hand side of (17.30) is also bounded above by $\varepsilon/3$. In summary, (17.30) reduces to

$$|\widehat{\mathcal{L}}_f^*(m)(g) - m(g)| \le \frac{\varepsilon}{3} + \frac{\varepsilon}{3} + 0 + 0 + \frac{\varepsilon}{3} = \varepsilon.$$

Letting $\varepsilon \to 0$, we deduce that $\widehat{\mathcal{L}}_f^*(m)(g) = m(g)$, i. e., $\mathcal{L}_f^*(m)(g) = e^{\mathrm{P}(f)}m(g)$. As $g \in C_b(E_A^\infty)$ was chosen arbitrarily, we conclude that $\mathcal{L}_f^*(m) = e^{\mathrm{P}(f)}m$. □

As an immediate consequence of Theorems 17.7.4, 17.6.5, 17.6.4, 17.4.6 and 17.5.1, we deduce the following fundamental result.

Corollary 17.7.5. *Suppose that a potential $f : E_A^\infty \to \mathbb{R}$ is summable and Hölder continuous on cylinders and that the incidence matrix A is finitely irreducible.*
(a) *There exists a unique Borel probability eigenmeasure m_f of the dual transfer operator $\mathcal{L}_f^*$ and its corresponding eigenvalue is $e^{\mathrm{P}(f)}$.*
(b) *The eigenmeasure m_f is a Gibbs state for f.*
(c) *The potential f has a unique σ-invariant Gibbs state μ_f, and this state is ergodic. If $\int_{E_A^\infty} f\,d\mu_f > -\infty$, then μ_f is also the unique equilibrium state for f. Finally, if A is finitely primitive, then μ_f is totally ergodic.*

The last result of this section explains the real dynamical meaning of the fixed points of the normalized transfer operator.

Proposition 17.7.6. *Suppose that a potential $f : E_A^\infty \to \mathbb{R}$ is summable and Hölder continuous on cylinders and that the incidence matrix A is finitely irreducible. Let m_f be the unique Borel probability measure fixed by the dual $\widehat{\mathcal{L}}_f^*$ of the normalized transfer operator $\widehat{\mathcal{L}}_f$. Let*

$$\mathrm{Fix}(\widehat{\mathcal{L}}_f) = \left\{g \in L^1(m_f) : \widehat{\mathcal{L}}_f(g) = g,\ \int_{E_A^\infty} g\,dm_f = 1,\ \text{and}\ g \geq 0\right\}$$

and

$$\mathrm{AI}(m_f) = \left\{g \in L^1(m_f) : gm_f \circ \sigma^{-1} = gm_f,\ \int_{E_A^\infty} g\,dm_f = 1,\ \text{and}\ g \geq 0\right\}.$$

Then $\mathrm{Fix}(\widehat{\mathcal{L}}_f) = \mathrm{AI}(m_f)$.

Proof. It follows from (17.29) that for every $e \in E$ and every $\omega \in E_A^\infty$ with $A_{e\omega_1} = 1$, we have

$$\frac{dm_f \circ e}{dm_f}(\omega) = \exp(f(e\omega) - \mathrm{P}(f)),$$

where the map $e : \{\omega \in E_A^\infty : A_{e\omega_1} = 1\} \to E_A^\infty$ on the left-hand side is defined by $e(\omega) := e\omega$. So if $g \in \mathrm{AI}(m_f)$, then

$$\frac{d(gm_f \circ e)}{dm_f}(\omega) = g(e\omega)\exp(f(e\omega) - \mathrm{P}(f)).$$

Therefore, $\widehat{\mathcal{L}}_f$ sends the Radon-Nikodym derivative of a measure m absolutely continuous with respect to m_f to the Radon-Nikodym derivative of the measure $m \circ \sigma^{-1}$ with respect to m_f and the result ensues. □

17.8 The invariant Gibbs state has positive entropy; pressure gap

First, we shall prove the following strengthening of Proposition 13.7.19.

Proposition 17.8.1. *Let (X, d) be a separable complete metric space and μ a Borel finite measure on X. For every $r > 0$, define*

$$S_\mu(r) := \sup\{\mu(B(x, r)) : x \in X\}.$$

If the measure μ is atomless, then

$$\lim_{r \to 0} S_\mu(r) = 0.$$

Proof. Fix $\varepsilon > 0$. Since μ is inner regular, there exists a compact set $K \subseteq X$ such that $\mu(X \backslash K) < \varepsilon/2$. It then follows from Proposition 13.7.19 that there is $r_\varepsilon > 0$ such that

$$\mu(B(z, 2r) \cap K) < \varepsilon/2, \quad \forall z \in K, \ \forall r \le r_\varepsilon.$$

Fix $x \in X$ and $0 < r \le r_\varepsilon$. If $B(x, r) \cap K = \emptyset$, then

$$\mu(B(x, r)) \le \mu(X \backslash K) < \varepsilon/2 < \varepsilon.$$

If $B(x, r) \cap K \ne \emptyset$, let $z \in B(x, r) \cap K$. Then $B(x, r) \subseteq B(z, 2r)$, whence

$$\mu(B(x, r)) = \mu(B(x, r) \cap K) + \mu(B(x, r) \backslash K) \le \mu(B(z, 2r) \cap K) + \mu(X \backslash K) < 2 \cdot \frac{\varepsilon}{2} = \varepsilon. \quad \square$$

Having this proposition, we can prove the following theorem in an entirely analogous way to Theorem 13.7.20.

Theorem 17.8.2. *If E is a countable set with at least two elements and $A : E \times E \to \{0, 1\}$ is a finitely irreducible incidence matrix such that E_A^∞ is not a single periodic orbit of the shift map $\sigma : E_A^\infty \to E_A^\infty$, then each potential $f : E_A^\infty \to \mathbb{R}$ which is summable and Hölder continuous on cylinders has a pressure gap with respect to the shift map $\sigma : E_A^\infty \to E_A^\infty$.*

Proof. Since the invariant Gibbs state μ_f is ergodic and has full support $\operatorname{supp}(\mu_f) = E_A^\infty$, and since E_A^∞ is infinite, we infer that μ_f is atomless. Let $C_{\mu_f} \ge 1$ be a constant C from (17.9) ascribed to μ_f and let $P = \mathrm{P}(f) = \mathrm{P}(\sigma, f)$. Since the metric space (E_A^∞, d_s) is separable and complete, it follows from Proposition 17.8.1 that there exists $k \in \mathbb{N}$ such that

$$\mu_f([\omega]) \leq \frac{1}{eC_{\mu_f}}, \qquad \forall \omega \in E_A^k.$$

Given any $\omega \in E_A^k$ and $\rho \in [\omega]$, it follows from this and (17.9) that

$$S_kf(\rho) - k\mathrm{P}(\sigma,f) \leq \log C_{\mu_f} + \log \mu_f([\omega]) \leq \log C_{\mu_f} + \log 1 - \log e - \log C_{\mu_f} = -1.$$

Using this and Exercise 17.9.9, we conclude that

$$S_kf(\rho) \leq k\mathrm{P}(\sigma,f) - 1 = \mathrm{P}(\sigma^k, S_kf) - 1$$

for all $\rho \in [\omega]$ and all $\omega \in E_A^k$. That is, there is a pressure gap. □

The existence of a pressure gap entails that the system has strictly positive entropy with respect to the unique σ-invariant Gibbs state μ_f.

Theorem 17.8.3. *Let E be a countable set with at least two elements and $A : E \times E \to \{0,1\}$ a finitely irreducible matrix such that E_A^∞ is not a single periodic orbit of the shift map $\sigma : E_A^\infty \to E_A^\infty$. If a potential $f : E_A^\infty \to \mathbb{R}$ is summable and Hölder continuous on cylinders and $\int_{E_A^\infty} f\, d\mu_f > -\infty$, then $\mathrm{h}_{\mu_f}(\sigma) > 0$.*

Proof. The proof is similar to the alternative argument given in the proof of Theorem 13.7.21. Use Corollary 17.7.5(c), the 3rd variational principle (Theorem 17.3.4), Exercise 17.9.9 and Theorem 17.8.2. □

17.9 Exercises

Exercise 17.9.1. Let E be a countably infinite alphabet and $A : E \times E \to \{0,1\}$ be a finitely irreducible incidence matrix. Prove that the symbolic space E_A^∞ is not locally compact nor even σ-compact.

Exercise 17.9.2. Show that any potential which is Hölder continuous on cylinders, is acceptable.

Exercise 17.9.3. Let $f : E_A^\infty \to \mathbb{R}$. Show that f is Hölder continuous if and only if it is Hölder continuous on cylinders and bounded.

Exercise 17.9.4. A sequence $(a_n)_{n=1}^\infty$ of positive real numbers is *boundedly submultiplicative* with constant $B > 0$ if

$$a_{m+n} \leq Ba_m a_n, \qquad \forall m,n \in \mathbb{N}.$$

(If $0 < B \le 1$, then the sequence is *submultiplicative*.) Prove that

$$\lim_{n\to\infty} \frac{1}{n} \log a_n = \inf_{n\in\mathbb{N}} \frac{1}{n} \log(Ba_n).$$

Show also that the above limit may be $-\infty$ but if $\inf_{n\in\mathbb{N}} a_n > 0$, then the limit is nonnegative. N. B.: This result is due to Fekete.

Exercise 17.9.5. A sequence $(a_n)_{n=1}^{\infty}$ of positive real numbers is *boundedly supermultiplicative* with constant $C > 0$ if

$$C^{-1} a_m a_n \le a_{m+n}, \quad \forall m, n \in \mathbb{N}.$$

(If $0 < C \le 1$, then the sequence is *supermultiplicative*.) Prove that

$$\lim_{n\to\infty} \frac{1}{n} \log a_n = \sup_{n\in\mathbb{N}} \frac{1}{n} \log(C^{-1} a_n).$$

Show also that the limit may be ∞ but if $\sup_{n\in\mathbb{N}} a_n < \infty$, then the limit is nonpositive.

Exercise 17.9.6. The *normalization* of a boundedly submultiplicative sequence $(a_n)_{n=1}^{\infty}$ is the sequence $(\tilde{a}_n)_{n=1}^{\infty}$, where

$$\tilde{a}_n = a_n e^{-Pn} \quad \text{and} \quad P = \lim_{n\to\infty} \frac{1}{n} \log a_n.$$

Prove that the normalization of a boundedly submultiplicative sequence with constant B is a boundedly submultiplicative sequence with the same constant B but that

$$\lim_{n\to\infty} \frac{1}{n} \log \tilde{a}_n = \inf_{n\in\mathbb{N}} \frac{1}{n} \log(B\tilde{a}_n) = 0.$$

Exercise 17.9.7. The normalization of a boundedly supermultiplicative sequence $(a_n)_{n=1}^{\infty}$ is the sequence $(\tilde{a}_n)_{n=1}^{\infty}$, where

$$\tilde{a}_n = a_n e^{-Pn} \quad \text{and} \quad P = \lim_{n\to\infty} \frac{1}{n} \log a_n.$$

Show that the normalization of a boundedly supermultiplicative sequence with constant C is a boundedly supermultiplicative sequence with the same constant C but

$$\lim_{n\to\infty} \frac{1}{n} \log \tilde{a}_n = \sup_{n\in\mathbb{N}} \frac{1}{n} \log(C^{-1}\tilde{a}_n) = 0.$$

Exercise 17.9.8. Prove that the normalization $(\tilde{a}_n)_{n=1}^{\infty}$ of a boundedly supermultiplicative and boundedly submultiplicative sequence $(a_n)_{n=1}^{\infty}$ of positive numbers such that $-\infty < P < \infty$ is a bounded sequence. More precisely, it is bounded from below by B^{-1}, where B is any constant of submultiplicativity, and from above by any constant C of supermultiplicativity.

Exercise 17.9.9. Suppose that $f : E_A^\infty \to \mathbb{R}$ is an acceptable potential and that A is a finitely irreducible matrix. Show that $\mathrm{P}(\sigma^k, S_k f) = k\mathrm{P}(\sigma, f)$ for any $k \in \mathbb{N}$. (This is a generalization of Theorem 11.1.22 to the countable state symbolic setting.)

Exercise 17.9.10. Let E be a countable alphabet and $A : E \times E \to \{0,1\}$ be a finitely irreducible matrix. Denote by $C_a(E_A^\infty)$ the set of all acceptable potentials on E_A^∞.
(a) Show that $C_a(E_A^\infty)$ is a convex set.
(b) Using the 1st variational principle (Theorem 17.3.2), prove that the pressure function $\mathrm{P}(\bullet) : C_a(E_A^\infty) \to \mathbb{R}$ is convex.
(c) Using rather the 3rd variational principle (Theorem 17.3.4), give an alternative demonstration of the convexity of the pressure function.

Exercise 17.9.11. Prove Proposition 17.4.3.

Exercise 17.9.12. Show that any potential $f : E_A^\infty \to \mathbb{R}$ that admits a Gibbs state must be summable.

Exercise 17.9.13. Prove Lemma 17.6.2.

Exercise 17.9.14. Prove that a Hölder continuous potential is not summable. Infer that such a potential does not admit a Gibbs state. Conclude that it is of no interest to study Hölder continuous potentials for countable state symbolic systems.

Exercise 17.9.15. Two functions $\varphi, \psi : E_A^\infty \to \mathbb{R}$ are said to be *cohomologous modulo a constant* (or, equivalently, $\varphi - \psi$ is cohomologous to a constant) in the additive group of bounded functions on E_A^∞ if there exist a bounded function $u : E_A^\infty \to \mathbb{R}$ and a constant $c \in \mathbb{R}$ such that

$$\varphi - \psi = u \circ \sigma - u + c.$$

(cf. Definition 13.2.5) The objective of this exercise is to establish results analogous to Lemma 13.2.6 and Exercise 12.4.5.

Suppose that $\varphi, \psi : E_A^\infty \to \mathbb{R}$ are acceptable potentials and that A is finitely irreducible.
(a) Show that $\mathrm{P}(\varphi) = \mathrm{P}(\psi) + c$.
(b) Prove that $\mathcal{G}_\varphi = \mathcal{G}_\psi$, i. e., φ and ψ share the same set of Gibbs states.
(c) Demonstrate that φ and ψ share the same set of equilibrium states.

Exercise 17.9.16. Repeat Exercise 13.11.13 but this time with a countable alphabet E.

Exercise 17.9.17. Repeat Exercise 13.11.14 but with a countable alphabet E.

18 Countable state thermodynamic formalism: finer properties

In this chapter, we continue the development of thermodynamic formalism for countable-alphabet subshifts of finite type $\sigma : E_A^\infty \to E_A^\infty$. This time, we deal with refined properties of σ-invariant Gibbs states μ_f for potentials $f : E_A^\infty \to \mathbb{R}$ that are summable and Hölder continuous on cylinders. We show that the measure-theoretic dynamical systems $(E_A^\infty, \sigma, \mu_f)$ deserve to be called random or chaotic by proving in Section 18.3 that they obey classical stochastic laws such as the exponential decay of correlations and the central limit theorem. Although we do not prove it, they also behave in accordance with the law of the iterated logarithm. Another property essential for randomness is the existence of asymptotic variances for Hölder continuous functions. We prove two very useful formulas for them. All of this is achieved via a refined study of transfer operators acting on the Banach space of Hölder continuous functions and especially by establishing their spectral properties, such as quasicompactness (spectral gap) and simplicity of the leading eigenvalue $e^{\mathrm{P}(f)}$. To set the table, we prove in Section 18.1 the famous Ionescu-Tulcea and Marinescu inequality. In Section 18.2, we establish the continuity (1) of the pressure function $f \mapsto \mathrm{P}(f)$; (2) of the unique σ-invariant Gibbs states $f \mapsto \mu_f$; (3) of the unique eigenmeasures of the dual operators $f \mapsto m_f$; and (4) of the Radon–Nikodym derivative of μ_f with respect to m_f. Finally, in Section 18.4 we characterize, primarily by means of cohomological equations, those potentials that have the same invariant Gibbs state.

18.1 Ionescu-Tulcea and Marinescu inequality

In this section, we prove the famous Ionescu-Tulcea and Marinescu inequality for countable state subshifts of finite type, with the aim of obtaining a spectral picture and the rate of convergence of the iterates of the transfer operator.

Unless otherwise stated, E is a countable alphabet and $A : E \times E \to \{0,1\}$ a finitely irreducible incidence matrix. Let $\beta \in (0,1]$. Denote by $\mathcal{K}_\beta(E_A^\infty)$ the set of all complex-valued functions that are Hölder continuous on cylinders with exponent β (cf. Definition 17.2.2). Denote by $\mathcal{K}_\beta(E_A^\infty, \mathbb{R})$ the subspace of $\mathcal{K}_\beta(E_A^\infty)$ consisting of those functions that are real valued. Set (cf. Definition 17.6.1)

$$\begin{aligned}\mathcal{K}_\beta^s(E_A^\infty) &:= \{f \in \mathcal{K}_\beta(E_A^\infty) : \mathrm{Re}(f) \text{ is summable}\} \\ &= \left\{f \in \mathcal{K}_\beta(E_A^\infty) : \sum_{e \in E} \exp(\sup \mathrm{Re}(f|_{[e]})) < \infty\right\}.\end{aligned}$$

Notice that the real part of the functions in $\mathcal{K}_\beta^s(E_A^\infty)$ is always bounded from above, but is not bounded from below when E is infinite.

https://doi.org/10.1515/9783110702699-018

Let $C_b(E_A^\infty)$ be the set of all bounded continuous complex-valued functions on E_A^∞ and

$$\mathcal{K}_\beta^b(E_A^\infty) := \mathcal{K}_\beta(E_A^\infty) \cap C_b(E_A^\infty).$$

Recall that a function $f : E_A^\infty \to \mathbb{C}$ is Hölder continuous if and only if it is Hölder continuous on cylinders and bounded. Thus, $\mathcal{K}_\beta^b(E_A^\infty)$ is the set of all complex-valued functions on E_A^∞ that are Hölder with exponent β. When endowed with the natural norm

$$\|f\|_\beta := v_\beta(f) + \|f\|_\infty, \tag{18.1}$$

the set $\mathcal{K}_\beta^b(E_A^\infty)$ becomes a Banach space. Note that $\mathcal{K}_\beta^b(E_A^\infty) = \mathcal{K}_\beta^s(E_A^\infty)$ when E is finite, while $\mathcal{K}_\beta^b(E_A^\infty) \cap \mathcal{K}_\beta^s(E_A^\infty) = \emptyset$ when E is infinite.

For any $f \in \mathcal{K}_\beta^b(E_A^\infty)$ and $F \subseteq E_A^\infty$, define

$$\|f\|_{\beta,F} = v_{\beta,F}(f) + \|f\|_{\infty,F},$$

where

$$v_{\beta,F}(f) := v_\beta(f|_F) = \sup_{\substack{\omega\neq\tau\in F:\\ \omega_1=\tau_1}} \frac{|f(\omega)-f(\tau)|}{d^\beta(\omega,\tau)} \quad \text{and} \quad \|f\|_{\infty,F} := \big\|f|_F\big\|_\infty = \sup_{\omega\in F}|f(\omega)|.$$

Clearly, $v_{\beta,F}(f) \le v_\beta(f)$ and $\|f\|_{\infty,F} \le \|f\|_\infty$, so that $\|f\|_{\beta,F} \le \|f\|_\beta$. It is further easy to see that (cf. Exercise 18.5.4)

$$\|fg\|_{\beta,F} \le \|f\|_{\beta,F}\|g\|_{\beta,F}, \quad \forall f,g \in \mathcal{K}_\beta^b(E_A^\infty). \tag{18.2}$$

We briefly look at the iterates of the normalized transfer operator.

Lemma 18.1.1. *Let E be a countable alphabet and $A : E\times E \to \{0,1\}$ a finitely irreducible incidence matrix. Let $f \in \mathcal{K}_\beta^s(E_A^\infty, \mathbb{R})$ be a potential and let C be a ratio bounding constant for a Gibbs state for f. Then*

$$\big\|\widehat{\mathcal{L}}_f^n(\mathbb{1})\big\|_\infty \le C, \quad \forall n\in\mathbb{N} \quad \textit{and hence} \quad \big\|\widehat{\mathcal{L}}_f^n(g)\big\|_\infty \le C\|g\|_\infty, \quad \forall n\in\mathbb{N},\ \forall g\in C_b(E_A^\infty).$$

Proof. First, a Gibbs state exists according to Corollary 17.7.5. Denote by m the Gibbs state with ratio bounding constant C. By (17.22) and (17.9), for every $\rho \in E_A^\infty$ we have

$$\widehat{\mathcal{L}}_f^n(\mathbb{1})(\rho) = \sum_{\omega\in E_A^n:\, A_{\omega_n\rho_1}=1} \exp(S_n f(\omega\rho) - \mathrm{P}(f)n) \le C \sum_{\omega\in E_A^n:\, A_{\omega_n\rho_1}=1} m([\omega]) \le C. \qquad \square$$

Lemma 18.1.2. *Suppose Λ is a finite set of words witnessing the finite irreducibility of the incidence matrix A. Let $|\Lambda|$ be the maximal length of words in Λ. If $f \in \mathcal{K}_\beta^s(E_A^\infty, \mathbb{R})$,* *then there exists a constant $D > 0$ such that*

$$\sum_{k=n}^{n+|\Lambda|} \widehat{\mathcal{L}}_f^k(\mathbb{1})(\omega) \geq D, \quad \forall n \in \mathbb{N},\ \forall \omega \in E_A^\infty.$$

Proof. As $f \in \mathcal{K}_\beta^s(E_A^\infty, \mathbb{R})$, it follows from Corollary 17.7.5 and (17.25) that $\int \widehat{\mathcal{L}}_f^k(\mathbb{1})\, dm_f = 1$ for all $k \in \mathbb{N}$. So, for every $k \in \mathbb{N}$ there is $\gamma(k) \in E_A^\infty$ such that $\widehat{\mathcal{L}}_f^k(\mathbb{1})(\gamma(k)) \geq 1$. Since Λ is finite and $f \in \mathcal{K}_\beta(E_A^\infty, \mathbb{R})$, we have

$$\ell := \min_{\alpha \in \Lambda} \exp\Big(\inf(S_{|\alpha|} f|_{[\alpha]}) - \mathrm{P}(f)|\alpha|\Big) > 0.$$

Let $n \in \mathbb{N}$ and $\tau \in E_A^\infty$. Then

$$\sum_{k=n}^{n+|\Lambda|} \widehat{\mathcal{L}}_f^k(\mathbb{1})(\tau) = \sum_{k=n}^{n+|\Lambda|} \sum_{\omega \in E_A^k : A_{\omega_k \tau_1} = 1} \exp(S_k f(\omega\tau) - \mathrm{P}(f)k).$$

For each $\omega \in E_A^n$, choose $\alpha(\omega, \tau) \in \Lambda$ such that $\omega\alpha(\omega,\tau)\tau \in E_A^\infty$. Then

$$\begin{aligned}
\sum_{k=n}^{n+|\Lambda|} \widehat{\mathcal{L}}_f^k(\mathbb{1})(\tau) &\geq \sum_{\omega \in E_A^n} \exp\Big(S_{n+|\alpha(\omega,\tau)|} f(\omega\alpha(\omega,\tau)\tau) - \mathrm{P}(f)(n + |\alpha(\omega,\tau)|)\Big) \\
&= \sum_{\omega \in E_A^n} \exp\Big(S_n f(\omega\alpha(\omega,\tau)\tau) - \mathrm{P}(f)n\Big) \\
&\quad \times \exp\Big(S_{|\alpha(\omega,\tau)|} f(\alpha(\omega,\tau)\tau) - \mathrm{P}(f)|\alpha(\omega,\tau)|\Big) \\
&\geq \ell \sum_{\omega \in E_A^n} \exp\Big(S_n f(\omega\alpha(\omega,\tau)\tau) - \mathrm{P}(f)n\Big).
\end{aligned}$$

If $A_{\omega_n \gamma(n)_1} = 1$, then by Lemma 17.2.3 we obtain that

$$\exp\Big(S_n f(\omega\alpha(\omega,\tau)\tau) - \mathrm{P}(f)n\Big) \geq e^{-\frac{v_\beta(f)}{e^\beta - 1}} \exp\Big(S_n f(\omega\gamma(n)) - \mathrm{P}(f)n\Big).$$

Therefore,

$$\begin{aligned}
\sum_{k=n}^{n+|\Lambda|} \widehat{\mathcal{L}}_f^k(\mathbb{1})(\tau) &\geq \ell e^{-\frac{v_\beta(f)}{e^\beta - 1}} \sum_{\omega \in E_A^n : A_{\omega_n \gamma(n)_1} = 1} \exp\Big(S_n f(\omega\gamma(n)) - \mathrm{P}(f)n\Big) \\
&\geq \ell e^{-\frac{v_\beta(f)}{e^\beta - 1}} \widehat{\mathcal{L}}_f^n(\mathbb{1})(\gamma(n)) \geq \ell e^{-\frac{v_\beta(f)}{e^\beta - 1}}.
\end{aligned}$$

□

Remark 18.1.3. If the underlying matrix A in Lemma 18.1.2 is finitely primitive, then there is a constant $D > 0$ such that $\widehat{\mathcal{L}}_f^n(\mathbb{1})(\omega) \geq D$ for all $n \in \mathbb{N}$ and all $\omega \in E_A^\infty$.

The main technical result of this section is the following.

Lemma 18.1.4 (Ionescu-Tulcea and Marinescu inequality). *Let E be a countable alphabet and $A : E \times E \to \{0,1\}$ a finitely irreducible incidence matrix. If $f \in \mathcal{K}^s_\beta(E_A^\infty, \mathbb{R})$, then the normalized transfer operator $\widehat{\mathcal{L}}_f : C_b(E_A^\infty) \to C_b(E_A^\infty)$ preserves the space $\mathcal{K}^b_\beta(E_A^\infty)$, i. e. $\widehat{\mathcal{L}}_f(\mathcal{K}^b_\beta(E_A^\infty)) \subseteq \mathcal{K}^b_\beta(E_A^\infty)$. Moreover, there exists a constant $Q > 1$ such that*

$$v_\beta(\widehat{\mathcal{L}}_f^n(g)) \le C(e^{-\beta n}\|g\|_\beta + (Q-1)\|g\|_\infty) \quad \text{and} \quad \|\widehat{\mathcal{L}}_f^n(g)\|_\infty \le C\|g\|_\infty$$

for all $n \in \mathbb{N}$ and all $g \in \mathcal{K}^b_\beta(E_A^\infty)$, where C is a ratio bounding constant for a Gibbs state for f and C, Q are independent of n and g. Consequently,

$$\|\widehat{\mathcal{L}}_f^n(g)\|_\beta \le C(e^{-\beta n}\|g\|_\beta + Q\|g\|_\infty), \quad \forall n \in \mathbb{N},\ \forall g \in \mathcal{K}^b_\beta(E_A^\infty).$$

Proof. Let $\rho, \tau \in E_A^\infty$ be such that $|\rho \wedge \tau| \ge 1$. For every $n \in \mathbb{N}$,

$$\begin{aligned}
\widehat{\mathcal{L}}_f^n(g)(\rho) - \widehat{\mathcal{L}}_f^n(g)(\tau) &= \sum_{\omega \in E_A^n : A_{\omega_n \rho_1}=1} g(\omega\rho)\exp(S_n f(\omega\rho) - \mathrm{P}(f)n) \\
&\quad - \sum_{\omega \in E_A^n : A_{\omega_n \rho_1}=1} g(\omega\tau)\exp(S_n f(\omega\tau) - \mathrm{P}(f)n) \\
&= \sum_{\omega \in E_A^n : A_{\omega_n \rho_1}=1} (g(\omega\rho) - g(\omega\tau))\exp(S_n f(\omega\rho) - \mathrm{P}(f)n) \\
&\quad + \sum_{\omega \in E_A^n : A_{\omega_n \rho_1}=1} g(\omega\tau)\Big[\exp(S_n f(\omega\rho) - \mathrm{P}(f)n) \\
&\qquad\qquad - \exp(S_n f(\omega\tau) - \mathrm{P}(f)n)\Big].
\end{aligned} \tag{18.3}$$

But

$$|g(\omega\rho) - g(\omega\tau)| \le v_\beta(g) d^\beta(\omega\rho, \omega\tau) = e^{-\beta n} v_\beta(g) d^\beta(\rho, \tau).$$

Using Lemma 18.1.1, we obtain that

$$\begin{aligned}
\sum_{\omega \in E_A^n : A_{\omega_n \rho_1}=1} |g(\omega\rho) - g(\omega\tau)| \exp(S_n f(\omega\rho) - \mathrm{P}(f)n) &\le e^{-\beta n} v_\beta(g) d^\beta(\rho,\tau)\,(\widehat{\mathcal{L}}_f^n(\mathbb{1}))(\rho) \\
&\le C e^{-\beta n}\|g\|_\beta d^\beta(\rho,\tau).
\end{aligned} \tag{18.4}$$

Notice that there exists a constant $Q_0 \ge 1$ such that $|1 - e^x| \le Q_0|x|$ for all x with $|x| \le \frac{v_\beta(f)}{e^\beta - 1}$. Employing Lemma 17.2.3, we get

$$\begin{aligned}
&|\exp(S_n f(\omega\rho) - \mathrm{P}(f)n) - \exp(S_n f(\omega\tau) - \mathrm{P}(f)n)| \\
&\quad = \exp(S_n f(\omega\rho) - \mathrm{P}(f)n)|1 - \exp(S_n f(\omega\tau) - S_n f(\omega\rho))| \\
&\quad \le Q_0 \exp(S_n f(\omega\rho) - \mathrm{P}(f)n)|S_n f(\omega\rho) - S_n f(\omega\tau)| \\
&\quad \le Q_0 \exp(S_n f(\omega\rho) - \mathrm{P}(f)n)\frac{v_\beta(f)}{e^\beta - 1} d^\beta(\rho,\tau).
\end{aligned}$$

Therefore, using Lemma 18.1.1, we have that

$$\sum_{\omega\in E_A^n:A_{\omega_n\rho_1}=1} |g(\omega\tau)|\Big|\exp(S_nf(\omega\rho)-\mathrm{P}(f)n)-\exp(S_nf(\omega\tau)-\mathrm{P}(f)n)\Big|$$

$$\le \|g\|_\infty Q_0\frac{\upsilon_\beta(f)}{e^\beta-1}d^\beta(\rho,\tau)\sum_{\omega\in E_A^n:A_{\omega_n\rho_1}=1}\exp(S_nf(\omega\rho)-\mathrm{P}(f)n)$$

$$=\|g\|_\infty Q_0\frac{\upsilon_\beta(f)}{e^\beta-1}d^\beta(\rho,\tau)\widehat{\mathcal{L}}_f^n(\mathbb{1})(\rho)\le CQ_0\frac{\upsilon_\beta(f)}{e^\beta-1}\|g\|_\infty d^\beta(\rho,\tau).$$

Combining this inequality with (18.4) and (18.3), we get

$$|\widehat{\mathcal{L}}_f^n(g)(\rho)-\widehat{\mathcal{L}}_f^n(g)(\tau)|\le\left[Ce^{-\beta n}\|g\|_\beta+CQ_0\frac{\upsilon_\beta(f)}{e^\beta-1}\|g\|_\infty\right]d^\beta(\rho,\tau),$$

i. e., $\upsilon_\beta(\widehat{\mathcal{L}}_f^n(g))\le Ce^{-\beta n}\|g\|_\beta+CQ_0\frac{\upsilon_\beta(f)}{e^\beta-1}\|g\|_\infty$. Joining this with Lemma 18.1.1, we conclude that

$$\|\widehat{\mathcal{L}}_f^n(g)\|_\beta\le Ce^{-\beta n}\|g\|_\beta+C\Big(Q_0\frac{\upsilon_\beta(f)}{e^\beta-1}+1\Big)\|g\|_\infty.$$

Set $Q=Q_0\frac{\upsilon_\beta(f)}{e^\beta-1}+1$. □

Remark 18.1.5. The proof of Lemma 18.1.4 used only a "weaker" property of Gibbs states, namely only the right inequality of (17.9).

Remark 18.1.6. If the unit ball in $\mathcal{K}_\beta^b(E_A^\infty)$ was a compact subset of the Banach space $(C_b(E_A^\infty),\|\cdot\|_\infty)$, we could use the famous Ionescu-Tulcea and Marinescu theorem (Theorem B.1.1 in Appendix B) to establish some useful spectral properties of the normalized transfer operator $\widehat{\mathcal{L}}_f$. But this ball is compact only in the topology of uniform convergence on compact subsets of E_A^∞ and we thus need to prove these spectral properties directly.

Next, we introduce an important fixed point ψ_f of $\widehat{\mathcal{L}}_f$.

Theorem 18.1.7. *Let E be a countable alphabet and $A:E\times E\to\{0,1\}$ a finitely irreducible incidence matrix. Let $f\in\mathcal{K}_\beta^s(E_A^\infty,\mathbb{R})$ be a potential and m_f be the unique Borel probability eigenmeasure m_f of the dual transfer operator $\mathcal{L}_f^*$ or, equivalently, the unique fixed point of $\widehat{\mathcal{L}}_f^*$ per Corollary 17.7.5. Then $\widehat{\mathcal{L}}_f$ has a fixed point $\psi_f\in\mathcal{K}_\beta^b(E_A^\infty)$ such that*

$$\int_{E_A^\infty}\psi_f\,dm_f=1\quad\text{and}\quad 0<\frac{D}{|\Lambda|+1}\le\psi_f\le C_f,\tag{18.5}$$

where C_f is the ratio bounding constant for the Gibbs state m_f from Theorem 17.6.4 *while D and $|\Lambda|$ are the constants from Lemma* 18.1.2. *Furthermore,*

$$v_\beta(\psi_f) \le C_f Q \quad \textit{and, consequently,} \quad \|\psi_f\|_\beta \le C_f(1+Q),$$

where Q is the constant from Lemma 18.1.4.

Proof. Lemma 18.1.4 reveals that $v_\beta(\widehat{\mathcal{L}}_f^n(\mathbb{1})) \le C_f Q$ and $\|\widehat{\mathcal{L}}_f^n(\mathbb{1})\|_\beta \le C_f(1+Q)$ for every $n \ge 0$. Then the Cesàro averages satisfy

$$v_\beta\left(\frac{1}{n}\sum_{j=0}^{n-1}\widehat{\mathcal{L}}_f^j(\mathbb{1})\right) \le C_f Q \quad \text{and} \quad \left\|\frac{1}{n}\sum_{j=0}^{n-1}\widehat{\mathcal{L}}_f^j(\mathbb{1})\right\|_\beta \le C_f(1+Q), \quad \forall n \in \mathbb{N}. \tag{18.6}$$

By the Arzelà–Ascoli theorem (Theorem A.2.3 in Appendix A), there then exists a strictly increasing sequence of positive integers $(n_k)_{k=1}^\infty$ such that the sequence $(\frac{1}{n_k}\sum_{j=0}^{n_k-1}\widehat{\mathcal{L}}_f^j(\mathbb{1}))_{k=1}^\infty$ converges uniformly on compact subsets of E_A^∞ to some function $\psi_f : E_A^\infty \to \mathbb{R}$. Clearly, $v_\beta(\psi_f) \le C_f Q$ and $\|\psi_f\|_\beta \le C_f(1+Q)$. In particular, $\psi_f \in \mathcal{K}_\beta^b(E_A^\infty)$.

Since m_f is a fixed point of the operator conjugate to $\widehat{\mathcal{L}}_f$, we know that $\int \widehat{\mathcal{L}}_f^j(\mathbb{1})\,dm_f = 1$ for every $j \in \mathbb{N}$. Consequently, $\int \frac{1}{n}\sum_{j=0}^{n-1}\widehat{\mathcal{L}}_f^j(\mathbb{1})\,dm_f = 1$ for every $n \in \mathbb{N}$. So, applying Lebesgue's dominated convergence theorem (Theorem A.1.38) along with Lemma 18.1.1, we conclude that $\int \psi_f\,dm_f = 1$ and $\psi_f \le C_f$.

Moreover, by uniquely writing $n_k = q_k(|\Lambda|+1)+r_k$, with $q_k, r_k \in \mathbb{Z}_+$ and $0 \le r_k \le |\Lambda|$, we deduce from Lemma 18.1.2 that

$$\begin{aligned}
\frac{1}{n_k}\sum_{j=0}^{n_k-1}\widehat{\mathcal{L}}_f^j(\mathbb{1}) &\ge \frac{1}{(q_k+1)(|\Lambda|+1)}\sum_{j=1}^{(q_k-1)(|\Lambda|+1)}\widehat{\mathcal{L}}_f^j(\mathbb{1}) \\
&= \frac{1}{(q_k+1)(|\Lambda|+1)}\sum_{q=0}^{q_k-2}\sum_{j=q(|\Lambda|+1)+1}^{(q+1)(|\Lambda|+1)}\widehat{\mathcal{L}}_f^j(\mathbb{1}) \\
&\ge \frac{1}{(q_k+1)(|\Lambda|+1)}\sum_{q=0}^{q_k-2} D = \frac{(q_k-1)D}{(q_k+1)(|\Lambda|+1)} \\
&\to \frac{D}{|\Lambda|+1} \quad \text{as } k \to \infty.
\end{aligned}$$

Thus, $\psi_f \ge \frac{D}{|\Lambda|+1}$.

It remains to show that $\widehat{\mathcal{L}}_f(\psi_f) = \psi_f$. Using Lemma 18.1.1, for every $k \in \mathbb{N}$ we get

$$\left\|\widehat{\mathcal{L}}_f\left(\frac{1}{n_k}\sum_{j=0}^{n_k-1}\widehat{\mathcal{L}}_f^j(\mathbb{1})\right) - \frac{1}{n_k}\sum_{j=0}^{n_k-1}\widehat{\mathcal{L}}_f^j(\mathbb{1})\right\|_\infty = \frac{1}{n_k}\|\widehat{\mathcal{L}}_f^{n_k}(\mathbb{1}) - \widehat{\mathcal{L}}_f^0(\mathbb{1})\|_\infty \le \frac{2C_f}{n_k}.$$

Therefore,

$$\widehat{\mathcal{L}}_f\left(\frac{1}{n_k}\sum_{j=0}^{n_k-1}\widehat{\mathcal{L}}_f^j(\mathbb{1})\right)-\frac{1}{n_k}\sum_{j=0}^{n_k-1}\widehat{\mathcal{L}}_f^j(\mathbb{1})\to 0 \quad \text{uniformly as } k\to\infty. \tag{18.7}$$

In order to infer from this that $\widehat{\mathcal{L}}_f(\psi_f)=\psi_f$, it suffices to show that if a sequence $(g_k)_{k=1}^\infty$ in $C_b(E_A^\infty)$ is uniformly bounded and converges uniformly on compact subsets of E_A^∞ to a function g, then $(\widehat{\mathcal{L}}_f(g_k))_{k=1}^\infty$ converges uniformly on compact subsets of E_A^∞ to $\widehat{\mathcal{L}}_f(g)$. To see this, first notice that $\|g\|_\infty\le B$, where B is an upper bound for the sequence $(\|g_k\|_\infty)_{k=1}^\infty$. Fix $\varepsilon>0$. As f is summable, the series $\sum_{e\in E}\exp(\sup(f|_{[e]})-\mathrm{P}(f))$ converges to, say, $S>0$. Consequently, there exists a finite set $E_\varepsilon\subseteq E$ such that

$$2B\sum_{e\in E\setminus E_\varepsilon}\exp(\sup(f|_{[e]})-\mathrm{P}(f))<\frac{\varepsilon}{2}. \tag{18.8}$$

Fix any compact set $K\subseteq E_A^\infty$. For every $e\in E$, the set $eK:=\{e\omega:\omega\in K \text{ and } A_{e\omega_1}=1\}$ is also compact and so is the set $\cup_{e\in E_\varepsilon}eK$. Since $(g_k)_{k=1}^\infty$ converges uniformly on compact sets to g, there exists $N\in\mathbb{N}$ such that

$$\|g_n-g\|_{\infty,\bigcup_{e\in E_\varepsilon}eK}\le\frac{\varepsilon}{2S},\quad \forall n\ge N. \tag{18.9}$$

Using (18.8)–(18.9), we get for every $n\ge N$ and every $\omega\in K$ that

$$\begin{aligned}
\left|\widehat{\mathcal{L}}_f(g)(\omega)-\widehat{\mathcal{L}}_f(g_n)(\omega)\right| &= \left|\widehat{\mathcal{L}}_f(g-g_n)(\omega)\right|\\
&\le \sum_{e\in E_\varepsilon:A_{e\omega_1}=1}\left|g_n(e\omega)-g(e\omega)\right|\exp(f(e\omega)-\mathrm{P}(f))\\
&\quad+\sum_{e\in E\setminus E_\varepsilon:A_{e\omega_1}=1}\left|g_n(e\omega)-g(e\omega)\right|\exp(f(e\omega)-\mathrm{P}(f))\\
&\le \sum_{e\in E_\varepsilon:A_{e\omega_1}=1}\frac{\varepsilon}{2S}\exp(f(e\omega)-\mathrm{P}(f))\\
&\quad+\sum_{e\in E\setminus E_\varepsilon:A_{e\omega_1}=1}2B\exp(f(e\omega)-\mathrm{P}(f))\\
&\le\frac{\varepsilon}{2}+\frac{\varepsilon}{2}=\varepsilon.
\end{aligned}$$

Hence, $(\widehat{\mathcal{L}}_f(g_k))_{k=1}^\infty$ converges uniformly on compact subsets of E_A^∞ to $\widehat{\mathcal{L}}_f(g)$ whenever a sequence $(g_k)_{k=1}^\infty$ in $C_b(E_A^\infty)$ is uniformly bounded and converges uniformly on compact subsets of E_A^∞ to a function g. As the sequence $(\frac{1}{n_k}\sum_{j=0}^{n_k-1}\widehat{\mathcal{L}}_f^j(\mathbb{1}))_{k\in\mathbb{N}}$ is uniformly bounded and converges uniformly on compact subsets of E_A^∞ to $\psi_f:E_A^\infty\to\mathbb{R}$, we conclude from (18.7) that $\widehat{\mathcal{L}}_f(\psi_f)=\psi_f$. □

Corollary 18.1.8. *Let E be a countable alphabet and $A : E \times E \to \{0,1\}$ a finitely irreducible incidence matrix. Let $f \in \mathcal{K}^s_\beta(E_A^\infty, \mathbb{R})$ be a potential. Per Corollary 17.7.5, let m_f be the unique Borel probability measure fixed by $\widehat{\mathcal{L}}_f^*$ and let μ_f be the unique σ-invariant Gibbs state for the potential f. Finally, let $\psi_f \in \mathcal{K}^b_\beta(E_A^\infty)$ be the fixed point of $\widehat{\mathcal{L}}_f$ identified in Theorem 18.1.7. Then*

$$\mu_f = \psi_f m_f.$$

Moreover, μ_f and m_f are equivalent with Radon–Nikodym derivatives bounded away from 0 and ∞. In addition, ψ_f is the unique $\psi \in C_b(E_A^\infty)$ which is fixed by $\widehat{\mathcal{L}}_f$, is nonnegative and such that $\int \psi \, dm_f = 1$.

Proof. The proof is in some respect similar to that of Theorem 13.7.7. However, we do know that f admits a unique σ-invariant Gibbs state, and that state is μ_f.

By virtue of its definition and (18.5) in Theorem 18.1.7, we know that $\psi_f m_f$ is a Borel probability measure and that it is equivalent to m_f with Radon–Nikodym derivatives bounded away from 0 and ∞.

Since m_f is a Gibbs state for f and since $\psi_f m_f$ is boundedly equivalent to m_f, we know by Proposition 17.4.2 that $\psi_f m_f$ is a Gibbs state for f, too.

Let us show that $\psi_f m_f$ is σ-invariant. Let $g \in C_b(E_A^\infty)$. Then

$$\begin{aligned}
\psi_f m_f(g) &= \int_{E_A^\infty} g \, d(\psi_f m_f) = \int_{E_A^\infty} g \cdot \psi_f \, dm_f = \int_{E_A^\infty} g \cdot \widehat{\mathcal{L}}_f(\psi_f) \, dm_f \\
&= \int_{E_A^\infty} \widehat{\mathcal{L}}_f((g \circ \sigma) \cdot \psi_f) \, dm_f = \int_{E_A^\infty} (g \circ \sigma) \cdot \psi_f \, d\widehat{\mathcal{L}}_f^*(m_f) \\
&= \int_{E_A^\infty} (g \circ \sigma) \cdot \psi_f \, dm_f = \int_{E_A^\infty} g \circ \sigma \, d(\psi_f m_f) = \psi_f m_f(g \circ \sigma).
\end{aligned}$$

As $C_b(E_A^\infty)$ is dense in $L^1(\psi_f m_f)$, we deduce that $\psi_f m_f$ is σ-invariant.

In summary, $\psi_f m_f$ is a σ-invariant Gibbs state and by the uniqueness of such a state we conclude that $\mu_f = \psi_f m_f$.

It remains to prove that ψ_f is unique. Suppose that two such functions $\psi_1, \psi_2 \in C_b(E_A^\infty)$ exist but that $\psi_1 \neq \psi_2$. By the uniqueness of the σ-invariant Gibbs state for f, we would then have that $\psi_1 m_f = \mu_f = \psi_2 m_f$. But there would also exist a point $\omega \in E_A^\infty$ and $\delta > 0$ such that $\psi_1(\omega) \le \psi_2(\omega) - 2\delta$ (without loss of generality; otherwise, interchange ψ_1 and ψ_2). By continuity, there would then exist $n \in \mathbb{N}$ such that $\psi_1(\tau) \le \psi_2(\tau) - \delta$ for all $\tau \in [\omega|_n]$. Upon integrating over $[\omega|_n]$, we would get

$$\mu_f([\omega|_n]) = \psi_1 m_f([\omega|_n]) \le (\psi_2 - \delta) m_f([\omega|_n]) = \mu_f([\omega|_n]) - \delta m_f([\omega|_n]) < \mu_f([\omega|_n]),$$

which is impossible. □

As ψ_f is uniformly bounded away from 0, we can mimick the main idea in Section 13.8.1 by introducing the function

$$\hat{f} = f + \log\psi_f - \log\psi_f \circ \sigma$$

and observing that the weighted normalized transfer operator $\widehat{\mathcal{L}}_{\hat{f}} : C_b(E_A^\infty) \to C_b(E_A^\infty)$ is well-defined and satisfies

$$\widehat{\mathcal{L}}_{\hat{f}}(g) = \frac{1}{\psi_f}\widehat{\mathcal{L}}_f(g\psi_f). \tag{18.10}$$

It is straightforward to check that $\widehat{\mathcal{L}}_{\hat{f}}(\mathcal{K}_\beta^b(E_A^\infty)) \subseteq \mathcal{K}_\beta^b(E_A^\infty)$, as $1/\psi_f \in \mathcal{K}_\beta^b(E_A^\infty)$ and the product of any two functions in $\mathcal{K}_\beta^b(E_A^\infty)$ is in $\mathcal{K}_\beta^b(E_A^\infty)$; see Exercise 18.5.4. The basic properties of the operators $\widehat{\mathcal{L}}_f$ and $\widehat{\mathcal{L}}_{\hat{f}}$, which follow from Lemma 18.1.4, Lemma 18.1.2 and Theorem 18.1.7, are listed in the next two theorems.

Theorem 18.1.9. *Let E be a countable alphabet and $A : E \times E \to \{0,1\}$ a finitely irreducible incidence matrix. Let $f \in \mathcal{K}_\beta^s(E_A^\infty, \mathbb{R})$ be a potential. The following statements hold:*

(a) $\widehat{\mathcal{L}}_f(\psi_f) = \psi_f$ *for some positive $\psi_f \in \mathcal{K}_\beta^b(E_A^\infty)$ whose values are uniformly bounded away from 0 and such that $m_f(\psi_f) = 1$.*

(b) $\widehat{\mathcal{L}}_{\hat{f}}^*(m_f) = m_f$, *where m_f is the unique Borel probability eigenmeasure m_f of the dual transfer operator $\mathcal{L}_f^*$ per Corollary 17.7.5.*

(c) *The vector subspaces $\mathbb{R}\psi_f$,*

$$C_b^{0,m_f}(E_A^\infty) := \{g \in C_b(E_A^\infty) : m_f(g) = 0\} \quad \textit{and} \quad \mathcal{K}_\beta^{b,0,m_f}(E_A^\infty) := \mathcal{K}_\beta(E_A^\infty) \cap C_b^{0,m_f}(E_A^\infty)$$

are $\widehat{\mathcal{L}}_f$-invariant. Moreover, $C_b^{0,m_f}(E_A^\infty)$ is closed in $(C_b(E_A^\infty), \|\cdot\|_\infty)$ while $\mathcal{K}_\beta^{b,0,m_f}(E_A^\infty)$ is closed in $(\mathcal{K}_\beta^b(E_A^\infty), \|\cdot\|_\beta)$.

(d) $\sup_{n\in\mathbb{N}} \|\widehat{\mathcal{L}}_f^n\|_\beta < \infty$.

(e) $\mathcal{K}_\beta^b(E_A^\infty) = \mathbb{R}\psi_f \bigoplus \mathcal{K}_\beta^{b,0,m_f}(E_A^\infty)$ *as $g = m_f(g)\psi_f + (g - m_f(g)\psi_f)$.*

(f) *If the spaces $C_b(E_A^\infty)$ and $\mathcal{K}_\beta(E_A^\infty)$ are extended to complex-valued functions in the obvious and natural way, then the above statements with $\mathbb{R}$ replaced by $\mathbb{C}$ hold. Consequently, the spectrum of $\widehat{\mathcal{L}}_f : \mathcal{K}_\beta^b(E_A^\infty) \to \mathcal{K}_\beta^b(E_A^\infty)$ is the union of the spectrum of $\widehat{\mathcal{L}}_f$ restricted to $\mathbb{C}\psi_f$ and of the spectrum of $\widehat{\mathcal{L}}_f$ restricted to the complex version of $\mathcal{K}_\beta^{b,0,m_f}(E_A^\infty)$.*

Theorem 18.1.10. *Let E be a countable alphabet and $A : E \times E \to \{0,1\}$ a finitely irreducible incidence matrix. Let $f \in \mathcal{K}_\beta^s(E_A^\infty, \mathbb{R})$ be a potential. For every $n \in \mathbb{N}$, define the* *balancing function* $u_n : E_A^\infty \to \mathbb{R}$ *by*

$$u_n(\omega) := \exp(S_n f(\omega) - \mathrm{P}(f)n)\frac{\psi_f(\omega)}{\psi_f(\sigma^n(\omega))}.$$

The following statements hold:

(a) *For all* $g \in \mathcal{K}_\beta^b(E_A^\infty)$ *and all* $n \in \mathbb{N}$*, we have*

$$\widehat{\mathcal{L}}_{\hat{f}}^n(g)(\omega) = \frac{1}{\psi_f(\omega)}\mathcal{L}_f^n(g\psi_f)(\omega) = \sum_{\tau\in E_A^n:\,A_{\tau_n\omega_1}=1} u_n(\tau\omega)g(\tau\omega).$$

(b) $\widehat{\mathcal{L}}_{\hat{f}}(\mathbb{1}) = \mathbb{1}$ *and* $\|\widehat{\mathcal{L}}_{\hat{f}}^n\|_\infty = 1$ *for all* $n \in \mathbb{N}$.

(c) $\widehat{\mathcal{L}}_{\hat{f}}^*(\mu_f) = \mu_f$*, where* μ_f *is the unique (ergodic)* σ*-invariant Gibbs state for* f*, per Corollary 17.7.5.*

(d) *The vector subspaces* $\mathbb{R}\mathbb{1}$,

$$C_b^{0,\mu_f}(E_A^\infty) := \{g \in C_b(E_A^\infty) : \mu_f(g) = 0\} \quad \textit{and} \quad \mathcal{K}_\beta^{b,0,\mu_f}(E_A^\infty) := \mathcal{K}_\beta(E_A^\infty) \cap C_b^{0,\mu_f}(E_A^\infty)$$

are $\widehat{\mathcal{L}}_{\hat{f}}$*–invariant. Moreover,* $C_b^{0,\mu_f}(E_A^\infty)$ *is closed in* $(C_b(E_A^\infty), \|\cdot\|_\infty)$ *while* $\mathcal{K}_\beta^{b,0,\mu_f}(E_A^\infty)$ *is closed in* $(\mathcal{K}_\beta^b(E_A^\infty), \|\cdot\|_\beta)$.

(e) $M := \sup_{n\in\mathbb{N}}\|\widehat{\mathcal{L}}_{\hat{f}}^n\|_\beta < \infty$.

(f) $\mathcal{K}_\beta^b(E_A^\infty) = \mathbb{R}\mathbb{1} \bigoplus \mathcal{K}_\beta^{b,0,\mu_f}(E_A^\infty)$ *as* $g = \mu_f(g)\mathbb{1} + (g - \mu_f(g)\mathbb{1})$.

(g) *If the spaces* $C_b(E_A^\infty)$ *and* $\mathcal{K}_\beta(E_A^\infty)$ *are extended to complex-valued functions in the obvious and natural way, then the above statements with* $\mathbb{R}$ *replaced by* $\mathbb{C}$ *hold. This implies that the spectrum of* $\widehat{\mathcal{L}}_{\hat{f}} : \mathcal{K}_\beta^b(E_A^\infty) \to \mathcal{K}_\beta^b(E_A^\infty)$ *is the union of the spectrum of* $\widehat{\mathcal{L}}_{\hat{f}}$ *restricted to* $\mathbb{C}\mathbb{1}$ *and of the spectrum of* $\widehat{\mathcal{L}}_{\hat{f}}$ *restricted to the complex version of* $\mathcal{K}_\beta^{b,0,\mu_f}(E_A^\infty)$.

Denote

$$\mathcal{K}_{\beta,1}^{b,0,\mu_f}(E_A^\infty) := \{g \in \mathcal{K}_\beta^{b,0,\mu_f}(E_A^\infty) : \|g\|_\beta \le 1\}.$$

We prove the following.

Lemma 18.1.11. *Let* E *be a countable alphabet and* $A : E\times E \to \{0,1\}$ *a finitely primitive incidence matrix. Let* $f \in \mathcal{K}_\beta^s(E_A^\infty, \mathbb{R})$ *be a potential. For each* $n \ge 0$*, let*

$$b_n = \sup\left\{\|\widehat{\mathcal{L}}_{\hat{f}}^n(g)\|_\beta : g \in \mathcal{K}_{\beta,1}^{b,0,\mu_f}(E_A^\infty)\right\}.$$

Then $\lim_{n\to\infty} b_n = 0$.

Proof. For every $n \geq 0$, define

$$a_n := \sup\left\{\|\widehat{\mathcal{L}}_f^n(g)\|_\infty : g \in \mathcal{K}_{\beta,1}^{b,0,\mu_f}(E_A^\infty)\right\} \geq 0.$$

It follows from Theorem 18.1.10(b) that the sequence $(a_n)_{n=1}^\infty$ is nonincreasing and $a_1 \leq 1$. We first show that $a := \lim_{n\to\infty} a_n = 0$.

Suppose on the contrary that $a > 0$. By Theorem 18.1.10(e), we deduce that $\sup\{v_\beta(\widehat{\mathcal{L}}_f^n(g)) : g \in \mathcal{K}_{\beta,1}^{b,0,\mu_f}(E_A^\infty),\ n \in \mathbb{N}\} \leq M < \infty$. Therefore, there exists $k \in \mathbb{N}$ such that

$$|\omega \wedge \tau| \geq k \implies |\widehat{\mathcal{L}}_f^n(g)(\omega) - \widehat{\mathcal{L}}_f^n(g)(\tau)| \leq a/2, \quad \forall g \in \mathcal{K}_{\beta,1}^{b,0,\mu_f}(E_A^\infty),\ \forall n \geq 0. \qquad (18.11)$$

Let μ_f be the unique ergodic σ-invariant Gibbs state for f, per Corollary 17.7.5. As μ_f is regular, there exists a compact set $K \subseteq E_A^\infty$ such that $\mu_f(K) \geq 1 - a/8$. Fix $g \in \mathcal{K}_{\beta,1}^{b,0,\mu_f}(E_A^\infty)$ and $n \geq 0$. Suppose that $\widehat{\mathcal{L}}_f^n(g)(\tau) > a/4$ for all $\tau \in K$. Since $\int \widehat{\mathcal{L}}_f^n(g)\,d\mu_f = 0$ by Theorem 18.1.10(d) and since $\|\widehat{\mathcal{L}}_f^n(g)\|_\infty \leq \|g\|_\infty \leq 1$, we find that

$$0 = \int_K \widehat{\mathcal{L}}_f^n(g)\,d\mu_f + \int_{E_A^\infty \setminus K} \widehat{\mathcal{L}}_f^n(g)\,d\mu_f \geq \frac{a}{4}\left(1 - \frac{a}{8}\right) - \frac{a}{8} = \frac{a}{8}\left(1 - \frac{a}{4}\right) > 0.$$

This contradiction shows that $\widehat{\mathcal{L}}_f^n(g)(\tau) \leq a/4$ for some $\tau \in K$. Fix this τ. Let $\Lambda \subseteq E_A^*$ be a finite set of finite words witnessing the finite primitivity of the incidence matrix A and let $|\Lambda|$ be the length of these words. For every $\alpha \in \Lambda$, pick a word $\overline{\alpha} \in E_A^\infty$ such that $\overline{\alpha}|_{|\alpha|} = \alpha$. For every $\omega \in E_A^\infty$, there exists $\alpha(\omega) \in \Lambda$ such that $\rho(\omega) := \tau|_k\alpha(\omega)\omega \in E_A^\infty$. Since $\rho(\omega)|_k = \tau|_k$, it follows from (18.11) that

$$\widehat{\mathcal{L}}_f^n(g)(\rho(\omega)) \leq \widehat{\mathcal{L}}_f^n(g)(\tau) + \frac{a}{2} \leq \frac{3}{4}a \leq a_n - \frac{a}{4}. \qquad (18.12)$$

Moreover, let $p := k + |\Lambda|$. Using Theorem 18.1.7 (with the notation $D_\Lambda := \frac{\eta}{|\Lambda|+1}$) and Lemma 17.2.3, we get

$$\begin{aligned}
\frac{u_p(\rho(\omega))}{u_p(\tau)} &= \exp(S_pf(\rho(\omega)) - S_pf(\tau))\frac{\psi_f(\rho(\omega))}{\psi_f(\sigma^p(\rho(\omega)))}\frac{\psi_f(\sigma^p(\tau))}{\psi_f(\tau)} \\
&\geq \exp(S_kf(\rho(\omega)) - S_kf(\tau))\exp\Big(S_{|\Lambda|}f(\sigma^k(\rho(\omega))) - S_{|\Lambda|}f(\sigma^k(\tau))\Big)\cdot\frac{D_\Lambda^2}{C_f^2} \\
&\geq e^{-\frac{v_\beta(f)}{e^\beta - 1}}\cdot\exp\Big(S_{|\Lambda|}f(\alpha(\omega)\omega) - |\Lambda|\sup(f)\Big)\cdot\frac{D_\Lambda^2}{C_f^2} \\
&\geq (D_\Lambda/C_f)^2 e^{-2\frac{v_\beta(f)}{e^\beta - 1}}\exp(S_{|\Lambda|}f(\overline{\alpha(\omega)}))\exp(-|\Lambda|\sup(f)) \\
&\geq (D_\Lambda/C_f)^2 e^{-2\frac{v_\beta(f)}{e^\beta - 1}}\exp(\min\{S_{|\Lambda|}f(\overline{\alpha}) : \alpha \in \Lambda\})\exp(-|\Lambda|\sup(f)).
\end{aligned}$$

Denote the last (constant) expression appearing in this formula by U. As u_p is positive and continuous and K is compact, we infer that $u_p(\rho(\omega)) \geq U \inf(u_p|_K) > 0$. Using Theorem 18.1.10(a,b) and (18.12), we deduce that

$$\begin{aligned}\widehat{\mathcal{L}}^p_{\hat{f}}(\widehat{\mathcal{L}}^n_{\hat{f}}(g))(\omega) &= u_p(\rho(\omega))\widehat{\mathcal{L}}^n_{\hat{f}}(g)(\rho(\omega)) + \sum_{\eta\in\sigma^{-p}(\omega)\setminus\{\rho(\omega)\}} u_p(\eta)\widehat{\mathcal{L}}^n_{\hat{f}}(g)(\eta) \\ &\leq \left(a_n - \frac{a}{4}\right)u_p(\rho(\omega)) + a_n \sum_{\eta\in\sigma^{-p}(\omega)\setminus\{\rho(\omega)\}} u_p(\eta) \\ &= a_n\widehat{\mathcal{L}}^n_{\hat{f}}(\mathbb{1})(\omega) - \frac{a}{4}u_p(\rho(\omega)) \leq a_n - \frac{a}{4}U\inf(u_p|_K).\end{aligned}$$

Similarly, we get $\widehat{\mathcal{L}}^p_{\hat{f}}(\widehat{\mathcal{L}}^n_{\hat{f}}(g))(\omega) \geq -a_n + (a/4)U\inf(u_p|_K)$. Consequently, we have $\|\widehat{\mathcal{L}}^{p+n}_{\hat{f}}(g)\|_\infty \leq a_n - (a/4)U\inf(u_p|_K)$ or, equivalently,

$$\|\widehat{\mathcal{L}}^n_{\hat{f}}(g)\|_\infty \leq a_{n-p} - \frac{a}{4}U\inf(u_p|_K), \quad \forall n \geq p.$$

Taking the supremum over all $g \in \mathcal{K}^{b,0,\mu_f}_{\beta,1}(E^\infty_A)$, we find $a_n \leq a_{n-p} - (a/4)U\inf(u_p|_K)$. So $a := \lim_{n\to\infty} a_n \leq \lim_{n\to\infty} a_{n-p} - (a/4)U\inf(u_p|_K) = a - (a/4)U\inf(u_p|_K) < a$. This contradiction shows that $a := \lim_{n\to\infty} a_n = 0$.

Now, fix $\varepsilon > 0$. Choose $N \in \mathbb{N}$ so large that $a_N \leq \varepsilon/(2C_fQ)$ and $C_fe^{-\beta N}M \leq \varepsilon/2$. By virtue of Lemma 18.1.4, for every $n \geq 2N$ and every $g \in \mathcal{K}^{b,0,\mu_f}_{\beta,1}(E^\infty_A)$ we get

$$\begin{aligned}\|\widehat{\mathcal{L}}^n_{\hat{f}}(g)\|_\beta = \|\widehat{\mathcal{L}}^{n-N}_{\hat{f}}(\widehat{\mathcal{L}}^N_{\hat{f}}(g))\|_\beta &\leq C_f\left(e^{-\beta(n-N)}\|\widehat{\mathcal{L}}^N_{\hat{f}}(g)\|_\beta + Q\|\widehat{\mathcal{L}}^N_{\hat{f}}(g)\|_\infty\right) \\ &\leq C_fe^{-\beta N}M + C_fQa_N \leq \varepsilon.\end{aligned}$$

So $b_n \leq \varepsilon$ for all $n \geq 2N$. The arbitrariness of ε implies that $\lim_{n\to\infty} b_n = 0$. □

We now describe the rates of convergence of the iterates of the normalized transfer operator and of its weighted counterpart.

Theorem 18.1.12. *Let E be a countable alphabet and $A : E \times E \to \{0,1\}$ a finitely primitive incidence matrix. Let $f \in \mathcal{K}^s_\beta(E^\infty_A, \mathbb{R})$ be a potential. Let m_f be the unique Borel probability measure fixed by the dual $\widehat{\mathcal{L}}^*_f$ of the normalized transfer operator $\widehat{\mathcal{L}}_f$ and let μ_f be the unique ergodic σ-invariant Gibbs state for f, per Corollary 17.7.5. Finally, let ψ_f be the fixed point of $\widehat{\mathcal{L}}_f$ identified in Theorem 18.1.7 and $\widehat{\mathcal{L}}_{\hat{f}}$ be the weighted normalized transfer operator defined by (18.10). Then there exist constants $c > 0$ and $0 < \zeta < 1$ such that for every $g \in \mathcal{K}^b_\beta(E^\infty_A)$ and every $n \geq 0$,*

(a)

$$\left\|\widehat{\mathcal{L}}^n_{\hat{f}}(g) - \left(\int_{E^\infty_A} g\,d\mu_f\right)\mathbb{1}\right\|_\beta \leq c\zeta^n\|g\|_\beta.$$

(b)

$$\left\| \widehat{\mathcal{L}}_f^n(g) - \left(\int_{E_A^\infty} g\, dm_f \right) \psi_f \right\|_\beta \le c\zeta^n \|g\|_\beta.$$

(c) *Upon complexifying $\mathcal{K}_\beta^b(E_A^\infty)$ in the natural way, the above relations also hold for all complex-valued functions $g \in \mathcal{K}_\beta^b(E_A^\infty)$. Additionally, the number 1 is a simple isolated eigenvalue of the operators $\widehat{\mathcal{L}}_f$ and $\widehat{\mathcal{L}}_{\hat{f}}$. More precisely, apart from number 1, the rest of the spectrum of each of these operators is contained in $\overline{B_{\mathbb{C}}(0,\zeta)}$.*

Proof. Lemma 18.1.11 says that $\lim_{n\to\infty} \|\widehat{\mathcal{L}}_{\hat{f}}|^n_{\mathcal{K}_\beta^{b,0,\mu_f}(E_A^\infty)}\|_\beta = 0$. Thus, there exists $q \in \mathbb{N}$ such that $\|\widehat{\mathcal{L}}_{\hat{f}}|^q_{\mathcal{K}_\beta^{b,0,\mu_f}(E_A^\infty)}\|_\beta \le 1/2$. By induction, we have $\|\widehat{\mathcal{L}}_{\hat{f}}|^{pq}_{\mathcal{K}_\beta^{b,0,\mu_f}(E_A^\infty)}\|_\beta \le (1/2)^p$ for all $p \in \mathbb{Z}_+$. Consider an arbitrary $n \ge 0$ and write $n = pq + r$, with $0 \le r < q$ and $p \in \mathbb{Z}_+$ uniquely determined. Using Theorem 18.1.10(d,e), we get for every $h \in \mathcal{K}_\beta^{b,0,\mu_f}(E_A^\infty)$ that

$$\|\widehat{\mathcal{L}}_{\hat{f}}^n(h)\|_\beta = \|\widehat{\mathcal{L}}_{\hat{f}}^{pq}(\widehat{\mathcal{L}}_{\hat{f}}^r(h))\|_\beta \le (1/2)^p \|\widehat{\mathcal{L}}_{\hat{f}}^r(h)\|_\beta \le (1/2)^{\frac{n-r}{q}} M\|h\|_\beta \le 2M(1/2)^{\frac{n}{q}}\|h\|_\beta.$$

Therefore, $\|\widehat{\mathcal{L}}_{\hat{f}}|^n_{\mathcal{K}_\beta^{b,0,\mu_f}(E_A^\infty)}\|_\beta \le 2M\zeta^n$ for every $n \ge 0$, where $\zeta = (1/2)^{1/q}$. If $g \in \mathcal{K}_\beta^b(E_A^\infty)$, then $g - \mu_f(g) \in \mathcal{K}_\beta^{b,0,\mu_f}(E_A^\infty)$ and $\|g - \mu_f(g)\|_\beta \le \|g\|_\beta + \|\mu_f(g)\|_\beta \le 2\|g\|_\beta$. Using also Theorem 18.1.10(b), we obtain that

$$\|\widehat{\mathcal{L}}_{\hat{f}}^n(g) - \mu_f(g)\mathbb{1}\|_\beta = \|\widehat{\mathcal{L}}_{\hat{f}}^n(g - \mu_f(g))\|_\beta \le \|\widehat{\mathcal{L}}_{\hat{f}}|^n_{\mathcal{K}_\beta^{b,0,\mu_f}(E_A^\infty)}\|_\beta \|g - \mu_f(g)\|_\beta \le 4M\zeta^n\|g\|_\beta$$

for every $n \ge 0$ and $g \in \mathcal{K}_\beta^b(E_A^\infty)$. This proves (a). Part (b) is a consequence of (a). For part (c), let us concentrate on the operator $\widehat{\mathcal{L}}_f$. A similar argument holds for $\widehat{\mathcal{L}}_{\hat{f}}$. By Theorem 18.1.9(f), the spectrum of $\widehat{\mathcal{L}}_f : \mathcal{K}_\beta^b(E_A^\infty) \to \mathcal{K}_\beta^b(E_A^\infty)$ is the union of the spectra of $\widehat{\mathcal{L}}_f$ restricted to $\mathbb{C}\psi_f$ and of $\widehat{\mathcal{L}}_f$ restricted to the complex version of $\mathcal{K}_\beta^{b,0,m_f}(E_A^\infty)$. The spectrum of the former is 1 by Theorem 18.1.9(a). The spectrum of the latter is contained in $\overline{B_{\mathbb{C}}(0,\zeta)}$ by the just proved part (b). It then clear that 1 is a simple isolated eigenvalue for $\widehat{\mathcal{L}}_f$. □

18.2 Continuity of Gibbs states

In this section, we shall prove continuity results analogous to those for open distance expanding maps from Subsection 13.8.3. However, the situation is now much more delicate for several reasons. First of all, when the alphabet is infinite the potentials are not bounded from below, and thus do not exhibit a natural metric space structure. Furthermore, the symbolic space E_A^∞ is not compact, so we must use Arzèla–Ascoli's theorem with special care. Moreover, the space $M(E_A^\infty)$ of all Borel probability measures on E_A^∞, endowed with the weak* topology, is no longer compact, thus we make use of the concept of tightness in a somewhat similar manner to the proof of Theorem 17.7.4.

Fix $\beta > 0$ and equip the set $\mathcal{K}_\beta^s(E_A^\infty, \mathbb{R})$ with the metric

$$\rho_\infty(f, g) := \begin{cases} \min\{1, \|f - g\|_\infty\} & \text{if } f - g \in \mathcal{K}_\beta^b(E_A^\infty, \mathbb{R}) \\ 1 & \text{if } f - g \notin \mathcal{K}_\beta^b(E_A^\infty, \mathbb{R}). \end{cases}$$

Also endow $\mathcal{K}_\beta^s(E_A^\infty, \mathbb{R})$ with the pseudometric

$$\rho_\beta(f, g) := v_\beta(f - g).$$

Nevertheless, convergence in $\mathcal{K}_\beta^s(E_A^\infty, \mathbb{R})$ and its subspaces will be solely determined by the metric ρ_∞. That is, convergence is uniform convergence. The pseudometric ρ_β will be strictly used to restrict the subspace over which the next result holds, namely, to concentrate on the subspace $\mathcal{K}_\beta^s(E_A^\infty, \mathbb{R}) \cap \overline{B_{\rho_\beta}(0, R)}$.

We shall prove the following.

Theorem 18.2.1. *Let E be a countable alphabet and $A : E \times E \to \{0, 1\}$ a finitely irreducible incidence matrix. For every $\beta > 0$ and every $R > 0$, the following statements hold:*

(a) *The function $\mathcal{K}_\beta^s(E_A^\infty, \mathbb{R}) \ni f \longmapsto \mathrm{P}(f) \in \mathbb{R}$ is continuous.*

(b) *The function $\mathcal{K}_\beta^s(E_A^\infty, \mathbb{R}) \cap \overline{B_{\rho_\beta}(0, R)} \ni f \longmapsto m_f \in M(E_A^\infty)$ is continuous, where m_f is the unique eigenmeasure of the dual transfer operator $\mathcal{L}_f^*$ (and also a Gibbs state), per Corollary* 17.7.5.

(c) *The function $\mathcal{K}_\beta^s(E_A^\infty, \mathbb{R}) \cap \overline{B_{\rho_\beta}(0, R)} \ni f \longmapsto \psi_f \in C_b(E_A^\infty)$ is continuous, where ψ_f is the fixed point of $\widehat{\mathcal{L}}_f$ identified in Theorem* 18.1.7 *(see also Corollary* 18.1.8*).*

(d) *The function $\mathcal{K}_\beta^s(E_A^\infty, \mathbb{R}) \cap \overline{B_{\rho_\beta}(0, R)} \ni f \longmapsto \mu_f \in M(E_A^\infty)$ is continuous, where μ_f is the unique ergodic σ-invariant Gibbs state for f, per Corollary* 17.7.5.

Proof. Part (a) follows from a straightforward calculation from the very definition of topological pressure; alternatively, it follows by an even easier calculation from Theorem 17.3.4 (3rd variational principle).

The proof of part (b) is motivated by and similar to that of Theorem 17.7.4. Let $(f_k)_{k=1}^\infty$ be a sequence of functions in $\mathcal{K}_\beta^s(E_A^\infty, \mathbb{R}) \cap \overline{B_{\rho_\beta}(0, R)}$ converging in the ρ_∞-metric (i. e. uniformly) to a function $f \in \mathcal{K}_\beta^s(E_A^\infty, \mathbb{R}) \cap \overline{B_{\rho_\beta}(0, R)}$. We must show that the sequence of measures $(m_{f_k})_{k=1}^\infty$ converges weak* to the measure m_f. Let $\varepsilon > 0$. Disregarding finitely many terms in the sequence $(f_k)_{k=1}^\infty$, we may assume without loss of generality that

$$\rho_\infty(f_k, f) \le 1/2, \quad \forall k \in \mathbb{N} \tag{18.13}$$

and, invoking part (a), that

$$|\mathrm{P}(f_k) - \mathrm{P}(f)| \le 1, \quad \forall k \in \mathbb{N}. \tag{18.14}$$

A somewhat long but natural series of estimates then shows that the constant C_{f_k} arising from Theorem 17.6.4 for the potential f_k satisfies

$$C_{f_k} \le e^{\frac{R}{e^{\beta}-1}} \max\Big\{1, \#\Lambda \exp\Big(2|\Lambda| \cdot (|\mathrm{P}(f)| + 1) + |\Lambda| - 2 \min_{\alpha\in\Lambda} \underline{S}_{|\alpha|} f([\alpha])\Big)\Big\} =: C_{f,R} \tag{18.15}$$

for every $k \in \mathbb{N}$. We first establish the following fact.

Claim 1. *The sequence $(f_k)_{k=1}^{\infty}$ is tight.*

Proof of Claim 1. Employing Corollary 17.7.5(b) and inequalities (18.13)–(18.15), we get for every $k \in \mathbb{N}$, every $n \in \mathbb{N}$, and every $e \in E = \mathbb{N}$, that

$$\begin{aligned}
m_{f_k}(\pi_n^{-1}(e)) &= \sum_{\omega\in E_A^n:\,\omega_n=e} m_{f_k}([\omega]) \le C_{f_k} \sum_{\omega\in E_A^n:\,\omega_n=e} \exp(\sup(S_n f_k|_{[\omega]}) - n\mathrm{P}(f_k)) \\
&\le C_{f,R} e^{-n\mathrm{P}(f_k)} \sum_{\omega\in E_A^n:\,\omega_n=e} \exp(\sup(S_{n-1} f_k|_{[\omega]}) + \sup(f_k|_{[e]})) \\
&\le C_{f,R} e^{-n(\mathrm{P}(f)-1)} \exp(\sup(f_k|_{[e]})) \sum_{\omega\in E_A^{n-1}} \exp\Bigg(\sum_{j=1}^{n-1} \sup(f_k|_{[\omega_j]})\Bigg) \\
&\le C_{f,R} e^{-n(\mathrm{P}(f)-1)} \exp(\sup(f_k|_{[e]})) \Bigg[\sum_{i\in E} \exp(\sup(f_k|_{[i]}))\Bigg]^{n-1} \\
&\le C_{f,R} e^{-n(\mathrm{P}(f)-3/2)} \exp(\sup(f|_{[e]})) \Bigg[\sum_{i\in E} \exp(\sup(f|_{[i]}))\Bigg]^{n-1},
\end{aligned}$$

which is finite since the potential f is summable. Therefore,

$$m_{f_k}(\pi_n^{-1}([e+1,\infty))) \le C_{f,R} e^{-n(\mathrm{P}(f)-3/2)} \Bigg[\sum_{i\in E} \exp(\sup(f|_{[i]}))\Bigg]^{n-1} \sum_{j=e+1}^{\infty} \exp(\sup(f|_{[j]})).$$

Since f is summable, there exists $q_n \in \mathbb{N}$ such that

$$C_{f,R} e^{-n(\mathrm{P}(f)-3/2)} \Bigg[\sum_{i\in E} \exp(\sup(f|_{[i]}))\Bigg]^{n-1} \sum_{j=q_n+1}^{\infty} \exp(\sup(f|_{[j]})) < \frac{\varepsilon}{2^n}.$$

So

$$m_{f_k}(\pi_n^{-1}([q_n+1,\infty))) \le \frac{\varepsilon}{2^n}, \quad \forall n \in \mathbb{N},\ \forall k \in \mathbb{N}.$$

Hence,

$$m_{f_k}\Big(E_A^\infty \cap \prod_{n=1}^{\infty}[1,q_n]\Big) \ge 1-\sum_{n=1}^{\infty} m_{f_k}(\pi_n^{-1}([q_n+1,\infty))) \ge 1-\sum_{n=1}^{\infty}\frac{\varepsilon}{2^n} = 1-\varepsilon$$

for all $k \in \mathbb{N}$. Since $E_A^\infty \cap \prod_{n=1}^{\infty}[1,q_n]$ is a compact subset of E_A^∞, Claim 1 ensues. ■

It follows from Claim 1 and Prokhorov's theorem that every subsequence of the sequence $(m_{f_k})_{k=1}^{\infty}$ has a weak* convergent subsequence. In order to prove part (b), it suffices to show that the limit m of any weak* convergent subsequence $(m_{f_{k_\ell}})_{\ell=1}^{\infty}$ of the sequence $(m_{f_k})_{k=1}^{\infty}$ is equal to m_f. An easy calculation shows the following.

Claim 2. *The function*

$$\mathcal{K}_\beta^s(E_A^\infty,\mathbb{R}) \ni g \longmapsto \mathcal{L}_g \in L(C_b(E_A^\infty))$$

is continuous.

As its immediate consequence, we get the following claim.

Claim 3. *The function*

$$\mathcal{K}_\beta^s(E_A^\infty,\mathbb{R}) \ni g \longmapsto \mathcal{L}_g^* \in L(C_b(E_A^\infty)^*)$$

is continuous.

Using part (a), we deduce that

$$m = \lim_{\ell\to\infty} m_{f_{k_\ell}} = \lim_{\ell\to\infty}\big(e^{-\mathrm{P}(f_{k_\ell})}\mathcal{L}_{f_{k_\ell}}^*(m_{f_{k_\ell}})\big) = e^{-\mathrm{P}(f)}\mathcal{L}_f^*(m).$$

Hence, invoking Theorem 17.6.5 (see also Corollary 17.7.5(a)), we infer that $m = m_f$. This proves part (b).

To demonstrate part (c), let $(f_k)_{k=1}^{\infty}$ be a sequence in $\mathcal{K}_\beta^s(E_A^\infty,\mathbb{R})\cap\overline{B_{\rho_\beta}(0,R)}$ converging in the ρ_∞-metric (i. e. uniformly) to a function $f \in \mathcal{K}_\beta^s(E_A^\infty,\mathbb{R})\cap\overline{B_{\rho_\beta}(0,R)}$. Disregarding finitely many terms, we may assume without loss of generality that

$$\rho_\infty(f_k,f) \le 1/2 \quad \text{and} \quad |\mathrm{P}(f_k)-\mathrm{P}(f)| \le 1, \quad \forall k \in \mathbb{N}.$$

In a similar way to (18.15), it follows from Theorems 17.6.4 and 18.1.7 that there exists a constant $C(f,R) \ge C_f$ such that

$$\|\psi_{f_k}\|_\beta \le C(f,R), \quad \forall k \in \mathbb{N}. \tag{18.16}$$

Take any subsequence of $(\psi_{f_k})_{k=1}^{\infty}$. By virtue of Arzelà–Ascoli's theorem (Theorem A.2.3 in Appendix A), there is a subsequence of that subsequence, which converges uniformly on compact subsets of E_A^∞ to some function $\psi \in B_{\mathcal{K}_\beta^b(E_A^\infty)}(0,C(f,R))$.

To lighten notation, this subsubsequence will be denoted like the sequence $(\psi_{f_k})_{k=1}^{\infty}$. Denote $\psi_k := \psi_{f_k}$ and $\mathcal{L}_k := \mathcal{L}_{f_k}$. Fix $\varepsilon > 0$. By Claim 2, there exists $N_1 \in \mathbb{N}$ such that

$$\|\mathcal{L}_k - \mathcal{L}_f\|_\infty \le \varepsilon, \quad \forall k \ge N_1. \tag{18.17}$$

Using (18.16)–(18.17), we get

$$\|\mathcal{L}_k\psi_k - \mathcal{L}_f\psi_k\|_\infty = \|(\mathcal{L}_k - \mathcal{L}_f)\psi_k\|_\infty \le \|\mathcal{L}_k - \mathcal{L}_f\|_\infty \cdot \|\psi_k\|_\infty \le C(f,R)\varepsilon, \quad \forall k \ge N_1. \tag{18.18}$$

Since the function f is summable, there exists $N_2 \in \mathbb{N}$ such that

$$\sum_{n=N_2+1}^{\infty} \exp(\sup(f|_{[n]})) \le \varepsilon. \tag{18.19}$$

Fix $\omega \in E_A^\infty$. Since the set $\{n\omega \in E_A^\infty : n \le N_2\}$ is finite and $\lim_{k\to\infty} \rho_\infty(\psi_k, \psi) = 0$, there is $N_3 \ge N_1$ such that

$$|\psi_k(\omega) - \psi(\omega)| < \varepsilon \tag{18.20}$$

and

$$\Bigg|\sum_{\substack{n\le N_2:\\ n\omega\in E_A^\infty}} e^{f(n\omega)}\psi_k(\omega) - \sum_{\substack{n\le N_2:\\ n\omega\in E_A^\infty}} e^{f(n\omega)}\psi(\omega)\Bigg| \le \sum_{\substack{n\le N_2:\\ n\omega\in E_A^\infty}} e^{f(n\omega)}|\psi_k(\omega) - \psi(\omega)| < \varepsilon \tag{18.21}$$

for every $k \ge N_3$. Per part (a), there exists $N_4 > N_3$ such that

$$|e^{\mathrm{P}(f)} - e^{\mathrm{P}(f_k)}| < \varepsilon, \quad \forall k \ge N_4. \tag{18.22}$$

Using (18.16), (18.18), (18.20) and (18.22), for all $k \ge N_4$ we get that

$$\begin{aligned}
&|e^{\mathrm{P}(f)}\psi(\omega) - \mathcal{L}_f\psi(\omega)| \\
&\quad = \Big|e^{\mathrm{P}(f)}(\psi(\omega) - \psi_k(\omega)) + (e^{\mathrm{P}(f)} - e^{\mathrm{P}(f_k)})\psi_k(\omega) \\
&\qquad + (e^{\mathrm{P}(f_k)}\psi_k(\omega) - \mathcal{L}_k\psi_k(\omega)) + (\mathcal{L}_k\psi_k(\omega) - \mathcal{L}_f\psi_k(\omega)) + (\mathcal{L}_f\psi_k(\omega) - \mathcal{L}_f\psi(\omega))\Big| \\
&\quad \le e^{\mathrm{P}(f)}|\psi(\omega) - \psi_k(\omega)| + |e^{\mathrm{P}(f)} - e^{\mathrm{P}(f_k)}| \cdot |\psi_k(\omega)| \\
&\qquad + |e^{\mathrm{P}(f_k)}\psi_k(\omega) - \mathcal{L}_k\psi_k(\omega)| + |\mathcal{L}_k\psi_k(\omega) - \mathcal{L}_f\psi_k(\omega)| + |\mathcal{L}_f\psi_k(\omega) - \mathcal{L}_f\psi(\omega)| \\
&\quad \le e^{\mathrm{P}(f)}\varepsilon + |e^{\mathrm{P}(f)} - e^{\mathrm{P}(f_k)}| \cdot \|\psi_k\|_\infty + 0 + \|\mathcal{L}_k\psi_k - \mathcal{L}_f\psi_k\|_\infty + |\mathcal{L}_f\psi_k(\omega) - \mathcal{L}_f\psi(\omega)| \\
&\quad \le e^{\mathrm{P}(f)}\varepsilon + \varepsilon \cdot C(f,R) + C(f,R)\varepsilon + |\mathcal{L}_f\psi_k(\omega) - \mathcal{L}_f\psi(\omega)| \\
&\quad \le (e^{\mathrm{P}(f)} + 2C(f,R))\varepsilon + |\mathcal{L}_f\psi_k(\omega) - \mathcal{L}_f\psi(\omega)|.
\end{aligned} \tag{18.23}$$

This last term can be estimated using (18.19) and (18.20). For all $k \geq \max\{N_2, N_4\}$, we obtain that

$$
\begin{aligned}
|\mathcal{L}_f\psi_k(\omega) - \mathcal{L}_f\psi(\omega)| &= \Bigg| \Bigg(\sum_{\substack{n\leq N_2:\\ n\omega\in E_A^\infty}} e^{f(n\omega)}\psi_k(\omega) + \sum_{\substack{n>N_2:\\ n\omega\in E_A^\infty}} e^{f(n\omega)}\psi_k(\omega)\Bigg) \\
&\quad - \Bigg(\sum_{\substack{n\leq N_2:\\ n\omega\in E_A^\infty}} e^{f(n\omega)}\psi(\omega) + \sum_{\substack{n>N_2:\\ n\omega\in E_A^\infty}} e^{f(n\omega)}\psi(\omega)\Bigg)\Bigg| \\
&\leq \Bigg| \sum_{\substack{n\leq N_2:\\ n\omega\in E_A^\infty}} e^{f(n\omega)}\psi_k(\omega) - \sum_{\substack{n\leq N_2:\\ n\omega\in E_A^\infty}} e^{f(n\omega)}\psi(\omega)\Bigg| + \sum_{\substack{n>N_2:\\ n\omega\in E_A^\infty}} e^{f(n\omega)}|\psi_k(\omega) - \psi(\omega)| \\
&\leq \varepsilon + \sum_{\substack{n>N_2:\\ n\omega\in E_A^\infty}} e^{f(n\omega)} \cdot \varepsilon \leq (1+\varepsilon)\varepsilon.
\end{aligned}
$$

Inserting this into (18.23), we have that

$$|e^{\mathrm{P}(f)}\psi(\omega) - \mathcal{L}_f\psi(\omega)| \leq (e^{\mathrm{P}(f)} + 2C(f,R) + 1 + \varepsilon)\varepsilon.$$

As $\varepsilon > 0$ is arbitrary, we deduce that $\mathcal{L}_f\psi(\omega) = e^{\mathrm{P}(f)}\psi(\omega)$. Thus,

$$\mathcal{L}_f\psi = e^{\mathrm{P}(f)}\psi.$$

Using part (b), we get that

$$\int_{E_A^\infty} \psi\, dm_f = \lim_{k\to\infty} \int_{E_A^\infty} \psi_k\, dm_{f_k} = \lim_{k\to\infty} 1 = 1.$$

Hence, invoking Corollary 18.1.8, we conclude that $\psi = \psi_f$. So every subsequence of $(\psi_{f_k})_{k=1}^\infty$ has itself a subsequence which converges uniformly on compact subsets of E_A^∞ to ψ_f. This implies that the sequence $(\psi_{f_k})_{k=1}^\infty$ converges uniformly on compact subsets of E_A^∞ to ψ_f. That is, part (c) is proved.

Finally, part (d) is a by-product of (b) and (c). Indeed, let $(f_k)_{k=1}^\infty$ be a sequence of functions in $\mathcal{K}_\beta^s(E_A^\infty, \mathbb{R}) \cap \overline{B_{\rho_\beta}(0,R)}$ converging to a function $f \in \mathcal{K}_\beta^s(E_A^\infty, \mathbb{R}) \cap \overline{B_{\rho_\beta}(0,R)}$. By (b), the sequence $(m_{f_k})_{k=1}^\infty$ converges to m_f in the weak* topology on $M(E_A^\infty)$. By (c), the sequence $(\psi_{f_k})_{k=1}^\infty$ converges to ψ_f in $C_b(E_A^\infty)$. For every function $g \in C_b(E_A^\infty)$, we then have

$$\lim_{k\to\infty} \int_{E_A^\infty} g\, d\mu_{f_k} = \lim_{k\to\infty} \int_{E_A^\infty} g \cdot \psi_{f_k}\, dm_{f_k} = \int_{E_A^\infty} g \cdot \psi_f\, dm_f = \int_{E_A^\infty} g\, d\mu_f.$$

As $C_b(E_A^\infty)$ is dense in $L^1(\mu_f)$, the above equality holds for all $g \in L^1(\mu_f)$. That is, the sequence $(\mu_{f_k})_{k=1}^\infty$ converges to μ_f in the weak* topology on $M(E_A^\infty)$. □

18.3 Stochastic laws

In this section, we prove the basic stochastic laws for countable state subshifts of finite type and Hölder-on-cylinders summable potentials. This section, its content and proofs, is analogous to Section 13.9. It relies on all that has been proved in Section 18.1, the main tool being Theorem 18.1.12.

18.3.1 Exponential decay of correlations

We start with a demonstration that, given a potential f which is summable and Hölder continuous on cylinders, for suitable functions g and h the correlations $C_n(g,h)$ decay exponentially with respect to the σ-invariant Gibbs state μ_f. This is a counterpart to Theorem 13.9.1.

Theorem 18.3.1. *Let E be a countable alphabet, let $A : E \times E \to \{0,1\}$ be a finitely primitive incidence matrix, and let $f \in \mathcal{K}_\beta^s(E_A^\infty, \mathbb{R})$ be a potential. Let μ_f be the unique σ-invariant Gibbs state for f, per Corollary 17.7.5. Then there exist constants $c \geq 0$ and $0 < \zeta < 1$ such that for all $g \in L^1(\mu_f)$ and all $h \in \mathcal{K}_\beta^b(E_A^\infty)$ we have*

$$C_{\mu_f,n}(g,h) \leq c\zeta^n \|g - E_{\mu_f}(g)\|_{L^1(\mu_f)} \|h - E_{\mu_f}(h)\|_\beta, \quad \forall n \in \mathbb{N}.$$

Proof. Let $g \in L^1(\mu_f)$ and $h \in \mathcal{K}_\beta^b(E_A^\infty)$. Replacing g by $g - E_{\mu_f}(g)$ and h by $h - E_{\mu_f}(h)$, we may assume without loss of generality that $E_{\mu_f}(g) = E_{\mu_f}(h) = 0$. Using (13.76) and Theorem 18.1.10(c), we obtain for every $n \in \mathbb{N}$ that

$$C_{\mu_f,n}(g,h) = \left| \int_{E_A^\infty} h \cdot (g \circ \sigma^n)\, d\mu_f \right| = \left| \int_{E_A^\infty} \hat{\mathcal{L}}_f^n(h \cdot (g \circ \sigma^n))\, d\mu_f \right| = \left| \int_{E_A^\infty} g \cdot \hat{\mathcal{L}}_f^n(h)\, d\mu_f \right|. \tag{18.24}$$

Applying Theorem 18.1.12(a), we have that

$$\|\hat{\mathcal{L}}_f^n(h)\|_\infty \leq \|\hat{\mathcal{L}}_f^n(h)\|_\beta \leq c\zeta^n \|h\|_\beta$$

for some $c \geq 0$ and $0 < \zeta < 1$ independent of n. Inserting this into (18.24), we conclude that

$$C_{\mu_f,n}(g,h) \leq \int_{E_A^\infty} |g \cdot \hat{\mathcal{L}}_f^n(h)|\, d\mu_f \leq \|\hat{\mathcal{L}}_f^n(h)\|_\infty \int_{E_A^\infty} |g|\, d\mu_f \leq c\zeta^n \|h\|_\beta \|g\|_{L^1(\mu_f)}. \qquad \square$$

18.3.2 Asymptotic variance

An immediate consequence of Theorems 18.3.1 and 13.9.5 is the following counterpart of Corollary 13.9.6.

Theorem 18.3.2. *Let E be a countable alphabet, let $A : E \times E \to \{0,1\}$ be a finitely primitive incidence matrix, and let a potential $f : E_A^\infty \to \mathbb{R}$ be summable and Hölder continuous on cylinders. Let μ_f be the unique σ-invariant Gibbs state for f, per Corollary 17.7.5. If $g : E_A^\infty \to \mathbb{R}$ is Hölder continuous, then the asymptotic variance $\sigma^2_{\mu_f,\infty}(g)$ exists and the following three conditions are equivalent:*
(a) $\sigma^2_{\mu_f,\infty}(g) = 0$.
(b) $\sup_{n\in\mathbb{N}} \|S_n(g - \mu_f(g)\mathbb{1})\|_{L^2(\mu_f)} < \infty$.
(c) *g is cohomologous to a constant in the Hilbert space $L^2(\mu_f)$.*

As a direct consequence of Theorem 18.3.1 and Lemma 13.9.9, we obtain the following generalization of Lemma 13.9.11 (cf. Lemma 13.9.13).

Lemma 18.3.3. *Let E be a countable alphabet, let $A : E\times E \to \{0,1\}$ be a finitely primitive incidence matrix, and let $f \in \mathcal{K}^s_\alpha(E_A^\infty, \mathbb{R})$ be a potential. Let μ_f be the unique σ-invariant Gibbs state for f, per Corollary 17.7.5. If $g \in \mathcal{K}^b_\beta(E_A^\infty, \mathbb{R})$ and $\sigma^2_{\mu_f,\infty}(g) = 0$, then there exists $u \in \mathcal{K}^b_{\min\{\alpha,\beta\}}(E_A^\infty, \mathbb{R})$ such that*

$$g = \mu_f(g) + u - u \circ \sigma,$$

i. e., g is cohomologous to the constant $\mu_f(g)$ in the additive group $\mathcal{K}^b_{\min\{\alpha,\beta\}}(E_A^\infty, \mathbb{R})$.

18.3.3 Central limit theorem

The celebrated central limit theorem also holds for countable state subshifts of finite type. This is the counterpart of Theorem 13.9.15.

Theorem 18.3.4 (Central limit theorem). *Let E be a countable alphabet, let $A : E \times E \to \{0,1\}$ be a finitely primitive incidence matrix, and let a potential $f : E_A^\infty \to \mathbb{R}$ be summable and Hölder continuous on cylinders. Let μ_f be the unique σ-invariant Gibbs state for f, per Corollary 17.7.5. If $g : E_A^\infty \to \mathbb{R}$ is Hölder continuous but is not cohomological to a constant in the class of Hölder continuous functions on E_A^∞, then for every Lebesgue measurable set $B \subseteq \mathbb{R}$ with boundary of Lebesgue measure zero we have that*

$$\begin{aligned}\lim_{n\to\infty} \mu_f\left(\left\{\omega \in E_A^\infty : \frac{S_n g(\omega) - \mu_f(g)n}{\sqrt{n}} \in B\right\}\right) &= \mathcal{N}_0(\sigma^2_{\mu_f,\infty}(g))(B)\\ &= \frac{1}{\sqrt{2\pi\sigma^2_{\mu_f,\infty}(g)}} \int_B \exp\left(\frac{-u^2}{2\sigma^2_{\mu_f,\infty}(g)}\right) du,\end{aligned}$$

where $\sigma^2_{\mu_f,\infty}(g)$ is the asymptotic variance of g whose existence is guaranteed by Theorem 18.3.2 and whose positivity follows from Lemma 18.3.3.

Proof. The proof is very similar to that of Theorem 13.9.15. It is based on Theorem 18.1.12 in place of Theorem 13.8.18. □

18.3.4 Law of the iterated logarithm

Finally, the law of the iterated logarithm also has a version for countable state subshifts of finite type. For the sake of completeness, we state it, though like its counterpart Theorem 13.9.17 a full proof necessitates deep results from probability theory and is too involved to be presented in this book. We instead refer the reader to [41, 98].

Theorem 18.3.5 (Law of the iterated logarithm). *Let E be a countable alphabet, let $A : E \times E \to \{0,1\}$ be a finitely primitive incidence matrix, and let a potential $f : E_A^\infty \to \mathbb{R}$ be summable and Hölder continuous on cylinders. Let μ_f be the unique σ-invariant Gibbs state for f, per Corollary 17.7.5. If $g : E_A^\infty \to \mathbb{R}$ is Hölder continuous but is not cohomological to a constant in the class of Hölder continuous functions on E_A^∞, then*

$$\limsup_{n\to\infty} \frac{S_n g(\omega) - n\int_{E_A^\infty} g\, d\mu_f}{\sqrt{n\log\log n}} = \sqrt{2\sigma^2_{\mu_f,\infty}(g)} \quad \text{for } \mu_f\text{-a.e. } \omega \in E_A^\infty.$$

18.4 Potentials with the same Gibbs states

We are ready to give a complete characterization of the potentials which have the same invariant Gibbs state. The next result is analogous to Theorem 13.9.14.

Theorem 18.4.1. *Let E be a countable alphabet and $A : E \times E \to \{0,1\}$ a finitely irreducible incidence matrix. If two potentials $\varphi, \psi : E_A^\infty \to \mathbb{R}$ are summable and Hölder continuous on cylinders, then the following statements are equivalent:*

(a) *The σ-invariant Gibbs states of φ and ψ coincide, i. e., $\mu_\varphi = \mu_\psi$.*

(b) *There exists a constant $R_1 \in \mathbb{R}$ such that*

$$S_n\varphi(\omega) - S_n\psi(\omega) = R_1 n$$

whenever $n \in \mathbb{N}$ and $\omega \in E_A^\infty$ is a periodic point of σ of period n.

(c) *There exist constants $R_2 \in \mathbb{R}$ and $C \in [0,\infty)$ such that*

$$|S_n\varphi(\omega) - S_n\psi(\omega) - R_2 n| \le C, \quad \forall n \in \mathbb{N},\ \forall \omega \in E_A^\infty.$$

(d) *$\varphi - \psi$ is cohomologous to a constant $R_3 \in \mathbb{R}$ in the additive group of bounded functions on E_A^∞.*

(e) *$\varphi - \psi$ is cohomologous to a constant $R_4 \in \mathbb{R}$ in the additive group of bounded continuous functions on E_A^∞.*

(f) $\varphi - \psi$ *is cohomologous to a constant* $R_5 \in \mathbb{R}$ *in the additive group of Hölder continuous functions on* E_A^∞.
(g) $\mathcal{G}_\varphi = \mathcal{G}_\psi$.
(h) $\mathcal{G}_\varphi \cap \mathcal{G}_\psi \neq \emptyset$.

If any of the above conditions holds and A is finitely primitive, then
(i) $\sigma^2_{\mu_f,\infty}(\varphi - \psi) = 0$ *for all potentials* $f : E_A^\infty \to \mathbb{R}$ *which are summable and Hölder continuous on cylinders.*

If, in addition to the hypotheses of the present theorem, the function $\varphi - \psi$ *is bounded and A is finitely primitive then*
(j) *the equality* $\sigma^2_{\mu_f,\infty}(\varphi - \psi) = 0$ *for some potential* $f : E_A^\infty \to \mathbb{R}$ *which is summable and Hölder continuous on cylinders, entails all of the above items* (a)–(g).

Finally, if any of the above conditions holds, then all the constants R_1*–*R_5 *are the same and equal to* $\mathrm{P}(\sigma, \varphi) - \mathrm{P}(\sigma, \psi)$.

Proof. Obviously, (a)⇒(g)⇒(h) and (f)⇒(e)⇒(d). It is also straightforward to see that (d)⇒(c).

To see that (c)⇒(b), take a periodic point ω of period n, write the formula in (c) with n replaced by kn for an arbitrary $k \in \mathbb{N}$, and let $k \to \infty$ to conclude that the formula in (b) holds. It follows in turn that $R_5 = R_4 = R_3 = R_2 = R_1$ (cf. Exercise 13.11.5).

The implication (d)⇒(a) is a consequence of two facts: (1) φ and ψ share the same set of Gibbs states according to Exercise 17.9.15(b); and (2) each potential has a unique σ-invariant Gibbs state by virtue of Corollary 17.7.5.

Let us now prove that (h)⇒(b). Let $\chi := [\varphi - \mathrm{P}(\sigma, \varphi)] - [\psi - \mathrm{P}(\sigma, \psi)] = [\varphi - \psi] - [\mathrm{P}(\sigma, \varphi) - \mathrm{P}(\sigma, \psi)]$. As there is a Gibbs state common to both φ and ψ (cf. (17.9)), there exists a constant $D \geq 1$ such that

$$D^{-1} \leq \exp([S_n\varphi(\tau) - n\mathrm{P}(\sigma, \varphi)] - [S_n\psi(\tau) - n\mathrm{P}(\sigma, \psi)]) = e^{S_n\chi(\tau)} \leq D \tag{18.25}$$

for all $n \in \mathbb{N}$ and all $\tau \in E_A^\infty$. In particular, if τ is a periodic point of period n then $S_n\chi(\tau) = 0$. Otherwise, we would have $S_{kn}\chi(\tau) = kS_n\chi(\tau)$ for all $k \in \mathbb{N}$ and (18.25) would not hold. Thus, item (b) holds with $R_1 = \mathrm{P}(\sigma, \varphi) - \mathrm{P}(\sigma, \psi)$.

With this value for R_1, let us establish that (b) implies (f). Keep χ the same as above. Since the shift map σ is transitive, let $\tau \in E_A^\infty$ be such that its ω-limit set is $\omega(\tau) = E_A^\infty$. Define $u(\sigma^n(\tau)) := S_n\chi(\tau)$ for every $n \geq 0$, where we define $S_0\chi \equiv 0$. This function u is well-defined on $\mathcal{O}_+(\tau)$. Indeed, if τ is periodic then u is well-defined since $S_N\chi(\tau) = 0$, where N is the prime period of τ. If τ is not periodic, then u is well defined simply because $\sigma^m(\tau) \neq \sigma^n(\tau)$ whenever $m \neq n$. Observe further that

$$\begin{aligned}\chi(\sigma^n(\tau)) &= S_{n+1}\chi(\tau) - S_n\chi(\tau) = u(\sigma^{n+1}(\tau)) - u(\sigma^n(\tau)) \\ &= u \circ \sigma(\sigma^n(\tau)) - u(\sigma^n(\tau)) \\ &= (u \circ \sigma - u)(\sigma^n(\tau))\end{aligned} \tag{18.26}$$

for all $n \ge 0$, that is, χ is cohomologous to 0 on $\mathcal{O}_+(\tau)$ with u as a coboundary. Equivalently, $\varphi - \psi$ is cohomologous to $\mathrm{P}(\sigma,\varphi) - \mathrm{P}(\sigma,\psi)$ on $\mathcal{O}_+(\tau)$ with u as a coboundary. If τ is periodic, then $\mathcal{O}_+(\tau) = E_A^\infty$ and u is the sought coboundary. If τ is not periodic, then $\mathcal{O}_+(\tau) \ne E_A^\infty$ and we need to extend u continuously to $\overline{\mathcal{O}_+(\tau)} = E_A^\infty$. In order to do this (and more), assume that $|\sigma^k(\tau) \wedge \sigma^\ell(\tau)| \ge 1$ for some $k < \ell$. Let $\eta := \tau|_k(\sigma^k(\tau)|_{\ell-k})^\infty \in E_A^\infty$. Then $\sigma^k(\eta) = (\sigma^k(\tau)|_{\ell-k})^\infty$ is a periodic point of period $\ell - k$. So $S_{\ell-k}\chi(\sigma^k(\eta)) = 0$. Hence,

$$\begin{aligned} u(\sigma^\ell(\tau)) - u(\sigma^k(\tau)) &= S_\ell\chi(\tau) - S_k\chi(\tau) = S_{\ell-k}\chi(\sigma^k(\tau)) \\ &= S_{\ell-k}\chi(\sigma^k(\tau)) - S_{\ell-k}\chi(\sigma^k(\eta)) \\ &= S_{\ell-k}\chi(\sigma^k(\tau)|_{\ell-k}\sigma^\ell(\tau)) - S_{\ell-k}\chi(\sigma^k(\eta)|_{\ell-k}\sigma^\ell(\eta)). \end{aligned}$$

Observe that $|\tau \wedge \eta| = \ell + |\sigma^k(\tau) \wedge \sigma^\ell(\tau)|$, and thus $|\sigma^\ell(\tau) \wedge \sigma^\ell(\eta)| = |\sigma^k(\tau) \wedge \sigma^\ell(\tau)|$. Let β be an Hölder-on-cylinders exponent of χ (see Exercise 18.5.5). Using Lemma 17.2.3 with $f = \chi$, $n = \ell - k$ and $\omega = \sigma^k(\tau)|_{\ell-k}$, we deduce that

$$\begin{aligned} |u(\sigma^\ell(\tau)) - u(\sigma^k(\tau))| &= \left|S_{\ell-k}\chi(\sigma^k(\tau)|_{\ell-k}\sigma^\ell(\tau)) - S_{\ell-k}\chi(\sigma^k(\eta)|_{\ell-k}\sigma^\ell(\eta))\right| \\ &\le \frac{v_\beta(\chi)}{e^\beta - 1}[d(\sigma^\ell(\tau),\sigma^\ell(\eta))]^\beta \\ &\le \frac{v_\beta(\varphi) + v_\beta(\psi)}{e^\beta - 1}[d(\sigma^k(\tau),\sigma^\ell(\tau))]^\beta. \end{aligned} \tag{18.27}$$

That is,

$$|u(\gamma) - u(\rho)| \le \frac{v_\beta(\varphi) + v_\beta(\psi)}{e^\beta - 1}[d(\gamma,\rho)]^\beta \tag{18.28}$$

whenever $\gamma, \rho \in \mathcal{O}_+(\tau)$ and $\gamma_1 = \rho_1$.

In particular, it follows from this that u is uniformly continuous on $\mathcal{O}_+(\tau)$. Since $\mathcal{O}_+(\tau)$ is a dense subset of E_A^∞, we therefore conclude that u has a unique continuous extension on E_A^∞, which we also denote by u. Moreover, (18.28) extends to all $\gamma, \rho \in E_A^\infty$. In particular, the function u is Hölder continuous on cylinders. Hence, (18.26) holds on all of E_A^∞, that is, χ is cohomologous to 0 with u as a coboundary.

We are left to prove that u is bounded. The proof is essentially identical to the end of the proof of Lemma 13.9.9. Let $\Lambda \subseteq E_A^*$ be a finite set witnessing the finite irreducibility of A. Let

$$|\Lambda| := \max\{|\alpha| : \alpha \in \Lambda\} \quad \text{and} \quad \Lambda_1 := \{\alpha_1 : \alpha \in \Lambda\}.$$

Since both functions χ and u are Hölder continuous on cylinders, $\sup(|\chi||_{[e]})$ and $\sup(|u||_{[e]})$ are finite for all $e \in E$. So

$$M_\chi := \max\{\sup(|\chi||_{[e]}) : e \in \Lambda_1\} < \infty \quad \text{and} \quad M_u := \max\{\sup(|u||_{[e]}) : e \in \Lambda_1\} < \infty.$$

Pick an arbitrary $\rho \in E_A^\infty$ and choose any $\hat{\rho} \in \Lambda$ such that $\hat{\rho}\rho \in E_A^\infty$. Let $k := |\hat{\rho}| \leq |\Lambda|$. As $\chi = u \circ \sigma - u$, it ensues that

$$S_k\chi(\hat{\rho}\rho) = u(\sigma^k(\hat{\rho}\rho)) - u(\hat{\rho}\rho) = u(\rho) - u(\hat{\rho}\rho).$$

Therefore,

$$|u(\rho)| \leq |S_k\chi(\hat{\rho}\rho)| + |u(\hat{\rho}\rho)| \leq S_k|\chi|(\hat{\rho}\rho) + \sup(|u||_{[\hat{\rho}_1]}) \leq kM_\chi + M_u \leq |\Lambda|M_\chi + M_u.$$

So $\|u\|_\infty \leq |\Lambda|M_\chi + M_u < \infty$, and the proof of the implication (b)⇒(f) is complete.

If any of the conditions (a)–(h) holds, then we already know that all of them hold, whence (i) directly follows from (h) under the additional assumption that A is finitely primitive according to Theorem 18.3.2. Obviously, (i)⇒(j). Finally, if (j) holds then (h) follows from Lemma 18.3.3. □

18.5 Exercises

Exercise 18.5.1. Let E be a countably infinite alphabet and $A : E \times E \to \{0,1\}$ be a finitely irreducible incidence matrix. Show that the closed unit ball in the Banach space $\mathcal{K}_\beta^b(E_A^\infty)$ is not compact when treated as a subset of $C_b(E_A^\infty)$.

Exercise 18.5.2. Let $f, g \in \mathcal{K}_\beta(E_A^\infty)$, $F \subseteq E_A^\infty$ and $c \in \mathbb{C}$. Prove the following statements:
(a) $v_{\beta,F}(f+g) \leq v_{\beta,F}(f) + v_{\beta,F}(g)$.
(b) $v_{\beta,F}(cf) = |c|\, v_{\beta,F}(f)$.
(c) If $m := \inf_{\omega\in F}|f| > 0$, then $v_{\beta,F}(1/f) \leq v_{\beta,F}(f)/m^2$.

Exercise 18.5.3. Let $f, g \in C_b(E_A^\infty)$, $F \subseteq E_A^\infty$ and $c \in \mathbb{C}$. Prove the following statements:
(a) $\|f+g\|_{\infty,F} \leq \|f\|_{\infty,F} + \|g\|_{\infty,F}$.
(b) $\|cf\|_{\infty,F} = |c|\,\|f\|_{\infty,F}$.
(c) $\|fg\|_{\infty,F} \leq \|f\|_{\infty,F}\|g\|_{\infty,F}$.
(d) If $m := \inf_{\omega\in F}|f| > 0$, then $\|1/f\|_{\infty,F} \leq 1/m$.

Exercise 18.5.4. Let $f, g \in \mathcal{K}_\beta^b(E_A^\infty)$, $F \subseteq E_A^\infty$ and $c \in \mathbb{C}$. Prove the following statements:
(a) $f+g \in \mathcal{K}_{\beta,F}^b(E_A^\infty)$ and $\|f+g\|_{\beta,F} \leq \|f\|_{\beta,F} + \|g\|_{\beta,F}$.
(b) $cf \in \mathcal{K}_{\beta,F}^b(E_A^\infty)$ and $\|cf\|_{\beta,F} = |c|\,\|f\|_{\beta,F}$.
(c) $v_{\beta,F}(fg) \leq \|f\|_{\infty,F}v_{\beta,F}(g) + v_{\beta,F}(f)\|g\|_{\infty,F}$.
(d) $fg \in \mathcal{K}_{\beta,F}^b(E_A^\infty)$ and $\|fg\|_{\beta,F} \leq \|f\|_{\beta,F}\|g\|_{\beta,F}$.
(e) If $m := \inf_{\omega\in F}|f| > 0$, then $1/f \in \mathcal{K}_{\beta,F}^b(E_A^\infty)$ and $\|1/f\|_{\beta,F} \leq \|f\|_{\beta,F}/m^2$.

Exercise 18.5.5. Let $\alpha, \beta \in (0,1]$ and $0 < \gamma \leq \min\{\alpha,\beta\}$. Let also $c \in \mathbb{C}$ and let $f \in \mathcal{K}_\alpha(E_A^\infty)$ while $g \in \mathcal{K}_\beta(E_A^\infty)$. Finally, let $F \subseteq E_A^\infty$. Prove the following statements:

(a) $f \in \mathcal{K}_\gamma(E_A^\infty)$ and $\upsilon_{\gamma,F}(f) \le \upsilon_{\alpha,F}(f)$.
(b) $\upsilon_{\gamma,F}(f+g) \le \upsilon_{\alpha,F}(f) + \upsilon_{\beta,F}(g)$.
(c) Deduce similar statements for cf, fg and $1/f$ using the previous exercises.

Exercise 18.5.6. For any $\psi \in \mathcal{K}_\beta(E_A^\infty)$ and $e \in E$, show that

$$\|\psi|_{[e]}\|_\infty \le |\sup \psi|_{[e]}| + \upsilon_\beta(\psi).$$

Exercise 18.5.7. Prove Theorems 18.1.9–18.1.10.

19 Conformal graph directed Markov systems

In this chapter, we describe a powerful method for constructing and studying the geometric and dynamical properties of quite general fractal sets. This method arises from the theory of countable-alphabet conformal iterated function systems, or more generally from the theory of countable-alphabet conformal graph directed Markov systems, as defined and developed in the paper [75] and the book [81]. Many sets (not only fractal ones) can be viewed as limit sets of conformal (or sometimes even more simply, of similarity) iterated function systems (IFSs). Among them are illustrious fractal sets such as the middle-third Cantor set, the Sierpiński triangle, the Sierpiński carpet, the Koch curve, various kinds of Menger sets and the limit set of real and complex continued fractions. Despite being less glamorous, the unit interval $[0, 1]$, every cube $[0, 1]^n$ and the set of irrational numbers in $[0, 1]$, are also limit sets of conformal IFSs. On the other hand, the unit circle $\mathbb{S}^1 \subseteq \mathbb{R}^2$ is not the limit set of any conformal IFS; it is, however, the limit set of a graph directed Markov system (GDMS). A GDMS is a generalization of an IFS, with a graph comprising vertices and edges and with an edge-transition matrix.

Historically, an iterative scheme for defining sets with desired topological and geometric properties, came into use in the 19th century works of Georg Cantor and others, and has been pretty common in topology ever since. It was formalized in John Hutchinson's groundbreaking paper [56]. Michael Barnsley popularized the scheme and coined its name in [4], and a flow of research ensued. At the early stage, the work on these so-called IFSs focused mainly on finite-alphabet systems consisting of similarities and satisfying the celebrated Open Set Condition (OSC). This condition was introduced in [56] where a formula, nowadays called Hutchinson's formula, was discovered. This formula describes the value of the Hausdorff dimension of the limit set/attractor of such an IFS solely in terms of the scaling factors of its generating similarities.

A leap forward in the advancement of the theory occurred with Rufus Bowen's seminal paper [12]. Although this article did not concern IFSs per se, an analogous formula was nevertheless proved in the context of quasi-Fuchsian groups, which are conformal (but nonsimilarity) IFSs. According to Bowen's formula (cf. Chapter 16), the Hausdorff dimension of the limit set of a conformal expanding repeller is equal to the unique zero of the pressure function for a particular, geometric potential. Following Bowen's paper, research on Bowen's formula and its variations in many settings, abounded. Bedford proved an appropriate counterpart for conformal IFSs in [5]. Mauldin and Williams introduced GDMSs consisting of similarities with finite sets of edges and vertices in [82]. A proper version of Hutchinson's formula was proved therein. Countable-alphabet conformal IFSs were introduced in Mauldin and Urbański's paper [75], where among the main results a suitable form of Bowen's formula was demonstrated. It is in that article that the intriguing possibility of a pressure function having no zero was first observed, leading to the concept of irregular systems.

https://doi.org/10.1515/9783110702699-019

It was further established that if h is the Hausdorff dimension of the limit set then, although this set's h-dimensional Hausdorff measure is always finite and this set's h-dimensional packing measure is usually positive (whenever the Strong Open Set Condition (SOSC) is satisfied), the former may vanish while the latter may be infinite. In the book [81], conformal GDMSs with a finite set of vertices but a countable set of edges, were introduced and systematically studied. Some elements of this theory were presented in the books [50] and [18]. The presentation in this chapter stems from [81] but develops and improves it in many respects.

Turning our attention to the contents of this chapter, the first five sections concentrate on the foundations of the theory of conformal graph directed Markov systems. In Section 19.1, we start with the introduction of basic notions and notation for GDMSs: in particular, the underlying symbolic space and the shift map acting on it, the limit set and the projection (coding map) from the symbolic space onto the limit set. Section 19.2 collects properties of conformal maps in Euclidean spaces that will be crucial in deriving in Section 19.3 the core properties of conformal GDMSs (CGDMSs). In Section 19.4, we define an appropriate version of the topological pressure, the finiteness parameter, and the Bowen parameter. Based on the behavior of the pressure function, we then classify in Section 19.5 CGDMSs as irregular or regular, and in this latter case a finer splitting into strongly regular and critically regular systems is described. Cofinite or hereditary regularity is also defined. In Section 19.6, a variation of Bowen's formula is derived. Among its many consequences, we get an almost cost-free, effective and nontrivial lower estimate for the Hausdorff dimension of the limit sets of finitely irreducible CGDMSs. We further deduce from it the classical version of Hutchinson's formula. A particularly important tool to study the fractal geometry of limit sets of CGDMSs is the concept of conformal measure. We introduce and study these measures at length in Section 19.8, and we give a formula, called volume lemma, for the Hausdorff dimension of projections of ergodic invariant measures from the underlying symbolic space. This formula is expressed in the familiar form of the ratio of the measure-theoretic entropy and the Lyapunov exponent. Beforehand, though, we examine in Section 19.7 different separation and geometric conditions: the Strong Open Set Condition (SOSC), the Boundary Separation Condition (BSC), the Strong Separation Condition (SSC) and the Cone Condition (CC). Any one of those, when satisfied, has a profound impact on conformal measures, invariant measures and the whole dynamics and geometry of CGDMSs. Finally, we provide a variety of examples in Section 19.9.

Returning to the all important topic of Section 19.8, conformal measures were first defined by Samuel Patterson in his influential paper [92] (see also [93]) in the context of Fuchsian groups. Dennis Sullivan extended this concept to all Kleinian groups in [122, 124, 126]. In the papers [123, 125], he then introduced and proved the existence of conformal measures for all rational functions of the Riemann sphere. Both Patterson and Sullivan created the concept of conformal measure to get an understanding of geometric measures, i. e., of Hausdorff and packing measures. Though

Sullivan already noticed that there are conformal measures for Kleinian groups that are not equal, nor even equivalent, to any Hausdorff or packing measure, their main purpose is to comprehend Hausdorff and packing measures.

Conformal measures for rational functions, in the sense of Sullivan, have been studied in greater detail by Denker and Urbański in [24] where, in particular, the structure of the set of their exponents was examined. Subsequently, the concept of conformal measure, still in the sense of Sullivan, was extended to countable-alphabet conformal IFSs (CIFSs) in [75] and to CGDMSs in [81].

Additional information about conformal graph directed Markov systems can be found in many papers and books, including [16–19, 36–39, 50, 73, 74, 76, 78–80, 83, 102, 132, 133].

19.1 Graph directed Markov systems

Graph directed Markov systems are based on a directed multigraph $\mathcal{G}$ and an incidence matrix A.

19.1.1 The underlying multigraph $\mathcal{G}$

The directed graph $\mathcal{G} = (V, E, i, t)$ consists of a **finite** set V of vertices, a **countable** (finite or infinite) set E of directed edges, and two functions $i, t : E \to V$ that indicate for each directed edge $e \in E$ its initial vertex $i(e)$ and its terminal vertex $t(e)$, respectively.

In symbolic terms, the set E of edges will play the role of the alphabet. Accordingly, we will use the terminology and notation from Chapter 17.

The functions i and t can naturally be extended to $E^*\backslash\{\epsilon\} = \bigcup_{n=1}^{\infty} E^n$ by defining $i(\omega) := i(\omega_1)$ and $t(\omega) := t(\omega_{|\omega|})$. The function i also extends to E^∞.

The directed graph $\mathcal{G}$ imposes some restrictions on the edges that can follow any given edge. To reflect these constraints, we set

$$E_{\mathcal{G}}^{\infty} := \{\omega \in E^\infty \mid t(\omega_j) = i(\omega_{j+1}), \forall j \in \mathbb{N}\}.$$

The elements of $E_{\mathcal{G}}^{\infty}$ are called infinite $\mathcal{G}$-admissible words. For each $n \geq 0$, let

$$E_{\mathcal{G}}^{n} := \{\omega \in E^n \mid t(\omega_j) = i(\omega_{j+1}), \forall 1 \leq j < n\}$$

be the set of all $\mathcal{G}$-admissible n-letter words (note: $E_{\mathcal{G}}^0 = \{\epsilon\}$ and $E_{\mathcal{G}}^1 = E$). Let

$$E_{\mathcal{G}}^{*} := \bigcup_{n=0}^{\infty} E_{\mathcal{G}}^{n}$$

be the set of all finite $\mathcal{G}$-admissible words. For every $\omega \in E_{\mathcal{G}}^*$, let

$$[\omega]_{\mathcal{G}} := \{\tau \in E_{\mathcal{G}}^\infty : \tau|_{|\omega|} = \omega\}.$$

These sets are called initial $\mathcal{G}$-cylinders. For every $\omega \in E_{\mathcal{G}}^*$, let also

$$E_{\mathcal{G}}^{\infty,\omega} := \{\gamma \in E_{\mathcal{G}}^\infty \mid \omega\gamma \in E_{\mathcal{G}}^\infty\} = \bigcup_{e\in E:\, t(\omega)=i(e)} [e]_{\mathcal{G}}$$

be the set of all infinite $\mathcal{G}$-admissible suffixes of ω, whereas

$$E_{\mathcal{G}}^{*,\omega} := \{\gamma \in E_{\mathcal{G}}^* \mid \omega\gamma \in E_{\mathcal{G}}^*\} = \{\gamma \in E_{\mathcal{G}}^* \mid t(\omega) = i(\gamma)\}$$

will denote the set of all finite $\mathcal{G}$-admissible suffixes of ω and

$$E_{\mathcal{G}}^{n,\omega} := \{\gamma \in E_{\mathcal{G}}^n \mid \omega\gamma \in E_{\mathcal{G}}^*\} = \{\gamma \in E_{\mathcal{G}}^n \mid t(\omega) = i(\gamma)\}$$

be the set of all $\mathcal{G}$-admissible n-letter suffixes of ω.

19.1.2 The underlying matrix *A*

The matrix $A : E\times E \to \{0,1\}$ is an edge incidence matrix, i. e., it dictates which edge(s) may follow any given edge. By definition, edge $e_2 \in E$ can follow edge $e_1 \in E$ when $A_{e_1e_2} = 1$ but not otherwise. The matrix A must further respect the graph $\mathcal{G}$, meaning that if $A_{e_1e_2} = 1$ then $t(e_1) = i(e_2)$.

More precisely, the incidence matrix $A : E\times E \to \{0,1\}$ must respect the underlying graph $\mathcal{G}$ (i. e., $A_{e_1e_2} = 0$ whenever $t(e_1) \neq i(e_2)$) but potentially imposes additional restrictions on the admissibility of words (i. e., it may be such that $A_{e_1e_2} = 0$ even though $t(e_1) = i(e_2)$). Accordingly,

$$E_A^\infty := \{\omega \in E^\infty \mid A_{\omega_j\omega_{j+1}} = 1, \forall j \in \mathbb{N}\} \subseteq E_{\mathcal{G}}^\infty.$$

Complete metrizability and separability of E_A^∞ can be achieved using, for instance, one of the classical metrics $d_s(\omega,\tau) = s^{|\omega\wedge\tau|}$, where $s \in (0,1)$ and $|\omega \wedge \tau|$ is the length of the initial block that ω and τ have in common (see Chapter 17). However, note that E_A^∞ is generally not compact when E is infinite.

For each $n \geq 0$, we also have that

$$E_A^n := \{\omega \in E^n \mid A_{\omega_j\omega_{j+1}} = 1, \forall 1 \leq j < n\} \subseteq E_{\mathcal{G}}^n,$$

and

$$E_A^* := \bigcup_{n=0}^{\infty} E_A^n \subseteq E_{\mathcal{G}}^*.$$

Similarly, for every $\omega \in E_A^*$ the following hold:

$$[\omega] = [\omega]_A := \{\tau \in E_A^\infty : \tau|_{|\omega|} = \omega\} = [\omega]_{\mathcal{G}} \cap E_A^\infty$$
$$E_A^{\infty,\omega} := \{\gamma \in E_A^\infty \mid \omega\gamma \in E_A^\infty\} = \bigcup_{e\in E:\, A_{\omega_{|\omega|}e}=1} [e] \subseteq E_{\mathcal{G}}^{\infty,\omega}$$
$$E_A^{*,\omega} := \{\gamma \in E_A^* \mid \omega\gamma \in E_A^*\} = \{\gamma \in E_A^* \mid A_{\omega_{|\omega|}\gamma_1} = 1\} \subseteq E_{\mathcal{G}}^{*,\omega}$$
$$E_A^{n,\omega} := \{\gamma \in E_A^n \mid \omega\gamma \in E_A^*\} = \{\gamma \in E_A^n \mid A_{\omega_{|\omega|}\gamma_1} = 1\} \subseteq E_{\mathcal{G}}^{n,\omega}.$$

19.1.3 The system itself

A *graph directed Markov system* (*GDMS*) Φ consists of a directed graph $\mathcal{G}$ and a matrix A as described above, together with a set of nonempty compact subsets $\{X_v\}_{v\in V}$ of a common Euclidean space $\mathbb{R}^d$ and a set of one-to-one contractions $\{\varphi_e : X_{t(e)} \to X_{i(e)}\}_{e\in E}$ with common contraction ratio $s < 1$. Note that the contraction φ_e "goes in the opposite direction" to the edge e. This is because each contraction is conceived as an inverse branch of a distance expanding dynamical system.

A GDMS is said to be finite if the set of edges E is finite. Otherwise, a GDMS is said to be infinite.

A GDMS is called a *graph directed system* (*GDS*) if the matrix A solely provides the information already carried by the graph $\mathcal{G}$, i. e., if the transition (passage) from one edge to another only depends on the graph, i. e., provided that $A_{e_1e_2} = 1$ if and only if $t(e_1) = i(e_2)$. Since the matrix A must always respect the graph $\mathcal{G}$, every GDMS is a subsystem of a unique GDS. This latter is called the underlying GDS.

A GDS is called an *iterated function system* (*IFS*) if the set V is a singleton. The hyperbolic Cantor sets studied in Section 4.1.2 (see also Example 16.2.2 and Remark 16.2.3) are classical examples of IFSs.

In a GDMS $\Phi = \{\varphi_e\}_{e\in E}$, the composition of maps is first subject to the graph admissibility of the underlying symbolic words. For $\omega \in E_{\mathcal{G}}^* \backslash \{\epsilon\}$, we define

$$\varphi_\omega := \varphi_{\omega_1} \circ \varphi_{\omega_2} \circ \cdots \circ \varphi_{\omega_{|\omega|}} : X_{t(\omega)} \to X_{i(\omega)}$$

and $\varphi_\epsilon = \mathrm{Id}_X$, where $X := \bigsqcup_{v\in V} X_v$ is the disjoint union of the compact sets X_v.

The main object of interest generated by a GDMS Φ is its limit set J. This set is the image of the symbolic subspace E_A^∞ under a coding map π. Indeed, given any $\omega \in E_A^\infty$, the sets $\varphi_{\omega|_n}(X_{t(\omega|_n)})$, $n \in \mathbb{N}$, form a descending sequence of nonempty compact sets. Their diameters do not exceed $s^n \max\{\mathrm{diam}(X_v) \mid v \in V\}$, and thus converge to zero as $n \to \infty$. Therefore, their intersection

$$\bigcap_{n=1}^{\infty} \varphi_{\omega|_n}(X_{t(\omega|_n)})$$

is a singleton, and we denote its element by $\pi(\omega)$. This defines the *coding map* (or projection) $\pi : E_A^\infty \to X$. The *limit set* of the GDMS Φ is

$$J := \pi(E_A^\infty) = \bigcup_{\omega \in E_A^\infty} \bigcap_{n=1}^{\infty} \varphi_{\omega|_n}(X_{t(\omega|_n)}).$$

Clearly, π is a continuous map when E_A^∞ is equipped with the topology generated by the initial A-cylinders $[e]$, $e \in E$. This is the case under any of the metrics $d_s(\omega,\tau) = s^{|\omega\wedge\tau|}$, where $s \in (0,1)$. Like E_A^∞, the limit set J is generally not compact when E is infinite. Nevertheless, since J is a continuous image of the Polish (i. e., the completely metrizable and separable) space E_A^∞, we have the following.

Theorem 19.1.1. *The limit set of any GDMS is an analytic set in the sense of descriptive set theory.*

For more information on analytic sets and descriptive set theory, see [120].

When a GDMS is pointwise finite, there is an alternative description of its limit set.

Definition 19.1.2. A GDMS $\Phi = \{\varphi_e : X_{t(e)} \to X_{i(e)}\}_{e\in E}$ is said to be *pointwise finite* if every point of the phase space belongs to finitely many first-level sets, i. e., if

$$\#\{e \in E : x \in \varphi_e(X_{t(e)})\} < \infty, \quad \forall x \in \mathbb{R}^d.$$

It is straightforward to see that the limit set of a pointwise finite GDMS is the intersection of its level sets.

Lemma 19.1.3. *If* $\Phi = \{\varphi_e\}_{e\in E}$ *is a pointwise finite GDMS, then*

$$J = \bigcap_{n=0}^{\infty} \bigcup_{\omega\in E_A^n} \varphi_\omega(X_{t(\omega)}).$$

In particular, J is a $F_{\sigma\delta}$–set.

The limit set J can also be defined in terms of "vertexwise components": for every $v \in V$, define

$$J_v := J \cap X_v$$

to be the section of the limit set at the vertex v. Observe that $J = \bigcup_{v\in V} J_v$.

It is further worth noticing that for any $\omega \in E_A^\infty$,

$$\begin{aligned}\{\pi(\sigma(\omega))\} &= \bigcap_{n=1}^{\infty} \varphi_{\sigma(\omega)|_n}(X_{t(\sigma(\omega)|_n)}) = \bigcap_{n=1}^{\infty} \varphi_{\omega_2\omega_3\cdots\omega_{n+1}}(X_{t(\omega_{n+1})}) = \bigcap_{n=1}^{\infty} \varphi_{\omega_2\omega_3\cdots\omega_n} \circ \varphi_{\omega_{n+1}}(X_{t(\omega_{n+1})})\\ &\subseteq \bigcap_{n=1}^{\infty} \varphi_{\omega_2\omega_3\cdots\omega_n}(X_{i(\omega_{n+1})}) = \bigcap_{n=1}^{\infty} \varphi_{\omega_2\omega_3\cdots\omega_n}(X_{t(\omega_n)}).\end{aligned}$$

In fact,

$$\{\pi(\sigma(\omega))\} = \bigcap_{n=1}^{\infty} \varphi_{\omega_2\omega_3\cdots\omega_n}(X_{t(\omega_n)})$$

since both sides are singleton sets. It follows that

$$\{\varphi_{\omega_1}(\pi(\sigma(\omega)))\} = \varphi_{\omega_1}\left(\bigcap_{n=1}^{\infty} \varphi_{\omega_2\omega_3\cdots\omega_n}(X_{t(\omega_n)})\right) \subseteq \bigcap_{n=1}^{\infty} \varphi_{\omega_1\omega_2\omega_3\cdots\omega_n}(X_{t(\omega_n)}) = \{\pi(\omega)\}.$$

Hence,

$$\varphi_{\omega_1}(\pi(\sigma(\omega))) = \pi(\omega).$$

By induction on $n \in \mathbb{N}$,

$$\varphi_{\omega|_n}(\pi(\sigma^n(\omega))) = \pi(\omega). \tag{19.1}$$

By the chain rule,

$$\varphi'_{\omega|_n}(\pi(\sigma^n(\omega))) = \varphi'_{\omega_1}(\pi(\sigma(\omega))) \circ \varphi'_{\omega_2}(\pi(\sigma^2(\omega))) \circ \cdots \circ \varphi'_{\omega_n}(\pi(\sigma^n(\omega))). \tag{19.2}$$

Finally, we introduce natural terminology.

Definition 19.1.4. A GDMS Φ is said to be *finitely irreducible* if its underlying matrix A is finitely irreducible (see Definition 17.1.2). In the same vein, a GDMS Φ is said to be *finitely primitive* if its underlying matrix A is finitely primitive (see Definition 17.1.4).

19.2 Properties of conformal maps in $\mathbb{R}^d$, $d \geq 2$

As we will later concentrate on GDMSs whose generators are conformal maps, in this section we shall study the analytic, geometric, and particularly the distortion properties of conformal maps in $\mathbb{R}^d$, $d \geq 2$.

Let U be a connected, open subset of $\mathbb{R}^d$. From Section 16.1, a C^1 diffeomorphism $\varphi : U \to \mathbb{R}^d$ is conformal if and only if its derivative at every point of U is a similarity map without a translation component. Recall that such maps satisfy properties (16.1)–(16.3).

When $d = 1$, every C^1 diffeomorphism is conformal since $T'(x) = \alpha(x)\cdot\mathrm{Id}_{\mathbb{R}}$ for some $\alpha(x) \in \mathbb{R}$. When $d = 2$, the conformal C^1 diffeomorphisms, as functions of a complex variable, are the holomorphic and antiholomorphic univalent functions. When $d \geq 3$, we have the following characterization.

Theorem 19.2.1 (Liouville's theorem). *If $d \geq 3$ and U is a connected, open subset of $\mathbb{R}^d$, then every conformal C^1 diffeomorphism $T : U \to \mathbb{R}^d$ is a Möbius transformation, i. e.,* *takes the form*

$$\varphi = \lambda A \circ i + b, \tag{19.3}$$

where $\lambda > 0$ *and* $b \in \mathbb{R}^d$ *are constants,* A *is an orthogonal matrix and* i *is either the identity map* $\mathrm{Id}_{\mathbb{R}^d}$ *or the inversion* $i_a(x) = \frac{x-a}{\|x-a\|^2}$ *in the unit sphere centered at* $a \in \mathbb{R}^d$.

The proof is long and we will not reproduce it here. For details, we refer the reader to Blair [10] and Benedetti and Petronio [6]. Details of the two-dimensional case can also be found there.

It is easy to compute that

$$|\varphi'(x)| = \frac{\lambda}{\|x-a\|^2} \quad \text{and} \quad \frac{\varphi'(x)}{|\varphi'(x)|} = \mathrm{Id}_{\mathbb{R}^d} - 2Q(x-a), \tag{19.4}$$

where $Q(x)$ is the matrix whose entries are

$$(Q(x))_{ij} = \frac{x_i x_j}{\|x\|^2} \tag{19.5}$$

and where

$$|\varphi'(x)| := \sup_{\|v\|\le 1} \|\varphi'(x)v\|$$

is the operator norm of the derivative $\varphi'(x)$.

In dimension $d = 2$, we will make extensive use of the part of Koebe distortion theorem about the derivative of univalent functions (see Theorems A.3.8–A.3.10 in Appendix A). We will also need the following reformulation of that theorem (for a proof, see the arguments leading to formula (17.4.21) on p. 353 of [53]).

Theorem 19.2.2. *Fix* $0 < \gamma \le 1/2$ *so small that*

$$\max\left\{\left[\frac{1-\gamma}{(1+\gamma)^3}\right]^{-1}, \frac{1+\gamma}{(1-\gamma)^3}\right\} \le \sqrt{2}.$$

There exists a constant $K_2 \ge 1$ *such that if* $x \in \mathbb{C}$, $R > 0$ *and* $\varphi : B(x,R) \to \mathbb{C}$ *is a univalent holomorphic function, then*

$$|\varphi'(y) - \varphi'(x)| \le \frac{K_2}{R}|\varphi'(x)| \cdot |y-x|, \quad \forall y \in B(x, \delta_2 R),$$

where $\delta_2 = \min\{\gamma, \pi/24\}$.

In dimension $d \ge 3$, we have the following corresponding result.

Theorem 19.2.3. *Suppose that* Y *is a bounded subset of* $\mathbb{R}^d$, *where* $d \ge 3$, *and that* W *is an open subset of* $\mathbb{R}^d$ *containing* $\overline{Y}$. *Then there exists a constant* $K_3 \ge 1$ *such that if*

$\varphi : \overline{\mathbb{R}^d} \to \overline{\mathbb{R}^d}$ is a conformal C^1 diffeomorphism with $\varphi(W) \subseteq \mathbb{R}^d$, then

$$\big||\varphi'(y)| - |\varphi'(x)|\big| \leq K_3 |\varphi'(x)| \cdot \|y - x\|, \quad \forall x \in Y, \forall y \in B(Y, \delta_3),$$

where $\delta_3 = \frac{1}{2}\min\{\operatorname{diam}(Y), \operatorname{dist}(Y, \partial W)\}$. In particular,

$$\frac{|\varphi'(y)|}{|\varphi'(x)|} \leq 1 + K_3 \operatorname{diam}(Y), \quad \forall x \in Y, \forall y \in B(Y, \delta_3).$$

Proof. In view of (19.3), there exist $\lambda > 0$, a linear isometry $A : \mathbb{R}^d \to \mathbb{R}^d$, an inversion $i = i_a$ or the identity map $i = \mathrm{Id}_{\mathbb{R}^d}$, and a vector $b \in \mathbb{R}^d$ such that $\varphi = \lambda A \circ i + b$. When i is the identity map, we have $|\varphi'(y)| = |\varphi'(x)| = \lambda$ for all $x, y \in \mathbb{R}^d$ and the statement of the theorem is obviously true. Assume now that i is an inversion. Since $\varphi(W) \subseteq \mathbb{R}^d$, we know that $a \notin W$. Therefore, for all $x \in Y$ and all $y \in B(Y, \delta_3)$ we have

$$\frac{\|x - a\|}{\|y - a\|} \leq \frac{\|x - y\| + \|y - a\|}{\|y - a\|} = 1 + \frac{\|x - y\|}{\|y - a\|} \leq 1 + \frac{\frac{3}{2}\operatorname{diam}(Y)}{\frac{1}{2}\operatorname{dist}(Y, \partial W)}. \tag{19.6}$$

Using (19.4), we deduce that

$$\frac{|\varphi'(y)|}{|\varphi'(x)|} = \frac{\|x - a\|^2}{\|y - a\|^2} \leq \left(1 + \frac{3\operatorname{diam}(Y)}{\operatorname{dist}(Y, \partial W)}\right)^2.$$

The second part of the statement of the theorem follows easily. In order to prove the first part, we may assume without losing generality that $|\varphi'(x)| \leq |\varphi'(y)|$. Using (19.4) and (19.6), we then get

$$\begin{aligned}
|\varphi'(y)| - |\varphi'(x)| &= |\varphi'(x)|\left(\frac{|\varphi'(y)|}{|\varphi'(x)|} - 1\right) = |\varphi'(x)|\left(\frac{\|x - a\|^2}{\|y - a\|^2} - 1\right) \\
&= |\varphi'(x)|\left(\frac{\|x - a\|}{\|y - a\|} + 1\right)\left(\frac{\|x - a\|}{\|y - a\|} - 1\right) \\
&\leq |\varphi'(x)|\left(2 + \frac{3\operatorname{diam}(Y)}{\operatorname{dist}(Y, \partial W)}\right)\frac{\|x - y\|}{\|y - a\|} \\
&\leq |\varphi'(x)|\left(2 + \frac{3\operatorname{diam}(Y)}{\operatorname{dist}(Y, \partial W)}\right)\frac{\|x - y\|}{\frac{1}{2}\operatorname{dist}(Y, \partial W)} \\
&= \left[\left(2 + \frac{3\operatorname{diam}(Y)}{\operatorname{dist}(Y, \partial W)}\right)\frac{2}{\operatorname{dist}(Y, \partial W)}\right]|\varphi'(x)| \cdot \|x - y\|. \qquad \square
\end{aligned}$$

As a consequence of Theorems 19.2.2 and 19.2.3, we obtain the following result on local bounded distortion.

Corollary 19.2.4. *Suppose that Y is a bounded subset of $\mathbb{R}^d$, $d \geq 2$, and W is an open subset of $\mathbb{R}^d$ containing $\overline{Y}$. For every $\varepsilon > 0$, there exists $0 < \delta < \operatorname{dist}(Y, \mathbb{R}^d \setminus W)$ such that* *if* $\varphi : \overline{\mathbb{R}^d} \to \overline{\mathbb{R}^d}$ *is a conformal* C^1 *diffeomorphism such that* $\varphi(W) \subseteq \mathbb{R}^d$*, then*

$$(1+\varepsilon)^{-1} \le \frac{|\varphi'(y)|}{|\varphi'(x)|} \le 1+\varepsilon, \quad \forall x \in Y, \forall y \in B(x,\delta).$$

19.3 Conformal graph directed Markov systems

Definition 19.3.1. A GDMS $\Phi = \{\varphi_e\}_{e\in E}$ is called *conformal* (and thereby a *CGDMS*) if its underlying GDS satisfies the following four conditions:

(a) For every vertex $v \in V$, the set X_v is a compact subset of $\mathbb{R}^d$ which is the closure of its interior, i. e., $X_v = \overline{\mathrm{Int}_{\mathbb{R}^d}(X_v)}$.

(b) (Open Set Condition (OSC)) For all edges $e_1, e_2 \in E$, $e_1 \ne e_2$,

$$\varphi_{e_1}(\mathrm{Int}(X_{t(e_1)})) \cap \varphi_{e_2}(\mathrm{Int}(X_{t(e_2)})) = \emptyset.$$

(c) For every vertex $v \in V$, there exists an open, connected and bounded set W_v such that $X_v \subseteq W_v \subseteq \mathbb{R}^d$ and such that for every $e \in E$ with $t(e) = v$, the map φ_e extends to a contracting conformal diffeomorphism of W_v into $\mathbb{R}^d$, and all of these extensions have a common contraction ratio $s < 1$.

(d) There exist constants $L \ge 1$ and $\alpha > 0$ such that

$$\big||\varphi_e'(y)| - |\varphi_e'(x)|\big| \le L|\varphi_e'(x)| \cdot \|y-x\|^{\alpha}$$

for every $e \in E$ and every pair of points $x, y \in W_{t(e)}$.

Remark 19.3.2. Let $\Phi = \{\varphi_e\}_{e\in E}$ be a CGDMS.

(a) For all incomparable words $\omega, \tau \in E_G^*$, it readily follows from the OSC (condition (b) of a CGDMS) that

$$\varphi_\omega(\mathrm{Int}(X_{t(\omega)})) \cap \varphi_\tau(\mathrm{Int}(X_{t(\tau)})) = \emptyset$$

and thus

$$\varphi_\omega(X_{t(\omega)}) \cap \varphi_\tau(X_{t(\tau)}) = \varphi_\omega(\partial X_{t(\omega)}) \cap \varphi_\tau(\partial X_{t(\tau)}).$$

(b) The boundedness of the sets $\{W_v\}_{v\in V}$ is required in condition (c) of a CGDMS to ensure that the system exhibits bounded distortion. If unbounded sets $\{W_v\}_{v\in V}$ otherwise satisfy condition (c), then there exist open, connected and bounded sets $\{W_v'\}_{v\in V}$ that fulfill condition (c). In fact, we have the following.

Claim. *For every vertex* $v \in V$*, there exists an open, connected and bounded set* W_v' *such that* $X_v \subseteq W_v' \subseteq \overline{W_v'} \subseteq W_v$*.*

Proof of Claim. Fix $v \in V$. Let $\delta_v := \operatorname{dist}(X_v, \mathbb{R}^d \backslash W_v) > 0$. The neighborhood $B(X_v, \delta_v/2) := \bigcup_{x \in X_v} B(x, \delta_v/2)$ of X_v is open and bounded but not necessarily connected. However, each open ball $B(x, \delta_v/2)$, $x \in X$, is connected, and thus every connected component of $B(X_v, \delta_v/2)$ contains at least one such ball. Since $B(X_v, \delta_v/2)$ is bounded and its connected components are mutually disjoint and each contains at least one ball of radius $\delta_v/2$, the neighborhood $B(X_v, \delta_v/2)$ has only finitely many connected components. Denote these components by $B_v^{(1)}, B_v^{(2)}, \ldots, B_v^{(k_v)}$. For each $1 \leq j \leq k_v$, choose $x_v^{(j)} \in B_v^{(j)} \cap X_v$. Since W_v is path-connected, for every $1 \leq j < k_v$ there is a piecewise C^1 curve $\gamma_v^{(j)}$ joining $x_v^{(j)}$ and $x_v^{(j+1)}$ in W_v. Since $\gamma_v^{(j)}$ is compact and W_v is open, there is $\delta_v^{(j)} > 0$ such that $B(\gamma_v^{(j)}, \delta_v^{(j)}) \subseteq W_v$. Set $W_v' := B(X_v, \delta_v/2) \cup \bigcup_{j=1}^{k_v-1} B(\gamma_v^{(j)}, \delta_v^{(j)}/2)$. □

(c) When $d \geq 2$, condition (d) of a CGDMS is respected with $\alpha = 1$ provided that the $\{W_v\}_{v \in V}$ are replaced by the $\{W_v'\}_{v \in V}$. In the case $d = 2$, this follows from condition (c), remark (b) and Theorem 19.2.2, by covering each W_v' with a finite chain of open balls of radius smaller than $\operatorname{dist}(W_v', \mathbb{R}^d \backslash W_v) > 0$. In the case $d \geq 3$, this follows from condition (c), remark (b) and Theorem 19.2.3 with $Y = W_v'$ and $W = W_v$.

(d) In [81], a cone condition was part of the definition of a CGDMS. However, this condition is not necessary for our purposes in most of this chapter. It will be described in Section 19.7 and used for establishing the existence of conformal measures and a volume lemma in Section 19.8. Conformality will play a central role in Chapter 21.

Remark 19.3.3. Let $\Phi = \{\varphi_e\}_{e \in E}$ be a CGDMS. By (19.2) and the conformality of the generators φ_e, $e \in E$, for every $\omega \in E_A^\infty$ and $n \in \mathbb{N}$ we have that

$$|\varphi'_{\omega|_n}(\pi(\sigma^n(\omega)))| = \prod_{j=1}^{n} |\varphi'_{\omega_j}(\pi(\sigma^j(\omega)))|.$$

As a consequence of condition (d) of a CGDMS, we get the following.

Lemma 19.3.4. *Let $\Phi = \{\varphi_e\}_{e \in E}$ be a CGDMS. For all $\omega \in E_\mathcal{G}^*$ and all $x, y \in W_{t(\omega)}$, we have*

$$\big|\log|\varphi'_\omega(y)| - \log|\varphi'_\omega(x)|\big| \leq \frac{L}{1 - s^\alpha} \|y - x\|^\alpha.$$

Proof. Let $\omega \in E_\mathcal{G}^*$, say $\omega \in E_\mathcal{G}^n$, and let $x, y \in W_{t(\omega)}$. For every $1 \leq k \leq n$ and $z \in W_{t(\omega)}$, set $z_k := \varphi_{\omega_{n-k+1}} \circ \varphi_{\omega_{n-k+2}} \circ \cdots \circ \varphi_{\omega_n}(z)$; set also $z_0 = z$. The chain rule states that

$$\varphi'_\omega(z) = \varphi'_{\omega_1}(z_{n-1}) \circ \varphi'_{\omega_2}(z_{n-2}) \circ \cdots \circ \varphi'_{\omega_n}(z_0).$$

The conformality of the generators implies that

$$|\varphi'_\omega(z)| = \prod_{k=1}^{n} |\varphi'_{\omega_k}(z_{n-k})|.$$

Thus,

$$\begin{aligned}\log|\varphi'_\omega(y)| - \log|\varphi'_\omega(x)| &= \sum_{k=1}^{n}\log|\varphi'_{\omega_k}(y_{n-k})| - \sum_{k=1}^{n}\log|\varphi'_{\omega_k}(x_{n-k})| \\ &= \sum_{k=1}^{n}\log\frac{|\varphi'_{\omega_k}(y_{n-k})|}{|\varphi'_{\omega_k}(x_{n-k})|} \\ &= \sum_{k=1}^{n}\log\left(1+\frac{|\varphi'_{\omega_k}(y_{n-k})| - |\varphi'_{\omega_k}(x_{n-k})|}{|\varphi'_{\omega_k}(x_{n-k})|}\right) \\ &\le \sum_{k=1}^{n}\log\left(1+\frac{\big||\varphi'_{\omega_k}(y_{n-k})| - |\varphi'_{\omega_k}(x_{n-k})|\big|}{|\varphi'_{\omega_k}(x_{n-k})|}\right) \\ &\le \sum_{k=1}^{n}\frac{\big||\varphi'_{\omega_k}(y_{n-k})| - |\varphi'_{\omega_k}(x_{n-k})|\big|}{|\varphi'_{\omega_k}(x_{n-k})|},\end{aligned}$$

where the second-last inequality is just a consequence of the increasing behavior of the logarithm function while the last inequality follows from the fact that $\log(1+t) \le t$ for all $t \ge 0$. Using condition (d) of a CGDMS and the uniform contraction ratio of the generators, we deduce that

$$\log|\varphi'_\omega(y)| - \log|\varphi'_\omega(x)| \le \sum_{k=1}^{n} L\|y_{n-k} - x_{n-k}\|^\alpha \le L\sum_{k=1}^{n} s^{\alpha(n-k)}\|y-x\|^\alpha \le \frac{L}{1-s^\alpha}\|y-x\|^\alpha.$$

Interchanging x and y leads to the conclusion. □

As an immediate consequence of Lemma 19.3.4, and thus as a by-product of condition (d) of a CGDMS, we get bounded distortion.

Corollary 19.3.5 (Bounded Distortion Property (BDP)). *Let $\Phi = \{\varphi_e\}_{e\in E}$ be a CGDMS. There exists a constant $K \ge 1$ such that for all $\omega \in E^*_{\mathcal{G}}$ and all $x, y \in W_{t(\omega)}$ we have*

$$|\varphi'_\omega(y)| \le K|\varphi'_\omega(x)|.$$

An immediate repercussion of BDP comes next.

Corollary 19.3.6. *Let $\Phi = \{\varphi_e\}_{e\in E}$ be a CGDMS. For all $\omega\tau \in E^*_{\mathcal{G}}$, the following holds:*

$$K^{-1}\|\varphi'_\omega\|_{X_{t(\omega)}}\|\varphi'_\tau\|_{X_{t(\tau)}} \le \|\varphi'_{\omega\tau}\|_{X_{t(\omega\tau)}} \le \|\varphi'_\omega\|_{X_{t(\omega)}}\|\varphi'_\tau\|_{X_{t(\tau)}}.$$

N. B.: The X_v's can be replaced by the W_v's in this relationship.

Proof. We will prove the lower bound and leave the upper bound to the reader. Let $\omega\tau \in E^*_{\mathcal{G}}$. Then

$$\begin{aligned}\|\varphi'_{\omega\tau}\|_{X_{t(\omega\tau)}} &= \sup_{x\in X_{t(\omega\tau)}}|\varphi'_{\omega\tau}(x)| \\ &= \sup_{x\in X_{t(\tau)}}|\varphi'_\omega(\varphi_\tau(x))\circ\varphi'_\tau(x)|\end{aligned}$$

$$\begin{aligned}
&= \sup_{x\in X_{t(\tau)}} |\varphi'_\omega(\varphi_\tau(x))|\cdot|\varphi'_\tau(x)| \\
&\geq \inf_{x\in X_{t(\tau)}} |\varphi'_\omega(\varphi_\tau(x))|\cdot \sup_{x\in X_{t(\tau)}} |\varphi'_\tau(x)| \\
&\geq \inf_{z\in X_{i(\tau)}} |\varphi'_\omega(z)|\cdot \|\varphi'_\tau\|_{X_{t(\tau)}} \\
&\geq K^{-1} \sup_{z\in X_{t(\omega)}} |\varphi'_\omega(z)|\cdot \|\varphi'_\tau\|_{X_{t(\tau)}} \quad \text{by BDP} \\
&= K^{-1}\|\varphi'_\omega\|_{X_{t(\omega)}} \|\varphi'_\tau\|_{X_{t(\tau)}}.
\end{aligned}$$

It is easy to see that the X_v's can be replaced by the W_v's. □

We shall now prove some geometric consequences of conditions (a)–(d) of a CGDMS and BDP. We first obtain a metric upper bound on the size of the image of any convex set, as well as a set-theoretic upper bound on the image of any ball.

Lemma 19.3.7. *Let $\Phi = \{\varphi_e\}_{e\in E}$ be a CGDMS. For all $\omega \in E^*_{\mathcal{G}}$ and convex sets $C \subseteq W_{t(\omega)}$, we have*

$$\operatorname{diam}(\varphi_\omega(C)) \leq \|\varphi'_\omega\|_{W_{t(\omega)}} \operatorname{diam}(C) \leq K\|\varphi'_\omega\|_{X_{t(\omega)}} \operatorname{diam}(C). \tag{19.7}$$

Moreover, for all $\omega \in E^*_{\mathcal{G}}$, all $x \in W_{t(\omega)}$ and all radii $0 < r \leq \operatorname{dist}(x, \partial W_{t(\omega)})$, we have

$$\varphi_\omega(B(x,r)) \subseteq B(\varphi_\omega(x), \|\varphi'_\omega\|_{W_{t(\omega)}} r) \subseteq B(\varphi_\omega(x), K|\varphi'_\omega(x)|r). \tag{19.8}$$

Proof. These results are simple consequences of the mean value theorem and the BDP (Corollary 19.3.5). □

We now aim at finding an upper bound on the size of the images of the X_v's. Since these sets need not be convex (not even connected), we show that points in a same X_v can be linked in W_v by piecewise C^1 curves of uniformly bounded lengths.

Lemma 19.3.8. *Let $\Phi = \{\varphi_e\}_{e\in E}$ be a CGDMS. There exists a constant $\ell > 0$ such that for every $v \in V$ and every $x, y \in X_v$ there is a piecewise C^1 curve $\gamma_{x,y}$ joining x and y in W_v whose length is at most ℓ.*

Proof. Since the set of vertices V is finite, it suffices to prove the statement for one vertex v. Fix $z \in W_v$. Let $\ell > 0$. Consider the set

$$W_v(\ell) := \left\{ w \in W_v \;\middle|\; \begin{array}{l}\exists \text{ a piecewise } C^1 \text{ curve } \gamma_{z,w} \text{ joining } z \text{ and } w \text{ in } W_v \\ \text{and whose length is smaller than } \ell\end{array} \right\}.$$

The set $W_v(\ell)$ is open and connected, and thus path-connected, since W_v is open and connected. Moreover, if $0 < \ell_1 \leq \ell_2$ then $W_v(\ell_1) \subseteq W_v(\ell_2)$ and $\bigcup_{\ell>0} W_v(\ell) = W_v \supseteq X_v$. Since X_v is compact, there exists $\ell > 0$ such that $W_v(\ell/2) \supseteq X_v$.

Now, let $x, y \in X_v \subseteq W_v(\ell/2)$. Then there are piecewise C^1 curves $\gamma_{z,x}$ joining z and x and $\gamma_{z,y}$ joining z and y in W_v whose lengths are smaller than $\ell/2$. Gluing those curves at z, we obtain a piecewise C^1 curve $\gamma_{x,y}$ joining x and y in W_v and whose length is smaller than ℓ. □

Lemma 19.3.9. *Let $\Phi = \{\varphi_e\}_{e\in E}$ be a CGDMS. There exists a constant $D \geq 1$ such that*

$$\operatorname{diam}(\varphi_\omega(X_{t(\omega)})) \leq D\|\varphi'_\omega\|_{W_{t(\omega)}} \leq KD\|\varphi'_\omega\|_{X_{t(\omega)}}, \quad \forall \omega \in E^*_{\mathcal{G}}. \tag{19.9}$$

Proof. Let $\omega \in E^*_{\mathcal{G}}$ and $x, y \in X_{t(\omega)}$. By Lemma 19.3.8, there exists a piecewise C^1 curve $\gamma_{x,y} : [0,1] \to W_{t(\omega)}$ joining x and y and such that $L(\gamma_{x,y}) \leq \ell$, where $L(\gamma)$ denotes the length of a curve γ. Then $\varphi_\omega(\gamma_{x,y})$ is a piecewise C^1 curve joining $\varphi_\omega(x)$ and $\varphi_\omega(y)$, and

$$\begin{aligned}\|\varphi_\omega(x) - \varphi_\omega(y)\| &\leq L(\varphi_\omega(\gamma_{x,y})) := \int_{[0,1]} |\varphi'_\omega(\gamma_{x,y}(t))|\, dt \\ &\leq \|\varphi'_\omega\|_{\gamma_{x,y}} \cdot L(\gamma_{x,y}) \leq \ell\|\varphi'_\omega\|_{W_{t(\omega)}} \leq \ell K\|\varphi'_\omega\|_{X_{t(\omega)}}.\end{aligned}$$

Let $D = \ell$. □

We shall now obtain lower bounds. First, we establish the counterpart of (19.8).

Lemma 19.3.10. *Let $\Phi = \{\varphi_e\}_{e\in E}$ be a CGDMS. For all $\omega \in E^*_{\mathcal{G}}$, all $x \in W_{t(\omega)}$ and all $0 < r \leq \operatorname{dist}(x, \partial W_{t(\omega)})$, we have*

$$\varphi_\omega(B(x,r)) \supseteq B(\varphi_\omega(x), K^{-1}\|\varphi'_\omega\|_{W_{t(\omega)}} r) \supseteq B(\varphi_\omega(x), K^{-1}|\varphi'_\omega(x)|r). \tag{19.10}$$

Proof. First, using BDP (Corollary 19.3.5), observe that for any $\omega \in E^*_{\mathcal{G}}$ and $z \in W_{t(\omega)}$,

$$|(\varphi_\omega^{-1})'(\varphi_\omega(z))|^{-1} = |\varphi'_\omega(z)| \geq K^{-1}\|\varphi'_\omega\|_{W_{t(\omega)}}.$$

Thus,

$$\|(\varphi_\omega^{-1})'\|_{\varphi_\omega(W_{t(\omega)})} \leq K\|\varphi'_\omega\|^{-1}_{W_{t(\omega)}}. \tag{19.11}$$

Now, fix ω, x and r as in the statement. Let $R > 0$ be the maximal radius such that

$$B(\varphi_\omega(x), R) \subseteq \varphi_\omega(B(x,r)). \tag{19.12}$$

By virtue of the mean value theorem and (19.11)–(19.12), we obtain

$$\varphi_\omega^{-1}(B(\varphi_\omega(x), R)) \subseteq B(x, \|(\varphi_\omega^{-1})'\|_{\varphi_\omega(W_{t(\omega)})} R) \subseteq B(x, K\|\varphi'_\omega\|^{-1}_{W_{t(\omega)}} R).$$

This implies that $B(\varphi_\omega(x), R) \subseteq \varphi_\omega(B(x, K\|\varphi'_\omega\|^{-1}_{W_{t(\omega)}} R))$. From the maximality of R in (19.12) and the openness of φ_ω, it ensues that $K\|\varphi'_\omega\|^{-1}_{W_{t(\omega)}} R \geq r$. That is, $R \geq K^{-1}\|\varphi'_\omega\|_{W_{t(\omega)}} r$ and (19.12) leads to (19.10). □

We shall now prove the counterpart of (19.9).

Lemma 19.3.11. *Let $\Phi = \{\varphi_e\}_{e\in E}$ be a CGDMS. There exists a constant $D \geq 1$ such that*

$$\operatorname{diam}(\varphi_\omega(X_{t(\omega)})) \geq D^{-1}\|\varphi'_\omega\|_{W_{t(\omega)}}, \quad \forall \omega \in E^*_{\mathcal{G}}. \tag{19.13}$$

Proof. Let $\overline{\delta} := \min\{\operatorname{dist}(X_v, \mathbb{R}^d\backslash W_v) : v \in V\}$. By condition (a) of a CGDMS, for each $v \in V$ there are $x_v \neq y_v \in X_v$ such that $y_v \in B(x_v, K^{-1}\overline{\delta})$. For every $\omega \in E^*_{\mathcal{G}}$, we have by (19.10) that

$$\varphi_\omega(B(x_{t(\omega)}, \overline{\delta})) \supseteq B(\varphi_\omega(x_{t(\omega)}), K^{-1}\|\varphi'_\omega\|_{W_{t(\omega)}}\overline{\delta}),$$

while we have by (19.8) in Lemma 19.3.7 that

$$\varphi_\omega(y_{t(\omega)}) \in B(\varphi_\omega(x_{t(\omega)}), K^{-1}\|\varphi'_\omega\|_{W_{t(\omega)}}\overline{\delta}).$$

Therefore, applying the mean value theorem to the map φ_ω^{-1} restricted to the convex set $B(\varphi_\omega(x_{t(\omega)}), K^{-1}\|\varphi'_\omega\|_{W_{t(\omega)}}\overline{\delta})$ followed by (19.11), we obtain

$$\begin{aligned}
\|y_{t(\omega)} - x_{t(\omega)}\| &= \|\varphi_\omega^{-1}(\varphi_\omega(y_{t(\omega)})) - \varphi_\omega^{-1}(\varphi_\omega(x_{t(\omega)}))\| \\
&\leq \|(\varphi_\omega^{-1})'\|_{\varphi_\omega(W_{t(\omega)})} \cdot \|\varphi_\omega(y_{t(\omega)}) - \varphi_\omega(x_{t(\omega)})\| \\
&\leq K\|\varphi'_\omega\|^{-1}_{W_{t(\omega)}}\|\varphi_\omega(y_{t(\omega)}) - \varphi_\omega(x_{t(\omega)})\|.
\end{aligned}$$

Thus,

$$\begin{aligned}
\operatorname{diam}(\varphi_\omega(X_{t(\omega)})) > \|\varphi_\omega(y_{t(\omega)}) - \varphi_\omega(x_{t(\omega)})\| &\geq K^{-1}\|\varphi'_\omega\|_{W_{t(\omega)}} \cdot \|y_{t(\omega)} - x_{t(\omega)}\| \\
&\geq K^{-1}\min\{\|y_v - x_v\| : v \in V\}\|\varphi'_\omega\|_{W_{t(\omega)}} = D^{-1}\|\varphi'_\omega\|_{W_{t(\omega)}},
\end{aligned}$$

where $D = K(\min\{\|y_v - x_v\| : v \in V\})^{-1}$. □

We now make a simple geometric observation which follows from the OSC (condition (b) of a CGDMS) and some of the previous lemmas.

Lemma 19.3.12. *Let $\Phi = \{\varphi_e\}_{e\in E}$ be a CGDMS. For all $0 < \kappa_1 \leq \kappa_2 < \infty$, all $r > 0$ and all $x \in X := \bigsqcup_{v\in V} X_v$, the cardinality of any collection of mutually incomparable words $\omega \in E^*_{\mathcal{G}}$ that satisfy the conditions*

$$\varphi_\omega(X_{t(\omega)}) \cap B(x, r) \neq \emptyset \quad \text{and} \quad \kappa_1 r \leq \operatorname{diam}(\varphi_\omega(X_{t(\omega)})) \leq \kappa_2 r,$$

is bounded above by the number

$$((1 + \kappa_2)KD(R\kappa_1)^{-1})^d,$$

where R is the radius of the largest ball that can be inscribed in all the sets X_v, $v \in V$.

Proof. Let $V_d = \lambda_d(B(0,1))$ be the Lebesgue measure of the unit ball in $\mathbb{R}^d$. For every $v \in V$, let $x_v \in X_v$ be such that $B(x_v, R) \subseteq X_v$. Fix $0 < \kappa_1 \leq \kappa_2 < \infty$, $r > 0$ and $x \in X$. Let W be a collection of incomparable words as described in the statement. Then for every $\omega \in W$, we have

$$\varphi_\omega(X_{t(\omega)}) \subseteq B(x, r + \operatorname{diam}(\varphi_\omega(X_{t(\omega)}))) \subseteq B(x, (1+\kappa_2)r).$$

As the sets $\{\varphi_\omega(\operatorname{Int}(X_{t(\omega)}))\}_{\omega\in W}$ are mutually disjoint due to the OSC, using Lemmas 19.3.10 and 19.3.9 successively we obtain that

$$\begin{aligned}
V_d(1+\kappa_2)^d r^d &= \lambda_d(B(x,(1+\kappa_2)r)) \\
&\geq \lambda_d\Big(\bigcup_{\omega\in W} \varphi_\omega(\operatorname{Int}(X_{t(\omega)}))\Big) \\
&= \sum_{\omega\in W} \lambda_d(\varphi_\omega(\operatorname{Int}(X_{t(\omega)}))) \\
&\geq \sum_{\omega\in W} \lambda_d(\varphi_\omega(B(x_{t(\omega)}, R))) \\
&\geq \sum_{\omega\in W} \lambda_d\Big(B(\varphi_\omega(x_{t(\omega)}), K^{-1}\|\varphi'_\omega\|_{W_{t(\omega)}} R)\Big) \\
&\geq \sum_{\omega\in W} \lambda_d\Big(B(\varphi_\omega(x_{t(\omega)}), K^{-1}D^{-1}\operatorname{diam}(\varphi_\omega(X_{t(\omega)}))R)\Big) \\
&\geq \sum_{\omega\in W} \lambda_d\Big(B(\varphi_\omega(x_{t(\omega)}), (KD)^{-1}\kappa_1 rR)\Big) \\
&= \#W((KD)^{-1}R\kappa_1 r)^d V_d.
\end{aligned}$$

Hence, $\#W \leq ((1+\kappa_2)(KD)(R\kappa_1)^{-1})^d$. □

We now look at the points in the limit set that lie in the boundary of the real space. First, we show that the set of admissible infinite words that generate points on the boundary of the real space is forward shift-invariant.

Lemma 19.3.13. *For any CGDMS $\Phi = \{\varphi_e\}_{e\in E}$,*

$$\sigma(\pi^{-1}(\partial X)) \subseteq \pi^{-1}(\partial X).$$

Proof. Since the φ_e's are open maps, $\varphi_e(\operatorname{Int}(X_{t(e)})) \subseteq \operatorname{Int}(X_{i(e)})$ for every $e \in E$. Because $\partial X_v \cap \operatorname{Int}(X_v) = \emptyset$ for all $v \in V$, it is then clear that

$$\varphi_e^{-1}(\partial X_{i(e)}) \cap X_{t(e)} \subseteq \partial X_{t(e)}.$$

If $\omega \in \pi^{-1}(\partial X)$, then $\pi(\omega) \in \partial X_{i(\omega_1)}$. As $\pi(\omega) = \varphi_{\omega_1}(\pi(\sigma(\omega)))$, we deduce that

$$\pi(\sigma(\omega)) = \varphi_{\omega_1}^{-1}(\pi(\omega)) \in \varphi_{\omega_1}^{-1}(\partial X_{i(\omega_1)}) \cap X_{t(\omega_1)} \subseteq \partial X_{t(\omega_1)}.$$

Hence, $\sigma(\omega) \in \pi^{-1}(\partial X_{t(\omega_1)}) \subseteq \pi^{-1}(\partial X)$. So $\sigma(\pi^{-1}(\partial X)) \subseteq \pi^{-1}(\partial X)$. □

Lemma 19.3.14. *Let $\Phi = \{\varphi_e\}_{e\in E}$ be a CGDMS. For every $\omega \in E_G^*$, we have*

$$\sigma^{|\omega|}(\pi^{-1}(\varphi_\omega(\partial X_{t(\omega)}))) \subseteq \pi^{-1}(\partial X).$$

Proof. Set $n = |\omega|$. Let $\rho \in \sigma^{|\omega|}(\pi^{-1}(\varphi_\omega(\partial X_{t(\omega)})))$. Then $\rho = \sigma^n(\gamma)$ for some $\gamma \in \pi^{-1}(\varphi_\omega(\partial X_{t(\omega)}))$. Thus,

$$\varphi_{\gamma|_n}(X_{t(\gamma|_n)}) \ni \varphi_{\gamma|_n}(\pi(\sigma^n(\gamma))) = \pi(\gamma) \in \varphi_\omega(\partial X_{t(\omega)}).$$

It follows from Remark 19.3.2(a) that $\pi(\gamma) \in \varphi_{\gamma|_n}(\partial X_{t(\gamma|_n)})$. Then $\pi(\rho) = \pi(\sigma^n(\gamma)) = \varphi_{\gamma|_n}^{-1}(\pi(\gamma)) \in \partial X_{t(\gamma_n)} \subseteq \partial X$. Hence, $\rho \in \pi^{-1}(\partial X)$. □

19.4 Topological pressure, finiteness parameter, and Bowen parameter for CGDMSs

Let $\Phi = \{\varphi_e\}_{e\in E}$ be a CGDMS. We will now impose on the symbolic space E_A^∞ a family of potentials $t\Xi$, $t \geq 0$, which is intimately related to the dynamics taking place on the limit set $J := \pi(E_A^\infty)$ of Φ.

Lemma 19.4.1. *Let $\Phi = \{\varphi_e\}_{e\in E}$ be a CGDMS. The potential $\Xi : E_A^\infty \to \mathbb{R}$ given by*

$$\Xi(\omega) := \log|\varphi'_{\omega_1}(\pi(\sigma(\omega)))| \tag{19.14}$$

is Hölder continuous on cylinders, and hence is acceptable. This potential induces the pressure function $\mathrm{P}(\cdot\,\Xi) : [0,\infty) \to \mathbb{R} \cup \{\infty\}$, where $\mathrm{P}(t\Xi) := \mathrm{P}(\sigma, t\Xi)$ is the topological pressure of the shift map $\sigma : E_A^\infty \to E_A^\infty$ subjected to the potential $t\Xi$, as described in Definition 17.2.7.

If Φ is finitely irreducible, then this pressure function satisfies

$$\mathrm{P}(t\Xi) = \lim_{n\to\infty} \frac{1}{n}\log Z_n(t\Xi) = \inf_{n\in\mathbb{N}} \frac{1}{n}\log Z_n(t\Xi),$$

with

$$K^{-t}\sum_{\omega\in E_A^n} \|\varphi'_\omega\|_{X_{t(\omega)}}^t \leq Z_n(t\Xi) \leq K^t \sum_{\omega\in E_A^n} \|\varphi'_\omega\|_{X_{t(\omega)}}^t, \quad \forall n \in \mathbb{N}.$$

Proof. Using Lemma 19.3.4, we deduce for any $\omega, \tau \in E_A^\infty$ with $|\omega \wedge \tau| \geq 1$ that

$$\begin{aligned}|\Xi(\omega) - \Xi(\tau)| &= \Big|\log|\varphi'_{\omega_1}(\pi(\sigma(\omega)))| - \log|\varphi'_{\omega_1}(\pi(\sigma(\tau)))|\Big| \\ &\leq \frac{L}{1-s^\alpha}\|\pi(\sigma(\omega)) - \pi(\sigma(\tau))\|^\alpha \\ &\leq \frac{L}{1-s^\alpha}s^{|\sigma(\omega)\wedge\sigma(\tau)|\alpha} = \frac{L}{1-s^\alpha}s^{(|\omega\wedge\tau|-1)\alpha} = \frac{Ls^{-\alpha}}{1-s^\alpha}[d_s(\omega,\tau)]^\alpha.\end{aligned}$$

This proves the Hölder continuity on cylinders. Then, by Theorem 17.2.8,

$$\mathrm{P}(t\Xi) = \lim_{n\to\infty} \frac{1}{n} \log Z_n(t\Xi) = \inf_{n\in\mathbb{N}} \frac{1}{n} \log Z_n(t\Xi_\Phi).$$

Using Remark 19.3.3, we observe that

$$\begin{aligned}
Z_n(t\Xi) &= \sum_{\omega\in E_A^n} \exp\Big(\overline{S}_n(t\Xi)([\omega])\Big) \\
&= \sum_{\omega\in E_A^n} \exp\Big(\sup_{\tau\in[\omega]} \sum_{j=0}^{n-1} t\Xi(\sigma^j(\tau))\Big) \\
&= \sum_{\omega\in E_A^n} \exp\Big(\sup_{\tau\in[\omega]} \sum_{j=0}^{n-1} t\log\Big|\varphi'_{(\sigma^j(\tau))_1}(\pi(\sigma(\sigma^j(\tau))))\Big|\Big) \\
&= \sum_{\omega\in E_A^n} \exp\Big(\sup_{\tau\in[\omega]} \log\prod_{j=0}^{n-1}\Big|\varphi'_{\tau_{j+1}}(\pi(\sigma^{j+1}(\tau)))\Big|^t\Big) \\
&= \sum_{\omega\in E_A^n} \exp\Big(\log\sup_{\tau\in[\omega]} \prod_{j=1}^{n}\Big|\varphi'_{\tau_j}(\pi(\sigma^j(\tau)))\Big|^t\Big) \\
&= \sum_{\omega\in E_A^n} \sup_{\tau\in[\omega]}\Big|\varphi'_{\tau|_n}(\pi(\sigma^n(\tau)))\Big|^t = \sum_{\omega\in E_A^n} \sup_{\tau\in[\omega]}\Big|\varphi'_{\omega}(\pi(\sigma^{|\omega|}(\tau)))\Big|^t \\
&\asymp \sum_{\omega\in E_A^n} \|\varphi'_\omega\|^t_{X_{t(\omega)}},
\end{aligned}$$

where a constant of comparability is K^t, the t-th power of the constant K of bounded distortion from Corollary 19.3.5. □

As a result, we make the following definition.

Definition 19.4.2. Let $n \in \mathbb{N}$. The nth *partition function* $Z_n : [0,\infty) \to [0,\infty]$ of a CGDMS $\Phi = \{\varphi_e\}_{e\in E}$ is defined at every parameter $t \geq 0$ by

$$Z_n(t) := \sum_{\omega\in E_A^n} \|\varphi'_\omega\|^t_{X_{t(\omega)}}.$$

Every partition function is nonincreasing and may take the value ∞ when $\#E = \infty$. Accordingly, for each $n \in \mathbb{N}$ let

$$\mathrm{Fin}_n := \{t \geq 0 : Z_n(t) < \infty\}$$

be the *finiteness parameter set* for the nth partition function Z_n. Thereafter, let

$$\theta_n := \inf \mathrm{Fin}_n$$

be the *finiteness parameter* of that function.

Definition 19.4.3. The topological pressure function $\mathrm{P} : [0,\infty) \to (-\infty,\infty]$ of a finitely irreducible CGDMS $\Phi = \{\varphi_e\}_{e\in E}$ is defined at the parameter $t \geq 0$ by

$$\mathrm{P}(t) := \lim_{n\to\infty} \frac{1}{n} \log Z_n(t).$$

Note that $K^{-t}Z_n(t) \leq Z_n(t\Xi) \leq K^t Z_n(t)$ for all $n \in \mathbb{N}$, according to Lemma 19.4.1. Thus, $\mathrm{P}(t) = \mathrm{P}(t\Xi)$ for all $t \geq 0$ and the pressure function is thereby well defined.

The finiteness parameter set for the topological pressure function

$$\mathrm{Fin} := \{t \geq 0 : \mathrm{P}(t) < \infty\}$$

is called the *finiteness parameter set* of the CGDMS Φ. The *finiteness parameter* θ of Φ is defined by

$$\theta := \inf \mathrm{Fin}.$$

The partition functions enjoy the following properties.

Proposition 19.4.4. *The partition functions $Z_n : [0,\infty) \to [0,\infty]$, $n \in \mathbb{N}$, of a finitely irreducible CGDMS $\Phi = \{\varphi_e\}_{e\in E}$ have the following properties:*

(a) *For every $t \geq 0$, the sequence $(Z_n(t))_{n=1}^{\infty}$ is submultiplicative and boundedly supermultiplicative.*
(b) *$\mathrm{Fin}_n = \mathrm{Fin}$ for all $n \in \mathbb{N}$. In particular, $\theta_n = \theta$ for all $n \in \mathbb{N}$.*
(c) *Z_n is nonincreasing on $[0,\infty)$, is strictly decreasing to $-\infty$ and convex on $[\theta,\infty)$, is continuous on (θ,∞), and is right-continuous at θ.*

Proof. (a) Fix $t \geq 0$. Since $Z_n(t)$ is comparable to $Z_n(t\Xi)$ according to Lemma 19.4.1 and since the sequence $(Z_n(t\Xi))_{n=1}^{\infty}$ is submultiplicative according to Lemma 17.2.5, the sequence $(Z_n(t))_{n=1}^{\infty}$ is boundedly submultiplicative, i. e., there exists a constant $B(t) > 0$ such that $Z_{m+n}(t) \leq B(t)Z_m(t)Z_n(t)$ for all $m,n \in \mathbb{N}$. One such constant is K^{3t} since $K^{-t}Z_n(t) \leq Z_n(t\Xi) \leq K^t Z_n(t)$. In fact, we even have submultiplicativity since

$$\begin{aligned} Z_{m+n}(t) &:= \sum_{\omega\in E_A^{m+n}} \|\varphi'_\omega\|_{X_{t(\omega)}}^t \leq \sum_{\omega\in E_A^{m+n}} \|\varphi'_{\omega|_m}\|_{X_{t(\omega|_m)}}^t \|\varphi'_{\omega|_{m+1}^{m+n}}\|_{X_{t(\omega|_{m+1}^{m+n})}}^t \\ &\leq \sum_{\tau\in E_A^m} \sum_{\rho\in E_A^n} \|\varphi'_\tau\|_{X_{t(\tau)}}^t \|\varphi'_\rho\|_{X_{t(\rho)}}^t = Z_m(t)Z_n(t). \end{aligned}$$

Similarly, as $Z_n(t)$ is comparable to $Z_n(t\Xi)$ and as the sequence $(Z_n(t\Xi))_{n=1}^{\infty}$ is boundedly supermultiplicative according to Lemma 17.2.5, the sequence $(Z_n(t))_{n=1}^{\infty}$ is boundedly supermultiplicative, i. e., there exists a constant $C(t) > 0$ such that $Z_{m+n}(t) \geq C(t)Z_m(t)Z_n(t)$ for all $m,n \in \mathbb{N}$.

(b) Since $Z_n(t)$ is comparable to $Z_n(t\Xi)$, Theorem 17.2.8 affirms that $\mathrm{Fin} = \mathrm{Fin}_n$ for all $n \in \mathbb{N}$. It follows immediately that $\theta = \theta_n$ for all $n \in \mathbb{N}$.

(c) The nonincreasing property of Z_n is clear. The strictly decreasing behavior of Z_n on $[\theta,\infty)$ can be more precisely described as follows. Let $\theta \le t_1 < t_2$. Then $Z_n(t_2) \le s^{n(t_2-t_1)} Z_n(t_1)$ for all $n \in \mathbb{N}$. The convexity of Z_n on Fin follows from the fact that it is the sum of the convex functions $t \mapsto \|\varphi'_\omega\|^t_{X_{t(\omega)}}$, $\omega \in E^n_A$. Its continuity on (θ,∞) is a direct consequence of its convexity on that interval. The right-continuity at θ follows immediately from the convexity on $[\theta,\infty)$ if $\theta \in$ Fin. However, if $\theta \notin$ Fin, then $Z_n(\theta) = \infty$, or equivalently, $Z_n(\theta\Xi) = \infty$. By Lemma 17.2.6, this means that $\sup\{Z_n(\theta\Xi, F) : F \subseteq E \text{ and } \#F < \infty\} = \infty$. If Z_n were not right-continuous at θ, then we would have $\sup_{t>\theta} Z_n(t) < \infty$. Therefore, there would exist a finite set $F \subseteq E$ such that $Z_n(\theta\Xi, F) > 2K^\theta \sup_{t>\theta} Z_n(t)$. Since $K^{-\theta} Z_n(\theta, F) \le Z_n(\theta\Xi, F) \le K^\theta Z_n(\theta, F)$, this implies that $Z_n(\theta, F) > 2\sup_{t>\theta} Z_n(t)$. As F is finite, $Z_n(\cdot, F)$ is strictly decreasing and continuous on $\mathbb{R}$ (recall that $Z_n(t_2, F) \le s^{n(t_2-t_1)} Z_n(t_1, F)$), and thus $\sup_{t>\theta} Z_n(t, F) = Z_n(\theta, F)$. Therefore, $\sup_{t>\theta} Z_n(t) \ge \sup_{t>\theta} Z_n(t, F) = Z_n(\theta, F) > 2\sup_{t>\theta} Z_n(t)$. This is a contradiction. So Z_n is right-continuous at θ. Finally, the convexity of Z_n on $[0,\infty)$ follows from its (already established) convexity on Fin and its right-continuity at θ. □

We now turn our attention to the topological pressure function. The following is a rewriting of Theorem 17.2.8 for $\mathrm{P}(t)$.

Proposition 19.4.5. *If $\Phi = \{\varphi_e\}_{e\in E}$ is a finitely irreducible CGDMS, then:*
(a) $\mathrm{P}(t) < \infty \iff Z_1(t) < \infty \iff Z_n(t) < \infty$ *for some* $n \in \mathbb{N} \iff Z_n(t) < \infty, \forall n \in \mathbb{N}$.
(b) $\mathrm{P}(t) = \infty \iff Z_1(t) = \infty \iff Z_n(t) = \infty$ *for some* $n \in \mathbb{N} \iff Z_n(t) = \infty, \forall n \in \mathbb{N}$.
(c) $\mathrm{P}(t) = \lim_{n\to\infty} \frac{1}{n} \log Z_n(t) = \inf_{n\in\mathbb{N}} \frac{1}{n} \log Z_n(t)$.

Proof. Statements (a) and (b) follow from the fact that $Z_n(t)$ is comparable to $Z_n(t\Xi)$. (Recall that $K^{-t} Z_n(t) \le Z_n(t\Xi) \le K^t Z_n(t)$.) Statement (c) is a consequence of (a) and (b), Proposition 19.4.4 and (17.5). □

It ensues that the pressure function has the following properties.

Proposition 19.4.6. *The pressure function $\mathrm{P} : [0,\infty) \to \mathbb{R} \cup \{\infty\}$ of a finitely irreducible CGDMS $\Phi = \{\varphi_e\}_{e\in E}$, exhibits the following basic properties:*
(a) $\mathrm{P}(t) = \mathrm{P}(t\Xi)$.
(b) $\mathrm{P}(t) = \sup\{\mathrm{P}_F(t) : F \subseteq E$ *and* $\#F < \infty\}$, *where $\mathrm{P}_F(t)$ is the pressure of the subsystem $\Phi|_F := \{\varphi_e\}_{e\in F}$.*
(c) P *is nonincreasing on $[0,\infty)$, strictly decreasing to $-\infty$ and convex on $[\theta,\infty)$, continuous on (θ,∞), and right-continuous at θ.*
(d) *If $\#E < \infty$ then $\mathrm{P}(t) < \infty$ for all $t \in [0,\infty)$. Furthermore, $\mathrm{P}(0) \le \log \#E$. On the other hand, if $\#E = \infty$ then $\mathrm{P}(0) = \infty$.*

Proof. (a) Since $Z_n(t)$ is comparable to $Z_n(t\Xi)$ (recall that $K^{-t} Z_n(t) \le Z_n(t\Xi) \le K^t Z_n(t)$), it follows from Theorem 17.2.8 that

$$\mathrm{P}(t\Xi) = \lim_{n\to\infty} \frac{1}{n} \log Z_n(t\Xi) = \lim_{n\to\infty} \frac{1}{n} \log Z_n(t) = \mathrm{P}(t).$$

(b) By (a), we obtain

$$\mathrm{P}(t) = \mathrm{P}(t\Xi) = \sup_{F \subseteq E \text{ finite}} \mathrm{P}_F(t\Xi_\Phi) = \sup_{F \subseteq E \text{ finite}} \mathrm{P}_F(t).$$

(c) The nonincreasing property of the pressure function P follows from the nonincreasing property of the partition functions Z_n, $n \in \mathbb{N}$. The strictly decreasing behavior of P on $[\theta, \infty)$ can be more specifically described as follows. Let $\theta \le t_1 < t_2$. Since $Z_n(t_2) \le s^{n(t_2-t_1)} Z_n(t_1)$ for all $n \in \mathbb{N}$, we deduce that $\mathrm{P}(t_2) \le (t_2 - t_1)\log s + \mathrm{P}(t_1)$. The convexity of the pressure on $[\theta, \infty)$ follows from the convexity of the partition functions on $[\theta, \infty)$. The continuity of P on (θ, ∞) and its right-continuity at θ is a direct consequence of its convexity, except at θ when $\theta \notin \mathrm{Fin}$, in which case an argument similar to that given in the proof of Proposition 19.4.4(c) can be formulated for the pressure.

(d) If $\#E < \infty$, then $\#(E_A^n) \le (\#E)^n < \infty$ for all $n \in \mathbb{N}$, and thus $Z_n(t) < \infty$ for all $n \in \mathbb{N}$ and all $t \ge 0$. It follows from Proposition 19.4.5 that $\mathrm{P}(t) < \infty$ for all $t \ge 0$. In particular, $Z_1(0) = \#E$, and thus $\mathrm{P}(0) \le \log \#E$. If $\#E = \infty$, then $Z_1(0) = \infty$ and $\mathrm{P}(0) = \infty$ by Proposition 19.4.5. □

Some parameters t have a special meaning. One such parameter is the finiteness parameter θ. Another one is the Bowen parameter h.

Definition 19.4.7. The number

$$h := \inf\{t \ge 0 : \mathrm{P}(t) \le 0\}$$

is called the *Bowen parameter* of the CGDMS Φ.

By virtue of Proposition 19.4.6, we know that $\theta \le h$ and that $\mathrm{P}(h) \le 0$. In Section 19.6, we will show that the Bowen parameter h of a finitely irreducible CGDMS Φ is equal to the Hausdorff dimension of the limit set J of Φ. This is accordingly called Bowen's formula.

19.5 Classification of CGDMSs

Infinite CGDMSs naturally break into two main classes: regular and irregular systems. This dichotomy is determined by the existence of a zero of the pressure function.

Definition 19.5.1. A CGDMS Φ is called *regular* if there is some $t \ge 0$ such that $\mathrm{P}(t) = 0$. Otherwise, the system is said to be *irregular*.

For finitely irreducible CGDMSs, it follows from the nonincrease of P and its strict decrease on $[\theta, \infty)$ (see Proposition 19.4.6) that such a t is unique (if it exists) and that we have the following.

Lemma 19.5.2. *A finitely irreducible CGDMS Φ is regular if and only if* $\mathrm{P}(h) = 0$.

Note that every finite CGDMS is regular. There are natural subclasses of regular systems.

Definition 19.5.3. A CGDMS Φ is said to be
(a) *strongly regular* if $0 < \mathrm{P}(t) < \infty$ for some $t \geq 0$;
(b) *critically regular* if $\mathrm{P}(\theta) = 0$. In this case, $\theta = h$.

It is an immediate consequence of the continuity and strict decrease of P on $[\theta, \infty)$ that the class of finitely irreducible, strongly regular systems consists of those systems whose Bowen parameter is strictly larger than their finiteness parameter.

Lemma 19.5.4. *A finitely irreducible CGDMS Φ is strongly regular if and only if $h > \theta$.*

Within the subclass of strongly regular systems can be found yet another subclass.

Definition 19.5.5. A CGDMS Φ is called *cofinitely regular* provided it has infinite pressure at its finiteness parameter, i. e., if $\mathrm{P}(\theta) = \infty$.

19.6 Bowen's formula for CGDMSs

We will now derive a formula for the Hausdorff dimension of the limit set of a finitely irreducible CGDMS. It is entirely expressed in dynamical terms. Since it stems from the seminal work of Bowen [12], it is referred to as Bowen's formula.

19.6.1 The finite case

Assume that E is a finite alphabet and that $A : E \times E \to \{0, 1\}$ is an irreducible matrix. Then the symbolic space E_A^∞ is compact due to the finiteness of E. Moreover, Theorem 3.2.14 states that the shift map $\sigma : E_A^\infty \to E_A^\infty$ is transitive precisely when A is irreducible. According to Theorem 3.2.12, the shift map is also open. Furthermore, as witnessed in Example 4.1.3, this map is distance expanding. (These properties can be proved directly.) Thus, the theory developed in Chapter 13 applies with $X = E_A^\infty$ and $T = \sigma$. Of particular interest are Sections 13.5, 13.6 and 13.7.

Lemma 19.6.1. *Let $\Phi = \{\varphi_e\}_{e\in E}$ be a finite and irreducible CGDMS. Let also $M(E_A^\infty)$ be the space of all Borel probability measures on E_A^∞. For every $t \geq 0$, there exists a unique σ-invariant Gibbs state and equilibrium state $\widetilde{m}_t \in M(E_A^\infty)$ for the Hölder continuous potential $f_t : E_A^\infty \to \mathbb{R}$ defined by $f_t = t\Xi$, where Ξ is the potential introduced in (19.14). The state $\widetilde{m}_t$ is also ergodic.*

In particular, there exists a constant $C(t) \geq 1$ such that

$$C(t)^{-1} \leq \frac{\widetilde{m}_t([\omega])}{e^{-|\omega|\,\mathrm{P}(t)}\|\varphi'_\omega\|^t_{X_{t(\omega)}}} \leq C(t), \quad \forall \omega \in E_A^* \tag{19.15}$$

and

$$\mathrm{P}(t) = \mathrm{h}_{\widetilde{m}_t}(\sigma) + t \int_{E_A^\infty} \log|\varphi'_{\rho_1}(\pi(\sigma(\rho)))|\, d\widetilde{m}_t(\rho). \tag{19.16}$$

Proof. Let $t \geq 0$. By Lemma 19.4.1, we know that f_t is Hölder continuous on cylinders since Ξ is. As E_A^∞ is compact, the potential f_t is bounded and it is then easy to see that Hölder continuity on cylinders is equivalent to global Hölder continuity. By Proposition 13.7.12, there thus exists a unique σ-invariant Gibbs state and equilibrium state $\widetilde{m}_t \in M(E_A^\infty)$ associated to f_t. This state is also ergodic.

Regarding the Gibbs property of $\widetilde{m}_t$ (cf. Definition 13.2.1), observe that the initial cylinders are the open balls of the symbolic space E_A^∞ (they are also the closed balls of that space). Let us use the metric $d_{1/2}(\omega,\tau) := (1/2)^{|\omega\wedge\tau|}$ on E_A^∞. As the image under the shift map of an initial cylinder of length k is an initial cylinder of length $k-1$ whenever $k > 2$, we can choose $\delta = \xi = 1/2$ in (13.1), (13.2) and (13.13). Let $\tau \in E_A^\infty$ and $n \in \mathbb{N}$. Then $B(\sigma^n(\tau),\xi) = [\tau_{n+1}\tau_{n+2}]$. Consequently, $\sigma_\tau^{-n}(B(\sigma^n(\tau),\xi)) = [\tau_1 \ldots \tau_n\tau_{n+1}\tau_{n+2}] = [\tau|_{n+2}]$. Moreover,

$$e^{S_n f_t(\tau)} = e^{tS_n\Xi(\tau)} = \exp\Big(t\log|\varphi'_{\tau|_n}(\pi(\sigma^n(\tau)))|\Big) = |\varphi'_{\tau|_n}(\pi(\sigma^n(\tau)))|^t \asymp \|\varphi'_{\tau|_n}\|^t_{X_{t(\tau|_n)}},$$

where a constant of comparability is K^t, the t-th power of the constant K of bounded distortion from Corollary 19.3.5. Finally, recall that $\mathrm{P}(t) = \mathrm{P}(f_t)$ according to Proposition 19.4.6(a). Rewriting (13.13) in our symbolic setting, we deduce that there exists a constant $\overline{C}(t) \geq 1$ such that

$$\overline{C}(t)^{-1} \leq \frac{\widetilde{m}_t([\tau|_{n+2}])}{e^{-n\mathrm{P}(t)}\|\varphi'_{\tau|_n}\|^t_{X_{t(\tau|_n)}}} \leq \overline{C}(t), \qquad \forall n \in \mathbb{N},\ \forall \tau \in E_A^\infty.$$

There is a mismatch between $n+2$ in the numerator and n in the denominator. Concerning the lower bound, observe that $\widetilde{m}_t([\tau|_n]) \geq \widetilde{m}_t([\tau|_{n+2}])$. So

$$\overline{C}(t)^{-1} \leq \frac{\widetilde{m}_t([\tau|_n])}{e^{-n\mathrm{P}(t)}\|\varphi'_{\tau|_n}\|^t_{X_{t(\tau|_n)}}}, \qquad \forall n \in \mathbb{N},\ \forall \tau \in E_A^\infty.$$

Deriving the upper bound is trickier. First, $e^{-(n+2)\mathrm{P}(t)} = e^{-n\mathrm{P}(t)}e^{-2\mathrm{P}(t)}$. Moreover, by Corollary 19.3.6,

$$\begin{aligned}
\|\varphi'_{\tau|_{n+2}}\|^t_{X_{t(\tau|_{n+2})}} &\geq \|\varphi'_{\tau|_n}\|^t_{X_{t(\tau|_n)}} \cdot K^{-t}\|\varphi'_{\tau_{n+1}}\|^t_{X_{t(\tau_{n+1})}} \cdot K^{-t}\|\varphi'_{\tau_{n+2}}\|^t_{X_{t(\tau_{n+2})}} \\
&\geq \|\varphi'_{\tau|_n}\|^t_{X_{t(\tau|_n)}} \cdot K^{-2t}\Big(\min_{e\in E}\|\varphi'_e\|_{X_{t(e)}}\Big)^{2t} \\
&= \widetilde{K}^{2t}\|\varphi'_{\tau|_n}\|^t_{X_{t(\tau|_n)}},
\end{aligned}$$

where $\widetilde{K} := K^{-1} \min_{e\in E} \|\varphi'_e\|_{X_{t(e)}}$. Therefore,

$$\frac{\widetilde{m}_t([\tau|_{n+2}])}{e^{-(n+2)\,\mathrm{P}(t)} \|\varphi'_{\tau|_{n+2}}\|^t_{X_{t(\tau|_{n+2})}}} \le \overline{C}(t) e^{2\,\mathrm{P}(t)} \widetilde{K}^{-2t}, \quad \forall n \in \mathbb{N}, \forall \tau \in E_A^\infty.$$

In summary, we have proved the lower bound in (19.15) for all cylinders and the upper bound for all cylinders of length at least 3. Since there are only finitely many cylinders of length smaller than 3, the upper bound holds for all cylinders by increasing the constant if necessary.

Finally, relation (19.16) is just the definition of an equilibrium state for the potential f_t (see Definition 12.2.1). □

Since the system Φ is finite, its Bowen parameter h satisfies the equation

$$\mathrm{P}(h) = 0. \tag{19.17}$$

The previous lemma has the following immediate ramification.

Corollary 19.6.2. *Let $\Phi = \{\varphi_e\}_{e\in E}$ be a finite and irreducible CGDMS, and h be its Bowen parameter. The unique ergodic σ-invariant Gibbs state and equilibrium state $\widetilde{m}_h \in M(E_A^\infty)$ for the Hölder continuous potential $f_h = h\Xi$ satisfies*

$$\widetilde{m}_h([\omega]) \asymp \|\varphi'_\omega\|^h_{X_{t(\omega)}}. \tag{19.18}$$

Now, let

$$m_h := \widetilde{m}_h \circ \pi^{-1}$$

be the pushdown onto the limit set J of the measure $\widetilde{m}_h$ under the coding map $\pi : E_A^\infty \to J$. We shall prove the following result, whose last part is Bowen's formula for finite and irreducible CGDMSs. In the case of finite-alphabet conformal IFSs, this result was first obtained by Bedford [5].

Theorem 19.6.3. *Let Φ be a finite and irreducible CGDMS, and h be its Bowen parameter. The following statements hold:*

(a) *m_h is a geometric measure with exponent h, i. e., there exists a constant $C(h) \ge 1$ such that*

$$C(h)^{-1} \le \frac{m_h(B(x,r))}{r^h} \le C(h)$$

for all $x \in J$ and all sufficiently small $r > 0$.

(b) *$0 < H_h(J), P_h(J) < \infty$, where $H_h(J)$ and $P_h(J)$ respectively denote the h-dimensional Hausdorff and packing measures of the limit set J.*

(c) (*Bowen's formula*) $\mathrm{HD}(J) = \mathrm{PD}(J) = \mathrm{BD}(J) = h$, *where* $\mathrm{HD}(J)$, $\mathrm{PD}(J)$ *and* $\mathrm{BD}(J)$ *respectively denote the Hausdorff, packing and box-counting dimensions of* J.

Proof. Let $x \in J$ and $0 < r < \frac{1}{2}\min\{\mathrm{diam}(X_v) : v \in V\}$. Then $x = \pi(\omega)$ for some $\omega \in E_A^\infty$. Take the smallest $n \in \mathbb{N}$ such that $\varphi_{\omega|_n}(X_{t(\omega_n)}) \subseteq B(x,r)$. By (19.18), we find that

$$m_h(B(x,r)) \geq \widetilde{m}_h \circ \pi^{-1}(\varphi_{\omega|_n}(X_{t(\omega_n)})) \geq \widetilde{m}_h([\omega|_n]) \geq \widetilde{C}(h)\|\varphi'_{\omega|_n}\|^h_{X_{t(\omega_n)}} \tag{19.19}$$

for some constant $\widetilde{C}(h)$ independent of ω and n, i. e., independent of x and r. Since $\#E < \infty$, we know that

$$\xi' := \min\{\|\varphi'_e\|_{X_{t(e)}} : e \in E\} > 0.$$

By the definition of n, by Lemma 19.3.9 and Corollary 19.3.6, we also have that

$$\begin{aligned} r \leq \mathrm{diam}(\varphi_{\omega|_{n-1}}(X_{t(\omega_{n-1})})) &\leq KD\|\varphi'_{\omega|_{n-1}}\|_{X_{t(\omega_{n-1})}} \\ &\leq KD\|\varphi'_{\omega|_{n-1}}\|_{X_{t(\omega_{n-1})}} \frac{\|\varphi'_{\omega_n}\|_{X_{t(\omega_n)}}}{\xi'} \\ &\leq (\xi')^{-1}K^2D\|\varphi'_{\omega|_n}\|_{X_{t(\omega_n)}}. \end{aligned}$$

Using (19.19), it ensues that

$$m_h(B(x,r)) \geq [\widetilde{C}(h)((\xi')^{-1}K^2D)^{-h}] \cdot r^h. \tag{19.20}$$

In order to prove a similar inequality in the opposite direction, let W be the family of all minimal (in the sense of length) words $\tau \in E_A^*$ such that

$$\varphi_\tau(X_{t(\tau)}) \cap B(x,r) \neq \emptyset \quad \text{and} \quad \mathrm{diam}(\varphi_\tau(X_{t(\tau)})) < 2r. \tag{19.21}$$

Let $\tau \in W$. Then $\mathrm{diam}(\varphi_{\tau|_{|\tau|-1}}(X_{t(\iota|_{|\tau|-1})})) \geq 2r$ and it follows from Lemma 19.3.11, Corollary 19.3.6 and Lemma 19.3.9 that

$$\begin{aligned} \mathrm{diam}(\varphi_\tau(X_{t(\tau)})) &\geq D^{-1}\|\varphi'_\tau\|_{X_{t(\tau)}} \\ &\geq (KD)^{-1}\|\varphi'_{\tau|_{|\tau|-1}}\|_{X_{t(\tau|_{|\tau|-1})}} \cdot \|\varphi'_{\tau_{|\tau|}}\|_{X_{t(\tau_{|\tau|})}} \\ &\geq (KD)^{-1} \cdot (KD)^{-1}\,\mathrm{diam}(\varphi_{\tau|_{|\tau|-1}}(X_{t(\tau|_{|\tau|-1})})) \cdot \xi' \\ &\geq 2(KD)^{-2}\xi' r. \end{aligned}$$

By definition, the family W consists of mutually incomparable words. Therefore, in virtue of Lemma 19.3.12 with $\kappa_1 = 2(KD)^{-2}\xi'$ and $\kappa_2 = 2$, we have

$$\#W \leq (3KD(R \cdot 2(KD)^{-2}\xi')^{-1})^d = (3(KD)^3(2\xi' R)^{-1})^d. \tag{19.22}$$

As $\pi^{-1}(B(x,r)) \subseteq \bigcup_{\tau\in W}[\tau]$, we get from (19.18), Lemma 19.3.11, (19.21) and (19.22) that

$$
\begin{aligned}
m_h(B(x,r)) &= \widetilde{m}_h \circ \pi^{-1}(B(x,r)) \le \widetilde{m}_h\Big(\bigcup_{\tau\in W}[\tau]\Big) \\
&\le \sum_{\tau\in W} \widetilde{m}_h([\tau]) \le \sum_{\tau\in W} \widetilde{C}(h)\|\varphi_\tau'\|_{X_{t(\tau)}}^h \\
&\le \widetilde{C}(h) \sum_{\tau\in W} (D\operatorname{diam}(\varphi_\tau(X_{t(\tau)})))^h \\
&\le \widetilde{C}(h) \sum_{\tau\in W} (D\cdot 2r)^h \\
&= \widetilde{C}(h)\cdot(2D)^h r^h \cdot \#W \\
&\le \big[\widetilde{C}(h)(2D)^h(3(KD)^3(2\xi' R)^{-1})^d\big]\cdot r^h.
\end{aligned}
$$

Along with (19.20), this proves that m_h is a geometric measure with exponent h, i. e., statement (a) is proved. Statement (b) is then an immediate consequence of Theorem 15.6.14(a), while statement (c) follows directly from Theorem 15.6.14(b). □

Using the communication classes introduced in Definition 3.2.23, a generalization of Bowen's formula for all finite CGDMSs was later derived by Ghenciu, Mauldin and Roy [36].

19.6.2 The general case

We are now ready to provide a short and simple proof of the main theorem of this chapter. The following characterization of the Hausdorff dimension $\mathrm{HD}(J)$ of the limit set J, which is a variation of the classical Bowen's formula, was proved for finitely irreducible CGDMSs by Mauldin and Urbański [81]. A further generalization of that formula was subsequently found by Roy [102].

For every $F \subseteq E$, we write $\Phi|_F$ for the subsystem $\{\varphi_e\}_{e\in F}$ of Φ, and J_F for the limit set of $\Phi|_F$.

Theorem 19.6.4 (Bowen's formula). *If Φ is a finitely irreducible CGDMS, then its Bowen parameter h satisfies*

$$h := \inf\{t \ge 0 : \mathrm{P}(t) \le 0\} = \mathrm{HD}(J) = \sup\{\mathrm{HD}(J_F) : F \subseteq E, \#F < \infty\} \ge \theta.$$

Moreover, if $\mathrm{P}(t) = 0$ then t is the only zero of the pressure function and $t = h = \mathrm{HD}(J)$.

Proof. Set $H = \mathrm{HD}(J)$ and $h_\infty = \sup\{\mathrm{HD}(J_F) : F \subseteq E, \#F < \infty\}$. Fix $t > h$. Then $\mathrm{P}(t) = \lim_{n\to\infty} \frac{1}{n}\log Z_n(t) < 0$. Therefore, for all $n \in \mathbb{N}$ large enough,

$$\sum_{\omega\in E_A^n} \|\varphi_\omega'\|_{X_{t(\omega)}}^t =: Z_n(t) \le \exp(n\,\mathrm{P}(t)/2).$$

Using Lemma 19.3.9, we obtain that

$$\sum_{\omega\in E_A^n}\left(\mathrm{diam}(\varphi_\omega(X_{t(\omega)}))\right)^t \le (KD)^t \sum_{\omega\in E_A^n}\|\varphi'_\omega\|^t_{X_{t(\omega)}} \le (KD)^t \exp(n\,\mathrm{P}(t)/2).$$

Given that the family $\{\varphi_\omega(X_{t(\omega)})\}_{\omega\in E_A^n}$ forms a cover of J of diameter at most $s^n \max\{\mathrm{diam}(X_v) : v \in V\}$, by letting $n \to \infty$ we get that $H_t(J) = 0$. This implies that $t \ge H$ for all $t > h$, and consequently $h \ge H$. Obviously $h_\infty \le H$, and hence

$$h_\infty \le H \le h.$$

We are thus left to show that $h \le h_\infty$. If F is a finite subset of E such that $A : F \times F \to \{0,1\}$ is irreducible, then $h_F = \mathrm{HD}(J_F)$ according to Theorem 19.6.3(c). So $h_F \le h_\infty$. Moreover, $\mathrm{P}_F(h_F) = 0$ by (19.17). Then, by virtue of Proposition 19.4.6(c), it follows that $\mathrm{P}_F(h_\infty) \le \mathrm{P}_F(h_F) = 0$. So, in view of Theorem 17.2.9, $\mathrm{P}(h_\infty) = \sup\{\mathrm{P}_F(h_\infty) : F \subseteq E, \#F < \infty\} \le 0$. Therefore, $h \le h_\infty$. □

Combining this theorem with Section 19.5, we deduce the following result.

Corollary 19.6.5. *Let Φ be a finitely irreducible CGDMS. Then:*
(a) *Φ is strongly regular if and only if* $\mathrm{HD}(J) = h > \theta$.
(b) *Φ is critically regular or irregular if and only if* $\mathrm{HD}(J) = h = \theta$.

Remark 19.6.6. The finiteness parameter θ is an important lower bound for the Hausdorff dimension of J since it requires no iteration: θ can be expressed in terms of $Z_1(t)$ alone.

19.6.3 Hutchinson's formula

From Bowen's formula (more precisely, Theorem 19.6.3(c)), we deduce Hutchinson's formula in its full generality. A special case of that formula was obtained in Section 16.3.1.

Theorem 19.6.7. *Let $\{F_e(x) = r_eA_e(x)+b_e\}_{e\in E}$ (with $2 \le \#E < \infty$) be an IFS of contracting similarities satisfying the open set condition. Then its limit set J has the property that* $\mathrm{HD}(J) = \mathrm{PD}(J) = \mathrm{BD}(J) = h > 0$, *where h is the unique number such that*

$$\sum_{e\in E} r_e^h = 1.$$

This equation is called Hutchinson's formula.

19.7 Other separation conditions and cone condition

The Open Set Condition (OSC) is one of the conditions defining a CGDMS (see Definition 19.3.1). Some CGDMSs satisfy stricter separation or other geometric conditions. We describe some of these conditions as well as some of their consequences in this section. In the next section, we will demonstrate that some of these conditions guarantee the existence and uniqueness of conformal measures, measures which will underpin the analysis of the local multifractal structure of the limit sets of CGDMSs in Chapter 21.

19.7.1 The strong open set condition

Definition 19.7.1. A CGDMS Φ satisfies the *strong open set condition* (*SOSC*) if $J \cap \mathrm{Int}(X) \neq \emptyset$, i. e., if $\bigcup_{v\in V}(J_v \cap \mathrm{Int}(X_v)) \neq \emptyset$.

All finite-alphabet CGDMSs satisfy the SOSC (see [116]). However, there are infinite-alphabet CGDMSs which do not fulfill the SOSC (see [129]).

The SOSC has several interesting and important consequences. First, it implies that the set of infinite A-admissible words that generate points on the boundary of the real space is negligible under any ergodic shift-invariant measure that has full topological support.

Theorem 19.7.2. *Let* $\Phi = \{\varphi_e\}_{e\in E}$ *be a CGDMS satisfying the SOSC. For any ergodic σ-invariant Borel probability measure μ on E_A^∞ such that* $\mathrm{supp}(\mu) = E_A^\infty$, *the following statements hold:*

(a) $\mu \circ \pi^{-1}(\partial X) = \mu(\bigcup_{n=0}^{\infty} \sigma^{-n}(\pi^{-1}(\partial X))) = 0$.

(b) $\mu \circ \pi^{-1}(\varphi_\omega(\partial X_{t(\omega)})) = 0$ *for all* $\omega \in E_A^*$.

(c) *If* $\omega, \tau \in E_A^*$ *are incomparable words, then*

$$\mu \circ \pi^{-1}(\varphi_\omega(X_{t(\omega)}) \cap \varphi_\tau(X_{t(\tau)})) = 0.$$

Proof. (a) Since Φ fulfills the SOSC, there exists $\omega \in E_A^\infty$ such that $\pi(\omega) \in \mathrm{Int}(X_{i(\omega_1)})$. Then there is $n \in \mathbb{N}$ such that $\varphi_{\omega|_n}(X_{t(\omega_n)}) \subseteq \mathrm{Int}(X_{i(\omega_1)})$. As $\pi([\omega|_n]) \subseteq \varphi_{\omega|_n}(X_{t(\omega_n)})$ and $\mathrm{supp}(\mu) = E_A^\infty$, we deduce that

$$\mu \circ \pi^{-1}(\mathrm{Int}(X)) \geq \mu \circ \pi^{-1}(\mathrm{Int}(X_{i(\omega_1)})) \geq \mu \circ \pi^{-1}(\varphi_{\omega|_n}(X_{t(\omega_n)})) \geq \mu([\omega|_n]) > 0.$$

Therefore, $\mu \circ \pi^{-1}(\partial X) < 1$. It follows from Lemma 19.3.13 and the ergodicity of μ (see Exercise 8.5.36) that $\mu \circ \pi^{-1}(\partial X) = \mu(\bigcup_{n=0}^{\infty} \sigma^{-n}(\pi^{-1}(\partial X))) = 0$.

(b) Let $\omega \in E_A^*$. It ensues from part (a), Lemma 19.3.14 and the σ-invariance of μ that

$$0 = \mu \circ \pi^{-1}(\partial X) \geq \mu(\sigma^{|\omega|}(\pi^{-1}(\varphi_\omega(\partial X_{t(\omega)})))) \geq \mu(\pi^{-1}(\varphi_\omega(\partial X_{t(\omega)}))).$$

(c) If $\omega, \tau \in E_A^*$ are incomparable words, then $\omega = \gamma a \alpha$ and $\tau = \gamma b \beta$ for some $\gamma, \alpha, \beta \in E_A^*$ and $a, b \in E$ with $a \neq b$. Using the OSC, we obtain that

$$\begin{aligned}\varphi_\omega(X_{t(\omega)}) \cap \varphi_\tau(X_{t(\tau)}) &\subseteq \varphi_\gamma(\varphi_a(X_{t(a)}) \cap \varphi_b(X_{t(b)})) \\ &= \varphi_\gamma(\varphi_a(\partial X_{t(a)}) \cap \varphi_b(\partial X_{t(b)})) \subseteq \varphi_{\gamma a}(\partial X_{t(a)}).\end{aligned}$$

So, by (b),

$$\mu \circ \pi^{-1}(\varphi_\omega(X_{t(\omega)}) \cap \varphi_\tau(X_{t(\tau)})) \leq \mu \circ \pi^{-1}(\varphi_{\gamma a}(\partial X_{t(a)})) = 0. \qquad \square$$

According to part (a) of the above result, the completely σ-invariant set

$$\begin{aligned}\{\omega \in E_A^\infty \mid \pi(\mathcal{O}_+(\omega)) \cap \partial X \neq \emptyset\} &= \{\omega \in E_A^\infty \mid \exists N \geq 0 \text{ with } \pi(\sigma^N(\omega)) \in \partial X_{i(\sigma^N(\omega))}\} \\ &= \{\omega \in E_A^\infty \mid \exists N \geq 0 \text{ with } \pi(\sigma^n(\omega)) \in \partial X_{i(\sigma^n(\omega))}, \forall n \geq N\}\end{aligned}$$

is negligible under any ergodic shift-invariant measure that has full topological support. There is an even bigger similar set with this property, namely

$$\left\{\omega \in E_A^\infty \;\middle|\; \lim_{n\to\infty} \operatorname{dist}(\pi(\sigma^n(\omega)), \partial X_{i(\sigma^n(\omega))}) = 0\right\}.$$

In Chapter 21, we will develop a multifractal analysis on the complement of this set, which can be described by

$$E_0^\infty := \left\{\omega \in E_A^\infty \;\middle|\; \limsup_{n\to\infty} \operatorname{dist}(\pi(\sigma^n(\omega)), \partial X_{i(\sigma^n(\omega))}) > 0\right\}. \tag{19.23}$$

Clearly, this set is completely σ-invariant. We now prove that it is of full measure under any ergodic shift-invariant measure with full topological support.

Theorem 19.7.3. *Let $\Phi = \{\varphi_e\}_{e\in E}$ be a CGDMS satisfying the SOSC. For every ergodic σ-invariant Borel probability measure μ on E_A^∞ such that $\operatorname{supp}(\mu) = E_A^\infty$, we have $\mu(E_0^\infty) = 1$.*

Proof. Since E_0^∞ is completely σ-invariant, the ergodicity of μ implies that $\mu(E_0^\infty)$ is 0 or 1. We shall now show that this latter possibility always prevails. Since Φ satisfies the SOSC, there exists $x \in J \cap \operatorname{Int}(X)$. Let $\omega \in E_A^\infty$ be such that $\pi(\omega) = x$. Let also $0 < r < \operatorname{dist}(x, \partial X)$. Recall that for any $\tau \in E_A^*$, we have $\pi([\tau]) \subseteq \varphi_\tau(X_{t(\tau)})$. Since $\varphi_{\omega|_k}(X_{t(\omega_k)}) \subseteq B(\pi(\omega), r)$ for all $k \in \mathbb{N}$ large enough, we obtain that $[\omega|_k] \subseteq \pi^{-1}(B(x,r))$ for some $k \in \mathbb{N}$. Then $\mu(\pi^{-1}(B(x,r))) \geq \mu([\omega|_k]) > 0$ since $\operatorname{supp}(\mu) = E_A^\infty$. It follows from Corollary 8.2.15 to Birkhoff's ergodic theorem (Theorem 8.2.11) that the set of infinite A-admissible words whose iterates' projections visit infinitely many times the ball $B(x,r)$ has measure 1, i. e.,

$$\mu\left(\{\rho \in E_A^\infty \mid \pi(\sigma^n(\rho)) \in B(x,r) \text{ for infinitely many } n\}\right) = 1.$$

Thus, $\mu(E_0^\infty) = 1$. (Note: Once we know that $\mu(E_0^\infty)$ is 0 or 1 and $\mu(\pi^{-1}(B(x,r))) > 0$, the conclusion can also be drawn by means of Poincaré's recurrence theorem (Theorem 8.1.2).) □

Under the OSC or even under the SOSC, the limit set J of a CGDMS Φ is generally only an analytic set in the sense of descriptive set theory (see Theorem 19.1.1 and [120]). If the alphabet E is finite, it is considerably more: J is compact.

To avoid problems with boundary intersections (and in particular with nonunique coding), we sometimes consider the Borel subsets

$$\tilde{J} := J \setminus \bigcup_{\omega \in E_A^*} \varphi_\omega(\partial X_{t(\omega)}) \quad \text{and} \quad \widetilde{E_A^\infty} := \pi^{-1}(\tilde{J}).$$

Observe that $\sigma(\widetilde{E_A^\infty}) \subseteq \widetilde{E_A^\infty}$. So let $\tilde{\sigma} := \sigma|_{\widetilde{E_A^\infty}}$. Notice also that for every $x \in \tilde{J}$ there exists a unique $\omega(x) \in E_A^\infty$ such that $x = \pi(\omega(x))$. Obviously, $\omega(x) \in \widetilde{E_A^\infty}$. Accordingly, let $\tilde{\pi} := \pi|_{\widetilde{E_A^\infty}}$. Then $\tilde{\pi} : \widetilde{E_A^\infty} \to \tilde{J}$ is a continuous bijection, with $\tilde{\pi}^{-1}(x) = \omega(x)$. The restricted shift map $\tilde{\sigma} : \widetilde{E_A^\infty} \to \widetilde{E_A^\infty}$ induces a map $\tilde{f} : \tilde{J} \to \tilde{J}$ via the formula

$$\tilde{f}(x) := \tilde{\pi} \circ \tilde{\sigma} \circ \tilde{\pi}^{-1}(x) = \varphi_{\omega(x)_1}^{-1}(x),$$

so that the diagram

$$\begin{array}{ccc} \widetilde{E_A^\infty} & \xrightarrow{\ \tilde{\sigma}\ } & \widetilde{E_A^\infty} \\ {\scriptstyle \tilde{\pi}}\downarrow & & \downarrow{\scriptstyle \tilde{\pi}} \\ \tilde{J} & \xrightarrow[\ \tilde{f}\ ]{} & \tilde{J} \end{array}$$

commutes. Let $\tilde{\mu} := \mu|_{\widetilde{E_A^\infty}}$. If (σ, μ) is a measure-preserving dynamical system, then so is the system $(\tilde{\sigma}, \tilde{\mu})$. By definition of $\tilde{f}$, it ensues immediately that $(\tilde{f}, \tilde{\mu} \circ \tilde{\pi}^{-1})$ is also measure-preserving. This discussion and Theorem 19.7.2 lead to the following result.

Theorem 19.7.4. *Let $\Phi = \{\varphi_e\}_{e \in E}$ be a CGDMS satisfying the SOSC. For any ergodic σ-invariant Borel probability measure μ on E_A^∞ such that $\operatorname{supp}(\mu) = E_A^\infty$, it turns out that $\mu \circ \pi^{-1}(\tilde{J}) = 1$ and the measure-preserving dynamical systems (σ, μ) and $(\tilde{f}, \mu \circ \pi^{-1})$ are isomorphic.*

Proof. We have already observed that the set $\widetilde{E_A^\infty} := \pi^{-1}(\tilde{J})$ is forward σ-invariant. As μ is σ-invariant and σ-ergodic, it ensues that $\mu(\widetilde{E_A^\infty}) \in \{0, 1\}$. Theorem 19.7.2(a) reveals that $\mu(\widetilde{E_A^\infty}) = 1$. The rest of the statement was proved in the discussion preceding this lemma. □

19.7.2 The boundary separation condition

A condition considerably stricter than the SOSC is the following.

Definition 19.7.5. A CGDMS Φ is said to satisfy the *Boundary Separation Condition* (*BSC*) if

$$\min_{v\in V} \operatorname{dist}\Bigg(\partial X_v, \overline{\bigcup_{\substack{e\in E:\\ i(e)=v}} \varphi_e(X_{t(e)})}\Bigg) > 0.$$

In other words, a CGDMS satisfies the BSC if its first-level sets stay uniformly away from the boundary of the phase space.

19.7.3 The strong separation condition

Definition 19.7.6. A CGDMS Φ is said to satisfy the *Strong Separation Condition* (*SSC*) if

$$\varphi_{e_1}(X_{t(e_1)}) \cap \varphi_{e_2}(X_{t(e_2)}) = \emptyset, \quad \forall e_1, e_2 \in E,\ e_1 \neq e_2.$$

Clearly, the SSC is a stricter requirement than the OSC. It does not necessarily imply the SOSC, though. The BSC and the SSC are related in the following way.

Lemma 19.7.7. *If a CGDMS $\Phi = \{\varphi_e : X_{t(e)} \to X_{i(e)}\}_{e\in E}$ satisfies the BSC, then the CGDMS $\Phi' = \{\varphi_e : X'_{t(e)} \to X'_{i(e)}\}_{e\in E}$ satisfies the SSC, where*

$$X'_v = X_v \setminus B(\partial X_v, \delta) \quad \text{for any } 0 < \delta < \min_{v\in V} \operatorname{dist}\Bigg(\partial X_v, \overline{\bigcup_{\substack{e\in E:\\ i(e)=v}} \varphi_e(X_{t(e)})}\Bigg).$$

The CGDMSs Φ and Φ' have the same coding map $\pi : E_A^\infty \to X$, and hence the same limit set J. Loosely speaking, we can thus say that a CGDMS that satisfies the BSC also satisfies the SSC.

Proof. The proof is left to the reader as an exercise. □

It is not difficult to observe the following.

Theorem 19.7.8. *If a CGDMS satisfies the BSC or the SSC, then the coding map $\pi : E_A^\infty \to J$ is one-to-one. In particular, if E is finite (so E_A^∞ is compact), then the coding map π is a homeomorphism. In this case, the limit set J, as a compact, perfect, totally disconnected, metrizable set, is a topological Cantor set, i. e., it is homeomorphic to the middle-third Cantor set.*

Proof. The proof is left to the reader as an exercise. □

19.7.4 The cone condition

Given a point $x \in \mathbb{R}^d$, an angle $\alpha \in (0, \pi/2)$ and a vector $u \in \mathbb{R}^d$, let

$$\begin{aligned}\operatorname{Con}(x, \alpha, u) &= \{y \in \mathbb{R}^d : \angle(y - x, u) \le \alpha \text{ and } \|y - x\| \le \|u\|\} \\ &= \{y \in \mathbb{R}^d : \langle y - x, u\rangle \ge \|y - x\| \cdot \|u\| \cos\alpha \text{ and } \|y - x\| \le \|u\|\},\end{aligned}$$

where $\langle v, u\rangle := \|v\| \cdot \|u\| \cos\angle(v, u)$ is the usual inner product in $\mathbb{R}^d$. The set $\operatorname{Con}(x, \alpha, u)$ is called the (closed) cone with vertex x, central angle α and direction vector u.

We aim to prove a geometrical fact about the images of cones under conformal maps, like we did for balls in Lemma 19.3.7. For this, we need two intermediate results. First, as an immediate consequence of Theorem 19.2.2 we obtain the following result.

Theorem 19.7.9. *If $x \in \mathbb{C}$, $R > 0$ and $\varphi : B(x, R) \to \mathbb{C}$ is a univalent holomorphic function, then*

$$|\varphi(y) - \varphi(x) - \varphi'(x)(y - x)| \le \frac{K_2}{R}|\varphi'(x)| \cdot |y - x|^2, \quad \forall y \in B(x, \delta_2 R),$$

where $\delta_2 = \min\{\gamma, \pi/24\}$.

Proof. In view of Theorem 19.2.2, we have

$$|\varphi(y) - \varphi(x) - \varphi'(x)(y - x)| = \left|\int_x^y (\varphi'(z) - \varphi'(x))\, dz\right| \le \frac{K_2}{R}|\varphi'(x)| \cdot |y - x| \cdot |y - x|. \quad \square$$

A similar result holds in higher-dimensional Euclidean spaces.

Theorem 19.7.10. *Suppose that Y is a bounded subset of $\mathbb{R}^d$, $d \ge 3$, and that W is an open subset of $\mathbb{R}^d$ containing $\overline{Y}$. Then there exists a constant $K_4 \ge 1$ such that if $\varphi : \overline{\mathbb{R}^d} \to \overline{\mathbb{R}^d}$ is a conformal C^1 diffeomorphism such that $\varphi(W) \subseteq \mathbb{R}^d$, then for all $x \in Y$ and all $y \in B(Y, \frac{1}{2}\operatorname{dist}(Y, \partial W))$ we have*

$$\|\varphi(y) - \varphi(x) - \varphi'(x)(y - x)\| \le K_4|\varphi'(x)| \cdot \|y - x\|^2.$$

Proof. Per Liouville's Theorem (Theorem 19.2.1), write $\varphi = \lambda A \circ i + b$. If i is the identity map, then the left-hand side of the claimed inequality is equal to 0 and we are done. If $i = i_a$, then setting $\nabla\varphi := \varphi(y) - \varphi(x) - \varphi'(x)(y - x)$ and using (19.4) we get

$$\begin{aligned}\nabla\varphi &= \lambda A \circ i(y) - \lambda A \circ i(x) - \lambda A \circ i'(x)(y - x) \\ &= \lambda A(i(y) - i(x) - i'(x)(y - x)) \\ &= \lambda A\left(\frac{y - a}{\|y - a\|^2} - \frac{x - a}{\|x - a\|^2} - \frac{1}{\|x - a\|^2}(\mathrm{Id}_{\mathbb{R}^d} - 2Q(x - a))(y - x)\right) =: \lambda A(\Delta). \quad (19.24)\end{aligned}$$

By (19.5), we have

$$Q(x-a)u = \frac{\langle x-a, u\rangle}{\|x-a\|^2}(x-a), \quad \forall u \in \mathbb{R}^d.$$

Hence,

$$\begin{aligned}\Delta &= \frac{y-x}{\|y-a\|^2} + \frac{x-a}{\|y-a\|^2} - \frac{x-a}{\|x-a\|^2} - \frac{y-x}{\|x-a\|^2} + 2\frac{\langle x-a, y-x\rangle}{\|x-a\|^4}(x-a) \\ &= \left(\frac{1}{\|y-a\|^2} - \frac{1}{\|x-a\|^2}\right)(y-x) + \left(\frac{1}{\|y-a\|^2} - \frac{1}{\|x-a\|^2} + 2\frac{\langle x-a, y-x\rangle}{\|x-a\|^4}\right)(x-a).\end{aligned}$$

Now decompose

$$x - a = \alpha(y-x) + \beta(y-x)^{\perp},$$

where $(y-x)^{\perp}$ is a vector perpendicular to $y-x$ with $\|(y-x)^{\perp}\| = \|y-x\|$. Then

$$\|x-a\|^2 = (\alpha^2+\beta^2)\|y-x\|^2, \quad \|y-a\|^2 = ((\alpha+1)^2+\beta^2)\|y-x\|^2$$
and
$$\langle x-a, y-x\rangle = \alpha\|y-x\|^2. \tag{19.25}$$

Using these expressions, we get

$$\begin{aligned}\Delta &= \left(\frac{1}{\|y-a\|^2} - \frac{1}{\|x-a\|^2}\right)(y-x) \\ &\quad + \alpha\left(\frac{1}{\|y-a\|^2} - \frac{1}{\|x-a\|^2} + \frac{2\alpha\|y-x\|^2}{\|x-a\|^4}\right)(y-x) \\ &\quad + \beta\left(\frac{1}{\|y-a\|^2} - \frac{1}{\|x-a\|^2} + \frac{2\alpha\|y-x\|^2}{\|x-a\|^4}\right)(y-x)^{\perp} \\ &= \frac{1}{\|x-a\|^2}\left[\left(\frac{(1+\alpha)(\alpha^2+\beta^2)}{(\alpha+1)^2+\beta^2} - (1+\alpha) + \frac{2\alpha^2}{\alpha^2+\beta^2}\right)(y-x)\right. \\ &\quad \left. + \beta\left(\frac{\alpha^2+\beta^2}{(\alpha+1)^2+\beta^2} - 1 + \frac{2\alpha}{\alpha^2+\beta^2}\right)(y-x)^{\perp}\right].\end{aligned} \tag{19.26}$$

Since $\varphi(W) \subseteq \mathbb{R}^d$, we know that $a \notin W$. A direct calculation using (19.25) shows that

$$\begin{aligned}\left|\frac{(1+\alpha)(\alpha^2+\beta^2)}{(\alpha+1)^2+\beta^2} - (1+\alpha) + \frac{2\alpha^2}{\alpha^2+\beta^2}\right| &= \frac{|\alpha^3+\alpha^2-3\alpha\beta^2-\beta^2|}{(\alpha^2+\beta^2)((\alpha+1)^2+\beta^2)} \\ &\le \frac{\|y-x\|^4}{\|x-a\|^2\|y-a\|^2}(|\alpha|\alpha^2+\alpha^2+3|\alpha|\beta^2+\beta^2) \\ &\le \frac{\|y-x\|^4}{\|x-a\|^2\|y-a\|^2}(3|\alpha|(\alpha^2+\beta^2)+(\alpha^2+\beta^2))\end{aligned}$$

$$\begin{aligned}
&\le \frac{\|y-x\|^4}{\|x-a\|^2\|y-a\|^2}\left(3\frac{\|x-a\|^3}{\|y-x\|^3}+\frac{\|x-a\|^2}{\|y-x\|^2}\right)\\
&= \frac{\|y-x\|}{\|y-a\|}\left(3\frac{\|x-a\|}{\|y-a\|}+\frac{\|y-x\|}{\|y-a\|}\right)\\
&\le \frac{\|y-x\|}{\|y-a\|}\left(3\frac{\|y-x\|+\|y-a\|}{\|y-a\|}+\frac{\|y-x\|}{\|y-a\|}\right)\\
&= \frac{\|y-x\|}{\|y-a\|}\left(4\frac{\|y-x\|}{\|y-a\|}+3\right)\\
&\le \frac{1}{R}\left(4\frac{\operatorname{diam}(Y)+R}{R}+3\right)\|y-x\|,
\end{aligned}$$

where $R=\frac{1}{2}\operatorname{dist}(Y,\partial W)$, while

$$\begin{aligned}
&\left|\beta\left(\frac{\alpha^2+\beta^2}{(\alpha+1)^2+\beta^2}-1+\frac{2\alpha}{\alpha^2+\beta^2}\right)\right|\\
&= \frac{|\beta|\cdot|3\alpha^2-\beta^2+2\alpha|}{(\alpha^2+\beta^2)((\alpha+1)^2+\beta^2)} \le \frac{|\beta|(3\alpha^2+\beta^2+2|\alpha|)}{\|x-a\|^2\|y-a\|^2}\|y-x\|^4\\
&\le \frac{1}{\|x-a\|^2\|y-a\|^2}\left(3\frac{\|x-a\|^3}{\|y-x\|^3}+2\frac{\|x-a\|^2}{\|y-x\|^2}\right)\|y-x\|^4\\
&= \frac{\|y-x\|}{\|y-a\|^2}(3\|x-a\|+2\|y-x\|) \le \frac{\|y-x\|}{\|y-a\|^2}(3\|y-a\|+5\|y-x\|)\\
&= \frac{\|y-x\|}{\|y-a\|}\left(3+5\frac{\|y-x\|}{\|y-a\|}\right) \le \frac{1}{R}\left(3+5\frac{\operatorname{diam}(Y)+R}{R}\right)\|y-x\|.
\end{aligned}$$

Combining the previous two estimates with (19.26), we find a constant $K_4>0$ such that

$$\|\Delta\| \le \frac{1}{\|x-a\|^2}K_4\|y-x\|^2.$$

It follows from this, along with (19.24) and (19.4), that

$$\|\nabla\varphi\| \le \frac{\lambda}{\|x-a\|^2}K_4\|y-x\|^2 = K_4|\varphi'(x)|\cdot\|y-x\|^2. \qquad \square$$

We can now derive the main fact we are after, which pertains to the images of cones under conformal maps.

Theorem 19.7.11. *Suppose that Y is a bounded subset of $\mathbb{R}^d$, $d\ge 2$, and that W is an open subset of $\mathbb{R}^d$ containing $\overline{Y}$. Then for every $\varepsilon\in(0,\pi/2)$ there exists $\delta>0$ such that if $x\in Y$, $\alpha\in(0,\pi/2-\varepsilon)$, $\|u\|\le\delta$, and $\varphi:\overline{\mathbb{R}^d}\to\overline{\mathbb{R}^d}$ is a conformal C^1 diffeomorphism such that $\varphi(W)\subseteq\mathbb{R}^d$, then*

$$\varphi(\operatorname{Con}(x,\alpha,u))\subseteq\operatorname{Con}(\varphi(x),\alpha+\varepsilon,2\varphi'(x)u).$$

Proof. Let $\varepsilon \in (0, \pi/2)$ and $\alpha \in (0, \pi/2 - \varepsilon)$. If $d = 2$, then φ is either holomorphic or antiholomorphic. Since complex conjugacy is an isometry, we may assume that φ is holomorphic when $d = 2$. In view of Theorems 19.7.9 and 19.7.10, setting $K_5 = \max\{K_4, 2K_2/\operatorname{dist}(Y, \partial W)\}$ we see that if $x \in Y$ and $\|y - x\| < \delta_2 \operatorname{dist}(Y, \partial W)$, then

$$\|\Delta\varphi - \varphi'(x)(y - x)\| \le K_5|\varphi'(x)| \cdot \|y - x\|^2, \tag{19.27}$$

where $\Delta\varphi := \varphi(y) - \varphi(x)$. Fix $\eta > 0$. By virtue of Corollary 19.2.4, there exists a constant $0 < \delta < \delta_2 \operatorname{dist}(Y, \partial W)$, independent of φ, such that

$$(1 + \eta)^{-1} \le \frac{|\varphi'(y)|}{|\varphi'(x)|} \le 1 + \eta, \quad \forall x \in Y, \forall y \in B(x, \delta). \tag{19.28}$$

Fix $u \in \mathbb{R}^d$ with $\|u\| \le \delta$, and let $y \in \operatorname{Con}(x, \alpha, u)$. This means that $\|y - x\| \le \|u\| \le \delta$ and $\langle y - x, u\rangle \ge \|y - x\| \cdot \|u\| \cos\alpha$. By the mean value theorem, (19.28) and (16.1), we get

$$\|\Delta\varphi\| \le (1 + \eta)|\varphi'(x)| \cdot \|y - x\| \le (1 + \eta)|\varphi'(x)| \cdot \|u\| = \|(1 + \eta)\varphi'(x)u\|. \tag{19.29}$$

Moreover,

$$\begin{aligned}\langle \Delta\varphi, (1 + \eta)\varphi'(x)u\rangle &= \langle \varphi'(x)(y - x), (1 + \eta)\varphi'(x)u\rangle \\ &\quad + \langle \Delta\varphi - \varphi'(x)(y - x), (1 + \eta)\varphi'(x)u\rangle.\end{aligned} \tag{19.30}$$

Using (16.1), (16.2) and (19.29), we obtain the following lower bound for the first term on the right-hand side of (19.30):

$$\begin{aligned}\langle \varphi'(x)(y - x), (1 + \eta)\varphi'(x)u\rangle &= (1 + \eta)|\varphi'(x)|^2\langle y - x, u\rangle \\ &\ge (1 + \eta)|\varphi'(x)|^2\|y - x\| \cdot \|u\| \cos\alpha \\ &\ge \|\Delta\varphi\| \cdot \|\varphi'(x)u\| \cos\alpha.\end{aligned} \tag{19.31}$$

Concerning the second term, it follows from (19.27) that

$$\begin{aligned}|\langle \Delta\varphi - \varphi'(x)(y - x), (1 + \eta)\varphi'(x)u\rangle| &\le \|\Delta\varphi - \varphi'(x)(y - x)\| \cdot \|(1 + \eta)\varphi'(x)u\| \\ &\le (1 + \eta)K_5|\varphi'(x)| \cdot \|y - x\|^2 \cdot \|\varphi'(x)u\|.\end{aligned} \tag{19.32}$$

Assuming $\delta > 0$ to be smaller than $(2K_5)^{-1}$, from (19.27) we have

$$\begin{aligned}\|\Delta\varphi\| &\ge |\varphi'(x)| \cdot \|y - x\| - K_5|\varphi'(x)| \cdot \|y - x\|^2 \\ &\ge |\varphi'(x)| \cdot \|y - x\| - K_5|\varphi'(x)| \cdot \|y - x\|(2K_5)^{-1} \\ &= \frac{1}{2}|\varphi'(x)| \cdot \|y - x\|.\end{aligned}$$

Consequently, we can continue (19.32) as follows and obtain an upper estimate of the absolute value of the second term on the right-hand side of (19.30):

$$\begin{aligned}|\langle\Delta\varphi-\varphi'(x)(y-x),(1+\eta)\varphi'(x)u\rangle| &\le (1+\eta)K_5\cdot 2\|\Delta\varphi\|\cdot\|y-x\|\cdot\|\varphi'(x)u\| \\ &\le 2(1+\eta)K_5\|\Delta\varphi\|\cdot\delta\|\varphi'(x)u\|. \qquad (19.33)\end{aligned}$$

Combining (19.31) with (19.33), we obtain a lower estimate of the left-hand side of (19.30):

$$\begin{aligned}\langle\Delta\varphi,(1+\eta)\varphi'(x)u\rangle &\ge (\cos\alpha-2(1+\eta)K_5\delta)\|\varphi'(x)u\|\cdot\|\Delta\varphi\| \\ &= ((1+\eta)^{-1}\cos\alpha-2K_5\delta)\cdot\|(1+\eta)\varphi'(x)u\|\cdot\|\Delta\varphi\|.\end{aligned}$$

Therefore,

$$\cos\angle(\Delta\varphi,(1+\eta)\varphi'(x)u)\ge(1+\eta)^{-1}\cos\alpha-2K_5\delta.$$

Taking $0<\eta\le 1$ and $\delta>0$ so small that $(1+\eta)^{-1}\cos\alpha-2K_5\delta\ge\cos(\alpha+\varepsilon)$, the above inequality implies that $\angle(\Delta\varphi,(1+\eta)\varphi'(x)u)\le\alpha+\varepsilon$ while (19.29) yields that $\|\Delta\varphi\|\le\|(1+\eta)\varphi'(x)u\|$. That is, $\varphi(y)\in\mathrm{Con}(\varphi(x),\alpha+\varepsilon,(1+\eta)\varphi'(x)u)$. Hence,

$$\varphi(\mathrm{Con}(x,\alpha,u))\subseteq\mathrm{Con}(\varphi(x),\alpha+\varepsilon,(1+\eta)\varphi'(x)u)\subseteq\mathrm{Con}(\varphi(x),\alpha+\varepsilon,2\varphi'(x)u).\qquad\square$$

Recall from Section 19.1.1 that

$$E_{\mathcal{G}}^*:=\{\omega\in E^*\mid t(\omega_j)=i(\omega_{j+1}),\forall 1\le j<|\omega|\}$$

is the set of all finite words that are permissible based solely on the underlying graph $\mathcal{G}$, i. e., making abstraction of the matrix A. Obviously, $E_{\mathcal{G}}^*\supseteq E_A^*$.

The next proposition follows immediately from Theorem 19.7.11 when $d\ge 2$ and is a direct consequence of the Bounded Distortion Property (BDP) when $d=1$ (see Corollary 19.3.5).

Proposition 19.7.12. *Let $\Phi=\{\varphi_e\}_{e\in E}$ be a CGDMS. For every $\varepsilon>0$, there exists $\delta>0$ such that if $\omega\in E_{\mathcal{G}}^*$, $x\in X_{t(\omega)}$, $\|u\|\le\delta$ and $\alpha\in(0,\pi/2-\varepsilon)$, then*

$$\varphi_\omega(\mathrm{Con}(x,\alpha,u))\subseteq\mathrm{Con}(\varphi_\omega(x),\alpha+\varepsilon,2\varphi_\omega'(x)u).$$

In the next section, we will see that the following purely geometric condition ensures the existence and uniqueness of a conformal measure.

Definition 19.7.13. A CGDMS Φ is said to satisfy the *Cone Condition* (*CC*) if there exist $\psi,\ell>0$ such that for every $v\in V$ and every $x\in X_v$ there is a cone with vertex x, central angle ψ, altitude ℓ and unit direction vector u_x such that $\mathrm{Con}(x,\psi,\ell u_x)\backslash\{x\}\subseteq\mathrm{Int}(X_v)$.

This condition has the following important geometrical repercussion.

Theorem 19.7.14. *Let Φ be a CGDMS satisfying the cone condition. There exists $\beta > 0$ such that if $D \geq 1$ is sufficiently large, then for all $\omega \in E_{\mathcal{G}}^*$ and $x \in X_{t(\omega)}$ we have*

$$\begin{aligned}\varphi_\omega(X_{t(\omega)}) &\supseteq \mathrm{Con}(\varphi_\omega(x), \beta, D^{-1}\varphi_\omega'(x)u_x)\\ &\supseteq \mathrm{Con}(\varphi_\omega(x), \beta, D^{-2}\,\mathrm{diam}(\varphi_\omega(X_{t(\omega)}))\overline{u}_x),\end{aligned} \tag{19.34}$$

where u_x is a unit direction vector from the cone condition and $\overline{u}_x = \frac{\varphi_\omega'(x)}{|\varphi_\omega'(x)|}u_x$.

Proof. When $d = 1$, the first inclusion is a consequence of Lemma 19.3.10, which clearly applies to all words in $E_{\mathcal{G}}^*$. So we may assume that $d \geq 2$ when proving this inclusion. Let $0 < \psi < \pi/2$ be an allowable central angle from the cone condition. Then there is $q \in \mathbb{N}$ such that for every $x \in X$ there exist unit vectors $u_1(x), \ldots, u_q(x) \in \mathbb{R}^d$ so that all q-tuples $(u_x, u_1(x), \ldots, u_q(x))$, $x \in X$, are linearly isometric to one another and

$$\mathrm{Con}(x, \psi, \mathbb{R}_+ u_x) \cup \bigcup_{j=1}^{q} \mathrm{Con}(x, \psi, \mathbb{R}_+ u_j(x)) = \mathbb{R}^d \tag{19.35}$$

while

$$\mathrm{Con}(x, \psi/2, \mathbb{R}_+ u_x) \cap \bigcup_{j=1}^{q} \mathrm{Con}(x, \psi, \mathbb{R}_+ u_j(x)) = \emptyset. \tag{19.36}$$

Then there exists $\varepsilon > 0$ such that

$$\mathrm{Con}(x, \psi/4, \mathbb{R}_+ \tilde{u}_x) \cap \bigcup_{j=1}^{q} \mathrm{Con}(x, \psi + \varepsilon, \mathbb{R}_+ \tilde{u}_j(x)) = \emptyset \tag{19.37}$$

for all $x \in X$ and all q-tuples $(\tilde{u}_x, \tilde{u}_1(x), \ldots, \tilde{u}_q(x))$ that are equivalent, via a similarity map, to $(u_x, u_1(x), \ldots, u_q(x))$. Let $\ell > 0$ be an allowable altitude according to the cone condition and let $0 < \delta < \min\{\ell, \mathrm{dist}(X, \partial W)\}$ be ascribed to the above ε by virtue of Proposition 19.7.12. In view of that proposition, relation (19.37) and the fact that $\varphi_\omega'(x)$ is a similarity map, we get

$$\mathrm{Con}(\varphi_\omega(x), \psi/4, \mathbb{R}_+ \varphi_\omega'(x)u_x) \cap \bigcup_{j=1}^{q} \varphi_\omega(\mathrm{Con}(x, \psi, \delta u_j(x))) = \emptyset \tag{19.38}$$

for all $\omega \in E_{\mathcal{G}}^*$ and $x \in X_{t(\omega)}$. In view of (19.35) and Lemma 19.3.10, which applies to all words in $E_{\mathcal{G}}^*$, we find that

$$\varphi_\omega(\mathrm{Con}(x, \psi, \delta u_x)) \cup \bigcup_{j=1}^{q} \varphi_\omega(\mathrm{Con}(x, \psi, \delta u_j(x))) \supseteq \varphi_\omega(B(x, \delta)) \supseteq B(\varphi_\omega(x), K^{-1}\|\varphi_\omega'\|_{W_{t(\omega)}}\delta).$$

This and (19.38) imply that

$$\begin{aligned}\varphi_\omega(\mathrm{Con}(x,\psi,\delta u_x)) &\supseteq \mathrm{Con}(\varphi_\omega(x),\psi/4,K^{-1}\|\varphi'_\omega\|_{W_{t(\omega)}}\delta|\varphi'_\omega(x)|^{-1}\varphi'_\omega(x)u_x)\\ &\supseteq \mathrm{Con}(\varphi_\omega(x),\psi/4,K^{-1}\delta\varphi'_\omega(x)u_x).\end{aligned}$$

Taking any $D \geq K\delta^{-1}$ and $\beta = \psi/4$ achieves the first inclusion in (19.34). The second inclusion holds if D is chosen large enough in (19.9) of Lemma 19.3.9, which applies to all words in $E^*_{\mathcal{G}}$. □

It ensues from this theorem that each point x of the space belongs to a finite number of n-th level sets, and that number is actually uniformly bounded in n and x.

Corollary 19.7.15. *If Φ is a CGDMS satisfying the cone condition, then*

$$\sup_{n\in\mathbb{N}}\sup_{x\in X}\#\{\omega\in E^n_{\mathcal{G}} : x\in\varphi_\omega(X_{t(\omega)})\} \leq \frac{\lambda_{d-1}(S^{d-1})}{\beta} < \infty,$$

where S^{d-1} is the $(d-1)$-dimensional unit sphere and λ_{d-1} is the $(d-1)$-dimensional Lebesgue measure on S^{d-1}, while β arises from Theorem 19.7.14.

Proof. Let $x \in X$ and $n \in \mathbb{N}$. By Theorem 19.7.14, any set $\varphi_\omega(X_{t(\omega)}) \ni x$ contains a cone with vertex x and central angle β. But by the OSC, the sets $\varphi_\tau(X_{t(\tau)})$, $\tau \in E^n_{\mathcal{G}}$, are nonoverlapping beyond their boundaries. Therefore the sets $\varphi_\omega(X_{t(\omega)})$, with $\omega \in E^n_{\mathcal{G}}$ and $x \in \varphi_\omega(X_{t(\omega)})$, contain mutually disjoint open cones with vertex x and central angle β. There are at most $\lambda_{d-1}(S^{d-1})/\beta$ such cones. □

We next establish a result similar to Lemma 19.3.12.

Lemma 19.7.16. *If Φ is a CGDMS satisfying the cone condition, then for every $\kappa > 0$, every $r > 0$, and every $x \in X$, the cardinality of any collection of mutually incomparable words $\omega \in E^*_{\mathcal{G}}$ that satisfy the conditions*

$$\varphi_\omega(X_{t(\omega)}) \cap B(x,r) \neq \emptyset \quad \text{and} \quad \kappa r \leq \mathrm{diam}(\varphi_\omega(X_{t(\omega)}))$$

is bounded above by the number

$$V_d\kappa^{-d}D^{2d}\beta^{-1}(1+\kappa D^{-2})^d,$$

where V_d is the volume of a unit ball in $\mathbb{R}^d$ while D and β originate from Theorem 19.7.14.

Proof. Fix $\kappa > 0$, $r > 0$ and $x \in X$, and let W be a collection as described in the statement of the lemma. For each $\omega \in W$, let $x_\omega \in X_{t(\omega)}$ be such that $\varphi_\omega(x_\omega) \in \varphi_\omega(X_{t(\omega)}) \cap B(x,r)$. It follows from Theorem 19.7.14 that

$$\varphi_\omega(X_{t(\omega)}) \supseteq \mathrm{Con}(\varphi_\omega(x_\omega),\beta,D^{-2}\kappa r\overline{u}_{x_\omega}) \subseteq B(x,(1+D^{-2}\kappa)r).$$

Due to the OSC (see Definition 19.3.1), the cones $\mathrm{Con}(\varphi_\omega(x_\omega), \beta, D^{-2}\kappa r\overline{u}_{x_\omega})$, $\omega \in W$, intersect at most on their boundaries. Therefore,

$$\lambda_d(B(x,(1+D^{-2}\kappa)r)) \geq \sum_{\omega\in W} \lambda_d(\mathrm{Con}(\varphi_\omega(x_\omega), \beta, D^{-2}\kappa r\overline{u}_{x_\omega})) = \#W \cdot \beta(D^{-2}\kappa r)^d.$$

So $\#W \leq \beta^{-1}(D^{-2}\kappa r)^{-d}\lambda_d(B(x,(1+D^{-2}\kappa)r)) = \beta^{-1}D^{2d}(\kappa r)^{-d}V_d(1+D^{-2}\kappa)^d r^d$. □

19.7.5 Conformal-likeness

Definition 19.7.17. Given $q \in \mathbb{N}$, we say that two distinct words $\omega, \tau \in E_G^*$ of equal length $n > q$ form a pair of q-codes for a point $x \in X$ if

$$x \in \varphi_\omega(X_{t(\omega)}) \cap \varphi_\tau(X_{t(\tau)}) \quad \text{and} \quad \omega|_{n-q} = \tau|_{n-q}.$$

A CGDMS Φ is said to be *conformal-like* if for every $q \in \mathbb{N}$ there is no point in J with arbitrarily long pairs of q-codes.

We now make a simple observation.

Lemma 19.7.18. *Each conformal-like CGDMS is pointwise finite, and hence its associated coding map $\pi : E_A^\infty \to J$ is bounded-to-one and its limit set J is a $F_{\sigma\delta}$–set.*

Proof. The reader can convince themself of this by looking at Definitions 19.7.17 and 19.1.2 as well as Lemma 19.1.3. □

Later on, we will observe that conformal-likeness implies the existence and uniqueness of a conformal measure. However, conformal-likeness can be challenging to demonstrate directly. We thus identify sufficient conditions for it to hold.

Proposition 19.7.19. *Every CGDMS satisfying the Boundary Separation Condition (BSC), the Strong Separation Condition (SSC) or the Cone Condition (CC) is conformal-like.*

Proof. Suppose that a CGDMS Φ satisfying the CC is not conformal-like. This means that there exist a point $x \in X$, an integer $q \in \mathbb{N}$ and a strictly increasing sequence $(n_k)_{k=1}^\infty$ along with words $\omega^{(k)} \neq \tau^{(k)} \in E_G^*$ of length n_k such that

$$x \in \varphi_{\omega^{(k)}}(X_{t(\omega^{(k)})}) \cap \varphi_{\tau^{(k)}}(X_{t(\tau^{(k)})}) \quad \text{and} \quad \omega^{(k)}|_{n_k-q} = \tau^{(k)}|_{n_k-q}.$$

Passing to a subsequence if necessary, we may assume that $n_{k+1} - n_k \geq q$ for every $k \in \mathbb{N}$. We shall construct by induction on k a sequence $(C_k)_{k-1}^\infty$ such that C_k consists of at least $k+1$ incomparable words from $\{\omega^{(j)}, \tau^{(j)} : j \leq k\}$. In particular, note that this will imply that $|C_k| := \max\{|\rho| : \rho \in C_k\} \leq n_k$. In fact, we will see that $|C_k| = n_k$. Set $C_1 = \{\omega^{(1)}, \tau^{(1)}\}$. Suppose now that C_k has been defined. If $\omega^{(k+1)}$ is comparable to a word $\rho \in C_k$, then ρ is unique. Indeed, if $\omega^{(k+1)}$ is comparable to $\rho_1, \rho_2 \in C_k$, then as

$|\omega^{(k+1)}| = n_{k+1} \geq n_k = |C_k| \geq \max\{|\rho_1|, |\rho_2|\}$ and as the words in C_k are incomparable, we deduce that $\rho_1 = \rho_2$. This means that $\omega^{(k+1)}$ (and similarly $\tau^{(k+1)}$) is comparable to at most one word in C_k. If $\omega^{(k+1)}$ is not comparable to any word in C_k, then we form C_{k+1} by simply adjoining $\omega^{(k+1)}$ to C_k. We can do a similar addition if $\tau^{(k+1)}$ is not comparable to any word in C_k. If $\omega^{(k+1)}$ and $\tau^{(k+1)}$ are respectively comparable to words ρ and η in C_k, then in fact ρ and η are both extended by $\omega^{(k+1)}$ since $\omega^{(k+1)}|_{n_{k+1}-q} = \tau^{(k+1)}|_{n_{k+1}-q}$ and $n_{k+1} - q \geq n_k = |C_k| \geq \max\{|\rho|, |\eta|\}$. As the words in C_k are incomparable, we deduce that $\rho = \eta$. In this case, we form C_{k+1} from C_k by taking away ρ and adding both $\omega^{(k+1)}$ and $\tau^{(k+1)}$. This completes the inductive proof of the existence of the sequence $(C_k)_{k=1}^{\infty}$. But C_k consists of at least $k+1$ incomparable words ρ such that $x \in \varphi_\rho(X_{t(\rho)})$. This contradicts Corollary 19.7.15.

The proof of the conformal-likeness of any CGDMS Φ satisfying the BSC or the SSC is left to the reader as an exercise. □

As an immediate consequence of Proposition 19.7.19 and Lemma 19.7.18, we obtain the following.

Corollary 19.7.20. *Every CGDMS satisfying the BSC, the SSC or the CC is pointwise finite and its limit set is an $F_{\sigma\delta}$–set.*

In Theorem 19.7.2(c), we saw that for every CGDMS satisfying the SOSC the intersection of any two level sets generated by incomparable words is negligible under any ergodic shift-invariant measure that has full topological support. For conformal-like CGDMSs, this is also true even when the measure is not ergodic or does not have full topological support.

Theorem 19.7.21. *Let $\Phi = \{\varphi_e\}_{e\in E}$ be a conformal-like CGDMS. For any σ-invariant Borel probability measure μ on E_A^∞, we have*

$$\mu \circ \pi^{-1}(\varphi_\omega(X_{t(\omega)}) \cap \varphi_\tau(X_{t(\tau)})) = 0$$

whenever $\omega, \tau \in E_A^*$ are incomparable words.

Proof. Suppose on the contrary that there are some incomparable words $\omega, \tau \in E_A^*$ such that $\mu \circ \pi^{-1}(\varphi_\omega(X_{t(\omega)}) \cap \varphi_\tau(X_{t(\tau)})) > 0$. Then $i(\omega) = i(\tau) =: v \in V$. We may assume without loss of generality that ω and τ are of the same length, say $q \in \mathbb{N}$. Let $B = \varphi_\omega(X_{t(\omega)}) \cap \varphi_\tau(X_{t(\tau)})$ and for every $n \in \mathbb{N}$ let

$$B_n = \bigcup_{\alpha\in E_A^n:\, t(\alpha)=v} \varphi_\alpha(B).$$

Since each element of B_n admits at least two q-codes of length $n+q$ (namely, $\alpha\omega$ and $\alpha\tau$), we deduce from the conformal-likeness of Φ (see Definition 19.7.17) that

$$J \cap \bigcap_{k=1}^{\infty} \bigcup_{n=k}^{\infty} B_n = \emptyset. \tag{19.39}$$

On the other hand, $\pi^{-1}(B_n) \supseteq \sigma^{-n}(\pi^{-1}(B))$ for all $n \in \mathbb{N}$. Therefore,

$$\mu(\pi^{-1}(B_n)) \geq \mu(\sigma^{-n}(\pi^{-1}(B))) = \mu(\pi^{-1}(B))$$

for all $n \in \mathbb{N}$, and hence

$$\mu \circ \pi^{-1}\left(J \cap \bigcap_{k=1}^{\infty} \bigcup_{n=k}^{\infty} B_n\right) = \lim_{k\to\infty} \mu \circ \pi^{-1}\left(\bigcup_{n=k}^{\infty} B_n\right) \geq \liminf_{k\to\infty} \mu(\pi^{-1}(B_k)) \geq \mu(\pi^{-1}(B)) > 0.$$

In particular, $J \cap \bigcap_{k=1}^{\infty} \bigcup_{n=k}^{\infty} B_n \neq \emptyset$. This contradicts (19.39). □

19.8 Hölder families of functions and conformal measures

We now turn our attention to families of functions on the real space. By convention, such families shall be denoted by uppercase letters, while their members and amalgamation will be denoted by the corresponding lowercase letters.

19.8.1 Basic definitions and properties

All of the notions that we introduce in this section for families of functions on the real space are counterparts to those that we studied earlier for potentials on the symbolic space. In fact, each family of functions on the real space generates a special potential on the symbolic space.

Definition 19.8.1. Let $\Phi = \{\varphi_e : X_{t(e)} \to X_{i(e)}\}_{e\in E}$ be a CGDMS and let $F = \{f_e : X_{t(e)} \to \mathbb{R}\}_{e\in E}$ be a family of real-valued functions. The potential $f : E_A^{\infty} \to \mathbb{R}$ defined by

$$f(\omega) := f_{\omega_1}(\pi(\sigma(\omega)))$$

is called the *amalgamated function* induced by the family F.

Given $\omega \in E_A^*$, define the function $S_\omega F : X_{t(\omega)} \to \mathbb{R}$ by declaring

$$S_\omega F := \sum_{j=1}^{|\omega|} f_{\omega_j} \circ \varphi_{\sigma^j(\omega)}.$$

By convention, the empty word ϵ is the only word of length 0 and $\varphi_\epsilon = \mathrm{Id}_X$.

Recall that

$$E_A^{\infty,\omega} = E_A^{\infty,\omega_{|\omega|}} = \{\rho \in E_A^{\infty} : \omega\rho \in E_A^{\infty}\}$$

is the set of all infinite A-admissible suffixes of ω. The values of $S_\omega F$ and $S_{|\omega|} f$ are related as follows.

Lemma 19.8.2. *Let $\Phi = \{\varphi_e : X_{t(e)} \to X_{i(e)}\}_{e\in E}$ be a CGDMS and let f be the amalgamated function induced by a family of functions $F = \{f_e : X_{t(e)} \to \mathbb{R}\}_{e\in E}$. For every $\omega \in E_A^*$ and every $\rho \in E_A^{\infty,\omega}$, we have*

$$S_\omega F(\pi(\rho)) = S_{|\omega|} f(\omega\rho).$$

Proof. Let $\omega \in E_A^*$, let $1 \le j \le |\omega|$ and $\rho \in E_A^{\infty,\omega}$. By (19.1),

$$\varphi_{\sigma^j(\omega)}(\pi(\rho)) = \varphi_{\sigma^j(\omega)}(\pi(\sigma^{|\omega|-j}(\sigma^j(\omega\rho)))) = \pi(\sigma^j(\omega\rho)).$$

Hence,

$$\begin{aligned} S_\omega F(\pi(\rho)) &= \sum_{j=1}^{|\omega|} f_{\omega_j} \circ \varphi_{\sigma^j(\omega)}(\pi(\rho)) = \sum_{j=1}^{|\omega|} f_{\omega_j}(\pi(\sigma(\sigma^{j-1}(\omega\rho)))) \\ &= \sum_{j=1}^{|\omega|} f(\sigma^{j-1}(\omega\rho)) = \sum_{j=0}^{|\omega|-1} f(\sigma^j(\omega\rho)) = S_{|\omega|} f(\omega\rho). \end{aligned}$$ □

Definition 19.8.3. Let $\Phi = \{\varphi_e : X_{t(e)} \to X_{i(e)}\}_{e\in E}$ be a CGDMS. A family of functions $F = \{f_e : X_{t(e)} \to \mathbb{R}\}_{e\in E}$ is called *summable* if

$$\sum_{e\in E} \|\exp(f_e)\|_{X_{t(e)}} < \infty.$$

The following fact is obvious (cf. Definition 17.6.1).

Lemma 19.8.4. *Let $\Phi = \{\varphi_e : X_{t(e)} \to X_{i(e)}\}_{e\in E}$ be a CGDMS. A family of functions $F = \{f_e : X_{t(e)} \to \mathbb{R}\}_{e\in E}$ is summable if and only if its amalgamated function f is summable.*

Now, for every $\beta > 0$ and $n \in \mathbb{N}$ let

$$v_{\beta,n}(F) := \sup_{\omega\in E_A^n} \sup_{x,y\in X_{t(\omega)}} |f_{\omega_1}(\varphi_{\sigma(\omega)}(x)) - f_{\omega_1}(\varphi_{\sigma(\omega)}(y))| e^{\beta n}.$$

Thus, $v_{\beta,1}(F) < \infty$ simply means that the diameters of the sets $f_e(X_{t(e)})$, $e \in E$, are uniformly bounded.

Definition 19.8.5. Let $\Phi = \{\varphi_e : X_{t(e)} \to X_{i(e)}\}_{e\in E}$ be a CGDMS. A family of functions $F = \{f_e : X_{t(e)} \to \mathbb{R}\}_{e\in E}$ is said to be *Hölder with exponent β* if

$$v_\beta(F) := \sup_{n\in\mathbb{N}} v_{\beta,n}(F) < \infty.$$

The following fact is fairly straightforward to establish and is left to the reader (cf. Definition 17.2.2).

Lemma 19.8.6. *Let $\Phi = \{\varphi_e\}_{e\in E}$ be a CGDMS. If $F = \{f_e\}_{e\in E}$ is a Hölder family with exponent β, then its amalgamated function f is Hölder continuous on cylinders with exponent β.*

Hölder families obey to a bounded variation principle (cf. Lemma 17.2.3).

Lemma 19.8.7 (Bounded variation principle). *Let $\Phi = \{\varphi_e\}_{e\in E}$ be a CGDMS and $F = \{f_e\}_{e\in E}$ a Hölder family of functions with exponent β. If $\omega \in E_A^*$ and $x, y \in X_{t(\omega)} \cap \varphi_\tau(X_{t(\tau)})$ for some $\tau \in E_A^*$, then*

$$|S_\omega F(x) - S_\omega F(y)| \le \frac{\upsilon_\beta(F)}{1 - e^{-\beta}} e^{-\beta|\tau|}.$$

Proof. There exist $\tilde{x}, \tilde{y} \in X_{t(\tau)}$ such that $x = \varphi_\tau(\tilde{x})$ and $y = \varphi_\tau(\tilde{y})$. Then

$$\begin{aligned}|S_\omega F(x) - S_\omega F(y)| &= \left|\sum_{j=1}^{|\omega|} f_{\omega_j} \circ \varphi_{\sigma^j(\omega)}(\varphi_\tau(\tilde{x})) - \sum_{j=1}^{|\omega|} f_{\omega_j} \circ \varphi_{\sigma^j(\omega)}(\varphi_\tau(\tilde{y}))\right| \\ &\le \sum_{j=1}^{|\omega|} |f_{(\omega\tau)_j} \circ \varphi_{\sigma^j(\omega\tau)}(\tilde{x}) - f_{(\omega\tau)_j} \circ \varphi_{\sigma^j(\omega\tau)}(\tilde{y})| \\ &\le \sum_{j=1}^{|\omega|} \upsilon_\beta(F) e^{-\beta(|\omega|+|\tau|-j)} \le \frac{\upsilon_\beta(F)}{1 - e^{-\beta}} e^{-\beta|\tau|}.\end{aligned}$$

□

Not surprisingly, the pressure of a family of functions is defined as follows.

Definition 19.8.8. Let $\Phi = \{\varphi_e\}_{e\in E}$ be a CGDMS. The topological pressure $\mathrm{P}(F)$ of a summable family of functions $F = \{f_e\}_{e\in E}$ is defined by

$$\mathrm{P}(F) := \lim_{n\to\infty} \frac{1}{n} \log \sum_{\omega\in E_A^n} \|\exp(S_\omega F)\|_{X_{t(\omega)}}.$$

The limit exists because the sequence $(\sum_{\omega\in E_A^n} \|\exp(S_\omega F)\|_{X_{t(\omega)}})_{n=1}^\infty$ is submultiplicative. Moreover, $\mathrm{P}(F) \le \sum_{e\in E} \|\exp(f_e)\|_{X_{t(e)}} < \infty$.

The next result asserts that the topological pressure of a family of functions coincides with the topological pressure of the amalgamated function it induces.

Proposition 19.8.9. *Let $\Phi = \{\varphi_e\}_{e\in E}$ be a CGDMS. If $F = \{f_e\}_{e\in E}$ is a summable Hölder family of functions and f is the amalgamated function induced by F, then $\mathrm{P}(F) = \mathrm{P}(f)$.*

Proof. Let $\omega \in E_A^*$. By taking the supremum over all $\rho \in E_A^{\infty,\omega}$ in Lemma 19.8.2, we get

$$\|\exp(S_\omega F)\|_{X_{t(\omega)}} \ge \exp(\sup\{S_{|\omega|} f(\omega\rho) : \rho \in E_A^{\infty,\omega}\}) = \exp(\overline{S}_{|\omega|} f([\omega])).$$

Consequently, $\mathrm{P}(F) \ge \mathrm{P}(f)$ (cf. Definitions 19.8.8 and 17.2.7).

On the other hand, let $\rho \in E_A^{\infty,\omega}$ and $x \in X_{t(\omega)}$. Set $V_\beta(F) = \exp(\frac{v_\beta(F)}{1-e^{-\beta}})$. By Lemmas 19.8.7 and 19.8.2, we obtain that

$$\exp(S_\omega F(x)) \leq V_\beta(F)\exp(S_\omega F(\pi(\rho))) = V_\beta(F)\exp(S_{|\omega|}f(\omega\rho)) \leq V_\beta(F)\exp(\overline{S}_{|\omega|}f([\omega])).$$

Thus, $\|\exp(S_\omega F)\|_{X_{t(\omega)}} \leq V_\beta(F)\exp(\overline{S}_{|\omega|}f([\omega]))$ and thereby $\mathrm{P}(F) \leq \mathrm{P}(f)$. □

One family of functions will play a special role for CGDMSs, as it induces the potential Ξ, which is at the heart of the definition of the topological pressure of a CGDMS (see Section 19.4).

Proposition 19.8.10. *For any CGDMS $\Phi = \{\varphi_e\}_{e\in E}$, the family $\mathrm{Log}_\Phi := \{\log|\varphi_e'|\}_{e\in E}$ has for amalgamated function the potential Ξ. The family Log_Φ is Hölder. Moreover, for any $t \in \mathrm{Fin}$, the family $t\,\mathrm{Log}_\Phi$ is summable and $\mathrm{P}(t\,\mathrm{Log}_\Phi) = \mathrm{P}(t)$.*

Proof. It is easy to see that the potential Ξ is induced by the Log_Φ family. This family is Hölder with exponent $-\alpha\log s$ by Lemma 19.3.4. For any $t \in \mathrm{Fin}$, the family $t\,\mathrm{Log}_\Phi$ is summable since $Z_1(t) < \infty$ according to Proposition 19.4.5(a). Finally, $\mathrm{P}(t\,\mathrm{Log}_\Phi) = \mathrm{P}(t\Xi) = \mathrm{P}(t)$ by Proposition 19.8.9 and the note subsequent to Definition 19.4.3. □

19.8.2 Conformal measures for summable Hölder families of functions

In Chapter 13, we witnessed the usefulness of conformal measures associated with potentials imposed on open, distance expanding systems. We now introduce this concept for families of functions imposed on CGDMSs (cf. Definition 13.6.12).

Definition 19.8.11. Let $\Phi = \{\varphi_e\}_{e\in E}$ be a CGDMS and $F = \{f_e\}_{e\in E}$ be a summable Hölder family of functions. A Borel probability measure m on X is said to be F-*conformal* provided it is supported on the limit set J and the following two conditions are satisfied:

(1) For every $e \in E$ and for every Borel set $B \subseteq \pi(E_A^{\infty,e})$, we have

$$m(\varphi_e(B)) = \int_B \exp(f_e - \mathrm{P}(F))\,dm. \tag{19.40}$$

(2) For all $e_1 \neq e_2 \in E$, we have

$$m(\varphi_{e_1}(X_{t(e_1)}) \cap \varphi_{e_2}(X_{t(e_2)})) = 0.. \tag{19.41}$$

A simple inductive argument shows that it follows from this that for every $\omega \in E_A^*$ and for every Borel set $B \subseteq \pi(E_A^{\infty,\omega})$ we get

$$m(\varphi_\omega(B)) = \int_B \exp(S_\omega F - |\omega|\,\mathrm{P}(F))\,dm, \tag{19.42}$$

and for all incomparable words $\omega, \tau \in E_A^*$,

$$m(\varphi_\omega(X_{t(\omega)}) \cap \varphi_\tau(X_{t(\tau)})) = 0. \tag{19.43}$$

This last property implies that for all incomparable words $\omega, \tau \in E_A^*$,

$$m(\varphi_\omega(X_{t(\omega)}) \cap \pi([\tau])) = 0 \tag{19.44}$$

and

$$m(\pi([\omega]) \cap \pi([\tau])) = 0. \tag{19.45}$$

It might be tempting to define conformality of a measure in terms of all Borel sets $B \subseteq X_{t(\omega)}$. However, it is subtle but crucial that conformality be defined solely in terms of the Borel subsets of $\pi(E_A^{\infty,\omega})$, as (19.42) does not generally hold for all Borel sets $B \subseteq X_{t(\omega)}$. Indeed, we have the following.

Lemma 19.8.12. *Let $\Phi = \{\varphi_e\}_{e\in E}$ be a CGDMS and $F = \{f_e\}_{e\in E}$ be a summable Hölder family of functions. If Φ admits an F-conformal measure m, then for every $\omega \in E_A^*$ and every Borel set $B \subseteq X_{t(\omega)}$ we have that*

$$m(\varphi_\omega(B)) = m(\varphi_\omega(B \cap \pi(E_A^{\infty,\omega}))) = \int_{B\cap\pi(E_A^{\infty,\omega})} \exp(S_\omega F - |\omega|\,\mathrm{P}(F))\,dm.$$

Proof. Recall that $E_A^{1,\omega} := \{e \in E : \omega e \in E_A^*\}$. Then

$$\begin{aligned}
m(\varphi_\omega(B)) &= \sum_{\tau\in E_A^{|\omega|}} m(\varphi_\omega(B) \cap \pi([\tau])) \quad \text{by (19.45) and Lemma A.1.19(g)}\\
&= m(\varphi_\omega(B) \cap \pi([\omega])) \quad \text{by (19.44)}\\
&= m\Big(\varphi_\omega(B) \cap \bigcup_{e\in E_A^{1,\omega}} \pi([\omega e])\Big)\\
&= \sum_{e\in E_A^{1,\omega}} m(\varphi_\omega(B) \cap \pi([\omega e])) \quad \text{by (19.45) and Lemma A.1.19(g)}\\
&= \sum_{e\in E_A^{1,\omega}} m(\varphi_\omega(B) \cap \varphi_\omega(\pi([e])))\\
&= \sum_{e\in E_A^{1,\omega}} m(\varphi_\omega(B \cap \pi([e]))) \quad \text{by injectivity of } \varphi_\omega\\
&= \int_{\bigcup_{e\in E_A^{1,\omega}} B\cap\pi([e])} \exp(S_\omega F - |\omega|\,\mathrm{P}(F))\,dm \quad \text{by (19.45) and Lemma A.1.19(g)}
\end{aligned}$$

$$= \int_{B\cap\pi(E_A^{\infty,\omega})} \exp(S_\omega F - |\omega|\,\mathrm{P}(F))\,dm$$

$$= m(\varphi_\omega(B\cap\pi(E_A^{\infty,\omega}))) \quad \text{by conformality of } m. \qquad \square$$

Remark 19.8.13. Let $\Phi = \{\varphi_e\}_{e\in E}$ be a CGDMS and $F = \{f_e\}_{e\in E}$ be a summable Hölder family of functions. Suppose that Φ admits an F-conformal measure m. Let $\omega \in E_A^*$ and $B \subseteq X_{t(\omega)}$ a Borel set. According to Lemma 19.8.12,

$$m(\varphi_\omega(B)) \le \int_B \exp(S_\omega F - |\omega|\,\mathrm{P}(F))\,dm$$

but the equality does not hold in general.

We now establish the existence and uniqueness of a conformal measure for any summable Hölder family of functions associated to a conformal-like CGDMS. We further describe its form.

Theorem 19.8.14. *Let $\Phi = \{\varphi_e\}_{e\in E}$ be a finitely irreducible, conformal-like CGDMS. For any summable Hölder family of functions $F = \{f_e\}_{e\in E}$, the CGDMS Φ has a unique F-conformal measure m_F. Moreover,*

$$m_F = m_f \circ \pi^{-1},$$

where f is the amalgamated function induced by F and m_f is the eigenmeasure of the dual transfer operator $\mathcal{L}_f^*$.

Proof. According to Lemmas 19.8.4 and 19.8.6, the amalgamated function $f : E_A^\infty \to \mathbb{R}$ induced by the family F is summable and Hölder continuous on cylinders. By Corollary 17.7.5(a), there then exists an eigenmeasure m_f of the dual operator $\mathcal{L}_f^*$ whose corresponding eigenvalue is $e^{\mathrm{P}(f)}$. Let $m_F = m_f \circ \pi^{-1}$. We shall show that m_F is an F-conformal measure.

First, we establish (19.43). Per Corollary 17.7.5(c) there exists an ergodic σ-invariant Gibbs state μ_f for f which, by Proposition 17.4.2(b), is boundedly equivalent with m_f. Therefore, $m_F = m_f \circ \pi^{-1}$ is boundedly equivalent with $\mu_f \circ \pi^{-1}$. As (19.43) holds for $\mu_f \circ \pi^{-1}$ according to Theorem 19.7.21, it must also hold for m_F.

It remains to prove (19.42). Fix $\omega \in E_A^*$, say $\omega \in E_A^n$, and a Borel set $B \subseteq \pi(E_A^{\infty,\omega})$. Set

$$\pi_\omega^{-1}(B) := \pi^{-1}(B) \cap E_A^{\infty,\omega} = \{\rho \in \pi^{-1}(B) \mid A_{\omega_n\rho_1} = 1\}.$$

Then

$$\pi(\pi_\omega^{-1}(B)) = B \tag{19.46}$$

and

$$\pi^{-1}(B)\setminus\pi_\omega^{-1}(B)\subseteq E_A^\infty\setminus E_A^{\infty,\omega}\subseteq\bigcup_{e\in E\setminus E_A^{1,\omega}}[e]=\bigcup_{e\in E:\,A_{\omega_n e}=0}[e].$$

So, using (19.46), we obtain

$$\begin{aligned}\pi(\pi^{-1}(B)\setminus\pi_\omega^{-1}(B))&\subseteq B\cap\pi\Big(\bigcup_{b\in E:\,A_{\omega_n b}=0}[b]\Big)\\&\subseteq\Big[\bigcup_{a\in E:\,A_{\omega_n a}=1}\varphi_a(X_{t(a)})\Big]\cap\Big[\bigcup_{b\in E:\,A_{\omega_n b}=0}\varphi_b(X_{t(b)})\Big]\\&\subseteq\bigcup_{a\in E:\,A_{\omega_n a}=1}\ \bigcup_{b\in E:\,A_{\omega_n b}=0}\varphi_a(X_{t(a)})\cap\varphi_b(X_{t(b)}).\end{aligned}$$

Since property (19.43) holds for $m_F:=m_f\circ\pi^{-1}$, we deduce that

$$m_F(\pi(\pi^{-1}(B)\setminus\pi_\omega^{-1}(B)))=0,$$

and hence

$$m_f(\pi^{-1}(B)\setminus\pi_\omega^{-1}(B))=0. \tag{19.47}$$

Now, $\rho\in\pi^{-1}(\varphi_\omega(B))$ if and only if $\pi(\rho)\in\varphi_\omega(B)$, i. e., $\varphi_{\rho|_n}(\pi(\sigma^n(\rho)))\in\varphi_\omega(B)$. Thus,

$$\omega\pi_\omega^{-1}(B)\subseteq\pi^{-1}(\varphi_\omega(B))\subseteq\omega\pi_\omega^{-1}(B)\cup\pi^{-1}\Big(\bigcup_{\tau\in E_A^n\setminus\{\omega\}}\varphi_\tau(X_{t(\tau)})\cap\varphi_\omega(X_{t(\omega)})\Big).$$

Using (19.43) and the definition of the measure m_F, it ensues that

$$m_f(\pi^{-1}(\varphi_\omega(B)))=m_f(\omega\pi_\omega^{-1}(B)).$$

Combining this equality with (17.26), Lemma 19.8.2, Proposition 19.8.9, (19.47) and (19.46), we get

$$\begin{aligned}m_F(\varphi_\omega(B))&=m_f\circ\pi^{-1}(\varphi_\omega(B))=m_f(\omega\pi_\omega^{-1}(B))\\&=\int_{\pi_\omega^{-1}(B)}\exp(S_nf(\omega\rho)-n\,\mathrm{P}(f))\,dm_f(\rho)\\&=\int_{\pi_\omega^{-1}(B)}\exp(S_\omega F(\pi(\rho))-n\,\mathrm{P}(F))\,dm_f(\rho)\\&=\int_{\pi^{-1}(B)}\exp(S_\omega F(\pi(\rho))-n\,\mathrm{P}(F))\,dm_f(\rho)\end{aligned}$$

$$= \int_B \exp(S_\omega F(x) - n\,\mathrm{P}(F))\, dm_f \circ \pi^{-1}(x)$$
$$= \int_B \exp(S_\omega F - n\,\mathrm{P}(F))\, dm_F.$$

This establishes (19.42) and completes the proof that m_F is an F-conformal measure.

To prove uniqueness, suppose that ν_F is an F-conformal measure. It is not difficult to show that $\nu := \nu_F \circ \pi$ is a Borel probability measure on E_A^∞. (The set of all unions of mutually disjoint initial cylinders forms an algebra on which ν can be shown to be countably additive using (19.43). Then Carathéodory's extension theorem (Theorem A.1.28) affirms that ν is a Borel probability measure.) Since π is surjective, note that $\nu_F = \nu \circ \pi^{-1}$.

Fix $n \in \mathbb{N}$, a word $\omega \in E_A^n$ and a Borel set $B \subseteq E_A^{\infty,\omega}$. Using (19.47) with m_f replaced by ν and B replaced by $\pi(B)$, observe that

$$\nu(B) = \nu_F(\pi(B)) = \nu(\pi^{-1}(\pi(B))) = \nu(\pi_\omega^{-1}(\pi(B))). \tag{19.48}$$

Then, using successively (19.42), Lemma 8.1.2, Lemma 19.8.2 and (19.48), we obtain that

$$\nu(\omega B) = \nu_F(\pi(\omega B)) = \nu_F(\varphi_\omega(\pi(B))) = \int_{\pi(B)} \exp(S_\omega F - n\,\mathrm{P}(F))\, d\nu \circ \pi^{-1}$$
$$= \int_{\pi^{-1}(\pi(B))} \exp(S_\omega F \circ \pi - n\,\mathrm{P}(F))\, d\nu$$
$$= \int_{\pi_\omega^{-1}(\pi(B))} \exp(S_\omega F \circ \pi - n\,\mathrm{P}(F))\, d\nu$$
$$= \int_B \exp(S_n f(\omega\rho) - n\,\mathrm{P}(f))\, d\nu(\rho).$$

Therefore, in view of Remark 17.6.3, ν is an eigenmeasure of the operator $\mathcal{L}_f^*$ corresponding to the eigenvalue $\exp(\mathrm{P}(f))$. Hence, $\nu = m_f$ by Corollary 17.7.5(a). Consequently, $\nu_F = \nu \circ \pi^{-1} = m_f \circ \pi^{-1} = m_F$. □

Remark 19.8.15. We would like to draw the reader's attention to the fact that the uniqueness part of the proof demonstrates an even stronger reality. Namely, if a measure m supported on J satisfies (19.43) and (19.42) with an arbitrary constant P replacing the constant $\mathrm{P}(F)$, then $m \circ \pi$ is an eigenmeasure of the dual transfer operator $\mathcal{L}_f^*$ corresponding to the eigenvalue $\exp(P)$. It then follows from Corollary 17.7.5(a) that $P = \mathrm{P}(f) = \mathrm{P}(F)$, and consequently m is F-conformal.

The following result is a direct consequence of Theorem 19.8.14 and Proposition 19.7.19.

Corollary 19.8.16. *Let $\Phi = \{\varphi_e\}_{e\in E}$ be a finitely irreducible CGDMS satisfying the Boundary Separation Condition (BSC), the Strong Separation Condition (SSC) or the Cone Condition (CC). For any summable Hölder family of functions $F = \{f_e\}_{e\in E}$, the CGDMS Φ has a unique F-conformal measure m_F. Moreover,*

$$m_F = m_f \circ \pi^{-1},$$

where f is the amalgamated function induced by F and m_f is the eigenmeasure of the dual transfer operator $\mathcal{L}_f^*$.

Conformal measures have nice properties. In the next few results, we describe some of them. First, conformal measures confer a positive measure to the image of any cylinder.

Lemma 19.8.17. *Let $\Phi = \{\varphi_e\}_{e\in E}$ be a finitely irreducible, conformal-like CGDMS. Let also $F = \{f_e\}_{e\in E}$ be a summable Hölder family of functions and m_F the unique F-conformal measure. Then $m_F(\pi([\omega])) > 0$ for every $\omega \in E_A^*$.*

Proof. First, observe that

$$1 = m_F(J) = m_F(\pi(E_A^\infty)) = m_F\Big(\bigcup_{e\in E} \pi([e])\Big) \le \sum_{e\in E} m_F(\pi([e])).$$

So there exists $e_0 \in E$ for which $m_F(\pi([e_0])) > 0$.

Now, let $\Lambda \subseteq E_A^*$ be a finite set of words witnessing the finite irreducibility of Φ. Pick any $\omega \in E_A^*$. Then there is $\tau = \tau(\omega, e_0) \in \Lambda$ such that $\omega\tau e_0 \in E_A^*$. It follows that $[e_0] \subseteq E_A^{\infty,\omega\tau}$ and it ensues from the conformality of m_F (more precisely, (19.42)) that

$$m_F(\pi([\omega])) \ge m_F(\pi([\omega\tau e_0])) = m_F(\varphi_{\omega\tau}(\pi([e_0]))) = \int_{\pi([e_0])} e^{S_{\omega\tau}F-|\omega\tau|\,\mathrm{P}(F)}\, dm_F.$$

This latter term is positive since the set $\pi([e_0])$ has positive m_F-measure and the integrand is positive and bounded below on $\pi([e_0])$. Indeed, $\mathrm{P}(F) < \infty$ since F is summable while $S_{\omega\tau}F$ is bounded below over $\pi([e_0])$ because F is Hölder. □

In fact, the set of followers to any finite word has positive measure, and that measure is uniformly bounded away from zero.

Lemma 19.8.18. *Let $\Phi = \{\varphi_e\}_{e\in E}$ be a finitely irreducible, conformal-like CGDMS. Let also $F = \{f_e\}_{e\in E}$ be a summable Hölder family of functions and m_F the unique F-conformal measure. Let $\Lambda \in E_A^*$ be a finite set of words witnessing the finite irreducibility of the system and denote by Λ_E the set of letters in words from Λ. Then*

$$\inf_{\omega\in E_A^*} m_F(\pi(E_A^{\infty,\omega})) \ge \min_{e\in\Lambda_E} m_F(\pi([e])).$$

Proof. For every $\omega \in E_A^*$, there is $e_\omega \in \Lambda_E$ such that $\omega e_\omega \in E_A^*$. Then $E_A^{\infty,\omega} \supseteq [e_\omega]$ and

$$m_F(\pi(E_A^{\infty,\omega})) \geq m_F(\pi([e_\omega])) \geq \min_{e\in\Lambda_E} m_F(\pi([e])).$$

The positivity of this last term is a consequence of Lemma 19.8.17. □

We now show that the measure of the closure of the image of a union of cylinder sets that are uniformly bounded in length, is equal to the measure of the image itself.

Lemma 19.8.19. *Let $\Phi = \{\varphi_e\}_{e\in E}$ be a finitely irreducible, conformal-like CGDMS satisfying the SOSC. Let also $F = \{f_e\}_{e\in E}$ be a summable Hölder family of functions and m_F the unique F-conformal measure. For any family $W \subseteq E_A^*$ of words of uniformly bounded lengths, we have*

$$m_F\Big(\overline{\bigcup_{\omega\in W} \pi([\omega])}\Big) = m_F\Big(\bigcup_{\omega\in W} \pi([\omega])\Big).$$

Proof. Let $n < \infty$ be the maximum length of words in W and let W' be the set of words of length n that are comparable to a word in W. Then $\bigcup_{\tau\in W'} \pi([\tau]) = \bigcup_{\omega\in W} \pi([\omega])$. We can thus assume without loss of generality that all words in W have the same length n. Let $x \in J \cap \overline{\bigcup_{\omega\in W} \pi([\omega])} \setminus \bigcup_{\omega\in W} \pi([\omega])$.

On one hand, as $x \in J \setminus \bigcup_{\omega\in W} \pi([\omega])$ and $J = \bigcup_{\rho\in E_A^n} \pi([\rho])$ we know that

$$x \in \bigcup_{\tau\in E_A^n\setminus W} \pi([\tau]) \subseteq \bigcup_{\tau\in E_A^n\setminus W} \varphi_\tau(X_{t(\tau)}).$$

So $x \in \varphi_\tau(X_{t(\tau)})$ for some $\tau \in E_A^n\setminus W$. In fact, $x \notin \mathrm{Int}(\varphi_\tau(X_{t(\tau)}))$. Otherwise, there would be $\varepsilon > 0$ such that $B(x,\varepsilon) \subseteq \mathrm{Int}(\varphi_\tau(X_{t(\tau)}))$. But since $x \in \overline{\bigcup_{\omega\in W} \pi([\omega])}$ there would also be $\omega \in W$ such that $B(x,\varepsilon) \cap \pi([\omega]) \neq \emptyset$. Therefore, $\mathrm{Int}(\varphi_\tau(X_{t(\tau)})) \cap \mathrm{Int}(\varphi_\omega(X_{t(\omega)})) \neq \emptyset$. This would mean that $\tau = \omega$, which is impossible. Hence,

$$x \in \varphi_\tau(X_{t(\tau)}) \setminus \mathrm{Int}(\varphi_\tau(X_{t(\tau)})) = \partial\varphi_\tau(X_{t(\tau)}) = \varphi_\tau(\partial X_{t(\tau)}).$$

We readily deduce that

$$J \cap \overline{\bigcup_{\omega\in W} \pi([\omega])} \setminus \bigcup_{\omega\in W} \pi([\omega]) \subseteq \bigcup_{\tau\in E_A^n\setminus W} \varphi_\tau(\partial X_{t(\tau)}).$$

Consequently,

$$\mu_F\Big(\overline{\bigcup_{\omega\in W} \pi([\omega])} \setminus \bigcup_{\omega\in W} \pi([\omega])\Big) \leq \sum_{\tau\in E_A^n\setminus W} \mu_f \circ \pi^{-1}(\varphi_\tau(\partial X_{t(\tau)})) = 0$$

by Theorem 19.7.2(b). The conclusion follows from the equivalence of m_F and μ_F. □

In the same vein, we have the following property.

Lemma 19.8.20. *Let $\Phi = \{\varphi_e\}_{e\in E}$ be a finitely irreducible, conformal-like CGDMS satisfying the SOSC. Let also $F = \{f_e\}_{e\in E}$ be a summable Hölder family of functions and m_F the unique F-conformal measure. For all $\omega \in E_A^*$,*

$$m_F(\overline{\pi(E_A^{\infty,\omega})}) = m_F(\pi(E_A^{\infty,\omega})).$$

Proof. Let $x \in J \cap \overline{\pi(E_A^{\infty,\omega})} \setminus \pi(E_A^{\infty,\omega})$.

On one hand, $x = \pi(\rho)$ for some $\rho \in E_A^\infty$ with $A_{\omega_{|\omega|}\rho_1} = 0$ since $x \in J \setminus \pi(E_A^{\infty,\omega})$. Then $x \in \varphi_{\rho_1}(X_{t(\rho_1)})$.

On the other hand, $x = \lim_{n\to\infty} \pi(\omega^{(n)})$ for some $\omega^{(n)} \in E_A^{\infty,\omega}$ since $x \in \overline{\pi(E_A^{\infty,\omega})}$. Then $x \in \overline{\bigcup_{n\in\mathbb{N}} \varphi_{\omega_1^{(n)}}(X_{t(\omega_1^{(n)})})}$. Also, $A_{\omega_{|\omega|}\omega_1^{(n)}} = 1$ and thus $\omega_1^{(n)} \neq \rho_1$ for all $n \in \mathbb{N}$.

In summary, $x \in \varphi_{\rho_1}(X_{t(\rho_1)}) \cap \overline{\bigcup_{e\in E\setminus\{\rho_1\}} \varphi_e(X_{t(e)})}$. The OSC then imposes that $x \in \partial\varphi_{\rho_1}(X_{t(\rho_1)})$. Therefore,

$$J \cap \overline{\pi(E_A^{\infty,\omega})} \setminus \pi(E_A^{\infty,\omega}) \subseteq \bigcup_{e\in E} \partial\varphi_e(X_{t(e)}).$$

Thus,

$$m_F\Big(J \cap \overline{\pi(E_A^{\infty,\omega})} \setminus \pi(E_A^{\infty,\omega})\Big) \le \sum_{e\in E} m_F(\partial\varphi_e(X_{t(e)})).$$

But in view of Corollary 17.7.5(c), there exists an ergodic σ-invariant Gibbs state μ_f for f which, by Proposition 17.4.2(b), is boundedly equivalent with m_f. By Theorem 19.7.2(b), it ensues that

$$m_F(\partial\varphi_e(X_{t(e)})) = m_f \circ \pi^{-1}(\partial\varphi_e(X_{t(e)})) \asymp \mu_f \circ \pi^{-1}(\partial\varphi_e(X_{t(e)})) = 0, \quad \forall e \in E.$$

Hence,

$$m_F\Big(J \cap \overline{\pi(E_A^{\infty,\omega})} \setminus \pi(E_A^{\infty,\omega})\Big) = 0.$$

Consequently,

$$m_F(\overline{\pi(E_A^{\infty,\omega})}) = m_F(J \cap \overline{\pi(E_A^{\infty,\omega})}) = m_F(\pi(E_A^{\infty,\omega})). \qquad \square$$

As the reader may recall from Lemma 13.6.15, if μ is a Borel probability measure on a compact metric space X and $\mathrm{supp}(\mu) = X$, then

$$\inf_{x\in X} \mu(B(x,r)) > 0$$

for any $r > 0$. We now show a refinement of this property for conformal measures.

Lemma 19.8.21. *Let $\Phi = \{\varphi_e\}_{e\in E}$ be a finitely irreducible, conformal-like CGDMS satisfying the SOSC. Let also $F = \{f_e\}_{e\in E}$ be a summable Hölder family of functions and m_F the unique F-conformal measure. For every $r > 0$, $\omega \in E_A^*$ and family $W \subseteq E_A^*$ of words uniformly bounded in length:*

(a) $\inf_{x\in\bigcup_{\omega\in W}\pi([\omega])} m_F\Big(B(x,r)\cap\bigcup_{\omega\in W}\pi([\omega])\Big) > 0.$

(b) $\inf_{x\in\pi(E_A^{\infty,\omega})} m_F(B(x,r)\cap\pi(E_A^{\infty,\omega})) > 0.$

Proof. As $\overline{\bigcup_{\omega\in W}\pi([\omega])}$ is closed, and thus compact ($\bigcup_{\omega\in W}\pi([\omega])$ might not be) and as $m_F(\overline{\bigcup_{\omega\in W}\pi([\omega])}) > 0$ according to Lemma 19.8.17, it follows from Lemma 13.6.15 that

$$\inf_{x\in\overline{\bigcup_{\omega\in W}\pi([\omega])}} m_F\Big(B(x,r)\cap\overline{\bigcup_{\omega\in W}\pi([\omega])}\Big) > 0.$$

However, it ensues from Lemma 19.8.19 that

$$m_F\Big(B(x,r)\cap\overline{\bigcup_{\omega\in W}\pi([\omega])}\Big) = m_F\Big(B(x,r)\cap\bigcup_{\omega\in W}\pi([\omega])\Big), \quad \forall x\in\overline{\bigcup_{\omega\in W}\pi([\omega])}.$$

From the above two relations, we deduce that

$$\inf_{x\in\bigcup_{\omega\in W}\pi([\omega])} m_F\Big(B(x,r)\cap\bigcup_{\omega\in W}\pi([\omega])\Big) \geq \inf_{x\in\overline{\bigcup_{\omega\in W}\pi([\omega])}} m_F\Big(B(x,r)\cap\overline{\bigcup_{\omega\in W}\pi([\omega])}\Big) > 0.$$

This establishes the first statement. Using Lemmas 19.8.20 and 19.8.18, we obtain along similar lines that

$$\inf_{x\in\pi(E_A^{\infty,\omega})} m_F(B(x,r)\cap\pi(E_A^{\infty,\omega})) \geq \inf_{x\in\overline{\pi(E_A^{\infty,\omega})}} m_F(B(x,r)\cap\overline{\pi(E_A^{\infty,\omega})}) > 0. \qquad \square$$

19.8.3 Conformal measures for CGDMSs

We have already seen that infinite CGDMSs naturally break into two main classes called irregular and regular systems. This dichotomy arises from the existence or nonexistence of a zero for the pressure function or, as we shall soon see, from the existence or nonexistence of a conformal measure (cf. Definitions 13.6.12 and 19.8.11).

Definition 19.8.22. Let $\Phi = \{\varphi_e\}_{e\in E}$ be a CGDMS. A Borel probability measure m on X is said to be *t-conformal* provided that $m(J) = 1$ and the following two conditions are satisfied:

(1) For every $e\in E$ and for every Borel set $B\subseteq\pi(E_A^{\infty,e})$, we have

$$m(\varphi_e(B)) = \int_B |\varphi_e'|^t\,dm. \tag{19.49}$$

(2) For all $e_1 \neq e_2 \in E$, we have

$$m(\varphi_{e_1}(X_{t(e_1)}) \cap \varphi_{e_2}(X_{t(e_2)})) = 0. \tag{19.50}$$

Note that the conformality of a measure depends on the measure of first-level sets and subsets. By induction and using the change-of-variables formula (13.20) in Proposition 13.3.1, the conformality cascades down to all levels. Indeed, a Borel probability measure m on X is t-conformal if and only if $m(J) = 1$, for every $\omega \in E_A^*$ and for every Borel set $B \subseteq \pi(E_A^{\infty,\omega})$

$$m(\varphi_\omega(B)) = \int_B |\varphi'_\omega|^t \, dm, \tag{19.51}$$

and for all incomparable words $\omega, \tau \in E_A^*$

$$m(\varphi_\omega(X_{t(\omega)}) \cap \varphi_\tau(X_{t(\tau)})) = 0. \tag{19.52}$$

The notion of t-conformality is reminiscent of that of F-conformality (compare Definitions 19.8.22 and 19.8.11). The two concepts are related. Indeed, in light of Proposition 19.8.10 and Theorem 19.8.14, any finitely irreducible, conformal-like CGDMS $\Phi = \{\varphi_e\}_{e\in E}$ admits for every $t \in$ Fin a unique $t\,\mathrm{Log}_\Phi$-conformal measure $m_{t\,\mathrm{Log}_\Phi}$. The family $t\,\mathrm{Log}_\Phi := \{t \log |\varphi'_e|\}_{e\in E}$ has for amalgamated function the potential $t\Xi$ and $\mathrm{P}(t\,\mathrm{Log}_\Phi) = \mathrm{P}(t)$. A simple calculation reveals that the measure $m_{t\,\mathrm{Log}_\Phi}$ is characterized by the following properties:

(1) For every $e \in E$ and for every Borel set $B \subseteq \pi(E_A^{\infty,e})$,

$$m_{t\,\mathrm{Log}_\Phi}(\varphi_e(B)) = e^{-\mathrm{P}(t)} \int_B |\varphi'_e|^t \, dm_{t\,\mathrm{Log}_\Phi}. \tag{19.53}$$

(2) For all $e_1 \neq e_2 \in E$,

$$m_{t\,\mathrm{Log}_\Phi}(\varphi_{e_1}(X_{t(e_1)}) \cap \varphi_{e_2}(X_{t(e_2)})) = 0. \tag{19.54}$$

It is therefore not surprising that the existence of a t-conformal measure is equivalent to t being a zero of the pressure function (compare (19.53) with (19.49)).

Theorem 19.8.23. *Let $\Phi = \{\varphi_e\}_{e\in E}$ be a finitely irreducible, conformal-like CGDMS. Then Φ is regular if and only if it admits a t-conformal measure. If such a measure exists, then it is unique, $\mathrm{P}(t) = 0$ and the t-conformal measure m is $t\,\mathrm{Log}_\Phi$-conformal, so $m = m_{t\,\mathrm{Log}_\Phi} = m_{t\Xi} \circ \pi^{-1}$.*

Proof. Recall that regularity means the existence of a parameter t such that $\mathrm{P}(t) = 0$.

Suppose that a t-conformal measure m exists. Then by Remark 19.8.15 and Proposition 19.8.10, we know that $\mathrm{P}(t) = 0$ and m is $t\,\mathrm{Log}_\Phi$-conformal. The uniqueness and the form of m are guaranteed by Theorem 19.8.14.

It therefore remains show that if $\mathrm{P}(t) = 0$, then a t-conformal measure exists. If $\mathrm{P}(t) = 0$, then $t \in \mathrm{Fin}$. By Proposition 19.8.10 and Theorem 19.8.14, there exists a unique $t\,\mathrm{Log}_\Phi$-conformal measure. Since $\mathrm{P}(t\,\mathrm{Log}_\Phi) = \mathrm{P}(t) = 0$, this measure is t-conformal. □

An immediate consequence of Theorem 19.8.23 and Proposition 19.7.19 is the following.

Corollary 19.8.24. *Let $\Phi = \{\varphi_e\}_{e\in E}$ be a finitely irreducible CGDMS satisfying the BSC, the SSC or the CC condition. Then Φ is regular if and only if it admits a t-conformal measure. If such a measure exists, then it is unique, $\mathrm{P}(t) = 0$ and the t-conformal measure m is $t\,\mathrm{Log}_\Phi$-conformal, so $m = m_{t\,\mathrm{Log}_\Phi} = m_{t\Xi} \circ \pi^{-1}$.*

For a general measure m and map f, the composition $m \circ f$ is not a measure even if m is. Nevertheless, by (19.50), we have the following.

Lemma 19.8.25. *Let $\Phi = \{\varphi_e\}_{e\in E}$ be a finitely irreducible CGDMS. If $t \geq 0$ and m is a t-conformal measure, then $m \circ \pi : E_A^\infty \to [0,1]$ is a measure such that $m = (m \circ \pi) \circ \pi^{-1}$.*

Proof. Because of the continuity of the map π, we know that $\pi(W)$ is a Lebesgue measurable set in X for any Borel set $W \subseteq E_A^\infty$. Let W_1 and W_2 be two disjoint Borel subsets of E_A^∞. If $x \in \pi(W_1) \cap \pi(W_2)$, then there are $w^{(1)} \in W_1$ and $w^{(2)} \in W_2$ such that $x = \pi(w^{(1)}) = \pi(w^{(2)})$. Let $n = |w^{(1)} \wedge w^{(2)}| + 1$. Then $x \in \varphi_{w^{(1)}|_n}(X_{t(w^{(1)}_n)}) \cap \varphi_{w^{(2)}|_n}(X_{t(w^{(2)}_n)})$. We infer from this that

$$\pi(W_1) \cap \pi(W_2) \subseteq \bigcup_{\substack{\omega,\tau\in E_A^* \\ \omega,\tau \text{ incomparable}}} \varphi_\tau(X_{t(\tau)}) \cap \varphi_\omega(X_{t(\omega)}).$$

By (19.52), we deduce that

$$m(\pi(W_1) \cap \pi(W_2)) \leq \sum_{\substack{\omega,\tau\in E_A^* \\ \omega,\tau \text{ incomparable}}} m(\varphi_\tau(X_{t(\tau)}) \cap \varphi_\omega(X_{t(\omega)})) = 0.$$

Consequently,

$$m \circ \pi(W_1 \cup W_2) = m \circ \pi(W_1) + m \circ \pi(W_2).$$

This argument easily generalizes to a sequence of disjoint sets $(W_k)_{k=1}^\infty$. So $m \circ \pi$ is a measure. □

We now uncover a counterpart of Lemma 13.6.13.

Lemma 19.8.26. *Let $\Phi = \{\varphi_e\}_{e\in E}$ be a finitely irreducible CGDMS. If m is a t-conformal measure for some $t \geq 0$, then $\mathrm{P}(t) = 0$ and $\mathcal{L}_t^*(m\circ\pi)(g\circ\pi) = m(g)$ for all $g \in C(X)$, where $\mathcal{L}_t := \mathcal{L}_{t\Xi}$.*

Proof. We have already established that $\mathrm{P}(t) = 0$ in Theorem 19.8.23 and the conformal-likeness was not used for this. A more direct way to prove this fact is, using the definition of t-conformality and the BDP (Corollary 19.3.5), to observe that

$$1 \leq \sum_{\omega\in E_A^n} \|\varphi_\omega'\|_{X_{t(\omega)}}^t \leq K^t, \quad \forall n \in \mathbb{N}.$$

So $\mathrm{P}(t) = 0$.

Now, let $g \in C(X)$. For every $e \in E$ and every Borel set $B \subseteq \pi(E_A^{\infty,e})$, it follows from (19.49) that

$$\int_{\varphi_e(B)} g\,dm = \int_B (g\circ\varphi_e)|\varphi_e'|^t\,dm. \tag{19.55}$$

It is clear that $\pi^{-1}(\pi(E_A^{\infty,e})) \supseteq E_A^{\infty,e}$. Moreover, if $\tau \in \pi^{-1}(\pi(E_A^{\infty,e})) \setminus E_A^{\infty,e}$ then $\pi(\tau) \in \pi(E_A^{\infty,e})$ and $A_{e\tau_1} = 0$, i. e., there is $\omega \in E_A^{\infty,e}$ such that $\pi(\tau) = \pi(\omega)$ and $A_{e\tau_1} = 0$. This means that $\pi(\tau) = \pi(\omega)$ with $A_{e\omega_1} = 1$ and $A_{e\tau_1} = 0$. So $\tau_1 \neq \omega_1$, and thus $\pi(\tau) = \pi(\omega) \in \varphi_{\tau_1}(X_{t(\tau_1)}) \cap \varphi_{\omega_1}(X_{t(\omega_1)})$. By (19.50), we deduce that

$$m\circ\pi(\pi^{-1}(\pi(E_A^{\infty,e})) \setminus E_A^{\infty,e}) \leq \sum_{a\neq b\in E} m(\varphi_a(X_{t(a)}) \cap \varphi_b(X_{t(b)})) = 0. \tag{19.56}$$

Using the t-conformality of m as well as (19.55) and (19.56), we obtain that

$$\begin{aligned}
m(g) &= \int_J g\,dm \\
&= \int_{\bigcup_{e\in E}\varphi_e(\pi(E_A^{\infty,e}))} g\,dm \\
&= \sum_{e\in E} \int_{\varphi_e(\pi(E_A^{\infty,e}))} g\,dm \\
&= \sum_{e\in E} \int_{\pi(E_A^{\infty,e})} g\circ\varphi_e(x)|\varphi_e'(x)|^t\,dm(x) \\
&= \sum_{e\in E} \int_{\pi^{-1}(\pi(E_A^{\infty,e}))} g\circ\varphi_e\circ\pi(\omega)|\varphi_e'\circ\pi(\omega)|^t\,dm\circ\pi(\omega)
\end{aligned}$$

$$= \sum_{e \in E} \int_{E_A^{\infty,e}} g \circ \varphi_e \circ \pi(\omega) |\varphi_e' \circ \pi(\omega)|^t \, dm \circ \pi(\omega)$$
$$= \int_{E_A^\infty} \sum_{e \in E: A_{e\omega_1}=1} g \circ \varphi_e \circ \pi(\omega) |\varphi_e' \circ \pi(\omega)|^t \, dm \circ \pi(\omega)$$
$$= \int_{E_A^\infty} \sum_{e \in E: A_{e\omega_1}=1} g \circ \pi(e\omega) \exp(t\Xi(e\omega)) \, dm \circ \pi(\omega)$$
$$= \int_{E_A^\infty} \mathcal{L}_{t\Xi}(g \circ \pi) \, dm \circ \pi$$
$$= \mathcal{L}_t^*(m \circ \pi)(g \circ \pi).$$ □

For CGDSs (conformal graph directed systems, i. e., when $A_{ab} = 1$ if and only if $t(a) = i(b)$), it is possible to define a transfer operator directly on the phase space $X := \bigsqcup_{v \in V} X_v$, namely $\widetilde{\mathcal{L}}_t : C(X) \to C(X)$, where

$$\widetilde{\mathcal{L}}_t(g)(x) := \sum_{e \in E: t(e)=v} g(\varphi_e(x)) |\varphi_e'(x)|^t, \quad \forall g \in C(X), \ \forall x \in X_v, \ \forall v \in V.$$

A closer analogue of Lemma 13.6.13 holds.

Lemma 19.8.27. *Let $\Phi = \{\varphi_e\}_{e \in E}$ be a CGDS. If m is a t-conformal measure for some $t \geq 0$, then $\mathrm{P}(t) = 0$ and $\widetilde{\mathcal{L}}_t^*(m) = m$.*

Proof. The proof is a little simpler than that of Lemma 19.8.26:

$$m(g) = \int_X g \, dm = \int_{\bigcup_{e \in E} \varphi_e(X_{t(e)})} g \, dm$$
$$= \sum_{e \in E} \int_{\varphi_e(X_{t(e)})} g \, dm = \sum_{e \in E} \int_{X_{t(e)}} g \circ \varphi_e |\varphi_e'|^t \, dm$$
$$= \sum_{v \in V} \sum_{e \in E: t(e)=v} \int_{X_v} g \circ \varphi_e |\varphi_e'|^t \, dm$$
$$= \sum_{v \in V} \int_{X_v} \sum_{e \in E: t(e)=v} g \circ \varphi_e |\varphi_e'|^t \, dm$$
$$= \int_X \widetilde{\mathcal{L}}_t(g) \, dm = \widetilde{\mathcal{L}}_t^*(m)(g).$$ □

19.8.4 Dimensions of measures for CGDMSs

The argument given to establish (19.43) in the proof of Theorem 19.8.14 also yields the following remarkable fact, which can be seen as a measure-theoretic open set condition.

Theorem 19.8.28. *Let $\Phi = \{\varphi_e\}_{e\in E}$ be a finitely irreducible, conformal-like CGDMS. If μ is a σ-invariant Borel probability measure on E_A^∞, then*

$$\mu \circ \pi^{-1}(\varphi_\omega(X_{t(\omega)}) \cap \varphi_\tau(X_{t(\tau)})) = 0 \tag{19.57}$$

for all incomparable words $\omega, \tau \in E_A^*$.

In turn, we observe the following interesting fact.

Corollary 19.8.29. *Let $\Phi = \{\varphi_e\}_{e\in E}$ be a finitely irreducible, conformal-like CGDMS. If μ is a σ-invariant Borel probability measure on E_A^∞, then*

$$\mu \circ \pi^{-1}(\varphi_\omega(X_{t(\omega)})) = \mu([\omega]), \quad \forall \omega \in E_A^*. \tag{19.58}$$

Proof. Clearly, $\pi^{-1}(\varphi_\omega(X_{t(\omega)})) \supseteq [\omega]$ and thus $\mu\circ\pi^{-1}(\varphi_\omega(X_{t(\omega)})) \geq \mu([\omega])$ for all $\omega \in E_A^*$. Suppose that $\mu \circ \pi^{-1}(\varphi_\tau(X_{t(\tau)})) > \mu([\tau])$ for some $\tau \in E_A^*$. Then $\mu \circ \pi^{-1}(\varphi_\tau(X_{t(\tau)})) = \mu([\tau]) + \varepsilon$ for some $\varepsilon > 0$. Using Theorem 19.8.28, we get

$$\begin{aligned} 1 = \mu\circ\pi^{-1}(J) &= \mu\circ\pi^{-1}\Bigg(\bigcup_{\omega\in E_A^{|\tau|}} \varphi_\omega(X_{t(\omega)})\Bigg) = \sum_{\omega\in E_A^{|\tau|}} \mu\circ\pi^{-1}(\varphi_\omega(X_{t(\omega)})) \\ &\geq \sum_{\omega\in E_A^{|\tau|}\setminus\{\tau\}} \mu([\omega]) + \mu([\tau]) + \varepsilon = \mu\Bigg(\bigcup_{\omega\in E_A^{|\tau|}} [\omega]\Bigg) + \varepsilon = \mu(E_A^\infty) + \varepsilon = 1 + \varepsilon. \end{aligned}$$

This contradiction shows that $\mu \circ \pi^{-1}(\varphi_\omega(X_{t(\omega)})) = \mu([\omega])$ for all $\omega \in E_A^*$. □

As before, the partition of E_A^∞ into its initial cylinders of length 1 is denoted by $\alpha = \{[e] : e \in E\}$. If μ is a σ-invariant Borel probability measure on E_A^∞, then the entropy of the partition α with respect to μ is denoted by $\mathrm{H}_\mu(\alpha)$, the entropy of μ with respect to the shift map $\sigma : E_A^\infty \to E_A^\infty$ by $\mathrm{h}_\mu(\sigma)$ and its (*characteristic*) *Lyapunov exponent* (cf. Definition 16.5.1) by

$$\chi_\mu(\sigma) := -\int_{E_A^\infty} \Xi_\Phi \, d\mu > 0,$$

where

$$\Xi(\omega) := \log|\varphi'_{\omega_1}(\pi(\sigma(\omega)))|$$

is the potential at the heart of the definition of the topological pressure of a CGDMS (see Section 19.4). This potential is also the amalgamated function induced by the family Log_Φ.

Recall from Definition 15.6.1 that the Hausdorff dimension $\mathrm{HD}(\mu\circ\pi^{-1})$ is the infimum of the Hausdorff dimensions of the sets of full $\mu\circ\pi^{-1}$-measure.

We now derive a volume lemma for measures associated with CGDMSs (cf. Lemma 16.5.2).

Theorem 19.8.30 (Volume lemma). *Let $\Phi = \{\varphi_e\}_{e\in E}$ be a finitely irreducible, conformal-like CGDMS. If μ is an ergodic σ-invariant Borel probability measure on E_A^∞ such that $\chi_\mu(\sigma) < \infty$, then*

$$\mathrm{HD}(\mu\circ\pi^{-1}) = \frac{\mathrm{h}_\mu(\sigma)}{\chi_\mu(\sigma)}.$$

Proof. By hypothesis, $\Xi \in L^1(\mu)$. Let $\alpha = \{[e] : e \in E\}$. Assume first that $\mathrm{H}_\mu(\alpha) < \infty$. Since α is a generating partition, we have that $\mathrm{h}_\mu(\sigma) = \mathrm{h}_\mu(\sigma,\alpha) \le \mathrm{H}_\mu(\alpha) < \infty$. Thus, in view of the ergodic case of Birkhoff's ergodic theorem (Corollary 8.2.14) and of Shannon–McMillan–Breiman's theorem (Corollary 9.5.5) there exists a set $Z \subseteq E_A^\infty$ such that $\mu(Z) = 1$ and for all $\omega \in Z$,

$$\lim_{n\to\infty}\frac{1}{n}\sum_{j=0}^{n-1}\Xi\circ\sigma^j(\omega) = -\chi_\mu(\sigma) \quad\text{and}\quad \lim_{n\to\infty}\frac{\log\mu([\omega|_n])}{n} = -\mathrm{h}_\mu(\sigma). \tag{19.59}$$

Fix $\omega \in Z$ and $0 < \varepsilon < \chi_\mu(\sigma)$. For each $r > 0$, let $n = n(\omega, r) \in \mathbb{N}$ be the least integer n such that $\varphi_{\omega|_n}(X_{t(\omega_n)}) \subseteq B(\pi(\omega), r)$. By (19.59), for every $r > 0$ small enough (which implies that $n = n(\omega, r)$ is large enough) we then have

$$\log\mu\circ\pi^{-1}(B(\pi(\omega),r)) \ge \log\mu\circ\pi^{-1}(\varphi_{\omega|_n}(X_{t(\omega_n)})) \ge \log\mu([\omega|_n]) \ge -(\mathrm{h}_\mu(\sigma)+\varepsilon)n$$

and

$$\mathrm{diam}(\varphi_{\omega|_{n-1}}(X_{t(\omega_{n-1})})) \ge r.$$

This last inequality, a slight adjustment to (19.9) in Lemma 19.3.9, and (19.59) imply that

$$\begin{aligned}\log r &\le \log\mathrm{diam}(\varphi_{\omega|_{n-1}}(X_{t(\omega_{n-1})})) \le \log(KD|\varphi'_{\omega|_{n-1}}(\pi(\sigma^{n-1}(\omega)))|)\\ &\le \log(KD) + \sum_{j=1}^{n-1}\log|\varphi'_{\omega_j}(\pi(\sigma^j(\omega)))| \le \log(KD) + (n-1)(-\chi_\mu(\sigma)+\varepsilon)\end{aligned}$$

for all $r > 0$ small enough (and corresponding n). Therefore, for r small enough we get

$$\frac{\log\mu\circ\pi^{-1}(B(\pi(\omega),r))}{\log r} \le \frac{-(\mathrm{h}_\mu(\sigma)+\varepsilon)n}{\log(KD)+(n-1)(-\chi_\mu(\sigma)+\varepsilon)} = \frac{\mathrm{h}_\mu(\sigma)+\varepsilon}{\frac{-\log(KD)}{n}+\frac{n-1}{n}(\chi_\mu(\sigma)-\varepsilon)}.$$

Letting $r \to 0$ (and consequently $n \to \infty$), the upper local dimension of $\mu \circ \pi^{-1}$ at $\pi(\omega)$ has for upper bound

$$\overline{d}_{\mu\circ\pi^{-1}}(\pi(\omega)) := \limsup_{r\to 0} \frac{\log \mu \circ \pi^{-1}(B(\pi(\omega), r))}{\log r} \leq \frac{\mathrm{h}_\mu(\sigma) + \varepsilon}{\chi_\mu(\sigma) - \varepsilon}.$$

As $\varepsilon > 0$ is arbitrary, we deduce that

$$\overline{d}_{\mu\circ\pi^{-1}}(\pi(\omega)) \leq \frac{\mathrm{h}_\mu(\sigma)}{\chi_\mu(\sigma)}, \quad \forall \omega \in Z.$$

Theorem 15.6.3(b) then asserts that $\mathrm{HD}(\pi(Z)) \leq \mathrm{h}_\mu(\sigma)/\chi_\mu(\sigma)$. As $\mu \circ \pi^{-1}(\pi(Z)) = 1$, we conclude that $\mathrm{HD}(\mu \circ \pi^{-1}) \leq \mathrm{h}_\mu(\sigma)/\chi_\mu(\sigma)$. If $\mathrm{h}_\mu(\sigma) = 0$, then this inequality suffices to establish the theorem.

To demonstrate the opposite inequality when $\mathrm{h}_\mu(\sigma) > 0$, let $J_1 \subseteq J$ be a Borel set such that $\mu \circ \pi^{-1}(J_1) > 0$. Fix $0 < \varepsilon < \mathrm{h}_\mu(\sigma)$. In view of (19.59) and Egorov's theorem (Theorem A.1.45), there exist $n_0 \in \mathbb{N}$ and a Borel set $\widetilde{J}_1 \subseteq \pi^{-1}(J_1)$ such that $\mu(\widetilde{J}_1) > \mu(\pi^{-1}(J_1))/2 > 0$,

$$\mu([\omega|_n]) \leq \exp((-\mathrm{h}_\mu(\sigma) + \varepsilon)n) \tag{19.60}$$

and $|\varphi'_{\omega|_n}(\pi(\sigma^n(\omega)))| \geq \exp((-\chi_\mu(\sigma)-\varepsilon)n)$ for all $\omega \in \widetilde{J}_1$ and all $n \geq n_0$. By Lemma 19.3.11, this last inequality implies that there exists $n_1 \geq n_0$ such that

$$\mathrm{diam}(\varphi_{\omega|_n}(X_{t(\omega_n)})) \geq D^{-1} e^{(-\chi_\mu(\sigma)-\varepsilon)n} \geq e^{-(\chi_\mu(\sigma)+2\varepsilon)n} \tag{19.61}$$

for all $n \geq n_1$ and all $\omega \in \widetilde{J}_1$. Given $0 < r < \exp(-(\chi_\mu(\sigma) + 2\varepsilon)n_1)$ and $\omega \in \widetilde{J}_1$, let $n(\omega, r)$ be the least integer n such that

$$\mathrm{diam}(\varphi_{\omega|_{n+1}}(X_{t(\omega_{n+1})})) < r. \tag{19.62}$$

Using (19.61), we know that $n(\omega, r) \geq n_1$ and

$$\mathrm{diam}(\varphi_{\omega|_{n(\omega,r)}}(X_{t(\omega_{n(\omega,r)})})) \geq r. \tag{19.63}$$

By virtue of Lemma 19.7.16 (with $\kappa = 1$), there is a universal constant $L \geq 1$ such that for every $\omega \in \widetilde{J}_1$ and $0 < r < \exp(-(\chi_\mu(\sigma) + 2\varepsilon)n_1)$ there exist words $\omega^{(1)}, \ldots, \omega^{(k)} \in \widetilde{J}_1$, where $k = k(\omega) \leq L$, such that

$$\pi(\widetilde{J}_1) \cap B(\pi(\omega), r) \subseteq \bigcup_{j=1}^{k} \varphi_{\omega^{(j)}|_{n(\omega^{(j)},r)}}(X_{t(\omega^{(j)}|_{n(\omega^{(j)},r)})}). \tag{19.64}$$

Let $\widetilde{\mu} := \mu|_{\widetilde{J}_1}$ be the restriction of the measure μ to the set $\widetilde{J}_1$. Using successively (19.64), (19.58), (19.60), (19.61) and (19.62) we get

$$\begin{aligned}
\widetilde{\mu} \circ \pi^{-1}(B(\pi(\omega), r)) &= \mu(\widetilde{J}_1 \cap \pi^{-1}(B(\pi(\omega), r))) \\
&\le \mu \circ \pi^{-1}(\pi(\widetilde{J}_1) \cap B(\pi(\omega), r)) \\
&\le \sum_{j=1}^{k} \mu \circ \pi^{-1}(\varphi_{\omega^{(j)}|_{n(\omega^{(j)},r)}}(X_{t(\omega^{(j)}|_{n(\omega^{(j)},r)})})) \\
&= \sum_{j=1}^{k} \mu([\omega^{(j)}|_{n(\omega^{(j)},r)}]) \le \sum_{j=1}^{k} \exp((-\mathrm{h}_\mu(\sigma) + \varepsilon) n(\omega^{(j)}, r)) \\
&= \sum_{j=1}^{k} \exp\left(-(\chi_\mu(\sigma) + 2\varepsilon)(n(\omega^{(j)}, r) + 1) \cdot \frac{n(\omega^{(j)}, r)}{n(\omega^{(j)}, r) + 1} \cdot \frac{-\mathrm{h}_\mu(\sigma) + \varepsilon}{-(\chi_\mu(\sigma) + 2\varepsilon)} \right) \\
&\le \sum_{j=1}^{k} \left(\operatorname{diam}(\varphi_{\omega^{(j)}|_{n(\omega^{(j)},r)+1}}(X_{t(\omega^{(j)}|_{n(\omega^{(j)},r)+1})})) \right)^{\frac{n(\omega^{(j)},r)}{n(\omega^{(j)},r)+1} \cdot \frac{\mathrm{h}_\mu(\sigma)-\varepsilon}{\chi_\mu(\sigma)+2\varepsilon}} \\
&\le \sum_{j=1}^{k} r^{\frac{n(\omega^{(j)},r)}{n(\omega^{(j)},r)+1} \cdot \frac{\mathrm{h}_\mu(\sigma)-\varepsilon}{\chi_\mu(\sigma)+2\varepsilon}} \le L r^{\frac{\mathrm{h}_\mu(\sigma)-2\varepsilon}{\chi_\mu(\sigma)+2\varepsilon}},
\end{aligned}$$

where the last inequality holds assuming n_1 to be so that $\frac{n_1}{n_1+1}(\mathrm{h}_\mu(\sigma) - \varepsilon) \ge \mathrm{h}_\mu(\sigma) - 2\varepsilon$. Consequently,

$$\frac{\log \widetilde{\mu} \circ \pi^{-1}(B(\pi(\omega), r))}{\log r} \ge \frac{\log L + \frac{\mathrm{h}_\mu(\sigma)-2\varepsilon}{\chi_\mu(\sigma)+2\varepsilon} \log r}{\log r} = \frac{\log L}{\log r} + \frac{\mathrm{h}_\mu(\sigma) - 2\varepsilon}{\chi_\mu(\sigma) + 2\varepsilon}.$$

Letting $r \to 0$, the lower local dimension of $\widetilde{\mu} \circ \pi^{-1}$ at $\pi(\omega)$ has for lower bound

$$\underline{d}_{\widetilde{\mu}\circ\pi^{-1}}(\pi(\omega)) := \liminf_{r\to 0} \frac{\log \widetilde{\mu} \circ \pi^{-1}(B(\pi(\omega), r))}{\log r} \ge \frac{\mathrm{h}_\mu(\sigma) - 2\varepsilon}{\chi_\mu(\sigma) + 2\varepsilon}.$$

Letting $\varepsilon \to 0$, we deduce that

$$\underline{d}_{\widetilde{\mu}\circ\pi^{-1}}(\pi(\omega)) \ge \frac{\mathrm{h}_\mu(\sigma)}{\chi_\mu(\sigma)}, \quad \forall \omega \in \widetilde{J}_1.$$

Theorem 15.6.3(a) then affirms that $\mathrm{HD}(J_1) \ge \mathrm{HD}(\pi(\widetilde{J}_1)) \ge \mathrm{h}_\mu(\sigma)/\chi_\mu(\sigma)$. Since J_1 is any Borel subset of J such that $\mu \circ \pi^{-1}(J_1) > 0$, we conclude that $\mathrm{HD}(\mu \circ \pi^{-1}) \ge \mathrm{h}_\mu(\sigma)/\chi_\mu(\sigma)$ and the proof is complete in the case where $\mathrm{H}_\mu(\alpha) < \infty$.

If $\mathrm{H}_\mu(\alpha) = \infty$, then the above considerations imply that $\mathrm{HD}(\mu \circ \pi^{-1}) = \infty$, which is impossible. □

A direct consequence of Theorem 19.8.30 (volume lemma) and Proposition 19.7.19 is the following.

Corollary 19.8.31 (Volume lemma). *Let $\Phi = \{\varphi_e\}_{e\in E}$ be a finitely irreducible CGDMS satisfying the BSC, the SSC or the CC condition. If μ is an ergodic σ-invariant Borel probability measure on E_A^∞ such that $\chi_\mu(\sigma) < \infty$, then*

$$\mathrm{HD}(\mu \circ \pi^{-1}) = \frac{\mathrm{h}_\mu(\sigma)}{\chi_\mu(\sigma)}.$$

19.9 Examples

Examples of CGDMSs abound. It is easy to construct them and the flexibility of some constructions can be enormous.

Example 19.9.1. Fix $d \in \mathbb{N}$. Let V be a finite set and $E = \mathbb{N}$. Choose a collection $\{X_v\}_{v\in V}$ of mutually disjoint compact subsets of $\mathbb{R}^d$ satisfying $X_v = \overline{\mathrm{Int}_{\mathbb{R}^d}(X_v)}$. Pick a collection $\{W_v\}_{v\in V}$ of mutually disjoint, open, connected and bounded subsets of $\mathbb{R}^d$ such that $X_v \subseteq W_v$ for all $v \in V$. Fix any countable set

$$\xi := \{\xi_n\}_{n\in\mathbb{N}} \subseteq \bigcup_{v\in V} X_v$$

such that for every $v \in V$ the set $X_v \cap \xi$ is discrete; it is straightforward to construct such a set inductively. For every $n \in \mathbb{N}$, let $u_n \in V$ be such that $\xi_n \in X_{u_n}$ and also select arbitrarily $v_n \in V$. Proceeding inductively with a different induction process, suppose that for some $n \in \mathbb{N}$ a closed ball $\overline{B(\xi_n, r_n)}$ has been chosen such that

$$\overline{B(\xi_n, 2r_n)} \cap \left(\bigcup_{k=1}^{n-1} \overline{B(\xi_k, 2r_k)} \cup \{\xi_\ell : \ell \geq n+1\} \right) = \emptyset.$$

Choose any $r_{n+1} > 0$ such that

$$\overline{B(\xi_{n+1}, 2r_{n+1})} \cap \left(\bigcup_{k=1}^{n} \overline{B(\xi_k, 2r_k)} \cup \{\xi_\ell : \ell \geq n+2\} \right) = \emptyset.$$

Then for every $n \in \mathbb{N}$ choose any conformal map

$$\varphi_n : W_{v_n} \to W_{u_n}$$

such that

$$\varphi_n(X_{v_n}) \subset X_{u_n} \quad \text{and} \quad \|\varphi_n'\|_{W_{v_n}} \leq 1/2.$$

For instance, the conformal map φ_n could simply be affine

$$\mathbb{R}^d \ni x \longmapsto a_n x + b_n,$$

with $a_n \in (0,1/2)$ and $b_n \in \mathbb{R}^d$ judiciously chosen. Then the maps $\{\varphi_n\}_{n\in\mathbb{N}}$ form a conformal IFS satisfying the Strong Open Set Condition (SOSC).

In the next few examples, we describe systems with specific features or of special interest.

Example 19.9.2. The limit set J is an $F_{\sigma\delta}$ but not a G_δ.

Denote by Q the set of all rational numbers in $[0,1]$. Let $X = [0,1] \times [0,1]$ and $\Delta = \{(x,x) : x \in [0,1]\}$ be the diagonal of X. Let $E = \{-1\} \cup Q$. Consider a conformal IFS $\{\varphi_e : X \to X\}_{e\in E}$ consisting of linear mappings such that:

(a) $\varphi_e(X) \cap \Delta = \{\varphi_e(0,1)\} = \{(e,e)\}$ for all $e \in Q$;
(b) $\varphi_{-1}(x,y) = (x/2, (y+1)/2)$;
(c) The sets $\varphi_e(X)$, $e \in E$, are mutually disjoint.

Then $J \cap \Delta = Q$ is not a G_δ–set, so neither is J. Note also that this system is pointwise finite but not locally finite (see Definition 19.1.2 and Exercise 19.10.18).

Example 19.9.3. An irregular similarity IFS.

Let

$$E = \{(n,k) : n \in \mathbb{N} \text{ and } 1 \le k \le 2^{n^2-1}\}.$$

Let $X = [0,1]$ and let $\Phi = \{\varphi_{n,k} : X \to X \mid (n,k) \in E\}$ be a system consisting of similarity maps $\varphi_{n,k}$ such that

$$\|\varphi'_{n,k}\| = 2^{-(n^2+n)}$$

and such that the intervals $\varphi_{n,k}(X)$ are mutually disjoint. This last requirement can be satisfied since

$$Z_1(1) = \sum_{(n,k)\in E} \|\varphi'_{n,k}\| = \sum_{n=1}^{\infty} 2^{n^2-1}2^{-(n^2+n)} = 1/2 < 1.$$

As the generators are similarities, this computation further reveals that $\mathrm{P}(1) = \log(1/2) < 0$. On the other hand, observe that

$$Z_1(t) = \sum_{n=1}^{\infty} 2^{n^2-1}2^{-(n^2+n)t} = \sum_{n=1}^{\infty} 2^{n^2(1-t)-nt-1} = \infty$$

for all $0 \le t < 1$. Thus, Φ is irregular and $\mathrm{HD}(J) = h = \theta = 1$ by Bowen's formula (Theorem 19.6.4 or Corollary 19.6.5).

Example 19.9.4. A critically regular similarity IFS.

This example is very similar to Example 19.9.3. The only difference in its construction is that we now take

$$E = \{(n,k) : n \in \mathbb{N} \text{ and } 1 \le k \le 2^{n^2}\}.$$

Then the same computations as in Example 19.9.3 show that $Z_1(1) = 1$, whence $\mathrm{P}(1) = 0$, and $Z_1(t) = \infty$ for all $0 \le t < 1$. Hence, Φ is critically regular, the only conformal measure is the Lebesgue measure on $[0,1]$, and $\mathrm{HD}(J) = h = \theta = 1$.

Example 19.9.5. Real continued fractions.

This conformal IFS (or CIFS), call it $\mathcal{G}$, is given by the maps $\varphi_n : [0,1] \to [0,1]$, $n \in \mathbb{N}$, defined by the formula

$$\varphi_n(x) = \frac{1}{x+n}.$$

It is easy to see that the corresponding coding map $\pi : \mathbb{N}^\infty \to [0,1]$ is given by the continued fractions formula

$$\pi(\omega) = \cfrac{1}{\omega_1 + \cfrac{1}{\omega_2 + \cfrac{1}{\omega_3 + \cdots}}} \tag{19.65}$$

and that its range, i. e., the limit set $J_{\mathcal{G}}$ of the system $\mathcal{G}$ is the set of all irrational numbers in $[0,1]$. In particular, $\mathrm{HD}(J_{\mathcal{G}}) = 1$ and the Lebesgue measure on $\mathbb{R}$ restricted to $J_{\mathcal{G}}$ is the 1-conformal measure for the system $\mathcal{G}$. Furthermore, since

$$|\varphi_n'(x)| = \frac{1}{(x+n)^2},$$

it is straightforward to see that $Z_1(t) < \infty$ for all $t > 1/2$ and $Z_1(t) = \infty$ for all $t \in [0,1/2]$. Thus, the finiteness parameter $\theta_{\mathcal{G}}$ is equal to $1/2$ and the system $\mathcal{G}$ is cofinitely regular. All theorems proved in this chapter for CGDMSs apply to $\mathcal{G}$ and all its subsystems.

Note that the maps $\varphi_n : [0,1] \to [0,1]$, $n \in \mathbb{N}$, are the continuous inverse branches of the Gauss map $G : [0,1) \to [0,1)$ given by the formula

$$G(x) := \frac{1}{x} - \left[\frac{1}{x}\right].$$

Its G-invariant Borel probability measure ν_G, absolutely continuous with respect to the Lebesgue measure on $\mathbb{R}$ restricted to J_G, is given by the formula

$$\nu_G(B) = \frac{1}{\log 2} \int_B \frac{1}{1+x}\, dx. \tag{19.66}$$

This measure is known as *Gauss measure*. It is the projection under the coding map π of the σ-invariant Gibbs state $\mu_{\mathcal{G}}$ of the system $\mathcal{G}$ corresponding to the potential

$$\mathbb{N}^\infty \ni \omega \longmapsto \Xi(\omega) := \log|\varphi'_{\omega_1}(\pi(\sigma(\omega)))| \in \mathbb{R},$$

i. e. $\nu_G = \mu_{\mathcal{G}} \circ \pi^{-1}$.

According to Theorem 19.7.4, the coding map $\pi : \mathbb{N}^\infty \to [0,1]$ is a measure-preserving isomorphism establishing a measure-theoretic conjugacy between the shift map $(\sigma : \mathbb{N}^\infty \to \mathbb{N}^\infty, \mu_{\mathcal{G}})$ and the Gauss map $(\tilde{f} = G : [0,1] \to [0,1], \mu_{\mathcal{G}} \circ \pi^{-1} = \nu_G)$. As an immediate consequence of this conjugacy and of Theorem 17.4.6, we deduce the following.

Theorem 19.9.6. *The Gauss measure ν_G, which is equivalent to the Lebesgue measure on $\mathbb{R}$ restricted to $J_{\mathcal{G}}$, is ergodic with respect to the Gauss map G.*

Moreover, all theorems proved in Chapter 17 for Gibbs measures and finitely primitive subshifts apply to $\mu_{\mathcal{G}}$ and $\mathcal{G}$.

Today, quite a lot is known about the CIFS $\mathcal{G}$, its subsystems, and the fractal geometry of their limit sets. It is impossible to cite all the relevant works here. We list only a few: [16, 17, 23, 49, 50, 58, 59, 76, 136].

Example 19.9.7. Complex continued fractions.

Let

$$E = \{m + ni \mid (m,n) \in \mathbb{N} \times \mathbb{Z}\}, \quad X = \overline{B(1/2, 1/2)} \subseteq \mathbb{C} \quad \text{and} \quad W = B(1/2, 3/4) \subseteq \mathbb{C}.$$

For every $e \in E$, define the map $\varphi_e : W \to W$ by

$$\varphi_e(z) := \frac{1}{e + z}.$$

Then $\varphi_e(X) \subseteq X$ for all $e \in E$ and the system $\mathbb{C}\mathcal{G} := \{\varphi_e\}_{e \in E}$ is a conformal IFS. It is again easy to see that the corresponding coding map $\pi : E^\infty \to X$ is given by the complex version of the continued fractions formula (19.65). Furthermore, since

$$|\varphi'_e(z)| = \frac{1}{|z + e|^2},$$

it is straightforward to see that $Z_1(t) < \infty$ for all $t > 1$ and $Z_1(t) = \infty$ for all $t \in [0,1]$. Thus, the finiteness parameter $\theta_{\mathbb{C}\mathcal{G}}$ is equal to 1 and the system $\mathbb{C}\mathcal{G}$, like its real counterpart, is cofinitely regular. In consequence, $\mathrm{HD}(J_{\mathbb{C}\mathcal{G}}) = h_{\mathbb{C}\mathcal{G}} > \theta_{\mathbb{C}\mathcal{G}} = 1$. All theorems proved in this chapter for CGDMSs apply to $\mathbb{C}\mathcal{G}$ and all its subsystems.

As for real continued fractions, a lot is known about complex continued fractions; see [16, 17, 31, 75].

19.10 Exercises

Exercise 19.10.1. Show that the unit circle $\mathbb{S}^1 \subseteq \mathbb{R}^2$ is not the limit set of any conformal IFS.

Exercise 19.10.2. Prove Lemma 19.1.3.

Exercise 19.10.3. Prove that if Φ is a conformal IFS, then

$$\{\mathrm{HD}(J_F) : F \subseteq E, \#F < \infty\} \supseteq [0, \theta].$$

(See [75].)

Exercise 19.10.4. Show that if Φ is a CGDMS, then the set

$$\{\mathrm{HD}(J_F) : F \subseteq E, \#F < \infty\}$$

is compact. (See [16, 75].)

Exercise 19.10.5. Build a CGDMS which satisfies the Strong Separation Condition (SSC) but not the Strong Open Set Condition (SOSC).

Exercise 19.10.6. Prove Lemma 19.7.7.

Exercise 19.10.7. Prove Theorem 19.7.8.

Exercise 19.10.8. Show that there is $q \in \mathbb{N}$ such that for every $x \in X$ there exist unit vectors $u_1(x), \dots, u_q(x) \in \mathbb{R}^d$ so that all q-tuples $(u_x, u_1(x), \dots, u_q(x))$, $x \in X$, are mutually linearly isometric and satisfy (19.35)–(19.36).

Exercise 19.10.9. Prove Proposition 19.7.19 in the case of a CGDMS satisfying the SSC. Deduce that the result also holds for any CGDMS satisfying the BSC.

Exercise 19.10.10. Prove Lemma 19.8.6.

Exercise 19.10.11. Prove that the pressure function is well defined in Definition 19.8.8.

Exercise 19.10.12. Using an inductive argument, prove that the pair of formulae (19.42) and (19.43) follow from the pair of formulae (19.40) and (19.41).

Exercise 19.10.13. Let Φ be a CGDMS and $f \in \mathcal{K}^s_\beta(E^\infty_A, \mathbb{R})$ be a potential such that $\chi_{\mu_f}(\sigma) < \infty$. Show that the following three conditions are equivalent:

(a) $\mathrm{HD}(\mu_f \circ \pi^{-1}) = \mathrm{HD}(J)$.
(b) $\mu_f = \mu_{\mathrm{HD}(J)\Xi}$.
(c) The potentials f and $\mathrm{HD}(J)\Xi$ are cohomologous modulo a constant in the class of continuous (equivalently, Hölder continuous) functions from E^∞_A to $\mathbb{R}$.

Exercise 19.10.14. Let Φ be a CGDMS. Strengthening Exercise 19.10.13, show that if $\mu \in M(\sigma)$ is such that $\chi_\mu(\sigma) < \infty$, then $\mathrm{HD}(\mu \circ \pi^{-1}) = \mathrm{HD}(J)$ if and only if $\mu = \mu_{\mathrm{HD}(J)\Xi}$.

Exercise 19.10.15. Let Φ be a CGDMS. Prove that the function $\mathcal{K}^s_\beta(E^\infty_A, \mathbb{R}) \ni f \mapsto \mathrm{HD}(\mu_f \circ \pi^{-1})$ is continuous.

Exercise 19.10.16. Let $\sigma : \mathbb{N}^\infty_A \to \mathbb{N}^\infty_A$ be a finitely irreducible subshift of finite type and $f \in \mathcal{K}^s_\beta(\mathbb{N}^\infty_A, \mathbb{R})$ a potential. Prove that the sequence

$$(\mu_{f|_{\{1,2,\dots,n\}^\infty_A}})_{n=1}^\infty$$

weak* converges to μ_f.

Exercise 19.10.17. Let E be a countable set and $p : E \to (0,1]$ be a function such that

$$\sum_{e\in E} p(e) = 1.$$

Show that the Bernoulli measure μ_p on E^∞ is the Gibbs state for the potential $E^\infty \ni \omega \mapsto \log p(\omega_1)$.

Exercise 19.10.18. A GDMS $\Phi = \{\varphi_e : X_{t(e)} \to X_{i(e)}\}_{e\in E}$ is said to be *locally finite* if every point of the phase space X has a neighbourhood that intersects only finitely many first-level sets, i. e., if for every $x \in X$ there is a neighbourhood V_x of x such that

$$\#\{e \in E : V_x \cap \varphi_e(X_{t(e)})\} < \infty.$$

(a) Prove that every locally finite GDMS is pointwise finite (cf. Definition 19.1.2).
(b) Show that the conformal IFS in Example 19.9.2 is pointwise finite but not locally finite.

Exercise 19.10.19. Prove that the Gauss measure ν_G in (19.66) is invariant under the Gauss map G. Then show that the Lebesgue measure on $[0,1]$ is not invariant under the Gauss map.

20 Real analyticity of topological pressure and Hausdorff dimension

The first three sections of this chapter are related to our considerations in Section 13.10. However, the forthcoming considerations are subtler and more technically involved. These sections are a continuation of Chapters 17 and 18, as they exclusively deal with finitely irreducible/primitive countable-alphabet subshifts of finite type. In these three sections, we show that under proper hypotheses the pressure function $\mathrm{P}(\varphi + g)$ depends real analytically on g (in an appropriately defined way). In all of these sections, the potential φ is assumed to be Hölder continuous on cylinders and strongly summable (see Definition 20.1.1). In Section 20.1, the Hölder-on-cylinders perturbation g, therein denoted by γ_z, need not be bounded but must fulfill several technical assumptions (see Definition 20.1.6). Section 20.2 is dedicated to the special case where g is of the form $z\gamma$. This case is very important, as it is frequently encountered in applications. Finally, in Section 20.3 we assume that the Hölder-on-cylinders perturbation g is essentially bounded. However, we impose no other technical condition. Each of these sections culminates with the establishment of the real-analytic dependence of the topological pressure $g \mapsto \mathrm{P}(\varphi + g)$. The methods of proof are of complex analytic and functional analytic character. The notorious difficulty in demonstrating real-analytic dependence of the topological pressure is that the limit of a uniformly convergent sequence of real-analytic functions need not be real analytic (look at the Stone–Weierstrass theorem, for example), and the pressure is defined as such a limit. Nevertheless, the limit of a uniformly convergent sequence of complex-analytic (i. e. holomorphic) functions is complex analytic. Accordingly, our strategy consists in complexifying potentials first, and thus Perron–Frobenius (transfer) operators, too. Then we prove that these complexified operators depend holomorphically on the perturbation g. This is where infinite-dimensional complex analysis on Banach spaces is used. Next enters functional analysis. Indeed, we know that the exponential of the pressure function is a simple isolated eigenvalue of the transfer operator. So the Kato–Rellich perturbation theorem yields holomorphic dependence of appropriate eigenvalues of the complexified transfer operators. Restricting these operators to the real realm gives real analyticity of the exponential of the pressure function, and hence of the pressure itself.

Once we know that the topological pressure is real analytic, it is natural to ask what its derivative is. Sections 20.2 and 20.3 provide an answer: the pressure's derivative is the integral of the perturbation against the appropriate Gibbs state.

The first results on the real analyticity of the topological pressure were obtained by David Ruelle in [111] and [113] for finite-alphabet subshifts of finite type and Hölder continuous potentials and perturbations. We followed here Ruelle's general strategy. In comparison to Section 13.10, the additional difficulties we face are caused by the fact that transfer operators are given as the sum of an infinite convergent series

https://doi.org/10.1515/9783110702699-020

(rather than a finite sum) and that neither the potentials nor the perturbations need be bounded.

In Section 20.4, we consider families of CGDMSs in the complex plane $\mathbb{C}$. We assume that these families depend in a suitable way on a complex parameter λ. In particular, we define a (somewhat technical) notion of strong parameter. We want to emphasize that at first we do not assume the open set condition to be satisfied when we show the real-analytic dependence on λ of the Bowen parameter h. This is achieved in two steps. First, we establish the real analyticity of the topological pressure by verifying that the conditions in Definition 20.1.6 are fulfilled. Then the use of the implicit function theorem yields the real analyticity of the Bowen parameter on some sufficiently small neighborhood of any strong parameter. As an immediate consequence, we conclude that if the open set condition holds, then the Hausdorff dimension of the limit set varies real analytically near any strong parameter. Our considerations up to that point in this section find their origin in the paper [134]. We somewhat drifted away from that paper, developing and improving it. The rest of the section is a detailed and improved account of [118]. It is devoted to providing easy-to-check sufficient conditions for an analytic family of CGDMSs to admit a strong parameter. For this, we first introduce the concept of Hölder stability of such families at their parameters and prove that this entails such parameters to be strong. Then we show that if such an analytic family is engendered by a holomorphic motion, then the parameters for which the corresponding CGDMSs are strongly regular, turn out to be parameters that are Hölder stable and hence strong. Finally, we describe a mild, easy-to-verify condition, called periodic separation, which guarantees that an analytic family of CGDMSs is generated by a holomorphic motion. However, this section is not the end of the story. In the paper [106], we proved a meaningful real-analytic dependence for CGDMSs living in higher-dimensional Euclidean spaces. Other analyticity and continuity results of a similar type can be found in [103, 104, 105].

20.1 Real analyticity of pressure: part I

Let E be a countable alphabet and $A : E \times E \to \{0, 1\}$ a finitely irreducible incidence matrix. Let also $\beta \in (0, 1]$. Recall that $\mathcal{K}_\beta(E_A^\infty)$ represents the set of all complex-valued functions that are Hölder continuous on cylinders with exponent β while its subset comprising those functions that are real-valued is denoted by $\mathcal{K}_\beta(E_A^\infty, \mathbb{R})$. Moreover, those functions whose real part is summable are designated by $\mathcal{K}_\beta^s(E_A^\infty)$. The real part of summable functions is always bounded from above, but it is not bounded from below when $\#E = \infty$.

Remember that $C_b(E_A^\infty)$ is the set of all bounded continuous complex-valued functions on E_A^∞ and the set $\mathcal{K}_\beta^b(E_A^\infty) := \mathcal{K}_\beta(E_A^\infty) \cap C_b(E_A^\infty)$ of all globally Hölder continuous functions, equipped with the norm $\|f\|_\beta := v_\beta(f) + \|f\|_\infty$, forms a Banach space. Recall that $\mathcal{K}_\beta^b(E_A^\infty) = \mathcal{K}_\beta^s(E_A^\infty)$ when $\#E < \infty$, while $\mathcal{K}_\beta^b(E_A^\infty) \cap \mathcal{K}_\beta^s(E_A^\infty) = \emptyset$ when $\#E = \infty$. (For more information, see the beginning of the Section 18.1.)

We now introduce a new subclass of $\mathcal{K}^s_\beta(E_A^\infty, \mathbb{R})$.

Definition 20.1.1. A function $\varphi \in \mathcal{K}^s_\beta(E_A^\infty, \mathbb{R})$ is *strongly summable* provided that there exists $v \in (0,1)$ such that $v\varphi$ is summable, i. e., $v\varphi \in \mathcal{K}^s_\beta(E_A^\infty, \mathbb{R})$.

We enumerate a few properties of summability and strong summability.

Lemma 20.1.2.
(a) *If $\varphi \in \mathcal{K}^s_\beta(E_A^\infty, \mathbb{R})$ and $t > 1$, then $t\varphi \in \mathcal{K}^s_\beta(E_A^\infty, \mathbb{R})$.*
(b) *If φ is strongly summable, then $u\varphi$ is strongly summable for any $u > 0$ large enough. More precisely, if $v\varphi \in \mathcal{K}^s_\beta(E_A^\infty, \mathbb{R})$ for some $v \in (0,1)$, then $u\varphi$ is strongly summable for any $u > v$.*
(c) *If $h \in \mathcal{K}^b_\beta(E_A^\infty, \mathbb{R})$ and $\varphi \in \mathcal{K}^s_\beta(E_A^\infty, \mathbb{R})$, then $\varphi + h \in \mathcal{K}^s_\beta(E_A^\infty, \mathbb{R})$.*
(d) *Let $h \in \mathcal{K}^b_\beta(E_A^\infty, \mathbb{R})$. If $\varphi \in \mathcal{K}^s_\beta(E_A^\infty, \mathbb{R})$ is strongly summable, then so is $\varphi + h$.*

We now turn our attention to the exponential of functions that are Hölder continuous on cylinders. We start with the following elementary observation.

Lemma 20.1.3. *For any $z, \zeta \in \mathbb{C}$, we have that*

$$|e^z - e^\zeta| \le \exp(\max\{\mathrm{Re}(z), \mathrm{Re}(\zeta)\})|z - \zeta|.$$

Proof. It follows from the mean value inequality that

$$|e^z - e^\zeta| \le \sup_{w \in [z,\zeta]} e^{\mathrm{Re}(w)} \cdot |z - \zeta| = \exp(\max\{\mathrm{Re}(z), \mathrm{Re}(\zeta)\})|z - \zeta|. \qquad \square$$

A fairly immediate application of the previous lemma leads to the following result.

Lemma 20.1.4. *If $f \in \mathcal{K}_\beta(E_A^\infty)$ is bounded above, then $e^f \in \mathcal{K}^b_\beta(E_A^\infty)$ and for any $F \subseteq E_A^\infty$,*

$$\|e^f\|_{\beta,F} \le (1 + v_{\beta,F}(f))e^{\sup \mathrm{Re}(f|_F)}.$$

Proof. It follows from the definition of $\mathcal{K}_\beta(E_A^\infty)$ and the upper boundedness of f that $e^f \in C_b(E_A^\infty)$ and

$$\|e^f\|_{\infty,F} \le e^{\sup \mathrm{Re}(f|_F)} < \infty. \tag{20.1}$$

If $\omega, \tau \in F$ and $\omega_1 = \tau_1$, we get from Lemma 20.1.3 that

$$|e^{f(\omega)} - e^{f(\tau)}| \le \exp\Big(\max\{\mathrm{Re}(f(\omega)), \mathrm{Re}(f(\tau))\}\Big)|f(\omega) - f(\tau)| \le e^{\sup \mathrm{Re}(f|_F)} v_{\beta,F}(f) d^\beta(\omega, \tau).$$

Therefore,

$$v_{\beta,F}(e^f) \le e^{\sup \mathrm{Re}(f|_F)} v_{\beta,F}(f). \tag{20.2}$$

Together, (20.1) and (20.2) yield the result. $\square$

Given $e \in E$ and $h : E_A^\infty \longrightarrow \mathbb{C}$, define the function $h \circ e : E_A^\infty \longrightarrow \mathbb{C}$ by

$$h \circ e(\omega) := \begin{cases} h(e\omega) & \text{if } A_{e\omega_1} = 1 \\ 0 & \text{if } A_{e\omega_1} = 0. \end{cases}$$

One can then rewrite the transfer operator $\mathcal{L}_f$ in the form (cf. (17.22))

$$\mathcal{L}_f(g)(\omega) = \sum_{e\in E} g \circ e(\omega) \exp(f \circ e(\omega)).$$

Moreover, we have the following basic properties.

Lemma 20.1.5. *For every $e \in E$, the following statements hold:*
(a) *If $f \in \mathcal{K}_\beta(E_A^\infty)$, then $f \circ e \in \mathcal{K}_\beta(E_A^\infty)$ and $v_\beta(f \circ e) = v_{\beta,[e]}(f) \le v_\beta(f)$.*
(b) *If $f \in \mathcal{K}_\beta^b(E_A^\infty)$, then $f \circ e \in \mathcal{K}_\beta^b(E_A^\infty)$ and $\|f \circ e\|_\beta = \|f\|_{\beta,[e]} \le \|f\|_\beta$.*
(c) *If $f, g : E_A^\infty \to \mathbb{C}$ and $h = f + g$, then $h \circ e = f \circ e + g \circ e$.*

In the sequel, we will make use of complex analysis on Banach spaces. For more information on this topic, we refer the reader to [25, 86, 110].

Definition 20.1.6. Consider the following conditions:
(A) E is a countable alphabet and $A : E \times E \to \{0, 1\}$ is a finitely irreducible matrix.
(B) $\varphi \in \mathcal{K}_\beta(E_A^\infty)$ for some $\beta \in (0, 1]$.
(C) $\psi \in \mathcal{K}_\beta^s(E_A^\infty, \mathbb{R})$ is strongly summable, i. e., there exists $\nu \in (0, 1)$ such that $\nu\psi \in \mathcal{K}_\beta^s(E_A^\infty, \mathbb{R})$.
(D) There are numbers $r_0 \in \mathbb{R}$ and $\delta > 0$, and a function

$$B_{\mathbb{C}}(r_0, 4\delta) \ni z \longmapsto (\gamma_z : E_A^\infty \to \mathbb{C})$$

with the following properties:
(D1) $\gamma_z \in \mathcal{K}_\beta(E_A^\infty)$ for every $z \in B_{\mathbb{C}}(r_0, 4\delta)$ and

$$Q := \sup\{v_\beta(\gamma_z) : z \in B_{\mathbb{C}}(r_0, 4\delta)\} < \infty.$$

(D2) For every $\omega \in E_A^\infty$, the function

$$B_{\mathbb{C}}(r_0, 4\delta) \ni z \longmapsto \gamma_z(\omega) \in \mathbb{C}$$

is holomorphic.
(E) There exists $\kappa \ge 1$ such that for every $z \in B_{\mathbb{C}}(r_0, 4\delta)$, the function γ_z satisfies

$$|\gamma_z| \le \kappa(|\psi| + 1).$$

(F) For every $z \in B_{\mathbb{C}}(r_0, 4\delta)$, the function $\varphi_z := \varphi + \gamma_z$ satisfies

$$\text{Re}(\varphi_z) \le \psi.$$

Under these conditions, the functions $\{\varphi_z\}_{z\in B_{\mathbb{C}}(r_0,4\delta)}$ are said to form a *complex-parametrized family of complex-valued potentials*.

As an immediate consequence of Cauchy's integral formula (Theorem A.3.2 in Appendix A), we get from conditions (D2) and (E) that

$$|\gamma_z'| \le \frac{\kappa}{\delta}(|\psi|+1), \quad \forall z \in B_{\mathbb{C}}(r_0,3\delta). \tag{20.3}$$

We shall now prove that

$$\sup\{v_\beta(\gamma_z') : z \in B_{\mathbb{C}}(r_0,3\delta)\} \le \frac{Q}{\delta} < \infty. \tag{20.4}$$

Indeed, condition (D1) guarantees that

$$|\omega \wedge \tau| \ge 1 \Longrightarrow |\gamma_z(\omega) - \gamma_z(\tau)| \le Qd^\beta(\omega,\tau), \quad \forall z \in B_{\mathbb{C}}(r_0,4\delta).$$

It follows from Cauchy's integral formula and condition (D2) that

$$|\omega \wedge \tau| \ge 1 \Longrightarrow |\gamma_z'(\omega) - \gamma_z'(\tau)| \le \frac{Q}{\delta}d^\beta(\omega,\tau), \quad \forall z \in B_{\mathbb{C}}(r_0,3\delta).$$

This means that $\gamma_z' \in \mathcal{K}_\beta(E_A^\infty)$ and $v_\beta(\gamma_z') \le Q/\delta$ for all $z \in B_{\mathbb{C}}(r_0,3\delta)$. Relation (20.4) ensues. Note that $\varphi_z' = \gamma_z'$ for all $z \in B_{\mathbb{C}}(r_0,4\delta)$. So (20.3)–(20.4) hold with γ_z' replaced by φ_z', i. e.,

$$\sup_{z\in B_{\mathbb{C}}(r_0,3\delta)} |\varphi_z'| \le \frac{\kappa}{\delta}(|\psi|+1) \quad \text{and} \quad \sup_{z\in B_{\mathbb{C}}(r_0,3\delta)} v_\beta(\varphi_z') \le \frac{Q}{\delta}. \tag{20.5}$$

For every $z \in B_{\mathbb{C}}(r_0,4\delta)$ and $e \in E$, define the linear operator $\mathcal{L}_{\varphi_z,e} : C_b(E_A^\infty) \longrightarrow C_b(E_A^\infty)$ by

$$\mathcal{L}_{\varphi_z,e}(g) := (g \circ e) \cdot \exp(\varphi_z \circ e). \tag{20.6}$$

As a direct repercussion of Lemmas 20.1.4–20.1.5, relation (18.2) with $F = [e]$, and conditions (D1) and (F), this operator has the following properties.

Lemma 20.1.7. *Let $\{\varphi_z\}_{z\in B_{\mathbb{C}}(r_0,4\delta)}$ be a complex-parametrized family of complex-valued potentials. For every $z \in B_{\mathbb{C}}(r_0,4\delta)$ and $e \in E$, the linear operator $\mathcal{L}_{\varphi_z,e}$ preserves the Banach space $\mathcal{K}_\beta^b(E_A^\infty)$, i. e.,*

$$\mathcal{L}_{\varphi_z,e}(\mathcal{K}_\beta^b(E_A^\infty)) \subseteq \mathcal{K}_\beta^b(E_A^\infty).$$

Furthermore, this operator is bounded and

$$\|\mathcal{L}_{\varphi_z,e}\|_\beta \le (1+v_\beta(\varphi_z))\exp(\sup \mathrm{Re}(\varphi_z|_{[e]})) \le (1+v_\beta(\varphi)+Q)\exp(\sup \psi|_{[e]}).$$

We now investigate the dependence on z of the operator $\mathcal{L}_{\varphi_z,e}$.

Lemma 20.1.8. *Let $\{\varphi_z\}_{z\in B_{\mathbb{C}}(r_0,4\delta)}$ be a complex-parametrized family of complex-valued potentials. For every $e \in E$, the function*

$$\overline{B_{\mathbb{C}}(r_0,2\delta)} \ni z \longmapsto \mathcal{L}_{\varphi_z,e} \in L(\mathcal{K}^b_\beta(E_A^\infty))$$

is continuous. In fact, there exists a constant $M \geq 1$ such that

$$\|\mathcal{L}_{\varphi_z,e} - \mathcal{L}_{\varphi_\zeta,e}\|_\beta \leq M\exp(v\sup\psi|_{[e]})|z-\zeta|$$

for every $e \in E$ and all $z,\zeta \in \overline{B_{\mathbb{C}}(r_0,2\delta)}$, where v comes from Definition 20.1.6(C).

Proof. By a standard calculus argument, there is a constant $D_\psi \geq 1$ such that

$$\frac{\kappa}{\delta}\max\{1, 2(v_\beta(\varphi)+Q)\}\cdot(|x|+v_\beta(\psi)+1)e^x \leq D_\psi e^{vx}, \quad \forall x \leq \sup\psi. \tag{20.7}$$

Fix $e \in E$ and $z,\zeta \in \overline{B_{\mathbb{C}}(r_0,2\delta)}$. We first seek an upper bound on the supremum norm of $e^{\varphi_z\circ e} - e^{\varphi_\zeta\circ e}$. Using successively Lemma 20.1.3, condition (F), the mean value inequality, relation (20.5), Exercise 18.5.6 and relation (20.7), we obtain that

$$\begin{aligned}
\|e^{\varphi_z\circ e} - e^{\varphi_\zeta\circ e}\|_\infty &\leq \exp\Big(\sup_{\omega\in[e]}\max\{\mathrm{Re}(\varphi_z(\omega)),\mathrm{Re}(\varphi_\zeta(\omega))\}\Big)\cdot\|\varphi_z-\varphi_\zeta\|_{\infty,[e]}\\
&\leq \exp(\sup\psi|_{[e]})\cdot\sup_{\xi\in\overline{B_{\mathbb{C}}(r_0,2\delta)}}\|\varphi'_\xi|_{[e]}\|_\infty\cdot|z-\zeta|\\
&\leq \frac{\kappa}{\delta}(\|\psi|_{[e]}\|_\infty+1)\exp(\sup\psi|_{[e]})|z-\zeta|\\
&\leq \frac{\kappa}{\delta}(|\sup\psi|_{[e]}|+v_\beta(\psi)+1)\exp(\sup\psi|_{[e]})|z-\zeta|\\
&\leq D_\psi\exp(v\sup\psi|_{[e]})|z-\zeta|.
\end{aligned} \tag{20.8}$$

We now find a Hölder constant for $e^{\varphi_z\circ e} - e^{\varphi_\zeta\circ e}$. For every $\omega \in E_A^\infty$, the fundamental theorem of calculus affirms that

$$e^{\varphi_z(\omega)} - e^{\varphi_\zeta(\omega)} = \int_\zeta^z \varphi'_\xi(\omega)e^{\varphi_\xi(\omega)}\,d\xi,$$

where the integration is along the closed line segment from ζ to z. Let $\omega,\tau \in E_A^\infty$. It ensues that

$$\begin{aligned}
(e^{\varphi_z(\omega)} - e^{\varphi_\zeta(\omega)}) - (e^{\varphi_z(\tau)} - e^{\varphi_\zeta(\tau)}) &= \int_\zeta^z\Big(\varphi'_\xi(\omega)e^{\varphi_\xi(\omega)} - \varphi'_\xi(\tau)e^{\varphi_\xi(\tau)}\Big)\,d\xi\\
&= \int_\zeta^z\Big[(\varphi'_\xi(\omega)-\varphi'_\xi(\tau))e^{\varphi_\xi(\omega)} + \varphi'_\xi(\tau)(e^{\varphi_\xi(\omega)} - e^{\varphi_\xi(\tau)})\Big]\,d\xi.
\end{aligned}$$

Assume that $\tau_1 = \omega_1$ and that $\psi(\omega) \le \psi(\tau)$. Using Lemma 20.1.3, conditions (F) and (D1), as well as relations (20.5) and (20.7), we get for every $\xi \in [\zeta, z]$ that

$$\begin{aligned}
&\left|(\varphi'_\xi(\omega) - \varphi'_\xi(\tau))e^{\varphi_\xi(\omega)} + \varphi'_\xi(\tau)(e^{\varphi_\xi(\omega)} - e^{\varphi_\xi(\tau)})\right| \\
&\quad \le |\varphi'_\xi(\omega) - \varphi'_\xi(\tau)|e^{\mathrm{Re}(\varphi_\xi(\omega))} + |\varphi'_\xi(\tau)| \cdot |e^{\varphi_\xi(\omega)} - e^{\varphi_\xi(\tau)}| \\
&\quad \le \upsilon_\beta(\varphi'_\xi)d^\beta(\omega,\tau)e^{\mathrm{Re}(\varphi_\xi(\omega))} \\
&\qquad + |\varphi'_\xi(\tau)| \cdot |\varphi_\xi(\omega) - \varphi_\xi(\tau)| \cdot \exp(\max\{\mathrm{Re}(\varphi_\xi(\omega)), \mathrm{Re}(\varphi_\xi(\tau))\}) \\
&\quad \le \upsilon_\beta(\varphi'_\xi)d^\beta(\omega,\tau)e^{\psi(\omega)} + |\varphi'_\xi(\tau)| \cdot \upsilon_\beta(\varphi_\xi)d^\beta(\omega,\tau) \cdot \exp(\max\{\psi(\omega), \psi(\tau)\}) \\
&\quad \le (\upsilon_\beta(\varphi'_\xi) + \upsilon_\beta(\varphi_\xi)|\varphi'_\xi(\tau)|)e^{\psi(\tau)}d^\beta(\omega,\tau) \\
&\quad \le \left(\frac{Q}{\delta} + \upsilon_\beta(\varphi_\xi) \cdot \frac{\kappa}{\delta}(|\psi(\tau)| + 1)\right)e^{\psi(\tau)}d^\beta(\omega,\tau) \\
&\quad \le \left(\frac{Q}{\delta} + \frac{\kappa}{\delta}(\upsilon_\beta(\varphi) + Q)(|\psi(\tau)| + 1)\right)e^{\psi(\tau)}d^\beta(\omega,\tau) \\
&\quad \le \frac{2\kappa}{\delta}(\upsilon_\beta(\varphi) + Q)(|\psi(\tau)| + 1)e^{\psi(\tau)}d^\beta(\omega,\tau) \\
&\quad \le D_\psi e^{\nu\psi(\tau)}d^\beta(\omega,\tau) \\
&\quad \le D_\psi \exp(\sup(\nu\psi|_{[\omega_1]}))d^\beta(\omega,\tau).
\end{aligned}$$

Hence,

$$|(e^{\varphi_z(\omega)} - e^{\varphi_\zeta(\omega)}) - (e^{\varphi_z(\tau)} - e^{\varphi_\zeta(\tau)})| \le D_\psi \exp(\sup(\nu\psi|_{[\omega_1]}))|z - \zeta|d^\beta(\omega,\tau).$$

Thus, for every $e \in E$ we deduce that

$$|(e^{\varphi_z\circ e(\omega)} - e^{\varphi_\zeta\circ e(\omega)}) - (e^{\varphi_z\circ e(\tau)} - e^{\varphi_\zeta\circ e(\tau)})| \le D_\psi \exp(\nu \sup \psi|_{[e]})|z - \zeta|d^\beta(\omega,\tau).$$

This holds for all $\omega, \tau \in E_A^\infty$ such that $\omega_1 = \tau_1$. Therefore,

$$\upsilon_\beta(e^{\varphi_z\circ e} - e^{\varphi_\zeta\circ e}) \le D_\psi \exp(\nu \sup \psi|_{[e]})|z - \zeta|.$$

Combining this with (20.8), we conclude that

$$\|e^{\varphi_z\circ e} - e^{\varphi_\zeta\circ e}\|_\beta \le 2D_\psi \exp(\nu \sup \psi|_{[e]})|z - \zeta|.$$

Applying Lemma 20.1.5(c) and (18.2) with $F = E_A^\infty$, we get for every $g \in \mathcal{K}_\beta^b(E_A^\infty)$ that

$$\begin{aligned}
\|\mathcal{L}_{\varphi_z,e}(g) - \mathcal{L}_{\varphi_\zeta,e}(g)\|_\beta &= \|(e^{\varphi_z\circ e}) \cdot (g \circ e) - (e^{\varphi_\zeta\circ e}) \cdot (g \circ e)\|_\beta \\
&= \|(e^{\varphi_z\circ e} - e^{\varphi_\zeta\circ e}) \cdot (g \circ e)\|_\beta \\
&\le 2D_\psi \exp(\nu \sup \psi|_{[e]})|z - \zeta| \cdot \|g\|_\beta.
\end{aligned}$$

Thus,

$$\|\mathcal{L}_{\varphi_z,e} - \mathcal{L}_{\varphi_\zeta,e}\|_\beta \le 2D_\psi \exp(\nu \sup \psi|_{[e]})|z - \zeta|.$$

Choose $M = 2D_\psi$. □

We now look at the operator which is the sum over $e \in E$ of the operators $\mathcal{L}_{\varphi_z,e}$.

Proposition 20.1.9. *Let $\{\varphi_z\}_{z\in B_{\mathbb{C}}(r_0,4\delta)}$ be a complex-parametrized family of complex-valued potentials.*

(a) *The series*

$$\overline{B_{\mathbb{C}}(r_0,2\delta)} \ni z \longmapsto \mathcal{L}_{\varphi_z} := \sum_{e\in E} \mathcal{L}_{\varphi_z,e}$$

converges absolutely uniformly in $\mathcal{K}^b_\beta(E^\infty_A)$.

(b) *In particular:*
   (b1) *The linear operator $\mathcal{L}_{\varphi_z}$ preserves the Banach space $\mathcal{K}^b_\beta(E^\infty_A)$, i. e.,*

$$\mathcal{L}_{\varphi_z}(\mathcal{K}^b_\beta(E^\infty_A)) \subseteq \mathcal{K}^b_\beta(E^\infty_A), \quad \forall z \in \overline{B_{\mathbb{C}}(r_0,2\delta)}.$$

   (b2) *The operator $\mathcal{L}_{\varphi_z} : \mathcal{K}^b_\beta(E^\infty_A) \longrightarrow \mathcal{K}^b_\beta(E^\infty_A)$ is bounded and*

$$\|\mathcal{L}_{\varphi_z}\|_\beta \le (1 + \upsilon_\beta(\varphi) + Q) \sum_{e\in E} \exp(\sup \psi|_{[e]}), \quad \forall z \in \overline{B_{\mathbb{C}}(r_0,2\delta)}.$$

   (b3) *The function $\overline{B_{\mathbb{C}}(r_0,2\delta)} \ni z \longmapsto \mathcal{L}_{\varphi_z} \in L(\mathcal{K}^b_\beta(E^\infty_A))$ is continuous.*

(c) *Moreover,*

$$\|\mathcal{L}_{\varphi_z} - \mathcal{L}_{\varphi_\zeta}\|_\beta \le M\Big(\sum_{e\in E} \exp(\nu \sup \psi|_{[e]})\Big)|z - \zeta|, \quad \forall z,\zeta \in \overline{B_{\mathbb{C}}(r_0,2\delta)}.$$

Proof. Items (a) and (b) are an immediate consequence of Lemma 20.1.7, condition (C) and the first part of Lemma 20.1.8. By invoking the second part of Lemma 20.1.8 and item (a), we get

$$\|\mathcal{L}_{\varphi_z} - \mathcal{L}_{\varphi_\zeta}\|_\beta \le \sum_{e\in E} \|\mathcal{L}_{\varphi_z,e} - \mathcal{L}_{\varphi_\zeta,e}\|_\beta \le M\Big(\sum_{e\in E} \exp(\nu \sup \psi|_{[e]})\Big)|z - \zeta|.$$

So item (c) is also established. □

More can be said about the function $z \longmapsto \mathcal{L}_{\varphi_z}$.

Lemma 20.1.10. *Let $\{\varphi_z\}_{z\in B_{\mathbb{C}}(r_0,4\delta)}$ be a complex-parametrized family of complex-valued potentials. Then the function*

$$B_{\mathbb{C}}(r_0,\delta) \ni z \longmapsto \mathcal{L}_{\varphi_z} \in L(\mathcal{K}^b_\beta(E_A^\infty))$$

is holomorphic.

Proof. Let $\Gamma \subseteq B_{\mathbb{C}}(r_0,\delta)$ be a closed rectifiable curve. Fix $g \in \mathcal{K}^b_\beta(E_A^\infty)$ and $\omega \in E_A^\infty$. As the function

$$B_{\mathbb{C}}(r_0,\delta) \ni z \longmapsto (\mathcal{L}_{\varphi_z,e}(g))(\omega) \in \mathbb{C}$$

is holomorphic for every $e \in E$, it follows from Proposition 20.1.9(a) that the function

$$B_{\mathbb{C}}(r_0,\delta) \ni z \longmapsto (\mathcal{L}_{\varphi_z}(g))(\omega) \in \mathbb{C}$$

is holomorphic. Hence, by Cauchy's theorem (Theorem A.3.1 in Appendix A),

$$\int_\Gamma (\mathcal{L}_{\varphi_z}(g))(\omega)\,dz = 0.$$

Since the function $B_{\mathbb{C}}(r_0,\delta) \ni z \longmapsto \mathcal{L}_{\varphi_z}(g) \in \mathcal{K}^b_\beta(E_A^\infty)$ is continuous according to Proposition 20.1.9(c), the integral $\int_\Gamma \mathcal{L}_{\varphi_z}(g)\,dz$ exists and

$$\int_\Gamma \mathcal{L}_{\varphi_z}(g)\,dz(\omega) := \int_\Gamma (\mathcal{L}_{\varphi_z}(g))(\omega)\,dz = 0, \quad \forall \omega \in E_A^\infty.$$

Thus,

$$\int_\Gamma \mathcal{L}_{\varphi_z}(g)\,dz = 0.$$

As the function $B_{\mathbb{C}}(r_0,\delta) \ni z \longmapsto \mathcal{L}_{\varphi_z} \in L(\mathcal{K}^b_\beta(E_A^\infty))$ is continuous by Proposition 20.1.9(c), the integral $\int_\Gamma \mathcal{L}_{\varphi_z}\,dz$ exists and

$$\int_\Gamma \mathcal{L}_{\varphi_z}\,dz(g) := \int_\Gamma \mathcal{L}_{\varphi_z}(g)\,dz = 0, \quad \forall g \in \mathcal{K}^b_\beta(E_A^\infty).$$

Therefore,

$$\int_\Gamma \mathcal{L}_{\varphi_z}\,dz = 0.$$

By Morera's theorem, the map $B_{\mathbb{C}}(r_0,\delta) \ni z \longmapsto \mathcal{L}_{\varphi_z} \in L(\mathcal{K}^b_\beta(E_A^\infty))$ is holomorphic. (Note: Morera's theorem for Banach space-valued functions is a direct consequence of the classical, complex-valued version of Morera's theorem (Theorem A.3.3) and the equivalence of weak and strong holomorphicity, which is for instance proved in Theorem 3.31 of Rudin [110]). □

As an immediate reverberation of this lemma and Hartogs' theorem (Theorem A.4.1 in Appendix A), we obtain the following.

Lemma 20.1.11. *Let $\{\varphi_z\}_{z\in B_{\mathbb{C}^d}(r_0,4\delta)}$ be a d-dimensional complex-parametrized family of complex-valued potentials with $\mathbb{C}$ replaced by $\mathbb{C}^d$ and $r_0 \in \mathbb{R}^d$. Then the function*

$$B_{\mathbb{C}^d}(r_0,\delta) \ni z \longmapsto \mathcal{L}_{\varphi_z} \in L(\mathcal{K}_\beta^b(E_A^\infty))$$

is holomorphic.

Finally, we can establish a first result on the real analyticity of the pressure function.

Theorem 20.1.12. *Let $\{\varphi_z\}_{z\in B_{\mathbb{C}^d}(r_0,4\delta)}$ be a d-dimensional complex-parametrized family of complex-valued potentials. If $r_0 \in \mathbb{R}^d$ and $\varphi_r \in \mathcal{K}_\beta(E_A^\infty, \mathbb{R})$ for all $r \in B_{\mathbb{R}^d}(r_0,\delta)$, then the function*

$$B_{\mathbb{R}^d}(r_0,\delta) \ni r \longmapsto \mathrm{P}(\varphi_r) \in \mathbb{R}$$

is real analytic.

Proof. Fix $u \in B_{\mathbb{R}^d}(r_0,\delta)$. Then there exists $R > 0$ such that

$$B_{\mathbb{C}^d}(u,R) \subseteq B_{\mathbb{C}^d}(r_0,\delta).$$

By Theorem 18.1.12, for all $r \in B_{\mathbb{R}^d}(u,R)$ the number $\exp(\mathrm{P}(\varphi_r))$ is a simple isolated eigenvalue of the operator $\mathcal{L}_{\varphi_r} \in L(\mathcal{K}_\beta^b(E_A^\infty))$. (For more information on the spectral theory of bounded linear operators, see [21, 26].) It follows from Lemma 20.1.11 and the Kato–Rellich theorem for perturbations of linear operators (Theorem A.2.5) that there exist $R' \in (0,R]$ and a holomorphic function

$$\rho : B_{\mathbb{C}^d}(u,R') \longrightarrow \mathbb{C}$$

such that $\rho(z)$ is an eigenvalue of the operator $\mathcal{L}_{\varphi_z}$ for all $z \in B_{\mathbb{C}^d}(u,R')$ and

$$\rho(r) = \exp(\mathrm{P}(\varphi_r)), \quad \forall r \in B_{\mathbb{R}^d}(u,R').$$

Consequently, the function

$$B_{\mathbb{R}^d}(u,R') \ni r \longmapsto \mathrm{P}(\varphi_r) \in \mathbb{R}$$

is real analytic. Since real analyticity is a local property, we are done. □

20.2 Real analyticity of pressure: part II

Let us look at the special case of a complex-parametrized family of complex-valued potentials $\{\varphi_z\}_{z\in B_{\mathbb{C}}(r_0,4\delta)}$ originating from a "linear" family

$$\varphi_z = \varphi + z\gamma \tag{20.9}$$

with φ_{r_0} strongly summable. The functions φ and γ are assumed to take the form

$$\varphi = b_\varphi + m_\varphi f \quad \text{and} \quad \gamma = b_\gamma + m_\gamma f, \tag{20.10}$$

where

$$b_\varphi, b_\gamma \in \mathcal{K}_\beta^b(E_A^\infty, \mathbb{R}), \quad m_\varphi, m_\gamma \in \mathbb{R}, \quad \text{and} \quad f \in \mathcal{K}_\beta(E_A^\infty, \mathbb{R}). \tag{20.11}$$

Under these assumptions, we will check that the family $\{\varphi_z\}_{z\in B_{\mathbb{C}}(r_0,4\delta)}$ fulfills conditions (A)–(F) in Definition 20.1.6 for appropriate choices of ψ, κ, and δ. Condition (A) is met de facto. From (20.10)–(20.11), it is immediate that $\varphi, \gamma \in \mathcal{K}_\beta(E_A^\infty, \mathbb{R})$. In particular, condition (B) holds.

Regarding condition (C), observe that

$$\varphi_r = (b_\varphi + rb_\gamma) + (m_\varphi + rm_\gamma)f, \quad \forall r \in B(r_0, 4\delta).$$

As $b_\varphi + rb_\gamma \in \mathcal{K}_\beta^b(E_A^\infty, \mathbb{R})$, Lemma 20.1.2(d) asserts that the strong summability of φ_r is equivalent to the strong summability of $(m_\varphi + rm_\gamma)f$. But then it follows from Lemma 20.1.2(b) that the strong summability of φ_{r_0} (or, equivalently, of $(m_\varphi + r_0 m_\gamma)f$) entails the existence of an open interval $I \subseteq \mathbb{R}$ centered on r_0 such that the function φ_r (or, equivalently, $(m_\varphi + rm_\gamma)f$) is strongly summable for every $r \in I$. If necessary, reduce $\delta > 0$ so that $B(r_0, 8\delta) \subseteq I$.

Let $m = \min\{m_\varphi + rm_\gamma : r \in B(r_0, 4\delta)\}$ and $M = \max\{m_\varphi + rm_\gamma : r \in B(r_0, 4\delta)\}$. By reducing δ if necessary, it is possible to make sure that $mM \neq 0$. Set

$$\psi = [\|b_\varphi\|_\infty + (|r_0| + 4\delta)\|b_\gamma\|_\infty] + \max\{mf, Mf\}.$$

By definition of m and M, the functions mf and Mf are both strongly summable. It readily ensues that their maximum is strongly summable. The term between square brackets being constant, it follows from Lemma 20.1.2(d) that the function ψ is strongly summable. Condition (C) is satisfied.

Concerning condition (D), by comparing the expression of φ_z in (20.9) with its definition in condition (F), we must set $\gamma_z = z\gamma$ in condition (D). As a multiple of $\gamma \in \mathcal{K}_\beta(E_A^\infty, \mathbb{R})$, it is clear that $\gamma_z \in \mathcal{K}_\beta(E_A^\infty)$ for every $z \in B_{\mathbb{C}}(r_0, 4\delta)$ and

$$Q := \sup_{z\in B_{\mathbb{C}}(r_0,4\delta)} v_\beta(\gamma_z) = \sup_{z\in B_{\mathbb{C}}(r_0,4\delta)} |z| \cdot v_\beta(\gamma) = (|r_0| + 4\delta)v_\beta(\gamma) < \infty.$$

Furthermore, for every $\omega \in E_A^\infty$ the function $B_{\mathbb{C}}(r_0, 4\delta) \ni z \longmapsto \gamma_z(\omega) = z\gamma(\omega) \in \mathbb{C}$ is obviously holomorphic. Thus, condition (D) is fulfilled.

For condition (E), let $z \in B_{\mathbb{C}}(r_0, 4\delta)$. First, note that the summability of ψ imposes that the set

$$E' := \left\{ e \in E : \psi|_{[e]} \le \frac{1}{2}\max\{mf|_{[e]}, Mf|_{[e]}\} < 0 \right\}$$

is cofinite. On $\bigcup_{e \in E'}[e]$, it turns out that

$$|\psi| \ge \frac{1}{2}|\max\{mf, Mf\}| \ge \frac{1}{2}|f|\min\{|m|, |M|\} > 0$$

and hence

$$\begin{aligned}
|\gamma_z| = |z||\gamma| &\le (|r_0| + 4\delta)|b_\gamma + m_\gamma f| \le (|r_0| + 4\delta)(\|b_\gamma\|_\infty + |m_\gamma||f|) \\
&\le (|r_0| + 4\delta)\left[\|b_\gamma\|_\infty + \frac{|m_\gamma|}{\frac{1}{2}\min\{|m|, |M|\}} \cdot \frac{1}{2}\min\{|m|, |M|\}|f| \right] \\
&\le (|r_0| + 4\delta)\left[\|b_\gamma\|_\infty + \frac{2|m_\gamma|}{\min\{|m|, |M|\}}|\psi| \right].
\end{aligned}$$

On the complement $\bigcup_{e \in E \setminus E'}[e]$, it is easy to see that $|\gamma_z| \le (|r_0| + 4\delta)U$, where $U := \sup_{e \in E \setminus E'} \|\gamma\|_{\infty,[e]} < \infty$. Choosing $\kappa = (|r_0| + 4\delta)\max\{U, \|b_\gamma\|_\infty, \frac{2|m_\gamma|}{\min\{|m|,|M|\}}\}$, condition (E) holds.

Finally, as $\mathrm{Re}(\varphi_z) = \varphi_{\mathrm{Re}(z)}$, condition (F) reduces to demonstrating that $\varphi_r \le \psi$ for every $r \in B(r_0, 4\delta)$. But

$$\psi \ge [\|b_\varphi\|_\infty + |r| \cdot \|b_\gamma\|_\infty] + (m_\varphi + rm_\gamma)f \ge (b_\varphi + rb_\gamma) + (m_\varphi + rm_\gamma)f = \varphi_r.$$

That is, condition (F) is respected.

Therefore, the following result is a consequence of Proposition 20.1.9(b).

Lemma 20.2.1. *Let $\sigma : E_A^\infty \to E_A^\infty$ be a finitely irreducible subshift of finite type over a countable alphabet E. Let $\{\varphi_z\}_{z \in B_{\mathbb{C}}(r_0, 4\delta)}$ be a complex-parametrized family of complex-valued potentials of the form (20.9)–(20.11). If the potential φ_{r_0} is strongly summable, then there exists $0 < \delta_{r_0} < \delta$ such that for every $z \in B_{\mathbb{C}}(r_0, 2\delta_{r_0})$ we have*

$$\mathcal{L}_{\varphi+z\gamma}(\mathcal{K}_\beta^b(E_A^\infty)) \subseteq \mathcal{K}_\beta^b(E_A^\infty) \tag{20.12}$$

and

$$\|\mathcal{L}_{\varphi+z\gamma}\|_\beta \le \left(1 + v_\beta(\varphi) + (|r_0| + 4\delta_{r_0})v_\beta(\gamma)\right) \sum_{e \in E} \exp(\sup \psi_{r_0}|_{[e]}) < \infty, \tag{20.13}$$

where

$$\psi_{r_0} = [\|b_\varphi\|_\infty + (|r_0| + 4\delta_{r_0})\|b_\gamma\|_\infty] + \max\{m_{r_0}f, M_{r_0}f\}$$

and $m_{r_0} = \min\{m_\varphi + rm_\gamma : r \in B(r_0, 4\delta_{r_0})\}$ *and* $M_{r_0} = \max\{m_\varphi + rm_\gamma : r \in B(r_0, 4\delta_{r_0})\}$.

Similarly, as a direct consequence of Lemma 20.1.10 we get the following.

Lemma 20.2.2. *Let $\sigma : E_A^\infty \to E_A^\infty$ be a finitely irreducible subshift of finite type over a countable alphabet E. Let $\{\varphi_z\}_{z\in B_{\mathbb{C}}(r_0,4\delta)}$ be a complex-parametrized family of complex-valued potentials of the form (20.9)–(20.11). If the potential φ_{r_0} is strongly summable, then there exists $0 < \delta_{r_0} < \delta$ such that the function*

$$B_{\mathbb{C}}(r_0, \delta_{r_0}) \ni z \longmapsto \mathcal{L}_{\varphi+z\gamma} \in L(\mathcal{K}_\beta^b(E_A^\infty))$$

is holomorphic.

Here is the second result on the real analyticity of pressure. It is a direct application of Theorem 20.1.12.

Theorem 20.2.3. *Let $\sigma : E_A^\infty \to E_A^\infty$ be a finitely irreducible subshift of finite type over a countable alphabet E. Let $\{\varphi_z\}_{z\in B_{\mathbb{C}}(r_0,4\delta)}$ be a complex-parametrized family of complex-valued potentials of the form (20.9)–(20.11). If the potential φ_{r_0} is strongly summable, then there exists an open real interval $I \ni r_0$ over which the function*

$$I \ni r \longmapsto \mathrm{P}(\varphi_r) = \mathrm{P}(\varphi + r\gamma) \in \mathbb{R}$$

is real analytic.

We now examine the derivative of the pressure function.

Theorem 20.2.4. *Let $\sigma : E_A^\infty \to E_A^\infty$ be a finitely irreducible subshift of finite type over a countable alphabet E. Let $\{\varphi_z\}_{z\in B_{\mathbb{C}}(r_0,4\delta)}$ be a complex-parametrized family of complex-valued potentials of the form (20.9)–(20.11). If the potential φ_{r_0} is strongly summable, then*

$$\left.\frac{d\mathrm{P}(\varphi + r\gamma)}{dr}\right|_{r=r_0} = \int_{E_A^\infty} \gamma \, d\mu_{\varphi+r_0\gamma},$$

where $\mu_{\varphi+r_0\gamma} = \mu_{\varphi_{r_0}}$ is the unique ergodic σ-invariant Gibbs state and the unique equilibrium state for φ_{r_0}.

Proof. The proof goes along similar lines to that of Theorem 16.4.10.

By Corollary 17.7.5, the potential φ_{r_0} has a unique σ-invariant Gibbs state $\mu_{\varphi_{r_0}}$. This state is ergodic. The strong summability of φ_{r_0} implies that $\int_{E_A^\infty} \varphi_r \, d\mu_{\varphi_{r_0}} > -\infty$ for all r in some open real interval containing r_0. In particular, $\int_{E_A^\infty} \varphi_{r_0} \, d\mu_{\varphi_{r_0}} > -\infty$ and

Corollary 17.7.5 guarantees that $\mu_{\varphi_{r_0}}$ is the unique equilibrium state for φ_{r_0}. The 3rd variational principle (Theorem 17.3.4) then entails that for every r in the said interval, we have

$$\begin{aligned}
\mathrm{P}(\varphi + r\gamma) &\geq \mathrm{h}_{\mu_{\varphi+r_0\gamma}}(\sigma) + \int_{E_A^\infty} (\varphi + r\gamma)\, d\mu_{\varphi+r_0\gamma} \\
&= \mathrm{h}_{\mu_{\varphi+r_0\gamma}}(\sigma) + \int_{E_A^\infty} (\varphi + r_0\gamma)\, d\mu_{\varphi+r_0\gamma} + (r - r_0)\int_{E_A^\infty} \gamma\, d\mu_{\varphi+r_0\gamma} \\
&= \mathrm{P}(\varphi + r_0\gamma) + (r - r_0)\int_{E_A^\infty} \gamma\, d\mu_{\varphi+r_0\gamma}.
\end{aligned}$$

On one hand,

$$\frac{d\mathrm{P}(\varphi + r\gamma)}{dr}(r_0) = \lim_{r\to r_0^+} \frac{\mathrm{P}(\varphi + r\gamma) - \mathrm{P}(\varphi + r_0\gamma)}{r - r_0} \geq \int_{E_A^\infty} \gamma\, d\mu_{\varphi+r_0\gamma}.$$

On the other hand,

$$\frac{d\mathrm{P}(\varphi + r\gamma)}{dr}(r_0) = \lim_{r\to r_0^-} \frac{\mathrm{P}(\varphi + r\gamma) - \mathrm{P}(\varphi + r_0\gamma)}{r - r_0} \leq \int_{E_A^\infty} \gamma\, d\mu_{\varphi+r_0\gamma}.$$

In conclusion, $\frac{d\mathrm{P}(\varphi+r\gamma)}{dr}(r_0) = \int_{E_A^\infty} \gamma\, d\mu_{\varphi+r_0\gamma}$.

An alternative proof analogous to that of Theorem 13.10.6 is left to the reader as an exercise. □

20.3 Real analyticity of pressure: part III

We now want to prove a result similar to Theorem 20.1.12, in a sense stronger but in another sense weaker. This time, the complex-parametrized family of complex-valued potentials $\{\varphi_z\}_{z\in B_{\mathbb{C}}(0,4\delta)}$ is assumed to originate from a "linear" family

$$\varphi_z = f + h + z\gamma, \tag{20.14}$$

with a strongly summable potential $f \in \mathcal{K}_\beta(E_A^\infty, \mathbb{R})$ and $h, \gamma \in \mathcal{K}_\beta^b(E_A^\infty)$.

Under these assumptions, we will check that the family $\{\varphi_z\}_{z\in B_{\mathbb{C}}(0,4\delta)}$ fulfills conditions (A)–(F) in Definition 20.1.6 for any $\delta > 0$ and appropriate choices of φ, ψ, and κ. Condition (A) is met de facto. Let

$$\varphi = f + h. \tag{20.15}$$

Then $\varphi \in \mathcal{K}_\beta(E_A^\infty)$ since $f, h \in \mathcal{K}_\beta(E_A^\infty)$. So condition (B) is satisfied.

Regarding condition (C), set

$$\psi = f + [\|h\|_\infty + 4\delta\|\gamma\|_\infty].$$

The term between square brackets being constant, it follows from Lemma 20.1.2(d) that the function ψ is strongly summable, like f. Condition (C) holds.

Concerning condition (D), by comparing the expression of φ_z in (20.14) with its definition in condition (F) and taking into account (20.15), we must set $\gamma_z = z\gamma$ in condition (D). As a multiple of $\gamma \in \mathcal{K}_\beta^b(E_A^\infty)$, it is clear that $\gamma_z \in \mathcal{K}_\beta^b(E_A^\infty)$ for every $z \in B_\mathbb{C}(0, 4\delta)$ and

$$Q := \sup_{z\in B_\mathbb{C}(0,4\delta)} \upsilon_\beta(\gamma_z) = \sup_{z\in B_\mathbb{C}(0,4\delta)} |z| \cdot \upsilon_\beta(\gamma) = 4\delta\upsilon_\beta(\gamma) < \infty.$$

Furthermore, for every $\omega \in E_A^\infty$ the function $B_\mathbb{C}(0, 4\delta) \ni z \longmapsto \gamma_z(\omega) = z\gamma(\omega) \in \mathbb{C}$ is obviously holomorphic. Thus, condition (D) is fulfilled.

For condition (E), let $z \in B_\mathbb{C}(0, 4\delta)$ and notice that $|\gamma_z| = |z||\gamma| \le 4\delta\|\gamma\|_\infty$. Choosing $\kappa = 4\delta\|\gamma\|_\infty$, condition (E) holds.

Finally, for any $z \in B_\mathbb{C}(0, 4\delta)$ we have

$$\psi = f + [\|h\|_\infty + 4\delta\|\gamma\|_\infty] \ge f + [|h| + |z||\gamma|] \ge f + \mathrm{Re}(h) + \mathrm{Re}(z\gamma) = \mathrm{Re}(\varphi_z).$$

That is, condition (F) is respected.

Therefore, as a direct consequence of Proposition 20.1.9(b) we obtain the following.

Lemma 20.3.1. *Let $\sigma : E_A^\infty \to E_A^\infty$ be a finitely irreducible subshift of finite type over a countable alphabet E. If $f \in \mathcal{K}_\beta(E_A^\infty, \mathbb{R})$ is strongly summable and $h, \gamma \in \mathcal{K}_\beta^b(E_A^\infty)$, then*

$$\mathcal{L}_{f+h+z\gamma}(\mathcal{K}_\beta^b(E_A^\infty)) \subseteq \mathcal{K}_\beta^b(E_A^\infty), \quad \forall z \in \mathbb{C} \tag{20.16}$$

and

$$\|\mathcal{L}_{f+h+z\gamma}\|_\beta \le (1 + \upsilon_\beta(f) + \upsilon_\beta(h) + 4\delta\upsilon_\beta(\gamma))e^{\|h\|_\infty + 4\delta\|\gamma\|_\infty} \sum_{e\in E} \exp(\sup f|_{[e]}) < \infty \tag{20.17}$$

for all $z \in B_\mathbb{C}(0, 2\delta)$.

Similarly, as a direct consequence of Lemma 20.1.10 we get the following.

Lemma 20.3.2. *Let $\sigma : E_A^\infty \to E_A^\infty$ be a finitely irreducible subshift of finite type over a countable alphabet E. If $f \in \mathcal{K}_\beta(E_A^\infty, \mathbb{R})$ is strongly summable and $h, \gamma \in \mathcal{K}_\beta^b(E_A^\infty)$, then the function*

$$\mathbb{C} \ni z \longmapsto \mathcal{L}_{f+h+z\gamma} \in L(\mathcal{K}_\beta^b(E_A^\infty))$$

is holomorphic. This means that the function

$$\mathcal{K}_\beta^b(E_A^\infty) \ni h \longmapsto \mathcal{L}_{f+h} \in L(\mathcal{K}_\beta^b(E_A^\infty))$$

is G (Goursat)-holomorphic in the sense of [86] *and* [25].

Here is the third result on the real analyticity of pressure.

Theorem 20.3.3. *Let $\sigma : E_A^\infty \to E_A^\infty$ be a finitely irreducible subshift of finite type over a countable alphabet E. If $f \in \mathcal{K}_\beta(E_A^\infty, \mathbb{R})$ is strongly summable, then the function*

$$\mathcal{K}_\beta^b(E_A^\infty, \mathbb{R}) \ni h \longmapsto \mathrm{P}(f+h) \in \mathbb{R}$$

is real analytic.

Proof. Fix $g \in \mathcal{K}_\beta^b(E_A^\infty, \mathbb{R})$. As real analyticity is a local property, it suffices to prove the real analyticity of the function $h \mapsto \mathrm{P}(f+h)$ at the point g. By Theorem 18.1.12, the number $\exp(\mathrm{P}(f+g))$ is a simple isolated eigenvalue of the operator $\mathcal{L}_{f+g} \in L(\mathcal{K}_\beta^b(E_A^\infty))$. So, it follows from Lemma 20.3.2 and the Kato–Rellich theorem for perturbations of linear operators (Theorem A.2.5) that there exist $r > 0$ and a holomorphic function

$$B_{\mathcal{K}_\beta^b(E_A^\infty)}(g,r) \ni h \longmapsto \rho_h \in \mathcal{K}_\beta^b(E_A^\infty)$$

such that ρ_h is an eigenvalue of the operator $\mathcal{L}_{f+h}$ for all $h \in B_{\mathcal{K}_\beta^b(E_A^\infty)}(g,r)$, and $\rho_h = \exp(\mathrm{P}(f+h))$ for those $h \in B_{\mathcal{K}_\beta^b(E_A^\infty,\mathbb{R})}(g,r)$. Therefore the function

$$B_{\mathcal{K}_\beta^b(E_A^\infty,\mathbb{R})}(g,r) \ni h \longmapsto \exp(\mathrm{P}(f+h)) \in \mathbb{R}$$

is real analytic, whence so is the function

$$B_{\mathcal{K}_\beta^b(E_A^\infty,\mathbb{R})}(g,r) \ni h \longmapsto \mathrm{P}(f+h) \in \mathbb{R}.$$

□

We now examine the derivative of the pressure function.

Theorem 20.3.4. *Let $\sigma : E_A^\infty \to E_A^\infty$ be a finitely irreducible subshift of finite type over a countable alphabet E. Let $h \in \mathcal{K}_\beta^b(E_A^\infty, \mathbb{R})$. Suppose that $q_0 \in \mathbb{R}$ and $f \in \mathcal{K}_\beta(E_A^\infty, \mathbb{R})$ are such that $q_0 f$ is strongly summable. Then*

$$\left.\frac{d\mathrm{P}(qf+h)}{dq}\right|_{q=q_0} = \int_{E_A^\infty} f \, d\mu_{q_0 f+h},$$

where $\mu_{q_0 f+h}$ is the unique ergodic σ-invariant Gibbs state and the unique equilibrium state for $q_0 f + h$.

Proof. The proof goes along the same lines as that of Theorem 20.2.4. □

20.4 Real analyticity of Hausdorff dimension for CGDMSs in $\mathbb{C}$

The results of this subsection form a far-reaching strengthening of existing theorems about real analyticity (see [106] and references therein) and even continuity (see [104]) of the Hausdorff dimension of limit sets of conformal graph directed Markov systems whose phase space is contained in $\mathbb{C}$.

It is worth emphasizing that most of the results in this subsection do not require that the Open Set Condition (OSC) hold, as they are primarily formulated in terms of the real analyticity of the Bowen parameter. Real analyticity of the Hausdorff dimension of limit sets will be obtained as an immediate corollary in the case where the OSC holds.

Let $\Lambda \subseteq \mathbb{C}^d$ be an open set. As in Chapter 19, let $\mathcal{G} = (V, E, i, t)$ be a directed multigraph and $A : E \times E \to \{0, 1\}$ an incidence matrix.

Let $\Phi = \{\Phi_\lambda\}_{\lambda\in\Lambda}$ be a family of CGDMSs (not necessarily satisfying the OSC; see Definition 19.3.1) with the following properties:

(i) Every CGDMS $\Phi_\lambda = \{\varphi_e^\lambda : X_{t(e)}^\lambda \to X_{i(e)}^\lambda \mid e \in E\}$ is generated over the (same) multigraph $\mathcal{G}$ and the (same) matrix A and compact sets $X_v^\lambda \subseteq \mathbb{C}$ for all $v \in V$ and $\lambda \in \Lambda$.

(ii) Though the seed sets $(X_v^\lambda)_{v\in V,\lambda\in\Lambda}$ may depend on λ, the open, connected and bounded sets $(W_v)_{v\in V}$ to which the generators $\{\varphi_e^\lambda\}_{e\in E,\lambda\in\Lambda}$ extend, are independent of λ.

(iii) $\inf\{\operatorname{dist}(X_v^\lambda, \partial W_v) \mid v \in V, \lambda \in \Lambda\} > 0$.

(iv) There exists $0 \le s < 1$ such that

$$\|(\varphi_e^\lambda)'\|_{W_{t(e)}} \le s, \quad \forall \lambda \in \Lambda, \forall e \in E.$$

Definition 20.4.1. A family $\Phi = \{\Phi_\lambda\}_{\lambda\in\Lambda}$ of CGDMSs in $\mathbb{C}$ (not necessarily satisfying the OSC) is called *analytic* if for any $e \in E$ and every $z \in W_{t(e)}$, the function $\Lambda \ni \lambda \longmapsto \varphi_e^\lambda(z) \in \mathbb{C}$ is holomorphic.

As in (19.14), for every $\lambda \in \Lambda$ let $f_\lambda := \Xi_{\Phi_\lambda} : E_A^\infty \to \mathbb{R}$ be the potential given by

$$f_\lambda(\omega) := \Xi_{\Phi_\lambda}(\omega) = \log|(\varphi_{\omega_1}^\lambda)'(\pi_\lambda(\sigma(\omega)))|, \quad \forall \omega \in E_A^\infty,$$

where $\pi_\lambda : E_A^\infty \to J_{\Phi_\lambda} =: J_\lambda$ is the canonical coding map induced by the CGDMS Φ_λ (cf. Subsection 19.1.3). For every $v \in V$, choose any $x_v \in W_v$. Because it is a uniform limit of the holomorphic functions $\lambda \longmapsto \varphi_{\omega|_n}^\lambda(x_{t(\omega_n)})$, the function $\lambda \longmapsto \pi_\lambda(\omega)$ is holomorphic for each $\omega \in E_A^\infty$.

Definition 20.4.2. An analytic family $\Phi = \{\Phi_\lambda\}_{\lambda\in\Lambda}$ of CGDMSs in $\mathbb{C}$ (not necessarily satisfying the OSC) is called *strong* at a parameter $\lambda_0 \in \Lambda$ if

(a) the CGDMS Φ_{λ_0} is strongly regular (then we also say that the parameter λ_0 is strongly regular); and

(b)

$$\sup\left\{\left|\frac{f_\lambda(\omega)}{f_{\lambda_0}(\omega)}\right| : \lambda \in \Lambda,\ \omega \in E_A^\infty\right\} < \infty.$$

Note that this latter condition is equivalent to

$$\sup\left\{\left|\frac{f_\lambda(\omega) - f_{\lambda_0}(\omega)}{f_{\lambda_0}(\omega)}\right| : \lambda \in \Lambda,\ \omega \in E_A^\infty\right\} < \infty. \tag{20.18}$$

For every $\lambda \in \Lambda$, let

$$\theta_\lambda := \inf\{t \geq 0 : \mathrm{P}_{\Phi_\lambda}(t) < \infty\} \quad \text{and} \quad h_\lambda := \inf\{t \geq 0 : \mathrm{P}_{\Phi_\lambda}(t) \leq 0\}$$

be respectively the finiteness parameter and the Bowen parameter of the CGDMS Φ_λ.

The main result of this section asserts that the Bowen parameter of an analytic family varies real analytically in the vicinity of any strong parameter.

Theorem 20.4.3. *Let $\Lambda \subseteq \mathbb{C}^d$ be an open set. If $\{\Phi_\lambda\}_{\lambda\in\Lambda}$ is an analytic family of finitely irreducible CGDMSs (not necessarily satisfying the OSC), then the Bowen parameter function $\Lambda \ni \lambda \longmapsto h_\lambda \in \mathbb{R}$ is real analytic on some neighborhood of every strong parameter.*

Proof. Let $\lambda_0 \in \Lambda$ be a strong parameter. For every $\omega \in E_A^\infty$, let $\eta_\omega : \Lambda \to \mathbb{C} \setminus \{0\}$ be the holomorphic function

$$\eta_\omega(\lambda) := \frac{(\varphi_{\omega_1}^\lambda)'(\pi_\lambda(\sigma(\omega)))}{(\varphi_{\omega_1}^{\lambda_0})'(\pi_{\lambda_0}(\sigma(\omega)))}.$$

Shrinking Λ if necessary, we may assume without loss of generality that Λ is simply connected. Then for every $\omega \in E_A^\infty$ the logarithm of η_ω has an holomorphic branch, which is denoted by $\log_\omega \eta_\omega$, such that $\log_\omega \eta_\omega(\lambda_0) = \log_\omega 1 = 0$. Define the holomorphic function $Z_\omega : \Lambda \to \mathbb{C}$ by

$$Z_\omega(\lambda) = \frac{\log_\omega \eta_\omega(\lambda)}{f_{\lambda_0}(\omega)}.$$

Claim 1. *For every $M > 0$, there exists $R_M > 0$ such that $B_{\mathbb{C}^d}(\lambda_0, R_M) \subseteq \Lambda$ and*

$$\sup\{|Z_\omega(\lambda)| : \omega \in E_A^\infty, \lambda \in B_{\mathbb{C}^d}(\lambda_0, R_M)\} < M.$$

Proof of Claim 1. Fix $M > 0$. Let $B > 0$ be the supremum in (20.18). For all $\omega \in E_A^\infty$ and all $\lambda \in \Lambda$, we then have that

$$|\exp(Z_\omega(\lambda))| \leq \exp|\mathrm{Re}(Z_\omega(\lambda))| = \exp\left|\frac{f_\lambda(\omega) - f_{\lambda_0}(\omega)}{f_{\lambda_0}(\omega)}\right| \leq e^B. \tag{20.19}$$

Fix an arbitrary $r > 0$ so small that $B_{\mathbb{C}^d}(\lambda_0, 2r) \subseteq \Lambda$ and let $\lambda \in B_{\mathbb{C}^d}(\lambda_0, r)$. Set

$$\Gamma_r = \{\gamma \in \Lambda : |\gamma_j - \lambda_j| = r, \ \forall 1 \le j \le d\}.$$

Using Cauchy's integral formula (Theorem A.3.2) and (20.19), we obtain that

$$\begin{aligned}\left|\frac{d}{d\lambda}\exp(Z_\omega(\lambda))\right| &= \left|\frac{1}{(2\pi i)^d}\int\limits_{\Gamma_r}\frac{\exp(Z_\omega(\gamma))}{(\gamma_1-\lambda_1)^2\cdots(\gamma_d-\lambda_d)^2}\,d\gamma_1\cdots d\gamma_d\right| \\ &\le \frac{1}{(2\pi)^d}\int\limits_{\Gamma_r}\frac{|\exp(Z_\omega(\gamma))|}{|\gamma_1-\lambda_1|^2\cdots|\gamma_d-\lambda_d|^2}|d\gamma_1|\cdots|d\gamma_d| \\ &= \frac{1}{(2\pi r^2)^d}\int\limits_{\Gamma_r}|\exp(Z_\omega(\gamma))||d\gamma_1|\cdots|d\gamma_d| \\ &\le \frac{e^B(2\pi r)^d}{(2\pi r^2)^d} = \frac{e^B}{r^d}.\end{aligned}$$

As $Z_\omega(\lambda_0) = 0$, we therefore get for all $\lambda \in B_{\mathbb{C}^d}(\lambda_0, r)$ that

$$|\exp(Z_\omega(\lambda)) - 1| = |\exp(Z_\omega(\lambda)) - \exp(Z_\omega(\lambda_0))| \le \frac{e^B}{r^d}|\lambda - \lambda_0|.$$

Taking $R_M \in (0, r]$ sufficiently small and recalling that $Z_\omega(\lambda_0) = 0$, we deduce that

$$\sup\{|Z_\omega(\lambda)| : \omega \in E_A^\infty, \lambda \in B_{\mathbb{C}^d}(\lambda_0, R_M)\} < M.$$

This completes the proof of Claim 1. ∎

Now, let $D_{\mathbb{C}^n}(z, r) := \prod_{k=1}^n B_{\mathbb{C}}(z_k, r)$ be the n-dimensional polydisk in $\mathbb{C}^n$ centered at $z = (z_1, \ldots, z_n)$ and of "radius" r, not to be confused with the open ball $B_{\mathbb{C}^n}(z, r)$ in $\mathbb{C}^n$ centered at z and of real radius r. Indeed, $B_{\mathbb{C}^n}(z, r) \subseteq D_{\mathbb{C}^n}(z, r) \subseteq B_{\mathbb{C}^n}(z, \sqrt{n}r)$ but the equality does not hold when $n > 1$ and $r > 0$.

Fix any $t > \theta_{\lambda_0}$. Pick any $0 < \varepsilon < t - \theta_{\lambda_0}$ and any $\theta_{\lambda_0} < \hat{t} < t - \varepsilon$. Thereafter take $M = 4^{-d}\min\{\hat{t}, t-\varepsilon-\hat{t}\}/(t+\varepsilon)$ and its associated $R_M > 0$ from Claim 1. For every $\omega \in E_A^\infty$, let $\widetilde{Z}_\omega : D_{\mathbb{C}^{2d}}(\widetilde{\lambda}_0, R_M/4) \to \mathbb{C}$ be the holomorphic function produced by Lemma 16.4.8 with $\psi = Z_\omega$. Obviously, we can always assume that $\lambda_0 = (\lambda_{0,1}, \ldots, \lambda_{0,d}) \in \mathbb{R}^d$, so that $\widetilde{\lambda}_0 = (\lambda_{0,1}, 0, \ldots, \lambda_{0,d}, 0) \in \mathbb{R}^{2d}$. Let $u \in B_{\mathbb{C}}(t, \varepsilon)$. We will show that $\{\varphi_{\widetilde{\lambda}}\}_{\widetilde{\lambda} \in B_{\mathbb{C}^{2d}}(r_0, 4\delta)} = \{\omega \mapsto uf_{\lambda_0}(\omega)(1 + \widetilde{Z}_\omega(\widetilde{\lambda}))\}_{\widetilde{\lambda} \in B_{\mathbb{C}^{2d}}(\widetilde{\lambda}_0, R_M/8)}$ is a complex-parametrized family of complex-valued potentials by checking that conditions (A)–(F) in Definition 20.1.6 are satisfied with $\mathbb{C}$ replaced by $\mathbb{C}^{2d}$ and

$$r_0 = \widetilde{\lambda}_0, \quad \delta = \frac{R_M}{32}, \quad \varphi(\omega) = uf_{\lambda_0}(\omega), \quad \gamma_{\widetilde{\lambda}}(\omega) = uf_{\lambda_0}(\omega)\widetilde{Z}_\omega(\widetilde{\lambda}), \quad \psi(\omega) = \hat{t}f_{\lambda_0}(\omega), \quad \kappa = 1.$$

Conditions (A) and (B) are immediately fulfilled. Condition (C) holds because $\widehat{t} > \theta_{\lambda_0}$. To verify condition (D), first note that (D2) is valid because all functions $\widetilde{Z}_\omega$, $\omega \in E_A^\infty$, are holomorphic and $D_{\mathbb{C}^{2d}}(\widetilde{\lambda}_0, R_M/4) \supseteq B_{\mathbb{C}^{2d}}(\widetilde{\lambda}_0, 4\delta)$. For (D1), fix $\omega, \tau \in E_A^\infty$ with $\omega_1 = \tau_1$. Define the holomorphic function $\Delta_{\omega,\tau} : \Lambda \to \mathbb{C}$ by

$$\Delta_{\omega,\tau}(\lambda) := \log_\omega \eta_\omega(\lambda) - \log_\tau \eta_\tau(\lambda). \tag{20.20}$$

Recall that the function $\lambda \mapsto \pi_\lambda(\omega)$ is holomorphic for each $\omega \in E_A^\infty$. Using Koebe distortion theorems (the reformulation in Theorem 19.2.2 and the version for the argument in Theorem A.3.11) and reducing R_M if needed, there exists a constant $K' > 0$ such that

$$|\Delta_{\omega,\tau}(\lambda)| \le K'\Big(|\pi_\lambda(\sigma(\omega)) - \pi_\lambda(\sigma(\tau))| + |\pi_{\lambda_0}(\sigma(\omega)) - \pi_{\lambda_0}(\sigma(\tau))|\Big), \quad \forall \lambda \in D_{\mathbb{C}^d}(\lambda_0, R_M/2).$$

So

$$|\Delta_{\omega,\tau}(\lambda)| \le K s^{|\omega\wedge\tau|-1}, \quad \forall \lambda \in D_{\mathbb{C}^d}(\lambda_0, R_M/2), \tag{20.21}$$

where $K = 2K'$. Let $\widetilde{\Delta}_{\omega,\tau} : D_{\mathbb{C}^{2d}}(\widetilde{\lambda}_0, R_M/8) \to \mathbb{C}$ be the holomorphic function produced by Lemma 16.4.8 with $\psi = \Delta_{\omega,\tau}$. By virtue of that lemma and (20.21), we have

$$|\widetilde{\Delta}_{\omega,\tau}(\widetilde{\lambda})| \le 4^d K s^{|\omega\wedge\tau|-1}, \quad \forall \widetilde{\lambda} \in D_{\mathbb{C}^{2d}}(\widetilde{\lambda}_0, R_M/8). \tag{20.22}$$

Since

$$\Delta_{\omega,\tau}(\lambda) = f_{\lambda_0}(\omega) Z_\omega(\lambda) - f_{\lambda_0}(\tau) Z_\tau(\lambda), \quad \forall \lambda \in D_{\mathbb{C}^d}(\lambda_0, R_M/2),$$

we conclude from the unique holomorphic extension theorem ($D_{\mathbb{C}^d}(\lambda_0, R_M/8) = D_{\mathbb{R}^{2d}}(\widetilde{\lambda}_0, R_M/8)$ is a real, not a complex, submanifold and the least complex submanifold containing $D_{\mathbb{R}^{2d}}(\widetilde{\lambda}_0, R_M/8)$ is $D_{\mathbb{C}^{2d}}(\widetilde{\lambda}_0, R_M/8)$) that

$$\widetilde{\Delta}_{\omega,\tau}(\widetilde{\lambda}) = f_{\lambda_0}(\omega)\widetilde{Z}_\omega(\widetilde{\lambda}) - f_{\lambda_0}(\tau)\widetilde{Z}_\tau(\widetilde{\lambda}), \quad \forall \widetilde{\lambda} \in D_{\mathbb{C}^{2d}}(\widetilde{\lambda}_0, R_M/8). \tag{20.23}$$

In particular, (20.22)–(20.23) hold for all $\widetilde{\lambda} \in B_{\mathbb{C}^{2d}}(\widetilde{\lambda}_0, 4\delta)$. These relations yield

$$|\gamma_{\widetilde{\lambda}}(\omega) - \gamma_{\widetilde{\lambda}}(\tau)| = |u\widetilde{\Delta}_{\omega,\tau}(\widetilde{\lambda})| \le (t+\varepsilon)\cdot 4^d K s^{-1} e^{-\beta|\omega\wedge\tau|}, \quad \forall \widetilde{\lambda} \in B_{\mathbb{C}^{2d}}(\widetilde{\lambda}_0, 4\delta),$$

where $\beta = -\log s$. This means that

$$v_\beta(\gamma_{\widetilde{\lambda}}) \le 4^d K s^{-1}(t+\varepsilon), \quad \forall \widetilde{\lambda} \in B_{\mathbb{C}^{2d}}(\widetilde{\lambda}_0, 4\delta), \tag{20.24}$$

whence condition (D1) is established.

By Lemma 16.4.8, Claim 1 and the choice of M, observe further that

$$|\gamma_{\tilde{\lambda}}(\omega)| = |uf_{\lambda_0}(\omega)\tilde{Z}_\omega(\tilde{\lambda})| \le (t+\varepsilon)|f_{\lambda_0}(\omega)| \cdot 4^d M \le \hat{t}|f_{\lambda_0}(\omega)| = |\psi(\omega)| \tag{20.25}$$

for all $\tilde{\lambda} \in D_{\mathbb{C}^{2d}}(\tilde{\lambda}_0, R_M/4) \supseteq B_{\mathbb{C}^{2d}}(\tilde{\lambda}_0, 4\delta)$. Condition (E) holds by setting $\kappa = 1$.

Finally, let us check condition (F). For every $\omega \in E_A^\infty$ and every $\tilde{\lambda} \in B_{\mathbb{C}^{2d}}(\tilde{\lambda}_0, 4\delta)$, we obtain from (20.25) and the choice of M that

$$\begin{aligned}
\operatorname{Re}(\varphi_{\tilde{\lambda}}(\omega)) &\le \operatorname{Re}(\varphi(\omega)) + |\gamma_{\tilde{\lambda}}(\omega)| \\
&\le \operatorname{Re}(u)f_{\lambda_0}(\omega) + 4^d M(t+\varepsilon)|f_{\lambda_0}(\omega)| \\
&\le (t-\varepsilon)f_{\lambda_0}(\omega) - 4^d M(t+\varepsilon)f_{\lambda_0}(\omega) \\
&\le \psi(\omega).
\end{aligned} \tag{20.26}$$

So condition (F) is fulfilled. Thus, we have verified all the conditions (A)–(F) and hence $\{\varphi_{\tilde{\lambda}}(\omega) = uf_{\lambda_0}(\omega)(1+\tilde{Z}_\omega(\tilde{\lambda}))\}_{\tilde{\lambda}\in B_{\mathbb{C}^{2d}}(\tilde{\lambda}_0,4\delta)}$ is a complex-parametrized family of complex-valued potentials. This holds for all $u \in B_{\mathbb{C}}(t,\varepsilon)$ and δ is independent of u.

For all $\omega \in E_A^\infty$ and $\tilde{\lambda} \in B_{\mathbb{C}^{2d}}(\tilde{\lambda}_0, 4\delta)$, set

$$\tilde{f}_{\tilde{\lambda}}(\omega) := f_{\lambda_0}(\omega)(1+\tilde{Z}_\omega(\tilde{\lambda})). \tag{20.27}$$

Per Lemma 20.1.11, we then know the following.

Claim 2. *If Λ is an open subset of $\mathbb{C}^d$ and $\Phi = \{\Phi_\lambda\}_{\lambda\in\Lambda}$ is an analytic family of CGDMSs (not necessarily satisfying the OSC) with a strong parameter $\lambda_0 \in \Lambda$, then for every $t > \theta_{\lambda_0}$ and every $0 < \varepsilon < t - \theta_{\lambda_0}$ there is $\delta > 0$ such that for every $u \in B_{\mathbb{C}}(t,\varepsilon)$ the function*

$$B_{\mathbb{C}^{2d}}(\tilde{\lambda}_0, \delta) \longmapsto \mathcal{L}_{\tilde{\lambda},u} \in L(\mathcal{K}_\beta^b(E_A^\infty))$$

is holomorphic, where $\mathcal{L}_{\tilde{\lambda},u}$ is the transfer operator corresponding to the potential $u\tilde{f}_{\tilde{\lambda}}$.

Now, using much of the work just done, it is not difficult to demonstrate that for every $\tilde{\lambda} \in B_{\mathbb{C}^{2d}}(\tilde{\lambda}_0, 4\delta)$ the family $\{\varphi_u\}_{u\in B_{\mathbb{C}}(t,\varepsilon)} = \{\omega \mapsto uf_{\lambda_0}(\omega)(1+\tilde{Z}_\omega(\tilde{\lambda}))\}_{u\in B_{\mathbb{C}}(t,\varepsilon)}$ is a complex-parametrized family of complex-valued potentials by checking that conditions (A)–(F) in Definition 20.1.6 are satisfied with

$$r_0 = t, \quad \delta = \varepsilon/4, \quad \varphi \equiv 0, \quad \gamma_u(\omega) = \varphi_u(\omega) = uf_{\lambda_0}(\omega)(1+\tilde{Z}_\omega(\tilde{\lambda})), \quad \psi = \hat{t}f_{\lambda_0}$$

and an appropriate κ. As before, conditions (A), (B), (C) and (D2) are fulfilled. For (D1), fix $\omega, \tau \in E_A^\infty$ with $\omega_1 = \tau_1$. Noting that $\gamma_u = uf_{\lambda_0} + \gamma_{\tilde{\lambda}}$, we deduce from (20.24) that

$$v_\beta(\gamma_u) \le |u|v_\beta(f_{\lambda_0}) + v_\beta(\gamma_{\tilde{\lambda}}) \le (v_\beta(f_{\lambda_0}) + 4^d Ks^{-1})(t+\varepsilon), \quad \forall u \in B_{\mathbb{C}}(t,\varepsilon),$$

whence condition (D) is established.

Using (20.25), we also get

$$|\gamma_u| \le |uf_{\lambda_0}| + |\gamma_{\tilde\lambda}| \le (t+\varepsilon)|f_{\lambda_0}| + 4^d M(t+\varepsilon)|f_{\lambda_0}| = \frac{(1+4^d M)(t+\varepsilon)}{\hat t}|\psi|, \quad \forall u \in B_{\mathbb{C}}(t,\varepsilon).$$

Condition (E) holds with $\kappa = (1+4^d M)(t+\varepsilon)/\hat t$.

Finally, condition (F) is respected because $\varphi_u = \varphi_{\tilde\lambda}$ and (20.26) holds.

Hence $\{\varphi_u(\omega) = uf_{\lambda_0}(\omega)(1+\tilde Z_\omega(\tilde\lambda))\}_{u\in B_{\mathbb{C}}(t,\varepsilon)}$ is a complex-parametrized family of potentials. This holds for all $\tilde\lambda \in B_{\mathbb{C}^{2d}}(\tilde\lambda_0, 4\delta)$ and ε is independent of $\tilde\lambda$.

Per Lemma 20.1.11, we then obtain the following result.

Claim 3. *If Λ is an open subset of $\mathbb{C}^d$ and $\Phi = \{\Phi_\lambda\}_{\lambda\in\Lambda}$ is an analytic family of CGDMSs (not necessarily satisfying the OSC) with a strong parameter $\lambda_0 \in \Lambda$, then for every $t > \theta_{\lambda_0}$ and every $0 < \varepsilon < t - \theta_{\lambda_0}$ there is $\delta > 0$ such that for every $\tilde\lambda \in B_{\mathbb{C}^{2d}}(\tilde\lambda_0, \delta)$ the function*

$$B_{\mathbb{C}}(t,\varepsilon) \ni u \longmapsto \mathcal{L}_{\tilde\lambda,u} \in L(\mathcal{K}^b_\beta(E^\infty_A))$$

is holomorphic.

Using Hartogs' theorem (Theorem A.4.1), we finally establish the following fact.

Claim 4. *If Λ is an open subset of $\mathbb{C}^d$ and $\Phi = \{\Phi_\lambda\}_{\lambda\in\Lambda}$ is an analytic family of CGDMSs (not necessarily satisfying the OSC) with a strong parameter $\lambda_0 \in \Lambda$, then for every $t > \theta_{\lambda_0}$ and every $0 < \varepsilon < t - \theta_{\lambda_0}$ there is $\delta > 0$ such that the function*

$$B_{\mathbb{C}^{2d}}(\tilde\lambda_0,\delta) \times B_{\mathbb{C}}(t,\varepsilon) \ni (\tilde\lambda, u) \longmapsto \mathcal{L}_{\tilde\lambda,u} \in L(\mathcal{K}^b_\beta(E^\infty_A))$$

is holomorphic, where $\mathcal{L}_{\tilde\lambda,u}$ is the transfer operator corresponding to the potential $u\tilde f_{\tilde\lambda}$.

We can now provide the concluding argument to the proof of the theorem. Looking at (20.26), observe that all the complex-valued potentials $u\tilde f_{\tilde\lambda}$, with $\tilde\lambda \in B_{\mathbb{C}^{2d}}(\tilde\lambda_0,\delta)$ and $u \in B_{\mathbb{C}}(t,\varepsilon)$, have a real part which is summable and thus even strongly summable (see Definition 20.1.1). In particular, this includes the real-valued potentials sf_λ, with $\lambda \in B_{\mathbb{C}^d}(\lambda_0,\delta)$ and $s \in B_{\mathbb{R}}(t,\varepsilon)$. These real-valued potentials are also Hölder on cylinders since $f_\lambda := \Xi_{\Phi_\lambda}$. Theorem 18.1.12 then ensures that the number $\exp(\mathrm{P}(sf_\lambda))$ is a simple isolated eigenvalue of the operator $\mathcal{L}_{\lambda,s}$ for all $\lambda \in B_{\mathbb{C}^d}(\lambda_0,\delta)$ and all $s \in B_{\mathbb{R}}(t,\varepsilon)$. It follows from Claim 4 and the Kato–Rellich theorem for perturbations of linear operators (Theorem A.2.5) that there exist $0 < \eta < \min\{\delta,\varepsilon\}$ and a holomorphic function

$$\rho : B_{\mathbb{C}^{2d}}(\tilde\lambda_0,\eta) \times B_{\mathbb{C}}(t,\eta) \to \mathbb{C}$$

such that $\rho(\tilde\lambda, u)$ is an eigenvalue of the operator $\mathcal{L}_{\tilde\lambda,u}$ for all $(\tilde\lambda,u) \in B_{\mathbb{C}^{2d}}(\tilde\lambda_0,\eta)\times B_{\mathbb{C}}(t,\eta)$ and

$$\rho(\lambda,s) = \exp(\mathrm{P}(sf_\lambda)), \quad \forall (\lambda,s) \in B_{\mathbb{C}^d}(\lambda_0,\eta) \times B_{\mathbb{R}}(t,\eta).$$

Hence, the function

$$B_{\mathbb{C}^d}(\lambda_0, \eta) \times B_{\mathbb{R}}(t, \eta) \ni (\lambda, s) \longmapsto \exp(\mathrm{P}(sf_\lambda)) \in \mathbb{R}$$

is real analytic. Consequently, the function

$$B_{\mathbb{C}^d}(\lambda_0, \eta) \times B_{\mathbb{R}}(t, \eta) \ni (\lambda, s) \longmapsto \mathrm{P}(sf_\lambda) \in \mathbb{R}$$

is real analytic. Moreover, it is easy to see that for each $\lambda \in B_{\mathbb{C}^d}(\lambda_0, \eta)$ and each $s \in B_{\mathbb{R}}(t, \eta)$ the family $\{\varphi_z\}_{z \in B_{\mathbb{C}}(s, \eta - |s-t|)} = \{\omega \mapsto zf_\lambda(\omega)\}_{z \in B_{\mathbb{C}}(s, \eta - |s-t|)}$ is a complex-parametrized family of complex-valued potentials of the form (20.9)–(20.11), with φ_s strongly summable for all $\lambda \in B_{\mathbb{C}^d}(\lambda_0, \eta)$. Then Theorem 20.2.3 asserts that $\mathrm{P}(sf_\lambda)$ depends real analytically on s for any λ and Theorem 20.2.4 affirms that $\frac{\partial}{\partial s}\mathrm{P}(sf_\lambda) = \int_{E_A^\infty} f_\lambda \, d\mu_{\lambda,s} < 0$ for all $(\lambda, s) \in B_{\mathbb{C}^d}(\lambda_0, \eta) \times B_{\mathbb{R}}(t, \eta)$. It follows from an application of the real-analytic implicit function theorem (a real-analytic version of Theorem A.3.16) to the equation $\mathrm{P}(sf_\lambda) = 0$ that the solution function, namely, the Bowen parameter function

$$B_{\mathbb{C}^d}(\lambda_0, \eta) \ni \lambda \longmapsto h_\lambda \in \mathbb{R}$$

is real analytic. □

Corollary 20.4.4. *Let $\Lambda \subseteq \mathbb{C}^d$ be an open set. If $\{\Phi_\lambda\}_{\lambda \in \Lambda}$ is an analytic family of finitely irreducible CGDMSs (satisfying the OSC), then the Hausdorff dimension function $\Lambda \ni \lambda \mapsto \mathrm{HD}(J_\lambda) \in \mathbb{R}$ is real analytic on some neighborhood of every strong parameter.*

Proof. By Theorem 19.6.4, the Bowen parameter of each CGDMS in this family is equal to the Hausdorff dimension of the corresponding limit set. So the result follows from Theorem 20.4.3. □

We now provide a useful, sufficient condition for a parameter to be strong.

Definition 20.4.5. An analytic family of CGDMSs $\{\Phi_\lambda\}_{\lambda \in \Lambda}$ (not necessarily satisfying the OSC) is *Hölder stable* at some parameter $\lambda_0 \in \Lambda$ if

(a) the CGDMS Φ_{λ_0} is strongly regular; and
(b) there are two constants $C > 0$ and $\beta \in (0, 1]$ such that for every $\lambda \in \Lambda$ there exists a homeomorphism $H_\lambda : J_{\lambda_0} \to J_\lambda$ with the properties

$$C^{-1}|z - w|^{1/\beta} \le |H_\lambda(z) - H_\lambda(w)| \le C|z - w|^\beta, \quad \forall z, w \in J_{\lambda_0}$$

and

$$\varphi_e^\lambda \circ H_\lambda = H_\lambda \circ \varphi_e^{\lambda_0}, \quad \forall e \in E.$$

Proposition 20.4.6. *If an analytic family of CGDMSs $\{\Phi_\lambda\}_{\lambda\in\Lambda}$ (not necessarily satisfying the OSC) is Hölder stable at some parameter $\lambda_0 \in \Lambda$, then λ_0 is a strong parameter for that family.*

Proof. It suffices to check condition (20.18). From condition (b) in Definition 20.4.5, it follows that

$$C^{-1}\operatorname{diam}^{1/\beta}(\varphi_e^{\lambda_0}(J_{\lambda_0})) \le \operatorname{diam}(\varphi_e^{\lambda}(J_\lambda)) \le C\operatorname{diam}^{\beta}(\varphi_e^{\lambda_0}(J_{\lambda_0})), \quad \forall e \in E,\ \forall \lambda \in \Lambda.$$

But by Koebe's distortion theorem (Theorem A.3.8 or A.3.9), the diameter of the image of a compact set (having at least two elements) under a univalent function is comparable to the derivative of the univalent function at any point of the domain multiplied by the diameter of the set. So there exists $\widehat{C} \ge 1$ such that

$$\widehat{C}^{-1}\big|(\varphi_e^{\lambda_0})'(z)\big|^{1/\beta} \le \big|(\varphi_e^{\lambda})'(w)\big| \le \widehat{C}\big|(\varphi_e^{\lambda_0})'(z)\big|^{\beta} \tag{20.28}$$

for all $z \in J_{\lambda_0}$, all $w \in J_\lambda$, all $e \in E$ and all $\lambda \in \Lambda$. Hence,

$$\begin{aligned} -\log\widehat{C} + \Big(\frac{1}{\beta} - 1\Big)\log\big|(\varphi_e^{\lambda_0})'(z)\big| &\le \log\big|(\varphi_e^{\lambda})'(w)\big| - \log\big|(\varphi_e^{\lambda_0})'(z)\big| \\ &\le \log\widehat{C} + (\beta - 1)\log\big|(\varphi_e^{\lambda_0})'(z)\big|. \end{aligned}$$

Therefore,

$$\left|\frac{\log|(\varphi_e^{\lambda})'(w)| - \log|(\varphi_e^{\lambda_0})'(z)|}{\log|(\varphi_e^{\lambda_0})'(z)|}\right| \le \max\Big\{1-\beta, \frac{1}{\beta} - 1\Big\} - \frac{\log\widehat{C}}{\log s} < \infty. \qquad \square$$

An immediate consequence of Theorem 20.4.3 and Proposition 20.4.6 is the following.

Corollary 20.4.7. *If an analytic family of finitely irreducible CGDMSs $\{\Phi_\lambda\}_{\lambda\in\Lambda}$ (not necessarily satisfying the OSC) is Hölder stable at some parameter $\lambda_0 \in \Lambda$, then the Bowen parameter function $\Lambda \ni \lambda \mapsto h_\lambda$ is real analytic in some neighborhood of λ_0.*

A straightforward repercussion of this and Theorem 19.6.4 is the following.

Corollary 20.4.8. *If an analytic family of finitely irreducible CGDMSs $\{\Phi_\lambda\}_{\lambda\in\Lambda}$ (satisfying the OSC) is Hölder stable at some parameter $\lambda_0 \in \Lambda$, then the Hausdorff dimension function $\Lambda \ni \lambda \mapsto \mathrm{HD}(J_\lambda) \in \mathbb{R}$ is real analytic in some neighborhood of λ_0.*

As the following result asserts, the existence of a holomorphic motion which is a conjugation of the generators of an analytic family, ensures the real analyticity of the Bowen parameter function in a neighborhood of any strongly regular system.

Theorem 20.4.9. *Let $\mathbb{D}$ be the unit disk in $\mathbb{C}$. Let $\{\Phi_\lambda\}_{\lambda\in\mathbb{D}}$ be an analytic family of finitely irreducible CGDMSs (not necessarily satisfying the OSC) such that Φ_0 is a strongly regular system. If there exists an holomorphic motion $H : \mathbb{D}\times\widehat{\mathbb{C}} \to \widehat{\mathbb{C}}$ such that*

$$\varphi_e^\lambda(H(\lambda,z)) = H(\lambda,\varphi_e^0(z)), \quad \forall\lambda\in\mathbb{D},\ \forall z\in J_0,$$

then the Bowen parameter function $\mathbb{D}\ni\lambda\to h_\lambda$ is real analytic on a neighborhood of 0.

Proof. By the λ-lemma (Theorem 16.4.4) and Remark 16.4.6, the analytic family $\{\Phi_\lambda\}_{\lambda\in\mathbb{D}}$ is Hölder stable at 0. A direct application of Corollary 20.4.7 yields the result. □

Below is a direct consequence of this and Theorem 19.6.4.

Corollary 20.4.10. *Let $\mathbb{D}$ be the unit disk in $\mathbb{C}$. Let $\{\Phi_\lambda\}_{\lambda\in\mathbb{D}}$ be an analytic family of finitely irreducible CGDMSs (satisfying the OSC). If Φ_0 is a strongly regular system and if there exists an holomorphic motion $H : \mathbb{D}\times\widehat{\mathbb{C}} \to \widehat{\mathbb{C}}$ such that*

$$\varphi_e^\lambda(H(\lambda,z)) = H(\lambda,\varphi_e^0(z)), \quad \forall\lambda\in\mathbb{D},\ \forall z\in J_0,$$

then the Hausdorff dimension function $\Lambda\ni\lambda\mapsto \mathrm{HD}(J_\lambda)\in\mathbb{R}$ is real analytic on a neighborhood of 0.

We now provide some mild, quite general, sufficient conditions for an analytic family of CGDMSs to admit a holomorphic motion.

Definition 20.4.11. Let $\Phi = \{\varphi_e\}_{e\in E}$ be a CGDMS (not necessarily satisfying the OSC) and

$$E_{\mathrm{per}}^* = \{\omega\in E_A^* : A_{\omega_{|\omega|}\omega_1} = 1 \text{ and } \omega\neq\tau^k \text{ for any } \tau\in E_A^* \text{ and } k\geq 2\}.$$

For every $\omega\in E_{\mathrm{per}}^*$, let $x_\omega\in X_{t(\omega)}$ be the fixed point of the map $\varphi_\omega : W_{t(\omega)}\to W_{t(\omega)}$. Then Φ is said to be *periodically separated* if $x_\omega\neq x_\tau$ for all $\omega,\tau\in E_{\mathrm{per}}^*$ such that $\omega\neq\tau$.

The following lemma asserts that periodic separation is a well-defined concept.

Lemma 20.4.12. *Let $\Phi = \{\varphi_e\}_{e\in E}$ be a CGDMS (not necessarily satisfying the OSC). Distinct words in E_{per}^* generate distinct periodic points.*

Proof. If $\omega,\chi\in E_{\mathrm{per}}^*$ are not comparable, then $\omega_i\neq\chi_i$ for some $i\leq\min\{|\omega|,|\chi|\}$ and hence $\omega^\infty\neq\chi^\infty$. So suppose that $\omega,\omega\rho\in E_{\mathrm{per}}^*$ with $\rho\neq\epsilon$. There are unique $m,n\in\mathbb{Z}_+$ with $n<|\omega|$ such that $|\rho| = m|\omega|+n$. In turn, there are unique $q\in\mathbb{N}$ and $0\leq r<n$ such that $|\omega| = qn+r$. If $\omega^\infty = (\omega\rho)^\infty$, then it can be shown that $\omega=\tau^k$ for some $\tau\in E_A^*$ with $|\tau| = \gcd(n,r)$. Obviously, $k = |\omega|/\gcd(n,r)\in\mathbb{N}\setminus\{1\}$. But this means that $\omega\notin E_{\mathrm{per}}^*$. Thus, $\omega^\infty\neq(\omega\rho)^\infty$. □

There is one obvious separation condition under which a CGDMS is periodically separated.

Proposition 20.4.13. *If a CGDMS $\Phi = \{\varphi_e\}_{e\in E}$ (not necessarily satisfying the OSC) satisfies the Strong Separation Condition (SSC), then Φ is periodically separated.*

Analytic families of periodically separated, finitely irreducible CGDMSs admit holomorphic motions.

Lemma 20.4.14. *If $\Lambda \subseteq \mathbb{C}$ is an open simply connected set and $\{\Phi_\lambda\}_{\lambda\in\Lambda}$ is an analytic family of periodically separated, finitely irreducible CGDMSs (not necessarily satisfying the OSC), then for every $\lambda_0 \in \Lambda$ there exists a holomorphic motion $H : \Lambda \times \overline{J_{\lambda_0}} \to \mathbb{C}$ such that*

$$\varphi_e^\lambda(H(\lambda,z)) = H(\lambda,\varphi_e^{\lambda_0}(z)), \quad \forall \lambda \in \Lambda,\ \forall z \in \overline{J_{\lambda_0}}.$$

In addition,

$$H(\{\lambda\}\times J_{\lambda_0}) = J_\lambda \quad \text{and} \quad H(\{\lambda\}\times \overline{J_{\lambda_0}}) = \overline{J_\lambda}, \quad \forall \lambda \in \Lambda.$$

Proof. Fix momentarily $\omega \in E^*_{\text{per}}$. Since the map

$$\Lambda \times W_{t(\omega)} \ni (\lambda,z) \longmapsto \varphi_\omega^\lambda(z) \in \mathbb{C}$$

is holomorphic and since $(\varphi_\omega^\lambda)'(z) \neq 1$ for all $(\lambda,z) \in \Lambda \times W_{t(\omega)}$, it follows from the implicit function theorem (Theorem A.3.16 with $F(z,\lambda) = \varphi_\omega^\lambda(z)-z$) that for every $\lambda_0 \in \Lambda$ and $r_{\lambda_0,\omega} > 0$ small enough there exists a unique holomorphic function

$$B(\lambda_0, r_{\lambda_0,\omega}) \ni \lambda \longmapsto x^\lambda_{\lambda_0,\omega} \in W_{t(\omega)}$$

such that $\varphi_\omega^\lambda(x^\lambda_{\lambda_0,\omega}) = x^\lambda_{\lambda_0,\omega}$, i. e., $x^\lambda_{\lambda_0,\omega}$ is the unique fixed point of $\varphi_\omega^\lambda : W_{t(\omega)} \to W_{t(\omega)}$. The uniqueness of that function imposes that if $\lambda \in B(\lambda_0', r_{\lambda_0',\omega}) \cap B(\lambda_0, r_{\lambda_0,\omega})$, then $x^\lambda_{\lambda_0',\omega} = x^\lambda_{\lambda_0,\omega}$ and thus all the maps $\{B(\lambda_0, r_{\lambda_0,\omega}) \ni \lambda \mapsto x^\lambda_{\lambda_0,\omega} \in W_{t(\omega)}\}_{\lambda_0\in\Lambda}$ glue together into a unique holomorphic function $\Lambda \ni \lambda \mapsto x^\lambda_\omega \in W_{t(\omega)}$ such that

$$\varphi_\omega^\lambda(x_\omega^\lambda) = x_\omega^\lambda, \quad \forall \lambda \in \Lambda. \tag{20.29}$$

We now consider the different $\omega \in E^*_{\text{per}}$ altogether. For each $\lambda \in \Lambda$, let

$$Y_\lambda = \{x_\omega^\lambda : \omega \in E^*_{\text{per}}\}.$$

Since the system Φ_λ is periodically separated, the map

$$\begin{aligned} F_\lambda : E^*_{\text{per}} &\longrightarrow Y_\lambda \\ \omega &\longmapsto F_\lambda(\omega) := x_\omega^\lambda \end{aligned}$$

is bijective. Consequently, so is the map

$$\begin{aligned} Y_{\lambda_0} &\longrightarrow Y_\lambda \\ z &\longmapsto F_\lambda \circ F_{\lambda_0}^{-1}(z) = x^\lambda_{F_{\lambda_0}^{-1}(z)}. \end{aligned}$$

Therefore, the map $H : \Lambda \times Y_{\lambda_0} \to \mathbb{C}$ given by

$$H(\lambda, z) = x^\lambda_{F_{\lambda_0}^{-1}(z)}$$

is a holomorphic motion. By the λ-lemma (Theorem 16.4.4), this map uniquely extends to a holomorphic motion $H : \Lambda \times \overline{Y_{\lambda_0}} \to \mathbb{C}$. Moreover, $\overline{Y_\lambda} = \overline{J_\lambda}$ for each $\lambda \in \Lambda$ due to the finite irreducibility of A. Furthermore, by continuity of H and (20.29) it ensues that

$$\varphi^\lambda_\omega(H(\lambda, z)) = H(\lambda, \varphi^{\lambda_0}_\omega(z)), \quad \forall \lambda \in \Lambda, \ \forall z \in \overline{J_{\lambda_0}}. \tag{20.30}$$

In addition, it follows from (20.30) that

$$\pi_\lambda(\omega) = H(\lambda, \pi_{\lambda_0}(\omega)), \quad \forall \lambda \in \Lambda, \ \forall \omega \in E^\infty_A.$$

Therefore,

$$J_\lambda = \pi_\lambda(E^\infty_A) = H(\{\lambda\} \times \pi_{\lambda_0}(E^\infty_A)) = H(\{\lambda\} \times J_{\lambda_0}), \quad \forall \lambda \in \Lambda.$$

By a standard argument, one deduces from all the above that

$$\overline{J_\lambda} = \overline{Y_\lambda} = \overline{H(\{\lambda\} \times Y_{\lambda_0})} = H(\{\lambda\} \times \overline{Y_{\lambda_0}}) = H(\{\lambda\} \times \overline{J_{\lambda_0}}), \quad \forall \lambda \in \Lambda. \qquad \square$$

Using this lemma and Slodkowski's theorem (Theorem A.3.18), we obtain the following.

Theorem 20.4.15. *Let $\Lambda \subseteq \mathbb{C}$ be an open simply connected set whose complement $\mathbb{C} \backslash \Lambda$ contains at least two points. If $\{\Phi_\lambda\}_{\lambda \in \Lambda}$ is an analytic family of periodically separated, finitely irreducible CGDMSs (not necessarily satisfying the OSC), then for every $\lambda_0 \in \Lambda$ there exists a holomorphic motion $H : \Lambda \times \widehat{\mathbb{C}} \to \widehat{\mathbb{C}}$ such that*

$$\varphi^\lambda_e(H(\lambda, z)) = H(\lambda, \varphi^{\lambda_0}_e(z)), \quad \forall \lambda \in \Lambda, \ \forall z \in \overline{J_{\lambda_0}}.$$

In addition,

$$H(\{\lambda\} \times J_{\lambda_0}) = J_\lambda \quad \text{and} \quad H(\{\lambda\} \times \overline{J_{\lambda_0}}) = \overline{J_\lambda} \quad \forall \lambda \in \Lambda.$$

Proof. By virtue of Lemma 20.4.14 there exists a holomorphic motion on $\lambda \times \overline{J_{\lambda_0}}$ satisfying the required properties. By Slodkowski's theorem, this motion can be extended to an holomorphic motion on $\Lambda \times \widehat{\mathbb{C}}$ with uniformly bounded dilatation. $\square$

The following result is an immediate consequence from Theorem 20.4.15 and Proposition 20.4.9.

Corollary 20.4.16. *Let $\Lambda \subseteq \mathbb{C}$ be an open simply connected set whose complement $\mathbb{C}\backslash\Lambda$ contains at least two points. Let $\{\Phi_\lambda\}_{\lambda\in\Lambda}$ be an analytic family of periodically separated, finitely irreducible CGDMSs (not necessarily satisfying the OSC). If the system Φ_{λ_0} is strongly regular, then the Bowen parameter function $\lambda \mapsto h_\lambda$ is real analytic on some neighborhood of λ_0.*

Corollary 20.4.17. *Let $\Lambda \subseteq \mathbb{C}$ be an open simply connected set whose complement $\mathbb{C}\backslash\Lambda$ contains at least two points. Let $\{\Phi_\lambda\}_{\lambda\in\Lambda}$ be an analytic family of periodically separated, finitely irreducible CGDMSs (satisfying the OSC). If the system Φ_{λ_0} is strongly regular, then the Hausdorff dimension function $\Lambda \ni \lambda \mapsto \mathrm{HD}(J_\lambda) \in \mathbb{R}$ is real analytic on some neighborhood of λ_0.*

20.5 Exercises

Exercise 20.5.1. Let Φ be a CGDMS and $f \in \mathcal{K}^s_\beta(E^\infty_A, \mathbb{R})$ a strongly summable potential. Show that the function $t \mapsto \mathrm{HD}(\mu_{tf} \circ \pi^{-1})$, defined on a sufficiently small neighborhood of 1, is real analytic.

Exercise 20.5.2. Let $\Lambda \subseteq \mathbb{C}^d$ be an open set. If $\{\Phi_\lambda\}_{\lambda\in\Lambda}$ is an analytic family of finitely irreducible CGDMSs (not necessarily satisfying the OSC) and if $f \in \mathcal{K}^s_\beta(E^\infty_A, \mathbb{R})$ is a potential such that $\chi_{\mu_f}(\sigma) < \infty$, prove that the function

$$\Lambda \ni \lambda \longmapsto \mathrm{HD}(\mu_f \circ \pi_\lambda^{-1})$$

is real analytic.

21 Multifractal analysis for conformal graph directed Markov systems

In this chapter, we perform a multifractal analysis of the Hausdorff dimension of the level sets of Gibbs states for any Hölder family of functions of the form $g_e + u \log|\varphi_e'|$, where $\{g_e\}_{e\in E}$ is a bounded Hölder family of functions and $\Phi = \{\varphi_e\}_{e\in E}$ is a CGDMS (conformal graph directed Markov system). This is a fertile field of applications of the theory of CGDMSs and more largely of this book. In addition to its intrinsic interest and importance, the multifractal analysis conducted in this chapter bears witness to the strength and sophistication of the tools and methods developed in this book.

To our best knowledge, rigorous mathematical research on multifractal analysis began in 1989 with David Rand's work [100] on cookie-cutter Cantor sets, which nowadays can be viewed as limit sets of finite(-alphabet) IFSs (iterated function systems) generated by similarities. Shortly afterwards, a multifractal analysis was done for general finite similarity IFSs by Cawley and Mauldin [15]. Roughly a decade later, a much more general and, we have to say, quite tedious multifractal analysis for countable(-alphabet) CIFSs (conformal IFSs) was achieved by Hanus, Mauldin and Urbański [46] and reproduced in [81] in the context of CGDMSs. That analysis relies on cylinders. Nearly a decade thereafter, a multifractal analysis was completed by Roy and Urbański [107] using balls. Our exposition stems from that paper but is more detailed and improved.

Recall that the Hausdorff dimension of a fractal set gives us an idea of the global structure of that set. Nevertheless, two fractal sets with similar global structures may have substantially different local structures. In Section 16.6, we saw that the pointwise or local dimension of a Borel measure seizes the local structure of a set and potentially helps in distinguishing that set from another one which may be globally similar. This is accomplished through a multifractal analysis, i. e., the examination of the multifractal spectrum, which describes the relative sizes of a multitude of fractal subsets which form a splitting of the original fractal set. More precisely, let μ be a Borel probability measure on a metric space X. For each number $\alpha \geq 0$, let $X_\mu(\alpha) := \{x \in X \mid d_\mu(x) = \alpha\}$ be the subset of all points $x \in X$ where the measure μ has local dimension α. Let also $X_\mu^c := \{x \in X \mid d_\mu(x)\nexists\} = X \setminus \bigcup_{\alpha\geq 0} X_\mu(\alpha)$ be the subset of all points $x \in X$ where the measure μ does not have a local dimension. In general, $X_\mu^c \neq \emptyset$ and a study of that set may be of interest. However, we will see that in our case $X_\mu^c = \emptyset$. Then $\bigcup_{\alpha\geq 0} X_\mu(\alpha) = X$. Let $d_\mu(\alpha) = \mathrm{HD}(X_\mu(\alpha))$ be the Hausdorff dimension of the fractal subset $X_\mu(\alpha)$. (Note: In Section 16.6, we used the more classical notation $F_\mu(\alpha)$. In this chapter, $d_\mu(\alpha)$ is more convenient since F will bear a different meaning.) The map $\alpha \mapsto d_\mu(\alpha)$ is the *(fine Hausdorff) multifractal spectrum* of the measure μ.

In [46, 81], the multifractal analysis used at each point of the limit set J the natural filtration generated by the initial cylinders associated to the word that encodes the point. In brief, the analysis was carried out using cylinders. Aiming to give the

https://doi.org/10.1515/9783110702699-021

multifractal analysis a transparent geometrical meaning, we derive in this chapter an analysis using as the filtration a base of balls centred at the given point. We will conduct the analysis of the conformal measure (or, equivalently, of the invariant measure) associated to a family of weights imposed upon a (finite or infinite) CGDMS. We will also conduct this analysis over the full set of parameters on which it can be expected to hold. However, we will restrict our attention to cofinitely regular, finitely irreducible, conformal-like CGDMSs that satisfy the SOSC (see Definitions 19.5.5, 17.1.2, 19.7.17 and 19.7.1, resp.) and to a large and dynamically significant subset J_0 of the limit set J, as opposed to the entire limit set, though the sets J_0 and J coincide under a mild boundary separation condition. An additional geometric flavor of this analysis results from the fact that we concentrate on a geometrically meaningful family of Hölder weights. The results apply to large classes of CGDMSs, one- and multi-dimensional alike, including real and complex continued fractions.

Let us describe the content of the chapter more precisely. In Section 21.1, we describe the Hölder families of weights F and $F_{q,t}$ we shall work with, and we study the properties of the pressure function $\mathrm{P}(q,t)$ and of the temperature function $T(q)$ they determine. Recall that the general properties of Hölder families of functions and the amalgamated function they induce were studied in Section 19.8. In Section 21.2, we carry out the multifractal analysis of the conformal measure m_F (or, equivalently, of the invariant measure μ_F) associated to the family of weights F. This analysis is done over a set $J_0 \subseteq J$ of full μ_F-measure and conducted by means of balls. In particular, we show that for each α there is an auxiliary measure that witnesses the Hausdorff dimension of the set $J_{0,\mu}(\alpha)$ and that the $d_{0,\mu}(\alpha)$ curve is the Legendre transform of the temperature function $T(q)$. Part of the analysis is restricted to a subclass called almost-CIFSs. In Section 21.3, we derive the multifractal analysis of m_F under additional conditions on the almost-CIFSs. In Subsection 21.3.1, we observe that $J_0 = J$ for all CGDMSs which satisfy the boundary separation condition. Real continued fractions with the digit 1 deleted are an important example of such systems. In Subsection 21.3.2, we derive the multifractal analysis over J under three conditions and show that there are families of one-dimensional CIFSs that meet these three conditions. Real continued fractions (with or without the digit 1) are a fundamental example of such families. Finally, in Section 21.4 we conduct a similar analysis on smaller subsets of points of the limit set J (they have a symbolic coding with at least one infinitely-seen letter from a finite subalphabet) but for all CGDMSs that are cofinitely regular, finitely irreducible, conformal-like, and satisfy the SOSC.

21.1 Pressure and temperature

In this section, we study the concept of temperature. This notion is defined in terms of the pressure of a two-parameter family of functions.

Let $\Phi = \{\varphi_e : X_{t(e)} \to X_{i(e)} \mid e \in E\}$ be a finitely irreducible CGDMS. Let $\mathrm{Fin}_\Phi := \{t \geq 0 : \mathrm{P}(t) < \infty\}$ and $\theta := \inf \mathrm{Fin}_\Phi$ be the finiteness set and finiteness parameter of Φ, respectively. Let $u \in \mathrm{Fin}_\Phi$. Let also

$$G = \{g_e : X_{t(e)} \to \mathbb{R} \mid e \in E\}$$

be a bounded Hölder family of functions, and let β be an Hölder exponent for that family. Bounded simply means that

$$\|G\| := \sup_{e\in E} \|g_e\|_{X_{t(e)}} < \infty.$$

The amalgamated function $g : E_A^\infty \to \mathbb{R}$ induced by the family G is clearly bounded, and Hölder continuous on cylinders with exponent β by Lemma 19.8.6.

Per Proposition 19.8.10, the family $\mathrm{Log}_\Phi := \{\log|\varphi_e'| : X_{t(e)} \to \mathbb{R} \mid e \in E\}$ is Hölder with exponent $-\alpha \log s$ and the family $t\,\mathrm{Log}_\Phi$ is summable for any $t \in \mathrm{Fin}_\Phi$. In particular, $u\,\mathrm{Log}_\Phi$ is summable and its amalgamated function $u\Xi_\Phi$, where Ξ_Φ was defined in (19.14), is summable and Hölder on cylinders with exponent $-\alpha \log s$.

The family $F := G + u\,\mathrm{Log}_\Phi$, i. e., $F = \{f_e : X_{t(e)} \to \mathbb{R} \mid e \in E\}$, where

$$f_e = g_e + u \log|\varphi_e'|,$$

is then summable and Hölder with exponent $\zeta := \min\{\beta, -\alpha \log s\}$. Its amalgamated function $f : E_A^\infty \to \mathbb{R}$ satisfies

$$f = g + u\Xi_\Phi$$

and is summable and Hölder continuous on cylinders with exponent ζ. Note also that F and f are bounded above by $\|G\|$.

Recall that the topological pressure $\mathrm{P}(F)$ of the family F is defined by

$$\mathrm{P}(F) := \lim_{n\to\infty} \frac{1}{n} \log \sum_{\omega\in E_A^n} \|\exp(S_\omega F)\|_{X_{t(\omega)}} < \infty.$$

Replacing the family G by $G - \mathrm{P}(F)$, we may assume without loss of generality that the family F is such that $\mathrm{P}(F) = 0$. **This will be a standing assumption throughout this chapter.** Equivalently, $\mathrm{P}(f) = 0$ by Proposition 19.8.9, where

$$\mathrm{P}(f) = \lim_{n\to\infty} \frac{1}{n} \log Z_n(f) = \lim_{n\to\infty} \frac{1}{n} \sum_{\omega\in E_A^n} \exp(\overline{S}_n f([\omega])).$$

Now, for every ordered pair of parameters $(q, t) \in \mathbb{R}^2$ define the family

$$F_{q,t} := qF + t\,\mathrm{Log}_\Phi = qG + (qu + t)\,\mathrm{Log}_\Phi$$

and its corresponding amalgamated function

$$f_{q,t} = qf + t\Xi_\Phi = qg + (qu + t)\Xi_\Phi.$$

Observe that $F_{1,0} = F$ and $f_{1,0} = f$. Note also that $F_{0,t} = t\,\mathrm{Log}_\Phi$ and $f_{0,t} = t\Xi_\Phi$. It is easy to see that all the $F_{q,t}$ are Hölder families of functions with exponent ζ (the exponent of F), and thus the $f_{q,t}$ are Hölder continuous on cylinders with exponent ζ (like f). By Proposition 19.8.10, $F_{q,t}$ and $f_{q,t}$ are summable if and only if $qu + t \in \mathrm{Fin}_\Phi$.

In particular, as Φ is finitely irreducible, Corollary 17.7.5 guarantees that for every $(q,t) \in \mathbb{R}^2$ such that $qu + t > \theta$ the dual transfer operator $\mathcal{L}^*_{f_{q,t}}$ has a unique eigenmeasure $m_{f_{q,t}}$. This measure is also a Gibbs state for $f_{q,t}$ and $f_{q,t}$ admits a unique σ-invariant Gibbs state $\mu_{f_{q,t}}$ according to that same corollary. As two Gibbs states for $f_{q,t}$, the measures $\mu_{f_{q,t}}$ and $m_{f_{q,t}}$ are boundedly equivalent. By the same corollary still, $\mu_{f_{q,t}}$ is ergodic, and even totally ergodic if Φ is finitely primitive. The measure $\mu_{f_{q,t}}$ is also the unique equilibrium state for $f_{q,t}$. Indeed, note that $f_{q,t} \in L^1(\mu_{f_{q,t}})$ since g is bounded and $\Xi_\Phi \in L^1(\mu_{f_{q,t}})$, as we will now show.

To shorten notation, we shall write $\|\varphi'_\omega\|$ for $\|\varphi'_\omega\|_{X_{t(\omega)}}$. Let $0 < \delta < (qu + t) - \theta$ and $I := \inf\{\|\varphi'_e\|^\delta \log\|\varphi'_e\| : e \in E\}$. Notice that $-\infty < I < 0$ because $\lim_{x\to 0^+} x^\delta \log x = 0^-$. Let $C_{q,t} \geq 1$ be a Gibbs constant for the Gibbs state $\mu_{f_{q,t}}$ and let $K \geq 1$ be a constant from the BDP (Corollary 19.3.5). Then

$$\begin{aligned}
\int_{E_A^\infty} |\Xi_\Phi|\,d\mu_{f_{q,t}} &= \sum_{e\in E}\int_{[e]} (-\Xi_\Phi)\,d\mu_{f_{q,t}} = \sum_{e\in E}\int_{[e]} -\log|\varphi'_e(\pi(\sigma(\omega)))|\,d\mu_{f_{q,t}}(\omega)\\
&\leq \sum_{e\in E}\int_{[e]} -\log(K^{-1}\|\varphi'_e\|)\,d\mu_{f_{q,t}}\\
&= \log K - \sum_{e\in E}\log\|\varphi'_e\|\mu_{f_{q,t}}([e])\\
&\leq \log K - C_{q,t}e^{-\mathrm{P}(f_{q,t})}\sum_{e\in E}\log\|\varphi'_e\|\exp(\sup(f_{q,t}|_{[e]}))\\
&\leq \log K - C_{q,t}e^{-\mathrm{P}(f_{q,t})}\sum_{e\in E}\log\|\varphi'_e\|\exp(\sup(qg_e) + (qu+t)\sup\log|\varphi'_e|)\\
&\leq \log K - C_{q,t}e^{-\mathrm{P}(f_{q,t})+|q|\cdot\|G\|}\sum_{e\in E}\log\|\varphi'_e\|\exp((qu+t)\log\|\varphi'_e\|)\\
&= \log K - C_{q,t}e^{-\mathrm{P}(f_{q,t})+|q|\cdot\|G\|}\sum_{e\in E}\|\varphi'_e\|^{qu+t}\log\|\varphi'_e\|\\
&= \log K - C_{q,t}e^{-\mathrm{P}(f_{q,t})+|q|\cdot\|G\|}\sum_{e\in E}\|\varphi'_e\|^{(qu+t)-\delta}\|\varphi'_e\|^\delta\log\|\varphi'_e\|\\
&\leq \log K - C_{q,t}e^{-\mathrm{P}(f_{q,t})+|q|\cdot\|G\|}I\sum_{e\in E}\|\varphi'_e\|^{(qu+t)-\delta}\\
&= \log K - C_{q,t}e^{-\mathrm{P}(f_{q,t})+|q|\cdot\|G\|}I\,Z_1((qu+t)-\delta)\\
&< \infty.
\end{aligned} \tag{21.1}$$

So Ξ_Φ, and hence $f_{q,t}$, are in $L^1(\mu_{f_{q,t}})$. By Corollary 17.7.5(c), the measure $\mu_{f_{q,t}}$ is then the unique equilibrium state for $f_{q,t}$.

Finally, for every $n \in \mathbb{N}$ observe that (cf. (17.1) and Definition 19.4.2)

$$\begin{aligned} Z_n(f_{q,t}) &= \sum_{\omega\in E_A^n} \sup_{\rho\in[\omega]} \Big[\exp(qS_n g(\rho)) \cdot |\varphi_\omega'(\pi(\sigma^n(\rho)))|^{qu+t}\Big] \\ &\geq \exp(-n|q| \cdot \|G\|) \cdot K^{-(qu+t)} Z_n(qu+t). \end{aligned}$$

Hence,

$$\mathrm{P}(f_{q,t}) \geq -|q| \cdot \|G\| + \mathrm{P}(qu+t).$$

A similar upper bound holds to yield

$$\mathrm{P}(qu+t) - |q| \cdot \|G\| \leq \mathrm{P}(f_{q,t}) \leq \mathrm{P}(qu+t) + |q| \cdot \|G\|. \tag{21.2}$$

We now state fundamental properties of the pressure as a function of the two variables q and t.

Theorem 21.1.1. *Let $\Phi = \{\varphi_e\}_{e\in E}$ be a finitely irreducible CGDMS. Let θ be the finiteness parameter of Φ and $u > \theta$. The pressure function $\mathbb{R}^2 \ni (q,t) \longmapsto \mathrm{P}(q,t) := \mathrm{P}(F_{q,t}) = \mathrm{P}(f_{q,t}) \in (-\infty,\infty]$ satisfies the following properties:*

(a) *The quantity $\mathrm{P}(q,t)$ has the same (finite or infinite) nature as $\mathrm{P}(qu+t)$. That is, $\mathrm{P}(q,t) < \infty$ if and only if $qu+t \in \mathrm{Fin}_\Phi$. Therefore, Φ is cofinitely regular if and only if $\mathrm{P}(q,t) = \infty$ whenever $qu+t = \theta$.*

(b) *If $q_1 \leq q_2$ and $t_1 \leq t_2$ are such that*

$$(q_2 - q_1)(\sup G + u\log s) + (t_2 - t_1)\log s \leq 0,$$

then

$$\mathrm{P}(q_2,t_2) \leq \mathrm{P}(q_1,t_1).$$

In particular, if $\sup G \leq -u\log s$, then $\mathrm{P}(q,t)$ is decreasing with respect to both variables q and t.

(c) *$t \longmapsto \mathrm{P}(q,t)$ is strictly decreasing on $(\theta - qu, \infty)$ for every $q \in \mathbb{R}$.*

(d) *$\lim_{t\to\infty} \mathrm{P}(q,t) = -\infty$ for every $q \in \mathbb{R}$.*

(e) *If Φ is cofinitely regular, then $\lim_{t\to(\theta-qu)^+} \mathrm{P}(q,t) = \infty$ for every $q \in \mathbb{R}$.*

(f) *$\dfrac{\partial P}{\partial q}(q,t) = \displaystyle\int_{E_A^\infty} f\, d\mu_{f_{q,t}} \leq 0$ for every $(q,t) \in \mathbb{R}^2$ such that $qu+t > \theta$.*

(g) *$\dfrac{\partial P}{\partial t}(q,t) = -\chi_{\mu_{f_{q,t}}}(\sigma) := \displaystyle\int_{E_A^\infty} \Xi_\Phi\, d\mu_{f_{q,t}} < 0$ for every $(q,t) \in \mathbb{R}^2$ such that $qu+t > \theta$.*

(h) *The pressure function* $t \longmapsto \mathrm{P}(q,t)$ *is convex on the interval* $(\theta - qu, \infty)$ *for every* $q \in \mathbb{R}$.

(i) *The pressure function* $(q,t) \longmapsto \mathrm{P}(q,t)$ *is convex on the half-plane* $\{(q,t) \in \mathbb{R}^2 : t > \theta - qu\}$.

Proof. (a) This follows from (21.2), the definition of Fin_Φ, and the fact that cofinite regularity is characterized by $\mathrm{P}(\theta) = \infty$.

(b) Let $q_1 \le q_2$ and $t_1 \le t_2$. If $q_1 u + t_1 \notin \mathrm{Fin}_\Phi$, then $\mathrm{P}(q_1, t_1) = \infty$ according to (a) and the statement thus holds. So suppose that $q_1 u + t_1 \in \mathrm{Fin}_\Phi$. Then $q_2 u + t_2 \in \mathrm{Fin}_\Phi$ and the nth partition function of f_{q_2,t_2} satisfies

$$
\begin{aligned}
Z_n(f_{q_2,t_2}) &= \sum_{\omega\in E_A^n} \exp\Big(\sup_{\rho\in[\omega]} S_n f_{q_2,t_2}(\rho)\Big) \\
&= \sum_{\omega\in E_A^n} \sup_{\rho\in[\omega]} \exp\Big(q_2 S_n g(\rho) + (q_2 u + t_2) S_n \Xi_\Phi(\rho)\Big) \\
&= \sum_{\omega\in E_A^n} \sup_{\rho\in[\omega]} \Big[\exp(q_2 S_n g(\rho)) \cdot |\varphi'_\omega(\pi(\sigma^n(\rho)))|^{q_2 u + t_2}\Big] \\
&\le \sum_{\omega\in E_A^n} \sup_{\rho\in[\omega]} \Big[\exp(q_2 S_n g(\rho))\Big] \cdot \|\varphi'_\omega\|_{X_{t(\omega)}}^{q_2 u + t_2} \\
&= \sum_{\omega\in E_A^n} \sup_{\rho\in[\omega]} \Big[\exp(q_1 S_n g(\rho) + (q_2 - q_1) S_n g(\rho))\Big] \\
&\qquad \cdot \|\varphi'_\omega\|_{X_{t(\omega)}}^{q_1 u + t_1} \cdot \|\varphi'_\omega\|_{X_{t(\omega)}}^{(q_2 - q_1)u + (t_2 - t_1)} \\
&\le \sum_{\omega\in E_A^n} \sup_{\rho\in[\omega]} \exp(q_1 S_n g(\rho)) \cdot e^{n(q_2 - q_1)\sup G} \\
&\qquad \cdot K^{q_1 u + t_1} \inf_{x\in X_{t(\omega)}} |\varphi'_\omega(x)|^{q_1 u + t_1} \cdot s^{n[(q_2 - q_1)u + (t_2 - t_1)]} \\
&\le e^{n(q_2 - q_1)\sup G} s^{n[(q_2 - q_1)u + (t_2 - t_1)]} K^{q_1 u + t_1} \\
&\qquad \cdot \sum_{\omega\in E_A^n} \sup_{\rho\in[\omega]} \Big[\exp(q_1 S_n g(\rho)) \cdot |\varphi'_\omega(\pi(\sigma^n(\rho)))|^{q_1 u + t_1}\Big] \\
&= e^{n(q_2 - q_1)\sup G} s^{n[(q_2 - q_1)u + (t_2 - t_1)]} K^{q_1 u + t_1} Z_n(f_{q_1,t_1}).
\end{aligned}
$$

Consequently,

$$
\mathrm{P}(q_2, t_2) \le (q_2 - q_1)\sup G + \big[(q_2 - q_1)u + (t_2 - t_1)\big] \log s + \mathrm{P}(q_1, t_1). \tag{21.3}
$$

Part (b) follows immediately.

(c) Letting $q_1 = q_2 = q$ and $t_1 < t_2$ in (21.3) yields (c).

(d) This also follows from (21.3) by setting $q_1 = q_2 = q$, $t_1 > \theta$ and $t_2 = t$, and then letting $t \to \infty$.

(e) Using (21.2) and the right-continuity of $\mathrm{P}(\bullet)$, we obtain that

$$\lim_{t\to(\theta-qu)^+} \mathrm{P}(q,t) \geq -|q|\|G\| + \mathrm{P}(\theta) = \infty$$

when Φ is cofinitely regular.

(f) First, note that $f \in L^1(\mu_{f_{q,t}})$ for all $(q,t) \in \mathbb{R}^2$ such that $qu + t > \theta$ since G is bounded, $u > \theta$ and $\Xi_\Phi \in L^1(\mu_{f_{q,t}})$ by (21.1). The statement then follows from Theorem 20.2.4, by replacing r with q and setting $\varphi = t\Xi_\Phi$ and $\gamma = f$, so that $\varphi_q = \varphi + q\gamma = qf + t\Xi_\Phi = f_{q,t}$. Furthermore, by the 2nd or 3rd variational principle (Theorem 17.3.3 or 17.3.4), observe that

$$\int_{E_A^\infty} f\,d\mu_{f_{q,t}} \leq \mathrm{P}(f) - \mathrm{h}_{\mu_{f_{q,t}}}(\sigma) = -\mathrm{h}_{\mu_{f_{q,t}}}(\sigma) \leq 0. \tag{21.4}$$

Recall that $\mathrm{P}(F) = \mathrm{P}(f) = 0$ is a standing assumption in this chapter.

(g) The proof goes along similar lines to (f), using the fact that $\chi_{\mu_{f_{q,t}}}(\sigma) < \infty$ whenever $qu + t > \theta$ according to (21.1). More precisely, in Theorem 20.2.4, replace r with t and set $\varphi = qf$ and $\gamma = \Xi_\Phi$, so that $\varphi_t = \varphi + t\gamma = qf + t\Xi_\Phi = f_{q,t}$.

(h,i) These statements follow from Exercise 17.9.10 and the fact that the family of potentials $f_{q,t}$ is convex on the half-plane $H = \{(q,t) \in \mathbb{R}^2 : t > \theta - qu\}$, i. e. for any $p \in [0,1]$ and $(q_1,t_1),(q_2,t_2) \in H$ we have $f_{pq_1+(1-p)q_2,\,pt_1+(1-p)t_2} = pf_{q_1,t_1} + (1-p)f_{q_2,t_2}$. □

Remark 21.1.2. In the proof of part (e), observe that $\mathrm{P}(\theta) \geq |q|\|G\|$ suffices to guarantee the existence of a zero for the pressure function $t \mapsto \mathrm{P}(q,t)$. The assumption of cofinite regularity on Φ ensures that the pressure function $t \mapsto \mathrm{P}(q,t)$ has a zero for every $q \in \mathbb{R}$.

Theorem 21.1.1 and Remark 21.1.2 lead to the next definition.

Definition 21.1.3. Let $\Phi = \{\varphi_e\}_{e\in E}$ be a cofinitely regular, finitely irreducible CGDMS. The function $T : \mathbb{R} \to \mathbb{R}$ defined for every $q \in \mathbb{R}$ as the unique $T(q) \in (\theta - qu, \infty)$ such that $\mathrm{P}(q, T(q)) = 0$, is called the *temperature function*.

In order to lighten notation, we shall henceforth let

$$f_q := f_{q,T(q)}, \quad F_q := F_{q,T(q)} \quad \text{and} \quad m_q := m_{f_{q,T(q)}}, \quad \mu_q := \mu_{f_{q,T(q)}}.$$

Let $q \in \mathbb{R}$. Recall that $\Xi_\Phi, f \in L^1(\mu_q)$. As $qu + T(q) > \theta$, Theorem 21.1.1(f,g) asserts that $\frac{\partial \mathrm{P}}{\partial q}(q,T(q)) = \int_{E_A^\infty} f\,d\mu_q \leq 0$ while $\frac{\partial \mathrm{P}}{\partial t}(q,T(q)) = -\chi_{\mu_q}(\sigma) = \int_{E_A^\infty} \Xi_\Phi\,d\mu_q < 0$. Consequently,

$$0 \leq \alpha(q) := \frac{\frac{\partial \mathrm{P}}{\partial q}(q,T(q))}{\frac{\partial \mathrm{P}}{\partial t}(q,T(q))} = \frac{\int_{E_A^\infty} f\,d\mu_q}{-\chi_{\mu_q}(\sigma)} < \infty. \tag{21.5}$$

We now enumerate basic features of the temperature function.

Theorem 21.1.4. *Let $\Phi = \{\varphi_e\}_{e\in E}$ be a cofinitely regular, finitely irreducible CGDMS. The temperature function $q \mapsto T(q)$ exhibits the following properties:*
(a) *The function $T : \mathbb{R} \to \mathbb{R}$ is real analytic.*
(b) *$T(0) = \mathrm{HD}(J)$ while $T(1) = 0$.*
(c) *$T'(q) = -\alpha(q) < 0$ for all $q \in \mathbb{R}$.*
(d) *The function $T : \mathbb{R} \to \mathbb{R}$ is convex, i. e. $T''(q) \geq 0$ for all $q \in \mathbb{R}$.*
(e) *If $\mu_f = \mu_{\mathrm{HD}(J)\Xi_\Phi}$ or, equivalently, if f and $\mathrm{HD}(J)\Xi_\Phi$ are cohomologous in the group of Hölder functions on E_A^∞, then T is linear. More precisely, $T(q) = \mathrm{HD}(J)(1-q)$.*

Proof. (a) By Theorem 21.1.1(g),

$$\frac{\partial P}{\partial t}(q, T(q)) = \int_{E_A^\infty} \Xi_\Phi \, d\mu_q = -\chi_{\mu_q}(\sigma) < 0, \quad \forall q \in \mathbb{R}.$$

Since $T(q)$ is uniquely determined by the condition $\mathrm{P}(q, T(q)) = 0$, it follows from the implicit function theorem and Theorem 20.2.3 (applied in the same setting as in the proof of Theorem 21.1.1(g)) that T is real analytic on $\mathbb{R}$.

(b) By definition, $T(0)$ is the unique number $t \in (\theta, \infty)$ such that $0 = \mathrm{P}(0,t) = \mathrm{P}(f_{0,t}) = \mathrm{P}(t\Xi_\Phi) = \mathrm{P}(t)$. However, $\mathrm{P}(\mathrm{HD}(J)) = 0$ since Φ is regular. So $T(0) = \mathrm{HD}(J)$.

Similarly, $T(1)$ is the unique number $t \in (\theta - u, \infty)$ such that $\mathrm{P}(1,t) = 0$. However, since $f_{1,0} = f$ and $\mathrm{P}(F) = \mathrm{P}(f) = 0$ by standing assumption, we deduce that $\mathrm{P}(1,0) = \mathrm{P}(f_{1,0}) = \mathrm{P}(f) = 0$. Therefore, $T(1) = 0$.

(c) It follows from the fact that $\mathrm{P}(q, T(q)) = 0$ for all $q \in \mathbb{R}$ and from Theorem 21.1.1(f,g) that

$$0 = \frac{dP}{dq}(q, T(q)) = \frac{\partial P}{\partial q}(q, T(q)) + \frac{\partial P}{\partial t}(q, T(q)) \cdot T'(q).$$

Hence,

$$T'(q) = -\frac{\frac{\partial P}{\partial q}(q, T(q))}{\frac{\partial P}{\partial t}(q, T(q))} = -\frac{\int_{E_A^\infty} f \, d\mu_q}{-\chi_{\mu_q}(\sigma)} = -\alpha(q). \tag{21.6}$$

Having already observed that $\alpha(q) \geq 0$, we thus know that $T'(q) \leq 0$. In order to prove that $T'(q) < 0$, we need to show that $\int_{E_A^\infty} f \, d\mu_q \neq 0$. But since $\int_{E_A^\infty} f \, d\mu_q \leq -\mathrm{h}_{\mu_q}(\sigma)$ by (21.4), it suffices to show that $\mathrm{h}_{\mu_q}(\sigma) > 0$. This follows immediately from Theorem 17.8.3 since $f_q \in \mathcal{K}_\zeta^s(E_A^\infty) \cap L^1(\mu_q)$. Also, E_A^∞ is not a single periodic orbit since Φ is cofinitely regular.

(d) The proof of this part essentially goes along the same lines as that of the last paragraph in Theorem 16.6.5(b,d). We recap part of the argument. The inequality $T''(q) \geq 0$ follows from the convexity of $\mathrm{P}(q,t)$ (see Theorem 21.1.1(i)). Indeed, the fact that the part of $\mathbb{R}^3$ above the graph of $\mathrm{P}(q,t)$ is convex implies that its intersection with the (q,t)-plane is also convex. Since $\frac{\partial \mathrm{P}(q,t)}{\partial t} = -\chi_{\mu_{f_{q,t}}}(\sigma) < 0$ by Theorem 21.1.1(g), this is the part of the plane above the graph of T. Hence, T is a convex function.

(e) If $\mu_f = \mu_{\mathrm{HD}(J)\Xi_\Phi}$ then Theorem 18.4.1 asserts that $f - \mathrm{HD}(J)\Xi_\Phi$ is cohomologous to a constant R via a Hölder coboundary $\upsilon : E_A^\infty \to \mathbb{R}$. By an analog of Lemma 13.2.6 (whose proof carries over to the current setting), Bowen's formula (Theorem 19.6.4) and our standing assumption on f, we deduce that $0 = \mathrm{P}(f) = \mathrm{P}(\mathrm{HD}(J)\Xi_\Phi) + R = R$. Consequently,

$$\begin{aligned}\mathrm{P}(q, T(q)) = 0 = \mathrm{P}(\mathrm{HD}(J)\Xi_\Phi) &= \mathrm{P}(q\upsilon - q\upsilon \circ \sigma + \mathrm{HD}(J)\Xi_\Phi)\\ &= \mathrm{P}\Big(q(\mathrm{HD}(J)\Xi_\Phi + \upsilon - \upsilon \circ \sigma) + \mathrm{HD}(J)(1-q)\Xi_\Phi\Big)\\ &= \mathrm{P}(qf + \mathrm{HD}(J)(1-q)\Xi_\Phi)\\ &= \mathrm{P}(q, \mathrm{HD}(J)(1-q)).\end{aligned}$$

By definition of $T(q)$, we infer that $T(q) = \mathrm{HD}(J)(1-q)$. □

21.2 Multifractal analysis of the conformal measure m_F over a subset of J

Let $\Phi = \{\varphi_e : X_{t(e)} \to X_{i(e)} \mid e \in E\}$ be a cofinitely regular, finitely irreducible, conformal-like CGDMS satisfying the Strong Open Set Condition (SOSC). Let θ be the finiteness parameter of Φ. Let also F be a family of functions of the form $F = G + u\,\mathrm{Log}_\Phi$ such that $\mathrm{P}(F) = 0$, where G is a bounded Hölder family of functions and $u > 0$.

If $qu + t \in \mathrm{Fin}_\Phi$, then $F_{q,t}$ and $f_{q,t}$ are summable and Theorem 19.8.14 asserts that there is a unique $F_{q,t}$-conformal measure $m_{F_{q,t}}$ supported on J. This measure is

$$m_{F_{q,t}} = m_{f_{q,t}} \circ \pi^{-1}.$$

Like $\mu_{f_{q,t}}$ and $m_{f_{q,t}}$, the measures $\mu_{F_{q,t}} := \mu_{f_{q,t}} \circ \pi^{-1}$ and $m_{F_{q,t}} = m_{f_{q,t}} \circ \pi^{-1}$ are boundedly equivalent. For this reason, $\mu_{F_{q,t}}$ is called the *Φ-invariant version* of $m_{F_{q,t}}$. To abridge notation, we will write

$$m_{F_q} := m_{F_{q,T(q)}} \quad \text{and} \quad \mu_{F_q} := \mu_{F_{q,T(q)}}.$$

We now develop a multifractal analysis of the conformal measure m_F (or, equivalently, of the invariant measure μ_F) associated to the family F. We shall conduct this analysis by means of balls and we will restrict ourselves to the subset $J_0 := \pi(E_0^\infty)$ of the limit set J of Φ. According to Theorem 19.7.3, the completely shift-invariant set

$$E_0^\infty := \Big\{\omega \in E_A^\infty \;\Big|\; \limsup_{n\to\infty} \mathrm{dist}(\pi(\sigma^n(\omega)), \partial X_{i(\sigma^n(\omega))}) > 0\Big\}$$

is of full measure under any ergodic σ-invariant Borel probability measure on E_A^∞ that has full topological support and we thus have the next result.

Corollary 21.2.1. *Let $\Phi = \{\varphi_e\}_{e\in E}$ be a finitely irreducible CGDMS satisfying the SOSC. For all $(q,t) \in \mathbb{R}^2$ such that $qu + t > \theta$, we have*

$$\mu_{F_{q,t}}(J_0) = \mu_{f_{q,t}}(E_0^\infty) = 1.$$

In particular,

$$\mu_F(J_0) = \mu_f(E_0^\infty) = 1,$$

i. e., the set J_0 is a set of full μ_F-measure.

If, additionally, Φ is cofinitely regular then for all $q \in \mathbb{R}$,

$$\mu_{F_q}(J_0) = \mu_q(E_0^\infty) = 1.$$

For every $\omega \in E_0^\infty$, let

$$r(\omega) := \limsup_{n\to\infty} \operatorname{dist}(\pi(\sigma^n(\omega)), \partial X_{i(\sigma^n(\omega))}) > 0. \tag{21.7}$$

For every $R > 0$, define the set

$$E_R^\infty := \{\omega \in E_0^\infty \mid r(\omega) \geq R\} \tag{21.8}$$

Observe that $E_0^\infty = \bigcup_{R>0} E_R^\infty$. Like E_0^∞, the sets E_R^∞ are completely σ-invariant. We now prove likewise that they are of full measure under any ergodic shift-invariant measure with full topological support, as long as R is small enough.

Theorem 21.2.2. *Let $\Phi = \{\varphi_e\}_{e\in E}$ be a CGDMS satisfying the SOSC. For every ergodic σ-invariant Borel probability measure μ on E_A^∞ such that $\operatorname{supp}(\mu) = E_A^\infty$, it turns out that*

$$\mu(E_R^\infty) = \mu \circ \pi^{-1}\left(\bigcup_{v\in V} X_v \setminus B(\partial X_v, R)\right) = 1$$

for all $0 < R < \sup_{x\in J} \operatorname{dist}(x, \partial X)$.

Proof. Let $0 < R < D := \sup_{x\in J} \operatorname{dist}(x,\partial X) = \sup_{v\in V}\sup_{x\in J_v} \operatorname{dist}(x,\partial X_v)$. The SOSC guarantees that $D > 0$, and thus there is $x \in J_{v^*} \setminus B(\partial X_{v^*}, (R+D)/2)$ for some $v^* \in V$. Let $\omega \in E_A^\infty$ be such that $\pi(\omega) = x$. Then there is $k \in \mathbb{N}$ such that

$$\pi([\omega|_k]) \subseteq \varphi_{\omega|_k}(X_{t(\omega_k)}) \subseteq B(\pi(\omega), (D-R)/2) \subseteq X_{v^*} \setminus B(\partial X_{v^*}, R).$$

Therefore,

$$\mu \circ \pi^{-1}\Big(\bigcup_{v \in V} X_v \backslash B(\partial X_v, R)\Big) \geq \mu \circ \pi^{-1}(X_{v^*} \backslash B(\partial X_{v^*}, R)) \geq \mu \circ \pi^{-1}(\pi([\omega|_k])) \geq \mu([\omega|_k]) > 0$$

since $\mathrm{supp}(\mu) = E_A^\infty$. Poincaré's recurrence theorem (Theorem 8.1.2) states that

$$\mu\Big(\Big\{\rho \in E_A^\infty \;\Big|\; \sigma^n(\rho) \in \pi^{-1}\Big(\bigcup_{v \in V} X_v \backslash B(\partial X_v, R)\Big) \text{ for infinitely many } n\Big\}\Big)$$
$$= \mu\Big(\pi^{-1}\Big(\bigcup_{v \in V} X_v \backslash B(\partial X_v, R)\Big)\Big) > 0.$$

But the set on the left-hand side of this relation is completely σ-invariant, so the ergodicity of μ imposes that its measure be 1. As that set is contained in E_R^∞, the result ensues. □

We now refine the sets we look at by restricting some of the symbols they can have. For every $\omega \in E_A^\infty$, every $R > 0$ and every $e \in E$, let $(n_j(\omega, R, e))_{j=1}^N$ be the sequence of all $n \geq 0$ such that

$$\sigma^n(\omega) \in \pi^{-1}\Big(\bigcup_{v \in V} X_v \backslash B(\partial X_v, R)\Big) \cap [e]. \tag{21.9}$$

This means that $n_j(\omega, R, e)$ is the jth time that an iterate of ω hits the set $\pi^{-1}(\bigcup_{v \in V} X_v \backslash B(\partial X_v, R)) \cap [e]$. In other words, $n_j(\omega, R, e)$ is the jth time that an iterate of ω starts with the letter e and is projected to a point that lies at a distance greater than or equal to R from the boundary of X. This sequence may be empty ($N = 0$), nonempty and finite ($N \in \mathbb{N}$), or infinite ($N = \infty$), depending on ω, R and e. To shorten notation, we shall usually write $n_j(\omega)$ instead of $n_j(\omega, R, e)$ when R and e are already fixed.

Define the sets

$$E_{R,e}^\infty := \Big\{\omega \in E_R^\infty \;\Big|\; \omega_{n_j(\omega)+1} = e, \forall j \in \mathbb{N},\ \lim_{j\to\infty} \frac{n_{j+1}(\omega)}{n_j(\omega)} = 1 \text{ and } \lim_{j\to\infty} \frac{S_{n_{j+1}(\omega)}\Xi_\Phi(\omega)}{S_{n_j(\omega)}\Xi_\Phi(\omega)} = 1\Big\}$$

and

$$E_{0,e}^\infty := \bigcup_{R>0} E_{R,e}^\infty \subseteq E_0^\infty.$$

Theorem 21.2.3. *Let $\Phi = \{\varphi_e\}_{e\in E}$ be a CGDMS satisfying the SOSC. For every $e \in E$, every $0 < R < \sup_{x\in J} \mathrm{dist}(x, \partial X)$ and every ergodic σ-invariant Borel probability measure μ on E_A^∞ such that $\mathrm{supp}(\mu) = E_A^\infty$ and $\int_{E_A^\infty} \Xi_\Phi \, d\mu > -\infty$, it holds that*

$$\mu(E_{R,e}^\infty) = 1, \quad \textit{and thus} \quad \mu(E_{0,e}^\infty) = 1.$$

Proof. Let $e \in E$ and $0 < R < D := \sup_{x \in J} \operatorname{dist}(x, \partial X)$. By Theorem 21.2.2, we know that $\mu \circ \pi^{-1}(\bigcup_{v \in V} X_v \backslash B(\partial X_v, R)) = 1$. Since $\operatorname{supp}(\mu) = E_A^\infty$, we also know that $\mu([e]) > 0$. Then

$$\mu\Big(\pi^{-1}\Big(\bigcup_{v \in V} X_v \backslash B(\partial X_v, R)\Big) \cap [e]\Big) = \mu([e]) > 0.$$

Applying twice the ergodic case of Birkhoff's ergodic theorem (Corollary 8.2.14), once with the indicator function of the set $\bigcup_{v \in V} \pi^{-1}(X_v \backslash B(\partial X_v, R)) \cap [e]$ and once with the potential function Ξ_Φ, we obtain that for μ-a. e. $\omega \in E_A^\infty$,

$$\begin{aligned} \lim_{n \to \infty} \frac{1}{n} \# \Big\{ 0 \le k \le n : \sigma^k(\omega) \in \pi^{-1}\Big(\bigcup_{v \in V} X_v \backslash B(\partial X_v, R)\Big) \cap [e] \Big\} \\ = \mu\Big(\pi^{-1}\Big(\bigcup_{v \in V} X_v \backslash B(\partial X_v, R)\Big) \cap [e]\Big) > 0 \end{aligned} \tag{21.10}$$

and

$$\lim_{n \to \infty} \frac{1}{n} S_n \Xi_\Phi(\omega) = -\chi_\mu(\sigma). \tag{21.11}$$

From (21.10), we deduce that the sequence $(n_j(\omega))_{j=1}^N$ is infinite, i. e., $N = \infty$, for μ-a. e. $\omega \in E_A^\infty$. Moreover,

$$\lim_{j \to \infty} \frac{1}{n_j(\omega)} j = \mu\Big(\pi^{-1}\Big(\bigcup_{v \in V} X_v \backslash B(\partial X_v, R)\Big) \cap [e]\Big) =: L > 0 \tag{21.12}$$

for μ-a. e. $\omega \in E_A^\infty$. Consequently,

$$\lim_{j \to \infty} \frac{n_{j+1}(\omega)}{n_j(\omega)} = \lim_{j \to \infty} \frac{j / n_j(\omega)}{(j+1)/n_{j+1}(\omega)} = \frac{L}{L} = 1 \tag{21.13}$$

for μ-a. e. $\omega \in E_A^\infty$. From (21.11), we infer that

$$\lim_{j \to \infty} \frac{1}{n_j(\omega)} S_{n_j(\omega)} \Xi_\Phi(\omega) = -\chi_\mu(\sigma) < 0 \tag{21.14}$$

for μ-a. e. $\omega \in E_A^\infty$. It follows from (21.14) and (21.13) that

$$\lim_{j \to \infty} \frac{S_{n_{j+1}(\omega)} \Xi_\Phi(\omega)}{S_{n_j(\omega)} \Xi_\Phi(\omega)} = \lim_{j \to \infty} \frac{S_{n_{j+1}(\omega)} \Xi_\Phi(\omega)/n_{j+1}(\omega)}{S_{n_j(\omega)} \Xi_\Phi(\omega)/n_j(\omega)} = \frac{-\chi_\mu(\sigma)}{-\chi_\mu(\sigma)} = 1. \tag{21.15}$$

Finally, by definition of $n_j(\omega)$, we have that $\sigma^{n_j(\omega)}(\omega) \in [e]$, i. e.,

$$\omega_{n_j(\omega)+1} = e, \quad \forall j \in \mathbb{N}. \tag{21.16}$$

By (21.13), (21.15) and (21.16), the set $E_{R,e}^\infty$ has full measure. □

As a direct consequence of this theorem, we get the following result.

Corollary 21.2.4. *Let $\Phi = \{\varphi_e\}_{e\in E}$ be a finitely irreducible CGDMS satisfying the SOSC. For every $e \in E$, every $0 < R < \sup_{x\in J} \operatorname{dist}(x, \partial X)$ and every $(q,t) \in \mathbb{R}^2$ such that $qu+t > \theta$, we have that*

$$\mu_{F_{q,t}}(\pi(E^\infty_{R,e})) = \mu_{f_{q,t}}(E^\infty_{R,e}) = 1 \quad \textit{and thus} \quad \mu_{F_{q,t}}(\pi(E^\infty_{0,e})) = \mu_{f_{q,t}}(E^\infty_{0,e}) = 1.$$

In particular,

$$\mu_F(\pi(E^\infty_{R,e})) = \mu_f(E^\infty_{R,e}) = 1 \quad \textit{and so} \quad \mu_F(\pi(E^\infty_{0,e})) = \mu_f(E^\infty_{0,e}) = 1.$$

If, additionally, Φ is cofinitely regular then for every $q \in \mathbb{R}$,

$$\mu_{F_q}(\pi(E^\infty_{R,e})) = \mu_q(E^\infty_{R,e}) = 1 \quad \textit{and hence} \quad \mu_{F_q}(\pi(E^\infty_{0,e})) = \mu_q(E^\infty_{0,e}) = 1.$$

We now recall a few basic definitions from multifractal analysis. While the Hausdorff dimension of a fractal set gives us an idea of the global structure of that set, the pointwise dimension of a measure on that set permits us to capture the local structure of the set.

Let μ be a Borel probability measure on

$$J_0 := \pi(E_0^\infty).$$

Per Definition 15.6.2, the *pointwise or local dimension* $d_\mu(x)$ of μ at a point $x \in J_0$ is the power law behavior (if any) of $\mu(B(x,r))$ for small $r > 0$, i. e.,

$$d_\mu(x) := \lim_{r\to 0} \frac{\log \mu(B(x,r))}{\log r}.$$

The local structure of J_0 is determined by the following partition of that set. For each number $\alpha \geq 0$, denote by $J_{0,\mu}(\alpha)$ the set of all points of J_0 at which the pointwise dimension of μ is equal to α, i. e.,

$$J_{0,\mu}(\alpha) := \{x \in J_0 \mid d_\mu(x) = \alpha\}.$$

Denote the Hausdorff dimension of $J_{0,\mu}(\alpha)$ by

$$d_{0,\mu}(\alpha) := \operatorname{HD}(J_{0,\mu}(\alpha)).$$

The multifractal analysis of J_0 consists in:

(1) determining the range of the function $x \mapsto d_\mu(x)$, and
(2) for each α in that range, determining $d_{0,\mu}(\alpha)$.

The map $\alpha \mapsto d_{0,\mu}(\alpha)$ is the (*fine Hausdorff*) *multifractal spectrum* of the measure μ. We now prove that for every $q \in \mathbb{R}$ the measure μ_{F_q} confers full measure to the subset of points of J_0 where the local dimension of the measure m_F (or, equivalently, μ_F) is $\alpha(q)$. In other terms, the measure μ_{F_q} is entirely distributed over the set $J_{0,m_F}(\alpha(q))$ and not at all over any of the other sets $J_{0,m_F}(\alpha)$, $\alpha \neq \alpha(q)$. Accordingly, the measure μ_{F_q} (and its sister m_{F_q}) will be used to determine the Hausdorff dimension of $J_{0,m_F}(\alpha(q))$.

Given $e \in E$, set

$$E_{0,e}^{\infty}(\alpha) := \left\{\omega \in E_{0,e}^{\infty} \;\middle|\; \lim_{n\to\infty} \frac{S_n f(\omega)}{S_n \Xi_{\Phi}(\omega)} = \alpha\right\} \subseteq E_{0,e}^{\infty} \subseteq E_0^{\infty}.$$

Theorem 21.2.5. *Let $\Phi = \{\varphi_e\}_{e\in E}$ be a cofinitely regular, finitely irreducible, conformal-like CGDMS satisfying the SOSC. For every $e \in E$, the following two statements hold:*
(a) $\pi(E_{0,e}^{\infty}(\alpha)) \subseteq J_{0,m_F}(\alpha)$ *for all* $\alpha \geq 0$.
(b) $\mu_q(E_{0,e}^{\infty}(\alpha(q))) = 1$ *for all* $q \in \mathbb{R}$.
Moreover,
(c) $\mu_{F_q}(J_{0,m_F}(\alpha(q))) = 1$ *for all* $q \in \mathbb{R}$.

Proof. (a) Let $x \in \pi(E_{0,e}^{\infty}(\alpha))$. Then there is some $\omega \in E_{0,e}^{\infty}(\alpha)$ such that $\pi(\omega) = x$. Therefore, $\omega \in E_{R,e}^{\infty}$ for some $R > 0$. Since the sets $E_{R,e}^{\infty}$ get bigger as R gets smaller, we may assume that $0 < R < \sup_{y\in J} \operatorname{dist}(y, \partial X)$. Let $(n_j)_{j=1}^{\infty} := (n_j(\omega, R, e))_{j=1}^{\infty}$ be the strictly increasing sequence of all $n \in \mathbb{N}$ defined in (21.9), i. e., of all $n \in \mathbb{N}$ such that

$$\pi(\sigma^n(\omega)) \in X_{t(\omega_n)} \setminus B(\partial X_{t(\omega_n)}, R) \quad \text{and} \quad \omega_{n+1} = e.$$

Let $0 < r \leq K^{-1}R|\varphi'_{\omega|_{n_1}}(\pi(\sigma^{n_1}(\omega)))|$. Let $j \in \mathbb{N}$ be the unique number such that

$$K^{-1}R|\varphi'_{\omega|_{n_{j+1}}}(\pi(\sigma^{n_{j+1}}(\omega)))| < r \leq K^{-1}R|\varphi'_{\omega|_{n_j}}(\pi(\sigma^{n_j}(\omega)))|. \tag{21.17}$$

Since $B(\pi(\sigma^{n_j}(\omega)), R) \subseteq X_{t(\omega_{n_j})}$, the right inequality in (21.17) and Lemma 19.3.10 yield

$$B(x, r) \subseteq B(\pi(\omega), K^{-1}R|\varphi'_{\omega|_{n_j}}(\pi(\sigma^{n_j}(\omega)))|) \subseteq \varphi_{\omega|_{n_j}}(B(\pi(\sigma^{n_j}(\omega)), R)). \tag{21.18}$$

Using successively (1) relation (21.18); (2) the conformality of m_F (by means of Lemma 19.8.12); (3) the fact that $\mathrm{P}(F) = 0$ by assumption; (4) Lemma 19.8.7 with $V_\gamma(F) := \frac{v_\gamma(F)}{1-e^{-\gamma}}$; and (5) Lemma 19.8.2, we obtain

$$\begin{aligned} m_F(B(x, r)) &\leq m_F\Big(\varphi_{\omega|_{n_j}}(B(\pi(\sigma^{n_j}(\omega)), R))\Big) \\ &\leq \exp\Big(\sup_{y\in X_{t(\omega_{n_j})}} S_{\omega|_{n_j}} F(y)\Big) \\ &\leq \exp\Big(S_{\omega|_{n_j}} F(\pi(\sigma^{n_j}(\omega))) + V_\gamma(F)\Big) \\ &= e^{V_\gamma(F)} \exp(S_{n_j} f(\omega)). \end{aligned} \tag{21.19}$$

On the other hand, by the left inequality in (21.17) and (19.8) in Lemma 19.3.7, we know that

$$B(x,r) \supseteq B(\pi(\omega), K^{-1}R|\varphi'_{\omega|_{n_{j+1}}}(\pi(\sigma^{n_{j+1}}(\omega)))|) \supseteq \varphi_{\omega|_{n_{j+1}}}(B(\pi(\sigma^{n_{j+1}}(\omega)), K^{-2}R)). \quad (21.20)$$

Using successively (1) relation (21.20); (2) the conformality of m_F; (3) $\mathrm{P}(F) = 0$; (4) Lemma 19.8.7; (5) the fact that $E_A^{\infty,\omega_{n_{j+1}}} \supseteq [\omega_{n_{j+1}+1}] = [e]$; and (6) Lemma 19.8.2, we get

$$\begin{aligned} m_F(B(x,r)) &\geq m_F\Big(\varphi_{\omega|_{n_{j+1}}}(B(\pi(\sigma^{n_{j+1}}(\omega)), K^{-2}R))\Big) \\ &\geq \exp\Big(\inf_{y\in X_{t(\omega_{n_{j+1}})}} S_{\omega|_{n_{j+1}}}F(y)\Big)m_F\Big(B(\pi(\sigma^{n_{j+1}}(\omega)), K^{-2}R) \cap \pi(E_A^{\infty,\omega_{n_{j+1}}})\Big) \\ &\geq \exp\Big(S_{\omega|_{n_{j+1}}}F(\pi(\sigma^{n_{j+1}}(\omega))) - V_\gamma(F)\Big) \\ &\quad \cdot m_F\Big(B(\pi(\sigma^{n_{j+1}}(\omega)), K^{-2}R) \cap \pi(E_A^{\infty,\omega_{n_{j+1}}})\Big) \\ &\geq e^{-V_\gamma(F)}\exp\Big(S_{\omega|_{n_{j+1}}}F(\pi(\sigma^{n_{j+1}}(\omega)))\Big)m_F\Big(B(\pi(\sigma^{n_{j+1}}(\omega)), K^{-2}R) \cap \pi([e])\Big) \\ &\geq e^{-V_\gamma(F)}\exp(S_{n_{j+1}}f(\omega))M_{F,e}(K^{-2}R), \end{aligned} \quad (21.21)$$

where

$$M_{F,e}(s) := \inf_{y\in\pi([e])} m_F(B(y,s) \cap \pi([e])) > 0. \quad (21.22)$$

The positivity of $M_{F,e}(s)$ is guaranteed by Lemma 19.8.21.

Moreover, we infer from (21.17) that

$$\log(K^{-1}R)+\log|\varphi'_{\omega|_{n_{j+1}}}(\pi(\sigma^{n_{j+1}}(\omega)))| < \log r \leq \log(K^{-1}R)+\log|\varphi'_{\omega|_{n_j}}(\pi(\sigma^{n_j}(\omega)))|. \quad (21.23)$$

We deduce from (21.21) and (21.23) that

$$\begin{aligned} \overline{d}_{m_F}(x) := \limsup_{r\to 0}\frac{\log m_F(B(x,r))}{\log r} &\leq \limsup_{j\to\infty}\frac{S_{n_{j+1}}f(\omega) + \log(M_{F,e}(K^{-2}R)e^{-V_\gamma(F)})}{\log(K^{-1}R) + \log|\varphi'_{\omega|_{n_j}}(\pi(\sigma^{n_j}(\omega)))|} \\ &= \limsup_{j\to\infty}\frac{S_{n_{j+1}}f(\omega) + \log(M_{F,e}(K^{-2}R)e^{-V_\gamma(F)})}{\log(K^{-1}R) + S_{n_j}\Xi_\Phi(\omega)} \\ &= \limsup_{j\to\infty}\frac{\frac{S_{n_{j+1}}f(\omega)}{S_{n_{j+1}}\Xi_\Phi(\omega)} + \frac{\log\big(M_{F,e}(K^{-2}R)e^{-V_\gamma(F)}\big)}{S_{n_{j+1}}\Xi_\Phi(\omega)}}{\frac{\log(K^{-1}R)}{S_{n_{j+1}}\Xi_\Phi(\omega)} + \frac{S_{n_j}\Xi_\Phi(\omega)}{S_{n_{j+1}}\Xi_\Phi(\omega)}} \\ &= \frac{\alpha+0}{0+1} = \alpha \end{aligned}$$

since $\omega \in E_{0,e}^{\infty}(\alpha) \cap E_{R,e}^{\infty}$ and $S_n\Xi_\Phi \le n \log s$ for all n. Similarly, we deduce from (21.19) and (21.23) that

$$\underline{d}_{m_F}(x) := \liminf_{r\to 0} \frac{\log m_F(B(x,r))}{\log r} \ge \liminf_{j\to\infty} \frac{S_{n_j}f(\omega) + V_\gamma(F)}{\log(K^{-1}R) + \log|\varphi'_{\omega|_{n_{j+1}}}(\pi(\sigma^{n_{j+1}}(\omega)))|} = \alpha.$$

Hence, $\alpha \le \underline{d}_{m_F}(x) \le \overline{d}_{m_F}(x) \le \alpha$, i. e., $d_{m_F}(x) = \alpha$. So $x \in J_{0,m_F}(\alpha)$.

(b) Let $q \in \mathbb{R}$. According to Corollary 21.2.4 and the ergodic case of Birkhoff's ergodic theorem (Corollary 8.2.14), there exists $E_q \subseteq E_{0,e}^{\infty}$ such that $\mu_q(E_q) = 1$ and so that for all $\omega \in E_q$ we have

$$\lim_{n\to\infty} \frac{1}{n} S_n f(\omega) = \int_{E_A^\infty} f \, d\mu_q \quad \text{and} \quad \lim_{n\to\infty} \frac{1}{n} S_n \Xi_\Phi(\omega) = \int_{E_A^\infty} \Xi_\Phi \, d\mu_q.$$

Using this and (21.5), we get

$$\lim_{n\to\infty} \frac{S_n f(\omega)}{S_n \Xi_\Phi(\omega)} = \frac{\int f \, d\mu_q}{\int \Xi_\Phi \, d\mu_q} = \alpha(q), \quad \forall \omega \in E_q.$$

Hence, $E_q \subseteq E_{0,e}^{\infty}(\alpha(q))$ and thus $\mu_q(E_{0,e}^{\infty}(\alpha(q))) \ge \mu_q(E_q) = 1$.

(c) Let $q \in \mathbb{R}$ and pick any $e \in E$. Using part (a) with $\alpha = \alpha(q)$ and part (b), we deduce that

$$\mu_{F_q}(J_{0,m_F}(\alpha(q))) \ge \mu_q \circ \pi^{-1}(\pi(E_{0,e}^{\infty}(\alpha(q)))) \ge \mu_q(E_{0,e}^{\infty}(\alpha(q))) = 1. \qquad \square$$

We shall now prove that $d_{0,m_F}(\alpha)$ and $T(q)$ form a Legendre transform pair under an additional restriction on the CGDMS. (For more information on the Legendre transform, see the discussion around Proposition 16.6.4.) This means that the Hausdorff dimension of the set of points of J_0 where the local dimension of the measure m_F is $\alpha(q)$ is equal to $q\alpha(q) + T(q)$.

Definition 21.2.6. A CGDMS $\Phi = \{\varphi_e\}_{e\in E}$ is said to be an *almost-CIFS* if $E = \mathbb{N}$ and there exists $P \in \mathbb{N}$ such that $A_{ab} = 1$ for all $a \in \mathbb{N}$ and all $b \ge P$.

It is obvious that any CIFS is finitely primitive. However, an almost-CIFS may not even be irreducible. Nevertheless, every irreducible almost-CIFS is finitely primitive.

Theorem 21.2.7. *Let $\Phi = \{\varphi_e\}_{e\in E}$ be a cofinitely regular, irreducible, conformal-like almost-CIFS satisfying the SOSC. For every $q \in \mathbb{R}$ we have $d_{0,m_F}(-T'(q)) = T(q) - qT'(q)$. Alternatively, $d_{0,m_F}(\alpha(q)) = q\alpha(q) + T(q)$.*

Proof. Using Theorems 21.2.5(a,b), 19.8.30 and 17.5.1 (which guarantees that μ_q is an (in fact, the unique) equilibrium state for f_q), and the fact that $0 = \mathrm{P}(q, T(q)) = \mathrm{P}(F_q) =$

$\mathrm{P}(f_q)$ by definition of $T(q)$, we obtain

$$\begin{aligned} d_{0,m_F}(\alpha(q)) &:= \mathrm{HD}(J_{0,m_F}(\alpha(q))) \geq \mathrm{HD}(\pi(E^\infty_{0,e}(\alpha(q)))) \text{ for any } e \in E \\ &\geq \mathrm{HD}(\mu_q \circ \pi^{-1}) = \frac{\mathrm{h}_{\mu_q}(\sigma)}{\chi_{\mu_q}(\sigma)} \\ &= \frac{\mathrm{P}(f_q) - \int f_q \, d\mu_q}{\chi_{\mu_q}(\sigma)} = \frac{-\int f_q \, d\mu_q}{-\int \Xi_\Phi \, d\mu_q} = \frac{\int (qf + T(q)\Xi_\Phi) \, d\mu_q}{\int \Xi_\Phi \, d\mu_q} \\ &= q \frac{\int f \, d\mu_q}{\int \Xi_\Phi \, d\mu_q} + T(q) = q\alpha(q) + T(q). \end{aligned}$$

To prove the opposite inequality, fix $x \in J_{0,m_F}(\alpha(q))$. Then there is $\omega \in E^\infty_0$ such that $\pi(\omega) = x$. Let $0 < R < \min\{K^{-1}, r(\omega)\}$, where $r(\omega)$ comes from (21.7). Let also $(n_j)_{j=1}^\infty$ be any subsequence of the strictly increasing sequence $(n_j(\omega, R))_{j=1}^\infty$ of all n such that $\pi(\sigma^n(\omega)) \in X_{t(\omega_n)} \backslash B(\partial X_{t(\omega_n)}, R)$. Relation (19.8) in Lemma 19.3.7 asserts that

$$\varphi_{\omega|_n}(B(\pi(\sigma^n(\omega)), R)) \subseteq B(\pi(\omega), KR|\varphi'_{\omega|_n}(\pi(\sigma^n(\omega)))|)$$

for every $n \in \mathbb{N}$. Since Φ is an almost-CIFS, we know that $E_A^{\infty,\omega_n} \supseteq [\omega_{n+1}] \cup \bigcup_{b \geq P}[b]$ for all $n \in \mathbb{N}$. In a similar way to (21.21), the conformality of m_{F_q} and the bounded variation principle for F_q give

$$\begin{aligned} m_{F_q}\Big(B(x, KR|\varphi'_{\omega|_n}(\pi(\sigma^n(\omega)))|)\Big) &\geq m_{F_q}\Big(\varphi_{\omega|_n}(B(\pi(\sigma^n(\omega)), R))\Big) \\ &\geq e^{-V_\gamma(F_q)} \exp(S_n f_q(\omega)) \\ &\quad \cdot m_{F_q}\Big(B(\pi(\sigma^n(\omega)), K^{-2}R) \cap \pi(E_A^{\infty,\omega_n})\Big) \\ &\geq e^{-V_\gamma(F_q)} \exp(S_n f_q(\omega)) \widetilde{M}_{F_q}(K^{-2}R), \end{aligned} \tag{21.24}$$

where

$$\widetilde{M}_{F_q}(s) := \min\left\{\min_{b<P} M_{F_q,b}(s), \inf_{x \in \bigcup_{b \geq P} \pi([b])} m_{F_q}\Big(B(x,s) \cap \bigcup_{b \geq P} \pi([b])\Big)\right\} > 0.$$

This last quantity is positive by Lemma 19.8.21 ($M_{F_q,b}(s)$ was defined in (21.22)). Hence,

$$\begin{aligned} \frac{\log m_{F_q}(B(x, KR|\varphi'_{\omega|_n}(\pi(\sigma^n(\omega)))|))}{\log(KR|\varphi'_{\omega|_n}(\pi(\sigma^n(\omega)))|)} &\leq \frac{\log(\widetilde{M}_{F_q}(K^{-2}R)e^{-V_\gamma(F_q)}) + S_n f_q(\omega)}{\log(KR) + S_n \Xi_\Phi(\omega)} \\ &= \frac{\frac{\log(\widetilde{M}_{F_q}(K^{-2}R)e^{-V_\gamma(F_q)})}{S_n \Xi_\Phi(\omega)} + q \frac{S_n f(\omega)}{S_n \Xi_\Phi(\omega)} + T(q)}{\frac{\log(KR)}{S_n \Xi_\Phi(\omega)} + 1} \end{aligned} \tag{21.25}$$

for all $n \in \mathbb{N}$. As in (21.19), we have

$$m_F(B(x, K^{-1}R|\varphi'_{\omega|_{n_j}}(\pi(\sigma^{n_j}(\omega)))|)) \le e^{V_\gamma(F)} \exp(S_{n_j} f(\omega)).$$

Thus,

$$\frac{\log m_F(B(x, K^{-1}R|\varphi'_{\omega|_{n_j}}(\pi(\sigma^{n_j}(\omega)))|))}{\log(K^{-1}R|\varphi'_{\omega|_{n_j}}(\pi(\sigma^{n_j}(\omega)))|)} \ge \frac{V_\gamma(F) + S_{n_j} f(\omega)}{\log(K^{-1}R) + S_{n_j}\Xi_\Phi(\omega)} \tag{21.26}$$

for all $j \in \mathbb{N}$ large enough. Using (21.25) and (21.26) as well as the facts that $S_n\Xi_\Phi \le n \log s$ for all n and that $x \in J_{0,m_F}(\alpha(q))$, we deduce that

$$\begin{aligned}
\underline{d}_{m_{F_q}}(x) &:= \liminf_{r\to 0} \frac{\log m_{F_q}(B(x,r))}{\log r} \\
&\le \liminf_{j\to\infty} \frac{\log m_{F_q}(B(x, KR|\varphi'_{\omega|_{n_j}}(\pi(\sigma^{n_j}(\omega)))|))}{\log(KR|\varphi'_{\omega|_{n_j}}(\pi(\sigma^{n_j}(\omega)))|)} \\
&\le q \liminf_{j\to\infty} \frac{S_{n_j} f(\omega)}{S_{n_j}\Xi_\Phi(\omega)} + T(q) \\
&= q \liminf_{j\to\infty} \frac{V_\gamma(F) + S_{n_j} f(\omega)}{\log(K^{-1}R) + S_{n_j}\Xi_\Phi(\omega)} + T(q) \\
&\le q \liminf_{j\to\infty} \frac{\log m_F(B(x, K^{-1}R|\varphi'_{\omega|_{n_j}}(\pi(\sigma^{n_j}(\omega)))|))}{\log(K^{-1}R|\varphi'_{\omega|_{n_j}}(\pi(\sigma^{n_j}(\omega)))|)} + T(q) \\
&= q\alpha(q) + T(q).
\end{aligned}$$

As $\underline{d}_{m_{F_q}}(x) \le q\alpha(q) + T(q)$ for every $x \in J_{0,m_F}(\alpha(q))$, we conclude by Theorem 15.6.3(b) that

$$d_{0,m_F}(\alpha(q)) := \mathrm{HD}(J_{0,m_F}(\alpha(q))) \le q\alpha(q) + T(q).$$

Finally, recall that $T'(q) = -\alpha(q)$ by Theorem 21.1.4. □

All of the above results give us an analog of Theorem 4.9.4 in [81] about the fine Hausdorff multifractal spectrum of the set J_0.

Theorem 21.2.8. *Let $\Phi = \{\varphi_e\}_{e\in E}$ be a cofinitely regular, irreducible, conformal-like almost-CIFS satisfying the SOSC. Then the following statements hold:*

(a) *For μ_F-a. e. $x \in J_0$,*

$$d_{\mu_F}(x) = \frac{\int_{E_A^\infty} f \, d\mu_f}{\int_{E_A^\infty} \Xi_\Phi \, d\mu_f} = \mathrm{HD}(\mu_F).$$

(b) *The temperature function* $T : \mathbb{R} \to \mathbb{R}$, *defined by* $\mathrm{P}(q, T(q)) = 0$ *for all* $q \in \mathbb{R}$, *is real analytic and such that*
 - $T(0) = \mathrm{HD}(J)$
 - $T(1) = 0$
 - $T'(q) = -\alpha(q) = -\frac{\int_{E_A^\infty} f\, d\mu_q}{\int_{E_A^\infty} \Xi_\Phi\, d\mu_q} < 0$
 - $T''(q) \geq 0$ *for all* $q \in \mathbb{R}$.

(c) *For every* $q \in \mathbb{R}$,

$$d_{0,\mu_F}(-T'(q)) = d_{0,\mu_F}(\alpha(q)) = q\alpha(q) + T(q) = T(q) - qT'(q).$$

That is, $d_{0,\mu_F}(\alpha)$ *and* $T(q)$ *form a Legendre transform pair of functions.*

(d) *If* $\mu_f = \mu_{\mathrm{HD}(J)\Xi_\Phi}$ *or, equivalently, if* f *and* $\mathrm{HD}(J)\Xi_\Phi$ *are cohomologous in the group of Hölder functions on* E_A^∞, *then* $T'(q) = -\mathrm{HD}(J)$ *for every* $q \in \mathbb{R}$ *and the domain and the range of the Hausdorff multifractal spectrum* $\alpha \mapsto d_{0,\mu_F}(\alpha)$ *are both reduced to the singleton* $\{\mathrm{HD}(J)\}$.

Proof. (a) Using Theorem 21.1.4(b), notice that $F_1 = F$ and $f_1 = f$. Thus, by Theorem 21.2.5(c), we have $\mu_F(J_{0,m_F}(\alpha(1))) = 1$. Since the measures μ_F and m_F are boundedly equivalent, we deduce that $\mu_F(J_{0,\mu_F}(\alpha(1))) = 1$, i. e., $d_{\mu_F}(x) = \alpha(1)$ for μ_F-a. e. $x \in J_0$. Since $\alpha(1) = \int f\, d\mu_f / \int \Xi_\Phi\, d\mu_f$, the proof of (a) is complete by calling upon Corollary 15.6.4. Part (b) is an excerpt of Theorem 21.1.4. Part (c) is simply Theorem 21.2.7. Finally, part (d) is an immediate consequence of Theorem 21.1.4(e) and part (c) of the present theorem. □

21.3 Multifractal analysis over J

21.3.1 Under the boundary separation condition

If an almost-CIFS Φ satisfies the boundary separation condition (BSC), then it is conformal-like by Proposition 19.7.19 and $E_0^\infty = E_A^\infty$. So $J_0 = J$ and Section 21.2 gives us the multifractal analysis over J. In particular, Theorem 21.2.8 reduces to the following.

Theorem 21.3.1. *Let* $\Phi = \{\varphi_e\}_{e\in E}$ *be a cofinitely regular, irreducible almost-CIFS satisfying the SOSC and the BSC. Then the following statements hold:*

(a) *For* μ_F*-a. e.* $x \in J$,

$$d_{\mu_F}(x) = \frac{\int_{E_A^\infty} f\, d\mu_f}{\int_{E_A^\infty} \Xi_\Phi\, d\mu_f} = \mathrm{HD}(\mu_F).$$

(b) *The temperature function* $T : \mathbb{R} \to \mathbb{R}$, *defined by* $\mathrm{P}(q, T(q)) = 0$ *for all* $q \in \mathbb{R}$, *is real analytic and such that* $T(0) = \mathrm{HD}(J)$, $T(1) = 0$, *and* $T'(q) = -\alpha(q) = -\frac{\int f\, d\mu_q}{\int \Xi_\Phi\, d\mu_q} < 0$ *while* $T''(q) \geq 0$ *for all* $q \in \mathbb{R}$.

(c) *For every* $q \in \mathbb{R}$, *we have* $d_{\mu_F}(-T'(q)) = d_{\mu_F}(\alpha(q)) = q\alpha(q) + T(q) = T(q) - qT'(q)$. *That is,* $d_{\mu_F}(\alpha)$ *and* $T(q)$ *form a Legendre transform pair of functions.*

(d) *If* $\mu_f = \mu_{\mathrm{HD}(J)\Xi_\Phi}$ *or, equivalently, if* f *and* $\mathrm{HD}(J)\Xi_\Phi$ *are cohomologous in the group of Hölder functions on* E_A^∞, *then* $T'(q) = -\mathrm{HD}(J)$ *for every* $q \in \mathbb{R}$ *and the domain and the range of the Hausdorff multifractal spectrum* $\alpha \mapsto d_{0,\mu_F}(\alpha)$ *are both reduced to the singleton* $\{\mathrm{HD}(J)\}$.

This theorem applies to an important CIFS which finds its origins in the real continued fractions with the digit 1 deleted (cf. Example 19.9.5).

Example 21.3.2. Let $-1/4 \leq \varepsilon < 0$. Set $X = [\varepsilon, 3/4]$. Let $\Phi = \{\varphi_n\}_{n \in \mathbb{N} \setminus \{1\}}$, where

$$\varphi_n(x) = \frac{1}{n + x}.$$

Then Φ is a cofinitely regular CIFS which satisfies both the SOSC and the BSC, and Theorem 21.3.1 applies.

21.3.2 Under other conditions

We shall now prove that $d_{m_F}(\alpha)$ and $T(q)$ form a Legendre transform pair under conditions other than the BSC.

Theorem 21.3.3. *Let* $\Phi = \{\varphi_e\}_{e \in E}$ *be a cofinitely regular, irreducible, conformal-like almost-CIFS satisfying the SOSC. Suppose that there exists a countable set* $J' \subseteq J$ *such that for every* $x \in J \setminus J'$ *there are* $\omega \in E_A^\infty$ *with* $\pi(\omega) = x$, *a constant* $C = C(x, \omega) > 0$, *a strictly increasing subsequence* $(n_j)_{j=1}^\infty = (n_j(x, \omega))_{j=1}^\infty$ *of natural numbers and a sequence* $(r_{n_j})_{j=1}^\infty = (r_{n_j}(x, \omega))_{j=1}^\infty$ *of positive real numbers such that*

(a) $m_F(B(x, r_{n_j})) \leq C \exp(S_{n_j} f(\omega))$ *for all* $j \in \mathbb{N}$;

(b) $1 > r_{n_j} \geq C \|\varphi'_{\omega|_{n_j}}\|_{X_{t(\omega_{n_j})}}$ *for all* $j \in \mathbb{N}$; *and*

(c) $\lim_{j \to \infty} r_{n_j} = 0$.

Then $d_{m_F}(\alpha(q)) = q\alpha(q) + T(q)$ *for all* $q \in \mathbb{R}$.

Proof. Clearly, $d_{m_F}(\alpha(q)) \geq d_{0,m_F}(\alpha(q)) = q\alpha(q) + T(q)$ using Theorem 21.2.7. We shall now prove the other inequality. Since J' is countable, according to Theorem 15.6.3(b) it suffices to show that $\underline{d}_{m_{F_q}}(x) \leq q\alpha(q) + T(q)$ for every $x \in J_{m_F}(\alpha(q)) \setminus J'$. Accordingly, fix $x \in J_{m_F}(\alpha(q)) \setminus J'$. Let ω, C, $(n_j)_{j=1}^\infty$ and $(r_{n_j})_{j=1}^\infty$ be as in the statement of the theorem.

By (19.8) in Lemma 19.3.7, we know that

$$\varphi_{\omega|_n}(B(\pi(\sigma^n(\omega)),R)) \subseteq B(\pi(\omega),KR|\varphi'_{\omega|_n}(\pi(\sigma^n(\omega)))|)$$

for every $n \in \mathbb{N}$ and $0 < R \leq \operatorname{dist}(X,\partial W) := \min\{\operatorname{dist}(X_v,\partial W_v) : v \in V\}$.

As in (21.24), the conformality of m_{F_q} and the bounded variation principle for F_q give

$$m_{F_q}\Big(B(x,KR|\varphi'_{\omega|_n}(\pi(\sigma^n(\omega)))|)\Big) \geq \widetilde{M}_{F_q}(K^{-2}R)e^{-V_\gamma(F_q)}\exp(S_nf_q(\omega)).$$

Hence, as in (21.25),

$$\frac{\log m_{F_q}(B(x,KR|\varphi'_{\omega|_n}(\pi(\sigma^n(\omega)))|))}{\log(KR|\varphi'_{\omega|_n}(\pi(\sigma^n(\omega)))|)} \leq \frac{\frac{\log(\widetilde{M}_{F_q}(K^{-2}R)e^{-V_\gamma(F_q)})}{S_n\Xi_\Phi(\omega)} + q\frac{S_nf(\omega)}{S_n\Xi_\Phi(\omega)} + T(q)}{\frac{\log(KR)}{S_n\Xi_\Phi(\omega)} + 1} \tag{21.27}$$

for all $n \in \mathbb{N}$ large enough. By assumptions (a) and (b), we obtain for all $j \in \mathbb{N}$,

$$\begin{aligned}\frac{\log m_F(B(x,r_{n_j}))}{\log r_{n_j}} &\geq \frac{\log C + S_{n_j}f(\omega)}{\log r_{n_j}}\\ &\geq \frac{\log C + S_{n_j}f(\omega)}{\log C + \log\|\varphi'_{\omega|_{n_j}}\|_{X_{t(\omega_{n_j})}}}\\ &\geq \frac{\log C + S_{n_j}f(\omega)}{\log C + \log|\varphi'_{\omega|_{n_j}}(\omega)|}\\ &= \frac{\log C + S_{n_j}f(\omega)}{\log C + S_{n_j}\Xi_\Phi(\omega)}.\end{aligned} \tag{21.28}$$

Using (21.27) and (21.28), we deduce that

$$\begin{aligned}\underline{d}_{m_{F_q}}(x) &:= \liminf_{r\to 0}\frac{\log m_{F_q}(B(x,r))}{\log r}\\ &\leq \liminf_{j\to\infty}\frac{\log m_{F_q}(B(x,KR|\varphi'_{\omega|_{n_j}}(\pi(\sigma^{n_j}(\omega)))|))}{\log(KR|\varphi'_{\omega|_{n_j}}(\pi(\sigma^{n_j}(\omega)))|)}\\ &\leq q\liminf_{j\to\infty}\frac{S_{n_j}f(\omega)}{S_{n_j}\Xi_\Phi(\omega)} + T(q)\\ &= q\liminf_{j\to\infty}\frac{\log C + S_{n_j}f(\omega)}{\log C + S_{n_j}\Xi_\Phi(\omega)} + T(q)\\ &\leq q\liminf_{j\to\infty}\frac{\log m_F(B(x,r_{n_j}))}{\log r_{n_j}} + T(q)\\ &= q\alpha(q) + T(q).\end{aligned}$$

As $\underline{d}_{m_{F_q}}(x) \le q\alpha(q) + T(q)$ for every $x \in J_{m_F}(\alpha(q)) \setminus J'$, we infer from Theorem 15.6.3(b) that $\mathrm{HD}(J_{m_F}(\alpha(q)) \setminus J') \le q\alpha(q) + T(q)$. Since J' is countable, it ensues that $d_{m_F}(\alpha(q)) := \mathrm{HD}(J_{m_F}(\alpha(q))) = \mathrm{HD}(J_{m_F}(\alpha(q)) \setminus J') \le q\alpha(q) + T(q)$. □

Theorem 21.3.4. *Let $\Phi = \{\varphi_e\}_{e\in E}$ be an almost-CIFS fulfilling the hypotheses of Theorem 21.3.3. Then the conclusion of Theorem 21.3.1 applies.*

There are interesting families of CIFSs which satisfy the conditions imposed in Theorem 21.3.3. Among others, let us mention real continued fractions over the unit interval (cf. Example 19.9.5).

Example 21.3.5. Let $X = [0,1]$. Let $\Phi = \{\varphi_n : X \to X \mid n \in \mathbb{N}\}$, where

$$\varphi_n(x) = \frac{1}{n+x}.$$

Then Φ is a cofinitely regular CIFS which satisfies the SOSC and the Cone Condition (CC). This latter implies that Φ is conformal-like, according to Proposition 19.7.19. This CIFS also satisfies conditions (a)–(c) of Theorem 21.3.3 but not the BSC. Moreover, writing

$$x = \frac{1}{x_1 + \frac{1}{x_2 + \dots}},$$

note that we can set $J' = \{x \in [0,1] : \liminf_{n\to\infty} x_n < \infty\}$, and thus $J \setminus J' = \{x \in [0,1] : \lim_{n\to\infty} x_n = \infty\}$.

In fact, there is a larger class of one-dimensional CIFSs for which the conditions of Theorem 21.3.3 are fulfilled and to which Theorem 21.3.4 thereby applies. By one-dimensional, we simply mean that X is a subinterval of $\mathbb{R}$. Every one-dimensional CIFS obeys the cone condition (CC), and hence is conformal-like by Proposition 19.7.19. Every one-dimensional CIFS also satisfies the SOSC.

Theorem 21.3.6. *Let $\Phi = \{\varphi_n : X \to X \mid n \in \mathbb{N}\}$ be a one-dimensional cofinitely regular CIFS satisfying the following properties:*

(a) *$\|\varphi_n'\|_X$ is comparable to $\|\varphi_{n+1}'\|_X$, i. e., there exists a constant $C \ge 1$ such that*

$$C^{-1} \le \frac{\|\varphi_{n+1}'\|_X}{\|\varphi_n'\|_X} \le C, \quad \forall n \in \mathbb{N}.$$

(b) *The generators φ_n of Φ either all preserve orientation (i. e., $\varphi_n' > 0$ for all $n \in \mathbb{N}$) or all reverse orientation (i. e., $\varphi_n' < 0$ for all $n \in \mathbb{N}$).*

(c) *Either $\varphi_{n+1} > \varphi_n$ for all $n \in \mathbb{N}$ or $\varphi_{n+1} < \varphi_n$ for all $n \in \mathbb{N}$.*

Then Φ satisfies the hypotheses of Theorem 21.3.3.

Proof. It follows from the bounded distortion property (BDP) and (a) that

$$(CK)^{-1} \le \frac{|\varphi_{n+1}(X)|}{|\varphi_n(X)|} \le CK, \quad \forall n \in \mathbb{N}, \tag{21.29}$$

where $|I|$ denotes the length of an interval I. Moreover, a rather straightforward calculation shows that the amalgamated function Ξ_Φ satisfies

$$\sup_{\omega,\tau \in E_A^\infty:\, |\omega_1-\tau_1|=1} |\Xi_\Phi(\omega) - \Xi_\Phi(\tau)| \le \log(CK) =: D.$$

The set $J' = \pi(\cup_{n=1}^\infty \sigma^{-n}(1^\infty))$ is clearly countable. Let $x \in J \setminus J'$ and $\omega \in \mathbb{N}^\infty$ such that $\pi(\omega) = x$. Let n_j be the position of the jth letter in ω which is not a 1. Let $r_{n_j} = |\varphi_{\omega_{n_j}}(X)|/(CK)$. By (21.29), we have

$$r_{n_j} \le \min\{ |\varphi_{\omega_{n_j}-1}(X)|, |\varphi_{\omega_{n_j}}(X)|, |\varphi_{\omega_{n_j}+1}(X)| \}$$

and hence

$$B(x, r_{n_j}) \cap J \subseteq \bigcup_{k=\omega_{n_j}-1}^{\omega_{n_j}+1} \varphi_{\omega|_{n_j-1}k}(X) \cap J.$$

Recall that $F = G + u\,\mathrm{Log}_\Phi$, so $f = g + u\Xi_\Phi$. Using the conformality of m_F, the standing assumption $\mathrm{P}(F) = 0$, and Lemmas 19.8.7 and 17.2.3 with $V_\gamma(h) := \frac{v_\gamma(h)}{1-e^{-\gamma}}$, we obtain that

$$\begin{aligned}
m_F(B(x, r_{n_j})) &\le \sum_{k=\omega_{n_j}-1}^{\omega_{n_j}+1} m_F(\varphi_{\omega|_{n_j-1}k}(X)) \\
&\le \sum_{k=\omega_{n_j}-1}^{\omega_{n_j}+1} \int_J \exp(S_{\omega|_{n_j-1}k}F(y) - n_j\mathrm{P}(F))\, dm_F(y) \\
&\le \sum_{k=\omega_{n_j}-1}^{\omega_{n_j}+1} \exp\Big(\sup_{y\in J} S_{\omega|_{n_j-1}k}F(y)\Big) \\
&\le \sum_{k=\omega_{n_j}-1}^{\omega_{n_j}+1} e^{V_\gamma(F)} \exp\Big(\inf_{\tau\in[\omega|_{n_j-1}k]} S_{n_j}f(\tau)\Big) \\
&\le e^{V_\gamma(F)} \sum_{k=\omega_{n_j}-1}^{\omega_{n_j}+1} \exp\Big(S_{n_j}f(\omega|_{n_j-1}k\omega|_{n_j+1}^\infty)\Big) \\
&= e^{V_\gamma(F)} \sum_{k=\omega_{n_j}-1}^{\omega_{n_j}+1} \exp\Big(S_{n_j-1}f(\omega|_{n_j-1}k\omega|_{n_j+1}^\infty) + f(k\omega|_{n_j+1}^\infty)\Big)
\end{aligned}$$

$$
\begin{aligned}
&= e^{V_\gamma(F)} \sum_{k=\omega_{n_j}-1}^{\omega_{n_j}+1} \exp\Big(S_{n_j-1}f(\omega|_{n_j-1}k\omega|_{n_j+1}^{\infty}) - S_{n_j-1}f(\omega)\Big) \\
&\quad \cdot \exp(S_{n_j-1}f(\omega)) \exp\Big(f(k\omega|_{n_j+1}^{\infty}) - f(\omega|_{n_j}^{\infty})\Big) \exp(f(\omega|_{n_j}^{\infty})) \\
&\le e^{V_\gamma(F)+V_\gamma(f)} \sum_{k=\omega_{n_j}-1}^{\omega_{n_j}+1} \exp(S_{n_j-1}f(\omega)) e^{2\|g\|_\infty + |k-\omega_{n_j}|uD} \exp(f(\omega|_{n_j}^{\infty})) \\
&\le 3e^{2\|g\|_\infty + uD} e^{V_\gamma(F)+V_\gamma(f)} \exp(S_{n_j}f(\omega)).
\end{aligned}
$$

Thus, condition (a) in Theorem 21.3.3 is satisfied. Condition (b) is fulfilled since

$$
r_{n_j} = \frac{1}{CK}|\varphi_{\omega_{n_j}}(X)| \ge \frac{1}{CK^2}\|\varphi'_{\omega_{n_j}}\|_X.
$$

Finally, condition (c) is obviously satisfied, as $\sum_{j=1}^{\infty}|\varphi_{\omega_{n_j}}(X)| \le |X| < \infty$, and hence $r_{n_j} \to 0$ as $j \to \infty$. □

21.4 Multifractal analysis over another subset of J

We will work in the general context of Section 21.2. Let $\Phi = \{\varphi_e : X_{t(e)} \to X_{i(e)} \mid e \in E\}$ be a cofinitely regular, finitely irreducible, conformal-like CGDMS satisfying the SOSC. Let θ be the finiteness parameter of Φ. Let $F = G + u\,\mathrm{Log}_\Phi$ be a family of functions such that $\mathrm{P}(F) = 0$, where G is a bounded Hölder family of functions and $u > \theta$.

For every $\omega \in E_A^\infty$, every $R > 0$, and every finite subalphabet $W \subseteq E$, let $(n_j(\omega, R, W))_{j=1}^{N}$ be the sequence of all $n \ge 0$ such that (cf. (21.9))

$$
\sigma^n(\omega) \in \pi^{-1}\Big(\bigcup_{v\in V} X_v \setminus B(\partial X_v, R)\Big) \cap \bigcup_{e\in W}[e]. \tag{21.30}
$$

This means that $n_j(\omega, R, W)$ is the jth time that an iterate of ω hits the set $\pi^{-1}(\bigcup_{v\in V} X_v \setminus B(\partial X_v, R)) \cap \bigcup_{e\in W}[e]$. This sequence may be empty ($N = 0$), nonempty and finite ($N \in \mathbb{N}$), or infinite ($N = \infty$), depending on ω, R and W. We will correspondingly write $N(\omega, R, W)$. To shorten notation, we shall usually write $n_j(\omega)$ instead of $n_j(\omega, R, W)$ and $N(\omega)$ in place of $N(\omega, R, W)$ when R and W are already fixed.

Recall the definition of E_R^∞ in (21.8). Define the sets

$$
\widehat{E}_{R,W}^{\infty} := \{\omega \in E_R^\infty \mid N(\omega, R, W) = \infty\}
$$

and

$$
E_{R,W}^{\infty} := \left\{\omega \in E_R^\infty \,\middle|\, \omega_{n_j(\omega)+1} \in W, \forall j \in \mathbb{N};\ \lim_{j\to\infty}\frac{n_{j+1}(\omega)}{n_j(\omega)} = 1;\ \lim_{j\to\infty}\frac{S_{n_{j+1}(\omega)}\Xi_\Phi(\omega)}{S_{n_j(\omega)}\Xi_\Phi(\omega)} = 1\right\} \subseteq \widehat{E}_{R,W}^{\infty}.
$$

Theorem 21.4.1. *Let $\Phi = \{\varphi_e\}_{e\in E}$ be a CGDMS satisfying the SOSC. For every $0 < R < \sup_{x\in J} \mathrm{dist}(x, \partial X)$, every finite subalphabet $W \subseteq E$ and every ergodic σ-invariant Borel probability measure μ on E_A^∞ such that $\mathrm{supp}(\mu) = E_A^\infty$ and $\int_{E_A^\infty} \Xi_\Phi \, d\mu > -\infty$, it holds that*

$$\mu(E_{R,W}^\infty) = 1 \quad \text{and thus} \quad \mu(\widehat{E}_{R,W}^\infty) = 1.$$

Proof. Simply replace $[e]$ by $\bigcup_{e\in W}[e]$ in the proof of Theorem 21.2.3. □

As a consequence of this theorem, we get the following result (cf. Corollary 21.2.4).

Corollary 21.4.2. *Let $\Phi = \{\varphi_e\}_{e\in E}$ be a finitely irreducible CGDMS satisfying the SOSC. For every $0 < R < \sup_{x\in J} \mathrm{dist}(x, \partial X)$, every finite subalphabet $W \subseteq E$, and every $(q,t) \in \mathbb{R}^2$ such that $qu + t > \theta$, we have that*

$$\mu_{F_{q,t}}(\pi(E_{R,W}^\infty)) = \mu_{f_{q,t}}(E_{R,W}^\infty) = 1 \quad \text{and thus} \quad \mu_{F_{q,t}}(\pi(\widehat{E}_{R,W}^\infty)) = \mu_{f_{q,t}}(\widehat{E}_{R,W}^\infty) = 1.$$

In particular,

$$\mu_F(\pi(E_{R,W}^\infty)) = \mu_f(E_{R,W}^\infty) = 1 \quad \text{and so} \quad \mu_F(\pi(\widehat{E}_{R,W}^\infty)) = \mu_f(\widehat{E}_{R,W}^\infty) = 1.$$

If, additionally, Φ is cofinitely regular then for every $q \in \mathbb{R}$,

$$\mu_{F_q}(\pi(E_{R,W}^\infty)) = \mu_q(E_{R,W}^\infty) = 1 \quad \text{and hence} \quad \mu_{F_q}(\pi(\widehat{E}_{R,W}^\infty)) = \mu_q(\widehat{E}_{R,W}^\infty) = 1.$$

Fix $0 < R < \sup_{x\in J} \mathrm{dist}(x, \partial X)$ and a finite subalphabet $W \subseteq E$. Let μ be a Borel probability measure on

$$J_{R,W} := \pi(\widehat{E}_{R,W}^\infty).$$

The local structure of $J_{R,W}$ is determined by the following partition of that set. For each number $\alpha \geq 0$, denote by $J_{R,W,\mu}(\alpha)$ the set of all points of $J_{R,W}$ at which the pointwise dimension of μ is equal to α, i. e.,

$$J_{R,W,\mu}(\alpha) := \{x \in J_{R,W} \mid d_\mu(x) = \alpha\}.$$

Denote the Hausdorff dimension of $J_{R,W,\mu}(\alpha)$ by

$$d_{R,W,\mu}(\alpha) := \mathrm{HD}(J_{R,W,\mu}(\alpha)).$$

The multifractal analysis of $J_{R,W}$ consists in:

(1) determining the range of the function $x \mapsto d_\mu(x)$, and
(2) for each α in that range, determining $d_{R,W,\mu}(\alpha)$.

The map $\alpha \mapsto d_{R,W,\mu}(\alpha)$ is the *(fine Hausdorff) multifractal spectrum* of the measure μ. We now prove that for every $q \in \mathbb{R}$ the measure μ_{F_q} confers full measure to the subset of points of $J_{R,W}$ where the local dimension of the measure m_F (or, equivalently, μ_F) is $\alpha(q)$. In other terms, the measure μ_{F_q} is entirely distributed over the set $J_{R,W,m_F}(\alpha(q))$ and not at all over any of the other sets $J_{R,W,m_F}(\alpha)$, $\alpha \neq \alpha(q)$. Accordingly, it is this measure (and its sister m_{F_q}), which will be used to determine the Hausdorff dimension of $J_{R,W,m_F}(\alpha(q))$.

Set

$$E^{\infty}_{R,W}(\alpha) := \left\{ \omega \in E^{\infty}_{R,W} \;\middle|\; \lim_{n\to\infty} \frac{S_n f(\omega)}{S_n \Xi_\Phi(\omega)} = \alpha \right\}.$$

Theorem 21.4.3. *Let $\Phi = \{\varphi_e\}_{e\in E}$ be a cofinitely regular, finitely irreducible, conformal-like CGDMS satisfying the SOSC. For every $0 < R < \sup_{x\in J} \operatorname{dist}(x, \partial X)$ and for every finite subalphabet $W \subseteq E$, the following statements hold:*

(a) $\pi(E^{\infty}_{R,W}(\alpha)) \subseteq J_{R,W,m_F}(\alpha)$ *for all* $\alpha \geq 0$.

(b) $\mu_q(E^{\infty}_{R,W}(\alpha(q))) = 1$ *for all* $q \in \mathbb{R}$.

(c) $\mu_{F_q}(J_{R,W,m_F}(\alpha(q))) = 1$ *for all* $q \in \mathbb{R}$.

Proof. The proof goes along similar lines to that of Theorem 21.2.5. □

Next, we establish that $d_{R,W,m_F}(\alpha)$ and $T(q)$ form a Legendre transform pair. This means that the Hausdorff dimension of the set of points of $J_{R,W}$ where the local dimension of the measure m_F is $\alpha(q)$ is equal to $q\alpha(q) + T(q)$.

Theorem 21.4.4. *Let $\Phi = \{\varphi_e\}_{e\in E}$ be a cofinitely regular, finitely irreducible, conformal-like CGDMS satisfying the SOSC. Fix $0 < R < \sup_{x\in J} \operatorname{dist}(x, \partial X)$ and a finite subalphabet $W \subseteq E$. For every $q \in \mathbb{R}$, we have $d_{R,W,m_F}(-T'(q)) = T(q) - qT'(q)$. Alternatively, $d_{R,W,m_F}(\alpha(q)) = q\alpha(q) + T(q)$.*

Proof. The proof goes along similar lines to that of Theorem 21.2.7. □

All of the above results give us an analog of Theorem 4.9.4 in [81] about the fine Hausdorff multifractal spectrum of the set $J_{R,W}$.

Theorem 21.4.5. *Let $\Phi = \{\varphi_e\}_{e\in E}$ be a cofinitely regular, finitely irreducible, conformal-like CGDMS satisfying the SOSC. For every $0 < R < \sup_{x\in J} \operatorname{dist}(x, \partial X)$ and for every finite subalphabet $W \subseteq E$, the following statements hold:*

(a) *For μ_F-a. e. $x \in J_{R,W}$,*

$$d_{\mu_F}(x) = \frac{\int_{E_A^\infty} f \, d\mu_f}{\int_{E_A^\infty} \Xi_\Phi \, d\mu_f} = \mathrm{HD}(\mu_F).$$

(b) *The temperature function $T : \mathbb{R} \to \mathbb{R}$, defined by $\mathrm{P}(q, T(q)) = 0$ for all $q \in \mathbb{R}$, is real analytic and such that*

- $T(0) = \mathrm{HD}(J)$
- $T(1) = 0$
- $T'(q) = -\alpha(q) = -\frac{\int_{E_A^\infty} f \, d\mu_q}{\int_{E_A^\infty} \Xi_\Phi \, d\mu_q} < 0$
- $T''(q) \geq 0$ *for all* $q \in \mathbb{R}$.

(c) *For every* $q \in \mathbb{R}$, *we have*

$$d_{R,W,\mu_F}(-T'(q)) = d_{R,W,\mu_F}(\alpha(q)) = q\alpha(q) + T(q) = T(q) - qT'(q).$$

That is, $d_{R,W,\mu_F}(\alpha)$ *and* $T(q)$ *form a Legendre transform pair of functions.*

(d) *If* $\mu_f = \mu_{\mathrm{HD}(J)\Xi_\Phi}$ *or, equivalently, if* f *and* $\mathrm{HD}(J)\Xi_\Phi$ *are cohomologous in the group of Hölder functions on* E_A^∞, *then* $T'(q) = -\mathrm{HD}(J)$ *for every* $q \in \mathbb{R}$ *and the domain and the range of the Hausdorff multifractal spectrum* $\alpha \mapsto d_{0,\mu_F}(\alpha)$ *are both reduced to the singleton* $\{\mathrm{HD}(J)\}$.

Proof. The proof goes along the same lines as that of Theorem 21.2.8. □

21.5 Exercises

Exercise 21.5.1. Let Φ be a CGDMS consisting of similarities, and F a Hölder family of functions all of which are constant (i. e., they depend on $e \in E$ but not on $x \in X$). Calculate the temperature function.

Exercise 21.5.2. Let Φ be a CGDMS consisting of similarities, and F a Hölder family of functions all of which are constant. Calculate the dimension spectrum of the measure μ_F.

Exercise 21.5.3. Prove that every irreducible almost-CIFS is finitely primitive.

Appendix A – A selection of classical results

This appendix lists classical definitions and results that will be used in this volume. Several of these results are stated without proofs. We sometimes complemented them with classical examples.

A.1 Measure theory

Let us begin by gathering together some of the standard results from measure theory that will be needed in this book. Measure theory is one of the main tools in ergodic theory, so it is important to be familiar with it. Proofs and further explanations of the results can be found in many books on measure theory, for instance, Billingsley [8, 9] and Rudin [109].

A.1.1 Collections of sets and measurable spaces

Given a set X, we shall denote the set of all subsets of X by $\mathcal{P}(X)$. Let us recall the definitions of some important collections of subsets of a set. The most basic collection is called a π-system.

Definition A.1.1. Let X be a set. A nonempty family $\mathcal{P} \subseteq \mathcal{P}(X)$ is a *π-system* on X if $P_1 \cap P_2 \in \mathcal{P}$ for all $P_1, P_2 \in \mathcal{P}$.

In other words, a π-system is a collection that is closed under finite intersections. For example, the family of open intervals $\{(a, \infty) : a \in \mathbb{R}\}$ constitutes a π-system on $\mathbb{R}$. So does the family of closed intervals $\{[a, \infty) : a \in \mathbb{R}\}$. Other examples are the families $\{(-\infty, b) : b \in \mathbb{R}\}$ and $\{(-\infty, b] : b \in \mathbb{R}\}$.

A "slightly" more complex collection is a semialgebra.

Definition A.1.2. Let X be a set. A family $\mathcal{S} \subseteq \mathcal{P}(X)$ is called a *semialgebra* on X if it satisfies the following three conditions:

(a) $\emptyset \in \mathcal{S}$.

(b) $\mathcal{S}$ is a π-system.

(c) If $S \in \mathcal{S}$, then $X \setminus S$ can be written as a finite union of mutually disjoint sets in $\mathcal{S}$. That is, $X \setminus S = \bigcup_{i=1}^{n} S_i$ for some $n \in \mathbb{N}$ and $S_1, S_2, \ldots, S_n \in \mathcal{S}$ with $S_i \cap S_j = \emptyset$ whenever $i \neq j$.

Every semialgebra is a π-system but the converse is not true in general. For instance, none of the π-systems described above is a semialgebra. However, the collection of all intervals forms a semialgebra on $\mathbb{R}$.

An even more intricate set collection is an algebra.

https://doi.org/10.1515/9783110702699-022

Definition A.1.3. Let X be a set. A family $\mathcal{A} \subseteq \mathcal{P}(X)$ is said to be an *algebra* on X if it satisfies the following three conditions:
(a) $\emptyset \in \mathcal{A}$.
(b) $\mathcal{A}$ is a π-system.
(c) If $A \in \mathcal{A}$, then $X \setminus A \in \mathcal{A}$.

Every algebra is a semialgebra, although the converse is not true in general. For instance, the semialgebra outlined earlier is not an algebra. Nevertheless, as we will observe in the next lemma, the collection of all subsets of $\mathbb{R}$ that can be expressed as a finite union of intervals is an algebra on $\mathbb{R}$.

The fact that an algebra is stable under finite intersections and complementation implies that an algebra is stable under finitely many set operations (e. g., unions, intersections, differences, symmetric differences, complementation and combinations thereof).

Note that $\{\emptyset, X\}$ and $\mathcal{P}(X)$ are trivial algebras on X. Since the intersection of any family of algebras is an algebra, the algebra $\mathcal{A}(\mathcal{C})$ generated by any collection $\mathcal{C}$ of subsets of X is well-defined as the smallest, in the sense of set inclusion, of all algebras on X that contain $\mathcal{C}$. If the collection $\mathcal{C}$ is a semialgebra, then it is easy to describe the algebra it generates.

Lemma A.1.4. *Let $\mathcal{S}$ be a semialgebra on X. The algebra $\mathcal{A}(\mathcal{S})$ generated by $\mathcal{S}$ consists of those subsets A of X which can be written as a finite union of mutually disjoint sets in $\mathcal{S}$, i. e., all sets $A \subseteq X$ such that $A = \bigcup_{i=1}^{n} S_i$ for some $S_1, S_2, \ldots, S_n \in \mathcal{S}$ with $S_i \cap S_j = \emptyset$ whenever $i \neq j$.*

Proof. Let

$$\mathcal{A} := \left\{ A \subseteq X \mid \exists S_1, \ldots, S_n \in \mathcal{S},\ S_i \cap S_j = \emptyset, \forall i \neq j \text{ such that } A = \bigcup_{i=1}^{n} S_i \right\}.$$

It is easy to see that $\mathcal{A}$ is an algebra containing $\mathcal{S}$. Therefore, $\mathcal{A} \supseteq \mathcal{A}(\mathcal{S})$. On the other hand, since any algebra is closed under finite unions, any algebra containing $\mathcal{S}$ must contain $\mathcal{A}$. Thus, $\mathcal{A}(\mathcal{S}) \supseteq \mathcal{A}$. Hence, $\mathcal{A} = \mathcal{A}(\mathcal{S})$. □

In measure theory, the most important type of collection of subsets of a given set is a σ-algebra.

Definition A.1.5. Let X be a set. A family $\mathcal{B} \subseteq \mathcal{P}(X)$ is called a *σ-algebra* on X if it satisfies the following three conditions:
(a) $\emptyset \in \mathcal{B}$.
(b) $\bigcap_{n=1}^{\infty} B_n \in \mathcal{B}$ for every sequence $(B_n)_{n=1}^{\infty}$ of sets in $\mathcal{B}$.
(c) If $B \in \mathcal{B}$, then $X \setminus B \in \mathcal{B}$.

Note that condition (b) can be replaced by:

(b′) $\bigcup_{n=1}^{\infty} B_n \in \mathcal{B}$ for every sequence $(B_n)_{n=1}^{\infty}$ of sets in $\mathcal{B}$.

A σ-algebra on X is thus a family of subsets of X which is closed under countably many set operations. Clearly, any σ-algebra is an algebra, though the converse is not true in general.

Note that $\{\emptyset, X\}$ and $\mathcal{P}(X)$ are trivial σ-algebras on X. They are respectively called the *indiscrete* and *discrete* σ-algebras. As the intersection of any family of σ-algebras is itself a σ-algebra, the σ-algebra $\sigma(\mathcal{C})$ generated by any collection $\mathcal{C}$ of subsets of X is well defined as the smallest σ-algebra that contains $\mathcal{C}$. In particular, if the collection $\mathcal{C}$ is finite then the algebra $\mathcal{A}(\mathcal{C})$ it generates is also finite, and thus $\sigma(\mathcal{C}) = \mathcal{A}(\mathcal{C})$ (see Exercise 8.5.1).

A set X equipped with a σ-algebra $\mathcal{B}$ is called a *measurable space* and the elements of $\mathcal{B}$ are accordingly called *measurable sets*.

Example A.1.6. Let X be a topological space and let $\mathcal{T}$ be the topology of X, i. e., the collection of all open subsets of X. Then $\sigma(\mathcal{T})$ is a σ-algebra on X called the *Borel σ-algebra* of X. Henceforth, we will denote this latter by $\mathcal{B}(X)$. In particular, $\mathcal{B}(X)$ contains all open sets and closed sets, as well as all countable unions of closed sets and all countable intersections of open sets, i. e., all F_σ- and G_δ-sets, respectively. Note that $\mathcal{T}$ is a π-system but not a semialgebra in general.

In the Euclidean space $\mathbb{R}$, the Borel σ-algebra $\mathcal{B}(\mathbb{R})$ is generated by the even simpler π-system of open intervals $\{(a,\infty) : a \in \mathbb{R}\}$. Similarly, it is generated by the semialgebra comprising all intervals.

Sometimes the functions considered take values in the extended real numbers $\overline{\mathbb{R}} := [-\infty,\infty]$. A base for the order topology of $\overline{\mathbb{R}}$ is the π-system of all open intervals, namely $\{[-\infty,b) : b \in \overline{\mathbb{R}}\} \cup \{(a,b) : a,b \in \overline{\mathbb{R}}\} \cup \{(a,\infty] : a \in \overline{\mathbb{R}}\} \cup \{[-\infty,\infty]\}$. The Borel σ-algebra $\mathcal{B}(\overline{\mathbb{R}})$ is generated by the even simpler π-system of open intervals $\{(a,\infty] : a \in \mathbb{R}\}$. Note that

$$\mathcal{B}(\overline{\mathbb{R}}) = \{B, \{-\infty\} \cup B, B \cup \{\infty\}, \{-\infty\} \cup B \cup \{\infty\} \mid B \in \mathcal{B}(\mathbb{R})\}.$$

More examples of algebras and σ-algebras are presented in Exercises 8.5.2–8.5.4.

We now introduce λ-systems, also called Dynkin systems. These collections of sets are closed under complementation and countable disjoint unions.

Definition A.1.7. Let X be a set. A family $\mathcal{L} \subseteq \mathcal{P}(X)$ is called a *λ-system* on X if it satisfies the following three conditions:

(a) $X \in \mathcal{L}$.
(b) $\bigcup_{n=1}^{\infty} L_n \in \mathcal{L}$ for every sequence $(L_n)_{n=1}^{\infty}$ of sets in $\mathcal{L}$ such that $L_n \cap L_m = \emptyset$ for all $m \neq n$.
(c) If $L \in \mathcal{L}$, then $X \setminus L \in \mathcal{L}$.

Note that condition (c) can be replaced by:

(c′) If $K, L \in \mathcal{L}$ and $K \subseteq L$, then $L \setminus K \in \mathcal{L}$.

Every σ-algebra is a λ-system but the converse is not true in general. Nevertheless, it is not difficult to see how these two concepts are related.

Lemma A.1.8. *A collection of sets forms a σ-algebra if and only if it is both a λ-system and a π-system.*

Proof. We have already observed that every σ-algebra is a λ-system and a π-system. So suppose that $\mathcal{B}$ is both a λ-system and a π-system on a set X. Since $\mathcal{B}$ is a π-system that enjoys properties (a) and (c) of a λ-system, it is clear that $\mathcal{B}$ is an algebra. Therefore, it just remains to prove that $\mathcal{B}$ satisfies condition (b') of Definition A.1.5. Let $(B_n)_{n=1}^{\infty}$ be a sequence of sets in $\mathcal{B}$. For every $n \in \mathbb{N}$, let $B'_n = \bigcup_{k=1}^{n} B_k$. As $\mathcal{B}$ is an algebra, $B'_n \in \mathcal{B}$ for all $n \in \mathbb{N}$. The sequence $(B'_n)_{n=1}^{\infty}$ is ascending and is such that $\bigcup_{n=1}^{\infty} B_n = \bigcup_{n=1}^{\infty} B'_n$. Thus, it suffices to prove condition (b') for ascending sequences in $\mathcal{B}$. Moreover, observe that $\bigcup_{n=1}^{\infty} B'_n = B'_1 \cup \bigcup_{n=1}^{\infty} (B'_{n+1} \setminus B'_n)$. By condition (c') of a λ-system, we know that $B'_{n+1} \setminus B'_n \in \mathcal{B}$ for each $n \in \mathbb{N}$. Furthermore, the sets B'_1 and $B'_{n+1} \setminus B'_n$, $n \in \mathbb{N}$, are mutually disjoint. By condition (b) of a λ-system, it follows that

$$\bigcup_{n=1}^{\infty} B_n = \bigcup_{n=1}^{\infty} B'_n = B'_1 \cup \bigcup_{n=1}^{\infty} (B'_{n+1} \setminus B'_n) \in \mathcal{B}.$$

Hence, $\mathcal{B}$ is an algebra satisfying condition (b') of Definition A.1.5. So $\mathcal{B}$ is a σ-algebra. □

The importance and usefulness of λ-systems mostly lie in the following theorem.

Theorem A.1.9 (Dynkin's π-λ theorem). *If $\mathcal{P}$ is a π-system and $\mathcal{L}$ is a λ-system such that $\mathcal{P} \subseteq \mathcal{L}$, then $\sigma(\mathcal{P}) \subseteq \mathcal{L}$.*

Proof. See Theorem 3.2 in Billingsley [8]. □

Furthermore, as the intersection of any family of λ-systems is a λ-system, the λ-system $\mathcal{L}(\mathcal{C})$ generated by any collection $\mathcal{C} \subseteq \mathcal{P}(X)$ is well defined as the intersection of all λ-systems that comprise $\mathcal{C}$. When $\mathcal{C}$ is a π-system, the λ-system and the σ-algebra that are generated by $\mathcal{C}$ are one and the same.

Corollary A.1.10. *If $\mathcal{P}$ is a π-system, then $\sigma(\mathcal{P}) = \mathcal{L}(\mathcal{P})$.*

Proof. This immediately follows from Lemma A.1.8 and Theorem A.1.9. □

Finally, let us recall yet another type of collection of sets named, for obvious reasons, a monotone class.

Definition A.1.11. Let X be a set. A family $\mathcal{M} \subseteq \mathcal{P}(X)$ is called a *monotone class* on X if it is stable under countable monotone unions and countable monotone intersections.

In other words,

$$\text{if } (M_n)_{n=1}^{\infty} \subseteq \mathcal{M} \text{ is such that } M_1 \subseteq M_2 \subseteq M_3 \subseteq \cdots, \quad \text{then } \bigcup_{n=1}^{\infty} M_n \in \mathcal{M}$$

and

$$\text{if } (M_n)_{n=1}^{\infty} \subseteq \mathcal{M} \text{ is such that } M_1 \supseteq M_2 \supseteq M_3 \supseteq \cdots, \quad \text{then } \bigcap_{n=1}^{\infty} M_n \in \mathcal{M}.$$

Every σ-algebra is a monotone class but the converse is not true in general. Nevertheless, a monotone class which is an algebra is a σ-algebra. There is an analogue of Dynkin's theorem for monotone classes.

Theorem A.1.12 (Halmos' monotone class theorem). *If $\mathcal{A}$ is an algebra and $\mathcal{M}$ is a monotone class such that $\mathcal{A} \subseteq \mathcal{M}$, then $\sigma(\mathcal{A}) \subseteq \mathcal{M}$.*

Proof. See Theorem 3.4 in Billingsley [8]. □

Because the intersection of any family of monotone classes is a monotone class, the monotone class $\mathcal{M}(\mathcal{C})$ generated by any collection $\mathcal{C} \subseteq \mathcal{P}(X)$ is well defined as the intersection of all monotone classes that comprise $\mathcal{C}$. When $\mathcal{C}$ is a semialgebra, the σ-algebra and the monotone class generated by $\mathcal{C}$ coincide.

Theorem A.1.13. *If $\mathcal{S}$ is a semialgebra, then $\sigma(\mathcal{S}) = \sigma(\mathcal{A}(\mathcal{S})) = \mathcal{M}(\mathcal{A}(\mathcal{S})) = \mathcal{M}(\mathcal{S})$.*

Proof. The equalities $\sigma(\mathcal{S}) = \sigma(\mathcal{A}(\mathcal{S}))$ and $\mathcal{M}(\mathcal{A}(\mathcal{S})) = \mathcal{M}(\mathcal{S})$ are obvious. Since a σ-algebra is a monotone class and $\sigma(\mathcal{A}(\mathcal{S})) \supseteq \mathcal{A}(\mathcal{S})$, it is evident that $\sigma(\mathcal{A}(\mathcal{S})) \supseteq \mathcal{M}(\mathcal{A}(\mathcal{S}))$. The opposite inclusion is the object of Halmos' monotone class theorem. □

A.1.2 Measurable transformations

We now look at maps between measurable spaces. Recall that a measurable space is a set X equipped with a σ-algebra $\mathcal{A}$. The elements of $\mathcal{A}$ are the measurable sets in that space.

Definition A.1.14. Let $(X, \mathcal{A})$ and $(Y, \mathcal{B})$ be measurable spaces. A transformation $T : X \to Y$ is said to be *measurable* provided that $T^{-1}(B) \in \mathcal{A}$ for every $B \in \mathcal{B}$.

We have earlier mentioned that σ-algebras are the most important collections of sets in measure theory. However, it is generally impossible to describe, in a simple form, the sets in a σ-algebra. Luckily, σ-algebras that are generated by smaller and simpler structures like π-systems, semialgebras or algebras, are much easier to cope with. In this situation, proving that some interesting property is satisfied for the sets in these smaller and simpler structures is often sufficient to guarantee that that property holds for all sets in the σ-algebra. This is the case for the measurability of transformations.

Theorem A.1.15. *Let $T : (X, \mathcal{A}) \to (Y, \mathcal{B})$ be a transformation. If $\mathcal{B} = \sigma(\mathcal{C})$ is a σ-algebra generated by a collection $\mathcal{C} \subseteq \mathcal{P}(Y)$, then T is measurable if and only if $T^{-1}(C) \in \mathcal{A}$ for all $C \in \mathcal{C}$.*

Proof. It is clear that if T is measurable, then $T^{-1}(C) \in \mathcal{A}$ for all $C \in \mathcal{C}$ since $\mathcal{C} \subseteq \sigma(\mathcal{C}) = \mathcal{B}$. Conversely, suppose that $T^{-1}(C) \in \mathcal{A}$ for all $C \in \mathcal{C}$. Consider the collection of sets $\mathcal{B}' = \{B \subseteq Y : T^{-1}(B) \in \mathcal{A}\}$. By assumption, $\mathcal{B}' \supseteq \mathcal{C}$. It is also easy to see that $\mathcal{B}'$ is a σ-algebra. Thus, $\mathcal{B}' \supseteq \sigma(\mathcal{C}) = \mathcal{B}$, and hence T is measurable. □

If the range Y of a transformation is a Borel subset of $\overline{\mathbb{R}}$, then unless otherwise stated Y will be assumed to be endowed with its Borel σ-algebra, which is just the projection of $\mathcal{B}(\overline{\mathbb{R}})$ onto Y (see Exercise 8.5.5). In this context, we will use the term function instead of transformation.

Example A.1.16. Let $(X, \mathcal{A})$ be a measurable space.

(a) Let $A \subseteq X$. The *indicator function* $\mathbb{1}_A : X \to \{0, 1\}$ (also called *characteristic function*) defined by

$$\mathbb{1}_A(x) := \begin{cases} 1 & \text{if } x \in A \\ 0 & \text{if } x \notin A \end{cases}$$

is measurable if and only if A is measurable, i. e., $A \in \mathcal{A}$.

(b) A function $s : X \to \mathbb{R}$ which takes only finitely many values is called a *simple function*. Such a function can be expressed in the form

$$s = \sum_{i=1}^{n} \alpha_i \mathbb{1}_{A_i},$$

where $A_i = \{x \in X : s(x) = \alpha_i\}$ and the α_i's are the values of the function s. Such a function is measurable if and only if each set A_i is measurable.

The following theorem shows the utility of simple functions. It states that any nonnegative measurable function is the pointwise limit of a nondecreasing sequence of nonnegative measurable simple functions.

Theorem A.1.17. *Let $(X, \mathcal{A})$ be a measurable space and $f : X \to [0, \infty]$ be a measurable function. Then there exists a sequence $(s_n)_{n=1}^{\infty}$ of measurable simple functions on X such that:*

(a) $0 \le s_1 \le s_2 \le \cdots \le f$.

(b) $\lim_{n\to\infty} s_n(x) = f(x)$, $\forall x \in X$.

Proof. See Theorem 1.17 in Rudin [109]. □

A.1.3 Measure spaces

The concept of measure is obviously central to measure theory.

Definition A.1.18. Let $(X, \mathcal{A})$ be a measurable space. A set function $\mu : \mathcal{A} \to [0, \infty]$ is said to be a *measure* on X provided that:

(a) $\mu(\emptyset) = 0$.

(b) μ is countably additive, i. e., for each sequence $(A_n)_{n=1}^{\infty}$ of pairwise disjoint sets belonging to $\mathcal{A}$, the function μ is such that

$$\mu\Big(\bigcup_{n=1}^{\infty} A_n\Big) = \sum_{n=1}^{\infty} \mu(A_n).$$

The triple $(X, \mathcal{A}, \mu)$ is called a *measure space*.

If $\mu(X) < \infty$, then μ is said to be a *finite measure*. If $\mu(X) = 1$, then μ is a *probability measure*. Finally, μ is said to be *σ-finite* if there exists a sequence $(A_n)_{n=1}^{\infty}$ of sets in $\mathcal{A}$ such that $\mu(A_n) < \infty$ for all $n \in \mathbb{N}$ and $\bigcup_{n=1}^{\infty} A_n = X$.

Here are a few basic properties of measures.

Lemma A.1.19. *Let $(X, \mathcal{A}, \mu)$ be a measure space and $A, B \in \mathcal{A}$.*

(a) *If $A \subset B$, then $\mu(A) \le \mu(B)$.*

(b) *If $A \subseteq B$ and $\mu(B \setminus A) < \infty$, then $\mu(A) = \mu(B) - \mu(B \setminus A)$.*

(c) *$\mu(A \cup B) \le \mu(A) + \mu(B)$.*

(d) *If $(A_n)_{n=1}^{\infty}$ is an ascending sequence in $\mathcal{A}$ (i. e., $A_n \subseteq A_{n+1}$, $\forall n \in \mathbb{N}$), then*

$$\mu\Big(\bigcup_{n=1}^{\infty} A_n\Big) = \lim_{n\to\infty} \mu(A_n) = \sup_{n\in\mathbb{N}} \mu(A_n).$$

(e) *If $(B_n)_{n=1}^{\infty}$ is a descending sequence in $\mathcal{A}$ (i. e., $A_n \supseteq A_{n+1}, \forall n \in \mathbb{N}$) and if $\mu(B_1) < \infty$, then*

$$\mu\Big(\bigcap_{n=1}^{\infty} B_n\Big) = \lim_{n\to\infty} \mu(B_n) = \inf_{n\in\mathbb{N}} \mu(B_n).$$

(f) *If $(C_n)_{n=1}^{\infty}$ is any sequence in $\mathcal{A}$, then*

$$\mu\Big(\bigcup_{n=1}^{\infty} C_n\Big) \le \sum_{n=1}^{\infty} \mu(C_n).$$

(g) *If* $(D_n)_{n=1}^{\infty}$ *is a sequence in* $\mathcal{A}$ *such that* $\mu(D_m \cap D_n) = 0$ *for all* $m \neq n$*, then*

$$\mu\Big(\bigcup_{n=1}^{\infty} D_n\Big) = \sum_{n=1}^{\infty} \mu(D_n).$$

Proof. The proof is left to the reader as an exercise. □

A more intricate property of measures deserves a special name.

Lemma A.1.20 (Borel–Cantelli lemma). *Let* $(X, \mathcal{A}, \mu)$ *be a measure space and* $(A_n)_{n=1}^{\infty}$ *a sequence in* $\mathcal{A}$. *If* $\sum_{n=1}^{\infty} \mu(A_n) < \infty$, *then* $\mu(\bigcap_{k=1}^{\infty} \bigcup_{n=k}^{\infty} A_n) = 0$.

Proof. The proof is left to the reader as an exercise. □

We now provide two simple examples of measures. The first of these may seem insignificant at first glance but turns out to be very useful in practice.

Example A.1.21. Let X be a nonempty set and $\mathcal{P}(X)$ be the discrete σ-algebra on X.

(a) Choose a point $x \in X$. Define the set function $\delta_x : \mathcal{P}(X) \to \{0, 1\}$ by setting

$$\delta_x(A) := \begin{cases} 1 & \text{if } x \in A \\ 0 & \text{if } x \notin A. \end{cases}$$

One readily verifies that δ_x is a probability measure. It is referred to as the *Dirac point mass* or *Dirac measure* concentrated at the point x.

(b) For any $A \subseteq X$, define $m(A)$ to be the number of elements in the set A if A is finite and set $m(A) = \infty$ if the set A is infinite. Then $m : \mathcal{P}(X) \to \mathbb{N} \cup \{\infty\}$ is called the *counting measure* on X.

There is a notion of completeness for measure spaces.

Definition A.1.22. A measure space $(X, \mathcal{A}, \mu)$ is said to be *complete* if every subset of a set of measure zero is measurable. That is, if $A \subseteq X$ and there is $B \in \mathcal{A}$ such that $A \subseteq B$ and $\mu(B) = 0$, then $A \in \mathcal{A}$.

Note that any measure space can be extended to a complete one (see Exercises 8.5.7–8.5.8).

In Example A.1.6, we introduced the concept of Borel σ-algebra. We now introduce Borel measures and describe different forms of regularity for these measures.

Definition A.1.23. Let X be a topological space and $\mathcal{B}(X)$ be the Borel σ-algebra on X.

(a) A *Borel measure* μ on X is a measure defined on the Borel σ-algebra $\mathcal{B}(X)$ of X. The resulting measure space $(X, \mathcal{B}(X), \mu)$ is called a *Borel measure space*. In particular, if μ is a probability measure then $(X, \mathcal{B}(X), \mu)$ is called a Borel probability space.

(b) A Borel measure μ is said to be *inner regular* if

$$\mu(B) = \sup\{\mu(K) : K \subseteq B,\ K \text{ compact}\}, \quad \forall B \in \mathcal{B}(X).$$

(c) A Borel measure μ is said to be *outer regular* if

$$\mu(B) = \inf\{\mu(G) : B \subseteq G,\ G \text{ open}\}, \quad \forall B \in \mathcal{B}(X).$$

(d) A Borel measure is called *regular* if it is both inner and outer regular.

Theorem A.1.24. *Every Borel probability measure on a separable, completely metrizable space is regular.*

Proof. See Theorem 17.11 in Kechris [61]. □

Example A.1.25. Let $k \in \mathbb{N}$. There exists a complete, regular measure λ defined on a σ-algebra $\mathcal{L}$ on $\mathbb{R}^k$ with the following properties:
(a) $\lambda(R) = \mathrm{Vol}(R)$ for every k-rectangle $R \subseteq \mathbb{R}^k$, where Vol denotes the usual k-dimensional volume in $\mathbb{R}^k$.
(b) $\mathcal{L}$ is the completion of the Borel σ-algebra $\mathcal{B}(\mathbb{R}^k)$; more precisely, $E \in \mathcal{L}$ if and only if there exist an F_σ-set F and a G_δ-set G such that $F \subseteq E \subseteq G$ and $\lambda(G \setminus F) = 0$.
(c) λ is *translation invariant*, i. e., $\lambda(E + x) = \lambda(E)$ for all $E \in \mathcal{L}$ and all $x \in \mathbb{R}^k$.

Moreover, up to a multiplicative constant, λ is the only translation-invariant Borel measure on $\mathbb{R}^k$ that gives finite measure to all compact sets.

The measure λ is called the *Lebesgue measure on* $\mathbb{R}^k$ and, accordingly, the sets in $\mathcal{L}$ are said to be *Lebesgue measurable*. Unlike in Example A.1.21, not all subsets of $\mathbb{R}^k$ are Lebesgue measurable. Indeed, Vitali (see 5.7, The Vitali Monsters on p. 120 of [51]) showed that it is impossible to construct a measure having properties (a)–(c) on the set of all subsets of $\mathbb{R}^k$.

The following lemma is one more eloquent manifestation of the relevance of structures simpler than σ-algebras.

Lemma A.1.26. *Let X be a set and $\mathcal{P}$ be a π-system on X. Let μ and ν be probability measures on $(X, \sigma(\mathcal{P}))$. Then*

$$\mu = \nu \quad \Longleftrightarrow \quad \mu(P) = \nu(P), \quad \forall P \in \mathcal{P}.$$

Proof. The implication $\Rightarrow$ is trivial. For the opposite one $\Leftarrow$, suppose that $\mu(P) = \nu(P)$ for all $P \in \mathcal{P}$. Consider the collection of sets $\mathcal{A} := \{A \subseteq X : \mu(A) = \nu(A)\}$. By assumption, $\mathcal{P} \subseteq \mathcal{A}$. It is also easy to see that $\mathcal{A}$ is a λ-system. Per Corollary A.1.10, it follows that $\sigma(\mathcal{P}) = \mathcal{L}(\mathcal{P}) \subseteq \mathcal{A}$. □

In summary, two probability measures that agree on a π-system are equal on the σ-algebra generated by that π-system. However, this result does not generally hold for infinite measures (see Exercise 8.5.9).

A.1.4 Extension of set functions to measures

The main shortcoming of the preceding lemma lies in the assumption that the set functions μ and ν are measures defined on a σ-algebra. In particular, this means that they are countably additive on that entire σ-algebra. However, we frequently define a set function $\mu : \mathcal{C} \to [0,\infty]$ on a collection $\mathcal{C}$ of subsets of a set X on which the values of μ are naturally determined but it may be unclear whether μ may be extended to a measure on $\sigma(\mathcal{C})$. The forthcoming results are extremely useful in that regard. The first one concerns the extension of a finitely/countably additive set function from a semialgebra to an algebra.

Theorem A.1.27. *Let $\mathcal{S}$ be a semialgebra on a set X and let $\mu : \mathcal{S} \to [0,\infty]$ be a finitely additive set function, i. e., a function such that*

$$\mu\Big(\bigcup_{i=1}^{n} S_i\Big) = \sum_{i=1}^{n} \mu(S_i)$$

for every finite family $(S_i)_{i=1}^{n}$ of mutually disjoint sets in $\mathcal{S}$ such that $\bigcup_{i=1}^{n} S_i \in \mathcal{S}$. Then there exists a unique finitely additive set function $\overline{\mu} : \mathcal{A}(\mathcal{S}) \to [0,\infty]$ which is an extension of μ to $\mathcal{A}(\mathcal{S})$, the algebra generated by $\mathcal{S}$. Moreover, the extension $\overline{\mu}$ is countably additive whenever the original set function μ is.

Proof. This directly follows from Lemma A.1.4. For more detail, see Theorems 3.4 and 3.5 in Kingman and Taylor [62]. □

The second result concerns the extension of a countably additive set function from an algebra to a σ-algebra. This time, though, the function is assumed to be finite.

Theorem A.1.28 (Carathéodory's extension theorem). *Let $\mathcal{A}$ be an algebra on a set X and let $\mu : \mathcal{A} \to [0,1]$ be a countably additive set function, i. e., a function such that*

$$\mu\Big(\bigcup_{i=1}^{\infty} A_i\Big) = \sum_{i=1}^{\infty} \mu(A_i)$$

for every sequence $(A_i)_{i=1}^{\infty}$ of mutually disjoint sets in $\mathcal{A}$ such that $\bigcup_{i=1}^{\infty} A_i \in \mathcal{A}$. Suppose also that $\mu(X) = 1$. Then there exists a unique probability measure $\overline{\mu} : \sigma(\mathcal{A}) \to [0,1]$ which is an extension of $\mu : \mathcal{A} \to [0,1]$.

Proof. See Theorem 3.1 in Billingsley [8] or Theorem 4.2 in Kingman and Taylor [62]. □

Carathéodory's extension theorem thus reduces the problem to demonstrating that a set function is countably additive on an algebra. This can be hard to prove, so we sometimes rely upon the following result.

Lemma A.1.29. *Let $\mathcal{A}$ be an algebra on a set X and $\mu : \mathcal{A} \to [0,\infty)$ be a finitely additive function. Then μ is countably additive if and only if*

$$\lim_{i\to\infty} \mu(A_i) = 0 \tag{A.1}$$

for every descending sequence $(A_i)_{i=1}^{\infty}$ of sets in $\mathcal{A}$ such that $\bigcap_{i=1}^{\infty} A_i = \emptyset$.

Furthermore, if (A.1) *holds, then μ has a unique extension to a σ-additive function (a measure) from $\sigma(\mathcal{A})$, the σ-algebra generated by $\mathcal{A}$, to $[0,\infty)$.*

Proof. Recall that measures enjoy property (d) of Lemma A.1.19. However, the set function μ considered here is not a measure since it is defined on an algebra rather than on a σ-algebra. Nevertheless, we will show that μ satisfies property (d) on the algebra $\mathcal{A}$ because of hypothesis (A.1). Let $(A_i)_{i=1}^{\infty}$ be an ascending sequence of sets in $\mathcal{A}$ such that $A := \bigcup_{i=1}^{\infty} A_i \in \mathcal{A}$. Since $\mathcal{A}$ is an algebra, we know that $A \setminus A_i \in \mathcal{A}$ for all $i \in \mathbb{N}$. Therefore, the sequence $(A \setminus A_i)_{i=1}^{\infty}$ is a descending sequence of sets in $\mathcal{A}$ such that

$$\bigcap_{i=1}^{\infty}(A \setminus A_i) = A \setminus \left(\bigcup_{i=1}^{\infty} A_i\right) = \emptyset.$$

By hypothesis (A.1), we infer that

$$\lim_{i\to\infty} \mu(A \setminus A_i) = 0.$$

Moreover, since $A = (A \setminus A_i) \cup A_i$ and μ is finitely additive on $\mathcal{A}$ and finite, we deduce that

$$\mu(A \setminus A_i) = \mu(A) - \mu(A_i), \quad \forall i \in \mathbb{N}.$$

Hence,

$$\lim_{i\to\infty} (\mu(A) - \mu(A_i)) = 0.$$

Thus,

$$\mu\left(\bigcup_{i=1}^{\infty} A_i\right) = \lim_{i\to\infty} \mu(A_i). \tag{A.2}$$

Now, let $(B_i)_{i=1}^{\infty}$ be any sequence of mutually disjoint sets in $\mathcal{A}$ such that $\bigcup_{i=1}^{\infty} B_i \in \mathcal{A}$. For every $i \in \mathbb{N}$, let $B_i' = \bigcup_{j=1}^{i} B_j$. As $\mathcal{A}$ is an algebra, $B_i' \in \mathcal{A}$ for all $i \in \mathbb{N}$. The sequence $(B_i')_{i=1}^{\infty}$ is ascending and is such that $\bigcup_{i=1}^{\infty} B_i' = \bigcup_{i=1}^{\infty} B_i \in \mathcal{A}$. By (A.2), we know that

$$\mu\left(\bigcup_{i=1}^{\infty} B_i\right) = \mu\left(\bigcup_{i=1}^{\infty} B_i'\right) = \lim_{i\to\infty} \mu(B_i').$$

Using this, the fact that the B_i's are mutually disjoint and that μ is finitely additive, we conclude that

$$\mu\Big(\bigcup_{i=1}^{\infty} B_i\Big) = \lim_{i\to\infty} \mu(B_i') = \lim_{i\to\infty} \mu\Big(\bigcup_{j=1}^{i} B_j\Big) = \lim_{i\to\infty} \sum_{j=1}^{i} \mu(B_j) = \sum_{i=1}^{\infty} \mu(B_i).$$

That is, μ is countably additive on $\mathcal{A}$.

The second part of the statement directly follows from the first one and Theorem A.1.28. □

As an immediate consequence of Lemma A.1.29 we get the following.

Lemma A.1.30. *Let $\mathcal{A}$ be an algebra on a set X, let $\nu : \sigma(\mathcal{A}) \to [0, \infty)$ be a finite measure, and let $\mu : \mathcal{A} \to [0, \infty)$ be a finitely additive function which is absolutely continuous with respect to ν, meaning that $\mu(A) = 0$ whenever $A \in \mathcal{A}$ and $\nu(A) = 0$. Then μ is countably additive and has a unique extension to a σ-additive function (a measure) from $\sigma(\mathcal{A})$ to $[0, \infty)$.*

Finally, as a straightforward consequence of Lemma A.1.29 we have the following.

Lemma A.1.31. *Let $\mathcal{A}$ be a σ-algebra on a set X, let $\nu : \mathcal{A} \to [0, \infty)$ be a finite measure, and let $\mu : \mathcal{A} \to [0, \infty)$ be a finitely additive function which is absolutely continuous with respect to ν on some algebra generating $\mathcal{A}$, meaning that $\mu(A) = 0$ whenever A belongs to this algebra and $\nu(A) = 0$. Then μ is a measure.*

Another feature of an algebra is that any element of a σ-algebra generated by an algebra can be approximated as closely as desired by an element of the algebra. Before stating the precise result, recall that the *symmetric difference* of two sets A and B is denoted by $A \triangle B$ and is the set of all points that belong to exactly one of those two sets. That is,

$$A \triangle B := (A \setminus B) \cup (B \setminus A) = (A \cup B) \setminus (A \cap B).$$

Properties of the symmetric difference are examined in Exercises 8.5.10–8.5.11.

Lemma A.1.32. *Let $\mathcal{A}$ be an algebra on a set X and μ be a probability measure on $(X, \sigma(\mathcal{A}))$. Then for every $\varepsilon > 0$ and $B \in \sigma(\mathcal{A})$ there exists some $A \in \mathcal{A}$ such that $\mu(A \triangle B) < \varepsilon$.*

Proof. See Theorem 4.4 in Kingman and Taylor [62]. □

A.1.5 Integration

Let us now briefly recollect some facts about integration. First, the definition of the integral of a measurable function with respect to a measure.

Definition A.1.33. Let $(X, \mathcal{A}, \mu)$ be a measure space and let $A \in \mathcal{A}$.

(a) If $s : X \to [0, \infty)$ is a measurable simple function of the form

$$s = \sum_{i=1}^{n} \alpha_i \mathbb{1}_{A_i},$$

then the integral of the function s over the set A with respect to the measure μ is defined as

$$\int_A s \, d\mu := \sum_{i=1}^{n} \alpha_i \mu(A_i \cap A).$$

We use the convention that $0 \cdot \infty = 0$ in case it happens that $\alpha_i = 0$ and $\mu(A_i \cap A) = \infty$ for some $1 \leq i \leq n$.

(b) If $f : X \to [0, \infty]$ is a measurable function, then the integral of the function f over the set A with respect to the measure μ is defined as

$$\int_A f \, d\mu := \sup \int_A s \, d\mu,$$

where the supremum is taken over all measurable simple functions $0 \leq s \leq f$. Note that if f is simple, then definitions (a) and (b) coincide.

(c) If $f : X \to \overline{\mathbb{R}}$ is a measurable function, then the integral of the function f over the set A with respect to the measure μ is defined as

$$\int_A f \, d\mu := \int_A f_+ \, d\mu - \int_A f_- \, d\mu,$$

as long as $\min\{\int_A f_+ \, d\mu, \int_A f_- \, d\mu\} < \infty$, where f_+ and f_- respectively denote the positive and negative parts of f. That is, $f_+(x) := \max\{f(x), 0\}$ whereas $f_-(x) := \max\{-f(x), 0\}$.

(d) A measurable function $f : X \to \overline{\mathbb{R}}$ is said to be *integrable* if $\int_X |f| \, d\mu < \infty$. We denote this by $f \in L^1(X, \mathcal{A}, \mu)$. If no confusion is possible, we simply write $f \in L^1(\mu)$.

(e) A property is said to hold μ-*almost everywhere* (sometimes abbreviated μ-a. e.) if the property holds on the entire space except possibly on a set of μ-measure zero.

The following properties follow from this definition.

Lemma A.1.34. *Let $(X, \mathcal{A}, \mu)$ be a measure space. Let $f, g \in L^1(X, \mathcal{A}, \mu)$, $A, B \in \mathcal{A}$ and $a, b \in \mathbb{R}$.*

(a) *If $f \leq g$ μ-a. e., then $\int_A f \, d\mu \leq \int_A g \, d\mu$. Also, if $f < g$ μ-a. e., then $\int_A f \, d\mu < \int_A g \, d\mu$.*

(b) *If $A \subseteq B$ and $0 \leq f$ μ-a. e., then $0 \leq \int_A f \, d\mu \leq \int_B f \, d\mu$.*

(c) *"Triangle inequality"*:

$$\left|\int_A f\,d\mu\right| \le \int_A |f|\,d\mu.$$

(d) *Linearity*:

$$\int_A (af + bg)\,d\mu = a\int_A f\,d\mu + b\int_A g\,d\mu.$$

(e) *If $(A_n)_{n=1}^{\infty}$ is a sequence of mutually disjoint measurable sets, then*

$$\int_{\bigcup_{n=1}^{\infty} A_n} f\,d\mu = \sum_{n=1}^{\infty}\int_{A_n} f\,d\mu.$$

(f) $f = g$ *μ-a. e.* $\Longleftrightarrow \int_A f\,d\mu = \int_A g\,d\mu,\ \forall A \in \mathcal{A}$.

(g) *The relation $f = g$ μ-a. e. is an equivalence relation on the set $L^1(X, \mathcal{A}, \mu)$. The equivalence classes generated by this relation form a Banach space also denoted by $L^1(X, \mathcal{A}, \mu)$ (or $L^1(\mu)$, for short) with norm*

$$\|f\|_1 := \int_X |f|\,d\mu < \infty.$$

A.1.6 Convergence theorems

In measure theory, there are fundamental theorems that are especially helpful for finding the integral of functions that are the pointwise limit of sequences of functions.

The first of these results applies to monotone sequences of functions. A sequence of functions $(f_n)_{n=1}^{\infty}$ is monotone if it is increasing pointwise ($f_{n+1}(x) \ge f_n(x)$ for all $x \in X$ and all $n \in \mathbb{N}$) or decreasing pointwise ($f_{n+1}(x) \le f_n(x)$ for all $x \in X$ and all $n \in \mathbb{N}$). We state the theorem for increasing sequences, but its counterpart for decreasing sequences can be easily deduced from it.

Theorem A.1.35 (Monotone convergence theorem). *Let $(X, \mathcal{A}, \mu)$ be a measure space. If $(f_n)_{n=1}^{\infty}$ is an increasing sequence of nonnegative measurable functions, then the integral of their pointwise limit is equal to the limit of their integrals, i. e.,*

$$\int_X \lim_{n\to\infty} f_n\,d\mu = \lim_{n\to\infty}\int_X f_n\,d\mu.$$

Proof. See Theorem 1.26 in Rudin [109]. □

Note that this theorem holds for almost everywhere increasing sequences of almost everywhere nonnegative measurable functions with an almost everywhere pointwise limit.

For general sequences of nonnegative functions, we have the following immediate consequence.

Lemma A.1.36 (Fatou's lemma). *Let $(X, \mathcal{A}, \mu)$ be a measure space. For any sequence $(f_n)_{n=1}^{\infty}$ of nonnegative measurable functions,*

$$\int_X \liminf_{n\to\infty} f_n \, d\mu \le \liminf_{n\to\infty} \int_X f_n \, d\mu.$$

Proof. For every $x \in X$ and $n \in \mathbb{N}$, define $g_n(x) = \inf\{f_i(x) : i \ge n\}$ and apply the monotone convergence theorem to the sequence $(g_n)_{n=1}^{\infty}$. For more detail, see Theorem 1.28 in Rudin [109]. □

The following lemma is another application of the monotone convergence theorem. It offers another way of integrating a nonnegative function.

Lemma A.1.37. *Let $(X, \mathcal{A}, \mu)$ be a measure space. Let f be a nonnegative measurable function and $A \in \mathcal{A}$. Then*

$$\int_A f \, d\mu = \int_0^{\infty} \mu(\{x \in A : f(x) > r\}) \, dr.$$

Proof. Suppose that $f = \mathbb{1}_B$ for some $B \in \mathcal{A}$. Then

$$\begin{aligned}\int_0^{\infty} \mu(\{x \in A : f(x) > r\}) \, dr &= \int_0^{1} \mu(\{x \in A : \mathbb{1}_B(x) > r\}) \, dr \\ &= \int_0^{1} \mu(A \cap B) \, dr = \mu(A \cap B) \\ &= \int_X \mathbb{1}_{A\cap B} \, d\mu = \int_X \mathbb{1}_A \cdot \mathbb{1}_B \, d\mu = \int_A f \, d\mu.\end{aligned}$$

So the equality holds for characteristic functions. We leave it to the reader to show that the equality prevails for all nonnegative measurable simple functions. If f is a general nonnegative measurable function, then by Theorem A.1.17 there exists an increasing sequence $(s_n)_{n=1}^{\infty}$ of nonnegative measurable simple functions such that $\lim_{n\to\infty} s_n(x) = f(x)$ for every $x \in X$. For every $r \ge 0$, let

$$\tilde{f}(r) = \mu(\{x \in A : f(x) > r\}).$$

This function is obviously nonnegative and decreasing. Hence, by Theorem A.1.15 it is Borel measurable since $\tilde{f}^{-1}((t,\infty])$ is an interval for all $t \in [0,\infty]$ and the sets $\{(t,\infty]\}_{t\in[0,\infty]}$ generate the Borel σ-algebra of $[0,\infty]$. Fix momentarily $r \geq 0$. Since $s_n \nearrow f$, the sets $(\{x \in A : s_n(x) > r\})_{n=1}^{\infty}$ form an ascending sequence such that

$$\bigcup_{n=1}^{\infty}\{x \in A : s_n(x) > r\} = \{x \in A : f(x) > r\}.$$

Then by Lemma A.1.19(d),

$$\tilde{f}(r) = \mu\Big(\bigcup_{n=1}^{\infty}\{x \in A : s_n(x) > r\}\Big) = \lim_{n\to\infty} \mu(\{x \in A : s_n(x) > r\}) = \lim_{n\to\infty} \tilde{s}_n(r).$$

Observe that $(\tilde{s}_n)_{n=1}^{\infty}$ is an increasing sequence of nonnegative decreasing functions. So $(\tilde{s}_n)_{n=1}^{\infty}$ is a sequence of nonnegative Borel measurable functions which increases pointwise to $\tilde{f}$. It follows from the monotone convergence theorem (Theorem A.1.35) that

$$\begin{aligned}\int_0^{\infty} \mu(\{x \in A : f(x) > r\})\, dr &= \int_0^{\infty} \tilde{f}(r)\, dr = \int_0^{\infty} \lim_{n\to\infty} \tilde{s}_n(r)\, dr\\ &= \lim_{n\to\infty}\int_0^{\infty} \tilde{s}_n(r)\, dr = \lim_{n\to\infty}\int_0^{\infty} \mu(\{x \in A : s_n(x) > r\})\, dr\\ &= \lim_{n\to\infty}\int_A s_n\, d\mu = \int_A \lim_{n\to\infty} s_n\, d\mu = \int_A f\, d\mu.\end{aligned}$$

□

Pointwise convergence of a sequence of integrable functions does not guarantee convergence in L^1 (see Exercise 8.5.14). However, under one relatively weak additional assumption, this becomes true. The second fundamental theorem of convergence applies to sequences of functions which have an almost everywhere pointwise limit and are dominated (i. e., uniformly bounded) almost everywhere by an integrable function.

Theorem A.1.38 (Lebesgue's dominated convergence theorem). *If a sequence of measurable functions $(f_n)_{n=1}^{\infty}$ on a measure space $(X, \mathcal{A}, \mu)$ converges pointwise μ-a. e. to a function f and if there exists $g \in L^1(\mu)$ such that $|f_n(x)| \leq g(x)$ for all $n \in \mathbb{N}$ and μ-a. e. $x \in X$, then $f \in L^1(\mu)$ and*

$$\lim_{n\to\infty} \|f_n - f\|_1 = 0 \quad \textit{and} \quad \lim_{n\to\infty}\int_X f_n\, d\mu = \int_X f\, d\mu.$$

Proof. Apply Fatou's lemma to $2g - |f_n - f| \geq 0$. See Theorem 1.34 in Rudin [109] for more detail. □

Note that

$$\lim_{n\to\infty} \|f_n - f\|_1 = 0 \quad \Longrightarrow \quad \lim_{n\to\infty} \|f_n\|_1 = \|f\|_1$$

since $\big|\|f_n\|_1 - \|f\|_1\big| \le \|f_n - f\|_1$ and

$$\lim_{n\to\infty} \|f_n - f\|_1 = 0 \quad \Longrightarrow \quad \lim_{n\to\infty} \int_X f_n \, d\mu = \int_X f \, d\mu$$

by applying Lemma A.1.34(c) to $f_n - f$. The opposite implications do not hold in general. Nevertheless, the following lemma states that any sequence of integrable functions $(f_n)_{n=1}^\infty$ that converges pointwise almost everywhere to an integrable function f will also converge to that function in L^1 if and only if their L^1 norms converge to the L^1 norm of f.

Lemma A.1.39 (Scheffé's lemma). *Let $(X, \mathcal{A}, \mu)$ be a measure space. If a sequence $(f_n)_{n=1}^\infty$ of functions in $L^1(\mu)$ converges pointwise μ-a. e. to a function $f \in L^1(\mu)$, then*

$$\lim_{n\to\infty} \|f_n - f\|_1 = 0 \quad \Longleftrightarrow \quad \lim_{n\to\infty} \|f_n\|_1 = \|f\|_1.$$

In particular, if $f_n \ge 0$ μ-a. e. for all $n \in \mathbb{N}$, then $f \ge 0$ μ-a. e. and

$$\lim_{n\to\infty} \|f_n - f\|_1 = 0 \quad \Longleftrightarrow \quad \lim_{n\to\infty} \int_X f_n \, d\mu = \int_X f \, d\mu.$$

Proof. The direct implication $\Rightarrow$ is trivial. For the converse implication, assume that $\lim_{n\to\infty} \|f_n\|_1 = \|f\|_1$. Suppose first that $f_n \ge 0$ for all $n \in \mathbb{N}$. Then $f \ge 0$, and hence our assumption reduces to $\lim_{n\to\infty} \int_X f_n \, d\mu = \int_X f \, d\mu$. Let $\ell_n = \min\{f, f_n\}$ and $u_n = \max\{f, f_n\}$. Then both $(\ell_n)_{n=1}^\infty$ and $(u_n)_{n=1}^\infty$ converge pointwise μ-a. e. to f. Also, $|\ell_n| = \ell_n \le f$ for all n, so Lebesgue's dominated convergence theorem asserts that

$$\lim_{n\to\infty} \int_X \ell_n \, d\mu = \int_X f \, d\mu.$$

Observing that $u_n = f + f_n - \ell_n$, we also get that

$$\lim_{n\to\infty} \int_X u_n \, d\mu = \int_X f \, d\mu + \lim_{n\to\infty} \int_X f_n \, d\mu - \lim_{n\to\infty} \int_X \ell_n \, d\mu = \int_X f \, d\mu.$$

Thus,

$$\lim_{n\to\infty} \|f_n - f\|_1 = \lim_{n\to\infty} \int_X |f_n - f| \, d\mu = \lim_{n\to\infty} \left(\int_X u_n \, d\mu - \int_X \ell_n \, d\mu \right) = 0.$$

So the implication $\Leftarrow$ holds for nonnegative functions. As $g = g_+ - g_-$ and $((f_n)_+)_{n=1}^{\infty}$ and $((f_n)_-)_{n=1}^{\infty}$ are sequences of functions in $L^1(\mu)$ converging pointwise μ-a. e. to $f_+ \in L^1(\mu)$ and $f_- \in L^1(\mu)$, respectively, it is easy to see that the general case follows from the case for nonnegative functions. □

If g is a L^1 function on a general measure space $(X, \mathcal{A}, \mu)$, then the sequence of nonnegative measurable functions $(g_M)_{M=1}^{\infty}$, where $g_M = |g| \cdot \mathbb{1}_{\{|g| \geq M\}}$, decreases to 0 pointwise and is dominated by $|g|$. Therefore, the monotone convergence theorem (or, alternatively, Lebesgue's dominated convergence theorem) affirms that

$$\lim_{M\to\infty} \int_{\{|g|\geq M\}} |g| \, d\mu = 0.$$

This suggests introducing the following concept.

Definition A.1.40. Let $(X, \mathcal{A}, \mu)$ be a measure space. A sequence of measurable functions $(f_n)_{n=1}^{\infty}$ is *uniformly integrable* if

$$\lim_{M\to\infty} \sup_{n\in\mathbb{N}} \int_{\{|f_n|\geq M\}} |f_n| \, d\mu = 0.$$

On finite measure spaces, there exists a generalization of Lebesgue's dominated convergence theorem (Theorem A.1.38).

Theorem A.1.41. *Let $(X, \mathcal{A}, \mu)$ be a finite measure space and $(f_n)_{n=1}^{\infty}$ a sequence of measurable functions that converges pointwise μ-a. e. to a function f.*

(a) *If $(f_n)_{n=1}^{\infty}$ is uniformly integrable, then $f_n \in L^1(\mu)$ for all $n \in \mathbb{N}$ and $f \in L^1(\mu)$. Moreover,*

$$\lim_{n\to\infty} \|f_n - f\|_1 = 0 \quad \textit{and} \quad \lim_{n\to\infty} \int_X f_n \, d\mu = \int_X f \, d\mu.$$

(b) *If $f, f_n \in L^1(\mu)$ and $f_n \geq 0$ μ-a. e. for all $n \in \mathbb{N}$, then $\lim_{n\to\infty} \int_X f_n \, d\mu = \int_X f \, d\mu$ implies that $(f_n)_{n=1}^{\infty}$ is uniformly integrable.*

Proof. See Theorem 16.14 in Billingsley [8]. □

Corollary A.1.42. *Let $(X, \mathcal{A}, \mu)$ be a finite measure space and $(f_n)_{n=1}^{\infty}$ a sequence of integrable functions that converges pointwise μ-a. e. to an integrable function f. Then the following conditions are equivalent:*

(a) *The sequence $(f_n)_{n=1}^{\infty}$ is uniformly integrable.*

(b) $\lim_{n\to\infty} \|f_n - f\|_1 = 0$.

(c) $\lim_{n\to\infty} \|f_n\|_1 = \|f\|_1$.

Proof. Part (a) of Theorem A.1.41 yields (a)⇒(b). That (b)⇒(c) follows from $\big|\|f_n\|_1 - \|f\|_1\big| \le \|f_n - f\|_1$. Finally, replacing f_n by $|f_n - f|$ and f by 0 in part (b) of Theorem A.1.41 gives (c)⇒(a). □

It is obvious that uniform convergence of a sequence of functions implies pointwise convergence of the sequence. The converse is generally not true but the following partial converse holds.

Theorem A.1.43 (Dini's theorem). *If X is a compact topological space and $(f_n)_{n=1}^\infty$ is an increasing sequence of real-valued continuous functions on X converging pointwise to a continuous function f, then the convergence is uniform. The same conclusion holds if $(f_n)_{n=1}^\infty$ is decreasing instead of increasing.*

It is also natural to ask whether, in some way, a pointwise convergent sequence converges "almost" uniformly.

Definition A.1.44. Let $(X, \mathcal{A}, \mu)$ be a measure space. A sequence $(f_n)_{n=1}^\infty$ of measurable functions on X is said to *converge μ-almost uniformly* to a function f if for every $\varepsilon > 0$ there exists $Y \in \mathcal{A}$ such that $\mu(Y) < \varepsilon$ and $(f_n)_{n=1}^\infty$ converges uniformly to f on $X \setminus Y$.

It is clear that almost uniform convergence implies almost everywhere pointwise convergence. The converse is not true in general but these two types of convergence are one and the same on any finite measure space.

Theorem A.1.45 (Egorov's theorem). *Let $(X, \mathcal{A}, \mu)$ be a finite measure space. A sequence $(f_n)_{n=1}^\infty$ of measurable functions on X converges pointwise μ-almost everywhere to a limit function f if and only if that sequence converges μ-almost uniformly to f.*

Proof. See Chapter 3, Exercise 16 in Rudin [109]. □

The reader ought to convince themself that this result does not generally hold on infinite spaces.

Convergence in measure is another interesting type of convergence.

Definition A.1.46. Let $(X, \mathcal{A}, \mu)$ be a measure space. A sequence $(f_n)_{n=1}^\infty$ of measurable functions *converges in measure* to a measurable function f provided that for each $\varepsilon > 0$,

$$\lim_{n\to\infty} \mu(\{x \in X : |f_n(x) - f(x)| > \varepsilon\}) = 0.$$

Lemma A.1.47. *Let $(X, \mathcal{A}, \mu)$ be a measure space. If a sequence $(f_n)_{n=1}^\infty$ of measurable functions converges in $L^1(\mu)$ to a measurable function f, then $(f_n)_{n=1}^\infty$ converges in measure to f.*

Proof. Let $\varepsilon > 0$. Then

$$\mu(\{x \in X : |f_n(x) - f(x)| > \varepsilon\}) \leq \int\limits_{\{x\in X : |f_n(x)-f(x)|>\varepsilon\}} \frac{|f_n - f|}{\varepsilon}\, d\mu \leq \frac{1}{\varepsilon}\|f_n - f\|_1.$$

Taking the limit of both sides as $n \to \infty$ completes the proof. □

When the measure is finite, there is a close relationship between pointwise convergence and convergence in measure.

Theorem A.1.48. *Let $(X, \mathcal{A}, \mu)$ be a finite measure space and $(f_n)_{n=1}^{\infty}$ a sequence of measurable functions.*

(a) *If $(f_n)_{n=1}^{\infty}$ converges pointwise μ-a. e. to a function f, then $(f_n)_{n=1}^{\infty}$ converges in measure to f.*
(b) *If $(f_n)_{n=1}^{\infty}$ converges in measure to a function f, then there exists a subsequence $(f_{n_k})_{k=1}^{\infty}$ which converges pointwise μ-a. e. to f.*
(c) *$(f_n)_{n=1}^{\infty}$ converges in measure to a function f if and only if each subsequence $(f_{n_k})_{k=1}^{\infty}$ admits a further subsequence $(f_{n_{k_l}})_{l=1}^{\infty}$ that converges pointwise μ-a. e. to f.*

Proof. See Theorem 20.5 in Billingsley [8]. □

The previous two results reveal that, on a finite measure space, a sequence of integrable functions that converges in L^1 to an integrable function admits a subsequence which converges pointwise almost everywhere to that function. In general, the sequence itself might not converge pointwise almost everywhere (see Exercise 8.5.17).

In some sense, the following result is a form of convergence theorem. It asserts that Borel measurable functions can be approximated by continuous functions on "arbitrarily large" portions of their domain.

Theorem A.1.49 (Luzin's theorem). *Let $(X, \mathcal{B}(X), \mu)$ be a finite Borel measure space and let $f : X \to \overline{\mathbb{R}}$ be a Borel measurable function. Given any $\varepsilon > 0$, for every $B \in \mathcal{B}(X)$ there is a closed set E with $\mu(B \setminus E) < \varepsilon$ such that $f|_E$ is continuous. If B is locally compact, then the set E can be chosen to be compact and then there is a continuous function $f_\varepsilon : X \to \overline{\mathbb{R}}$ with compact support that coincides with f on E and such that $\sup_{x\in X} |f_\varepsilon(x)| \leq \sup_{x\in X} |f(x)|$.*

A.1.7 Mutual singularity, absolute continuity and equivalence of measures

We now leave aside convergence of sequences of functions and recall the definitions of mutually singular, absolutely continuous and equivalent measures.

Definition A.1.50. Let $(X, \mathcal{A})$ be a measurable space, and μ and ν be two measures on $(X, \mathcal{A})$.

(a) The measures μ and ν are said to be *mutually singular*, denoted by $\mu \perp \nu$, if there exist disjoint sets $X_\mu, X_\nu \in \mathcal{A}$ such that $\mu(X \setminus X_\mu) = 0 = \nu(X \setminus X_\nu)$.
(b) The measure μ is said to be *absolutely continuous with respect to* ν, denoted $\mu \ll \nu$, if $\nu(A) = 0 \Longrightarrow \mu(A) = 0$.
(c) The measures μ and ν are said to be *equivalent* if $\mu \ll \nu$ and $\nu \ll \mu$.

The Radon–Nikodym theorem provides a characterization of absolute continuity. Though it is valid for σ-finite measures, the following version for finite measures is sufficient for our purposes.

Theorem A.1.51 (Radon–Nikodym theorem). *Let $(X, \mathcal{A})$ be a measurable space and let μ and ν be two finite measures on $(X, \mathcal{A})$. The following statements are equivalent:*
(a) $\mu \ll \nu$.
(b) *For every $\varepsilon > 0$, there exists $\delta > 0$ such that $\nu(A) < \delta \Longrightarrow \mu(A) < \varepsilon$.*
(c) *There exists a ν-a. e. unique function $f \in L^1(\nu)$ such that $f \geq 0$ and*

$$\mu(A) = \int_A f \, d\nu, \quad \forall A \in \mathcal{A}.$$

Proof. See relation (32.4) and Theorem 32.2 in Billingsley [8]. □

Remark A.1.52. The function f is often denoted by $\frac{d\mu}{d\nu}$ and called the *Radon–Nikodym derivative* of μ with respect to ν.

Per Lemma A.1.34(f), two integrable functions are equal almost everywhere if and only if their integrals are equal over every measurable set. When the measure is finite, we can restrict our attention to any generating π-system.

Corollary A.1.53. *Let $(X, \mathcal{A}, \nu)$ be a finite measure space and suppose that $\mathcal{A} = \sigma(\mathcal{P})$ for some π-system $\mathcal{P}$. Let $f, g \in L^1(\nu)$. Then*

$$f = g \ \nu\text{-a. e.} \quad \Longleftrightarrow \quad \int_P f \, d\nu = \int_P g \, d\nu, \quad \forall P \in \mathcal{P}.$$

Proof. The direct implication $\Rightarrow$ is obvious. So let us assume that $\int_P f \, d\nu = \int_P g \, d\nu$ for all $P \in \mathcal{P}$. The measures $\mu_f(A) := \int_A f \, d\nu$ and $\mu_g(A) := \int_A g \, d\nu$ are equal on the π-system $\mathcal{P}$. According to Lemma A.1.26, this implies that $\mu_f = \mu_g$. It follows from the uniqueness part of the Radon–Nikodym theorem that $f = g$ ν-almost everywhere. □

A.1.8 The space $C(X)$, its dual $C(X)^*$ and the subspace $M(X)$

Another important result is Riesz representation theorem. Before stating it, we first establish some notation. Let X be a compact metrizable space. Let $C(X)$ be the set of all continuous real-valued functions on X. This set becomes a normed vector space when endowed with the supremum norm

$$\|f\|_{\infty} := \sup\{|f(x)| : x \in X\}. \tag{A.3}$$

This norm defines a metric on $C(X)$ in the usual way:

$$d_{\infty}(f, g) := \|f - g\|_{\infty}.$$

The topology induced by the metric d_{∞} on $C(X)$ is called the *topology of uniform convergence* on X. Indeed, $\lim_{n\to\infty} d_{\infty}(f_n, f) = 0$ if and only if the sequence $(f_n)_{n=1}^{\infty}$ converges to f uniformly on X. It is not hard to see that $C(X)$ is a separable Banach space (i. e., a separable and complete normed vector space).

Let $C(X)^*$ denote the *dual space* of $C(X)$, i. e.,

$$C(X)^* := \{F : C(X) \to \mathbb{R} \mid F \text{ is continuous and linear}\}.$$

Recall that a real-valued function F defined on $C(X)$ is called a *functional* on $C(X)$. It is well known that a linear functional F is continuous if and only if it is *bounded*, i. e., if and only if its operator norm $\|F\|$ (sometimes denoted $\|F\|_{\infty}$) is finite, where

$$\|F\| := \sup\{|F(f)| : f \in C(X) \text{ and } \|f\|_{\infty} \le 1\}. \tag{A.4}$$

So $C(X)^*$ can also be described as the normed vector space of all bounded linear functionals on $C(X)$. The operator norm defines a metric on $C(X)^*$ in the usual manner:

$$d(F, G) := \|F - G\|.$$

The topology induced by the metric d on $C(X)^*$ is called the *operator norm topology*, or strong topology, on $C(X)^*$. It is not difficult to see that $C(X)^*$ is a separable Banach space.

Furthermore, a linear functional F is said to be *normalized* if $F(1) = 1$ and is called *positive* if $F(f) \ge 0$ whenever $f \ge 0$.

Finally, we denote the set of all Borel probability measures on X by $M(X)$. This set is clearly convex and can be characterized as follows.

Theorem A.1.54 (Riesz representation theorem). *Let X be a compact metrizable space, and let F be a normalized and positive linear functional on $C(X)$. Then there exists a unique $\mu \in M(X)$ such that*

$$F(f) = \int_X f\, d\mu, \quad \forall f \in C(X). \tag{A.5}$$

Conversely, any $\mu \in M(X)$ defines a normalized positive linear functional on $C(X)$ via formula (A.5). *This linear functional is bounded.*

Proof. The converse statement is straightforward to check. For the other direction, see Theorem 2.14 in Rudin [109]. □

It immediately follows from Riesz' representation theorem that every Borel probability measure on a compact metrizable space is uniquely determined by the way it integrates continuous functions on that space.

Corollary A.1.55. *If μ and ν are two Borel probability measures on a compact metrizable space X, then*

$$\mu = \nu \quad \Longleftrightarrow \quad \int_X f\,d\mu = \int_X f\,d\nu, \quad \forall f \in C(X).$$

Let us now discuss the *weak* topology* on the set $M(X)$. Recall that if Z is a set and $(Z_\alpha)_{\alpha\in A}$ is a family of topological spaces, then the weak topology induced on Z by a collection of maps $\{\psi_\alpha : Z \to Z_\alpha \mid \alpha \in A\}$ is the smallest topology on Z that makes each ψ_α continuous. Evidently, the sets $\psi_\alpha^{-1}(U_\alpha)$, for U_α open in Z_α, constitute a subbase for the weak topology. The weak* topology on $M(X)$ is the weak topology induced by $C(X)$ on its dual space $C(X)^*$, where measures in $M(X)$ and normalized positive linear functionals in $C(X)^*$ are identified via the Riesz representation theorem. Note that $M(X)$ is metrizable, although $C(X)^*$ with the weak* topology usually is not. Indeed, both $C(X)$ and its subspace $C(X,[0,1])$ of continuous functions on X taking values in $[0,1]$, are separable since X is a compact metrizable space. Then for any dense subset $\{f_n\}_{n=1}^\infty$ of $C(X,[0,1])$, a metric on $M(X)$ is

$$d(\mu,\nu) = \sum_{n=1}^\infty \frac{1}{2^n}\left|\int_X f_n\,d\mu - \int_X f_n\,d\nu\right|.$$

In this book, we will sometimes denote the convergence of a sequence of measures $(\mu_n)_{n=1}^\infty$ to a measure μ in the weak* topology of $M(X)$ by $\mu_n \xrightarrow{*} \mu$.

Remark A.1.56. Note that this notion is often presented as “weak convergence” of measures. This can be slightly confusing at first sight, but it helps to bear in mind that, as we have seen above, the weak* topology is just one instance of a weak topology.

The following theorem gives several equivalent characterizations of weak* convergence of Borel probability measures.

Theorem A.1.57 (Portmanteau theorem). *Let $(\mu_n)_{n=1}^\infty$ and μ be Borel probability measures on a compact metrizable space X. The following statements are equivalent:*

(a) $\mu_n \xrightarrow{*} \mu$.

(b) *For all continuous functions $f : X \to \mathbb{R}$,*

$$\lim_{n\to\infty} \int_X f\,d\mu_n = \int_X f\,d\mu.$$

(c) *For all closed sets $F \subseteq X$,*

$$\limsup_{n\to\infty} \mu_n(F) \le \mu(F).$$

(d) *For all open sets $G \subseteq X$,*

$$\liminf_{n\to\infty} \mu_n(G) \le \mu(G).$$

(e) *For all sets $A \in \mathcal{B}(X)$ such that $\mu(\partial A) = 0$,*

$$\lim_{n\to\infty} \mu_n(A) = \mu(A).$$

Proof. See Theorem 2.1 in Billingsley [9]. □

For us, the most important result concerning weak* convergence of measures is that the set $M(X)$ of all Borel probability measures on a compact metrizable space X is a compact and convex set in the weak* topology.

In order to establish this, we need to remember Banach–Alaoglu's theorem. In this theorem, note that the boundedness and closedness are with respect to the operator norm on the dual space while the compactness is with respect to the weak* topology on the dual space.

Theorem A.1.58 (Banach–Alaoglu's theorem). *The closed unit ball in the dual space B^* of a Banach space B is compact in the* weak* *topology on B^*. Furthermore, every closed, bounded subset of B^* is compact in the* weak* *topology on B^*.*

Proof. See Theorem V.4.2 and Corollary V.4.3 in Dunford and Schwartz [26]. □

Theorem A.1.59. *Let X be a compact metrizable space. The set $M(X)$ is compact and convex in the weak* topology of $C(X)^*$.*

Proof. The set $M(X)$ is closed with respect to the operator norm topology on $C(X)^*$. Indeed, suppose that $(\mu_n)_{n=1}^{\infty}$ is a sequence in $M(X)$ which converges to a $F \in C(X)^*$ in the operator norm topology of $C(X)^*$. In other words, suppose that $\lim_{n\to\infty} \|\mu_n - F\| = 0$. By definition of the operator norm (see (A.4)) and thanks to the linearity of F, this implies that

$$F(f) = \lim_{n\to\infty} \int_X f\, d\mu_n, \quad \forall f \in C(X).$$

In particular, F is normalized (since $F(1) = 1$) and positive (as $F(f) \ge 0$ for all $f \ge 0$). By Riesz representation theorem (Theorem A.1.54), there is $\mu \in M(X)$ that represents F. So $F \in M(X)$, and thus $M(X)$ is closed in the operator norm topology of $C(X)^*$.

The set $M(X)$ is also bounded in that topology. Indeed, if $\mu \in M(X)$ then

$$\|\mu\| \leq \sup\left\{\int_X |f|\,d\mu : f \in C(X),\ \|f\|_\infty \leq 1\right\} = 1.$$

Since X is a compact metrizable space, the space $C(X)$ is a Banach space, as earlier mentioned. We can then infer from Banach–Alaoglu's theorem that the set $M(X)$ is compact in the weak* topology.

The convexity of $M(X)$ is obvious. Indeed, if $\mu, \nu \in M(X)$ so is any convex combination $m = \alpha\mu + (1-\alpha)\nu$, where $\alpha \in [0,1]$. □

A.1.9 Expected values and conditional expectation functions

The mean or expected value of a function over a set is a straightforward generalization of the mean value of a real-valued function defined on an interval of the real line.

Definition A.1.60. Let $(X, \mathcal{A}, \mu)$ be a probability space and let $\varphi \in L^1(\mu)$. The *mean* or *expected value* $E(\varphi|A)$ of the function φ over the set $A \in \mathcal{A}$ is defined to be

$$E(\varphi|A) := \begin{cases} \frac{1}{\mu(A)}\int_A \varphi\,d\mu & \text{if } \mu(A) > 0 \\ 0 & \text{if } \mu(A) = 0. \end{cases}$$

Given that μ is a probability measure, the expected value of φ over the entire space is simply given by

$$E(\varphi) := E(\varphi|X) = \int_X \varphi\,d\mu =: \mu(\varphi).$$

Our next goal is to give the definition of the conditional expectation of a function with respect to a σ-algebra. Let $(X, \mathcal{A}, \mu)$ be a probability space and $\mathcal{B}$ be a sub-σ-algebra of $\mathcal{A}$. Let also $\varphi \in L^1(X, \mathcal{A}, \mu)$. Notice that $\varphi : X \to \mathbb{R}$ is not necessarily measurable if X is endowed with the sub-σ-algebra $\mathcal{B}$ instead of the σ-algebra $\mathcal{A}$. In short, we say that φ is $\mathcal{A}$-measurable but not necessarily $\mathcal{B}$-measurable. We aim to find a function $E(\varphi|\mathcal{B}) \in L^1(X, \mathcal{B}, \mu)$ such that

$$\int_B E(\varphi|\mathcal{B})\,d\mu = \int_B \varphi\,d\mu, \quad \forall B \in \mathcal{B}. \tag{A.6}$$

This condition means that the function $E(\varphi|\mathcal{B})$ has the same expected value as φ on every measurable set belonging to the sub-σ-algebra $\mathcal{B}$. Accordingly, $E(\varphi|\mathcal{B})$ is called the *conditional expectation of φ with respect to $\mathcal{B}$.*

We now demonstrate the existence and μ-a. e. uniqueness of the conditional expectation. Let us begin with the existence of that function. Suppose first that the function φ is nonnegative. If $\varphi = 0$ μ-a. e., then simply set $E(\varphi|\mathcal{B}) = 0$. If $\varphi \neq 0$ μ-a. e., then the set function $\nu(A) := \int_A \varphi\, d\mu$ defines a finite measure on $(X, \mathcal{A})$ which is absolutely continuous with respect to μ. The restriction of ν to $\mathcal{B}$ also determines a finite measure on $(X, \mathcal{B})$ which is absolutely continuous with respect to the restriction of μ to $\mathcal{B}$. So by the Radon–Nikodym theorem (Theorem A.1.51), there exists a μ-a. e. unique nonnegative function $\widehat{\varphi} \in L^1(X, \mathcal{B}, \mu)$ such that $\nu(B) = \int_B \widehat{\varphi}\, d\mu$ for every $B \in \mathcal{B}$. Then

$$\int_B \widehat{\varphi}\, d\mu = \nu(B) = \int_B \varphi\, d\mu, \quad \forall B \in \mathcal{B}.$$

The point here is that although it may look as if we have not really achieved anything, we have actually gained that $\widehat{\varphi}$ is $\mathcal{B}$-measurable, whereas φ may not be. Therefore, $\widehat{\varphi}$ is the sought-after conditional expectation $E(\varphi|\mathcal{B})$ of φ with respect to $\mathcal{B}$.

If φ takes both negative and positive values, write $\varphi = \varphi_+ - \varphi_-$, where $\varphi_+(x) := \max\{\varphi(x), 0\}$ is the positive part of φ and $\varphi_-(x) := \max\{-\varphi(x), 0\}$ is the negative part of φ. Then define the conditional expectation linearly, i. e., set

$$E(\varphi|\mathcal{B}) := E(\varphi_+|\mathcal{B}) - E(\varphi_-|\mathcal{B}).$$

This proves the existence of the conditional expectation function. Its μ-a. e. uniqueness follows from its defining property (A.6) and Lemma A.1.34(f).

The conditional expectation exhibits several natural properties. We mention a few of them in the next proposition.

Proposition A.1.61. *Let $(X, \mathcal{A}, \mu)$ be a probability space, let $\mathcal{B}$ and $\mathcal{C}$ denote sub-σ-algebras of $\mathcal{A}$ and let $\varphi \in L^1(X, \mathcal{A}, \mu)$.*
(a) *If $\varphi \geq 0$ μ-a. e., then $E(\varphi|\mathcal{B}) \geq 0$ μ-a. e.*
(b) *If $\varphi_1 \geq \varphi_2$ μ-a. e., then $E(\varphi_1|\mathcal{B}) \geq E(\varphi_2|\mathcal{B})$ μ-a. e.*
(c) $|E(\varphi|\mathcal{B})| \leq E(|\varphi| \,|\, \mathcal{B})$.
(d) *The functional $E(\cdot|\mathcal{B})$ is linear, i. e. for any $c_1, c_2 \in \mathbb{R}$ and $\varphi_1, \varphi_2 \in L^1(X, \mathcal{A}, \mu)$,*

$$E(c_1\varphi_1 + c_2\varphi_2|\mathcal{B}) = c_1 E(\varphi_1|\mathcal{B}) + c_2 E(\varphi_2|\mathcal{B}).$$

(e) *If φ is $\mathcal{B}$-measurable, then $E(\varphi|\mathcal{B}) = \varphi$. In particular, $E(\,E(\varphi|\mathcal{B}) \,|\, \mathcal{B}) = E(\varphi|\mathcal{B})$. Also, if $\varphi \equiv c \in \mathbb{R}$ is a constant function, then $E(\varphi|\mathcal{B}) = \varphi \equiv c$.*
(f) *If $\mathcal{C} \subseteq \mathcal{B}$, then $E(\varphi|\mathcal{C}) = E(\,E(\varphi|\mathcal{B}) \,|\, \mathcal{C})$.*

Proof. This is left as an exercise to the reader. □

We will now determine the conditional expectation of an arbitrary integrable function φ with respect to various sub-σ-algebras of particular interest.

Example A.1.62. Let $(X, \mathcal{A}, \mu)$ be a probability space. The family $\mathcal{N}$ of all measurable sets that are either of null or of full measure constitutes a sub-σ-algebra of $\mathcal{A}$. Let $\varphi \in L^1(X, \mathcal{A}, \mu)$. Then the function $E(\varphi|\mathcal{N})$ has to belong to $L^1(X, \mathcal{N}, \mu)$ and must satisfy condition (A.6). In particular, $E(\varphi|\mathcal{N})$ must be $\mathcal{N}$-measurable. This means that for each Borel subset R of $\mathbb{R}$, the function $E(\varphi|\mathcal{N})$ must be such that $E(\varphi|\mathcal{N})^{-1}(R) \in \mathcal{N}$. Among others, for every $t \in \mathbb{R}$ we must have $E(\varphi|\mathcal{N})^{-1}(\{t\}) \in \mathcal{N}$; in other words, for each $t \in \mathbb{R}$ the set $E(\varphi|\mathcal{N})^{-1}(\{t\})$ must be of measure zero or of measure one. Also bear in mind that

$$X = E(\varphi|\mathcal{N})^{-1}(\mathbb{R}) = \bigcup_{t\in\mathbb{R}} E(\varphi|\mathcal{N})^{-1}(\{t\}).$$

Since the above union consists of mutually disjoint sets of measure zero and one, it follows that only one of these sets can be of measure one. In other words, there exists a unique $t \in \mathbb{R}$ such that $E(\varphi|\mathcal{N})^{-1}(\{t\}) = A$ for some $A \in \mathcal{A}$ with $\mu(A) = 1$. Because the function $E(\varphi|\mathcal{N})$ is unique up to a set of measure zero, we may assume without loss of generality that $A = X$. Hence, $E(\varphi|\mathcal{N})$ is a constant function. More specifically, its value is

$$E(\varphi|\mathcal{N}) = \int_X E(\varphi|\mathcal{N})\, d\mu = \int_X \varphi\, d\mu.$$

Example A.1.63. Let $(X, \mathcal{A})$ be a measurable space and let $\alpha = \{A_n\}_{n=1}^{\infty}$ be a countable measurable partition of X. That is, each $A_n \in \mathcal{A}$, $A_i \cap A_j = \emptyset$ for all $i \neq j$ and $X = \bigcup_{n=1}^{\infty} A_n$. The sub-$\sigma$-algebra of $\mathcal{A}$ generated by α is the family of all sets which can be written as a union of elements of α, i. e.,

$$\sigma(\alpha) = \left\{A \subseteq X : A = \bigcup_{j\in J} A_j \text{ for some } J \subseteq \mathbb{N}\right\}.$$

When α is finite, so is $\sigma(\alpha)$. When α is countably infinite, $\sigma(\alpha)$ is uncountable. Let μ be a probability measure on $(X, \mathcal{A})$. Let $\varphi \in L^1(X, \mathcal{A}, \mu)$ and set $\mathcal{B} = \sigma(\alpha)$. Then the conditional expectation $E(\varphi|\mathcal{B}) : X \to \mathbb{R}$ has to be a $L^1(X, \mathcal{B}, \mu)$ function that satisfies condition (A.6). In particular, $E(\varphi|\mathcal{B})$ must be $\mathcal{B}$-measurable. Thus, for any $t \in \mathbb{R}$ we must have $E(\varphi|\mathcal{B})^{-1}(\{t\}) \in \mathcal{B}$, i. e., the set $E(\varphi|\mathcal{B})^{-1}(\{t\})$ must be a union of elements of α. This means that the conditional expectation function $E(\varphi|\mathcal{B})$ is constant on each element of α. Let $A_n \in \alpha$. If $\mu(A_n) = 0$, then $E(\varphi|\mathcal{B})|_{A_n} = 0$. Otherwise,

$$E(\varphi|\mathcal{B})|_{A_n} = \frac{1}{\mu(A_n)} \int_{A_n} E(\varphi|\mathcal{B})\, d\mu = \frac{1}{\mu(A_n)} \int_{A_n} \varphi\, d\mu = E(\varphi|A_n).$$

In summary, the conditional expectation $E(\varphi|\mathcal{B})$ of a function φ with respect to a sub-σ-algebra generated by a countable measurable partition is constant on each element of that partition. More precisely, on any given element of the partition, $E(\varphi|\mathcal{B})$ is equal to the mean value of φ on that element.

The next result is a special case of a theorem originally due to Doob and called the martingale convergence theorem. But, first, let us define the martingale itself.

Definition A.1.64. Let $(X, \mathcal{A}, \mu)$ be a probability space. Let $(\mathcal{A}_n)_{n=1}^{\infty}$ be a sequence of sub-σ-algebras of $\mathcal{A}$. Let also $(\varphi_n : X \to \mathbb{R})_{n=1}^{\infty}$ be a sequence of $\mathcal{A}$-measurable functions. The sequence $((\varphi_n, \mathcal{A}_n))_{n=1}^{\infty}$ is called a *martingale* if the following conditions are satisfied:
(a) $(\mathcal{A}_n)_{n=1}^{\infty}$ is an ascending sequence, i. e., $\mathcal{A}_n \subseteq \mathcal{A}_{n+1}$ for all $n \in \mathbb{N}$.
(b) φ_n is $\mathcal{A}_n$-measurable for all $n \in \mathbb{N}$.
(c) $\varphi_n \in L^1(\mu)$ for all $n \in \mathbb{N}$.
(d) $E(\varphi_{n+1}|\mathcal{A}_n) = \varphi_n$ μ-a. e. for all $n \in \mathbb{N}$.

Theorem A.1.65 (Martingale convergence theorem). *Let $(X, \mathcal{A}, \mu)$ be a probability space. If $((\varphi_n, \mathcal{A}_n))_{n=1}^{\infty}$ is a martingale such that*

$$\sup_{n\in\mathbb{N}} \|\varphi_n\|_1 < \infty,$$

then there exists $\widehat{\varphi} \in L^1(X, \mathcal{A}, \mu)$ such that

$$\lim_{n\to\infty} \varphi_n(x) = \widehat{\varphi}(x) \quad \textit{for } \mu\textit{-a. e. } x \in X \quad \textit{and} \quad \|\widehat{\varphi}\|_1 \le \sup_{n\in\mathbb{N}} \|\varphi_n\|_1.$$

Proof. See Theorem 35.5 in Billingsley [8]. □

One natural martingale is formed by the conditional expectations of a function with respect to an ascending sequence of sub-σ-algebras.

Example A.1.66. Let $(X, \mathcal{A}, \mu)$ be a probability space and $(\mathcal{A}_n)_{n=1}^{\infty}$ be an ascending sequence of sub-σ-algebras of $\mathcal{A}$. For any $\varphi \in L^1(X, \mathcal{A}, \mu)$, the sequence $((E(\varphi|\mathcal{A}_n), \mathcal{A}_n))_{n=1}^{\infty}$ is a martingale. Indeed, set $\varphi_n = E(\varphi|\mathcal{A}_n)$ for all $n \in \mathbb{N}$. Condition (a) in Definition A.1.64 is automatically fulfilled. Conditions (b) and (c) follow from the very definition of the conditional expectation function. Regarding condition (d), a straightforward application of Proposition A.1.61(f) gives

$$E(\varphi_{n+1}|\mathcal{A}_n) = E\big(E(\varphi|\mathcal{A}_{n+1}) \mid \mathcal{A}_n\big) = E(\varphi|\mathcal{A}_n) = \varphi_n \ \mu\text{-a. e.}, \quad \forall n \in \mathbb{N}.$$

So $((E(\varphi|\mathcal{A}_n), \mathcal{A}_n))_{n=1}^{\infty}$ is a martingale. Using Proposition A.1.61(c), note that

$$\sup_{n\in\mathbb{N}} \|\varphi_n\|_1 = \sup_{n\in\mathbb{N}} \int_X |E(\varphi|\mathcal{A}_n)|\, d\mu \le \sup_{n\in\mathbb{N}} \int_X E(|\varphi| \mid \mathcal{A}_n)\, d\mu = \int_X |\varphi|\, d\mu = \|\varphi\|_1 < \infty.$$

According to Theorem A.1.65, there thus exists $\widehat{\varphi} \in L^1(X, \mathcal{A}, \mu)$ such that

$$\lim_{n\to\infty} E(\varphi|\mathcal{A}_n)(x) = \widehat{\varphi}(x) \quad \text{for } \mu\text{-a. e. } x \in X \quad \text{and} \quad \|\widehat{\varphi}\|_1 \le \|\varphi\|_1.$$

What is $\widehat{\varphi}$? This is the question we will answer in Theorem A.1.68.

Beforehand, we establish the uniform integrability of this martingale (see Definition A.1.40).

Lemma A.1.67. *Let $(X, \mathcal{A}, \mu)$ be a probability space and let $(\mathcal{A}_n)_{n=1}^\infty$ be a sequence of sub-σ-algebras of $\mathcal{A}$. For any $\varphi \in L^1(X, \mathcal{A}, \mu)$, the sequence $(E(\varphi|\mathcal{A}_n))_{n=1}^\infty$ is uniformly integrable.*

Proof. Without loss of generality, we may assume that $\varphi \ge 0$. Let $\varepsilon > 0$. Since $\nu(A) = \int_A \varphi\, d\mu$ is absolutely continuous with respect to μ, it follows from the Radon–Nikodym theorem (Theorem A.1.51) that there exists $\delta > 0$ such that

$$A \in \mathcal{A},\ \mu(A) < \delta \quad \Longrightarrow \quad \int_A \varphi\, d\mu < \varepsilon. \tag{A.7}$$

Set $M > \int_X \varphi\, d\mu/\delta$. For each $n \in \mathbb{N}$, let

$$X_n(M) = \{x \in X : E(\varphi|\mathcal{A}_n)(x) \ge M\}.$$

Observe that $X_n(M) \in \mathcal{A}_n$ since $E(\varphi|\mathcal{A}_n)$ is $\mathcal{A}_n$-measurable. Therefore,

$$\mu(X_n(M)) \le \frac{1}{M} \int_{X_n(M)} E(\varphi|\mathcal{A}_n)\, d\mu = \frac{1}{M} \int_{X_n(M)} \varphi\, d\mu \le \frac{1}{M} \int_X \varphi\, d\mu < \delta$$

for all $n \in \mathbb{N}$. Consequently, by (A.7),

$$\int_{X_n(M)} E(\varphi|\mathcal{A}_n)\, d\mu = \int_{X_n(M)} \varphi\, d\mu < \varepsilon$$

for all $n \in \mathbb{N}$. Thus,

$$\sup_{n\in\mathbb{N}} \int_{\{E(\varphi|\mathcal{A}_n)\ge M\}} E(\varphi|\mathcal{A}_n)\, d\mu \le \varepsilon.$$

Since this holds for large enough M's and since $\varepsilon > 0$ is arbitrary, we have

$$\lim_{M\to\infty} \sup_{n\in\mathbb{N}} \int_{\{E(\varphi|\mathcal{A}_n)\ge M\}} E(\varphi|\mathcal{A}_n)\, d\mu = 0,$$

i. e., the sequence $(E(\varphi|\mathcal{A}_n))_{n=1}^\infty$ is uniformly integrable. □

Theorem A.1.68 (Martingale convergence theorem for conditional expectations). *Let $(X, \mathcal{A}, \mu)$ be a probability space and $\varphi \in L^1(X, \mathcal{A}, \mu)$. Let $(\mathcal{A}_n)_{n=1}^{\infty}$ be an ascending sequence of sub-σ-algebras of $\mathcal{A}$ and*

$$\mathcal{A}_\infty := \sigma\Big(\bigcup_{n=1}^{\infty} \mathcal{A}_n\Big).$$

Then

$$\lim_{n\to\infty} \|E(\varphi|\mathcal{A}_n) - E(\varphi|\mathcal{A}_\infty)\|_1 = 0 \quad \textit{and} \quad \lim_{n\to\infty} E(\varphi|\mathcal{A}_n) = E(\varphi|\mathcal{A}_\infty) \;\; \mu\textit{-a. e. on } X.$$

Proof. Let $\varphi_n = E(\varphi|\mathcal{A}_n)$. In Example A.1.66 and Lemma A.1.67, we have seen that $((\varphi_n, \mathcal{A}_n))_{n=1}^{\infty}$ is a uniformly integrable martingale such that

$$\lim_{n\to\infty} \varphi_n = \widehat{\varphi} \quad \mu\text{-a. e. on } X$$

for some $\widehat{\varphi} \in L^1(X, \mathcal{A}, \mu)$. For all $n \in \mathbb{N}$ the function φ_n is $\mathcal{A}_\infty$-measurable since it is $\mathcal{A}_n$-measurable and $\mathcal{A}_n \subseteq \mathcal{A}_\infty$. Thus, $\widehat{\varphi}$ is $\mathcal{A}_\infty$-measurable, too. Moreover, it follows from Theorem A.1.41 that

$$\lim_{n\to\infty} \|\varphi_n - \widehat{\varphi}\|_1 = 0 \quad \text{and} \quad \lim_{n\to\infty} \int_A \varphi_n \, d\mu = \int_A \widehat{\varphi} \, d\mu, \qquad \forall A \in \mathcal{A}.$$

Therefore, it just remains to show that $\widehat{\varphi} = E(\varphi|\mathcal{A}_\infty)$.

Let $k \in \mathbb{N}$ and $A \in \mathcal{A}_k$. If $n \geq k$, then $A \in \mathcal{A}_n \subseteq \mathcal{A}_\infty$, and thus

$$\int_A \varphi_n \, d\mu = \int_A E(\varphi|\mathcal{A}_n) \, d\mu = \int_A \varphi \, d\mu = \int_A E(\varphi|\mathcal{A}_\infty) \, d\mu.$$

Letting $n \to \infty$ yields

$$\int_A \widehat{\varphi} \, d\mu = \int_A E(\varphi|\mathcal{A}_\infty) \, d\mu, \quad \forall A \in \mathcal{A}_k.$$

As k is arbitrary,

$$\int_B \widehat{\varphi} \, d\mu = \int_B E(\varphi|\mathcal{A}_\infty) \, d\mu, \quad \forall B \in \bigcup_{k=1}^{\infty} \mathcal{A}_k.$$

Since $\bigcup_{k=1}^{\infty} \mathcal{A}_k$ is a π-system generating $\mathcal{A}_\infty$ and since both $\widehat{\varphi}$ and $E(\varphi|\mathcal{A}_\infty)$ are $\mathcal{A}_\infty$-measurable, Corollary A.1.53 affirms that $\widehat{\varphi} = E(\varphi|\mathcal{A}_\infty)$ μ-a. e. □

There is also a counterpart of this theorem for descending sequences of σ-algebras.

Theorem A.1.69 (Reversed martingale convergence theorem for conditional expectations). *Let $(X, \mathcal{A}, \mu)$ be a probability space and $\varphi \in L^1(X, \mathcal{A}, \mu)$. If $(\mathcal{A}_n)_{n=1}^{\infty}$ is a descending sequence of sub-σ-algebras of $\mathcal{A}$, then*

$$\lim_{n\to\infty} \left\| E(\varphi|\mathcal{A}_n) - E\left(\varphi \,\middle|\, \bigcap_{n=1}^{\infty} \mathcal{A}_n\right) \right\|_1 = 0 \quad \textit{and} \quad \lim_{n\to\infty} E(\varphi|\mathcal{A}_n) = E\left(\varphi \,\middle|\, \bigcap_{n=1}^{\infty} \mathcal{A}_n\right) \ \mu\text{-a. e.}$$

Proof. See Theorem 35.9 of Billingsley [8]. □

Theorems A.1.68/A.1.69 are especially useful for the calculation of the conditional expectation of a function with respect to a sub-σ-algebra generated by an uncountable measurable partition which can be approached by an ascending/descending sequence of sub-σ-algebras generated by countable measurable partitions. See Exercise 8.5.22.

They can also be used to approximate a measurable set by one from a generating sequence of sub-σ-algebras.

Corollary A.1.70. *Let $(X, \mathcal{A}, \mu)$ be a probability space. Let $(\mathcal{A}_n)_{n=1}^{\infty}$ be an ascending sequence of sub-σ-algebras of $\mathcal{A}$ and set $\mathcal{A}_\infty = \sigma(\bigcup_{n=1}^{\infty} \mathcal{A}_n)$. Let $B \in \mathcal{A}_\infty$. For every $\varepsilon > 0$, there exists $A \in \bigcup_{n=1}^{\infty} \mathcal{A}_n$ such that $\mu(A \triangle B) < \varepsilon$.*

Proof. Let $B \in \mathcal{A}_\infty$. It ensues from Theorem A.1.68 that

$$\lim_{n\to\infty} E(\mathbb{1}_B|\mathcal{A}_n) = E(\mathbb{1}_B|\mathcal{A}_\infty) = \mathbb{1}_B \quad \mu\text{-a. e. on } X.$$

By Theorem A.1.48, we deduce that

$$\lim_{n\to\infty} \mu\left(\left\{x \in X : \left|E(\mathbb{1}_B|\mathcal{A}_n)(x) - \mathbb{1}_B(x)\right| \geq \frac{1}{8}\right\}\right) = 0.$$

For every $n \in \mathbb{N}$, let

$$B_n := \left\{x \in X : \left|E(\mathbb{1}_B|\mathcal{A}_n)(x) - \mathbb{1}_B(x)\right| \geq \frac{1}{8}\right\}.$$

Then there exists $N = N(\varepsilon) \in \mathbb{N}$ such that

$$\mu(B_n) \leq \varepsilon/2, \quad \forall n \geq N.$$

For every $n \in \mathbb{N}$, let

$$A_n := \left\{x \in X : \left|E(\mathbb{1}_B|\mathcal{A}_n)(x) - 1\right| \leq \frac{1}{4}\right\} \in \mathcal{A}_n.$$

On one hand,

$$x \in B \setminus A_n \implies \left|E(\mathbb{1}_B|\mathcal{A}_n)(x) - \mathbb{1}_B(x)\right| > \frac{1}{4} \implies x \in B_n.$$

This means that

$$B \setminus A_n \subseteq B_n.$$

On the other hand,

$$x \in A_n \setminus B \implies \left|E(\mathbb{1}_B|\mathcal{A}_n)(x) - \mathbb{1}_B(x)\right| = \left|E(\mathbb{1}_B|\mathcal{A}_n)(x)\right| \geq \frac{3}{4} \implies x \in B_n.$$

This means that

$$A_n \setminus B \subseteq B_n.$$

Therefore,

$$\mu(A_n \triangle B) = \mu(A_n \setminus B) + \mu(B \setminus A_n) \leq 2\mu(B_n) \leq \varepsilon, \quad \forall n \geq N.$$

As $A_n \in \mathcal{A}_n$, we have found some $A \in \bigcup_{n=1}^{\infty} \mathcal{A}_n$ with $\mu(A \triangle B) < \varepsilon$. □

We will now give a proof of Lemma A.1.32 in the case where the algebra is countable.

Proof. Let $\mathcal{A} = \{A_n\}_{n=1}^{\infty}$ be a countable algebra on a set X and let μ be a probability measure on $(X, \sigma(\mathcal{A}))$. Set $\mathcal{A}'_n = \{A_k\}_{k=1}^{n}$. Since $\mathcal{A}'_n$ is a finite set, the algebra $\mathcal{A}(\mathcal{A}'_n)$ it generates is also finite, and thus $\sigma(\mathcal{A}'_n) = \mathcal{A}(\mathcal{A}'_n)$. Moreover, since $\mathcal{A}$ is an algebra, $\sigma(\mathcal{A}'_n) = \mathcal{A}(\mathcal{A}'_n) \subseteq \mathcal{A}$. It follows that

$$\sigma(\mathcal{A}) = \sigma\left(\bigcup_{n=1}^{\infty} \sigma(\mathcal{A}'_n)\right).$$

Let $B \in \sigma(\mathcal{A})$ and $\varepsilon > 0$. Since $(\sigma(\mathcal{A}'_n))_{n=1}^{\infty}$ is an ascending sequence of sub-σ-algebras of $\sigma(\mathcal{A})$ such that $\sigma(\mathcal{A}) = \sigma(\bigcup_{n=1}^{\infty} \sigma(\mathcal{A}'_n))$, it ensues from Corollary A.1.70 that there exists $A \in \bigcup_{n=1}^{\infty} \sigma(\mathcal{A}'_n)$ with $\mu(A \triangle B) < \varepsilon$. Since $\bigcup_{n=1}^{\infty} \sigma(\mathcal{A}'_n) \subseteq \mathcal{A}$, we have found some $A \in \mathcal{A}$ such that $\mu(A \triangle B) < \varepsilon$. □

Finally, we introduce the concept of conditional measure and relate it to the concept of expected value.

Definition A.1.71. Let $(X, \mathcal{A}, \mu)$ be a probability space and let $B \in \mathcal{A}$ be such that $\mu(B) > 0$. The set function $\mu_B : \mathcal{A} \to [0,1]$ defined by setting

$$\mu_B(A) := \frac{\mu(A \cap B)}{\mu(B)}, \quad \forall A \in \mathcal{A}$$

is a probability measure on $(X, \mathcal{A})$ called the *conditional measure of μ on B.*

Note that for every $\varphi \in L^1(X, \mathcal{A}, \mu)$,

$$\int_X \varphi\, d\mu_B = \int_B \varphi\, d\mu_B + \int_{X\setminus B} \varphi\, d\mu_B = \frac{1}{\mu(B)} \int_B \varphi\, d\mu + 0 = E(\varphi|B).$$

A.1.10 Characteristic functions and distributions

Definition A.1.72. If μ is a Borel probability measure on $\mathbb{R}$, then the function

$$\mathbb{R} \ni t \longmapsto \widehat{\mu}(t) := \int_{\mathbb{R}} e^{its}\, d\mu(s)$$

is called the *characteristic function* (commonly in probability theory) or *Fourier transform* (in analysis) of the measure μ.

The first classical result about this notion is the following.

Theorem A.1.73. *Two Borel probability measures μ and ν on $\mathbb{R}$ are equal if and only if their characteristic functions are equal, i. e., $\mu = \nu$ if and only if $\widehat{\mu} = \widehat{\nu}$.*

The second classical result goes as follows.

Theorem A.1.74. *A sequence $(\mu_n)_{n=1}^{\infty}$ of Borel probability measures on $\mathbb{R}$ weak* converges to a Borel probability measure μ on $\mathbb{R}$ if and only if the sequence $(\widehat{\mu}_n)_{n=1}^{\infty}$ of their characteristic functions converges pointwise to $\widehat{\mu}$, the characteristic function of μ, i. e., if and only if*

$$\lim_{n\to\infty} \widehat{\mu}_n(t) = \mu(t), \quad \forall t \in \mathbb{R}.$$

Definition A.1.75. Let $(X, \mathcal{A}, \mu)$ be a probability space and let $g : X \to \mathbb{R}$ be a function in $L^1(\mu)$. The function

$$\mathbb{R} \ni t \longmapsto \widehat{\mu \circ g^{-1}}(t) = \int_{\mathbb{R}} e^{its}\, d\mu \circ g^{-1}(s) = \int_X e^{itg(x)}\, d\mu(x) \tag{A.8}$$

is called the *characteristic function* (in probability theory) or *Fourier transform* (in analysis) of g.

The Borel probability measure $\mu \circ g^{-1}$ on $\mathbb{R}$ is called the *probability distribution* of the function (random variable) g with respect to the measure μ.

Note that it follows from (A.8) that the characteristic function of g depends only on the distribution of g.

As a direct consequence of Theorem A.1.73, we obtain the following.

Theorem A.1.76. *Any two random variables (not necessarily defined on the same probability space) have the same distribution if and only if they have the same characteristic function.*

Definition A.1.77. A sequence $(g_n)_{n=1}^{\infty}$ of random variables (not necessarily defined on the same probability space) is said to *converge in distribution* to a random variable g if the sequence $(\mu \circ g_n^{-1})_{n=1}^{\infty}$ of their probability distributions weak* converges to $\mu \circ g^{-1}$, the probability distribution of g.

As an immediate repercussion of Theorem A.1.74, we get the following.

Theorem A.1.78. *A sequence $(g_n)_{n=1}^{\infty}$ of random variables (not necessarily defined on the same probability space) converges in distribution to a random variable g if and only if the sequence $(\widehat{\mu \circ g_n^{-1}})_{n=1}^{\infty}$ of their characteristic functions converges pointwise to the characteristic function $\widehat{\mu \circ g^{-1}}$ of g.*

A.2 Functional analysis

Arzelà–Ascoli's theorem (cf. Theorem IV.6.7 in Dunford and Schwartz [26]) is a fundamental theorem in functional analysis and topology. Before stating it, let us first recall the following two definitions.

Definition A.2.1. Let X be a set and $\mathbb{R}^X$ be the set of real-valued functions on X. A subset F of $\mathbb{R}^X$ is said to be *pointwise bounded* if

$$\sup\{|f(x)| : f \in F\} < \infty, \quad \forall x \in X.$$

Definition A.2.2. Let X be a topological space, and let $C(X)$ be the space of real-valued continuous functions on X. A subset F of $C(X)$ is said to be *equicontinuous* if for every $x \in X$ and every $\varepsilon > 0$, there is a neighborhood $U(x, \varepsilon)$ of x such that

$$(y, f) \in U(x, \varepsilon) \times F \quad \Longrightarrow \quad |f(y) - f(x)| < \varepsilon.$$

The Arzelà–Ascoli theorem gives a characterization of the relative compactness of sets of continuous functions.

Theorem A.2.3 (Arzelà–Ascoli theorem). *Let X be a compact Hausdorff space. A subset F of $C(X)$ is relatively compact (i. e., has compact closure in the topology of uniform convergence induced by the supremum norm $\|\cdot\|_\infty$) if and only if F is equicontinuous and pointwise bounded.*

Theorem A.2.4 (Inverse function theorem). *Let F and G be Banach spaces and $u_0 \in F$. If $f : F \to G$ is a C^1 (i. e., continuously differentiable) function on some neighborhood of*

u_0 such that $f'(u_0)$ is invertible, then there exists an open neighborhood U of u_0 such that $f(U)$ is open in G and $f|_U : U \to f(U)$ is a bijection. Furthermore, the inverse function $f|_U^{-1} : f(U) \to U$ is C^1.

Let F be a complex Banach space and $L(F)$ be the Banach space of bounded (i. e., continuous) $\mathbb{C}$-linear operators from F to itself. For any operator $L \in L(F)$, recall that the operator norm of L is defined by

$$\|L\| := \inf\{c \geq 0 : \|L(v)\| \leq c\|v\|,\ \forall v \in F\} = \sup\left\{\frac{\|L(v)\|}{\|v\|} : v \in F \setminus \{0\}\right\}.$$

If $L_0 \in L(F)$ and $\delta > 0$, then $B_{L(F)}(L_0, \delta) = \{L \in L(F) : \|L - L_0\| < \delta\}$.

The following is a result on perturbations of linear operators.

Theorem A.2.5 (Kato–Rellich perturbation theorem). *Let F be a complex Banach space. If $L_0 \in L(F)$ has a simple isolated eigenvalue $\alpha(L_0)$, then for all $\varepsilon > 0$ there exists $\delta > 0$ such that every $L \in B_{L(F)}(L_0, \delta)$ has a simple isolated eigenvalue $\alpha(L)$ and there are two operators $Q_L, S_L \in L(F)$ such that:*

(a) *$\{\alpha(L)\} = \sigma(L) \cap B_{\mathbb{C}}(\alpha(L_0), \varepsilon)$ for all $L \in B_{L(F)}(L_0, \delta)$, where $\sigma(L)$ is the spectrum of L.*
(b) *The mapping $B_{L(F)}(L_0, \delta) \ni L \mapsto \alpha(L) \in \mathbb{C}$ is holomorphic.*
(c) *$L = \alpha(L)Q_L + S_L$, $Q_L S_L = S_L Q_L = 0$, $LQ_L = Q_L L$, $LS_L = S_L L$, $Q_L^2 = Q_L$, and $\dim(Q_L(F)) = 1$.*
(d) *The mappings $B_{L(F)}(L_0, \delta) \ni L \mapsto Q_L, S_L \in L(F)$ are holomorphic.*
(e) *Consequently, decreasing $\delta > 0$ if necessary, for every vector $v_0 \in F \setminus \mathrm{Ker}(L_0)$, the vector $Q_L(v_0)$ is an eigenvector of the operator L corresponding to the eigenvalue $\alpha(L)$ and the function $B_{L(F)}(L_0, \delta) \ni L \mapsto Q_L(v_0) \in F$ is holomorphic.*

There is another Riesz's Theorem that we will use in this book. This one is about the extension of a linear functional.

Theorem A.2.6. *Let E be a real vector space, F be a vector subspace of E, and K be a convex cone in E. A linear functional $L : F \to \mathbb{R}$ is called K-positive if $L|_{F \cap K} \geq 0$. A linear functional $\overline{L} : E \to \mathbb{R}$ is said to be a K-positive extension of L if $\overline{L}|_F = L$ and $\overline{L}|_K \geq 0$. If $E \subseteq F + K$, i. e. if for every $y \in E$ there exists $x \in F$ such that $y - x \in K$, then every K-positive linear functional on F can be extended to a K-positive linear functional on E.*

Note that, in general, a K-positive linear functional on F cannot be extended to a K-positive linear functional on E. Already in two dimensions there is a counterexample. Let $E = \mathbb{R}^2$, $F = \mathbb{R} \times \{0\}$, and $K = \{(x, y) : y > 0\} \cup \{(x, 0) : x > 0\}$. The K-positive functional $L(x, 0) = x$ can not be extended to a K-positive functional on E.

A.3 Complex analysis in one variable

Theorem A.3.1 (Cauchy's theorem). *Let $D \subseteq \mathbb{C}$ be a connected open set with piecewise smooth boundary ∂D. If f is an analytic function on D that extends smoothly to ∂D, then*

$$\int_{\partial D} f(z)\,dz = 0.$$

Theorem A.3.2 (Cauchy's integral formula). *Let $D \subseteq \mathbb{C}$ be a connected open set with piecewise smooth boundary ∂D. If f is an analytic function on D that extends smoothly to ∂D, then f has derivatives of all orders on D, which are given by*

$$f^{(n)}(z) = \frac{n!}{2\pi i} \int_{\partial D} \frac{f(w)}{(w-z)^{n+1}}\,dw, \quad \forall z \in D, \ \forall n \geq 0.$$

Theorem A.3.3 (Morera's theorem). *Let $D \subseteq \mathbb{C}$ be a connected open set. If a continuous function $f : D \to \mathbb{C}$ is such that $\int_{\partial R} f(z)\,dz = 0$ for every closed rectangle $R \subseteq D$ with sides parallel to the coordinate axes, then f is analytic on D.*

Theorem A.3.4 (Monodromy theorem). *Let $D \subseteq \mathbb{C}$ be a connected open set. Let $f : D \to \mathbb{C}$ be holomorphic at a point $z_0 \in D$, and let $\gamma_0, \gamma_1 : [0,1] \to D$ be two homotopic paths from z_0 to a point w. If f can be continued holomorphically along any path in D, then the holomorphic continuations of f along γ_0 and γ_1 coincide at w.*

Lemma A.3.5 (Schwarz lemma). *Let $\mathbb{D} = \{z \in \mathbb{C} : |z| < 1\}$. If $f : \mathbb{D} \to \mathbb{C}$ is an holomorphic function such that $f(0) = 0$ and $f(\mathbb{D}) \subseteq \mathbb{D}$, then*

$$|f'(0)| \leq 1 \quad \text{and} \quad |f(z)| \leq |z|, \quad \forall z \in \mathbb{D}.$$

If $|f'(0)| = 1$ or $|f(z)| = |z|$ for some $z \neq 0$ then $f(z) = az$ for some $a \in \mathbb{C}$ with $|a| = 1$.

Theorem A.3.6 (Schwarz reflection principle). *Let D be a connected open subset of $\mathbb{C}$ that is symmetric with respect to the real axis, and let $D^+ := D \cap \{\mathrm{Im}(z) > 0\}$. Let f be an holomorphic function on D^+ such that $\mathrm{Im}(f(z)) \to 0$ as $z \in D^+$ tends to $\overline{D} \cap \mathbb{R}$. Then f has an analytic extension to D, and that extension satisfies*

$$f(\overline{z}) = \overline{f(z)}, \quad \forall z \in D.$$

Theorem A.3.7 (Koebe quarter theorem). *Let $\mathbb{D} = \{z \in \mathbb{C} : |z| < 1\}$. If $f : \mathbb{D} \to \mathbb{C}$ is a univalent (i. e., one-to-one) holomorphic function, then*

$$f(\mathbb{D}) \supseteq D\Big(f(0), \frac{1}{4}|f'(0)|\Big),$$

where $D(w,r) = B_{\mathbb{C}}(w,r)$ is the open disc/ball centered at z and of radius r.

It is worth pointing out that this result is sharp, i. e., the constant 1/4 cannot be improved (read increased).

A repercussion of this theorem is the following well-known distortion theorem.

Theorem A.3.8 (Koebe distortion theorem). *Let $\mathbb{D} = \{z \in \mathbb{C} : |z| < 1\}$. If $f : \mathbb{D} \to \mathbb{C}$ is a univalent holomorphic function such that $f(0) = 0$ and $f'(0) = 1$, then*

$$\frac{|z|}{(1+|z|)^2} \le |f(z)| \le \frac{|z|}{(1-|z|)^2},$$
$$\frac{1-|z|}{(1+|z|)^3} \le |f'(z)| \le \frac{1+|z|}{(1-|z|)^3},$$

and

$$\frac{1-|z|}{1+|z|} \le \left|z\frac{f'(z)}{f(z)}\right| \le \frac{1+|z|}{1-|z|},$$

with equality if and only if f is a Koebe function, i. e., for some $\theta > 0$

$$f(z) = \frac{z}{(1-e^{i\theta}z)^2}.$$

The following is a variation of the part of the Koebe distortion theorem about the derivative of univalent functions.

Theorem A.3.9. *Let $w \in \mathbb{C}$ and $\varphi : D(w,R) \to \mathbb{C}$ be a univalent holomorphic function. Then for all $z \in D(w,R)$,*

$$\frac{1-\frac{|z-w|}{R}}{(1+\frac{|z-w|}{R})^3} \le \frac{|\varphi'(z)|}{|\varphi'(w)|} \le \frac{1+\frac{|z-w|}{R}}{(1-\frac{|z-w|}{R})^3}.$$

In this book, we will overwhelmingly use Koebe's distortion theorem in the following form.

Theorem A.3.10. *Let $w \in \mathbb{C}$ and $\varphi : D(w,R) \to \mathbb{C}$ be a univalent holomorphic function. There exists a function $K : [0,1) \to [1,\infty)$ such that*

$$\sup\{|\varphi'(z)| : z \in D(w,tR)\} \le K(t)\inf\{|\varphi'(z)| : z \in D(z,tR)\}.$$

We finish with a Koebe distortion theorem for the argument of univalent functions.

Theorem A.3.11 (Koebe distortion theorem for argument). *Let $\mathbb{D} = \{z \in \mathbb{C} : |z| < 1\}$. If $f : \mathbb{D} \to \mathbb{C}$ is a univalent holomorphic function such that $f(0) = 0$ and $f'(0) = 1$, then*

$$|\arg(f'(z))| \le 2\log\frac{1+r}{1-r}, \quad \forall |z| \le r < 1.$$

For more information on the Koebe distortion theorem, see [22, 53].

The next two results are exhibits of the general fact that the number of zeros of an analytic function does not change under small analytic perturbations.

Theorem A.3.12 (Rouché's theorem). *Let $D \subseteq \mathbb{C}$ be a connected open set with piecewise smooth boundary ∂D. If f and g are analytic functions on $D \cup \partial D$ such that $|g(z)| < |f(z)|$ for all $z \in \partial D$, then f and $f+g$ have the same number of zeros in D, counting multiplicities.*

Theorem A.3.13 (Hurwitz's theorem). *Let $D \subseteq \mathbb{C}$ be a connected open set. Suppose that $(f_n)_{n=1}^{\infty}$ is a sequence of holomorphic functions on D that converges normally on D (i. e. converges uniformly on compact subsets of D) to a function f, and suppose that f has a zero of order N at $z_0 \in D$. Then there exists $\delta > 0$ such that for all sufficiently large n, the functions f_n have each exactly N zeros in the disc $D(z_0, \delta)$, counting multiplicities, and these zeros converge to z_0 as $n \to \infty$. (Note: The normal convergence imposes that f be holomorphic.)*

Many theorems are attributed to Hurwitz, including the next one.

Theorem A.3.14 (Hurwitz's theorem II). *Let $D \subseteq \mathbb{C}$ be a connected open set. Suppose that $(f_n)_{n=1}^{\infty}$ is a sequence of univalent (i. e., one-to-one) holomorphic functions on D that converges normally on D to a function f. Then f is either constant or univalent on D.*

A partial converse to this theorem is the following.

Theorem A.3.15 (Converse to Hurwitz's theorem). *Let $D \subseteq \mathbb{C}$ be a connected open set. Suppose that a sequence $(f_n)_{n=1}^{\infty}$ of holomorphic functions on D converges normally on D to a univalent function f. Then for any compact set $K \subseteq D$, all but finitely many of the functions f_n are univalent on K.*

Theorem A.3.16 (Implicit function theorem). *Let $F(z,w)$ be a continuous function of z and w that depends analytically on z for each fixed w. Suppose that $F(z_0, w_0) = 0$ and $\frac{\partial F}{\partial z}(z_0, w_0) \neq 0$. Let $\delta > 0$ be such that $F(z, w_0) \neq 0$ for all $z \in \overline{D(z_0, \delta)} \setminus \{z_0\}$. Then there exists $\varepsilon > 0$ such that for every $w \in D(w_0, \varepsilon)$, there is a unique $z = g(w) \in D(z_0, \delta)$ such that $F(z, w) = 0$. In fact,*

$$g(w) = \frac{1}{2\pi i} \int_{|\zeta - z_0| = \delta} \frac{\zeta \frac{\partial F}{\partial z}(\zeta, w)}{F(\zeta, w)}\, d\zeta, \quad \forall w \in D(w_0, \varepsilon),$$

and $g : D(w_0, \varepsilon) \to D(z_0, \delta)$ is thus continuous. If, additionally, $F(z, w)$ is analytic in w for each fixed z, then g is analytic and

$$g'(w) = -\frac{\frac{\partial F}{\partial w}(g(w), w)}{\frac{\partial F}{\partial z}(g(w), w)}.$$

The following theorem is a corollary to the previous one.

Theorem A.3.17 (Inverse function theorem). *Suppose that an holomorphic function $f : D(z_0, \delta) \to \mathbb{C}$ is such that $f'(z_0) \neq 0$ and $f(z) \neq f(z_0)$ for all $z \in \overline{D(z_0, \delta)} \setminus \{z_0\}$. Let $\varepsilon > 0$ be such that $f(\partial D(z_0, \delta)) \subseteq \mathbb{C} \setminus D(f(z_0), \varepsilon)$. Then for every $w \in D(f(z_0), \varepsilon)$, there is a unique $z \in D(z_0, \delta)$ such that $f(z) = w$ and, writing $z = f^{-1}(w)$, the inverse function $f^{-1} : D(f(z_0), \varepsilon) \to D(z_0, \delta)$ satisfies*

$$f^{-1}(w) = \frac{1}{2\pi i} \int_{|\zeta - z_0| = \delta} \frac{\zeta f'(\zeta)}{f(\zeta) - w} \, d\zeta, \quad \forall w \in D(f(z_0), \varepsilon),$$

and f^{-1} is thus holomorphic.

The next theorem concerns extensions of holomorphic motions.

Theorem A.3.18 (Slodkowski's theorem). *Let $\mathbb{D} = \{z \in \mathbb{C} : |z| < 1\}$. Every holomorphic motion $f : \mathbb{D} \times A \to \mathbb{C}$ of any set $A \subseteq \mathbb{C}$ can be extended to a holomorphic motion $F : \mathbb{D} \times \mathbb{C} \to \mathbb{C}$ of $\mathbb{C}$.*

A.4 Complex analysis in several variables

Theorem A.4.1 (Hartogs' theorem). *Let Λ be an open subset of $\mathbb{C}^n$ and F be a complex Banach space. If a function $f : \Lambda \to F$ is separately holomorphic on Λ, then it is holomorphic on Λ.*

Appendix B – The Ionescu-Tulcea and Marinescu theorem

In this Appendix, we present a functional-analytic result originally due to Ionescu-Tulcea and Marinescu from their acclaimed paper [57]. We slightly extend the theorem to serve our purposes in Section 13.8.

Let E be a complex linear (i. e. vector) space and B a linear subspace of E. Suppose that E and B are Banach spaces when equipped with the norms $|\cdot|$ and $\|\cdot\|$, respectively. Suppose also that:

(a1) If a sequence $(x_n)_{n=1}^{\infty}$ in B is such that there exists a constant $K > 0$ with

$$\lim_{n\to\infty} |x_n - x| = 0 \text{ for some } x \in E \quad \text{and} \quad \|x_n\| \le K \text{ for all } n \in \mathbb{N},$$

then

$$x \in B \quad \text{and} \quad \|x\| \le K.$$

Let $C(B,E)$ be the subclass of all bounded linear operators $T : B \to B$ satisfying the following conditions:

(b1) There exists a constant $H > 0$ such that $|T^n| \le H$ for all $n \in \mathbb{N}$.

(c1) There exist $N \in \mathbb{N}$ and constants $0 < r < 1$ and $R > 0$ such that

$$\|T^N(x)\| \le r\|x\| + R|x|, \quad \forall x \subset B.$$

(d1) Every bounded subset of B is mapped by T to a relatively compact subset of E. That is, the operator $T : (B, \|\cdot\|) \to (E, |\cdot|)$ is compact.

Given the above notation, definitions and assumptions, the remainder of this Appendix will be dedicated to proving the following theorem.

Theorem B.1.1 (Ionescu-Tulcea and Marinescu theorem). *Let $T \in C(B,E)$. Then:*

(a2) *There are only a finite number of eigenvalues of $T : B \to B$ with modulus equal to 1. We denote them by $\lambda_1, \dots, \lambda_p \in \mathbb{S}^1 := \{\lambda \in \mathbb{C} : |\lambda| = 1\}$.*

(b2) *There exist operators $T_1, \dots, T_p, S \in C(B,E)$ such that*

$$T^n = \sum_{i=1}^{p} \lambda_i^n T_i + S^n, \quad \forall n \in \mathbb{N},$$

where $\dim(T_i(B)) < \infty$ for every $1 \le i \le p$.

(c2) *$T_i^2 = T_i$, $T_i \circ T_j = 0$ $(i \ne j)$ and $T_i \circ S = S \circ T_i = 0$ for each $1 \le i,j \le p$.*

(d2) *$T \circ T_i = T_i \circ T = \lambda_i T_i$ for every $1 \le i \le p$.*

(e2) *$r(S) < 1$, where $r(S)$ is the spectral radius of $S : B \to B$.*

https://doi.org/10.1515/9783110702699-023

(f2) *Addendum:* $\sigma(T) \setminus \{\lambda_1, \ldots, \lambda_p\} \subseteq \sigma(S) \subseteq \overline{B(0, r(S))} \subseteq \mathbb{C}$, *where* $\sigma(T)$ *denotes the spectrum of* T. *This property is called the spectral gap of* T.

Remark B.1.2. Because $T : B \to B$ is a bounded operator, instead of condition (d1) it suffices that each bounded subset of $(B, \|\cdot\|)$ be relatively compact in $(E, |\cdot|)$.

We will not prove Theorem B.1.1 directly; rather, the proof will result from the combination of a series of lemmas. First of all, let us observe that we may replace the norms $\|\cdot\|$ and $|\cdot|$ by any equivalent norms to prove this theorem.

Lemma B.1.3. *The norm* $\|\cdot\|_*$ *on* B *defined by*

$$\|x\|_* := \sum_{n=0}^{N-1} \|T^n(x)\|$$

is equivalent to $\|\cdot\|$ *and there exists a constant* $0 < \tilde{r} < 1$ *such that*

$$\|T(x)\|_* \le \tilde{r}\|x\|_* + R|x|, \quad \forall x \in B.$$

Proof. Let $x \in B$. Notice that

$$\|x\| \le \|x\|_* \le \sum_{n=0}^{N-1} \|T\|^n \cdot \|x\| =: m\|x\|.$$

Now, let $\tilde{r} = (m + r - 1)/m$. Clearly, $0 < \tilde{r} < 1$. Moreover,

$$\begin{aligned}
\|T(x)\|_* &= \sum_{n=0}^{N-1} \|T^{n+1}(x)\| = \sum_{n=1}^{N-1} \|T^n(x)\| + \|T^N(x)\| \\
&\le \sum_{n=1}^{N-1} \|T^n(x)\| + r\|x\| + R|x| \quad \text{by (c1)} \\
&= \tilde{r} \sum_{n=0}^{N-1} \|T^n(x)\| + (1-\tilde{r}) \sum_{n=0}^{N-1} \|T^n(x)\| + (r-1)\|x\| + R|x| \\
&\le \tilde{r}\|x\|_* + (1-\tilde{r})m\|x\| + (r-1)\|x\| + R|x| \\
&= \tilde{r}\|x\|_* + (m + r - 1 - \tilde{r}m)\|x\| + R|x| \\
&= \tilde{r}\|x\|_* + R|x|.
\end{aligned}$$

□

Remark B.1.4. This lemma means that we may, without loss of generality, assume that condition (c1) is satisfied with $N = 1$.

Lemma B.1.5. *There exists a constant* $\tilde{R} > 0$ *such that*

$$\|T^n(x)\| \le r^n\|x\| + \tilde{R}|x|, \quad \forall x \in B, \forall n \in \mathbb{N}.$$

Proof. Let $x \in B$ and $n \in \mathbb{N}$. By condition (c1) with $N = 1$,

$$\|T^n(x)\| = \|T(T^{n-1}(x))\| \le r\|T^{n-1}(x)\| + R|T^{n-1}(x)|.$$

From condition (b1), it follows that

$$|T^{n-1}(x)| \le |T^{n-1}| \cdot |x| \le H|x|.$$

Thus,

$$\|T^n(x)\| \le r\|T^{n-1}(x)\| + A|x|,$$

where $A = RH$. Observe that we can write $A|x| = (\frac{A}{1-r} - r\frac{A}{1-r})|x|$, so

$$\|T^n(x)\| - \frac{A}{1-r}|x| \le r\left(\|T^{n-1}(x)\| - \frac{A}{1-r}|x|\right).$$

Repeating this $n-1$ more times, we deduce that

$$\|T^n(x)\| - \frac{A}{1-r}|x| \le r^n\left(\|x\| - \frac{A}{1-r}|x|\right).$$

Hence,

$$\|T^n(x)\| \le r^n\|x\| + \frac{1-r^n}{1-r}A|x| \le r^n\|x\| + \frac{A}{1-r}|x|.$$

Set $\widetilde{R} = A/(1-r)$. □

Lemma B.1.6. *The norms $\|T^n\|$, $n \in \mathbb{N}$, are uniformly bounded. In other words,*

$$M := \sup\{\|T^n\| : n \in \mathbb{N}\} < \infty.$$

Proof. Let $x \in B$. Recalling that $r < 1$, Lemma B.1.5 implies that

$$\sup\{\|T^n(x)\| : n \in \mathbb{N}\} \le \|x\| + \widetilde{R}|x| < \infty.$$

The desired result follows upon applying the Banach–Steinhaus theorem (also known as the uniform boundedness principle; see Theorem 5.8 in Rudin [109]). □

For every complex number λ, define $B(\lambda) := \mathrm{Ker}(T - \lambda \cdot \mathrm{Id}) \subseteq B$. In other words, if $B(\lambda) \ne \{0\}$ then λ is an eigenvalue of the operator T and $B(\lambda)$ is the eigenspace corresponding to λ. It is easy to see that for each $\lambda \in \mathbb{C}$, the set $B(\lambda)$ is a linear subspace of B. Let us now proceed with the next lemma.

Lemma B.1.7. *If $\lambda \in \mathbb{S}^1$, then $B(\lambda)$ is finite dimensional.*

Proof. Let X denote the closed (with respect to $|\cdot|$) unit ball of $B(\lambda)$, i. e., let

$$X := \{x \in B(\lambda) : |x| \leq 1\}.$$

Clearly, X is a closed set in E. Let $x \in X$. Then $x = \lambda^{-1}T(x)$ and, therefore, $\|x\| = \|T(x)\| \leq r\|x\| + R|x| \leq r\|x\| + R$ by condition (c1). Hence,

$$\|x\| \leq \frac{R}{1-r}.$$

Thus, X is a bounded subset of B. This allows us to infer from condition (d1) that $\overline{T(X)}$ is compact in E. Since $X = \lambda^{-1}T(X)$, we deduce that $\overline{X} = \lambda^{-1}\overline{T(X)}$ is compact in E. Because X is closed in E, it follows that $X = \overline{X}$ is compact in E. As a Banach space is finite dimensional if and only if its unit ball is compact, the compactness in E of the unit ball X of $B(\lambda)$ implies that

$$\dim(B(\lambda)) < \infty. \qquad \square$$

Lemma B.1.8. *Let $T \in C(B,E)$. Then T has only a finite number of eigenvalues with modulus 1.*

Proof. Suppose, by way of contradiction, that there exists an infinite sequence of distinct eigenvalues $(\lambda_i)_{i=1}^{\infty}$ with $|\lambda_i| = 1$ for all $i \in \mathbb{N}$. For each λ_i choose any eigenvector $x_i \in B(\lambda_i) \setminus \{0\}$. Then the eigenvectors $(x_i)_{i=1}^{\infty}$ are linearly independent. For every $n \in \mathbb{N}$, let $X(n)$ be the linear span of $(x_1, \ldots, x_n)$, i. e., the set of all vectors of the form $a_1x_1 + \cdots + a_nx_n$, where $a_1, \ldots, a_n \in \mathbb{C}$. Notice that $X(n)$ is a proper closed linear subspace of $X(n+1)$. A straightforward calculation shows that

$$cT(X(n)) \subseteq X(n), \quad \forall n \in \mathbb{N},\ \forall c \in \mathbb{C}. \tag{B.1}$$

Indeed, if $x \in X(n)$,

$$cT(x) = c\sum_{i=1}^{n} a_iT(x_i) = \sum_{i=1}^{n} ca_i\lambda_ix_i \in X(n).$$

The remainder of the proof of this lemma depends on the following two claims.

Claim 1. *If $x \in X(n)$, then for every $k \in \mathbb{N}$ we have that*

$$T^k(x) - \lambda_n^k x \in X(n-1).$$

Proof of Claim 1. We will prove this claim by induction. Let $x \in X(n)$ and observe that

$$T(x) - \lambda_n x = \sum_{i=1}^{n} a_i\lambda_ix_i - \lambda_n\sum_{i=1}^{n} a_ix_i = \sum_{i=1}^{n-1}(a_i\lambda_i - a_i\lambda_n)x_i \in X(n-1).$$

Assume now that $T^{k-1}(x) - \lambda_n^{k-1}x \in X(n-1)$. Since T is a linear operator, it follows that

$$\begin{aligned} T^k(x) - \lambda_n^k x &= T(T^{k-1}(x) - \lambda_n^{k-1}x) + \lambda_n^{k-1}(T(x) - \lambda_n x) \\ &\in T(X(n-1)) + X(n-1) \\ &\subseteq X(n-1) + X(n-1) \subseteq X(n-1). \end{aligned}$$

This completes the proof of the first claim. ■

Claim 2. *There exists a sequence $(y_n)_{n=1}^\infty$ such that $y_n \in X(n)$, $|y_n| = 1$ and $|y_n - x| \geq \frac{1}{2}$ for all $x \in X(n-1)$.*

Proof of Claim 2. Choose $z \in X(n) \setminus X(n-1)$. As $X(n-1)$ is closed, we have

$$\beta := \operatorname{dist}(z, X(n-1)) = \inf\{|z - x| : x \in X(n-1)\} > 0.$$

So there exists some $\widehat{x} \in X(n-1)$ with the property that $|z - \widehat{x}| \leq 2\beta$. Let $y_n := (z - \widehat{x})/|z - \widehat{x}|$. Clearly, $y_n \in X(n)$ and $|y_n| = 1$. Let $x \in X(n-1)$. Then

$$|y_n - x| = \left| \frac{z - \widehat{x}}{|z - \widehat{x}|} - x \right| = \frac{1}{|z - \widehat{x}|} |z - (\widehat{x} + x|z - \widehat{x}|)| \geq \frac{1}{2\beta}\beta = \frac{1}{2}.$$

This ends the proof of the second claim. ■

In light of Lemma B.1.5, for all $j \in \mathbb{N}$ we have that

$$\|\lambda_j^{-n} T^n(y_j)\| = \|T^n(y_j)\| \leq r^n \|y_j\| + \widetilde{R}$$

for all $n \in \mathbb{N}$. Since $r < 1$, for each $j \in \mathbb{N}$ there exists $n_j \in \mathbb{N}$ such that for all $n \geq n_j - 1$,

$$\|\lambda_j^{-(n+1)} T^n(y_j)\| = \|\lambda_j^{-n} T^n(y_j)\| \leq 1 + \widetilde{R}.$$

Thus, the set $\{\lambda_j^{-n_j} T^{n_j - 1}(y_j) : j \in \mathbb{N}\}$ is bounded in B. By condition (d1), the closure of the set $\{\lambda_j^{-n_j} T^{n_j}(y_j) : j \in \mathbb{N}\}$ is compact in E. It ensues that the sequence $(\lambda_j^{-n_j} T^{n_j}(y_j))_{j=1}^\infty$ contains a subsequence that converges in E. To finish the proof of the lemma, consider

$$D_j := |\lambda_{j+1}^{-n_{j+1}} T^{n_{j+1}}(y_{j+1}) - \lambda_j^{-n_j} T^{n_j}(y_j)| = |\lambda_{j+1}^{-n_{j+1}} T^{n_{j+1}}(y_{j+1}) - y_{j+1} + y_{j+1} - \lambda_j^{-n_j} T^{n_j}(y_j)| = |y_{j+1} - x_j|,$$

where $x_j := \lambda_j^{-n_j} T^{n_j}(y_j) - (\lambda_{j+1}^{-n_{j+1}} T^{n_{j+1}}(y_{j+1}) - y_{j+1})$. According to Claim 1,

$$\lambda_{j+1}^{-n_{j+1}} T^{n_{j+1}}(y_{j+1}) - y_{j+1} \in X(j + 1 - 1) = X(j).$$

Also, $\lambda_j^{-n_j} T^{n_j}(y_j) \in X(j)$ by (B.1). Therefore, $x_j \in X(j)$. Finally, by the choice of the sequence $(y_j)_{j=1}^\infty$, we obtain from Claim 2 that $D_j \geq 1/2$ for all $j \in \mathbb{N}$. This contradicts the fact that $(\lambda_j^{-n_j} T^{n_j}(y_j))_{j=1}^\infty$ contains a convergent subsequence. □

In order to proceed further, we need the following theorem due to Kakutani and Yosida [60]. Here, we use the notation $\mathcal{B}(X)$ for the set of bounded linear operators on a Banach space $(X, |\cdot|)$.

Theorem B.1.9. *Let $T \in \mathcal{B}(X)$. Assume that:*
(a) *There exists $C \geq 0$ such that*

$$|T^n| \leq C, \quad \forall n \geq 0.$$

(b) *For each $x \in X$, the sequence $(\frac{1}{n}\sum_{j=1}^{n} T^j(x))_{n=1}^{\infty}$ contains a subsequence converging to a point denoted by $T_1(x)$.*

Then the sequence $(\frac{1}{n}\sum_{j=1}^{n} T^j(x))_{n=1}^{\infty}$ actually converges to $T_1(x)$ for each $x \in X$. Furthermore, $T_1 : X \to X$ is a bounded linear operator such that

$$T \circ T_1 = T_1 \circ T = T_1^2 = T_1 \quad \text{and} \quad |T_1| \leq C.$$

Proof. Fix $x \in X$ and, to shorten the notation, set

$$x_n := \frac{1}{n}\sum_{j=1}^{n} T^j(x).$$

By (b), let $\overline{x}$ be an accumulation point of the sequence $(x_n)_{n=1}^{\infty}$. We shall first prove that

$$T(\overline{x}) = \overline{x}. \tag{B.2}$$

Indeed, using (a),

$$|T(x_n) - x_n| = \left|\frac{1}{n}(T^{n+1}(x) - T(x))\right| \leq \frac{1}{n}(|T^{n+1}(x)| + |T(x)|) \leq \frac{2C}{n}|x|.$$

So, if $(x_{n_k})_{k=1}^{\infty}$ is a subsequence such that $x_{n_k} \to \overline{x}$, then

$$|T(x_{n_k}) - x_{n_k}| \leq \frac{2C}{n_k}|x|.$$

Upon letting $k \to \infty$, we obtain that $T(\overline{x}) = \overline{x}$. Now we shall prove that the entire sequence $(x_n)_{n=1}^{\infty}$ converges to $\overline{x}$. Write $x = \overline{x} + (x - \overline{x})$. By (B.2), we have

$$x_n = \overline{x} + \frac{1}{n}\sum_{j=1}^{n} T^j(x - \overline{x}).$$

Thus, in order to prove that $(x_n)_{n=1}^{\infty}$ converges to $\overline{x}$, it suffices to prove that

$$\lim_{n\to\infty} \frac{1}{n}\sum_{j=1}^{n} T^j(x - \overline{x}) = 0. \tag{B.3}$$

To that end, let the set R be defined by $R := (\mathrm{Id} - T)(X)$. We then have that

$$\Big(\mathrm{Id} - \frac{1}{n_k}\sum_{j=1}^{n_k} T^j\Big)(x) = (\mathrm{Id} - T)\Big(\mathrm{Id} + \frac{n_k - 1}{n_k}T + \frac{n_k - 2}{n_k}T^2 + \cdots + \frac{n_k - (n_k - 1)}{n_k}T^{n_k - 1}\Big)(x) \in R$$

for all $k \in \mathbb{N}$. Letting $k \to \infty$, we deduce that

$$x - \overline{x} \in \overline{R}.$$

To obtain (B.3), it suffices to prove that for all $y \in \overline{R}$,

$$\lim_{n\to\infty} \frac{1}{n}\sum_{j=1}^{n} T^j(y) = 0. \tag{B.4}$$

In order to show this, consider first $y \in R$. In that case, $y = z - T(z)$ for some $z \in X$ and so, using (a) again,

$$\Big|\frac{1}{n}\sum_{j=1}^{n} T^j(y)\Big| = \Big|\frac{1}{n}\sum_{j=1}^{n} T^j(z) - \frac{1}{n}\sum_{j=2}^{n+1} T^j(z)\Big| = \frac{1}{n}\big|T(z) - T^{n+1}(z)\big| \le \frac{2C}{n}|z|.$$

This last quantity tends to zero as $n \to \infty$.

On the other hand, if $y \in \overline{R} \setminus R$ then fix $\varepsilon > 0$ and pick $y' \in R$ such that $|y - y'| < \varepsilon$. Using (a) once more, we obtain that

$$\begin{aligned}\Big|\frac{1}{n}\sum_{j=1}^{n} T^j(y)\Big| &= \Big|\frac{1}{n}\sum_{j=1}^{n} T^j(y') + \frac{1}{n}\sum_{j=1}^{n} T^j(y - y')\Big| \\ &\le \Big|\frac{1}{n}\sum_{j=1}^{n} T^j(y')\Big| + \frac{1}{n}\sum_{j=1}^{n}\big|T^j(y - y')\big| \\ &\le \Big|\frac{1}{n}\sum_{j=1}^{n} T^j(y')\Big| + C|y - y'| \le \Big|\frac{1}{n}\sum_{j=1}^{n} T^j(y')\Big| + C\varepsilon.\end{aligned}$$

Letting $n \to \infty$, we deduce that

$$\limsup_{n\to\infty}\Big|\frac{1}{n}\sum_{j=1}^{n} T^j(y)\Big| \le C\varepsilon.$$

Letting $\varepsilon \to 0$, we conclude that

$$\lim_{n\to\infty} \frac{1}{n}\sum_{j=1}^{n} T^j(y) = 0.$$

We have now established (B.4) for any $y \in \overline{R}$. This completes the proof that the sequence $(x_n)_{n=1}^{\infty}$ converges to $\overline{x}$.

So it makes sense to define the operator T_1 by setting $T_1(x) := \overline{x}$. It is clear that $T_1 : X \to X$ is linear and bounded with $|T_1| \le C$. Furthermore, (B.2) is equivalent to the statement

$$T \circ T_1 = T_1.$$

Thus, for each $n \in \mathbb{N}$,

$$T^n \circ T_1 = T_1 \quad \text{and hence} \quad \frac{1}{n}\sum_{j=1}^{n} T^j \circ T_1 = T_1.$$

Upon passing to the limit $n \to \infty$, this implies that

$$T_1^2 = T_1.$$

Finally, we have that

$$\left|\frac{1}{n}\sum_{j=1}^{n} T^j(T(x)) - \frac{1}{n}\sum_{j=1}^{n} T^j(x)\right| = \left|\frac{1}{n}(T^{n+1}(x) - T(x))\right| \le \frac{2C}{n}|x|.$$

Letting $n \to \infty$, it follows that $T_1(T(x)) = T_1(x)$ for every $x \in X$. This completes the proof of Theorem B.1.9. □

Theorem B.1.10. *If $T \in \mathcal{B}(X)$ satisfies the hypotheses of Theorem* B.1.9, *then so does $\lambda^{-1}T$ for any $\lambda \in \mathbb{S}^1$ and there thus exists an operator $T_\lambda \in \mathcal{B}(X)$ such that:*
(a) $T_\lambda = \lim_{n\to\infty} \frac{1}{n}\sum_{j=1}^{n} \lambda^{-j}T^j$.
(b) $T_\lambda \circ T = T \circ T_\lambda = \lambda T_\lambda$, $T_\lambda^2 = T_\lambda$ *and* $|T_\lambda| \le C$.
(c) *If* $\lambda, \mu \in \mathbb{S}^1$ *with* $\lambda \ne \mu$, *then* $T_\lambda \circ T_\mu = 0$.
(d) $T(x) = \lambda x$ *if and only if* $T_\lambda(x) = x$.

Furthermore, if $S = T - \sum_{i=1}^{k} \lambda_i T_{\lambda_i}$ for some $\lambda_1, \ldots, \lambda_k \in \mathbb{S}^1$, then:
(e) $T_{\lambda_i} \circ S = S \circ T_{\lambda_i} = 0$ *for all* $1 \le i \le k$.
(f) $T \circ S = S \circ T = S^2$.
(g) $T^n = \sum_{i=1}^{k} \lambda_i^n T_{\lambda_i} + S^n$ *for all* $n \in \mathbb{N}$.
(h) λ *is an eigenvalue of S if and only if λ is an eigenvalue of T and $\lambda \ne \lambda_i$ for all* $1 \le i \le k$.

Proof. Part (a), as well as $T_\lambda^2 = T_\lambda$ and $|T_\lambda| \le C$ in part (b), follow directly from Theorem B.1.9 with T_λ being the operator T_1 associated to $\lambda^{-1}T$. Hence,

$$T_\lambda \circ T = \lambda T_\lambda \circ (\lambda^{-1}T) = \lambda T_\lambda = \lambda(\lambda^{-1}T) \circ T_\lambda = T \circ T_\lambda.$$

Alternatively,

$$T_\lambda \circ T = T \circ T_\lambda = \lim_{n\to\infty} \frac{1}{n}\sum_{j=1}^{n} \lambda^{-j} T^{j+1} = \lambda \lim_{n\to\infty} \frac{1}{n}\sum_{j=1}^{n} \lambda^{-(j+1)} T^{j+1} = \lambda \lim_{n\to\infty} \frac{1}{n+1}\sum_{j=1}^{n+1} \lambda^{-j} T^{j} = \lambda T_\lambda.$$

Part (b) is thus proved. Moreover, by induction,

$$T_\lambda \circ T^n = T^n \circ T_\lambda = \lambda^n T_\lambda, \quad \forall n \in \mathbb{N}. \tag{B.5}$$

In order to prove part (c), applying T_μ to part (a) from the right-hand side and using (B.5) for T_μ, we see that

$$T_\lambda \circ T_\mu = \lim_{n\to\infty} \frac{1}{n}\sum_{j=1}^{n} \lambda^{-j} T^{j} \circ T_\mu = \lim_{n\to\infty} \frac{1}{n}\sum_{j=1}^{n} \frac{1}{\lambda^j}\mu^j T_\mu = \lim_{n\to\infty} \frac{1}{n}\frac{1-(\frac{\mu}{\lambda})^n}{1-(\frac{\mu}{\lambda})}\frac{\mu}{\lambda} T_\mu = 0.$$

Thus, part (c) is proved.

For part (d), first notice that if $\lambda^{-1}T(x) = x$, then

$$T_\lambda(x) = \lim_{n\to\infty} \frac{1}{n}\sum_{j=1}^{n} \lambda^{-j} T^{j}(x) = \lim_{n\to\infty} \frac{1}{n}\sum_{j=1}^{n} x = x.$$

Also, if $T_\lambda(x) = x$, it follows from (B.5) that $T(x) = T \circ T_\lambda(x) = \lambda T_\lambda(x) = \lambda x$.

The proofs of parts (e) and (f) follow from the definition of S and parts (b) and (c):

$$T_{\lambda_i} \circ S = S \circ T_{\lambda_i} = T \circ T_{\lambda_i} - \sum_{j=1}^{k} \lambda_j T_{\lambda_j} \circ T_{\lambda_i} = \lambda_i T_{\lambda_i} - \lambda_i T_{\lambda_i}^2 = 0.$$

Therefore,

$$S^2 = \Big(T - \sum_{i=1}^{k} \lambda_i T_{\lambda_i}\Big) \circ S = T \circ S - \sum_{i=1}^{k} \lambda_i T_{\lambda_i} \circ S = T \circ S = S \circ T.$$

The proof of part (g) follows by induction from parts (b) and (f):

$$T^{n+1} = T \circ T^n = T \circ \Big(\sum_{i=1}^{k} \lambda_i^n T_{\lambda_i} + S^n\Big) = \sum_{i=1}^{k} \lambda_i^n T \circ T_{\lambda_i} + T \circ S \circ S^{n-1} = \sum_{i=1}^{k} \lambda_i^{n+1} T_{\lambda_i} + S^{n+1}.$$

It only remains to demonstrate part (h). To that end, suppose that λ is an eigenvalue of T and that $\lambda \neq \lambda_i$ for all $1 \le i \le k$. Then there exists $x \neq 0$ such that $T(x) = \lambda x$ and, therefore, in light of part (b),

$$\lambda_i T_{\lambda_i}(x) = T_{\lambda_i}(T(x)) = \lambda T_{\lambda_i}(x), \quad \forall 1 \le i \le k.$$

Since $\lambda \neq \lambda_i$, it follows that $T_{\lambda_i}(x) = 0$ for all $1 \le i \le k$. Hence,

$$\lambda x = T(x) = \sum_{i=1}^{k} \lambda_i T_{\lambda_i}(x) + S(x) = S(x).$$

In other words, λ is an eigenvalue of S. For the converse, suppose that λ is an eigenvalue of S. Then there exists $x \neq 0$ such that $S(x) = \lambda x$. Thus,

$$T(x) = T\Big(\frac{1}{\lambda}S(x)\Big) = \frac{1}{\lambda}T(S(x)) = \frac{1}{\lambda}S^2(x) = \frac{1}{\lambda}\lambda^2 x = \lambda x.$$

So λ is also an eigenvalue of T. By way of contradiction, suppose that $\lambda = \lambda_i$ for some $1 \le i \le k$. Then

$$T_{\lambda_i}(x) = \lim_{n\to\infty} \frac{1}{n}\sum_{j=1}^{n} \lambda_i^{-j} T^j(x) = \lim_{n\to\infty} \frac{1}{n}\sum_{j=1}^{n} \lambda_i^{-j}\lambda_i^j x = x \neq 0.$$

However,

$$T_{\lambda_i}(x) = T_{\lambda_i}\Big(\frac{1}{\lambda}S(x)\Big) = \frac{1}{\lambda}T_{\lambda_i} \circ S(x) = 0.$$

This contradiction finishes the proof of part (h), which completes the proof of Theorem B.1.10. □

Let us now return to our particular situation.

Lemma B.1.11. *Let $T \in C(B,E)$. For each $\lambda \in \mathbb{S}^1$ and each $x \in B$, there exists a unique $\overline{x} \in B$ such that*

$$\lim_{n\to\infty} \left| \frac{1}{n}\sum_{j=1}^{n} \lambda^{-j} T^j(x) - \overline{x} \right| = 0.$$

Proof. Let $\lambda \in \mathbb{S}^1$ and $x \in B$. First, note that

$$\sum_{j=1}^{n} \lambda^{-j} T^j(x) = T\Big(\sum_{j=1}^{n} \lambda^{-j} T^{j-1}(x)\Big)$$

and, by Lemma B.1.6,

$$\Big\| \sum_{j=1}^{n} \lambda^{-j} T^{j-1}(x) \Big\| \le \sum_{j=1}^{n} |\lambda^{-j}| \cdot M\|x\| = Mn\|x\|. \tag{B.6}$$

Consequently, by (d1), the sequence

$$\left(\frac{1}{n}\sum_{j=1}^{n} \lambda^{-j} T^j(x) = T\Big(\frac{1}{n}\sum_{j=1}^{n} \lambda^{-j} T^{j-1}(x)\Big)\right)_{n=1}^{\infty}$$

contains a convergent subsequence in E.

Let $\overline{B}$ denote the closure of B in E. Due to (b1), each map $T^n : B \to B$ extends continuously to a map $\widehat{T^n} = \widehat{T}^n : \overline{B} \to \overline{B}$ and

$$\left|\widehat{T}^n\right| \le H, \quad \forall n \in \mathbb{N}. \tag{B.7}$$

In order to shorten the notation, set

$$R := \lambda^{-1}\widehat{T} : \overline{B} \to \overline{B} \quad \text{and so} \quad R^n = \lambda^{-n}\widehat{T}^n : \overline{B} \to \overline{B}.$$

Take $x \in \overline{B}$. For all $k \in \mathbb{N}$ there exists $x_k \in B$ such that

$$|x - x_k| < \frac{1}{k}. \tag{B.8}$$

Then

$$\left|R^i(x - x_k)\right| \le H|x - x_k| < \frac{H}{k}, \quad \forall k \in \mathbb{N},\ \forall i \in \mathbb{N}. \tag{B.9}$$

Suppose inductively that an increasing subsequence $(n_j^{(k-1)})_{j=1}^{\infty}$ of positive integers has been constructed for some $k - 1 \in \mathbb{N}$. Then, just as argued above, the sequence

$$\left(\frac{1}{n_j^{(k-1)}} \sum_{i=1}^{n_j^{(k-1)}} R^i(x_k)\right)_{j=1}^{\infty}$$

contains a subsequence which converges to, say, $\overline{x}_k$ in E. Denote by $(n_j^{(k)})_{j=1}^{\infty}$ the corresponding subsequence of $(n_j^{(k-1)})_{j=1}^{\infty}$. We may assume, without loss of generality, that

$$\left|\frac{1}{n_j^{(k)}} \sum_{i=1}^{n_j^{(k)}} R^i(x_k) - \overline{x}_k\right| < \frac{1}{k}, \quad \forall j, k \in \mathbb{N}.$$

From this, (B.8) and (B.9), we deduce for all $k \ge l \in \mathbb{N}$ that

$$\begin{aligned}
|\overline{x}_k - \overline{x}_l| &\le \left|\overline{x}_k - \frac{1}{n_j^{(k)}} \sum_{i=1}^{n_j^{(k)}} R^i(x_k)\right| + \left|\frac{1}{n_j^{(k)}} \sum_{i=1}^{n_j^{(k)}} R^i(x_k - x_l)\right| + \left|\frac{1}{n_j^{(k)}} \sum_{i=1}^{n_j^{(k)}} R^i(x_l) - \overline{x}_l\right| \\
&\le \frac{1}{k} + \left|\frac{1}{n_j^{(k)}} \sum_{i=1}^{n_j^{(k)}} R^i(x_k - x)\right| + \left|\frac{1}{n_j^{(k)}} \sum_{i=1}^{n_j^{(k)}} R^i(x - x_l)\right| + \left|\frac{1}{n_{j'}^{(l)}} \sum_{i=1}^{n_{j'}^{(l)}} R^i(x_l) - \overline{x}_l\right| \\
&< \frac{1}{k} + \frac{H}{k} + \frac{H}{l} + \frac{1}{l} = \left(\frac{1}{k} + \frac{1}{l}\right)(1 + H),
\end{aligned}$$

where $j' \in \mathbb{N}$ is such that $n_{j'}^{(l)} = n_j^{(k)}$ for each $j \in \mathbb{N}$. The above calculation shows that $(\overline{x}_n)_{n=1}^{\infty}$ is a Cauchy sequence in $\overline{B}$ and, therefore, converges to a point in $\overline{B}$. Denote its limit by $\overline{x} \in \overline{B}$.

Now, consider the Cantor diagonal sequence $(n_k^{(k)})_{k=1}^{\infty}$. Then

$$\left| \frac{1}{n_k^{(k)}} \sum_{i=1}^{n_k^{(k)}} R^i(x_k) - \overline{x}_k \right| < \frac{1}{k}, \quad |\overline{x}_k - \overline{x}| \le \frac{1}{k}(1 + H),$$

and

$$|R^\ell(x_k) - R^\ell(x)| \le H|x_k - x| \le \frac{H}{k}, \quad \forall k, \ell \in \mathbb{N}.$$

Combining all of these observations, we obtain that

$$\begin{aligned} \left| \frac{1}{n_k^{(k)}} \sum_{i=1}^{n_k^{(k)}} R^i(x) - \overline{x} \right| &= \left| \frac{1}{n_k^{(k)}} \sum_{i=1}^{n_k^{(k)}} (R^i(x) - R^i(x_k)) + \frac{1}{n_k^{(k)}} \sum_{i=1}^{n_k^{(k)}} (R^i(x_k) - \overline{x}_k) + (\overline{x}_k - \overline{x}) \right| \\ &\le \left| \frac{1}{n_k^{(k)}} \sum_{i=1}^{n_k^{(k)}} (R^i(x) - R^i(x_k)) \right| + \left| \frac{1}{n_k^{(k)}} \sum_{i=1}^{n_k^{(k)}} (R^i(x_k) - \overline{x}_k) \right| + |\overline{x}_k - \overline{x}| \\ &\le \frac{1}{n_k^{(k)}} n_k^{(k)} \cdot \frac{H}{k} + \frac{1}{k} + \frac{1}{k}(1 + H) = \frac{2}{k}(1 + H). \end{aligned}$$

Hence, in E,

$$\lim_{k \to \infty} \frac{1}{n_k^{(k)}} \sum_{i=1}^{n_k^{(k)}} R^i(x) = \overline{x}.$$

Applying Theorem B.1.9 with R in place of T and $(E, |\cdot|)$ in place of $(X, |\cdot|)$, we conclude that, in E,

$$\lim_{n \to \infty} \frac{1}{n} \sum_{i=1}^{n} R^i(x) = \overline{x}.$$

Moreover, if $x \in B$ then $(\frac{1}{n} \sum_{i=1}^{n} R^i(x))_{n=1}^{\infty}$ is a sequence in B which converges to $\overline{x}$ in E. By (B.6), this sequence is bounded in B. Consequently, $\overline{x} \in B$ by condition (a1). □

Lemma B.1.12. *Let $T \in C(B, E)$ and $\lambda \in \mathbb{S}^1$. Let $\overline{x}$ be the unique element of B associated to x by Lemma* B.1.11. *Set $T_\lambda(x) = \overline{x}$ for all $x \in B$. Then $T_\lambda : B \to E$ is a linear operator which enjoys the following properties:*

(a) $T_\lambda(B) \subseteq B$.

(b) $|T_\lambda^n| \le H$ *and* $\|T_\lambda^n\| \le M$ *for all $n \in \mathbb{N}$, where H comes from the standing hypothesis* (b1) *while M arises from Lemma* B.1.6.

(c) $T_\lambda^2 = T_\lambda$ *and* $T_\alpha \circ T_\beta = 0$ *if* $\alpha, \beta \in \mathbb{S}^1$ *are such that* $\alpha \neq \beta$.
(d) $T_\lambda(B) = B(\lambda)$, *where* $B(\lambda) := \mathrm{Ker}(T - \lambda \cdot \mathrm{Id}) \subseteq B$.
(e) *If* $\lambda_1, \ldots, \lambda_p$ *are the eigenvalues of* T *with modulus 1, then*

$$T^n = \sum_{i=1}^{p} \lambda_i^n T_i + S^n, \quad \forall n \in \mathbb{N},$$

where $T_i := T_{\lambda_i} : B \to B$ *and* $S : B \to B$ *is defined by* $S = T - \sum_{i=1}^{p} \lambda_i T_i$.
(f) $T_i \circ T = T \circ T_i = \lambda_i T_i$ *for all* $1 \le i \le p$.
(g) S *has no eigenvalue of modulus 1 and* $T_i \circ S = S \circ T_i = 0$ *for all* $1 \le i \le p$.
(h) $|S^n| \le H(p+1)$ *and* $\|S^n\| \le M(p+1)$ *for every* $n \in \mathbb{N}$.

Proof. The linearity of T_λ and properties (a)–(g) follow directly from Lemma B.1.11, relations (B.7) and (B.6), Theorem B.1.10 and Lemma B.1.8. Part (h) can be deduced by combining parts (b) and (e), relation (B.7) and Lemma B.1.6. □

We now seek to deepen our understanding of the operator S. To lighten notation, set $T_* := \sum_{i=1}^{p} \lambda_i T_i : B \to B$. Then $|T_*| \le pH$, and the codomain of T_* is a finite-dimensional linear subspace of B by Lemma B.1.7. Recalling that all norms are equivalent on a finite-dimensional vector space, there is $c \ge 0$ such that $c^{-1}|y| \le \|y\| \le c|y|$ for all $y \in T_*(B)$. Hence, for all $x \in B$ we have

$$\|T_*(x)\| \le c|T_*(x)| \le cpH|x| = A|x|,$$

where $A = cpH$. So, by definition of S, condition (c1) and Remark B.1.4,

$$\|S(x)\| \le \|T(x)\| + \|T_*(x)\| \le r\|x\| + (R + A)|x|. \tag{B.10}$$

Lemma B.1.13. *The operator* $S : B \to E$ *is compact.*

Proof. Since $T : B \to E$ is compact by (d1) and $T_* : B \to E$ is compact as it is of finite-dimensional range, the operator $S = T - T_*$ is compact, too. (Note: All operators with finite-dimensional range are compact; this follows from the Bolzano–Weierstrass theorem; see Theorem 2.42 in Rudin [108].) □

Using the previous observations, we can prove the following lemma.

Lemma B.1.14. *If* $\lambda \in \mathbb{S}^1$, *then*

$$(\lambda \,\mathrm{Id} - S)(B) = B.$$

Proof. Recall that the spectral radius of S satisfies $r(S) = \lim_{n\to\infty} \|S^n\|^{1/n}$. It follows from Lemma B.1.12(h) that $r(S) \le 1$. So if $|\zeta| > 1$, then $(\zeta \,\mathrm{Id} - S) : B \to B$ is invertible. In particular,

$$(\zeta \operatorname{Id} - S)(B) = B.$$

Fix $\lambda \in \mathbb{S}^1$. Take an arbitrary sequence $(\zeta_n)_{n=1}^\infty$ in $\mathbb{C}$ such that

$$\lim_{n\to\infty} \zeta_n = \lambda \quad \text{and} \quad |\zeta_n| > 1, \quad \forall n \in \mathbb{N}.$$

Fix an arbitrary $y \in B$. Then, for all $n \in \mathbb{N}$ there exists $x_n \in B$ such that

$$(\zeta_n \operatorname{Id} - S)(x_n) = y. \tag{B.11}$$

Claim. *There exists a constant $D \geq 0$ such that*

$$|x_n| \leq D, \quad \forall n \in \mathbb{N}.$$

Proof of Claim. Suppose that the claim is not true. By passing to a subsequence if necessary, we may assume without loss of generality that each $|x_n| > 0$ and

$$\lim_{n\to\infty} |x_n| = \infty.$$

Set $x_n' := x_n/|x_n|$. Since

$$(\zeta_n \operatorname{Id} - S)(x_n') = \frac{y}{|x_n|}, \tag{B.12}$$

we obtain that

$$\zeta_n x_n' = S(x_n') + \frac{y}{|x_n|}. \tag{B.13}$$

Hence, using (B.10),

$$\|x_n'\| < \|\zeta_n x_n'\| \leq \|S(x_n')\| + \frac{\|y\|}{|x_n|} \leq r\|x_n'\| + (R + A)|x_n'| + \frac{\|y\|}{|x_n|} \leq r\|x_n'\| + R + A + \frac{\|y\|}{|x_n|}.$$

Thus,

$$\|x_n'\| < \frac{R + A + \frac{\|y\|}{|x_n|}}{1 - r}. \tag{B.14}$$

As $\lim_{n\to\infty} |x_n| = \infty$, we deduce that the sequence $(\|x_n'\|)_{n=1}^\infty$ is bounded. By Lemma B.1.13, the sequence $(S(x_n'))_{n=1}^\infty$ contains a convergent subsequence in the $|\cdot|$ norm. By (B.13), so does the sequence $(\zeta_n x_n')_{n=1}^\infty$. In turn, this implies that the sequence $(x_n')_{n=1}^\infty$ contains a subsequence $(x_{n_j}')_{j=1}^\infty$ convergent in E. Set

$$x' := \lim_{j\to\infty} x_{n_j}' \in E \quad \text{(i. e., in the } |\cdot| \text{ norm)}.$$

Using (a1) and the fact that $(\|x_n'\|)_{n=1}^{\infty}$ is a bounded sequence, we know that $x' \in B$. We also have that $|x'| = 1$, so $x' \neq 0$. According to Lemma B.1.12(h), the operator $S : (B, |\cdot|) \to (B, |\cdot|)$ is bounded. Since every bounded operator on a Banach space is continuous, $S : (B, |\cdot|) \to (B, |\cdot|)$ is continuous. We thus obtain from (B.12) that

$$(\lambda \operatorname{Id} - S)(x') = 0.$$

But this means that λ is an eigenvalue of S of modulus 1, which contradicts Lemma B.1.12(g). This completes the proof of the claim. ■

Continuing the proof of Lemma B.1.14, it follows from (B.14) and from the claim that

$$\|x_n\| = \|x_n'\| \cdot |x_n| \leq \frac{(R+A)|x_n| + \|y\|}{1-r} \leq \frac{(R+A)D + \|y\|}{1-r}.$$

So the sequence $(\|x_n\|)_{n=1}^{\infty}$ is bounded. Thus, in light of Lemma B.1.13, the sequence $(S(x_n))_{n=1}^{\infty}$ contains a convergent subsequence in $|\cdot|$. It follows from (B.11) that $(\zeta_n x_n)_{n=1}^{\infty}$ also contains a convergent subsequence, as, therefore, does the sequence $(x_n)_{n=1}^{\infty}$. Denote this latter convergent subsequence by $(x_{n_j})_{j=1}^{\infty}$ and set

$$x := \lim_{j\to\infty} x_{n_j} \in E.$$

Using (a1) and the fact that $(\|x_n\|)_{n=1}^{\infty}$ is bounded, we know that $x \in B$. Since the operator $S : (B, |\cdot|) \to (B, |\cdot|)$ is continuous, we deduce from (B.11) that

$$(\lambda \operatorname{Id} - S)(x) = y.$$

Since $y \in B$ is arbitrary, we conclude that $(\lambda \operatorname{Id} - S)(B) = B$. □

We now turn our attention to the spectral radius of S.

Lemma B.1.15. $r(S) < 1$.

Proof. Lemma B.1.14 states that the operator $(\lambda \operatorname{Id} - S) : B \to B$ is surjective for all $\lambda \in \mathbb{S}^1$. It follows from the fact that $S : B \to B$ has no eigenvalue of modulus 1 that $(\lambda \operatorname{Id} - S) : B \to B$ is also injective. Thus, the inverse $(\lambda \operatorname{Id} - S)^{-1}$ exists for all $\lambda \in \mathbb{S}^1$. Since the set of invertible operators on a Banach space is open, we deduce that S cannot have eigenvalues with moduli arbitrarily close to 1. □

Finally, we describe a relation between the spectra of S and T.

Lemma B.1.16. $\sigma(T) \setminus \{\lambda_1, \ldots, \lambda_p\} \subseteq \sigma(S) \subseteq \overline{B(0, r(S))} \subseteq \mathbb{C}$.

Proof. Recall that $\lambda \in \sigma(T)$ if and only if the operator $T - \lambda \operatorname{Id}$ is not invertible. Notice that if a Banach space B can be written as the direct sum $B_1 \bigoplus B_2$ of two Banach spaces

B_1 and B_2 and if $T : B \to B$ is a bounded linear operator under which B_1 and B_2 are invariant, i. e., $T(B_1) \subseteq B_1$ and $T(B_2) \subseteq B_2$, then

$$\sigma(T) = \sigma(T|_{B_1}) \cup \sigma(T|_{B_2}).$$

Indeed, in such circumstances, $T - \lambda\,\mathrm{Id}$ preserves B_1 and B_2, and hence $T - \lambda\,\mathrm{Id}$ is noninvertible on B if and only if $T - \lambda\,\mathrm{Id}$ is noninvertible on one of B_1 or B_2, i. e., if and only if $\lambda \in \sigma(T|_{B_1}) \cup \sigma(T|_{B_2})$.

Using Lemma B.1.12, split B into $B_1 := \bigoplus_{i=1}^p T_i(B)$ and $B_2 := (\mathrm{Id} - \sum_{i=1}^p T_i)(B)$. Observe that

$$T(T_i(B)) = T_i \circ T(B) \subseteq T_i(B), \quad \forall 1 \le i \le p,$$

and

$$T \circ \Big(\mathrm{Id} - \sum_{i=1}^p T_i\Big) = T - \sum_{i=1}^p T \circ T_i = T - \sum_{i=1}^p T_i \circ T = \Big(\mathrm{Id} - \sum_{i=1}^p T_i\Big) \circ T.$$

So T preserves each $T_i(B)$, as well as B_1 and B_2. We must now show that $B = B_1 \bigoplus B_2$, i. e., that any $x \in B$ can be uniquely written as $x_1 + x_2$ where $x_1 \in B_1$ and $x_2 \in B_2$. Clearly,

$$x = \sum_{i=1}^p T_i(x) + \Big(\mathrm{Id} - \sum_{i=1}^p T_i\Big)(x).$$

If this decomposition were not unique, then there would be $x_1 \neq x_2 \in B$ such that

$$\sum_{i=1}^p T_i(x_1) + \Big(\mathrm{Id} - \sum_{i=1}^p T_i\Big)(x_1) = \sum_{i=1}^p T_i(x_2) + \Big(\mathrm{Id} - \sum_{i=1}^p T_i\Big)(x_2).$$

Equivalently,

$$\sum_{i=1}^p T_i(x_1 - x_2) = \Big(\mathrm{Id} - \sum_{i=1}^p T_i\Big)(x_2 - x_1).$$

Suppose that

$$\sum_{i=1}^p T_i(z) = \Big(\mathrm{Id} - \sum_{i=1}^p T_i\Big)(y)$$

for some $y, z \in B$. Then $\sum_{i=1}^p T_i(z + y) = y$. Applying any T_j to both sides would result in

$$T_j(y) = T_j \circ \sum_{i=1}^p T_i(z + y) = T_j \circ T_j(z + y) = T_j(z + y).$$

Hence, $T_j(z) = 0$ for all $1 \le j \le p$ and $\sum_{i=1}^p T_i(z) = 0 = (\mathrm{Id} - \sum_{i=1}^p T_i)(y)$. In particular, this would mean that $T_j(x_1 - x_2) = 0$ for all $1 \le j \le p$, i. e. $T_j(x_1) = T_j(x_2)$ for all $1 \le j \le p$ and the decomposition is unique. Thus, we have shown that

$$B = B_1 \bigoplus B_2 = \bigoplus_{i=1}^{p} T_i(B) \bigoplus B_2$$

and that T preserves each subspace in these direct sums. Consequently,

$$\sigma(T) = \bigcup_{i=1}^{p} \sigma(T|_{T_i(B)}) \bigcup \sigma(T|_{B_2}). \tag{B.15}$$

We now examine the pointwise action of T on the subspace $T_i(B)$ for any fixed $1 \le i \le p$. If $x \in T_i(B)$, then $x = T_i(y)$ for some $y \in B$. Therefore,

$$T(x) = \sum_{j=1}^{p} \lambda_j T_j(T_i(y)) + S(T_i(y)) = \lambda_i T_i^2(y) = \lambda_i T_i(y) = \lambda_i x.$$

So $T|_{T_i(B)} = \lambda_i \,\mathrm{Id}\,|_{T_i(B)}$, i. e., T acts on $T_i(B)$ as the multiplication by λ_i, and hence $\sigma(T|_{T_i(B)}) = \{\lambda_i\}$.

Let us now investigate the pointwise action of T on B_2. Let $x \in B_2$, i. e., $x = (\mathrm{Id} - \sum_{i=1}^p T_i)(y)$ for some $y \in B$. Then

$$T(x) = \sum_{j=1}^{p} \lambda_j T_j\left(\left(\mathrm{Id} - \sum_{i=1}^{p} T_i\right)(y)\right) + S(x) = \sum_{j=1}^{p} \lambda_j T_j(y) - \sum_{i=1}^{p} \lambda_i T_i(y) + S(x) - S(x).$$

Thus, $\sigma(T|_{B_2}) = \sigma(S|_{B_2})$.

By (B.15), we know that $\sigma(T) = \{\lambda_1, \dots, \lambda_p\} \cup \sigma(S|_{B_2})$. By Lemma B.1.15 and the fact that all $\lambda_i \in \mathbb{S}^1$, we deduce that $\sigma(T) \setminus \{\lambda_1, \dots, \lambda_p\} = \sigma(S|_{B_2}) \subseteq \sigma(S) \subseteq \overline{B(0, r(S))}$. □

Proof of Theorem B.1.1. Lemma B.1.8 establishes (a2). Properties (b2), (c2) and (d2) are proved in Lemma B.1.12, in combination with Lemma B.1.7. Property (e2) is the object of Lemma B.1.15 while property (f2) is demonstrated in Lemma B.1.16. □

Bibliography

[1] Ahlfors, L., Lectures on Quasiconformal Mappings, 2nd edn. University Lecture Series, vol. **38**. American Mathematical Society (2006).

[2] Akter, H., Urbański, M., Real analyticity of Hausdorff dimension of Julia sets of parabolic polynomials $f_\lambda(z) = z(1 - z - \lambda z^2)$. Illinois J. Math. **55**(1), 157–184 (2011).

[3] Astala, K., Iwaniec, T., Martin, G., Elliptic partial differential equations and quasiconformal mappings in the plane. Princeton Mathematical Series, vol. **48**. Princeton University Press, Princeton (2009).

[4] Barnsley, M., Fractals Everywhere. Academic Press (1988).

[5] Bedford, T., Hausdorff dimension and box-dimension in self-similar sets. In: Proc. Topology and Measure V. Ernst-Moritz-Arndt Universität Greisfwald (1988).

[6] Benedetti, R., Petronio, C., Lectures on Hyperbolic Geometry. Springer, Berlin (1992).

[7] Billingsley, P., Ergodic Theory and Information. R. E. Krieger Pub. Co (1978).

[8] Billingsley, P., Probability and Measure, 3rd edn. John Wiley and Sons, New York (1995)..

[9] Billingsley, P., Convergence of Probability Measures, 2nd edn. John Wiley and Sons, New York (1999).

[10] Blair, D., Inversion Theory and Conformal Mapping. Student Mathematical Library, vol. **9**. American Mathematical Society (2000).

[11] Bowen, R., Equilibrium States and the Ergodic Theory of Anosov Diffeomorphisms. Lect. Notes in Math., vol. **470**. Springer (1975).

[12] Bowen, R., Hausdorff dimension of quasi-circles. Publ. Math. Inst. Hautes Études Sci. **50**, 11–25 (1979).

[13] Boyarsky, A., Góra, P., Laws of Chaos, Invariant Measures and Dynamical Systems in One Dimension. Birkhäuser (1997).

[14] Branner, B., Fagella, N., Quasiconformal Surgery in Holomorphic Dynamics. Cambridge Studies in Advanced Mathematics, vol. **141**. Cambridge University Press, New York (2014).

[15] Cawley, R., Mauldin, R. D., Multifractal decompositions of Moran fractals. Adv. Math. **92**, 196–236 (1992).

[16] Chousionis, V., Leykekhman, D., Urbanski, M., The dimension spectrum of graph directed Markov systems. Selecta Math. (N.S.) **25**(3), 40 (2019).

[17] Chousionis, V., Leykekhman, D., Urbanski, M., On the dimension spectrum of infinite subsystems of continued fractions. Trans. Amer. Math. Soc. **373**, 1009–1042 (2020).

[18] Chousionis, V., Tyson, J., Urbański, M., Conformal graph directed Markov systems on Carnot groups. Mem. Amer. Math. Soc. **266**(1291), 1–155 (2020).

[19] Chousionis, V., Urbański, M., Porosity in conformal dynamical systems, preprint (2019).

[20] Cieśla, T., On lifting invariant probability measures. Bull. Pol. Acad. Sci. **68**, 69–74 (2020).

[21] Conway, J. B., A Course in Functional Analysis, 2nd edn. Springer, New York (1990).

[22] Conway, J. B., Functions of One Complex Variable II. Graduate Texts in Mathematics. Springer, New York (1995).

[23] Das, T., Fishman, L., Simmons, D., Urbański, M., Hausdorff dimension of perturbations of a conformal iterated function system via thermodynamic formalism, preprint (2021).

[24] Denker, M., Urbański, M., On Sullivan's conformal measures for rational maps of the Riemann sphere. Nonlinearity **4**, 365–384 (1991).

[25] Dineen, S., Complex Analysis on Infinite Dimensional Spaces. Springer Monographs in Mathematics. Springer, London (1999).

[26] Dunford, N., Schwartz, J. T., Linear Operators, Part I: General Theory. John Wiley and Sons, New York (1964).

[27] Eggleston, H., Sets of fractional dimensions which occur in some problems of number theory. Proc. Lond. Math. Soc. **54**, 42–93 (1952).

https://doi.org/10.1515/9783110702699-024

[28] Falconer, K. J., The Geometry of Fractal Sets. Cambridge University Press (1985).

[29] Falconer, K. J., Techniques in Fractal Geometry. John Wiley and Sons, Chichester (1997).

[30] Falconer, K. J., Fractal Geometry: Mathematical Foundations and Applications, 2nd edn. John Wiley and Sons, Chichester (2003).

[31] Falk, R. S., Nussbaum, R., A new approach to numerical computation of Hausdorff dimension of iterated function systems: Applications to complex continued fractions. Integral Equations Operator Theory **90**(5), 61 (2018).

[32] Federer, H., Geometric Measure Theory. Die Grundlehren der mathematischen Wissenschaften, vol. **153**. Springer, New York (1969).

[33] Friedman, N. A., Ornstein, D. S., On isomorphisms of weak Bernoulli transformations. Adv. Math. **5**, 365–394 (1971).

[34] Frisch, U., Parisi, G., On the singularity structure of fully developed turbulence, appendix to U. Frisch, Fully developed turbulence and intermittency. In: Turbulence and Predictability in Geophysical Fluid Dynamics and Climate Dynamics (Proc. Int. Sch. Phys., "Enrico Fermi," Course LXXXVIII), North-Holland (Amsterdam, 1985), pp. 84–88 (1985).

[35] Gardiner, F. P., Lakic, N., Quasiconformal Teichmüller Theory. Mathematical Surveys and Monographs, vol. **76**. American Mathematical Society (1999).

[36] Ghenciu, A., Mauldin, R. D., Roy, M., Conformal graph directed Markov systems: beyond finite irreducibility. J. Fractal Geom. **3**(3), 217–243 (2016).

[37] Ghenciu, A., Roy, M., Bowen's formula for shift-generated finite conformal recursive constructions. Real Anal. Exchange **40**(1), 99–112 (2015).

[38] Ghenciu, A., Roy, M., Gibbs states for non-irreducible countable Markov shifts. Fund. Math. **221**(3), 231–265 (2013).

[39] Ghenciu, A., Munday, S., Roy, M., The Hausdorff dimension spectrum of conformal graph directed Markov systems and applications to nearest integer continued fractions. J. Number Theory **175**, 223–249 (2017).

[40] Gordin, M. I., The central limit theorem for stationary processes. Dokl. Akad. Nauk SSSR **188**, 1174–1176 (1969).

[41] Gouezel, S., Almost sure invariance principle for dynamical systems by spectral methods. Ann. Probab. **38**, 1639–1671 (2010).

[42] Grassberger, P., Procaccia, I., Characterization of strange attractors. Phys. Lett. **50**, 346–349 (1983).

[43] Gurewic, B. M., Shift entropy and Markov measures in the path space of a denumerable graph. Dokl. Akad. Nauk SSSR 192 (1970); English trans. in Sov. Math., Dokl. 11 744–747 (1970).

[44] Gurewic, B. M., Savchenko, S. V., Thermodynamic formalism for countable symbolic Markov chains. Russ. Math. Nauk **53**(2), 245–344 (1998).

[45] Halsey, T. C., Jensen, M. H., Kadanoff, L. P., Procaccia, I., Shraiman, B. J., Fractal measures and their singularities: the characterization of strange sets. Phys. Rev. A **33**, 1141–1151 (1986).

[46] Hanus, P., Mauldin, R. D., Urbański, M., Thermodynamic formalism and multifractal analysis of conformal infinite iterated function systems. Acta Math. Hungar. **96**, 27–98 (2002).

[47] Hausdorff, F., Dimension und äußeres Maß. Math. Ann. **79**, 157–179 (1919).

[48] Heinonen, J., Lectures on Analysis on Metric Spaces. Springer Universitext. Springer, New York (2001).

[49] Hensley, D., The Hausdorff dimensions of some continued fraction Cantor sets. J. Number Theory **33**(2), 182–198 (1989).

[50] Hensley, D., Continued Fractions. World Scientific (2006).

[51] Herrlich, H., Axiom of Choice. Lect. Notes in Math., vol. **1876**. Springer, Berlin Heidelberg (2006).

[52] Hewitt, E., Stromberg, K., Real and Abstract Analysis. Springer, New York (1965).

[53] Hille, E., Analytic Function Theory, vol. II. Introduction to Higher Mathematics. Ginn and Company (1962).
[54] Hofbauer, F., Keller, G., Ergodic properties of invariant measures for piecewise monotonic transformations. Math. Z. **180**, 119–140 (1982).
[55] Hubbard, J., Teichmüller Theory and Applications to Geometry, Topology, and Dynamics, vol. **1**. Matrix Press (2006).
[56] Hutchinson, J., Fractals and self-similarity. Indiana Univ. Math. J. **30**(5), 713–747 (1981).
[57] Ionescu-Tulcea, C. T., Marinescu, G., Théorie ergodique pour des classes d'opérations non complêtement continues. Ann. Math. **52**(1), 140–147 (1950).
[58] Iosifescu, M., Kraaikamp, C., Matrical theory of Continued Fractions. Kluwer (2002).
[59] Jenkinson, O., Pollicott, M., Rigorous effective bounds on the Hausdorff dimension of continued fraction Cantor sets: A hundred decimal digits for the dimension of E2. Adv. Math. **325**, 87–115 (2018).
[60] Kakutani, S., Yosida, K., Operator-theoretical treatment of Markoff process and the mean ergodic theorem. Ann. Math. **42**, 188–228 (1941).
[61] Kechris, A., Classical Descriptive Set Theory. Graduate Texts in Mathematics. Springer, New York (1995).
[62] Kingman, J. F. C., Taylor, S. J., Introduction to Measure and Probability. Cambridge University Press, London (1966).
[63] Kotus, J., Urbański, M., Ergodic Theory, Geometric Measure Theory, Conformal Measures and the Dynamics of Elliptic Functions, preprint. http://arxiv.org/pdf/2007.13235.pdf.
[64] Krzyżewski, K., Szlenk, W., On invariant measures for expanding differentiable mappings. Studia Math. **33**, 83–92 (1969).
[65] Lasota, A., Yorke, J. A., On the existence of invariant measures for piecewise monotonic transformations. Trans. Amer. Math. Soc. **186**, 481–488 (1973).
[66] Lehto, O., Univalent functions and Teichmüller spaces. Graduate Texts in Mathematics, vol. **109**. Springer, New York (1987).
[67] Lehto, O., Virtanen, K. I., In: Quasiconformal mappings in the plane, 2nd edn. Springer Series: Grundlehren der mathematischen Wissenschaften in Einzeldarstellungen mit besonderer Berücksichtigung der Anwendungsgebiete, vol. **126**. Springer, Berlin (1973).
[68] Mandelbrot, B., Intermittent turbulence in self-similar cascades: divergence of high moments and dimension of the carrier. J. Fluid Mech. **62**, 331–358 (1974).
[69] Mañé, R., The Hausdorff dimension of invariant probabilities of rational maps. Lecture Notes in Math. **1331**, 86–117 (1988).
[70] Mañé, R., Sad, P., Sullivan, D., On the dynamics of rational maps. Ann. Sci. Éc. Norm. Supér. **16**(2), 193–217 (1983).
[71] Manning, A., McCluskey, H., Hausdorff dimension for horseshoes. Ergodic Theory Dynam. Systems **3**, 251–260 (1983).
[72] Mattila, P., Geometry of Sets and Measures in Euclidean Spaces. Cambridge University Press, Cambridge (1995).
[73] Mauldin, R. D., Przytycki, F., Urbański, M., Rigidity of conformal iterated function systems. Compos. Math. **129**, 273–299 (2001).
[74] Mauldin, R. D., Szarek, T., Urbański, M., Graph directed Markov systems on Hilbert spaces. Math. Proc. Cambridge Philos. Soc. **147**, 455–488 (2009).
[75] Mauldin, R. D., Urbański, M., Dimensions and measures in infinite iterated function systems. Proc. Lond. Math. Soc. **73**, 105–154 (1996).
[76] Mauldin, R. D., Urbański, M., Conformal iterated function systems with applications to the geometry of continued fractions. Trans. Amer. Math. Soc. **351**, 4995–5025 (1999).
[77] Mauldin, R. D., Urbański, M., Gibbs states on the symbolic space over an infinite alphabet. Israel J. Math. **125**, 93–130 (2001).

[78] Mauldin, R. D., Urbański, M., Parabolic iterated function systems. Ergodic Theory Dynam. Systems **20**, 1423–1447 (2000).

[79] Mauldin, R. D., Urbański, M., Gibbs states on the symbolic space over an infinite alphabet. Israel J. Math. **125**, 93–130 (2001).

[80] Mauldin, R. D., Urbański, M., Fractal measures for parabolic IFS. Adv. Math. **168**, 225–253 (2002).

[81] Mauldin, R. D., Urbański, M., Graph Directed Markov Systems: Geometry and Dynamics of Limit Sets. Cambridge Tracts in Mathematics, vol. **148**. Cambridge University Press, Cambridge (2003).

[82] Mauldin, R. D., Williams, S. C., Hausdorff dimension in graph directed constructions. Trans. Amer. Math. Soc. **309**, 811–829 (1988).

[83] Mayer, V., Urbański, M., Finer geometric rigidity of limit sets of conformal IFS. Proc. Amer. Math. Soc. **131**, 3695–3702 (2003).

[84] Mayer, V., Urbański, M., Thermodynamical Formalism and Multifractal Analysis for Meromorphic Functions of Finite Order. Mem. Amer. Math. Soc. **203**(o), 954 (2010).

[85] Mihailescu, M., Urbański, M., Skew product Smale endomorphisms over countable shifts of finite type. Ergodic Theory Dynam. Systems **40**, 3105–3149 (2020).

[86] Mujica, J., Complex Analysis in Banach Spaces: Holomorphic Functions and Domains of Holomorphy in Finite and Infinite Dimensions. North-Holland Mathematics Studies, vol. **120**. Elsevier Science Pub. Co., New York (1986).

[87] Olsen, L., A multifractal formalism. Adv. Math. **116**, 82–196 (1995).

[88] Olsen, L., Multifractal geometry. In: Bandt, C., Graf, S., Zähle, M. (eds.) Proceedings Fractal Geometry and Stochastics II (Greifswald, Germany, August 1998). Progress in Probability, vol. **46**, pp. 1–37. Birkhäuser (2000).

[89] Olsen, L., Hausdorff and packing measure functions of self-similar sets: continuity and measurability. Ergodic Theory Dynam. Systems **28**, 1635–1655 (2008).

[90] Ornstein, D., Bernoulli shifts with the same entropy are isomorphic. Adv. Math. **4**, 337–352 (1970).

[91] Parry, W., Pollicott, M., Zeta functions and the periodic orbit structure of hyperbolic dynamics. Astérisque **187–188** (1990).

[92] Patterson, S. J., The limit set of a Fuchsian group. Acta Math. **136**, 241–273 (1976).

[93] Patterson, S. J., Lectures on measures on limit sets of Kleinian groups. In: Analytical and geometric aspects of hyperbolic space. London Math. Soc., Lecture Notes, vol. **111**. Cambridge Univ. Press (1987).

[94] Patzschke, N., Self-conformal multifractal measures. Adv. in Appl. Math. **19**, 486–513 (1997).

[95] Pesin, Y. B., Dimension Theory in Dynamical Systems. Chicago Lectures in Mathematics. University of Chicago Press, Chicago (1997).

[96] Pesin, Y. B., Weiss, H., A multifractal analysis of equilibrium measures for conformal expanding maps and Markov Moran geometric constructions. J. Stat. Phys. **86**, 233–275 (1997).

[97] Przytycki, F., Hausdorff dimension of harmonic measure on the boundary of an attractive basin for a holomorphic map. Invent. Math. **80**, 169–171 (1985).

[98] Przytycki, F., Urbański, M., Conformal Fractals: Ergodic Theory Methods. London Mathematical Society Lecture Note Series, vol. **371**. Cambridge University Press (2010).

[99] Przytycki, F., Urbański, M., Zdunik, A., Harmonic, Gibbs and Hausdorff measures on repellers for holomorphic maps I. Ann. Math. **130**, 1–40 (1989).

[100] Rand, D. A., The singularity spectrum for cookie-cutters. Ergodic Theory Dynam. Systems **9**, 527–541 (1989).

[101] Rogers, C. A., Hausdorff Measures. Cambridge University Press (1970, 1998).

[102] Roy, M., A new variation of Bowen's formula for graph directed Markov systems. Discrete Contin. Dyn. Syst. Ser. A **32**(7), 2533–2551 (2012).

[103] Roy, M., Sumi, H., Urbański, M., Analytic families of holomorphic iterated function systems. Nonlinearity **21**, 2255–2279 (2008).

[104] Roy, M., Sumi, H., Urbański, M., λ-topology vs. pointwise topology. Ergodic Theory Dynam. Systems **29**(2), 685–713 (2009).

[105] Roy, M., Urbański, M., Regularity properties of Hausdorff dimension in conformal infinite IFS. Ergodic Theory Dynam. Systems **25**, 1961–1983 (2005).

[106] Roy, M., Urbański, M., Real analyticity of Hausdorff dimension for higher dimensional hyperbolic graph directed Markov systems. Math. Z. **260**(1), 153–175 (2008).

[107] Roy, M., Urbański, M., Multifractal analysis for conformal graph directed Markov systems. Discrete Contin. Dyn. Syst. Ser. A **25**(2), 627–650 (2009).

[108] Rudin, W., Principles of Mathematical Analysis, 3rd edn. International Series in Pure and Applied Mathematics. McGraw-Hill, Singapore (1976).

[109] Rudin, W., Real and Complex Analysis, 3rd edn. McGraw-Hill, Singapore (1987).

[110] Rudin, W., Functional Analysis, 2nd edn. McGraw-Hill, Singapore (1991).

[111] Ruelle, D., Statistical Mechanics. Rigorous Results. Benjamin, New York (1969).

[112] Ruelle, D., Statistical mechanics on a compact set with $\mathbb{Z}^\nu$ action satisfying expansiveness and specification. Bull. Amer. Math. Soc. **78**, 988–991 (1972); Trans. Amer. Math. Soc. **185**, 237–251 (1973).

[113] Ruelle, D., Thermodynamic formalism. Encyclopedia of Math. and Appl., vol. **5**. Addison-Wesley, Reading, Mass. (1976).

[114] Rychlik, M., Bounded variation and invariant measures. Studia Math. **76**, 69–80 (1983).

[115] Sarig, O., Theormodynamic formalism for countable Markov shifts. Ergodic Theory Dynam. Systems **62**, 1565–1593 (1999).

[116] Schief, A., Separation properties for self-similar sets. Proc. Amer. Math. Soc. **122**(1), 111–115 (1994).

[117] Sinai, Ya.G., Gibbs Measures in Ergodic Theory. Russian Math. Surveys **27**, 21–70 (1972).

[118] Skorulski, B., Urbański, M., Finer fractal geometry for analytic families of conformal dynamical systems. Dyn. Syst. **29**, 369–398 (2014).

[119] Slodkowski, Z., Holomorphic motions and polynomial hulls. Proc. Amer. Math. Soc. **111**(2), 347–355 (1991).

[120] Srivastava, S. M., A Course on Borel Sets. Graduate Texts in Mathematics, vol. **180**. Springer, New York (1998).

[121] Stratmann, B. O., Urbański, M., Real analyticity of topological pressure for parabolically semihyperbolic generalized polynomial-like maps. Indag. Math. (N.S.) **14**, 119–134 (2003).

[122] Sullivan, D., The density at infinity of a discrete group. Publ. Math. Inst. Hautes Études Sci. **50** (1979).

[123] Sullivan, D., Seminar on conformal and hyperbolic geometry, preprint IHES (1982).

[124] Sullivan, D., Disjoint spheres, approximation by imaginary quadratic numbers and the logarithmic law for geodesics. Acta Math. **149**, 215–237 (1982).

[125] Sullivan, D., Conformal dynamical systems. In: Geometric dynamics. Lect. Notes in Math., vol. **1007**, pp. 725–752. Springer (1983).

[126] Sullivan D., E., Hausdorff measures old and new, and the limit set of geometrically finite Kleinian groups. Acta Math. **153**, 259–277 (1984).

[127] Sumi, H., Urbański, M., Real analyticity of Hausdorff dimension for expanding rational semigroups. Ergodic Theory Dynam. Systems **30**, 601–633 (2010).

[128] Szarek, T., Urbański, M., Zdunik, A., Continuity of Hausdorff measure for conformal dynamical systems. Discrete Contin. Dyn. Syst. Ser. A **33**, 4647–4692 (2013).

[129] Szarek, T., Wedrychowicz, S., The OSC does not imply the SOSC for infinite iterated function systems. Proc. Amer. Math. Soc. **133**(2), 437–440 (2004).

[130] Taylor, S. J., Tricot, C., Packing measure, and its evaluation for a Brownian path. Trans. Amer. Math. Soc. **288**, 679–699 (1985).

[131] Tricot, C., Two definitions of fractional dimension. Math. Proc. Cambridge Philos. Soc. **91**, 57–74 (1982).

[132] Urbański, M., Porosity in conformal infinite iterated function systems. J. Number Theory **88**, 283–312 (2001).

[133] Urbański, M., Diophantine approximation of self-conformal measures. J. Number Theory **110**, 219–235 (2005).

[134] Urbański, M., Analytic families of semihyperbolic generalized polynomial-like mappings. Monatsh. Math. **159**, 133–162 (2010).

[135] Urbański, M., Zdunik, A., Real analyticity of Hausdorff dimension of finer Julia sets of exponential family. Ergodic Theory Dynam. Systems **24**, 279–315 (2004).

[136] Urbański, M., Zdunik, A., Continuity of the Hausdorff measure of continued fractions and countable alphabet iterated function systems. J. Théor. Nombres Bordeaux **28**, 261–286 (2016).

[137] Urbański, M., Zinsmeister, M., Geometry of hyperbolic Julia-Lavaurs sets. Indag. Math. (N.S.) **12**, 273–292 (2001).

[138] Williams, D., Probability with Martingales. Cambridge University Press, Cambridge (1991).

[139] Young, L.-S., Dimension, entropy and Lyapunov exponents. Ergodic Theory Dynam. Systems **2**, 109–124 (1982).

Index

https://doi.org/10.1515/9783110702699-025

De Gruyter Expositions in Mathematics

Volume 70
Aron Simis, Zaqueu Ramos
Graded Algebras in Algebraic Geometry, 2022
ISBN 978-3-11-063754-0, e-ISBN 978-3-11-064069-4

Volume 69/1
Mariusz Urbański, Mario Roy, Sara Munday
Non-Invertible Dynamical Systems. Volume 1: Ergodic Theory – Finite and Infinite, Thermodynamic Formalism, Symbolic Dynamics and Distance Expanding Maps, 2021
ISBN 978-3-11-070264-4, e-ISBN 978-3-11-070268-2

Volume 68
Yuri A. Bahturin
Identical Relations in Lie Algebras, 2021
ISBN 978-3-11-056557-7, e-ISBN 978-3-11-056665-9

Volume 34
Marek Jarnicki, Peter Pflug
Extension of Holomorphic Functions, 2nd Edition, 2020
ISBN 978-3-11-062766-4, e-ISBN 978-3-11-063027-5

Volume 67
Alcides Lins Neto, Bruno Scárdua
Complex Algebraic Foliations, 2020
ISBN 978-3-11-060107-7, e-ISBN 978-3-11-060205-0

Volume 66
Anke Kalauch, Onno van Gaans
Pre-Riesz Spaces, 2018
ISBN 978-3-11-047539-5, e-ISBN 978-3-11-047629-3

Volume 65
Yakov Berkovich, Zvonimir Janko
Groups of Prime Power Order. Volume 6, 2018
ISBN 978-3-11-053097-1, e-ISBN 978-3-11-053314-9

Volume 64
Yakov Berkovich, Lev S. Kazarin, Emmanuel M. Zhmud'
Characters of Finite Groups. Volume 2, 2019
ISBN 978-3-11-022408-5, e-ISBN 978-3-11-022409-2

www.ingramcontent.com/pod-product-compliance
Ingram Content Group UK Ltd.
Pitfield, Milton Keynes, MK11 3LW, UK
UKHW051029041026
15470UKWH00016B/1817

9 783110 700619